Bayesian
Data Analysis

SECOND EDITION

CHAPMAN & HALL/CRC
Texts in Statistical Science Series

Series Editors
Chris Chatfield, *University of Bath, UK*
Martin Tanner, *Northwestern University, USA*
Jim Zidek, *University of British Columbia, Canada*

Bayesian Data Analysis

SECOND EDITION

Andrew Gelman
Columbia University, New York

John B. Carlin
University of Melbourne, Australia

Hal S. Stern
University of California, Irvine

Donald B. Rubin
Harvard University, Cambridge, Massachusetts

CHAPMAN & HALL/CRC

A CRC Press Company
Boca Raton   London   New York   Washington, D.C.

Library of Congress Cataloging-in-Publication Data

Bayesian data analysis / Andrew Gelman ... [et al.].—2nd ed.
   p. cm. — (Texts in statistical science)
   Includes bibliographical references and index.
   ISBN 1-58488-388-X (alk. paper)
   1. Bayesian statistical decision theory. I. Gelman, Andrew. II. Series.

  QA279.5.B386 2003
  519.5'42—dc21
                                    2003051474

Visit the CRC Press Web site at www.crcpress.com

© 2004 by Chapman & Hall/CRC

No claim to original U.S. Government works
International Standard Book Number 1-58488-388-X
Library of Congress Card Number 2003051474
Printed in the United States of America   6 7 8 9 0
Printed on acid-free paper

Contents

List of models

List of examples

Preface

This book is intended to have three roles and to serve three associated audiences: an introductory text on Bayesian inference starting from first principles, a graduate text on effective current approaches to Bayesian modeling and computation in statistics and related fields, and a handbook of Bayesian methods in applied statistics for general users of and researchers in applied statistics. Although introductory in its early sections, the book is definitely not elementary in the sense of a first text in statistics. The mathematics used in our book is basic probability and statistics, elementary calculus, and linear algebra. A review of probability notation is given in Chapter 1 along with a more detailed list of topics assumed to have been studied. The practical orientation of the book means that the reader's previous experience in probability, statistics, and linear algebra should ideally have included strong computational components.

To write an introductory text alone would leave many readers with only a taste of the conceptual elements but no guidance for venturing into genuine practical applications, beyond those where Bayesian methods agree essentially with standard non-Bayesian analyses. On the other hand, given the continuing scarcity of introductions to applied Bayesian statistics either in books or in statistical education, we feel it would be a mistake to present the advanced methods without first introducing the basic concepts from our data-analytic perspective. Furthermore, due to the nature of applied statistics, a text on current Bayesian methodology would be incomplete without a variety of worked examples drawn from real applications. To avoid cluttering the main narrative, *there are bibliographic notes at the end of each chapter* and references at the end of the book.

Examples of real statistical analyses are found throughout the book, and we hope thereby to give a genuine applied flavor to the entire development. Indeed, given the conceptual simplicity of the Bayesian approach, it is only in the intricacy of specific applications that novelty arises. Non-Bayesian approaches to inference have dominated statistical theory and practice for most of the past century, but the last two decades or so have seen a reemergence of the Bayesian approach. This has been driven more by the availability of new computational techniques than by what many would see as the philosophical and logical advantages of Bayesian thinking.

We hope that the publication of this book will enhance the spread of ideas that are currently trickling through the scientific literature. The models and methods developed recently in this field have yet to reach their largest possible audience, partly because the results are scattered in various journals and proceedings volumes. We hope that this book will help a new generation of statisticians and users of statistics to solve complicated problems with greater understanding.

Progress in Bayesian data analysis

Bayesian methods have matured and improved in several ways in the eight years since the first edition of this book appeared.

- Successful applications of Bayesian data analysis have appeared in many different fields, including business, computer science, economics, educational research, environmental science, epidemiology, genetics, geography, imaging, law, medicine, political science, psychometrics, public policy, sociology, and sports. In the social sciences, Bayesian ideas often appear in the context of multilevel modeling.

- New computational methods generalizing the Gibbs sampler and Metropolis algorithm, including some methods from the physics literature, have been adapted to statistical problems. Along with improvements in computing speed, these have made it possible to compute Bayesian inference for more complicated models on larger datasets.

- In parallel with the theoretical improvements in computation, the software package Bugs has allowed nonexperts in statistics to fit complex Bayesian models with minimal programming. Hands-on experience has convinced many applied researchers of the benefits of the Bayesian approach.

- There has been much work on model checking and comparison, from many perspectives, including predictive checking, cross-validation, Bayes factors, model averaging, and estimates of predictive errors and model complexity.

- In sample surveys and elsewhere, multiple imputation has become a standard method of capturing uncertainty about missing data. This has motivated ongoing work into more flexible models for multivariate distributions.

- There has been continuing progress by various researchers in combining Bayesian inference with existing statistical approaches from other fields, such as instrumental variables analysis in economics, and with nonparametric methods such as classification trees, splines, and wavelets.

- In general, work in Bayesian statistics now focuses on applications, computations, and models. Philosophical debates, abstract optimality criteria, and asymptotic analyses are fading to the background. It is now possible to do serious applied work in Bayesian inference without the need to debate foundational principles of inference.

Changes for the second edition

The major changes for the second edition of this book are:

- Reorganization and expansion of Chapters 6 and 7 on model checking and data collection;

- Revision of Part III on computation;

- New chapters on nonlinear models and decision analysis;
- An appendix illustrating computation using the statistical packages R and Bugs,
- New applied examples throughout, including:
  - Census record linkage, a data-based assignment of probability distributions (Section 1.7),
  - Cancer mapping, demonstrating the role of the prior distribution on data with different sample sizes (Section 2.8),
  - Psychological measurement data and the use of graphics in model checking (Section 6.4),
  - Survey of adolescent smoking, to illustrate numerical predictive checks (Section 6.5),
  - Two surveys using cluster sampling (Section 7.4),
  - Experiment of vitamin A intake, with noncompliance to assigned treatment (Section 7.7),
  - Factorial data on internet connect times, summarized using the analysis of variance (Section 15.6),
  - Police stops, modeled with hierarchical Poisson regressions (Section 16.5),
  - State-level opinions from national polls, using hierarchical modeling and poststratification (Section 16.6),
  - Serial dilution assays, as an example of a nonlinear model (Section 20.2),
  - Data from a toxicology experiment, analyzed with a hierarchical nonlinear model (Section 20.3),
  - Pre-election polls, with multiple imputation of missing data (Section 21.2),
  - Incentives for telephone surveys, a meta-analysis for a decision problem (Section 22.2),
  - Medical screening, an example of a decision analysis (Section 22.3),
  - Home radon measurement and remediation decisions, analyzed using a hierarchical model (Section 22.4).

We have added these examples because our readers have told us that one thing they liked about the book was the presentation of realistic problem-solving experiences. As in the first edition, we have included many applications from our own research because we know enough about these examples to convey the specific challenges that arose in moving from substantive goals to probability modeling and, eventually, to substantive conclusions. Also as before, some of the examples are presented schematically and others in more detail.

We changed the computation sections out of recognition that our earlier recommendations were too rigid: Bayesian computation is currently at a stage where there are many reasonable ways to compute any given posterior distribution, and the best approach is not always clear in advance. Thus we have moved to a more pluralistic presentation—we give advice about performing computations from many perspectives, including approximate computation, mode-finding, and simulations, while making clear, especially in the discussion of individual models in the later parts of the book, that it is important to be aware of the different ways of implementing any given iterative simulation computation. We briefly discuss some recent ideas in Bayesian computation but devote most of Part III to the practical issues of implementing the Gibbs sampler and the Metropolis algorithm. Compared to the first edition, we deemphasize approximations based on the normal distribution and the posterior mode, treating these now almost entirely as techniques for obtaining starting points for iterative simulations.

Contents

Part I introduces the fundamental Bayesian principle of treating all unknowns as random variables and presents basic concepts, standard probability models, and some applied examples. In Chapters 1 and 2, simple familiar models using the normal, binomial, and Poisson distributions are used to establish this introductory material, as well as to illustrate concepts such as conjugate and noninformative prior distributions, including an example of a nonconjugate model. Chapter 3 presents the Bayesian approach to multiparameter problems. Chapter 4 introduces large-sample asymptotic results that lead to normal approximations to posterior distributions.

Part II introduces more sophisticated concepts in Bayesian modeling and model checking. Chapter 5 introduces hierarchical models, which reveal the full power and conceptual simplicity of the Bayesian approach for practical problems. We illustrate issues of model construction and computation with a relatively complete Bayesian analysis of an educational experiment and of a meta-analysis of a set of medical studies. Chapter 6 discusses the key practical concerns of model checking, sensitivity analysis, and model comparison, illustrating with several examples. Chapter 7 discusses how Bayesian data analysis is influenced by data collection, including the topics of ignorable and nonignorable data collection rules in sample surveys and designed experiments, and specifically the topic of randomization, which is presented as a device for increasing the robustness of posterior inferences. This a difficult chapter, because it presents important ideas that will be unfamiliar to many readers. Chapter 8 discusses connections to non-Bayesian statistical methods, emphasizing common points in practical applications and current challenges in implementing Bayesian data analysis. Chapter 9 summarizes some of the key ideas of Bayesian modeling, inference, and model checking, illustrating issues with some relatively simple examples that highlight potential pitfalls in trying to fit models automatically.

Part III covers Bayesian computation, which can be viewed as a highly specialized branch of numerical analysis: given a posterior distribution function (possibly implicitly defined), how does one extract summaries such as quantiles, moments, and modes, and draw random samples of values? We emphasize iterative methods—the Gibbs sampler and Metropolis algorithm—for drawing random samples from the posterior distribution.

Part IV discusses regression models, beginning with a Bayesian treatment of classical regression illustrated using an example from the study of elections that has both causal and predictive aspects. The subsequent chapters give general principles and examples of hierarchical linear models, generalized linear models, and robust models.

Part V presents a range of other Bayesian probability models in more detail, with examples of multivariate models, mixtures, and nonlinear models. We conclude with methods for missing data and decision analysis, two practical concerns that arise implicitly or explicitly in many statistical problems.

Throughout, we illustrate in examples the three steps of Bayesian statistics: (1) setting up a full probability model using substantive knowledge, (2) conditioning on observed data to form a posterior inference, and (3) evaluating the fit of the model to substantive knowledge and observed data.

Appendixes provide a list of common distributions with their basic properties, a sketch of a proof of the consistency and limiting normality of Bayesian posterior distributions, and an extended example of Bayesian computation in the statistical packages Bugs and R.

Most chapters conclude with a set of exercises, including algebraic derivations, simple algebraic and numerical examples, explorations of theoretical topics covered only briefly in the text, computational exercises, and data analyses. The exercises in the later chapters tend to be more difficult; some are suitable for term projects.

One-semester or one-quarter course

This book began as lecture notes for a graduate course. Since then, we have attempted to create an advanced undergraduate text, a graduate text, and a reference work all in one, and so the instructor of any course based on this book must be selective in picking out material.

Chapters 1–6 should be suitable for a one-semester course in Bayesian statistics for advanced undergraduates, although these students might also be interested in the introduction to Markov chain simulation in Chapter 11.

Part I has many examples and algebraic derivations that will be useful for a lecture course for undergraduates but may be left to the graduate students to read at home (or conversely, the lectures can cover the examples and leave the theory for homework). The examples of Part II are crucial, however, since these ideas will be new to most graduate students as well. We see the first two chapters of Part III as essential for understanding modern Bayesian computation and the first three chapters of Part IV as basic to any graduate course because they take the student into the world of standard applied models; the remaining material in Parts III–V can be covered as time permits.

This book has been used as the text for one-semester and one-quarter courses for graduate students in statistics at many universities. We suggest the following syllabus for an intense fifteen-week course.

1. Setting up a probability model, Bayes' rule, posterior means and variances, binomial model, proportion of female births (Chapter 1, Sections 2.1–2.5).

2. Standard univariate models including the normal and Poisson models, cancer rate example, noninformative prior distributions (Sections 2.6–2.9).

3. Multiparameter models, normal with unknown mean and variance, the multivariate normal distribution, multinomial models, election polling, bioassay. Computation and simulation from arbitrary posterior distributions in two parameters (Chapter 3).

4. Inference from large samples and comparison to standard non-Bayesian methods (Chapter 4).

5. Hierarchical models, estimating population parameters from data, rat tumor rates, SAT coaching experiments, meta-analysis (Chapter 5).

6. Model checking, posterior predictive checking, sensitivity analysis, model comparison and expansion, checking the analysis of the SAT coaching experiments (Chapter 6).

7. Data collection—ignorability, surveys, experiments, observational studies, unintentional missing data (Chapter 7).

8. General advice, connections to other statistical methods, examples of potential pitfalls of Bayesian inference (Chapters 8 and 9).

9. Computation: overview, uses of simulations, Gibbs sampling (Chapter 10, Sections 11.1–11.3).

10. Markov chain simulation (Sections 11.4–11.10, Appendix C).

11. Normal linear regression from a Bayesian perspective, incumbency advantage in Congressional elections (Chapter 14).

12. Hierarchical linear models, selection of explanatory variables, forecasting Presidential elections (Chapter 15).

13. Generalized linear models, police stops example, opinion polls example (Chapter 16).

14. Final weeks: topics from remaining chapters (including advanced computational methods, robust inference, mixture models, multivariate models, nonlinear models, missing data, and decision analysis).

Computer sites and contact details

Additional materials, including the data used in the examples, solutions to many of the end-of-chapter exercises, and any errors found after the book goes to press, are posted at http://www.stat.columbia.edu/~gelman/. Please send any comments to us at gelman@stat.columbia.edu, sternh@uci.edu, jbcarlin@unimelb.edu.au, or rubin@stat.harvard.edu.

Acknowledgments

We thank Stephen Ansolabehere, Adriano Azevedo, Jarrett Barber, Tom Belin, Suzette Blanchard, Brad Carlin, Alicia Carriquiry, Samantha Cook, Victor De Oliveira, David Draper, John Emerson, Steve Fienberg, Yuri Goegebeur, Daniel Gianola, David Hammill, Chuanpu Hu, Zaiying Huang, Yoon-Sook Jeon, Shane Jensen, Jay Kadane, Jouni Kerman, Gary King, Lucien Le Cam, Rod Little, Tom Little, Chuanhai Liu, Xuecheng Liu, Peter McCullagh, Mary Sara McPeek, Xiao-Li Meng, Baback Moghaddam, Olivier Nimeskern, Ali Rahimi, Thomas Richardson, Scott Schmidler, Andrea Siegel, Sandip Sinharay, Elizabeth Stuart, Andrew Swift, Francis Tuerlinckx, Iven Van Mechelen, Rob Weiss, Alan Zaslavsky, several reviewers, many other colleagues, and the students in Statistics 238, 242A, and 260 at Berkeley, Statistics 36-724 at Carnegie Mellon, Statistics 320 at Chicago, Statistics 220 at Harvard, Statistics 544 at Iowa State, and Statistics 6102 at Columbia, for helpful discussions, comments, and corrections. We especially thank Phillip Price and Radford Neal for their thorough readings of different parts of this book. John Boscardin deserves special thanks for implementing many of the computations for Sections 5.5, 6.8, 15.2, and 17.4. We also thank Chad Heilig for help in preparing tables, lists, and indexes. The National Science Foundation provided financial support through a postdoctoral fellowship and grants SBR-9223637, 9708424, DMS-9404305, 9457824, 9796129, and SES-9987748, 0084368. The computations and figures were done using the S, S-Plus, R, and Bugs computer packages (see Appendix C).

Many of the examples in this book have appeared elsewhere, in books and articles by ourselves and others, as we indicate in the bibliographic notes and exercises in the chapters where they appear. (In particular: Figures 1.3–1.5 are adapted from the *Journal of the American Statistical Association* **90** (1995), pp. 696, 702, and 703, and are reprinted with permission of the American Statistical Association. Figures 2.7 and 2.8 come from Gelman, A., and Nolan, D., *Teaching Statistics: A Bag of Tricks*, Oxford University Press (1992), pp. 14 and 15, and are reprinted with permission of Oxford University Press. Figures 20.8–20.10 come from the *Journal of the American Statistical Association* **91** (1996), pp. 1407 and 1409, and are reprinted with permission of the American Statistical Association. Table 20.1 comes from Berry, D., *Statistics: A Bayesian Perspective*, Duxbury Press (1996), p. 81, and is reprinted with permission of Brooks/Cole, a division of Thomson Learning. Figures 21.1 and 21.2 come from the *Journal of the American Statistical Association* **93** (1998) pp. 851 and 853, and are reprinted with permission of the American Statistical Association. Figures 22.1–22.3 are adapted from the *Journal of Business and Economic Statistics* **21** (2003), pp. 219 and 223, and are reprinted with permission of the American Statistical Association.)

Finally, we thank our spouses, Caroline, Nancy, Hara, and Kathryn, for their love and support during the writing and revision of this book.

Part I: Fundamentals of Bayesian Inference

Bayesian inference is the process of fitting a probability model to a set of data and summarizing the result by a probability distribution on the parameters of the model and on unobserved quantities such as predictions for new observations. In Chapters 1–3, we introduce several useful families of models and illustrate their application in the analysis of relatively simple data structures. Some mathematics arises in the analytical manipulation of the probability distributions, notably in transformation and integration in multiparameter problems. We differ somewhat from other introductions to Bayesian inference by emphasizing stochastic simulation, and the combination of mathematical analysis and simulation, as general methods for summarizing distributions. Chapter 4 outlines the fundamental connections between Bayesian inference, other approaches to statistical inference, and the normal distribution. The early chapters focus on simple examples to develop the basic ideas of Bayesian inference; examples in which the Bayesian approach makes a practical difference relative to more traditional approaches begin to appear in Chapter 3. The major practical advantages of the Bayesian approach appear in hierarchical models, as discussed in Chapter 5 and thereafter.

CHAPTER 1

Background

1.1 Overview

By Bayesian data analysis, we mean practical methods for making inferences from data using probability models for quantities we observe and for quantities about which we wish to learn. The essential characteristic of Bayesian methods is their explicit use of probability for quantifying uncertainty in inferences based on statistical data analysis.

The process of Bayesian data analysis can be idealized by dividing it into the following three steps:

1. Setting up a *full probability model*—a joint probability distribution for all observable and unobservable quantities in a problem. The model should be consistent with knowledge about the underlying scientific problem and the data collection process.

2. Conditioning on observed data: calculating and interpreting the appropriate *posterior distribution*—the conditional probability distribution of the unobserved quantities of ultimate interest, given the observed data.

3. Evaluating the fit of the model and the implications of the resulting posterior distribution: does the model fit the data, are the substantive conclusions reasonable, and how sensitive are the results to the modeling assumptions in step 1? If necessary, one can alter or expand the model and repeat the three steps.

Great advances in all these areas have been made in the last forty years, and many of these are reviewed and used in examples throughout the book. Our treatment covers all three steps, the second involving computational methodology and the third a delicate balance of technique and judgment, guided by the applied context of the problem. The first step remains a major stumbling block for much Bayesian analysis: just where do our models come from? How do we go about constructing appropriate probability specifications? We provide some guidance on these issues and illustrate the importance of the third step in retrospectively evaluating the fit of models. Along with the improved techniques available for computing conditional probability distributions in the second step, advances in carrying out the third step alleviate to some degree the need for completely correct model specification at the first attempt. In particular, the much-feared dependence of conclusions on 'subjective' prior distributions can be examined and explored.

A primary motivation for believing Bayesian thinking important is that it facilitates a common-sense interpretation of statistical conclusions. For instance, a Bayesian (probability) interval for an unknown quantity of interest can be directly regarded as having a high probability of containing the unknown quantity, in contrast to a frequentist (confidence) interval, which may strictly be interpreted only in relation to a sequence of similar inferences that might be made in repeated practice. Recently in applied statistics, increased emphasis has been placed on interval estimation rather than hypothesis testing, and this provides a strong impetus to the Bayesian viewpoint, since it seems likely that most users of standard confidence intervals give them a common-sense Bayesian interpretation. One of our aims in this book is to indicate the extent to which Bayesian interpretations of common simple statistical procedures are justified.

Rather than engage in philosophical debates about the foundations of statistics, however, we prefer to concentrate on the pragmatic advantages of the Bayesian framework, whose flexibility and generality allow it to cope with very complex problems. The central feature of Bayesian inference, the direct quantification of uncertainty, means that there is no impediment in principle to fitting models with many parameters and complicated multilayered probability specifications. In practice, the problems are ones of setting up and computing with large models, and a large part of this book focuses on recently developed and still developing techniques for handling these modeling and computational challenges. The freedom to set up complex models arises in large part from the fact that the Bayesian paradigm provides a conceptually simple method for coping with multiple parameters, as we discuss in detail from Chapter 3 on.

1.2 General notation for statistical inference

Statistical inference is concerned with drawing conclusions, from numerical data, about quantities that are not observed. For example, a clinical trial of a new cancer drug might be designed to compare the five-year survival probability in a population given the new drug with that in a population under standard treatment. These survival probabilities refer to a large *population* of patients, and it is neither feasible nor ethically acceptable to experiment with an entire population. Therefore inferences about the true probabilities and, in particular, their differences must be based on a *sample* of patients. In this example, even if it were possible to expose the entire population to one or the other treatment, it is obviously never possible to expose anyone to both treatments, and therefore statistical inference would still be needed to assess the *causal inference*—the comparison between the observed outcome in each patient and that patient's unobserved outcome if exposed to the other treatment.

We distinguish between two kinds of *estimands*—unobserved quantities for which statistical inferences are made—first, potentially observable quantities, such as future observations of a process, or the outcome under the treatment not received in the clinical trial example; and second, quantities that are not directly observable, that is, parameters that govern the hypothetical process leading to the observed data (for example, regression coefficients). The distinction between these two kinds of estimands is not always precise, but is generally useful as a way of understanding how a statistical model for a particular problem fits into the real world.

Parameters, data, and predictions

As general notation, we let θ denote unobservable vector quantities or population *parameters* of interest (such as the probabilities of survival under each treatment for randomly chosen members of the population in the example of the clinical trial), y denote the observed data (such as the numbers of survivors and deaths in each treatment group), and $\tilde{y}$ denote unknown, but potentially observable, quantities (such as the outcomes of the patients under the other treatment, or the outcome under each of the treatments for a new patient similar to those already in the trial). In general these symbols represent multivariate quantities. We generally use Greek letters for parameters, lower-case Roman letters for observed or observable scalars and vectors (and sometimes matrices), and upper-case Roman letters for observed or observable matrices. When using matrix notation, we consider vectors as column vectors throughout; for example, if u is a vector with n components, then $u^T u$ is a scalar and $u u^T$ an $n \times n$ matrix.

Observational units and variables

In many statistical studies, data are gathered on each of a set of n objects or *units*, and we can write the data as a vector, $y = (y_1, \ldots, y_n)$. In the clinical trial example, we might label y_i as 1 if patient i is alive after five years or 0 if the patient dies. If several variables are measured on each unit, then each y_i is actually a vector, and the entire dataset y is a matrix (usually taken to have n rows). The y variables are called the 'outcomes' and are considered 'random' in the sense that, when making inferences, we wish to allow for the possibility that the observed values of the variables could have turned out otherwise, due to the sampling process and the natural variation of the population.

Exchangeability

The usual starting point of a statistical analysis is the (often tacit) assumption that the n values y_i may be regarded as *exchangeable*, meaning that the joint probability density $p(y_1, \ldots, y_n)$ should be invariant to permutations of the indexes. A nonexchangeable model would be appropriate if information relevant to the outcome were conveyed in the unit indexes rather than by explanatory variables (see below). The idea of exchangeability is fundamental to statistics, and we return to it repeatedly throughout the book.

Generally, it is useful and appropriate to model data from an exchangeable distribution as independently and identically distributed (*iid*) given some unknown parameter vector θ with distribution $p(\theta)$. In the clinical trial example, we might model the outcomes y_i as iid, given θ, the unknown probability of survival.

Explanatory variables

It is common to have observations on each unit that we do not bother to model as random. In the clinical trial example, such variables might include the age and previous health status of each patient in the study. We call this second class of variables *explanatory variables*, or *covariates*, and label them x. We use X to denote the entire set of explanatory variables for all n units; if there are k explanatory variables, then X is a matrix with n rows and k columns. Treating X as random, the notion of exchangeability can be extended to require the distribution of the n values of $(x, y)_i$ to be unchanged by arbitrary permutations of the indexes. It is *always* appropriate to assume an exchangeable model after incorporating sufficient relevant information in X that the indexes can be thought of as randomly assigned. It follows from the assumption of exchangeability that the distribution of y, given x, is the same for all units in the study in the sense that if two units have the same value of x, then their distributions of y are the same. Any of the explanatory variables x can of course be moved into the y category if we wish to model them. We discuss the role of explanatory variables (also called predictors) in detail in Chapter 7 in the context of analyzing surveys, experiments, and observational studies, and in Parts IV and V in the context of regression models.

Hierarchical modeling

In Chapter 5 and subsequent chapters, we focus on *hierarchical models* (also called *multilevel models*), which are used when information is available on several different levels of observational units. In a hierarchical model, it is possible to speak of exchangeability at each level of units. For example, suppose two medical treatments are applied, in separate randomized experiments, to patients in several different cities. Then, if no other information were available, it would be reasonable to treat the patients within each city as exchangeable and also treat the results from different cities as themselves exchangeable. In practice it would make sense to include, as explanatory variables at the city level, whatever relevant information we have on each city, as well as the explanatory variables mentioned before at the individual level, and then the conditional distributions given these explanatory variables would be exchangeable.

1.3 Bayesian inference

Bayesian statistical conclusions about a parameter θ, or unobserved data $\tilde{y}$, are made in terms of *probability* statements. These probability statements are conditional on the observed value of y, and in our notation are written simply as $p(\theta|y)$ or $p(\tilde{y}|y)$. We also implicitly condition on the known values of any covariates, x. It is at the fundamental level of conditioning on observed data that Bayesian inference departs from the approach to statistical inference described in many textbooks, which is based on a retrospective evaluation of the procedure used to estimate θ (or $\tilde{y}$) over the distribution of possible y values conditional on the true unknown value of θ. Despite this difference, it will be seen that in many simple analyses, superficially similar conclusions result from the two approaches to statistical inference. However, analyses obtained using Bayesian methods can be easily extended to more complex problems. In this section, we present the basic mathematics and notation of Bayesian inference, followed in the next section by an example from genetics.

Probability notation

Some comments on notation are needed at this point. First, $p(\cdot|\cdot)$ denotes a conditional probability density with the arguments determined by the context, and similarly for $p(\cdot)$, which denotes a marginal distribution. We use the terms 'distribution' and 'density' interchangeably. The same notation is used for continuous density functions and discrete probability mass functions. Different distributions in the same equation (or expression) will each be denoted by $p(\cdot)$, as in (1.1) below, for example. Although an abuse of standard mathematical notation, this method is compact and similar to the standard practice of using $p(\cdot)$ for the probability of any discrete event, where the sample space is also suppressed in the notation. Depending on context, to avoid confusion, we may use the notation $\Pr(\cdot)$ for the probability of an event; for example, $\Pr(\theta > 2) = \int_{\theta>2} p(\theta)d\theta$. When using a standard distribution, we use a notation based on the name of the distribution; for example, if θ has a normal distribution with mean μ and variance σ^2, we write $\theta \sim N(\mu, \sigma^2)$ or $p(\theta) = N(\theta|\mu, \sigma^2)$ or, to be even more explicit, $p(\theta|\mu, \sigma^2) = N(\theta|\mu, \sigma^2)$. Throughout, we use notation such as $N(\mu, \sigma^2)$ for random variables and $N(\theta|\mu, \sigma^2)$ for density functions. Notation and formulas for several standard distributions appear in Appendix A.

We also occasionally use the following expressions for all-positive random variables θ: the *coefficient of variation* (CV) is defined as $sd(\theta)/E(\theta)$, the *geometric mean* (GM) is $\exp(E[\log(\theta)])$, and the geometric standard deviation (GSD) is $\exp(sd[\log(\theta)])$.

Bayes' rule

In order to make probability statements about θ given y, we must begin with a *model* providing a *joint probability distribution* for θ and y. The joint probability mass or density function can be written as a product of two densities that are often referred to as the *prior distribution* $p(\theta)$ and the *sampling distribution* (or *data distribution*) $p(y|\theta)$ respectively:

$$p(\theta, y) = p(\theta)p(y|\theta).$$

Simply conditioning on the known value of the data y, using the basic property of conditional probability known as Bayes' rule, yields the *posterior* density:

$$p(\theta|y) = \frac{p(\theta, y)}{p(y)} = \frac{p(\theta)p(y|\theta)}{p(y)}, \tag{1.1}$$

where $p(y) = \sum_{\theta} p(\theta)p(y|\theta)$, and the sum is over all possible values of θ (or $p(y) = \int p(\theta)p(y|\theta)d\theta$ in the case of continuous θ). An equivalent form of (1.1) omits the factor $p(y)$, which does not depend on θ and, with fixed y, can thus be considered a constant, yielding the *unnormalized posterior density* , which is the right side of (1.2):

$$p(\theta|y) \propto p(\theta)p(y|\theta). \tag{1.2}$$

These simple expressions encapsulate the technical core of Bayesian inference: the primary task of any specific application is to develop the model $p(\theta, y)$ and perform the necessary computations to summarize $p(\theta|y)$ in appropriate ways.

Prediction

To make inferences about an unknown observable, often called predictive inferences, we follow a similar logic. Before the data y are considered, the distribution of the unknown but observable y is

$$p(y) = \int p(y, \theta)d\theta = \int p(\theta)p(y|\theta)d\theta. \tag{1.3}$$

This is often called the marginal distribution of y, but a more informative name is the *prior predictive distribution*: prior because it is not conditional on a previous observation of the process, and predictive because it is the distribution for a quantity that is observable.

After the data y have been observed, we can predict an unknown observable, $\tilde{y}$, from the same process. For example, $y = (y_1, \ldots, y_n)$ may be the vector of recorded weights of an object weighed n times on a scale, $\theta = (\mu, \sigma^2)$ may be the unknown true weight of the object and the measurement variance of the scale, and $\tilde{y}$ may be the yet to be recorded weight of the object in a planned new weighing. The distribution of $\tilde{y}$ is called the *posterior predictive distribution*, posterior because it is conditional on the observed y and predictive because it is a prediction for an observable $\tilde{y}$:

$$\begin{aligned} p(\tilde{y}|y) &= \int p(\tilde{y}, \theta|y)d\theta \\ &= \int p(\tilde{y}|\theta, y)p(\theta|y)d\theta \\ &= \int p(\tilde{y}|\theta)p(\theta|y)d\theta. \end{aligned} \tag{1.4}$$

The second and third lines display the posterior predictive distribution as an average of conditional predictions over the posterior distribution of θ. The last equation follows because y and $\tilde{y}$ are conditionally independent given θ in this model.

Likelihood

Using Bayes' rule with a chosen probability model means that the data y affect the posterior inference (1.2) *only* through the function $p(y|\theta)$, which, when regarded as a function of θ, for fixed y, is called the *likelihood function*. In this way Bayesian inference obeys what is sometimes called the *likelihood principle*, which states that for a given sample of data, any two probability models $p(y|\theta)$ that have the same likelihood function yield the same inference for θ.

The likelihood principle is reasonable, but only within the framework of the model or family of models adopted for a particular analysis. In practice, one can rarely be confident that the chosen model is *the* correct model. We shall see in Chapter 6 that sampling distributions (imagining repeated realizations of our data) can play an important role in checking model assumptions. In fact, our view of an applied Bayesian statistician is one who is willing to apply Bayes' rule under a variety of possible models.

Likelihood and odds ratios

The ratio of the posterior density $p(\theta|y)$ evaluated at the points θ_1 and θ_2 under a given model is called the posterior *odds* for θ_1 compared to θ_2. The most familiar application of this concept is with discrete parameters, with θ_2 taken to be the complement of θ_1. Odds provide an alternative representation of probabilities and have the attractive property that Bayes' rule takes a particularly simple form when expressed in terms of them:

$$\frac{p(\theta_1|y)}{p(\theta_2|y)} = \frac{p(\theta_1)p(y|\theta_1)/p(y)}{p(\theta_2)p(y|\theta_2)/p(y)} = \frac{p(\theta_1)}{p(\theta_2)} \frac{p(y|\theta_1)}{p(y|\theta_2)}. \tag{1.5}$$

In words, the posterior odds are equal to the prior odds multiplied by the *likelihood ratio*, $p(y|\theta_1)/p(y|\theta_2)$.

1.4 Example: inference about a genetic probability

The following example is not typical of *statistical* applications of the Bayesian method, because it deals with a very small amount of data and concerns a single individual's state (gene carrier or not) rather than with the estimation of a parameter that describes an entire population. Nevertheless it is a real example of the very simplest type of Bayesian calculation, where the estimand and the individual item of data each have only two possible values.

Human males have one X-chromosome and one Y-chromosome, whereas females have two X-chromosomes, each chromosome being inherited from one parent. Hemophilia is a disease that exhibits X-chromosome-linked recessive inheritance, meaning that a male who inherits the gene that causes the disease on the X-chromosome is affected, whereas a female carrying the gene on only one of her two X-chromosomes is not affected. The disease is generally fatal for women who inherit two such genes, and this is very rare, since the frequency of occurrence of the gene is low in human populations.

The prior distribution

Consider a woman who has an affected brother, which implies that her mother must be a carrier of the hemophilia gene with one 'good' and one 'bad' hemophilia gene. We are also told that her father is not affected; thus the woman herself has a fifty-fifty chance of having the gene. The unknown quantity of interest, the state of the woman, has just two values: the woman is either a carrier of the gene ($\theta = 1$) or not ($\theta = 0$). Based on the information provided thus far, the prior distribution for the unknown θ can be expressed simply as $\Pr(\theta = 1) = \Pr(\theta = 0) = \frac{1}{2}$.

The model and likelihood

The data used to update this prior information consist of the affection status of the woman's sons. Suppose she has two sons, neither of whom is affected. Let $y_i = 1$ or 0 denote an affected or unaffected son, respectively. The outcomes of the two sons are exchangeable and, conditional on the unknown θ, are independent; we assume the sons are not identical twins. The two items of independent data generate the following likelihood function:

$$\Pr(y_1 = 0, y_2 = 0 \,|\, \theta = 1) \quad = \quad (0.5)(0.5) = 0.25$$
$$\Pr(y_1 = 0, y_2 = 0 \,|\, \theta = 0) \quad = \quad (1)(1) = 1.$$

These expressions follow from the fact that if the woman is a carrier, then each of her sons will have a 50% chance of inheriting the gene and so being affected, whereas if she is not a carrier then there is a probability very close to 1 that a son of hers will be unaffected. (In fact, there is a nonzero probability of being affected even if the mother is not a carrier, but this risk—the mutation rate—is very small and can be ignored for this example.)

The posterior distribution

Bayes' rule can now be used to combine the information in the data with the prior probability; in particular, interest is likely to focus on the posterior probability that the woman is a carrier. Using y to denote the joint data (y_1, y_2), this is simply

$$\Pr(\theta = 1|y) \quad = \quad \frac{p(y|\theta = 1)\Pr(\theta = 1)}{p(y|\theta = 1)\Pr(\theta = 1) + p(y|\theta = 0)\Pr(\theta = 0)}$$
$$= \quad \frac{(0.25)(0.5)}{(0.25)(0.5) + (1.0)(0.5)} = \frac{0.125}{0.625} = 0.20.$$

Intuitively it is clear that if a woman has unaffected children, it is less probable that she is a carrier, and Bayes' rule provides a formal mechanism for determining the extent of the correction. The results can also be described in terms of prior and posterior odds. The prior odds of the woman being a carrier are $0.5/0.5 = 1$. The likelihood ratio based on the information about her two unaffected sons is $0.25/1 = 0.25$, so the posterior odds are obtained very simply as 0.25. Converting back to a probability, we obtain $0.25/(1+0.25) = 0.2$, just as before.

Adding more data

A key aspect of Bayesian analysis is the ease with which sequential analyses can be performed. For example, suppose that the woman has a third son, who is also unaffected. The entire calculation does not need to be redone; rather we use the previous posterior distribution as the new prior distribution, to obtain:

$$\Pr(\theta = 1|y_1, y_2, y_3) = \frac{(0.5)(0.20)}{(0.5)(0.20) + (1)(0.8)} = 0.111.$$

Alternatively, if we suppose that the third son is affected, it is easy to check that the posterior probability of the woman being a carrier becomes 1 (again ignoring the possibility of a mutation).

1.5 Probability as a measure of uncertainty

We have already used concepts such as probability density, and indeed we assume that the reader has a fair degree of familiarity with basic probability theory (although in Section 1.8 we provide a brief technical review of some probability calculations that often arise in Bayesian analysis). But since the uses of probability within a Bayesian framework are much broader than within non-Bayesian statistics, it is important to consider at least briefly the foundations of the concept of probability before considering more detailed statistical examples. We take for granted a common understanding on the part of the reader of the mathematical definition of probability: that probabilities are numerical quantities, defined on a set of 'outcomes,' that are nonnegative, additive over mutually exclusive outcomes, and sum to 1 over all possible mutually exclusive outcomes.

In Bayesian statistics, probability is used as the fundamental measure or yardstick of uncertainty. Within this paradigm, it is equally legitimate to discuss the probability of 'rain tomorrow' or of a Brazilian victory in the soccer World Cup as it is to discuss the probability that a coin toss will land heads. Hence, it becomes as natural to consider the probability that an unknown estimand lies in a particular range of values as it is to consider the probability that the mean of a random sample of 10 items from a known fixed population of size 100 will lie in a certain range. The first of these two probabilities is of more interest after data have been acquired whereas the second is more relevant beforehand. Bayesian methods enable statements to be made about the partial knowledge available (based on data) concerning some situation or 'state of nature' (unobservable or as yet unobserved) in a systematic way, using probability as the yardstick. The guiding principle is that the state of knowledge about anything unknown is described by a probability distribution.

What is meant by a numerical measure of uncertainty? For example, the probability of 'heads' in a coin toss is widely agreed to be $\frac{1}{2}$. Why is this so? Two justifications seem to be commonly given:

1. Symmetry or exchangeability argument:

$$\text{probability} = \frac{\text{number of favorable cases}}{\text{number of possibilities}},$$

   assuming equally likely possibilities. For a coin toss this is really a physical argument, based on assumptions about the forces at work in determining the manner in which the coin will fall, as well as the initial physical conditions of the toss.

2. Frequency argument: probability = relative frequency obtained in a very long sequence of tosses, assumed to be performed in an identical manner, physically independently of each other.

Both the above arguments are in a sense subjective, in that they require judgments about the nature of the coin and the tossing procedure, and both involve semantic arguments about the meaning of equally likely events, identical measurements, and independence. The frequency argument may be perceived to have certain special difficulties, in that it involves the hypothetical notion of a very long sequence of identical tosses. If taken strictly, this point of view does not allow a statement of probability for a single coin toss that does not happen to be embedded, at least conceptually, in a long sequence of identical events.

The following examples illustrate how probability judgments can be increasingly subjective. First, consider the following modified coin experiment. Suppose that a particular coin is stated to be either double-headed *or* double-tailed, with no further information provided. Can one still talk of the probability of heads? It seems clear that in common parlance one certainly can. It is less clear, perhaps, how to assess this new probability, but many would agree on the same value of $\frac{1}{2}$, perhaps based on the exchangeability of the labels 'heads' and 'tails.'

Now consider some further examples. Suppose Colombia plays Brazil in soccer tomorrow: what is the probability of Colombia winning? What is the probability of rain tomorrow? What is the probability that Colombia wins, if it rains tomorrow? What is the probability that the next space shuttle launched will explode? Although each of these questions seems reasonable in a common-sense way, it is difficult to contemplate strong frequency interpretations for the probabilities being referenced. Frequency interpretations can usually be *constructed*, however, and this is an extremely useful tool in statistics. For example, we can consider the next space shuttle launch as a sample from the population of potential space shuttle launches, and look at the frequency of past shuttle launches that have exploded (see the bibliographic note at the end of this chapter for more details on this example). Doing this sort of thing scientifically means creating a probability model (or, at the very least, a 'reference set' of comparable events), and this brings us back to a situation analogous to the simple coin toss, where we must consider the outcomes in question as exchangeable and thus equally likely.

Why is probability a reasonable way of quantifying uncertainty? The following reasons are often advanced.

1. By analogy: physical randomness induces uncertainty, so it seems reasonable to describe uncertainty in the language of random events. Common speech uses many terms such as 'probably' and 'unlikely,' and it appears consistent with such usage to extend a more formal probability calculus to problems of scientific inference.

2. Axiomatic or normative approach: related to decision theory, this approach places all statistical inference in the context of decision-making with gains and losses. Then reasonable axioms (ordering, transitivity, and so on) imply that uncertainty *must* be represented in terms of probability. We view this normative rationale as suggestive but not compelling.

3. Coherence of bets. *Define* the probability p attached (by you) to an event E as the fraction ($p \in [0, 1]$) at which you would exchange (that is, bet) \$$p$ for a return of \$1 if E occurs. That is, if E occurs, you gain \$$(1 - p)$; if the complement of E occurs, you lose \$$p$. For example:

   • Coin toss: thinking of the coin toss as a fair bet suggests even odds corresponding to $p = \frac{1}{2}$.

   • Odds for a game: if you are willing to bet on team A to win a game at 10 to 1 odds against team B (that is, you bet 1 to win 10), your 'probability' for team A winning is at least $1/11$.

The principle of coherence of probabilities states that your assignment of probabilities to all possible events should be such that it is not possible to make a definite gain by betting with you. It can be proved that probabilities constructed under this principle must satisfy the basic axioms of probability theory.

The betting rationale has some fundamental difficulties:

• Exact odds are required, on which you would be willing to bet in either direction, for all events. How can you assign exact odds if you are not sure?

• If a person is willing to bet with you, and has information you do not, it might not be wise for you to take the bet. In practice, probability is an incomplete (necessary but not sufficient) guide to betting.

All of these considerations suggest that probabilities may be a reasonable approach to summarizing uncertainty in applied statistics, but the ultimate proof is in the success of the applications. The remaining chapters of this book demonstrate that probability provides a rich and flexible framework for handling uncertainty in statistical applications.

Subjectivity and objectivity

All statistical methods that use probability are subjective in the sense of relying on mathematical idealizations of the world. Bayesian methods are sometimes said to be especially subjective because of their reliance on a prior distribution, but in most problems, scientific judgment is necessary to specify both the 'likelihood' and the 'prior' parts of the model. For example, linear regression models are generally at least as suspect as any prior distribution that might be assumed about the regression parameters. A general principle is at work here: whenever there is replication, in the sense of many exchangeable units observed, there is scope for estimating features of a probability distribution from data and thus making the analysis more 'objective.' If an experiment as a whole is replicated several times, then the parameters of the prior distribution can themselves be estimated from data, as discussed in Chapter 5. In any case, however, certain elements requiring scientific judgment will remain, notably the choice of data included in the analysis, the parametric forms assumed for the distributions, and the ways in which the model is checked.

1.6 Example of probability assignment: football point spreads

As an example of how probabilities might be assigned using empirical data and plausible substantive assumptions, we consider methods of estimating the probabilities of certain outcomes in professional (American) football games. This is an example only of probability assignment, not of Bayesian inference. A number of approaches to assigning probabilities for football game outcomes are illustrated: making subjective assessments, using empirical probabilities based on observed data, and constructing a parametric probability model.

Football point spreads and game outcomes

Football experts provide a *point spread* for every football game as a measure of the difference in ability between the two teams. For example, team A might be a 3.5-point favorite to defeat team B. The implication of this point spread is that the proposition that team A, the favorite, defeats team B, the underdog, by 4 or more points is considered a fair bet; in other words, the probability that A wins by more than 3.5 points is $\frac{1}{2}$. If the point spread is an integer, then the implication is that team A is as likely to win by more points than the point spread as it is to win by fewer points than the point spread (or to lose); there is positive probability that A will win by exactly the point spread, in which case neither side is paid off. The assignment of point spreads is itself an interesting exercise in probabilistic reasoning; one interpretation is that the

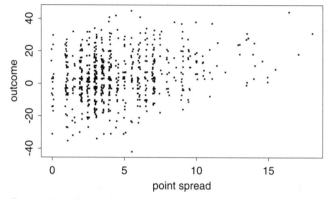

Figure 1.1 *Scatterplot of actual outcome vs. point spread for each of 672 professional football games. The x and y coordinates are jittered by adding uniform random numbers to each point's coordinates (between −0.1 and −0.1 for the x coordinate; between −0.2 and 0.2 for the y coordinate) in order to display multiple values but preserve the discrete-valued nature of each.*

point spread is the median of the distribution of the gambling population's beliefs about the possible outcomes of the game. For the rest of this example, we treat point spreads as given and do not worry about how they were derived.

The point spread and actual game outcome for 672 professional football games played during the 1981, 1983, and 1984 seasons are graphed in Figure 1.1. (Much of the 1982 season was canceled due to a labor dispute.) Each point in the scatterplot displays the point spread, x, and the actual outcome (favorite's score minus underdog's score), y. (In games with a point spread of zero, the labels 'favorite' and 'underdog' were assigned at random.) A small random jitter is added to the x and y coordinate of each point on the graph so that multiple points do not fall exactly on top of each other.

Assigning probabilities based on observed frequencies

It is of interest to assign probabilities to particular events: Pr(favorite wins), Pr(favorite wins | point spread is 3.5 points), Pr(favorite wins by more than the point spread), Pr(favorite wins by more than the point spread | point spread is 3.5 points), and so forth. We might report a subjective probability based on informal experience gathered by reading the newspaper and watching football games. The probability that the favored team wins a game should certainly be greater than 0.5, perhaps between 0.6 and 0.75? More complex events require more intuition or knowledge on our part. A more systematic approach is to assign probabilities based on the data in Figure 1.1. Counting a tied game as one-half win and one-half loss, and ignoring games for which the point spread is zero (and thus there is no favorite), we obtain empirical estimates such as:

- Pr(favorite wins) = $\frac{410.5}{655}$ = 0.63

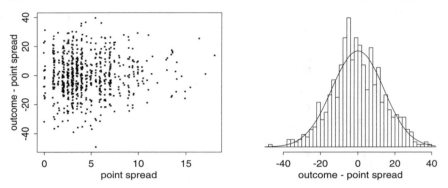

Figure 1.2 *(a) Scatterplot of (actual outcome − point spread) vs. point spread for each of 672 professional football games (with uniform random jitter added to x and y coordinates). (b) Histogram of the differences between the game outcome and the point spread, with the $N(0, 14^2)$ density superimposed.*

- $\Pr(\text{favorite wins} \mid x = 3.5) = \frac{36}{59} = 0.61$
- $\Pr(\text{favorite wins by more than the point spread}) = \frac{308}{655} = 0.47$
- $\Pr(\text{favorite wins by more than the point spread} \mid x = 3.5) = \frac{32}{59} = 0.54.$

These empirical probability assignments all seem sensible in that they match the intuition of knowledgeable football fans. However, such probability assignments are problematic for events with few directly relevant data points. For example, 8.5-point favorites won five out of five times during this three-year period, whereas 9-point favorites won thirteen out of twenty times. However, we realistically expect the probability of winning to be greater for a 9-point favorite than for an 8.5-point favorite. The small sample size with point spread 8.5 leads to very imprecise probability assignments. We consider an alternative method using a parametric model.

A parametric model for the difference between game outcome and point spread

Figure 1.2a displays the differences $y − x$ between the observed game outcome and the point spread, plotted versus the point spread, for the games in the football dataset. (Once again, random jitter was added to both coordinates.) This plot suggests that it may be roughly reasonable to model the distribution of $y − x$ as independent of x. (See Exercise 6.16.) Figure 1.2b is a histogram of the differences $y − x$ for all the football games, with a fitted normal density superimposed. This plot suggests that it may be reasonable to approximate the marginal distribution of the random variable $d = y − x$ by a normal distribution. The sample mean of the 672 values of d is 0.07, and the sample standard deviation is 13.86, suggesting that the results of football games are approximately normal with mean equal to the point spread and standard deviation nearly 14 points (two converted touchdowns). For the remainder of the discussion we take the distribution of d to be independent of x and normal with mean zero and standard deviation 14 for each x; that is,

$$d|x \sim N(0, 14^2),$$

as displayed in Figure 1.2b. We return to this example in Sections 2.7 and 3.4 to estimate the parameters of this normal distribution using Bayesian methods. The assigned probability model is not perfect: it does not fit the data exactly, and, as is often the case with real data, neither football scores nor point spreads are continuous-valued quantities.

Assigning probabilities using the parametric model

Nevertheless, the model provides a convenient approximation that can be used to assign probabilities to events. If d has a normal distribution with mean zero and is independent of the point spread, then the probability that the favorite wins by more than the point spread is $\frac{1}{2}$, conditional on any value of the point spread, and therefore unconditionally as well. Denoting probabilities obtained by the normal model as $\Pr_{\text{norm}}$, the probability that an x-point favorite wins the game can be computed, assuming the normal model, as follows:

$$\Pr_{\text{norm}}(y > 0 \,|\, x) = \Pr_{\text{norm}}(d > -x \,|\, x) = 1 - \Phi\left(-\frac{x}{14}\right),$$

where Φ is the standard normal cumulative distribution function. For example,

- $\Pr_{\text{norm}}(\text{favorite wins} \,|\, x = 3.5) = 0.60$
- $\Pr_{\text{norm}}(\text{favorite wins} \,|\, x = 8.5) = 0.73$
- $\Pr_{\text{norm}}(\text{favorite wins} \,|\, x = 9.0) = 0.74$.

The probability for a 3.5-point favorite agrees with the empirical value given earlier, whereas the probabilities for 8.5- and 9-point favorites make more intuitive sense than the empirical values based on small samples.

1.7 Example of probability assignment: estimating the accuracy of record linkage

We emphasize the essentially empirical (not 'subjective' or 'personal') nature of probabilities with another example in which they are estimated from data.

Record linkage refers to the use of an algorithmic technique to identify records from different databases that correspond to the same individual. Record-linkage techniques are used in a variety of settings. The work described here was formulated and first applied in the context of record linkage between the U.S. Census and a large-scale post-enumeration survey, which is the first step of an extensive matching operation conducted to evaluate census coverage for subgroups of the population. The goal of this first step is to declare as many records as possible 'matched' by computer without an excessive rate of error, thereby avoiding the cost of the resulting manual processing for all records not declared 'matched.'

Existing methods for assigning scores to potential matches

Much attention has been paid in the record-linkage literature to the problem of assigning 'weights' to individual fields of information in a multivariate record and obtaining a composite 'score,' which we call y, that summarizes the closeness of agreement between two records. Here, we assume that this step is complete in the sense that these rules have been chosen. The next step is the assignment of candidate matched pairs, where each pair of records consists of the best potential match for each other from the respective data bases. The specified weighting rules then order the candidate matched pairs. In the motivating problem at the Census Bureau, a binary choice is made between the alternatives 'declare matched' vs. 'send to followup,' where a cutoff score is needed above which records are declared matched. The false-match rate is then defined as the number of falsely matched pairs divided by the number of declared matched pairs.

Particularly relevant for any such decision problem is an accurate method for assessing the probability that a candidate matched pair is a correct match as a function of its score. Simple methods exist for converting the scores into probabilities, but these lead to extremely inaccurate, typically grossly optimistic, estimates of false-match rates. For example, a manual check of a set of records with nominal false-match probabilities ranging from 10^{-3} to 10^{-7} (that is, pairs deemed almost certain to be matches) found actual false-match rates closer to the 1% range. Records with nominal false-match probabilities of 1% had an actual false-match rate of 5%.

We would like to use Bayesian methods to recalibrate these to obtain objective probabilities of matching for a given decision rule—in the same way that in the football example, we used past data to estimate the probabilities of different game outcomes conditional on the point spread. Our approach is to work with the scores y and empirically estimate the probability of a match as a function of y.

Estimating match probabilities empirically

We obtain accurate match probabilities using mixture modeling, a topic we discuss in detail in Chapter 18. The distribution of previously-obtained scores for the candidate matches is considered a 'mixture' of a distribution of scores for true matches and a distribution for non-matches. The parameters of the mixture model are estimated from the data. The estimated parameters allow us to calculate an estimate of the probability of a false match (a pair declared matched that is not a true match) for any given decision threshold on the scores. In the procedure that was actually used, some elements of the mixture model (for example, the optimal transformation required to allow a mixture of normal distributions to apply) were fit using 'training' data with known match status (separate from the data to which we apply our calibration procedure), but we do not describe those details here. Instead we focus on how the method would be used with a set of data with unknown match status.

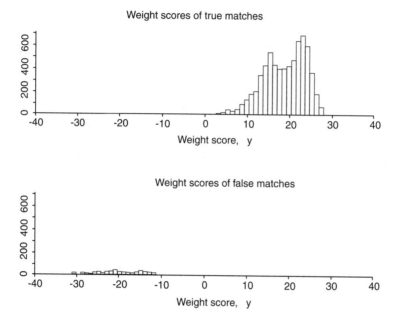

Figure 1.3 *Histograms of weight scores y for true and false matches in a sample of records from the 1988 test Census. Most of the matches in the sample are true (because a pre-screening process has already picked these as the best potential match for each case), and the two distributions are mostly, but not completely, separated.*

Support for this approach is provided in Figure 1.3, which displays the distribution of scores for the matches and non-matches in a particular data set obtained from 2300 records from a 'test Census' survey conducted in a single local area two years before the 1990 Census. The two distributions, $p(y|\text{match})$ and $p(y|\text{non-match})$, are mostly distinct—meaning that in most cases it is possible to identify a candidate as a match or not given the score alone—but with some overlap.

In our application dataset, we do not know the match status. Thus we are faced with a single combined histogram from which we estimate the two component distributions and the proportion of the population of scores that belong to each component. Under the mixture model, the distribution of scores can be written as,

$$p(y) = \Pr(\text{match})\, p(y|\text{match}) + \Pr(\text{non-match})\, p(y|\text{non-match}). \quad (1.6)$$

The mixture probability ($\Pr(\text{match})$) and the parameters of the distributions of matches ($p(y|\text{match})$) and non-matches ($p(y|\text{non-match})$) are estimated using the mixture model approach (as described in Chapter 18) applied to the combined histogram from the data with unknown match status.

To use the method to make record-linkage decisions, we construct a curve giving the false-match rate as a function of the decision threshold, the score

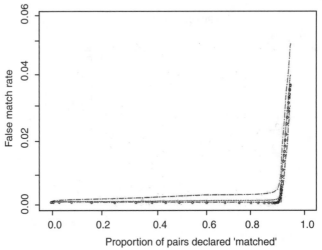

Figure 1.4 *Lines show expected false-match rate (and 95% bounds) as a function of the proportion of cases declared matches, based on the mixture model for record linkage. Dots show the actual false-match rate for the data.*

above which pairs will be 'declared' a match. For a given decision threshold, the probability distributions in (1.6) can be used to estimate the probability of a false match, a score y above the threshold originating from the distribution $p(y|\text{non-match})$. The lower the threshold, the more pairs we will declare as matches. As we declare more matches, the proportion of errors increases. The approach described here should provide an objective error estimate for each threshold. (See the validation in the next paragraph.) Then a decision maker can determine the threshold that provides an acceptable balance between the goals of declaring more matches automatically (thus reducing the clerical labor) and making fewer mistakes.

External validation of the probabilities using test data

The approach described above was externally validated using data for which the match status is known. The method was applied to data from three different locations of the 1988 test Census, and so three tests of the methods were possible. We provide detailed results for one; results for the other two were similar. The mixture model was fitted to the scores of all the candidate pairs at a test site. Then the estimated model was used to create the lines in Figure 1.4, which show the expected false-match rate (and uncertainty bounds) in terms of the proportion of cases declared matched, as the threshold varies from very high (thus allowing no matches) to very low (thus declaring almost all the candidate pairs to be matches). The false-match proportion is an increasing function of the number of declared matches, which makes sense: as

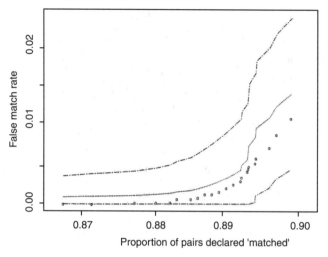

Figure 1.5 *Expansion of Figure 1.4 in the region where the estimated and actual match rates change rapidly. In this case, it would seem a good idea to match about 88% of the cases and send the rest to followup.*

we move rightward on the graph, we are declaring weaker and weaker cases to be matches.

The lines on Figure 1.4 display the expected proportion of false matches and 95% posterior bounds for the false-match rate as estimated from the model. (These bounds give the estimated range within which there is 95% posterior probability that the false-match rate lies. The concept of posterior intervals is discussed in more detail in the next chapter.) The dots in the graph display the actual false-match proportions, which track well with the model. In particular, the model would suggest a recommendation of declaring something less than 90% of cases as matched and giving up on the other 10% or so, so as to avoid most of the false matches, and the dots show a similar pattern.

It is clearly possible to match large proportions of the files with little or no error. Also, the quality of candidate matches becomes dramatically worse at some point where the false-match rate accelerates. Figure 1.5 takes a magnifying glass to the previous display to highlight the behavior of the calibration procedure in the region of interest where the false-match rate accelerates. The predicted false-match rate curves bend upward, close to the points where the observed false-match rate curves rise steeply, which is a particularly encouraging feature of the calibration method. The calibration procedure performs well from the standpoint of providing predicted probabilities that are close to the true probabilities and interval estimates that are informative and include the true values. By comparison, the original estimates of match probabilities, constructed by multiplying weights without empirical calibration, were highly inaccurate.

1.8 Some useful results from probability theory

We assume the reader is familiar with elementary manipulations involving probabilities and probability distributions. In particular, basic probability background that must be well understood for key parts of the book includes the manipulation of joint densities, the definition of simple moments, the transformation of variables, and methods of simulation. In this section we briefly review these assumed prerequisites and clarify some further notational conventions used in the remainder of the book. Appendix A provides information on some commonly used probability distributions.

As introduced in Section 1.3, we generally represent joint distributions by their joint probability mass or density function, with dummy arguments reflecting the name given to each variable being considered. Thus for two quantities u and v, we write the joint density as $p(u, v)$; if specific values need to be referenced, this notation will be further abused as with, for example, $p(u, v = 1)$.

In Bayesian calculations relating to a joint density $p(u, v)$, we will often refer to a *conditional* distribution or density function such as $p(u|v)$ and a *marginal* density such as $p(u) = \int p(u, v)dv$. In this notation, either or both u and v can be vectors. Typically it will be clear from the context that the range of integration in the latter expression refers to the entire range of the variable being integrated out. It is also often useful to *factor* a joint density as a product of marginal and conditional densities; for example, $p(u, v, w) = p(u|v, w)p(v|w)p(w)$.

Some authors use different notations for distributions on parameters and observables—for example, $\pi(\theta), f(y|\theta)$—but this obscures the fact that all probability distributions have the same *logical* status in Bayesian inference. We must always be careful, though, to indicate appropriate conditioning; for example, $p(y|\theta)$ is different from $p(y)$. In the interests of conciseness, however, our notation hides the conditioning on hypotheses that hold throughout—no probability judgments can be made in a vacuum—and to be more explicit one might use a notation such as the following:

$$p(\theta, y|H) = p(\theta|H)p(y|\theta, H),$$

where H refers to the set of hypotheses or assumptions used to define the model. Also, we sometimes suppress explicit conditioning on known explanatory variables, x.

We use the standard notations, $E(\cdot)$ and $var(\cdot)$, for mean and variance, respectively:

$$E(u) = \int up(u)du, \quad var(u) = \int (u - E(u))^2 p(u)du.$$

For a vector parameter u, the expression for the mean is the same, and the covariance matrix is defined as

$$var(u) = \int (u - E(u))(u - E(u))^T p(u)du,$$

where u is considered a column vector. (We use the terms 'variance matrix' and 'covariance matrix' interchangeably.) This notation is slightly imprecise, because $E(u)$ and $var(u)$ are really functions of the distribution function, $p(u)$, not of the variable u. In an expression involving an expectation, any variable that does not appear explicitly as a conditioning variable is assumed to be integrated out in the expectation; for example, $E(u|v)$ refers to the conditional expectation of u with v held fixed—that is, the conditional expectation as a function of v—whereas $E(u)$ is the expectation of u, averaging over v (as well as u).

Modeling using conditional probability

Useful probability models often express the distribution of observables conditionally or hierarchically rather than through more complicated unconditional distributions. For example, suppose y is the height of a university student selected at random. The marginal distribution $p(y)$ is (essentially) a mixture of two approximately normal distributions centered around 160 and 175 centimeters. A more useful description of the distribution of y would be based on the joint distribution of height and sex: $p(\text{male}) \approx p(\text{female}) \approx \frac{1}{2}$, along with the conditional specifications that $p(y|\text{female})$ and $p(y|\text{male})$ are each approximately normal with means 160 and 175 cm, respectively. If the conditional variances are not too large, the marginal distribution of y is bimodal. In general, we prefer to model complexity with a hierarchical structure using additional variables rather than with complicated marginal distributions, even when the additional variables are unobserved or even unobservable; this theme underlies mixture models, as discussed in Chapter 18. We repeatedly return to the theme of conditional modeling throughout the book.

Means and variances of conditional distributions

It is often useful to express the mean and variance of a random variable u in terms of the conditional mean and variance given some related quantity v. The mean of u can be obtained by averaging the conditional mean over the marginal distribution of v,

$$E(u) = E(E(u|v)), \qquad (1.7)$$

where the inner expectation averages over u, conditional on v, and the outer expectation averages over v. Identity (1.7) is easy to derive by writing the expectation in terms of the joint distribution of u and v and then factoring the joint distribution:

$$E(u) = \int\int u p(u,v) du dv = \int\int u\, p(u|v) du\, p(v) dv = \int E(u|v) p(v) dv.$$

The corresponding result for the variance includes two terms, the mean of the conditional variance and the variance of the conditional mean:

$$\text{var}(u) = \text{E}(\text{var}(u|v)) + \text{var}(\text{E}(u|v)). \tag{1.8}$$

This result can be derived by expanding the terms on the right side of (1.8):

$$
\begin{aligned}
\text{E}[\text{var}(u|v)] &+ \text{var}[\text{E}(u|v)] \\
&= \text{E}[\text{E}(u^2|v) - (\text{E}(u|v))^2] + \text{E}[(\text{E}(u|v))^2] - (\text{E}[\text{E}(u|v)])^2 \\
&= \text{E}(u^2) - \text{E}[(\text{E}(u|v))^2] + \text{E}[(\text{E}(u|v))^2] - (\text{E}(u))^2 \\
&= \text{E}(u^2) - (\text{E}(u))^2 = \text{var}(u).
\end{aligned}
$$

Identities (1.7) and (1.8) also hold if u is a vector, in which case $\text{E}(u)$ is a vector and $\text{var}(u)$ a matrix.

Transformation of variables

It is common to transform a probability distribution from one parameterization to another. We review the basic result here for a probability density on a transformed space. For clarity, we use subscripts here instead of our usual generic notation, $p(\cdot)$. Suppose $p_u(u)$ is the density of the vector u, and we transform to $v = f(u)$, where v has the same number of components as u.

If p_u is a discrete distribution, and f is a one-to-one function, then the density of v is given by

$$p_v(v) = p_u(f^{-1}(v)).$$

If f is a many-to-one function, then a sum of terms appears on the right side of this expression for $p_v(v)$, with one term corresponding to each of the branches of the inverse function.

If p_u is a continuous distribution, and $v = f(u)$ is a one-to-one transformation, then the joint density of the transformed vector is

$$p_v(v) = |J| \, p_u(f^{-1}(v))$$

where $|J|$ is the determinant of the Jacobian of the transformation $u = f^{-1}(v)$ as a function of v; the Jacobian J is the square matrix of partial derivatives (with dimension given by the number of components of u), with the (i,j)th entry equal to $\partial u_i / \partial v_j$. Once again, if f is many-to-one, then $p_v(v)$ is a sum or integral of terms.

In one dimension, we commonly use the logarithm to transform the parameter space from $(0, \infty)$ to $(-\infty, \infty)$. When working with parameters defined on the open unit interval, $(0, 1)$, we often use the logistic transformation:

$$\text{logit}(u) = \log \left(\frac{u}{1-u} \right), \tag{1.9}$$

whose inverse transformation is

$$\text{logit}^{-1}(v) = \frac{e^v}{1 + e^v}.$$

Another common choice is the probit transformation, $\Phi^{-1}(u)$, where Φ is the standard normal cumulative distribution function, to transform from $(0,1)$ to $(-\infty, \infty)$.

1.9 Summarizing inferences by simulation

Simulation forms a central part of much applied Bayesian analysis, because of the relative ease with which samples can often be generated from a probability distribution, even when the density function cannot be explicitly integrated. In performing simulations, it is helpful to consider the duality between a probability density function and a histogram of a set of random draws from the distribution: given a large enough sample, the histogram can provide practically complete information about the density, and in particular, various sample moments, percentiles, and other summary statistics provide estimates of any aspect of the distribution, to a level of precision that can be estimated. For example, to estimate the 95th percentile of the distribution of θ, draw a random sample of size L from $p(\theta)$ and use the $0.95L$th order statistic. For most purposes, $L = 1000$ is adequate for estimating the 95th percentile in this way.

Another advantage of simulation is that extremely large or small simulated values often flag a problem with model specification or parameterization (for example, see Figure 4.2) that might not be noticed if estimates and probability statements were obtained in analytic form.

Generating values from a probability distribution is often straightforward with modern computing techniques based on (pseudo)random number sequences. A well-designed pseudorandom number generator yields a deterministic sequence that appears to have the same properties as a sequence of independent random draws from the uniform distribution on $[0,1]$. Appendix A describes methods for drawing random samples from some commonly used distributions.

Sampling using the inverse cumulative distribution function.

As an introduction to the ideas of simulation, we describe a method for sampling from discrete and continuous distributions using the inverse cumulative distribution function. The *cumulative distribution function*, or *cdf*, F, of a one-dimensional distribution, $p(v)$, is defined by

$$F(v_*) \;=\; \Pr(v \leq v_*)$$

$$= \begin{cases} \sum_{v \leq v_*} p(v) & \text{if } p \text{ is discrete} \\ \int_{-\infty}^{v_*} p(v)dv & \text{if } p \text{ is continuous.} \end{cases}$$

The inverse cdf can be used to obtain random samples from the distribution p, as follows. First draw a random value, U, from the uniform distribution on $[0,1]$, using a table of random numbers or, more likely, a random number function on the computer. Now let $v = F^{-1}(U)$. The function F is not necessarily one-to-one—certainly not if the distribution is discrete—but $F^{-1}(U)$

Simulation draw	Parameters			Predictive quantities		
	θ_1	$\ldots$	θ_k	$\tilde{y}_1$	$\ldots$	$\tilde{y}_n$
1	θ_1^1	$\ldots$	θ_k^1	$\tilde{y}_1^1$	$\ldots$	$\tilde{y}_n^1$
$\vdots$	$\vdots$	$\ddots$	$\vdots$	$\vdots$	$\ddots$	$\vdots$
L	θ_1^L	$\ldots$	θ_k^L	$\tilde{y}_1^L$	$\ldots$	$\tilde{y}_n^L$

Table 1.1 *Structure of posterior and posterior predictive simulations. The super-scripts are indexes, not powers.*

is unique with probability 1. The value v will be a random draw from p, and is easy to compute as long as $F^{-1}(U)$ is simple. For a discrete distribution, F^{-1} can simply be tabulated.

For a continuous example, suppose v has an exponential distribution with parameter λ (see Appendix A); then its cdf is $F(v) = 1 - \exp(-\lambda v)$, and the value of v for which $U = F(v)$ is $v = -\log(1 - U)/\lambda$. Of course $1 - U$ also has the uniform distribution on $[0, 1]$, so we can obtain random draws from the exponential distribution as $-(\log U)/\lambda$. We discuss other methods of simulation in Part III of the book and Appendix A.

Simulation of posterior and posterior predictive quantities

In practice, we are most often interested in simulating draws from the posterior distribution of the model parameters θ, and perhaps from the posterior predictive distribution of unknown observables $\tilde{y}$. Results from a set of L simulation draws can be stored in the computer in an array, as illustrated in Table 1.1. We use the notation $l = 1, \ldots, L$ to index simulation draws; $(\theta^l, \tilde{y}^l)$ is the corresponding joint draw of parameters and predicted quantities from their joint posterior distribution.

From these simulated values, we can estimate the posterior distribution of any quantity of interest, such as θ_1/θ_3, by just computing a new column in Table 1.1 using the existing L draws of $(\theta, \tilde{y})$. We can estimate the posterior probability of any event, such as $\Pr(\tilde{y}_1 + \tilde{y}_2 > \exp(\theta_1))$, by the proportion of the L simulations for which it is true. We are often interested in posterior intervals; for example, the central 95% posterior interval $[a, b]$ for the parameter θ_j, for which $\Pr(\theta_j < a) = 0.025$ and $\Pr(\theta_j > b) = 0.025$. These values can be directly estimated by the appropriate simulated values of θ_j, for example, the 25th and 976th order statistics if $L = 1000$. We commonly summarize inferences by 50% and 95% intervals.

We return to the accuracy of simulation inferences in Section 10.2 after we have gained some experience using simulations of posterior distributions in some simple examples.

1.10 Computation and software

At the time of writing, the authors rely primarily on the statistical package R for graphs and basic simulations, fitting of classical simple models (including regression, generalized linear models, and nonparametric methods such as locally-weighted regression), optimization, and some simple programming.

We use the Bayesian inference package Bugs (using WinBugs run directly from within R; see Appendix C) as a first try for fitting most models. If there are difficulties in setting up the model or achieving convergence of the simulations, we explore the model more carefully in R and, if necessary for computational speed, a lower-level language such as Fortran or C (both of which can be linked from R). In any case, we typically work within R to plot and transform the data before model fitting, and to display inferences and model checks afterwards.

Specific computational tasks that arise in Bayesian data analysis include:

- Vector and matrix manipulations (see Table 1.1)

- Computing probability density functions (see Appendix A)

- Drawing simulations from probability distributions (see Appendix A for standard distributions and Exercise 1.9 for an example of a simple stochastic process)

- Structured programming (including looping and customized functions)

- Calculating the linear regression estimate and variance matrix (see Chapter 14)

- Graphics, including scatterplots with overlain lines and multiple graphs per page (see Chapter 6 for examples).

Our general approach to computation is to fit many models, gradually increasing the complexity. We do *not* recommend the strategy of writing a model and then letting the computer run overnight to estimate it perfectly. Rather, we prefer to fit each model relatively quickly, using inferences from the previously-fitted simpler models as starting values, and displaying inferences and comparing to data before continuing.

We discuss computation in detail in Part III of this book after first introducing the fundamental concepts of Bayesian modeling, inference, and model checking. Appendix C illustrates how to perform computations in R and Bugs in several different ways for a single example.

1.11 Bibliographic note

Several good introductory books have been written on Bayesian statistics, beginning with Lindley (1965). Berry (1996) presents, from a Bayesian perspective, many of the standard topics for an introductory statistics textbook. Congdon (2001, 2003) and Gill (2002) are recent introductory books on applied Bayesian statistics that use the statistical package Bugs. Carlin and

Louis (2001) cover the theory and applications of Bayesian inference, focusing on biological applications and connections to classical methods.

The bibliographic notes at the ends of the chapters in this book refer to a variety of specific applications of Bayesian data analysis. Several review articles in the statistical literature, such as Breslow (1990) and Racine et al. (1986), have appeared that discuss, in general terms, areas of application in which Bayesian methods have been useful. The volumes edited by Gatsonis et al. (1993–2002) are collections of Bayesian analyses, including extensive discussions about choices in the modeling process and the relations between the statistical methods and the applications.

The foundations of probability and Bayesian statistics are an important topic that we treat only very briefly. Bernardo and Smith (1994) give a thorough review of the foundations of Bayesian models and inference with a comprehensive list of references. Jeffreys (1961) is a self-contained book about Bayesian statistics that comprehensively presents an inductive view of inference; Good (1950) is another important early work. Jaynes (1983) is a collection of reprinted articles that present a deductive view of Bayesian inference, which we believe is quite similar to ours. Both Jeffreys and Jaynes focus on applications in the physical sciences. Jaynes (1996) focuses on connections between statistical inference and the philosophy of science and includes several examples of physical probability.

De Finetti (1974) is an influential work that focuses on the crucial role of exchangeability. More approachable discussions of the role of exchangeability in Bayesian inference are provided by Lindley and Novick (1981) and Rubin (1978a, 1987a). The non-Bayesian article by Draper et al. (1993) makes an interesting attempt to explain how exchangeable probability models can be justified in data analysis. Berger and Wolpert (1984) give a comprehensive discussion and review of the likelihood principle, and Berger (1985, Sections 1.6, 4.1, and 4.12) reviews a range of philosophical issues from the perspective of Bayesian decision theory.

Pratt (1965) and Rubin (1984) discuss the relevance of Bayesian methods for applied statistics and make many connections between Bayesian and non-Bayesian approaches to inference. Further references on the foundations of statistical inference appear in Shafer (1982) and the accompanying discussion. Kahneman, Slovic, and Tversky (1982) present the results of various psychological experiments that assess the meaning of 'subjective probability' as measured by people's stated beliefs and observed actions. Lindley (1971a) surveys many different statistical ideas, all from the Bayesian perspective. Box and Tiao (1973) is an early book on applied Bayesian methods. They give an extensive treatment of inference based on normal distributions, and their first chapter, a broad introduction to Bayesian inference, provides a good counterpart to Chapters 1 and 2 of this book.

The iterative process involving modeling, inference, and model checking that we present in Section 1.1 is discussed at length in the first chapter of Box and Tiao (1973) and also in Box (1980). Cox and Snell (1981) provide a more introductory treatment of these ideas from a less model-based perspective.

Many good books on the mathematical aspects of probability theory are available, such as Feller (1968) and Ross (1983); these are useful when constructing probability models and working with them. O'Hagan (1988) has written an interesting introductory text on probability from an explicitly Bayesian point of view.

Physical probability models for coin tossing are discussed by Keller (1986), Jaynes (1996), and Gelman and Nolan (2002b). The football example of Section 1.6 is discussed in more detail in Stern (1991); see also Harville (1980) and Glickman (1993) and Glickman and Stern (1998) for analyses of football scores not using the point spread. Related analyses of sports scores and betting odds appear in Stern (1997, 1998). For more background on sports betting, see Snyder (1975) and Rombola (1984).

An interesting real-world example of probability assignment arose with the explosion of the Challenger space shuttle in 1986; Martz and Zimmer (1992), Dalal, Fowlkes, and Hoadley (1989), and Lavine (1991) present and compare various methods for assigning probabilities for space shuttle failures. (At the time of writing we are not aware of similar contributions relating to the latest space shuttle tragedy in 2003.) The record-linkage example in Section 1.7 appears in Belin and Rubin (1995b), who discuss the mixture models and calibration techniques in more detail. The Census problem that motivated the record linkage is described by Hogan (1992).

In all our examples, probabilities are assigned using statistical modeling and estimation, not by 'subjective' assessment. Dawid (1986) provides a general discussion of probability assignment, and Dawid (1982) discusses the connections between calibration and Bayesian probability assignment.

The graphical method of jittering, used in Figures 1.1 and 1.2 and elsewhere in this book, is discussed in Chambers et al. (1983). For information on the statistical packages R and BUGS, see Becker, Chambers, and Wilks (1988), R Project (2002), Fox (2002), Venables and Ripley (2002), and Spiegelhalter et al. (1994, 2003).

1.12 Exercises

1. Conditional probability: suppose that if $\theta = 1$, then y has a normal distribution with mean 1 and standard deviation σ, and if $\theta = 2$, then y has a normal distribution with mean 2 and standard deviation σ. Also, suppose $\Pr(\theta = 1) = 0.5$ and $\Pr(\theta = 2) = 0.5$.

   (a) For $\sigma = 2$, write the formula for the marginal probability density for y and sketch it.

   (b) What is $\Pr(\theta = 1 | y = 1)$, again supposing $\sigma = 2$?

   (c) Describe how the posterior density of θ changes in shape as σ is increased and as it is decreased.

2. Conditional means and variances: show that (1.7) and (1.8) hold if u is a vector.

3. Probability calculation for genetics (from Lindley, 1965): suppose that in each individual of a large population there is a pair of genes, each of which can be either x or X, that controls eye color: those with xx have blue eyes, while heterozygotes (those with Xx or xX) and those with XX have brown eyes. The proportion of blue-eyed individuals is p^2 and of heterozygotes is $2p(1-p)$, where $0 < p < 1$. Each parent transmits one of its own genes to the child; if a parent is a heterozygote, the probability that it transmits the gene of type X is $\frac{1}{2}$. Assuming random mating, show that among brown-eyed children of brown-eyed parents, the expected proportion of heterozygotes is $2p/(1+2p)$. Suppose Judy, a brown-eyed child of brown-eyed parents, marries a heterozygote, and they have n children, all brown-eyed. Find the posterior probability that Judy is a heterozygote and the probability that her first grandchild has blue eyes.

4. Probability assignment: we will use the football dataset to estimate some conditional probabilities about professional football games. There were twelve games with point spreads of 8 points; the outcomes in those games were: $-7, -5, -3, -3, 1, 6, 7, 13, 15, 16, 20, 21$, with positive values indicating wins by the favorite and negative values indicating wins by the underdog. Consider the following conditional probabilities:

$$\Pr(\text{favorite wins} \,|\, \text{point spread} = 8),$$
$$\Pr(\text{favorite wins by at least } 8 \,|\, \text{point spread} = 8),$$
$$\Pr(\text{favorite wins by at least } 8 \,|\, \text{point spread} = 8 \text{ and favorite wins}).$$

   (a) Estimate each of these using the relative frequencies of games with a point spread of 8.

   (b) Estimate each using the normal approximation for the distribution of (outcome − point spread).

5. Probability assignment: the 435 U.S. Congress members are elected to two-year terms; the number of voters in an individual Congressional election varies from about 50,000 to 350,000. We will use various sources of information to estimate roughly the probability that at least one Congressional election is tied in the next national election.

   (a) Use any knowledge you have about U.S. politics. Specify clearly what information you are using to construct this conditional probability, even if your answer is just a guess.

   (b) Use the following information: in the period 1900–1992, there were 20,597 Congressional elections, out of which 6 were decided by fewer than 10 votes and 49 decided by fewer than 100 votes.

   See Gelman, King, and Boscardin (1998), Mulligan and Hunter (2001), and Gelman, Katz, and Tuerlinckx (2002) for more on this topic.

6. Conditional probability: approximately 1/125 of all births are fraternal twins and 1/300 of births are identical twins. Elvis Presley had a twin brother (who died at birth). What is the probability that Elvis was an identical twin? (You may approximate the probability of a boy or girl birth as $\frac{1}{2}$.)

7. Conditional probability: the following problem is loosely based on the television game show *Let's Make a Deal*. At the end of the show, a contestant is asked to choose one of three large boxes, where one box contains a fabulous prize and the other two boxes contain lesser prizes. After the contestant chooses a box, Monty Hall, the host of the show, opens one of the two boxes containing smaller prizes. (In order to keep the conclusion suspenseful, Monty does not open the box selected by the contestant.) Monty offers the contestant the opportunity to switch from the chosen box to the remaining unopened box. Should the contestant switch or stay with the original choice? Calculate the probability that the contestant wins under each strategy. This is an exercise in being clear about the information that should be conditioned on when constructing a probability judgment. See Selvin (1975) and Morgan et al. (1991) for further discussion of this problem.

8. Subjective probability: discuss the following statement. 'The probability of event E is considered "subjective" if two rational persons A and B can assign unequal probabilities to E, $P_A(E)$ and $P_B(E)$. These probabilities can also be interpreted as "conditional": $P_A(E) = P(E|I_A)$ and $P_B(E) = P(E|I_B)$, where I_A and I_B represent the knowledge available to persons A and B, respectively.' Apply this idea to the following examples.

   (a) The probability that a '6' appears when a fair die is rolled, where A observes the outcome of the die roll and B does not.

   (b) The probability that Brazil wins the next World Cup, where A is ignorant of soccer and B is a knowledgeable sports fan.

9. Simulation of a queuing problem: a clinic has three doctors. Patients come into the clinic at random, starting at 9 a.m., according to a Poisson process with time parameter 10 minutes: that is, the time after opening at which the first patient appears follows an exponential distribution with expectation 10 minutes and then, after each patient arrives, the waiting time until the next patient is independently exponentially distributed, also with expectation 10 minutes. When a patient arrives, he or she waits until a doctor is available. The amount of time spent by each doctor with each patient is a random variable, uniformly distributed between 5 and 20 minutes. The office stops admitting new patients at 4 p.m. and closes when the last patient is through with the doctor.

   (a) Simulate this process once. How many patients came to the office? How many had to wait for a doctor? What was their average wait? When did the office close?

(b) Simulate the process 100 times and estimate the median and 50% interval for each of the summaries in (a).

CHAPTER 2

Single-parameter models

Our first detailed discussion of Bayesian inference is in the context of statistical models where only a single scalar parameter is to be estimated; that is, the estimand θ is one-dimensional. In this chapter, we consider four fundamental and widely used one-dimensional models—the binomial, normal, Poisson, and exponential—and at the same time introduce important concepts and computational methods for Bayesian data analysis.

2.1 Estimating a probability from binomial data

In the simple binomial model, the aim is to estimate an unknown population proportion from the results of a sequence of 'Bernoulli trials'; that is, data $y_1, \ldots, y_n$, each of which is either 0 or 1. This problem provides a relatively simple but important starting point for the discussion of Bayesian inference. By starting with the binomial model, our discussion also parallels the very first published Bayesian analysis by Thomas Bayes in 1763, and his seminal contribution is still of interest.

The binomial distribution provides a natural model for data that arise from a sequence of n exchangeable trials or draws from a large population where each trial gives rise to one of two possible outcomes, conventionally labeled 'success' and 'failure.' Because of the exchangeability, the data can be summarized by the total number of successes in the n trials, which we denote here by y. Converting from a formulation in terms of exchangeable trials to one using independent and identically distributed random variables is achieved quite naturally by letting the parameter θ represent the proportion of successes in the population or, equivalently, the probability of success in each trial. The binomial sampling model states that

$$p(y|\theta) = \text{Bin}(y|n, \theta) = \binom{n}{y} \theta^y (1 - \theta)^{n-y}, \tag{2.1}$$

where on the left side we suppress the dependence on n because it is regarded as part of the experimental design that is considered fixed; all the probabilities discussed for this problem are assumed to be conditional on n.

Example. Estimating the probability of a female birth

As a specific application of the binomial model, we consider the estimation of the sex ratio within a population of human births. The proportion of births that are female has long been a topic of interest both scientifically and to the lay public. Two hundred years ago it was established that the proportion of female births in European populations was less than 0.5 (see Historical Note below), while in this century interest has focused on factors that may influence the sex ratio. The currently accepted value of the proportion of female births in very large European-race populations is 0.485.

For this example we define the parameter θ to be the proportion of female births, but an alternative way of reporting this parameter is as a ratio of male to female birth rates, $\phi = (1 - \theta)/\theta$.

Let y be the number of girls in n recorded births. By applying the binomial model (2.1), we are assuming that the n births are conditionally independent given θ, with the probability of a female birth equal to θ for all cases. This modeling assumption is motivated by the exchangeability that may be judged to arise when we have no explanatory information (for example, distinguishing multiple births or births within the same family) that might affect the sex of the baby.

To perform Bayesian inference in the binomial model, we must specify a prior distribution for θ. We will discuss issues associated with specifying prior distributions many times throughout this book, but for simplicity at this point, we assume that the prior distribution for θ is uniform on the interval $[0, 1]$.

Elementary application of Bayes' rule as displayed in (1.2), applied to (2.1), then gives the posterior density for θ as

$$p(\theta|y) \propto \theta^y (1 - \theta)^{n-y}. \tag{2.2}$$

With fixed n and y, the factor $\binom{n}{y}$ does not depend on the unknown parameter θ, and so it can be treated as a constant when calculating the posterior distribution of θ. As is typical of many examples, the posterior density can be written immediately in closed form, up to a constant of proportionality. In single-parameter problems, this allows immediate graphical presentation of the posterior distribution. For example, in Figure 2.1, the unnormalized density (2.2) is displayed for several different experiments, that is, different values of n and y. Each of the four experiments has the same proportion of successes, but the sample sizes vary. In the present case, the form of the unnormalized posterior density is recognizable as a *beta* distribution (see Appendix A),

$$\theta|y \sim \text{Beta}(y + 1, n - y + 1). \tag{2.3}$$

Historical note: Bayes and Laplace

Many early writers on probability dealt with the elementary binomial model. The first contributions of lasting significance, in the 17th and early 18th centuries, concentrated on the 'pre-data' question: given θ, what are the probabilities of the various possible outcomes of the random variable y? For example, the 'weak law of large numbers' of Jacob Bernoulli states that if $y \sim \text{Bin}(n, \theta)$, then $\Pr(|\frac{y}{n} - \theta| > \epsilon \mid \theta) \to 0$ as $n \to \infty$, for any θ and any fixed value of $\epsilon > 0$. The Reverend Thomas Bayes, an English part-time mathematician whose work was unpublished during his lifetime, and Pierre Simon Laplace, an inventive and productive mathematical scientist whose massive output spanned the Napoleonic era in France, receive independent credit as the first to *invert* the probability statement and obtain probability statements about θ, *given* observed y.

In his famous paper, published in 1763, Bayes sought, in our notation, the prob-

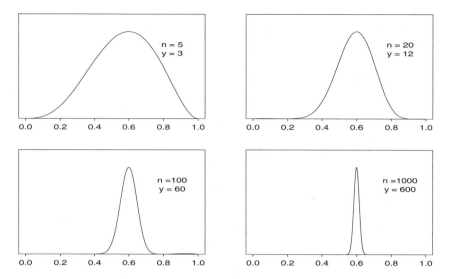

Figure 2.1 *Unnormalized posterior density for binomial parameter θ, based on uniform prior distribution and y successes out of n trials. Curves displayed for several values of n and y.*

ability $\Pr(\theta \in (\theta_1, \theta_2)|y)$; his solution was based on a physical analogy of a probability space to a rectangular table (such as a billiard table):

1. (Prior distribution) A ball W is randomly thrown (according to a uniform distribution on the table). The horizontal position of the ball on the table is θ, expressed as a fraction of the table width.

2. (Likelihood) A ball O is randomly thrown n times. The value of y is the number of times O lands to the right of W.

Thus, θ is assumed to have a (prior) *uniform distribution* on $[0, 1]$. Using elementary rules of probability theory, which he derived in the paper, Bayes then obtained

$$
\begin{aligned}
\Pr(\theta \in (\theta_1, \theta_2)|y) &= \frac{\Pr(\theta \in (\theta_1, \theta_2), y)}{p(y)} \\[2mm]
&= \frac{\int_{\theta_1}^{\theta_2} p(y|\theta)p(\theta)d\theta}{p(y)} \\[2mm]
&= \frac{\int_{\theta_1}^{\theta_2} \binom{n}{y}\theta^y(1-\theta)^{n-y}d\theta}{p(y)}.
\end{aligned}
\tag{2.4}
$$

Bayes succeeded in evaluating the denominator, showing that

$$
\begin{aligned}
p(y) &= \int_0^1 \binom{n}{y}\theta^y(1-\theta)^{n-y}d\theta \\[2mm]
&= \frac{1}{n+1} \quad \text{for } y = 0, \ldots, n.
\end{aligned}
\tag{2.5}
$$

This calculation shows that all possible values of y are equally likely *a priori*.

The numerator of (2.4) is an incomplete beta integral with no closed-form expression for large values of y and $(n - y)$, a fact that apparently presented some difficulties for Bayes.

Laplace, however, independently 'discovered' Bayes' theorem, and developed new analytic tools for computing integrals. For example, he expanded the function $\theta^y(1 - \theta)^{n-y}$ around its maximum at $\theta = y/n$ and evaluated the incomplete beta integral using what we now know as the normal approximation.

In analyzing the binomial model, Laplace also used the uniform prior distribution. His first serious application was to estimate the proportion of female births in a population. A total of 241,945 girls and 251,527 boys were born in Paris from 1745 to 1770. Letting θ be the probability that any birth is female, Laplace showed that

$$\Pr(\theta \geq 0.5 | y = 241{,}945, n = 251{,}527 + 241{,}945) \approx 1.15 \times 10^{-42},$$

and so he was 'morally certain' that $\theta < 0.5$.

Prediction

In the binomial example with the uniform prior distribution, the prior predictive distribution can be evaluated explicitly, as we have already noted in (2.5). Under the model, all possible values of y are equally likely, *a priori*. For posterior prediction from this model, we might be more interested in the outcome of one new trial, rather than another set of n new trials. Letting $\tilde{y}$ denote the result of a new trial, exchangeable with the first n,

$$
\begin{aligned}
\Pr(\tilde{y} = 1 | y) &= \int_0^1 \Pr(\tilde{y} = 1 | \theta, y) p(\theta | y) d\theta \\
&= \int_0^1 \theta p(\theta | y) d\theta = \mathrm{E}(\theta | y) = \frac{y+1}{n+2},
\end{aligned}
\tag{2.6}
$$

from the properties of the beta distribution (see Appendix A). It is left as an exercise to reproduce this result using direct integration of (2.6). This result, based on the uniform prior distribution, is known as 'Laplace's law of succession.' At the extreme observations $y = 0$ and $y = n$, Laplace's law predicts probabilities of $\frac{1}{n+2}$ and $\frac{n+1}{n+2}$, respectively.

2.2 Posterior distribution as compromise between data and prior information

The process of Bayesian inference involves passing from a prior distribution, $p(\theta)$, to a posterior distribution, $p(\theta | y)$, and it is natural to expect that some general relations might hold between these two distributions. For example, we might expect that, because the posterior distribution incorporates the information from the data, it will be less variable than the prior distribution. This notion is formalized in the second of the following expressions:

$$E(\theta) = E(E(\theta|y)) \tag{2.7}$$

and

$$\text{var}(\theta) = E(\text{var}(\theta|y)) + \text{var}(E(\theta|y)), \tag{2.8}$$

which are obtained by substituting (θ, y) for the generic (u, v) in (1.7) and (1.8). The result expressed by equation (2.7) is scarcely surprising: the prior mean of θ is the average of all possible posterior means over the distribution of possible data. The variance formula (2.8) is more interesting because it says that *the posterior variance is on average smaller than the prior variance*, by an amount that depends on the variation in posterior means over the distribution of possible data. The greater the latter variation, the more the potential for reducing our uncertainty with regard to θ, as we shall see in detail for the binomial and normal models in the next chapter. Of course, the mean and variance relations only describe expectations, and in particular situations the posterior variance can be similar to or even larger than the prior variance (although this can be an indication of conflict or inconsistency between the sampling model and prior distribution).

In the binomial example with the uniform prior distribution, the prior mean is $\frac{1}{2}$, and the prior variance is $\frac{1}{12}$. The posterior mean, $\frac{y+1}{n+2}$, is a compromise between the prior mean and the sample proportion, $\frac{y}{n}$, where clearly the prior mean has a smaller and smaller role as the size of the data sample increases. This is a very general feature of Bayesian inference: the posterior distribution is centered at a point that represents a compromise between the prior information and the data, and the compromise is controlled to a greater extent by the data as the sample size increases.

2.3 Summarizing posterior inference

The posterior probability distribution contains all the current information about the parameter θ. Ideally one might report the entire posterior distribution $p(\theta|y)$; as we have seen in Figure 2.1, a graphical display is useful. In Chapter 3, we use contour plots and scatterplots to display posterior distributions in multiparameter problems. A key advantage of the Bayesian approach, as implemented by simulation, is the flexibility with which posterior inferences can be summarized, even after complicated transformations. This advantage is most directly seen through examples, some of which will be presented shortly.

For many practical purposes, however, various numerical summaries of the distribution are desirable. Commonly used summaries of location are the mean, median, and mode(s) of the distribution; variation is commonly summarized by the standard deviation, the interquartile range, and other quantiles. Each summary has its own interpretation: for example, the mean is the posterior expectation of the parameter, and the mode may be interpreted as the single 'most likely' value, given the data (and, of course, the model). Furthermore, as we shall see, much practical inference relies on the use of normal approximations, often improved by applying a symmetrizing transformation

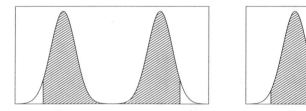

Figure 2.2 *Hypothetical posterior density for which the 95% central interval and 95% highest posterior density region dramatically differ: (a) central posterior interval, (b) highest posterior density region.*

to θ, and here the mean and the standard deviation play key roles. The mode is important in computational strategies for more complex problems because it is often easier to compute than the mean or median.

When the posterior distribution has a closed form, such as the beta distribution in the current example, summaries such as the mean, median, and standard deviation of the posterior distribution are often available in closed form. For example, applying the distributional results in Appendix A, the mean of the beta distribution in (2.3) is $\frac{y+1}{n+2}$, and the mode is $\frac{y}{n}$, which is well known from different points of view as the maximum likelihood and (minimum variance) unbiased estimate of θ.

Posterior quantiles and intervals

In addition to point summaries, it is nearly always important to report posterior uncertainty. Our usual approach is to present quantiles of the posterior distribution of estimands of interest or, if an interval summary is desired, a central interval of posterior probability, which corresponds, in the case of a $100(1-\alpha)\%$ interval, to the range of values above and below which lies exactly $100(\alpha/2)\%$ of the posterior probability. Such interval estimates are referred to as *posterior intervals*. For simple models, such as the binomial and normal, posterior intervals can be computed directly from cumulative distribution functions, often using calls to standard computer functions, as we illustrate in Section 2.5 with the example of the human sex ratio. In general, intervals can be computed using computer simulations from the posterior distribution, as described at the end of Section 1.9.

A slightly different method of summarizing posterior uncertainty is to compute a region of highest posterior density: the region of values that contains $100(1-\alpha)\%$ of the posterior probability and also has the characteristic that the density within the region is never lower than that outside. Obviously, such a region is identical to a central posterior interval if the posterior distribution is unimodal and symmetric. In general, we prefer the central posterior interval to the highest posterior density region because the former has a direct interpretation as the posterior $\alpha/2$ and $1-\alpha/2$ quantiles, is invariant to one-to-one transformations of the estimand, and is usually easier to compute.

An interesting comparison is afforded by the hypothetical bimodal posterior density pictured in Figure 2.2; the 95% central interval includes the area of zero probability in the center of the distribution, whereas the 95% highest posterior density region comprises two disjoint intervals. In this situation, the highest posterior density region is more cumbersome but conveys more information than the central interval; however, it is probably better not to try to summarize this bimodal density by any single interval. The central interval and the highest posterior density region can also differ substantially when the posterior density is highly skewed.

2.4 Informative prior distributions

In the binomial example, we have so far considered only the uniform prior distribution for θ. How can this specification be justified, and how in general do we approach the problem of constructing prior distributions?

We consider two basic interpretations that can be given to prior distributions. In the *population* interpretation, the prior distribution represents a population of possible parameter values, from which the θ of current interest has been drawn. In the more subjective *state of knowledge* interpretation, the guiding principle is that we must express our knowledge (and uncertainty) about θ as if its value could be thought of as a random realization from the prior distribution. For many problems, such as estimating the probability of failure in a new industrial process, there is no perfectly relevant population of θ's from which the current θ has been drawn, except in hypothetical contemplation. Typically, the prior distribution should include all plausible values of θ, but the distribution need not be realistically concentrated around the true value, because often the information about θ contained in the data will far outweigh *any* reasonable prior probability specification.

In the binomial example, we have seen that the uniform prior distribution for θ implies that the prior predictive distribution for y (given n) is uniform on the discrete set $\{0, 1, \ldots, n\}$, giving equal probability to the $n + 1$ possible values. In his original treatment of this problem (described in the Historical Note in Section 2.1), Bayes' justification for the uniform prior distribution appears to have been based on this observation; the argument is appealing because it is expressed entirely in terms of the *observable* quantities y and n. Laplace's rationale for the uniform prior density was less clear, but subsequent interpretations ascribe to him the so-called 'principle of insufficient reason,' which claims that if nothing is known about θ then a uniform specification is appropriate. We shall discuss in Section 2.9 the weaknesses of the principle of insufficient reason as a general approach for assigning probability distributions.

At this point, we discuss some of the issues that arise in assigning a prior distribution that reflects substantive information.

Binomial example with different prior distributions

We first pursue the binomial model in further detail using a parametric family of prior distributions that includes the uniform as a special case. For mathematical convenience, we construct a family of prior densities that lead to simple posterior densities.

Considered as a function of θ, the likelihood (2.1) is of the form,

$$p(y|\theta) \propto \theta^a (1 - \theta)^b.$$

Thus, if the prior density is of the same form, with its own values a and b, then the posterior density will also be of this form. We will parameterize such a prior density as

$$p(\theta) \propto \theta^{\alpha-1}(1 - \theta)^{\beta-1},$$

which is a beta distribution with parameters α and β: $\theta \sim \text{Beta}(\alpha, \beta)$. Comparing $p(\theta)$ and $p(y|\theta)$ suggests that this prior density is equivalent to $\alpha - 1$ prior successes and $\beta - 1$ prior failures. The parameters of the prior distribution are often referred to as *hyperparameters*. The beta prior distribution is indexed by two hyperparameters, which means we can specify a particular prior distribution by fixing two features of the distribution, for example its mean and variance; see (A.3) on page 580.

For now, assume that we can select reasonable values α and β. Appropriate methods for working with unknown hyperparameters in certain problems are described in Chapter 5. The posterior density for θ is

$$\begin{aligned} p(\theta|y) \quad &\propto \quad \theta^y (1 - \theta)^{n-y} \theta^{\alpha-1} (1 - \theta)^{\beta-1} \\ &= \quad \theta^{y+\alpha-1}(1 - \theta)^{n-y+\beta-1} \\ &= \quad \text{Beta}(\theta|\alpha + y, \beta + n - y). \end{aligned}$$

The property that the posterior distribution follows the same parametric form as the prior distribution is called *conjugacy*; the beta prior distribution is a *conjugate family* for the binomial likelihood. The conjugate family is mathematically convenient in that the posterior distribution follows a known parametric form. Of course, if information is available that contradicts the conjugate parametric family, it may be necessary to use a more realistic, if inconvenient, prior distribution (just as the binomial likelihood may need to be replaced by a more realistic likelihood in some cases).

To continue with the binomial model with beta prior distribution, the posterior mean of θ, which may be interpreted as the posterior probability of success for a future draw from the population, is now

$$\text{E}(\theta|y) = \frac{\alpha + y}{\alpha + \beta + n},$$

which always lies between the sample proportion, y/n, and the prior mean, $\alpha/(\alpha + \beta)$; see Exercise 2.5b. The posterior variance is

$$\text{var}(\theta|y) = \frac{(\alpha + y)(\beta + n - y)}{(\alpha + \beta + n)^2(\alpha + \beta + n + 1)} = \frac{E(\theta|y)[1 - E(\theta|y)]}{\alpha + \beta + n + 1}.$$

As y and $n - y$ become large with fixed α and β, $\mathrm{E}(\theta|y) \approx y/n$ and $\mathrm{var}(\theta|y) \approx \frac{1}{n}\frac{y}{n}(1 - \frac{y}{n})$, which approaches zero at the rate $1/n$. Clearly, in the limit, the parameters of the prior distribution have no influence on the posterior distribution.

In fact, as we shall see in more detail in Chapter 4, the central limit theorem of probability theory can be put in a Bayesian context to show:

$$\left(\left.\frac{\theta - \mathrm{E}(\theta|y)}{\sqrt{\mathrm{var}(\theta|y)}}\right| y\right) \to \mathrm{N}(0, 1).$$

This result is often used to justify approximating the posterior distribution with a normal distribution. For the binomial parameter θ, the normal distribution is a more accurate approximation in practice if we transform θ to the logit scale; that is, performing inference for $\log(\theta/(1 - \theta))$ instead of θ itself, thus expanding the probability space from $[0, 1]$ to $(-\infty, \infty)$, which is more fitting for a normal approximation.

Conjugate prior distributions

Conjugacy is formally defined as follows. If $\mathcal{F}$ is a class of sampling distributions $p(y|\theta)$, and $\mathcal{P}$ is a class of prior distributions for θ, then the class $\mathcal{P}$ is *conjugate* for $\mathcal{F}$ if

$$p(\theta|y) \in \mathcal{P} \text{ for all } p(\cdot|\theta) \in \mathcal{F} \text{ and } p(\cdot) \in \mathcal{P}.$$

This definition is formally vague since if we choose $\mathcal{P}$ as the class of all distributions, then $\mathcal{P}$ is always conjugate no matter what class of sampling distributions is used. We are most interested in *natural* conjugate prior families, which arise by taking $\mathcal{P}$ to be the set of all densities having the same functional form as the likelihood.

Conjugate prior distributions have the practical advantage, in addition to computational convenience, of being interpretable as additional data, as we have seen for the binomial example and will also see for the normal and other standard models in Sections 2.6 and 2.7.

Nonconjugate prior distributions

The basic justification for the use of conjugate prior distributions is similar to that for using standard models (such as binomial and normal) for the likelihood: it is easy to understand the results, which can often be put in analytic form, they are often a good approximation, and they simplify computations. Also, they will be useful later as building blocks for more complicated models, including in many dimensions, where conjugacy is typically impossible. For these reasons, conjugate models can be good starting points; for example, mixtures of conjugate families can sometimes be useful when simple conjugate distributions are not reasonable (see Exercise 2.4).

Although they can make interpretations of posterior inferences less transparent and computation more difficult, nonconjugate prior distributions do not pose any new conceptual problems. In practice, for complicated models, conjugate prior distributions may not even be possible. Section 2.5 and exercises 2.10 and 2.11 present examples of nonconjugate computation; a more extensive nonconjugate example, an analysis of a bioassay experiment, appears in Section 3.7.

Conjugate prior distributions, exponential families, and sufficient statistics

We close this section by relating conjugate families of distributions to the classical concepts of exponential families and sufficient statistics. Readers who are unfamiliar with these concepts can skip ahead to Section 2.5 with no loss.

Probability distributions that belong to an *exponential family* have natural conjugate prior distributions, so we digress at this point to review the definition of exponential families; for complete generality in this section, we allow data points y_i and parameters θ to be multidimensional. The class $\mathcal{F}$ is an exponential family if all its members have the form,

$$p(y_i|\theta) = f(y_i)g(\theta)e^{\phi(\theta)^T u(y_i)}.$$

The factors $\phi(\theta)$ and $u(y_i)$ are, in general, vectors of equal dimension to that of θ. The vector $\phi(\theta)$ is called the 'natural parameter' of the family $\mathcal{F}$. The likelihood corresponding to a sequence $y = (y_1, \ldots, y_n)$ of iid observations is

$$p(y|\theta) = \left[\prod_{i=1}^{n} f(y_i)\right] g(\theta)^n \exp\left(\phi(\theta)^T \sum_{i=1}^{n} u(y_i)\right).$$

For all n and y, this has a fixed form (as a function of θ):

$$p(y|\theta) \propto g(\theta)^n e^{\phi(\theta)^T t(y)}, \quad \text{where } t(y) = \sum_{i=1}^{n} u(y_i).$$

The quantity $t(y)$ is said to be a *sufficient statistic* for θ, because the likelihood for θ depends on the data y only through the value of $t(y)$. Sufficient statistics are useful in algebraic manipulations of likelihoods and posterior distributions. If the prior density is specified as

$$p(\theta) \propto g(\theta)^\eta e^{\phi(\theta)^T \nu},$$

then the posterior density is

$$p(\theta|y) \propto g(\theta)^{\eta+n} e^{\phi(\theta)^T(\nu+t(y))},$$

which shows that this choice of prior density is conjugate. It has been shown that, in general, the exponential families are the only classes of distributions that have natural conjugate prior distributions, since, apart from certain irregular cases, the only distributions having a fixed number of sufficient statistics for all n are of the exponential type. We have already discussed the binomial distribution, where for the likelihood $p(y|\theta, n) = \text{Bin}(y|n, \theta)$ with n known, the conjugate prior distributions on θ are beta distributions. It is left as an exercise to show that the binomial is an exponential family with natural parameter logit(θ).

2.5 Example: estimating the probability of a female birth given placenta previa

As a specific example of a factor that may influence the sex ratio, we consider the maternal condition *placenta previa*, an unusual condition of pregnancy in which the placenta is implanted very low in the uterus, obstructing the fetus from a normal vaginal delivery. An early study concerning the sex of placenta previa births in Germany found that of a total of 980 births, 437 were female. How much evidence does this provide for the claim that the proportion of female births in the population of placenta previa births is less than 0.485, the proportion of female births in the general population?

Analysis using a uniform prior distribution

Assuming a uniform prior distribution for the probability of a female birth, the posterior distribution is Beta(438, 544). Exact summaries of the posterior distribution can be obtained from the properties of the beta distribution (Appendix A): the posterior mean of θ is 0.446 and the posterior standard deviation is 0.016. Exact posterior quantiles can be obtained using numerical integration of the beta density, which in practice we perform by a computer function call; the median is 0.446 and the central 95% posterior interval is [0.415, 0.477]. This 95% posterior interval matches, to three decimal places, the interval that would be obtained by using a normal approximation with the calculated posterior mean and standard deviation. Further discussion of the approximate normality of the posterior distribution is given in Chapter 4.

In many situations it is not feasible to perform calculations on the posterior density function directly. In such cases it can be particularly useful to use simulation from the posterior distribution to obtain inferences. The first histogram in Figure 2.3 shows the distribution of 1000 draws from the Beta(438, 544) posterior distribution. An estimate of the 95% posterior interval, obtained by taking the 25th and 976th of the 1000 ordered draws, is [0.415, 0.476], and the median of the 1000 draws from the posterior distribution is 0.446. The sample mean and standard deviation of the 1000 draws are 0.445 and 0.016, almost identical to the exact results. A normal approximation to the 95% posterior interval is $[0.445 \pm 1.96(.016)] = [0.414, 0.476]$. Because of the large sample and the fact that the distribution of θ is concentrated away from zero and one, the normal approximation is quite good in this example.

As already noted, when estimating a proportion, the normal approximation is generally improved by applying it to the logit transform, $\log(\frac{\theta}{1-\theta})$, which transforms the parameter space from the unit interval to the real line. The second histogram in Figure 2.3 shows the distribution of the transformed

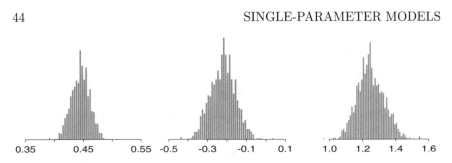

Figure 2.3 *Draws from the posterior distribution of (a) the probability of female birth,*
θ; *(b) the logit transform,* $\mathrm{logit}(\theta)$; *(c) the male-to-female sex ratio,* $\phi = (1-\theta)/\theta$.

Parameters of the prior distribution		Summaries of the posterior distribution	
$\frac{\alpha}{\alpha+\beta}$	$\alpha+\beta$	Posterior median of θ	95% posterior interval for θ
0.500	2	0.446	$[0.415, 0.477]$
0.485	2	0.446	$[0.415, 0.477]$
0.485	5	0.446	$[0.415, 0.477]$
0.485	10	0.446	$[0.415, 0.477]$
0.485	20	0.447	$[0.416, 0.478]$
0.485	100	0.450	$[0.420, 0.479]$
0.485	200	0.453	$[0.424, 0.481]$

Table 2.1 *Summaries of the posterior distribution of* θ, *the probability of a girl birth given placenta previa, under a variety of conjugate prior distributions.*

draws. The estimated posterior mean and standard deviation on the logit scale based on 1000 draws are -0.220 and 0.065. A normal approximation to the 95% posterior interval for θ is obtained by inverting the 95% interval on the logit scale $[-0.220 \pm (1.96)(0.065)]$, which yields $[0.414, 0.477]$ on the original scale. The improvement from using the logit scale is most noticeable when the sample size is small or the distribution of θ includes values near zero or one.

In any real data analysis, it is important to keep the applied context in mind. The parameter of interest in this example is traditionally expressed as the 'sex ratio,' $(1-\theta)/\theta$, the ratio of male to female births. The posterior distribution of the ratio is illustrated in the third histogram. The posterior median of the sex ratio is 1.24, and the 95% posterior interval is $[1.10, 1.41]$. The posterior distribution is concentrated on values far above the usual European-race sex ratio of 1.06, implying that the probability of a female birth given placenta previa is less than in the general population.

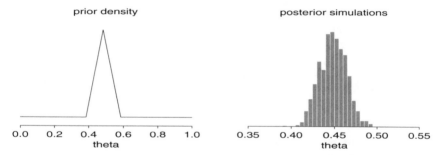

Figure 2.4 *(a) Prior density for θ in nonconjugate analysis of birth ratio example; (b) histogram of 1000 draws from a discrete approximation to the posterior density. Figures are plotted on different scales.*

Analysis using different conjugate prior distributions

The sensitivity of posterior inference about θ to the proposed prior distribution is exhibited in Table 2.1. The first row corresponds to the uniform prior distribution, $\alpha = 1$, $\beta = 1$, and subsequent rows of the table use prior distributions that are increasingly concentrated around 0.485, the mean proportion for female births in the general population. The first column shows the prior mean for θ, and the second column indexes the amount of prior information, as measured by $\alpha + \beta$; recall that $\alpha + \beta - 2$ is, in some sense, equivalent to the number of prior observations. Posterior inferences based on a large sample are not particularly sensitive to the prior distribution. Only at the bottom of the table, where the prior distribution contains information equivalent to 100 or 200 births, are the posterior intervals pulled noticeably toward the prior distribution, and even then, the 95% posterior intervals still exclude the prior mean.

Analysis using a nonconjugate prior distribution

As an alternative to the conjugate beta family for this problem, we might prefer a prior distribution that is centered around 0.485 but is flat far away from this value to admit the possibility that the truth is far away. The piecewise linear prior density in Figure 2.4a is an example of a prior distribution of this form; 40% of the probability mass is outside the interval [0.385, 0.585]. This prior distribution has mean 0.493 and standard deviation 0.21, similar to the standard deviation of a beta distribution with $\alpha + \beta = 5$. The unnormalized posterior distribution is obtained at a grid of θ values, $(0.000, 0.001, \ldots, 1.000)$, by multiplying the prior density and the binomial likelihood at each point. Samples from the posterior distribution can be obtained by normalizing the distribution on the discrete grid of θ values. Figure 2.4b is a histogram of 1000 draws from the discrete posterior distribution. The posterior median is 0.448, and the 95% central posterior interval is [0.419, 0.480]. Because the prior distribution is overwhelmed by the data, these results match those in Table 2.1

based on beta distributions. In taking the grid approach, it is important to avoid grids that are too coarse and distort a significant portion of the posterior mass.

2.6 Estimating the mean of a normal distribution with known variance

The normal distribution is fundamental to most statistical modeling. The central limit theorem helps to justify using the normal likelihood in many statistical problems, as an approximation to a less analytically convenient actual likelihood. Also, as we shall see in later chapters, even when the normal distribution does not itself provide a good model fit, it can be useful as a component of a more complicated model involving Student-t or finite mixture distributions. For now, we simply work through the Bayesian results assuming the normal model is appropriate. We derive results first for a single data point and then for the general case of many data points.

Likelihood of one data point

As the simplest first case, consider a single scalar observation y from a normal distribution parameterized by a mean θ and variance σ^2, where for this initial development we assume that σ^2 is known. The sampling distribution is

$$p(y|\theta) = \frac{1}{\sqrt{2\pi}\sigma} e^{-\frac{1}{2\sigma^2}(y-\theta)^2}.$$

Conjugate prior and posterior distributions

Considered as a function of θ, the likelihood is an exponential of a quadratic form in θ, so the family of conjugate prior densities looks like

$$p(\theta) = e^{A\theta^2 + B\theta + C}.$$

We parameterize this family as

$$p(\theta) \propto \exp\left(-\frac{1}{2\tau_0^2}(\theta - \mu_0)^2\right);$$

that is, $\theta \sim N(\mu_0, \tau_0^2)$, with hyperparameters μ_0 and τ_0^2. As usual in this preliminary development, we assume that the hyperparameters are known.

The conjugate prior density implies that the posterior distribution for θ is the exponential of a quadratic form and thus normal, but some algebra is required to reveal its specific form. In the posterior density, all variables except θ are regarded as constants, giving the conditional density,

$$p(\theta|y) \propto \exp\left(-\frac{1}{2}\left[\frac{(y-\theta)^2}{\sigma^2} + \frac{(\theta - \mu_0)^2}{\tau_0^2}\right]\right).$$

Expanding the exponents, collecting terms and then completing the square in

θ (see Exercise 2.14(a) for details) gives

$$p(\theta|y) \propto \exp\left(-\frac{1}{2\tau_1^2}(\theta - \mu_1)^2\right),$$ (2.9)

that is, $\theta|y \sim N(\mu_1, \tau_1^2)$, where

$$\mu_1 = \frac{\frac{1}{\tau_0^2}\mu_0 + \frac{1}{\sigma^2}y}{\frac{1}{\tau_0^2} + \frac{1}{\sigma^2}} \quad \text{and} \quad \frac{1}{\tau_1^2} = \frac{1}{\tau_0^2} + \frac{1}{\sigma^2}.$$ (2.10)

Precisions of the prior and posterior distributions. In manipulating normal distributions, the inverse of the variance plays a prominent role and is called the *precision*. The algebra above demonstrates that for normal data and normal prior distribution (each with known precision), *the posterior precision equals the prior precision plus the data precision.*

There are several different ways of interpreting the form of the posterior mean, μ_1. In (2.10), *the posterior mean is expressed as a weighted average of the prior mean and the observed value, y, with weights proportional to the precisions.* Alternatively, we can express μ_1 as the prior mean adjusted toward the observed y,

$$\mu_1 = \mu_0 + (y - \mu_0)\frac{\tau_0^2}{\sigma^2 + \tau_0^2},$$

or as the data 'shrunk' toward the prior mean,

$$\mu_1 = y - (y - \mu_0)\frac{\sigma^2}{\sigma^2 + \tau_0^2}.$$

Each formulation represents the posterior mean as a compromise between the prior mean and the observed value.

In extreme cases, the posterior mean equals the prior mean or the observed value:

$$\mu_1 = \mu_0 \quad \text{if} \quad y = \mu_0 \text{ or } \tau_0^2 = 0;$$
$$\mu_1 = y \quad \text{if} \quad y = \mu_0 \text{ or } \sigma^2 = 0.$$

If $\tau_0^2 = 0$, the prior distribution is infinitely more precise than the data, and so the posterior and prior distributions are identical and concentrated at the value μ_0. If $\sigma^2 = 0$, the data are perfectly precise, and the posterior distribution is concentrated at the observed value, y. If $y = \mu_0$, the prior and data means coincide, and the posterior mean must also fall at this point.

Posterior predictive distribution

The posterior predictive distribution of a future observation, $\tilde{y}$, $p(\tilde{y}|y)$, can be calculated directly by integration, using (1.4):

$$p(\tilde{y}|y) = \int p(\tilde{y}|\theta)p(\theta|y)d\theta$$

$$\propto \int \exp\left(-\frac{1}{2\sigma^2}(\tilde{y} - \theta)^2\right) \exp\left(-\frac{1}{2\tau_1^2}(\theta - \mu_1)^2\right) d\theta.$$

The first line above holds because the distribution of the future observation, $\tilde{y}$, given θ, does not depend on the past data, y. We can determine the distribution of $\tilde{y}$ more easily using the properties of the bivariate normal distribution. The product in the integrand is the exponential of a quadratic function of $(\tilde{y}, \theta)$; hence $\tilde{y}$ and θ have a joint normal posterior distribution, and so the marginal posterior distribution of $\tilde{y}$ is normal.

We can determine the mean and variance of the posterior predictive distribution using the knowledge from the posterior distribution that $E(\tilde{y}|\theta) = \theta$ and $\text{var}(\tilde{y}|\theta) = \sigma^2$, along with identities (2.7) and (2.8):

$$E(\tilde{y}|y) = E(E(\tilde{y}|\theta, y)|y) = E(\theta|y) = \mu_1,$$

and

$$
\begin{aligned}
\text{var}(\tilde{y}|y) &= E(\text{var}(\tilde{y}|\theta, y)|y) + \text{var}(E(\tilde{y}|\theta, y)|y) \\
&= E(\sigma^2|y) + \text{var}(\theta|y) \\
&= \sigma^2 + \tau_1^2.
\end{aligned}
$$

Thus, the posterior predictive distribution of the unobserved $\tilde{y}$ has mean equal to the posterior mean of θ and two components of variance: the predictive variance σ^2 from the model and the variance τ_1^2 due to posterior uncertainty in θ.

Normal model with multiple observations

This development of the normal model with a single observation can be easily extended to the more realistic situation where a sample of independent and identically distributed observations $y = (y_1, \ldots, y_n)$ is available. Proceeding formally, the posterior density is

$$
\begin{aligned}
p(\theta|y) &\propto p(\theta)p(y|\theta) \\
&= p(\theta) \prod_{i=1}^{n} p(y_i|\theta) \\
&\propto \exp\left(-\frac{1}{2\tau_0^2}(\theta - \mu_0)^2\right) \prod_{i=1}^{n} \exp\left(-\frac{1}{2\sigma^2}(y_i - \theta)^2\right) \\
&\propto \exp\left(-\frac{1}{2}\left[\frac{1}{\tau_0^2}(\theta - \mu_0)^2 + \frac{1}{\sigma^2}\sum_{i=1}^{n}(y_i - \theta)^2\right]\right).
\end{aligned}
$$

Algebraic simplification of this expression (along similar lines to those used in the single observation case, as explicated in Exercise 2.14(b)) shows that the posterior distribution depends on y only through the sample mean, $\bar{y} = \frac{1}{n}\sum_i y_i$; that is, $\bar{y}$ is a *sufficient statistic* in this model. In fact, since $\bar{y}|\theta, \sigma^2 \sim N(\theta, \sigma^2/n)$, the results derived for the single normal observation apply immediately (treating $\bar{y}$ as the single observation) to give

$$p(\theta|y_1 \ldots, y_n) = p(\theta|\bar{y}) = \mathrm{N}(\theta|\mu_n, \tau_n^2), \tag{2.11}$$

where

$$\mu_n = \frac{\frac{1}{\tau_0^2}\mu_0 + \frac{n}{\sigma^2}\bar{y}}{\frac{1}{\tau_0^2} + \frac{n}{\sigma^2}} \quad \text{and} \quad \frac{1}{\tau_n^2} = \frac{1}{\tau_0^2} + \frac{n}{\sigma^2}. \tag{2.12}$$

Incidentally, the same result is obtained by adding information for the data points $y_1, y_2, \ldots, y_n$ one point at a time, using the posterior distribution at each step as the prior distribution for the next (see Exercise 2.14(c)).

In the expressions for the posterior mean and variance, the prior precision, $\frac{1}{\tau_0^2}$, and the data precision, $\frac{n}{\sigma^2}$, play equivalent roles, so if n is at all large, the posterior distribution is largely determined by σ^2 and the sample value $\bar{y}$. For example, if $\tau_0^2 = \sigma^2$, then the prior distribution has the same weight as one extra observation with the value μ_0. More specifically, as $\tau_0 \to \infty$ with n fixed, *or* as $n \to \infty$ with τ_0^2 fixed, we have:

$$p(\theta|y) \approx \mathrm{N}(\theta|\bar{y}, \sigma^2/n), \tag{2.13}$$

which is, in practice, a good approximation whenever prior beliefs are relatively diffuse over the range of θ where the likelihood is substantial.

2.7 Other standard single-parameter models

Recall that, in general, the posterior density, $p(\theta|y)$, has no closed-form expression; the normalizing constant, $p(y)$, is often especially difficult to compute due to the integral (1.3). Much formal Bayesian analysis concentrates on situations where closed forms are available; such models are sometimes unrealistic, but their analysis often provides a useful starting point when it comes to constructing more realistic models.

The standard distributions—binomial, normal, Poisson, and exponential—have natural derivations from simple probability models. As we have already discussed, the binomial distribution is motivated from counting exchangeable outcomes, and the normal distribution applies to a random variable that is the sum of a large number of exchangeable or independent terms. We will also have occasion to apply the normal distribution to the logarithm of all-positive data, which would naturally apply to observations that are modeled as the product of many independent multiplicative factors. The Poisson and exponential distributions arise as the number of counts and the waiting times, respectively, for events modeled as occurring exchangeably in all time intervals; that is, independently in time, with a constant rate of occurrence. We will generally construct realistic probability models for more complicated outcomes by combinations of these basic distributions. For example, in Section 18.4, we model the reaction times of schizophrenic patients in a psychological experiment as a binomial mixture of normal distributions on the logarithmic scale.

Each of these standard models has an associated family of conjugate prior distributions, which we discuss in turn.

Normal distribution with known mean but unknown variance

The normal model with known mean θ and unknown variance is an important example, not necessarily for its direct applied value, but as a building block for more complicated, useful models, most immediately the normal distribution with unknown mean and variance, which we cover in Sections 3.2–3.4. In addition, the normal distribution with known mean but unknown variance provides an introductory example of the estimation of a scale parameter.

For $p(y|\theta, \sigma^2) = N(y|\theta, \sigma^2)$, with θ known and σ^2 unknown, the likelihood for a vector y of n iid observations is

$$
\begin{aligned}
p(y|\sigma^2) &\propto \sigma^{-n} \exp\left(-\frac{1}{2\sigma^2} \sum_{i=1}^{n}(y_i - \theta)^2\right) \\
&= (\sigma^2)^{-n/2} \exp\left(-\frac{n}{2\sigma^2}v\right).
\end{aligned}
$$

The sufficient statistic is

$$
v = \frac{1}{n} \sum_{i=1}^{n}(y_i - \theta)^2.
$$

The corresponding conjugate prior density is the inverse-gamma,

$$
p(\sigma^2) \propto (\sigma^2)^{-(\alpha+1)} e^{-\beta/\sigma^2},
$$

which has hyperparameters (α, β). A convenient parameterization is as a scaled inverse-χ^2 distribution with scale σ_0^2 and ν_0 degrees of freedom (see Appendix A); that is, the prior distribution of σ^2 is taken to be the distribution of $\sigma_0^2 \nu_0 / X$, where X is a $\chi_{\nu_0}^2$ random variable. We use the convenient but nonstandard notation, $\sigma^2 \sim \text{Inv-}\chi^2(\nu_0, \sigma_0^2)$.

The resulting posterior density for σ^2 is

$$
\begin{aligned}
p(\sigma^2|y) &\propto p(\sigma^2)p(y|\sigma^2) \\
&\propto \left(\frac{\sigma_0^2}{\sigma^2}\right)^{\nu_0/2+1} \exp\left(-\frac{\nu_0 \sigma_0^2}{2\sigma^2}\right) \cdot (\sigma^2)^{-n/2} \exp\left(-\frac{n}{2}\frac{v}{\sigma^2}\right) \\
&\propto (\sigma^2)^{-((n+\nu_0)/2+1)} \exp\left(-\frac{1}{2\sigma^2}(\nu_0 \sigma_0^2 + nv)\right).
\end{aligned}
$$

Thus,

$$
\sigma^2|y \sim \text{Inv-}\chi^2\left(\nu_0 + n, \frac{\nu_0 \sigma_0^2 + nv}{\nu_0 + n}\right),
$$

which is a scaled inverse-χ^2 distribution with scale equal to the degrees-of-freedom-weighted average of the prior and data scales and degrees of freedom equal to the sum of the prior and data degrees of freedom. The prior distribution can be thought of as providing the information equivalent to ν_0 observations with average squared deviation σ_0^2.

Example. Football scores and point spreads

We illustrate the problem of estimating an unknown normal variance using the football data presented in Section 1.6. As shown in Figure 1.2b, the 672 values of d, the difference between the game outcome and the point spread, have an approximate normal distribution with mean approximately zero. From prior knowledge, it may be reasonable to consider the values of d as samples from a distribution whose true median is zero—on average, neither the favorite nor the underdog should be more likely to beat the point spread. Based on the histogram, the normal distribution fits the data fairly well, and, as discussed in Section 1.6, the variation in the data is large enough that we will ignore the discreteness and pretend that d is continuous-valued. Finally, the data come from $n = 672$ different football games, and it seems reasonable to assume exchangeability of the d_i's given the point spreads. In summary, we assume the data points $d_i, i = 1, \ldots, n,$ are independent samples from a $\mathrm{N}(0, \sigma^2)$ distribution.

To estimate σ^2 as above, we need a prior distribution. For simplicity, we use a conjugate prior density corresponding to zero 'prior observations'; that is, $\nu_0 = 0$ in the scaled inverse-χ^2 density. This yields the 'prior density,' $p(\sigma^2) \propto \sigma^{-2}$, which is not integrable. However, we can formally combine this prior density with the likelihood to yield an acceptable posterior inference; Section 2.9 provides more discussion of this point in the general context of 'noninformative' prior densities. The prior density and posterior density for σ^2 are displayed in Figure 2.5. Algebraically, the posterior distribution for σ^2 can be written as $\sigma^2|d \sim$ Inv-$\chi^2(n, v)$, where $v = \frac{1}{n}\sum_{i=1}^n d_i^2$, since we are assuming that the mean of the distribution of d is zero.

For the football data, $n = 672$ and $v = 13.85^2$. Using values computed with a suitable software package, we find that the 2.5% and 97.5% points of the χ_{672}^2 distribution are 602.1 and 745.7, respectively, and so the central 95% posterior interval for σ^2 is $[nv/745.7, nv/602.1] = [172.8, 214.1]$. The interval for the standard deviation, σ, is thus $[13.1, 14.6]$. Alternatively, we could sample 1000 draws from a χ_{672}^2 distribution and use those to create 1000 draws from the posterior distribution of σ^2 and thus of σ. We did this, and the median of the 1000 draws of σ was 13.9, and the estimated 95% interval—given by the 25th and 976th largest values of σ in the simulation—was $[13.2, 14.7]$.

Poisson model

The Poisson distribution arises naturally in the study of data taking the form of counts; for instance, a major area of application is epidemiology, where the incidence of diseases is studied.

If a data point y follows the Poisson distribution with rate θ, then the probability distribution of a single observation y is

$$p(y|\theta) = \frac{\theta^y e^{-\theta}}{y!}, \quad \text{for } y = 0, 1, 2, \ldots,$$

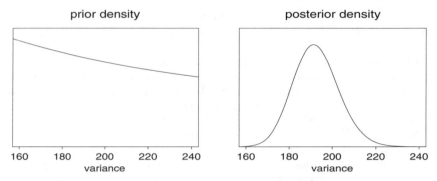

Figure 2.5 *(a) Prior density and (b) posterior density for σ^2 for the football example, graphed in the range over which the posterior density is substantial.*

and for a vector $y = (y_1, \ldots, y_n)$ of iid observations, the likelihood is

$$p(y|\theta) = \prod_{i=1}^{n} \frac{1}{y_i!} \theta^{y_i} e^{-\theta}$$

$$\propto \theta^{t(y)} e^{-n\theta},$$

where $t(y) = \sum_{i=1}^{n} y_i$ is the sufficient statistic. We can rewrite the likelihood in exponential family form as

$$p(y|\theta) \propto e^{-n\theta} e^{t(y) \log \theta},$$

revealing that the natural parameter is $\phi(\theta) = \log \theta$, and the natural conjugate prior distribution is

$$p(\theta) \propto (e^{-\theta})^\eta e^{\nu \log \theta},$$

indexed by hyperparameters (η, ν). To put this argument another way, the likelihood is of the form $\theta^a e^{-b\theta}$, and so the conjugate prior density must be of the form $p(\theta) \propto \theta^A e^{-B\theta}$. In a more conventional parameterization,

$$p(\theta) \propto e^{-\beta\theta} \theta^{\alpha-1},$$

which is a gamma density with parameters α and β, Gamma(α, β); see Appendix A. Comparing $p(y|\theta)$ and $p(\theta)$ reveals that the prior density is, in some sense, equivalent to a total count of $\alpha - 1$ in β prior observations. With this conjugate prior distribution, the posterior distribution is

$$\theta|y \sim \text{Gamma}(\alpha + n\bar{y}, \beta + n).$$

The negative binomial distribution. With conjugate families, the known form of the prior and posterior densities can be used to find the marginal distribution, $p(y)$, using the formula

$$p(y) = \frac{p(y|\theta)p(\theta)}{p(\theta|y)}.$$

For instance, the Poisson model for a single observation, y, has prior predictive distribution

$$p(y) = \frac{\text{Poisson}(y|\theta)\text{Gamma}(\theta|\alpha, \beta)}{\text{Gamma}(\theta|\alpha + y, 1 + \beta)}$$

$$= \frac{\Gamma(\alpha + y)\beta^\alpha}{\Gamma(\alpha)y!(1 + \beta)^{\alpha + y}},$$

which reduces to

$$p(y) = \binom{\alpha + y - 1}{y} \left(\frac{\beta}{\beta + 1}\right)^\alpha \left(\frac{1}{\beta + 1}\right)^y,$$

which is known as the *negative binomial* density:

$$y \sim \text{Neg-bin}(\alpha, \beta).$$

The above derivation shows that the negative binomial distribution is a *mixture* of Poisson distributions with rates, θ, that follow the gamma distribution:

$$\text{Neg-bin}(y|\alpha, \beta) = \int \text{Poisson}(y|\theta)\text{Gamma}(\theta|\alpha, \beta)d\theta.$$

We return to the negative binomial distribution in Section 17.2 as a robust alternative to the Poisson distribution.

Poisson model parameterized in terms of rate and exposure

In many applications, it is convenient to extend the Poisson model for data points $y_1, \ldots, y_n$ to the form

$$y_i \sim \text{Poisson}(x_i\theta), \tag{2.14}$$

where the values x_i are known positive values of an explanatory variable, x, and θ is the unknown parameter of interest. In epidemiology, the parameter θ is often called the *rate*, and x_i is called the *exposure* of the ith unit. This model is not exchangeable in the y_i's but is exchangeable in the pairs $(x, y)_i$. The likelihood for θ in the extended Poisson model is

$$p(y|\theta) \propto \theta^{\left(\sum_{i=1}^n y_i\right)} e^{-\left(\sum_{i=1}^n x_i\right)\theta}$$

(ignoring factors that do not depend on θ), and so the gamma distribution for θ is conjugate. With prior distribution

$$\theta \sim \text{Gamma}(\alpha, \beta),$$

the resulting posterior distribution is

$$\theta|y \sim \text{Gamma}\left(\alpha + \sum_{i=1}^n y_i, \beta + \sum_{i=1}^n x_i\right). \tag{2.15}$$

Estimating a rate from Poisson data: an idealized example

Suppose that causes of death are reviewed in detail for a city in the United States for a single year. It is found that 3 persons, out of a population of 200,000, died

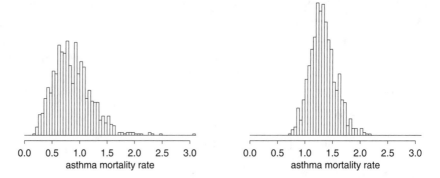

Figure 2.6 *Posterior density for θ, the asthma mortality rate in cases per 100,000 persons per year, with a Gamma(3.0, 5.0) prior distribution: (a) given y = 3 deaths out of 200,000 persons; (b) given y = 30 deaths in 10 years for a constant population of 200,000. The histograms appear jagged because they are constructed from only 1000 random draws from the posterior distribution in each case.*

of asthma, giving a crude estimated asthma mortality rate in the city of 1.5 cases per 100,000 persons per year. A Poisson sampling model is often used for epidemiological data of this form. The Poisson model derives from an assumption of exchangeability among all small intervals of exposure. Under the Poisson model, the sampling distribution of y, the number of deaths in a city of 200,000 in one year, may be expressed as Poisson(2.0θ), where θ represents the true underlying long-term asthma mortality rate in our city (measured in cases per 100,000 persons per year). In the above notation, $y = 3$ is a single observation with exposure $x = 2.0$ (since θ is defined in units of 100,000 people) and unknown rate θ. We can use knowledge about asthma mortality rates around the world to construct a prior distribution for θ and then combine the datum $y = 3$ with that prior distribution to obtain a posterior distribution.

Setting up a prior distribution. What is a sensible prior distribution for θ? Reviews of asthma mortality rates around the world suggest that mortality rates above 1.5 per 100,000 people are rare in Western countries, with typical asthma mortality rates around 0.6 per 100,000. Trial-and-error exploration of the properties of the gamma distribution, the conjugate prior family for this problem, reveals that a Gamma(3.0, 5.0) density provides a plausible prior density for the asthma mortality rate in this example if we assume exchangeability between this city and other cities and this year and other years. The mean of this prior distribution is 0.6 (with a mode of 0.4), and 97.5% of the mass of the density lies below 1.44. In practice, specifying a prior mean sets the ratio of the two gamma parameters, and then the shape parameter can be altered by trial and error to match the prior knowledge about the tail of the distribution.

Posterior distribution. The result in (2.15) shows that the posterior distribution of θ for a Gamma(α, β) prior distribution is Gamma($\alpha + y, \beta + x$) in this case. With the prior distribution and data described, the posterior distribution for θ is Gamma(6.0, 7.0), which has mean 0.86—substantial shrinkage has occurred toward the prior distribution. A histogram of 1000 draws from the posterior distribution for θ is shown as Figure 2.6a. For example, the posterior probability that the long-term death rate from asthma in our city is more than 1.0 per 100,000 per year, computed from the gamma posterior density, is 0.30.

Posterior distribution with additional data. To consider the effect of additional data, suppose that ten years of data are obtained for the city in our example, instead of just one, and it is found that the mortality rate of 1.5 per 100,000 is maintained; we find $y = 30$ deaths over 10 years. Assuming the population is constant at 200,000, and assuming the outcomes in the ten years are independent with constant long-term rate θ, the posterior distribution of θ is then Gamma(33.0, 25.0); Figure 2.6b displays 1000 draws from this distribution. The posterior distribution is much more concentrated than before, and it still lies between the prior distribution and the data. After ten years of data, the posterior mean of θ is 1.32, and the posterior probability that θ exceeds 1.0 is 0.93.

Exponential model

The exponential distribution is commonly used to model 'waiting times' and other continuous, positive, real-valued random variables, usually measured on a time scale. The sampling distribution of an outcome y, given parameter θ, is

$$p(y|\theta) = \theta \exp(-y\theta), \text{ for } y > 0,$$

and $\theta = 1/E(y|\theta)$ is called the 'rate.' Mathematically, the exponential is a special case of the gamma distribution with the parameters $(\alpha, \beta) = (1, \theta)$. In this case, however, it is being used as a sampling distribution for an outcome y, not a prior distribution for a parameter θ, as in the Poisson example.

The exponential distribution has a 'memoryless' property that makes it a natural model for survival or lifetime data; the probability that an object survives an additional length of time t is independent of the time elapsed to this point: $\Pr(y > t+s \,|\, y > s, \theta) = \Pr(y > t \,|\, \theta)$ for any s, t. The conjugate prior distribution for the exponential parameter θ, as for the Poisson mean, is Gamma$(\theta|\alpha, \beta)$ with corresponding posterior distribution Gamma$(\theta|\alpha+1, \beta+y)$. The sampling distribution of n independent exponential observations, $y = (y_1, \ldots, y_n)$, with constant rate θ is

$$p(y|\theta) = \theta^n \exp(-n\bar{y}\theta), \text{ for } \bar{y} \geq 0,$$

which when viewed as the likelihood of θ, for fixed y, is proportional to a Gamma$(n+1, n\bar{y})$ density. Thus the Gamma(α, β) prior distribution for θ can be viewed as $\alpha - 1$ exponential observations with total waiting time β (see Exercise 2.21).

2.8 Example: informative prior distribution and multilevel structure for estimating cancer rates

Section 2.5 considered the effect of the prior distribution on inference given a fixed quantity of data. Here, in contrast, we consider a large set of inferences,

Highest kidney cancer death rates

Figure 2.7 *The counties of the United States with the highest 10% age-standardized death rates for cancer of kidney/ureter for U.S. white males, 1980–1989. Why are most of the shaded counties in the middle of the country? See Section 2.8 for discussion.*

each based on different data but with a common prior distribution. In addition to illustrating the role of the prior distribution, this example introduces hierarchical modeling, to which we return in Chapter 5.

A puzzling pattern in a map

Figure 2.7 shows the counties in the United States with the highest kidney cancer death rates during the 1980s.* The most noticeable pattern in the map is that many of the counties in the Great Plains in the middle of the country, but relatively few counties near the coasts, are shaded.

When shown the map, people come up with many theories to explain the disproportionate shading in the Great Plains: perhaps the air or the water is polluted, or the people tend not to seek medical care so the cancers get detected too late to treat, or perhaps their diet is unhealthy ... These conjectures may all be true but they are not actually needed to explain the patterns in Figure 2.7. To see this, look at Figure 2.8, which plots the 10% of counties with the *lowest* kidney cancer death rates. These are also mostly in the middle of the country. So now we need to explain why these areas have the lowest, as well as the highest, rates.

The issue is sample size. Consider a county of population 1000. Kidney cancer is a rare disease, and, in any ten-year period, a county of 1000 will probably have zero kidney cancer deaths, so that it will be tied for the lowest rate in the country and will be shaded in Figure 2.8. However, there is a chance

---

* The rates are age-adjusted and restricted to white males, issues which need not concern us here.

Lowest kidney cancer death rates

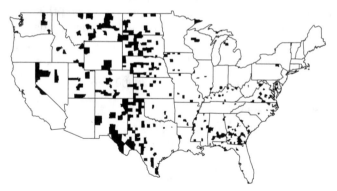

Figure 2.8 *The counties of the United States with the lowest 10% age-standardized death rates for cancer of kidney/ureter for U.S. white males, 1980–1989. Surprisingly, the pattern is somewhat similar to the map of the highest rates, shown in Figure 2.7.*

the county will have one kidney cancer death during the decade. If so, it will have a rate of 1 per 10,000 per year, which is high enough to put it in the top 10% so that it will be shaded in Figure 2.7. The Great Plains has many low-population counties, and so it is disproportionately represented in both maps. There is no evidence from these maps that cancer rates are particularly high there.

Bayesian inference for the cancer death rates

The misleading patterns in the maps of raw rates suggest that a model-based approach to estimating the true underlying rates might be helpful. In particular, it is natural to estimate the underlying cancer death rate in each county j using the model

$$y_j \sim \text{Poisson}(10n_j\theta_j), \qquad (2.16)$$

where y_j is the number of kidney cancer deaths in county j from 1980–1989, n_j is the population of the county, and θ_j is the underlying rate in units of deaths per person per year. (Here we are ignoring the age-standardization, although a generalization of the model to allow for this would be possible.)

This model differs from (2.14) in that θ_j varies between counties, so that (2.16) is a separate model for each of the counties in the U.S. We use the subscript j (rather than i) in (2.16) to emphasize that these are separate parameters, each being estimated from its own data. Were we performing inference for just one of the counties, we would simply write $y \sim \text{Poisson}(10n\theta)$.

To perform Bayesian inference, we need a prior distribution for the unknown rate θ_j. For convenience we use a gamma distribution, which is conjugate to the Poisson. As we shall discuss later, a gamma distribution with parameters $\alpha = 20$ and $\beta = 430{,}000$ is a reasonable prior distribution for underlying kidney cancer death rates in the counties of the U.S. during this period. This prior distribution has a mean of $\alpha/\beta = 4.65 \times 10^{-5}$ and standard deviation $\sqrt{\alpha}/\beta = 1.04 \times 10^{-5}$.

The posterior distribution of θ_j is then,

$$\theta_j | y_j \sim \text{Gamma}(20 + y_j, \, 430000 + 10n_j),$$

which has mean and variance,

$$\mathrm{E}(\theta_j | y_j) \;=\; \frac{20 + y_j}{430{,}000 + 10n_j}$$

$$\text{var}(\theta_j | y_j) \;=\; \frac{20 + y_j}{(430{,}000 + 10n_j)^2}.$$

The posterior mean can be viewed as a weighted average of the raw rate, $y_j/(10n_j)$, and the prior mean, $\alpha/\beta = 4.65 \times 10^{-5}$. (For a similar calculation, see Exercise 2.5.)

Relative importance of the local data and the prior distribution

Inference for a small county.   The relative weighting of prior information and data depends on the population size n_j. For example, consider a small county with $n_j = 1000$:

- For this county, if $y_j = 0$, then the raw death rate is 0 but the posterior mean is $20/440{,}000 = 4.55 \times 10^{-5}$.
- If $y_j = 1$, then the raw death rate is 1 per 1000 per 10 years, or 10^{-4} per person-year (about twice as high as the national mean), but the posterior mean is only $21/440{,}000 = 4.77 \times 10^{-5}$.
- If $y_j = 2$, then the raw death rate is an extremely high 2×10^{-4} per person-year, but the posterior mean is still only $22/440{,}000 = 5.00 \times 10^{-5}$.

With such a small population size, the data are dominated by the prior distribution.

But how likely, *a priori*, is it that y_j will equal 0, 1, 2, and so forth, for this county with $n_j = 1000$? This is determined by the predictive distribution, the marginal distribution of y_j, averaging over the prior distribution of θ_j. As discussed in Section 2.7, the Poisson model with gamma prior distribution has a negative binomial predictive distribution:

$$y_j \sim \text{Neg-bin}\left(\alpha, \frac{\beta}{10n_j}\right).$$

It is perhaps even simpler to simulate directly the predictive distribution of y_j as follows: (1) draw 500 (say) values of θ_j from the Gamma$(20, 430000)$ distribution; (2) for each of these, draw one value y_j from the Poisson distribution with parameter $100000\theta_j$. Of 500 simulations of y_j produced in this way, 319 were 0's, 141 were 1's, 33 were 2's, and 5 were 3's.

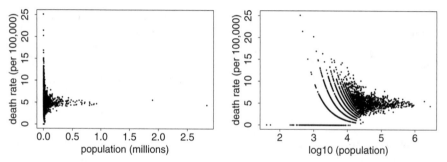

Figure 2.9 *(a) Kidney cancer death rates $y_j/(10n_j)$ vs. population size n_j. (b) Replotted on the scale of $\log_{10}$ population to see the data more clearly. The patterns come from the discreteness of the data ($n_j = 0, 1, 2, \ldots$).*

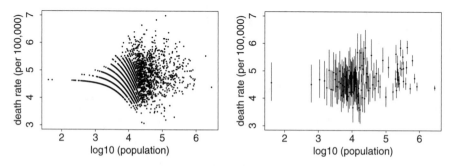

Figure 2.10 *(a) Bayes-estimated posterior mean kidney cancer death rates, $E(\theta_j|y_j) = \frac{20+y_j}{430\,000+10n_j}$ vs. logarithm of population size n_j, the 3071 counties in the U.S. (b) Posterior medians and 50% intervals for θ_j for a sample of 100 counties j. The scales on the y-axes differ from the plots in Figure 2.9b.*

Inference for a large county. Now consider a large county with $n_j = 1$ million. How many cancer deaths y_j might we expect to see in a ten-year period? Again we can use the Gamma(20, 430000) and Poisson($10^7 \cdot \theta_j$) distributions to simulate 500 values of y_j from the predictive distribution. Doing this we found a median of 473 and a 50% interval of [393, 545]. The raw death rate is then as likely or not to fall between 3.93×10^{-5} and 5.45×10^{-5}.

What about the Bayesianly estimated or 'Bayes-adjusted' death rate? For example, if y_j takes on the low value of 393, then the raw death rate is 3.93×10^{-5} and the posterior mean of θ_j is $(20 + 393)/(10,430,000) = 3.96 \times 10^{-5}$, and if $y_j = 545$, then the raw rate is 5.45×10^{-5} and the posterior mean is 5.41×10^{-5}. In this large county, the data dominate the prior distribution.

Comparing counties of different sizes. In the Poisson model (2.16), the variance of y_j is inversely proportional to the exposure parameter n_j, which can thus be considered a 'sample size' for county j. Figure 2.9 shows how the raw kidney cancer death rates vary by population. The extremely high and extremely low rates are all in low-population counties. By comparison, Figure 2.10a shows that the Bayes-estimated rates are much less variable. Finally, Figure 2.10b displays 50% interval estimates for a sample of counties (chosen because it would be hard to display all 3071 in a single plot). The smaller counties supply less information and thus have wider posterior intervals.

Constructing a prior distribution

We now step back and discuss where we got the Gamma$(20, 430000)$ prior distribution for the underlying rates. As we discussed when introducing the model, we picked the gamma distribution for mathematical convenience. We now explain how the two parameters α, β can be estimated from data to match the distribution of the observed cancer death rates $y_j/(10n_j)$. It might seem inappropriate to use the data to set the prior distribution, but we view this as a useful approximation to our preferred approach of hierarchical modeling (introduced in Chapter 5), in which distributional parameters such as α, β in this example are treated as unknowns to be estimated.

Under the model, the observed count y_j for any county j comes from the predictive distribution, $p(y_j) = \int p(y_j|\theta_j)p(\theta_j)d\theta_j$, which in this case is Neg-bin$(\alpha, \beta/(10n_j))$. From Appendix A, we can find the mean and variance of this distribution:

$$
\begin{aligned}
\mathrm{E}(y_j) &= 10n_j\frac{\alpha}{\beta} \\
\mathrm{var}(y_j) &= 10n_j\frac{\alpha}{\beta} + (10n_j)^2\frac{\alpha}{\beta^2}.
\end{aligned}
\tag{2.17}
$$

These can also be derived directly using the mean and variance formulas (1.7) and (1.8); see Exercise 2.6.

Matching the observed mean and variance to their expectations and solving for α and β yields the parameters of the prior distribution. The actual computation is more complicated because it is better to deal with the mean and variance of the rates $y_j/(10n_j)$ than the y_j and we must also deal with the age adjustment, but the basic idea of matching moments presented here illustrates that the information is present to estimate α and β from the data of the 3071 counties.

Figure 2.11 shows the empirical distribution of the raw cancer rates, along with the estimated Gamma$(20, 430000)$ prior distribution for the underlying cancer rates θ_j. The distribution of the raw rates is much broader, which makes sense since they include the Poisson variability as well as the variation between counties.

Our prior distribution is reasonable in this example, but this method of constructing it—by matching moments—is somewhat sloppy and can be difficult to apply in general. In Chapter 5, we discuss how to estimate this and other prior distributions in a more direct Bayesian manner, in the context of hierarchical models.

A more important way this model could be improved is by including infor-

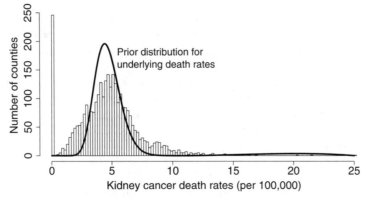

Figure 2.11 *Empirical distribution of the age-adjusted kidney cancer death rates,* $y_j/(10n_j)$, *for the 3071 counties in the U.S., along with the Gamma*$(20, 430000)$ *prior distribution for the underlying cancer rates* θ_j.

mation at the county level that could predict variation in the cancer rates. This would move the model toward a hierarchical Poisson regression of the sort discussed in Chapter 16.

2.9 Noninformative prior distributions

When prior distributions have no population basis, they can be difficult to construct, and there has long been a desire for prior distributions that can be guaranteed to play a minimal role in the posterior distribution. Such distributions are sometimes called 'reference prior distributions,' and the prior density is described as vague, flat, diffuse or *noninformative*. The rationale for using noninformative prior distributions is often said to be 'to let the data speak for themselves,' so that inferences are unaffected by information external to the current data.

Proper and improper prior distributions

We return to the problem of estimating the mean θ of a normal model with known variance σ^2, with a $N(\mu_0, \tau_0^2)$ prior distribution on θ. If the prior precision, $1/\tau_0^2$, is small relative to the data precision, n/σ^2, then the posterior distribution is approximately as if $\tau_0^2 = \infty$:

$$p(\theta|y) \approx N(\theta|\bar{y}, \sigma^2/n).$$

Putting this another way, the posterior distribution is approximately that which would result from assuming $p(\theta)$ is proportional to a constant for $\theta \in (-\infty, \infty)$. Such a distribution is not strictly possible, since the integral of the assumed $p(\theta)$ is infinity, which violates the assumption that probabilities sum to 1. In general, we call a prior density $p(\theta)$ *proper* if it does not depend on data and integrates to 1. (If $p(\theta)$ integrates to any positive finite value, it is called an *unnormalized density* and can be renormalized—multiplied by a constant—to integrate to 1.) Despite the impropriety of the prior distribution in this example, the posterior distribution is proper, given at least one data point.

As a second example of a noninformative prior distribution, consider the normal model with known mean but unknown variance, with the conjugate scaled inverse-χ^2 prior distribution. If the prior degrees of freedom, ν_0, are small relative to the data degrees of freedom, n, then the posterior distribution is approximately as if $\nu_0 = 0$:

$$p(\sigma^2|y) \approx \text{Inv-}\chi^2(\sigma^2|n, v).$$

This limiting form of the posterior distribution can also be derived by defining the prior density for σ^2 as $p(\sigma^2) \propto 1/\sigma^2$, which is improper, having an infinite integral over the range $(0, \infty)$.

Improper prior distributions can lead to proper posterior distributions

In neither of the above two examples does the prior density combine with the likelihood to define a proper joint probability model, $p(y, \theta)$. However, we can proceed with the algebra of Bayesian inference and define an unnormalized posterior density function by

$$p(\theta|y) \propto p(y|\theta)p(\theta).$$

In the above examples (but not always!), the posterior density is in fact proper; that is, $\int p(\theta|y)d\theta$ is finite for all y. Posterior distributions obtained from improper prior distributions must be interpreted with great care—one must always check that the posterior distribution has a finite integral and a sensible form. Their most reasonable interpretation is as approximations in situations where the likelihood dominates the prior density. We discuss this aspect of Bayesian analysis more completely in Chapter 4.

Jeffreys' invariance principle

One approach that is sometimes used to define noninformative prior distributions was introduced by Jeffreys, based on considering one-to-one transformations of the parameter: $\phi = h(\theta)$. By transformation of variables, the prior density $p(\theta)$ is equivalent, in terms of expressing the same beliefs, to the following prior density on ϕ:

$$p(\phi) = p(\theta)\left|\frac{d\theta}{d\phi}\right| = p(\theta)|h'(\theta)|^{-1}. \tag{2.18}$$

Jeffreys' general principle is that any rule for determining the prior density $p(\theta)$ should yield an equivalent result if applied to the transformed parameter; that is, $p(\phi)$ computed by determining $p(\theta)$ and applying (2.18) should match the distribution that is obtained by determining $p(\phi)$ directly using the transformed model, $p(y, \phi) = p(\phi)p(y|\phi)$.

Jeffreys' principle leads to defining the noninformative prior density as $p(\theta) \propto [J(\theta)]^{1/2}$, where $J(\theta)$ is the *Fisher information* for θ:

$$J(\theta) = \text{E}\left[\left(\frac{d\log p(y|\theta)}{d\theta}\right)^2 \middle| \theta\right] = -\text{E}\left[\frac{d^2 \log p(y|\theta)}{d\theta^2} \middle| \theta\right]. \qquad (2.19)$$

To see that Jeffreys' prior model is invariant to parameterization, evaluate $J(\phi)$ at $\theta = h^{-1}(\phi)$:

$$\begin{aligned} J(\phi) &= -\text{E}\left[\frac{d^2 \log p(y|\phi)}{d\phi^2}\right] \\ &= -\text{E}\left[\frac{d^2 \log p(y|\theta = h^{-1}(\phi))}{d\theta^2} \left|\frac{d\theta}{d\phi}\right|^2\right] \\ &= J(\theta)\left|\frac{d\theta}{d\phi}\right|^2; \end{aligned}$$

thus, $J(\phi)^{1/2} = J(\theta)^{1/2}\left|\frac{d\theta}{d\phi}\right|$, as required.

Jeffreys' principle can be extended to multiparameter models, but the results are more controversial. Simpler approaches based on assuming independent noninformative prior distributions for the components of the vector parameter θ can give different results than are obtained with Jeffreys' principle. When the number of parameters in a problem is large, we find it useful to abandon pure noninformative prior distributions in favor of hierarchical models, as we discuss in Chapter 5.

Various noninformative prior distributions for the binomial parameter

Consider the binomial distribution: $y \sim \text{Bin}(n, \theta)$, which has log-likelihood

$$\log p(y|\theta) = \text{constant} + y \log \theta + (n - y)\log(1 - \theta).$$

Routine evaluation of the second derivative and substitution of $\text{E}(y|\theta) = n\theta$ yields the Fisher information:

$$J(\theta) = -\text{E}\left[\frac{d^2 \log p(y|\theta)}{d\theta^2} \middle| \theta\right] = \frac{n}{\theta(1 - \theta)}.$$

Jeffreys' prior density is then $p(\theta) \propto \theta^{-1/2}(1 - \theta)^{-1/2}$, which is a Beta$(\frac{1}{2}, \frac{1}{2})$ density. By comparison, recall the Bayes–Laplace uniform prior density, which can be expressed as $\theta \sim \text{Beta}(1, 1)$. On the other hand, the prior density that is uniform in the natural parameter of the exponential family representation of the distribution is $p(\text{logit}(\theta)) \propto \text{constant}$ (see Exercise 2.7), which corresponds to the improper Beta$(0, 0)$ density on θ. In practice, the difference between these alternatives is often small, since to get from $\theta \sim \text{Beta}(0, 0)$ to $\theta \sim \text{Beta}(1, 1)$ is equivalent to passing from prior to posterior distribution given one more success and one more failure, and usually 2 is a small fraction of the total number of observations. But one must be careful with the improper

Beta$(0, 0)$ prior distribution—if $y = 0$ or n, the resulting posterior distribution is improper!

Pivotal quantities

For the binomial and other single-parameter models, different principles give (slightly) different noninformative prior distributions. But for two cases— location parameters and scale parameters—all principles seem to agree.

1. If the density of y is such that $p(y - \theta|\theta)$ is a function that is free of θ and y, say, $f(u)$, where $u = y - \theta$, then $y - \theta$ is a *pivotal quantity*, and θ is called a pure *location parameter*. In such a case, it is reasonable that a noninformative prior distribution for θ would give $f(y - \theta)$ for the posterior distribution, $p(y - \theta|y)$. That is, under the posterior distribution, $y - \theta$ should still be a pivotal quantity, whose distribution is free of both θ and y. Under this condition, using Bayes' rule, $p(y - \theta|y) \propto p(\theta)p(y - \theta|\theta)$, thereby implying that the noninformative prior density is uniform on θ, $p(\theta) \propto$ constant over the range $(-\infty, \infty)$.

2. If the density of y is such that $p(y/\theta|\theta)$ is a function that is free of θ and y— say, $g(u)$, where $u = y/\theta$—then $u = y/\theta$ is a pivotal quantity and θ is called a pure *scale parameter*. In such a case, it is reasonable that a noninformative prior distribution for θ would give $g(y/\theta)$ for the posterior distribution, $p(y/\theta|y)$. By transformation of variables, the conditional distribution of y given θ can be expressed in terms of the distribution of u given θ,

$$p(y|\theta) = \frac{1}{\theta}p(u|\theta),$$

and similarly,

$$p(\theta|y) = \frac{y}{\theta^2}p(u|y).$$

After letting both $p(u|\theta)$ and $p(u|y)$ equal $g(u)$, we have the identity $p(\theta|y) = \frac{y}{\theta}p(y|\theta)$. Thus, in this case, the reference prior distribution is $p(\theta) \propto 1/\theta$ or, equivalently, $p(\log \theta) \propto 1$ or $p(\theta^2) \propto 1/\theta^2$.

This approach, in which the sampling distribution of the pivot is used as its posterior distribution, can be applied to sufficient statistics in more complicated examples, such as hierarchical normal models.

Even these principles can be misleading in some problems, in the critical sense of suggesting prior distributions that can lead to improper posterior distributions. For example, the uniform prior distribution should *not* be applied to the logarithm of a hierarchical variance parameter, as we discuss in Section 5.4.

Difficulties with noninformative prior distributions

The search for noninformative prior distributions has several problems, including:

1. Searching for a prior distribution that is always vague seems misguided: if the likelihood is truly dominant in a given problem, then the choice among a range of relatively flat prior densities cannot matter. Establishing a particular specification as *the* reference prior distribution seems to encourage its automatic, and possibly inappropriate, use.

2. For many problems, there is no clear choice for a vague prior distribution, since a density that is flat or uniform in one parameterization will not be in another. This is the essential difficulty with Laplace's principle of insufficient reason—on what scale should the principle apply? For example, the 'reasonable' prior density on the normal mean θ above is uniform, while for σ^2, the density $p(\sigma^2) \propto 1/\sigma^2$ seems reasonable. However, if we define $\phi = \log \sigma^2$, then the prior density on ϕ is

$$p(\phi) = p(\sigma^2) \left| \frac{d\sigma^2}{d\phi} \right| \propto \frac{1}{\sigma^2} \sigma^2 = 1;$$

that is, uniform on $\phi = \log \sigma^2$. With discrete distributions, there is the analogous difficulty of deciding how to subdivide outcomes into 'atoms' of equal probability.

3. Further difficulties arise when averaging over a set of competing models that have improper prior distributions, as we discuss in Section 6.7.

Nevertheless, noninformative and reference prior densities are often useful when it does not seem to be worth the effort to quantify one's real prior knowledge as a probability distribution, as long as one is willing to perform the mathematical work to check that the posterior density is proper and to determine the sensitivity of posterior inferences to modeling assumptions of convenience.

   In almost every real problem, the data analyst will have more information than can be conveniently included in the statistical model. This is an issue with the likelihood as well as the prior distribution. In practice, there is always compromise for a number of reasons: to describe the model more conveniently; because it may be difficult to express knowledge accurately in probabilistic form; to simplify computations; or perhaps to avoid using a possibly unreliable source of information. Except for the last reason, these are all arguments for convenience and are best justified by the claim, 'the answer would not have changed much had we been more accurate.' If so few data are available that the choice of noninformative prior distribution makes a difference, one should put relevant information into the prior distribution, perhaps using a hierarchical model, as we discuss in Chapter 5. We return to the issue of accuracy vs. convenience in likelihoods and prior distributions in the examples of the later chapters.

2.10 Bibliographic note

A fascinating detailed account of the early development of the idea of 'inverse probability' (Bayesian inference) is provided in the book by Stigler (1986), on which our brief accounts of Bayes' and Laplace's solutions to the problem of estimating an unknown proportion are based. Bayes' famous 1763 essay in the *Philosophical Transactions of the Royal Society of London* has been reprinted as Bayes (1763).

Introductory textbooks providing complementary discussions of the simple models covered in this chapter were listed at the end of Chapter 1. In particular, Box and Tiao (1973) provide a detailed treatment of Bayesian analysis with the normal model and also discuss highest posterior density regions in some detail. The theory of conjugate prior distributions was developed in detail by Raiffa and Schlaifer (1961). An interesting account of inference for prediction, which also includes extensive details of particular probability models and conjugate prior analyses, appears in Aitchison and Dunsmore (1975).

Noninformative and reference prior distributions have been studied by many researchers. Jeffreys (1961) and Hartigan (1964) discuss invariance principles for noninformative prior distributions. Chapter 1 of Box and Tiao (1973) presents a straightforward and practically-oriented discussion, a brief but detailed survey is given by Berger (1985), and the article by Bernardo (1979) is accompanied by a wide-ranging discussion. Bernardo and Smith (1994) give an extensive treatment of this topic along with many other matters relevant to the construction of prior distributions. Barnard (1985) discusses the relation between pivotal quantities and noninformative Bayesian inference. Kass and Wasserman (1996) provide a review of many approaches for establishing noninformative prior densities based on Jeffreys' rule, and they also discuss the problems that may arise from uncritical use of purportedly noninformative prior specifications. Dawid, Stone, and Zidek (1973) discuss some difficulties that can arise with noninformative prior distributions; also see Jaynes (1980).

Jaynes (1983) discusses in several places the idea of objectively constructing prior distributions based on invariance principles and *maximum entropy*. Appendix A of Bretthorst (1988) outlines an objective Bayesian approach to assigning prior distributions, as applied to the problem of estimating the parameters of a sinusoid from time series data. More discussions of maximum entropy models appear in Jaynes (1982), Skilling (1989), and Gull (1989a); see Titterington (1984) and Donoho et al. (1992) for other views.

The data for the placenta previa example come from a study from 1922 reported in James (1987). The Bayesian analysis of age-adjusted kidney cancer death rates in Section 2.8 is adapted from Manton et al. (1989); see also Gelman and Nolan (2002a) for more on this particular example and Bernardinelli, Clayton, and Montomoli (1995) for a general discussion of prior distributions for disease mapping. Gelman and Price (1999) discuss artifacts in maps of parameter estimates, and Louis (1984), Shen and Louis (1998), and Louis and Shen (1999) analyze the general problem of estimation of ensembles of parameters, a topic to which we return in Chapter 5.

2.11 Exercises

1. Posterior inference: suppose there is Beta$(4, 4)$ prior distribution on the probability θ that a coin will yield a 'head' when spun in a specified manner. The coin is independently spun ten times, and 'heads' appear fewer than 3 times. You are not told how many heads were seen, only that the number is less than 3. Calculate your exact posterior density (up to a proportionality constant) for θ and sketch it.

2. Predictive distributions: consider two coins, C_1 and C_2, with the following characteristics: $\Pr(\text{heads}|C_1) = 0.6$ and $\Pr(\text{heads}|C_2) = 0.4$. Choose one of the coins at random and imagine spinning it repeatedly. Given that the first two spins from the chosen coin are tails, what is the expectation of the number of additional spins until a head shows up?

3. Predictive distributions: let y be the number of 6's in 1000 rolls of a fair die.

   (a) Sketch the approximate distribution of y, based on the normal approximation.

   (b) Using the normal distribution table, give approximate 5%, 25%, 50%, 75%, and 95% points for the distribution of y.

4. Predictive distributions: let y be the number of 6's in 1000 independent rolls of a particular real die, which may be unfair. Let θ be the probability that the die lands on '6.' Suppose your prior distribution for θ is as follows:

$$\begin{aligned}
\Pr(\theta = 1/12) &= 0.25, \\
\Pr(\theta = 1/6) &= 0.5, \\
\Pr(\theta = 1/4) &= 0.25.
\end{aligned}$$

   (a) Using the normal approximation for the conditional distributions, $p(y|\theta)$, sketch your approximate prior predictive distribution for y.

   (b) Give approximate 5%, 25%, 50%, 75%, and 95% points for the distribution of y. (Be careful here: y does not have a normal distribution, but you can still use the normal distribution as part of your analysis.)

5. Posterior distribution as a compromise between prior information and data: let y be the number of heads in n spins of a coin, whose probability of heads is θ.

   (a) If your prior distribution for θ is uniform on the range $[0, 1]$, derive your prior predictive distribution for y,

$$\Pr(y = k) = \int_0^1 \Pr(y = k|\theta)d\theta,$$

   for each $k = 0, 1, \ldots, n$.

   (b) Suppose you assign a Beta(α, β) prior distribution for θ, and then you observe y heads out of n spins. Show algebraically that your posterior mean of θ always lies between your prior mean, $\frac{\alpha}{\alpha+\beta}$, and the observed relative frequency of heads, $\frac{y}{n}$.

(c) Show that, if the prior distribution on θ is uniform, the posterior variance of θ is always less than the prior variance.

(d) Give an example of a Beta(α, β) prior distribution and data y, n, in which the posterior variance of θ is higher than the prior variance.

6. Predictive distributions: Derive the mean and variance (2.17) of the negative binomial predictive distribution for the cancer rate example, using the mean and variance formulas (1.7) and (1.8).

7. Noninformative prior densities:

(a) For the binomial likelihood, $y \sim \text{Bin}(n, \theta)$, show that $p(\theta) \propto \theta^{-1}(1 - \theta)^{-1}$ is the uniform prior distribution for the natural parameter of the exponential family.

(b) Show that if $y = 0$ or n, the resulting posterior distribution is improper.

8. Normal distribution with unknown mean: a random sample of n students is drawn from a large population, and their weights are measured. The average weight of the n sampled students is $\bar{y} = 150$ pounds. Assume the weights in the population are normally distributed with unknown mean θ and known standard deviation 20 pounds. Suppose your prior distribution for θ is normal with mean 180 and standard deviation 40.

(a) Give your posterior distribution for θ. (Your answer will be a function of n.)

(b) A new student is sampled at random from the same population and has a weight of $\tilde{y}$ pounds. Give a posterior predictive distribution for $\tilde{y}$. (Your answer will still be a function of n.)

(c) For $n = 10$, give a 95% posterior interval for θ and a 95% posterior predictive interval for $\tilde{y}$.

(d) Do the same for $n = 100$.

9. Setting parameters for a beta prior distribution: suppose your prior distribution for θ, the proportion of Californians who support the death penalty, is beta with mean 0.6 and standard deviation 0.3.

(a) Determine the parameters α and β of your prior distribution. Sketch the prior density function.

(b) A random sample of 1000 Californians is taken, and 65% support the death penalty. What are your posterior mean and variance for θ? Draw the posterior density function.

10. Discrete sample spaces: suppose there are N cable cars in San Francisco, numbered sequentially from 1 to N. You see a cable car at random; it is numbered 203. You wish to estimate N. (See Goodman, 1952, for a discussion and references to several versions of this problem, and Jeffreys, 1961, Lee, 1989, and Jaynes, 1996, for Bayesian treatments.)

Year	Fatal accidents	Passenger deaths	Death rate
1976	24	734	0.19
1977	25	516	0.12
1978	31	754	0.15
1979	31	877	0.16
1980	22	814	0.14
1981	21	362	0.06
1982	26	764	0.13
1983	20	809	0.13
1984	16	223	0.03
1985	22	1066	0.15

Table 2.2 *Worldwide airline fatalities, 1976–1985. Death rate is passenger deaths per 100 million passenger miles. Source:* Statistical Abstract of the United States.

(a) Assume your prior distribution on N is geometric with mean 100; that is,

$$p(N) = (1/100)(99/100)^{N-1}, \quad \text{for } N = 1, 2, \ldots.$$

What is your posterior distribution for N?

(b) What are the posterior mean and standard deviation of N? (Sum the infinite series analytically or approximate them on the computer.)

(c) Choose a reasonable 'noninformative' prior distribution for N and give the resulting posterior distribution, mean, and standard deviation for N.

11. Computing with a nonconjugate single-parameter model: suppose $y_1, \ldots, y_5$ are independent samples from a Cauchy distribution with unknown center θ and known scale 1: $p(y_i|\theta) \propto 1/(1 + (y_i - \theta)^2)$. Assume, for simplicity, that the prior distribution for θ is uniform on $[0, 1]$. Given the observations $(y_1, \ldots, y_5) = (-2, -1, 0, 1.5, 2.5)$:

(a) Compute the unnormalized posterior density function, $p(\theta)p(y|\theta)$, on a grid of points $\theta = 0, \frac{1}{m}, \frac{2}{m}, \ldots, 1$, for some large integer m. Using the grid approximation, compute and plot the normalized posterior density function, $p(\theta|y)$, as a function of θ.

(b) Sample 1000 draws of θ from the posterior density and plot a histogram of the draws.

(c) Use the 1000 samples of θ to obtain 1000 samples from the predictive distribution of a future observation, y_6, and plot a histogram of the predictive draws.

12. Jeffreys' prior distributions: suppose $y|\theta \sim \text{Poisson}(\theta)$. Find Jeffreys' prior density for θ, and then find α and β for which the Gamma(α, β) density is a close match to Jeffreys' density.

13. Discrete data: Table 2.2 gives the number of fatal accidents and deaths on scheduled airline flights per year over a ten-year period. We use these data as a numerical example for fitting discrete data models.

   (a) Assume that the numbers of fatal accidents in each year are independent with a Poisson(θ) distribution. Set a prior distribution for θ and determine the posterior distribution based on the data from 1976 through 1985. Under this model, give a 95% predictive interval for the number of fatal accidents in 1986. You can use the normal approximation to the gamma and Poisson or compute using simulation.

   (b) Assume that the numbers of fatal accidents in each year follow independent Poisson distributions with a constant rate and an exposure in each year proportional to the number of passenger miles flown. Set a prior distribution for θ and determine the posterior distribution based on the data for 1976–1985. (Estimate the number of passenger miles flown in each year by dividing the appropriate columns of Table 2.2 and ignoring round-off errors.) Give a 95% predictive interval for the number of fatal accidents in 1986 under the assumption that 8×10^{11} passenger miles are flown that year.

   (c) Repeat (a) above, replacing 'fatal accidents' with 'passenger deaths.'

   (d) Repeat (b) above, replacing 'fatal accidents' with 'passenger deaths.'

   (e) In which of the cases (a)–(d) above does the Poisson model seem more or less reasonable? Why? Discuss based on general principles, without specific reference to the numbers in Table 2.2.

   Incidentally, in 1986, there were 22 fatal accidents, 546 passenger deaths, and a death rate of 0.06 per 100 million miles flown. We return to this example in Exercises 3.12, 6.2, 6.3, and 7.14.

14. Algebra of the normal model:

   (a) Fill in the steps to derive (2.9)–(2.10), and (2.11)–(2.12).

   (b) Derive (2.11) and (2.12) by starting with a N(μ_0, τ_0^2) prior distribution and adding data points one at a time, using the posterior distribution at each step as the prior distribution for the next.

15. Beta distribution: assume the result, from standard advanced calculus, that

$$\int_0^1 u^{\alpha-1}(1-u)^{\beta-1}du = \frac{\Gamma(\alpha)\Gamma(\beta)}{\Gamma(\alpha+\beta)}.$$

   If Z has a beta distribution with parameters α and β, find $E[Z^m(1-Z)^n]$ for any nonnegative integers m and n. Hence derive the mean and variance of Z.

16. Beta-binomial distribution and Bayes' prior distribution: suppose y has a binomial distribution for given n and unknown parameter θ, where the prior distribution of θ is Beta(α, β).

(a) Find $p(y)$, the *marginal distribution* of y, for $y = 0, \ldots, n$ (unconditional on θ). This discrete distribution is known as the *beta-binomial*, for obvious reasons.

(b) Show that if the beta-binomial probability is constant in y, then the prior distribution has to have $\alpha = \beta = 1$.

17. Informative prior distribution: as a modern-day Laplace, you have more definite beliefs about the ratio of male to female births than reflected by his uniform prior distribution. In particular, if θ represents the proportion of female births in a given population, you are willing to place a Beta(100, 100) prior distribution on θ.

(a) Show that this means you are more than 95% sure that θ is between 0.4 and 0.6, although you are ambivalent as to whether it is greater or less than 0.5.

(b) Now you observe that out of a random sample of 1,000 births, 511 are boys. What is your posterior probability that $\theta > 0.5$?

Compute using the exact beta distribution or the normal approximation.

18. Predictive distribution and tolerance intervals: a scientist using an apparatus of known standard deviation 0.12 takes nine independent measurements of some quantity. The measurements are assumed to be normally distributed, with the stated standard deviation and unknown mean θ, where the scientist is willing to place a vague prior distribution on θ. If the sample mean obtained is 17.653, obtain limits between which a tenth measurement will lie with 99% probability. This is called a 99% *tolerance interval*.

19. Posterior intervals: unlike the central posterior interval, the highest posterior interval is *not* invariant to transformation. For example, suppose that, given σ^2, the quantity nv/σ^2 is distributed as χ^2_n, and that σ has the (improper) noninformative prior density $p(\sigma) \propto \sigma^{-1}, \sigma > 0$.

(a) Prove that the corresponding prior density for σ^2 is $p(\sigma^2) \propto \sigma^{-2}$.

(b) Show that the 95% highest posterior density region for σ^2 is not the same as the region obtained by squaring a posterior interval for σ.

20. Poisson model: derive the gamma posterior distribution (2.15) for the Poisson model parameterized in terms of rate and exposure with conjugate prior distribution.

21. Exponential model with conjugate prior distribution:

(a) Show that if $y|\theta$ is exponentially distributed with rate θ, then the gamma prior distribution is conjugate for inferences about θ given an iid sample of y values.

(b) Show that the equivalent prior specification for the mean, $\phi = 1/\theta$, is inverse-gamma. (That is, derive the latter density function.)

(c) The length of life of a light bulb manufactured by a certain process has an exponential distribution with unknown rate θ. Suppose the prior distribution for θ is a gamma distribution with coefficient of variation 0.5. (The *coefficient of variation* is defined as the standard deviation divided by the mean.) A random sample of light bulbs is to be tested and the lifetime of each obtained. If the coefficient of variation of the distribution of θ is to be reduced to 0.1, how many light bulbs need to be tested?

(d) In part (c), if the coefficient of variation refers to ϕ instead of θ, how would your answer be changed?

22. Censored and uncensored data in the exponential model:

(a) Suppose $y|\theta$ is exponentially distributed with rate θ, and the marginal (prior) distribution of θ is Gamma(α, β). Suppose we observe that $y \geq 100$, but do not observe the exact value of y. What is the posterior distribution, $p(\theta|y \geq 100)$, as a function of α and β? Write down the posterior mean and variance of θ.

(b) In the above problem, suppose that we are now told that y is exactly 100. Now what are the posterior mean and variance of θ?

(c) Explain why the posterior variance of θ is higher in part (b) even though more information has been observed. Why does this not contradict identity (2.8) on page 37?

23. Conjugate prior distributions for the normal variance: we will analyze the football example of Section 2.7 using informative conjugate prior distributions for σ^2.

(a) Express your own prior knowledge about σ^2 (before seeing the data) in terms of a prior mean and standard deviation for σ. Use the expressions for the mean and standard deviation of the inverse-gamma distribution in Appendix A to find the parameters of your conjugate prior distribution for σ^2.

(b) Suppose that instead, from our prior knowledge, we are 95% sure that σ falls between 3 points and 20 points. Find a conjugate inverse-gamma prior distribution for σ whose 2.5% and 97.5% quantiles are approximately 3 and 20. (You can do this using trial and error on the computer.)

(c) Reexpress each of the two prior distributions above in terms of the scaled inverse-χ^2 parameterization; that is, give σ_0 and ν_0 for each case.

(d) For each of the two prior distributions above, determine the posterior distribution for σ^2. Graph the posterior density and give the central posterior 95% interval for σ in each case.

(e) Do the answers for this problem differ much from the results using the noninformative prior distribution in Section 2.7? Discuss.

CHAPTER 3

Introduction to multiparameter models

Virtually every practical problem in statistics involves more than one unknown or unobservable quantity. It is in dealing with such problems that the simple conceptual framework of the Bayesian approach reveals its principal advantages over other methods of inference. Although a problem can include several parameters of interest, conclusions will often be drawn about one, or only a few, parameters at a time. In this case, the ultimate aim of a Bayesian analysis is to obtain the *marginal* posterior distribution of the particular parameters of interest. In principle, the route to achieving this aim is clear: we first require the *joint* posterior distribution of *all* unknowns, and then we integrate this distribution over the unknowns that are not of immediate interest to obtain the desired marginal distribution. Or equivalently, using simulation, we draw samples from the joint posterior distribution and then look at the parameters of interest and ignore the values of the other unknowns. In many problems there is no interest in making inferences about many of the unknown parameters, although they are required in order to construct a realistic model. Parameters of this kind are often called *nuisance parameters*. A classic example is the scale of the random errors in a measurement problem.

We begin this chapter with a general treatment of nuisance parameters and then cover the normal distribution with unknown mean and variance in Sections 3.2–3.4. Sections 3.5 and 3.6 present inference for the multinomial and multivariate normal distributions—the simplest models for discrete and continuous multivariate data, respectively. The chapter concludes with an analysis of a nonconjugate logistic regression model, using numerical computation of the posterior density on a grid.

3.1 Averaging over 'nuisance parameters'

To express the ideas of joint and marginal posterior distributions mathematically, suppose θ has two parts, each of which can be a vector, $\theta = (\theta_1, \theta_2)$, and further suppose that we are only interested (at least for the moment) in inference for θ_1, so θ_2 may be considered a 'nuisance' parameter. For instance, in the simple example,

$$y|\mu, \sigma^2 \sim \mathrm{N}(\mu, \sigma^2),$$

in which both μ ($=$'θ_1') and σ^2 ($=$'θ_2') are unknown, interest commonly centers on μ.

We seek the conditional distribution of the parameter of interest given the observed data; in this case, $p(\theta_1|y)$. This is derived from the *joint posterior density,*

$$p(\theta_1, \theta_2 | y) \propto p(y | \theta_1, \theta_2) p(\theta_1, \theta_2),$$

by averaging over θ_2:

$$p(\theta_1 | y) = \int p(\theta_1, \theta_2 | y) d\theta_2.$$

Alternatively, the joint posterior density can be factored to yield

$$p(\theta_1 | y) = \int p(\theta_1 | \theta_2, y) p(\theta_2 | y) d\theta_2, \qquad (3.1)$$

which shows that the posterior distribution of interest, $p(\theta_1 | y)$, is a *mixture* of the conditional posterior distributions given the nuisance parameter, θ_2, where $p(\theta_2 | y)$ is a weighting function for the different possible values of θ_2. The weights depend on the posterior density of θ_2 and thus on a combination of evidence from data and prior model. The averaging over nuisance parameters θ_2 can be interpreted very generally; for example, θ_2 can include a discrete component representing different possible sub-models.

We rarely evaluate the integral (3.1) explicitly, but it suggests an important practical strategy for both constructing and computing with multiparameter models. Posterior distributions can be computed by marginal and conditional simulation, first drawing θ_2 from its marginal posterior distribution and then θ_1 from its conditional posterior distribution, given the drawn value of θ_2. In this way the integration embodied in (3.1) is performed indirectly. A canonical example of this form of analysis is provided by the normal model with unknown mean and variance, to which we now turn.

3.2 Normal data with a noninformative prior distribution

As the prototype example of estimating the mean of a population from a sample, we consider a vector y of n iid observations from a univariate normal distribution, $N(\mu, \sigma^2)$; the generalization to the multivariate normal distribution appears in Section 3.6. We begin by analyzing the model under a noninformative prior distribution, with the understanding that this is no more than a convenient assumption for the purposes of exposition and is easily extended to informative prior distributions.

A noninformative prior distribution

We saw in Chapter 2 that a sensible vague prior density for μ and σ, assuming prior independence of location and scale parameters, is uniform on $(\mu, \log \sigma)$ or, equivalently,

$$p(\mu, \sigma^2) \propto (\sigma^2)^{-1}.$$

The joint posterior distribution, $p(\mu, \sigma^2|y)$

Under this conventional improper prior density, the joint posterior distribution is proportional to the likelihood function multiplied by the factor $1/\sigma^2$:

$$p(\mu, \sigma^2|y) \quad \propto \quad \sigma^{-n-2} \exp\left(-\frac{1}{2\sigma^2} \sum_{i=1}^{n} (y_i - \mu)^2\right)$$

$$= \quad \sigma^{-n-2} \exp\left(-\frac{1}{2\sigma^2} \left[\sum_{i=1}^{n} (y_i - \bar{y})^2 + n(\bar{y} - \mu)^2\right]\right)$$

$$= \quad \sigma^{-n-2} \exp\left(-\frac{1}{2\sigma^2}[(n-1)s^2 + n(\bar{y} - \mu)^2]\right), \quad (3.2)$$

where

$$s^2 = \frac{1}{n-1} \sum_{i=1}^{n} (y_i - \bar{y})^2$$

is the sample variance of the y_i's. The sufficient statistics are $\bar{y}$ and s^2.

The conditional posterior distribution, $p(\mu|\sigma^2, y)$

In order to factor the joint posterior density as in (3.1), we consider first the conditional posterior density, $p(\mu|\sigma^2, y)$, and then the marginal posterior density, $p(\sigma^2|y)$. To determine the posterior distribution of μ, given σ^2, we simply use the result derived in Section 2.6 for the mean of a normal distribution with *known* variance and a uniform prior distribution:

$$\mu|\sigma^2, y \sim N(\bar{y}, \sigma^2/n). \quad (3.3)$$

The marginal posterior distribution, $p(\sigma^2|y)$

To determine $p(\sigma^2|y)$, we must average the joint distribution (3.2) over μ:

$$p(\sigma^2|y) \propto \int \sigma^{-n-2} \exp\left(-\frac{1}{2\sigma^2}[(n-1)s^2 + n(\bar{y} - \mu)^2]\right) d\mu.$$

Integrating this expression over μ requires evaluating the integral of the factor $\exp\left(-\frac{1}{2\sigma^2}n(\bar{y} - \mu)^2\right)$; which is a simple normal integral; thus,

$$p(\sigma^2|y) \quad \propto \quad \sigma^{-n-2} \exp\left(-\frac{1}{2\sigma^2}(n-1)s^2\right) \sqrt{2\pi\sigma^2/n}$$

$$\propto \quad (\sigma^2)^{-(n+1)/2} \exp\left(-\frac{(n-1)s^2}{2\sigma^2}\right), \quad (3.4)$$

which is a scaled inverse-χ^2 density:

$$\sigma^2|y \sim \text{Inv-}\chi^2(n-1, s^2). \quad (3.5)$$

We have thus factored the joint posterior density (3.2) as the product of conditional and marginal posterior densities: $p(\mu, \sigma^2|y) = p(\mu|\sigma^2, y)p(\sigma^2|y)$.

This marginal posterior distribution for σ^2 has a remarkable similarity to the analogous sampling theory result: conditional on σ^2 (and μ), the distribution of the appropriately scaled sufficient statistic, $\frac{(n-1)s^2}{\sigma^2}$, is χ^2_{n-1}. Considering our derivation of the reference prior distribution for the scale parameter in Section 2.9, however, this result is not surprising.

Sampling from the joint posterior distribution

It is easy to draw samples from the joint posterior distribution: first draw σ^2 from (3.5), then draw μ from (3.3). We also derive some analytical results for the posterior distribution, since this is one of the few multiparameter problems simple enough to solve in closed form.

Analytic form of the marginal posterior distribution of μ

The population mean, μ, is typically the estimand of interest, and so the objective of the Bayesian analysis is the marginal posterior distribution of μ, which can be obtained by integrating σ^2 out of the joint posterior distribution. The representation (3.1) shows that the posterior distribution of μ can be regarded as a mixture of normal distributions, mixed over the scaled inverse-χ^2 distribution for the variance, σ^2. We can derive the marginal posterior density for μ by integrating the joint posterior density over σ^2:

$$p(\mu|y) = \int_0^\infty p(\mu, \sigma^2|y)d\sigma^2.$$

This integral can be evaluated using the substitution

$$z = \frac{A}{2\sigma^2}, \quad \text{where } A = (n-1)s^2 + n(\mu - \bar{y})^2,$$

and recognizing that the result is an unnormalized gamma integral:

$$p(\mu|y) \propto A^{-n/2} \int_0^\infty z^{(n-2)/2} \exp(-z)dz$$

$$\propto [(n-1)s^2 + n(\mu - \bar{y})^2]^{-n/2}$$

$$\propto \left[1 + \frac{n(\mu - \bar{y})^2}{(n-1)s^2}\right]^{-n/2}.$$

This is the $t_{n-1}(\bar{y}, s^2/n)$ density (see Appendix A).

To put it another way, we have shown that, under the noninformative uniform prior distribution on $(\mu, \log\sigma)$, the posterior distribution of μ has the form

$$\left.\frac{\mu - \bar{y}}{s/\sqrt{n}}\right|y \sim t_{n-1},$$

where t_{n-1} denotes the standard Student-t density (location 0, scale 1) with $n-1$ degrees of freedom. This marginal posterior distribution provides another interesting comparison with sampling theory. Under the sampling distribution, $p(y|\mu, \sigma^2)$, the following relation holds:

$$\left. \frac{\bar{y} - \mu}{s/\sqrt{n}} \right| \mu, \sigma^2 \sim t_{n-1}.$$

The sampling distribution of the *pivotal quantity* $(\bar{y} - \mu)/(s/\sqrt{n})$ does not depend on the nuisance parameter σ^2, and its posterior distribution does not depend on data. In general, a pivotal quantity for the estimand is defined as a nontrivial function of the data and the estimand whose sampling distribution is independent of all parameters and data.

Posterior predictive distribution for a future observation

The posterior predictive distribution for a future observation, $\tilde{y}$, can be written as a mixture, $p(\tilde{y}|y) = \int\int p(\tilde{y}|\mu, \sigma^2, y) p(\mu, \sigma^2|y) d\mu d\sigma^2$. The first of the two factors in the integral is just the normal distribution for the future observation given the values of (μ, σ^2), and in fact does not depend on y at all. To draw from the posterior predictive distribution, first draw μ, σ^2 from their joint posterior distribution and then simulate $\tilde{y} \sim N(\mu, \sigma^2)$.

In fact, the posterior predictive distribution of $\tilde{y}$ is a Student-t distribution with location $\bar{y}$, scale $(1 + \frac{1}{n})^{1/2} s$, and $n-1$ degrees of freedom. This analytic form is obtained using the same techniques as in the derivation of the posterior distribution of μ. Specifically, the distribution can be obtained by integrating out the parameters μ, σ^2 according to their joint posterior distribution. We can identify the result more easily by noticing that the factorization $p(\tilde{y}|\sigma^2, y) = \int p(\tilde{y}|\mu, \sigma^2, y) p(\mu|\sigma^2, y) d\mu$ leads to $p(\tilde{y}|\sigma^2, y) = N(\tilde{y}|\bar{y}, (1 + \frac{1}{n})\sigma^2)$, which is the same, up to a changed scale factor, as the distribution of $\mu|\sigma^2, y$.

Example. Estimating the speed of light
Simon Newcomb set up an experiment in 1882 to measure the speed of light. Newcomb measured the amount of time required for light to travel a distance of 7442 meters. A histogram of Newcomb's 66 measurements is shown in Figure 3.1. There are two unusually low measurements and then a cluster of measurements that are approximately symmetrically distributed. We (inappropriately) apply the normal model, assuming that all 66 measurements are independent draws from a normal distribution with mean μ and variance σ^2. The main substantive goal is posterior inference for μ. The outlying measurements do not fit the normal model; we discuss Bayesian methods for measuring the lack of fit for these data in Section 6.3. The mean of the 66 measurements is $\bar{y} = 26.2$, and the sample standard deviation is $s = 10.8$. Assuming the noninformative prior distribution $p(\mu, \sigma^2) \propto (\sigma^2)^{-1}$, a 95% central posterior interval for μ is obtained from the t_{65} marginal posterior distribution of μ as $[\bar{y} \pm 1.997s/\sqrt{66}] = [23.6, 28.8]$.

The posterior interval can also be obtained by simulation. Following the factorization of the posterior distribution given by (3.5) and (3.3), we first draw a random value of $\sigma^2 \sim \text{Inv-}\chi^2(65, s^2)$ as $65s^2$ divided by a random draw from the χ_{65}^2 distribution (see Appendix A). Then given this value of σ^2, we draw μ from its conditional posterior distribution, $N(26.2, \sigma^2/66)$. Based on 1000 simulated

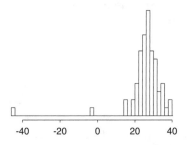

Figure 3.1 *Histogram of Simon Newcomb's measurements for estimating the speed of light, from Stigler (1977). The data are recorded as deviations from 24,800 nanoseconds.*

values of (μ, σ^2), we estimate the posterior median of μ to be 26.2 and a 95% central posterior interval for μ to be $[23.6, 28.9]$, quite close to the analytically calculated interval.

Incidentally, based on the currently accepted value of the speed of light, the 'true value' for μ in Newcomb's experiment is 33.0, which falls outside our 95% interval. This reinforces the fact that posterior inferences are only as good as the model and the experiment that produced the data.

3.3 Normal data with a conjugate prior distribution

A family of conjugate prior distributions

A first step toward a more general model is to assume a conjugate prior distribution for the two-parameter univariate normal sampling model in place of the noninformative prior distribution just considered. The form of the likelihood displayed in (3.2) and the subsequent discussion shows that the conjugate prior density must also have the product form $p(\sigma^2)p(\mu|\sigma^2)$, where the marginal distribution of σ^2 is scaled inverse-χ^2 and the conditional distribution of μ given σ^2 is normal (so that marginally μ has a Student-t distribution). A convenient parameterization is given by the following specification:

$$\mu|\sigma^2 \sim \text{N}(\mu_0, \sigma^2/\kappa_0)$$
$$\sigma^2 \sim \text{Inv-}\chi^2(\nu_0, \sigma_0^2),$$

which corresponds to the joint prior density

$$p(\mu, \sigma^2) \propto \sigma^{-1}(\sigma^2)^{-(\nu_0/2+1)} \exp\left(-\frac{1}{2\sigma^2}[\nu_0\sigma_0^2 + \kappa_0(\mu_0 - \mu)^2]\right). \qquad (3.6)$$

We label this the N-Inv-$\chi^2(\mu_0, \sigma_0^2/\kappa_0; \nu_0, \sigma_0^2)$ density; its four parameters can be identified as the location and scale of μ and the degrees of freedom and scale of σ^2, respectively.

The appearance of σ^2 in the conditional distribution of $\mu|\sigma^2$ means that μ and σ^2 are necessarily dependent in their joint conjugate prior density: for example, if σ^2 is large, then a high-variance prior distribution is induced on μ. This dependence is notable, considering that conjugate prior distributions are used largely for convenience. Upon reflection, however, it often makes sense for the prior variance of the mean to be tied to σ^2, which is the sampling variance of the observation y. In this way, prior belief about μ is calibrated by the scale of measurement of y and is equivalent to κ_0 prior measurements on this scale. In the next section, we consider a prior distribution in which μ and σ^2 are independent.

The joint posterior distribution, $p(\mu, \sigma^2 | y)$

Multiplying the prior density (3.6) by the normal likelihood yields the posterior density

$$
\begin{aligned}
p(\mu, \sigma^2 | y) \quad \propto \quad & \sigma^{-1}(\sigma^2)^{-(\nu_0/2+1)} \exp\left(-\frac{1}{2\sigma^2}[\nu_0\sigma_0^2 + \kappa_0(\mu - \mu_0)^2]\right) \times \\
& \times (\sigma^2)^{-n/2} \exp\left(-\frac{1}{2\sigma^2}[(n-1)s^2 + n(\bar{y} - \mu)^2]\right) \qquad (3.7) \\
= \quad & \text{N-Inv-}\chi^2(\mu_n, \sigma_n^2/\kappa_n; \nu_n, \sigma_n^2),
\end{aligned}
$$

where, after some algebra (see Exercise 3.9), it can be shown that

$$
\begin{aligned}
\mu_n &= \frac{\kappa_0}{\kappa_0 + n}\mu_0 + \frac{n}{\kappa_0 + n}\bar{y} \\
\kappa_n &= \kappa_0 + n \\
\nu_n &= \nu_0 + n \\
\nu_n\sigma_n^2 &= \nu_0\sigma_0^2 + (n-1)s^2 + \frac{\kappa_0 n}{\kappa_0 + n}(\bar{y} - \mu_0)^2.
\end{aligned}
$$

The parameters of the posterior distribution combine the prior information and the information contained in the data. For example μ_n is a weighted average of the prior mean and the sample mean, with weights determined by the relative precision of the two pieces of information. The posterior degrees of freedom, ν_n, is the prior degrees of freedom plus the sample size. The posterior sum of squares, $\nu_n\sigma_n^2$, combines the prior sum of squares, the sample sum of squares, and the additional uncertainty conveyed by the difference between the sample mean and the prior mean.

The conditional posterior distribution, $p(\mu | \sigma^2, y)$

The conditional posterior density of μ, given σ^2, is proportional to the joint posterior density (3.7) with σ^2 held constant,

$$
\begin{aligned}
\mu | \sigma^2, y \quad &\sim \quad N(\mu_n, \sigma^2/\kappa_n) \\
&= \quad N\left(\frac{\frac{\kappa_0}{\sigma^2}\mu_0 + \frac{n}{\sigma^2}\bar{y}}{\frac{\kappa_0}{\sigma^2} + \frac{n}{\sigma^2}}, \frac{1}{\frac{\kappa_0}{\sigma^2} + \frac{n}{\sigma^2}}\right), \qquad (3.8)
\end{aligned}
$$

which agrees, as it must, with the analysis in Section 2.6 of μ with σ considered fixed.

The marginal posterior distribution, $p(\sigma^2|y)$

The marginal posterior density of σ^2, from (3.7), is scaled inverse-χ^2:

$$\sigma^2|y \sim \text{Inv-}\chi^2(\nu_n, \sigma_n^2). \tag{3.9}$$

Sampling from the joint posterior distribution

To sample from the joint posterior distribution, just as in the previous section, we first draw σ^2 from its marginal posterior distribution (3.9), then draw μ from its normal conditional posterior distribution (3.8), using the simulated value of σ^2.

Analytic form of the marginal posterior distribution of μ

Integration of the joint posterior density with respect to σ^2, in a precisely analogous way to that used in the previous section, shows that the marginal posterior density for μ is

$$\begin{aligned}
p(\mu|y) \quad &\propto \quad \left[1 + \frac{\kappa_n(\mu - \mu_n)^2}{\nu_n \sigma_n^2}\right]^{-(\nu_n+1)/2} \\
&= \quad t_{\nu_n}(\mu|\mu_n, \sigma_n^2/\kappa_n).
\end{aligned}$$

3.4 Normal data with a semi-conjugate prior distribution

A family of prior distributions

Another approach to constructing a prior distribution for the mean and variance of a normal distribution is to specify the prior distributions for μ and σ^2 independently. If we try to follow the conjugate forms as closely as possible, we have

$$\begin{aligned}
\mu|\sigma^2 \quad &\sim \quad N(\mu_0, \tau_0^2) \\
\sigma^2 \quad &\sim \quad \text{Inv-}\chi^2(\nu_0, \sigma_0^2).
\end{aligned} \tag{3.10}$$

This prior independence—that is, the assumption that the distribution of μ, given σ^2, does not depend on σ^2—is attractive in problems for which the prior information on μ does *not* take the form of a fixed number of observations with variance σ^2. For example, suppose μ is the unknown weight of a particular student, the data y are weighings on a particular scale with unknown variance, and the prior information consists of a visual inspection: the student looks to weigh about 150 pounds, with a subjective 95% probability that the weight is in the range $[150 \pm 20]$. Then it would be reasonable to express the prior information from the visual inspection as $\mu \sim N(150, 10^2)$, independent of the unknown value σ^2. Prior information about the scale's measurement variance, σ^2, could be expressed as a scaled inverse-χ^2 density as above, possibly fitting the parameters ν_0 and σ_0^2 to a 95% prior interval for σ^2. For a noninformative prior distribution for the variance, set $\nu_0 = 0$; that is, $p(\sigma^2) \propto \sigma^{-2}$.

Semi-conjugacy

The joint prior distribution (3.10) is *not* a conjugate family for the normal likelihood on (μ, σ^2): in the resulting posterior distribution, μ and σ^2 are dependent, and the posterior density does not follow any standard parametric form. However, we can obtain useful results by considering the conditional posterior distribution of μ, given σ^2, then averaging over σ^2.

The conditional posterior distribution, $p(\mu|\sigma^2, y)$

Given σ^2, we just have normal data with a normal prior distribution, so the posterior distribution is normal:

$$\mu|\sigma^2, y \sim N(\mu_n, \tau_n^2),$$

where

$$\mu_n = \frac{\frac{1}{\tau_0^2}\mu_0 + \frac{n}{\sigma^2}\bar{y}}{\frac{1}{\tau_0^2} + \frac{n}{\sigma^2}} \quad \text{and} \quad \tau_n^2 = \frac{1}{\frac{1}{\tau_0^2} + \frac{n}{\sigma^2}}, \tag{3.11}$$

and we must remember that both μ_n and τ_n^2 depend on σ^2.

Calculating the marginal posterior distribution, $p(\sigma^2|y)$, using a basic identity of conditional probability

The posterior density of σ^2 can be determined by integrating the joint posterior density over μ:

$$p(\sigma^2|y) \propto \int N(\mu|\mu_0, \tau_0^2) \text{Inv-}\chi^2(\sigma^2|\nu_0, \sigma_0^2) \prod_{i=1}^{n} N(y_i|\mu, \sigma^2) d\mu.$$

This integral can be solved in closed form because the integrand, considered as a function of μ, is proportional to a normal density. A convenient way to keep track of the constants of integration involving σ^2 is to use the following identity based on the definition of conditional probability:

$$p(\sigma^2|y) = \frac{p(\mu, \sigma^2|y)}{p(\mu|\sigma^2, y)}. \tag{3.12}$$

Using the result derived above for $p(\mu|\sigma^2, y)$, we obtain

$$p(\sigma^2|y) \propto \frac{N(\mu|\mu_0, \tau_0^2) \text{Inv-}\chi^2(\sigma^2|\nu_0, \sigma_0^2) \prod_{i=1}^{n} N(y_i|\mu, \sigma^2)}{N(\mu|\mu_n, \tau_n^2)}, \tag{3.13}$$

which has no simple conjugate form but can easily be computed on a discrete grid of values of σ^2, where μ_n and τ_n^2 are given by (3.11). Notationally, the

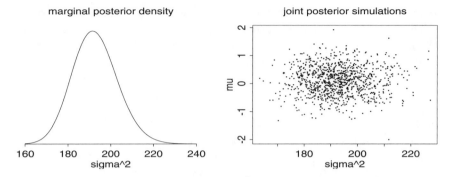

Figure 3.2 *(a) Marginal posterior density of σ^2 for football scores data with semiconjugate prior distribution; (b) 1000 simulations from the joint posterior distribution of μ, σ^2.*

right side of (3.13) depends on μ, but from (3.12), we know it cannot depend on μ in a real sense. That is, the factors that depend on μ in the numerator and denominator must cancel. Thus, we can compute (3.13) by inserting any value of μ into the expression; a convenient choice, from the standpoint of computational simplicity and stability, is to set $\mu = \mu_n$, which simplifies the denominator of (3.13):

$$p(\sigma^2|y) \propto \tau_n N(\mu_n|\mu_0, \tau_0^2) \text{Inv-}\chi^2(\sigma^2|\nu_0, \sigma_0^2) \prod_{i=1}^{n} N(y_i|\mu_n, \sigma^2). \tag{3.14}$$

Sampling from the joint posterior distribution

As with the conjugate prior distribution, the easiest way to draw (μ, σ^2) from their joint posterior distribution, $p(\mu, \sigma^2|y)$, is to draw σ^2 from its marginal posterior density and then μ from its conditional posterior density, given the drawn value σ^2. The first step—drawing σ^2—must be done numerically, for example using the inverse cdf method based on a computation of the posterior density (3.14) on a discrete grid of values σ^2. The second step is immediate: draw $\mu \sim N(\mu_n, \tau_n^2)$, with μ_n and τ_n^2 from (3.11).

Example. Football scores and point spreads

We return to the normal distribution model described in Section 2.7 for the football data introduced in Section 1.6. The differences between the game outcomes and the point spreads, d_i, $i = 1, \ldots, 672$, are modeled as $N(\mu, \sigma^2)$ random variables. The variance σ^2 was estimated in Section 2.7 under a noninformative prior distribution with the rather strong assumption that $\mu = 0$. Here, we incorporate a $N(0, 2^2)$ prior distribution for μ, consistent with our belief that the point spread is approximately the mean of the game outcome but with nonzero variance indicating some degree of uncertainty (approximately 95% prior probability that μ falls in the range $[-4, 4]$). We continue to use a noninformative prior distribution for the variance, $p(\sigma^2) \propto \sigma^{-2}$. The marginal posterior distribution of σ^2, calculated from (3.14), is evaluated for a grid of values on the interval $[150, 250]$

and displayed in Figure 3.2a. The posterior median and a 95% posterior interval can be computed directly from the grid approximation to the marginal distribution or by drawing a random sample from the grid approximation. The posterior median of σ based on 1000 draws is 13.86, and the posterior 95% interval is [13.19, 14.62]. Figure 3.2b displays the 1000 draws from the joint posterior distribution of (μ, σ^2). The posterior distribution of μ is concentrated between -1 and 1.

3.5 The multinomial model

The binomial distribution that was emphasized in Chapter 2 can be generalized to allow more than two possible outcomes. The multinomial sampling distribution is used to describe data for which each observation is one of k possible outcomes. If y is the vector of counts of the number of observations of each outcome, then

$$p(y|\theta) \propto \prod_{j=1}^{k} \theta_j^{y_j},$$

where the sum of the probabilities, $\sum_{j=1}^{k} \theta_j$, is 1. The distribution is typically thought of as implicitly conditioning on the number of observations, $\sum_{j=1}^{k} y_j = n$. The conjugate prior distribution is a multivariate generalization of the beta distribution known as the Dirichlet,

$$p(\theta|\alpha) \propto \prod_{j=1}^{k} \theta_j^{\alpha_j - 1},$$

where the distribution is restricted to nonnegative θ_j's with $\sum_{j=1}^{k} \theta_j = 1$; see Appendix A for details. The resulting posterior distribution for the θ_j's is Dirichlet with parameters $\alpha_j + y_j$.

The prior distribution is mathematically equivalent to a likelihood resulting from $\sum_{j=1}^{k} \alpha_j$ observations with α_j observations of the jth outcome category. As in the binomial there are several plausible noninformative Dirichlet prior distributions. A uniform density is obtained by setting $\alpha_j = 1$ for all j; this distribution assigns equal density to any vector θ satisfying $\sum_{j=1}^{k} \theta_j = 1$. Setting $\alpha_j = 0$ for all j results in an improper prior distribution that is uniform in the $\log(\theta_j)$'s. The resulting posterior distribution is proper if there is at least one observation in each of the k categories, so that each component of y is positive. The bibliographic note at the end of this chapter points to other suggested noninformative prior distributions for the multinomial model.

Example. Pre-election polling

For a simple example of a multinomial model, we consider a sample survey question with three possible responses. In late October, 1988, a survey was conducted by CBS News of 1447 adults in the United States to find out their preferences in the upcoming Presidential election. Out of 1447 persons, $y_1 = 727$ supported George Bush, $y_2 = 583$ supported Michael Dukakis, and $y_3 = 137$ supported other

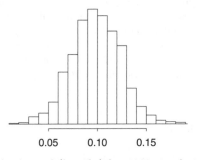

Figure 3.3 *Histogram of values of* $(\theta_1 - \theta_2)$ *for 1000 simulations from the posterior distribution for the election polling example.*

candidates or expressed no opinion. Assuming no other information on the respondents, the 1447 observations are exchangeable. If we also assume simple random sampling (that is, 1447 names 'drawn out of a hat'), then the data (y_1, y_2, y_3) follow a multinomial distribution, with parameters $(\theta_1, \theta_2, \theta_3)$, the proportions of Bush supporters, Dukakis supporters, and those with no opinion in the survey population. An estimand of interest is $\theta_1 - \theta_2$, the population difference in support for the two major candidates.

With a noninformative uniform prior distribution on θ, $\alpha_1 = \alpha_2 = \alpha_3 = 1$, the posterior distribution for $(\theta_1, \theta_2, \theta_3)$ is Dirichlet(728, 584, 138). We could compute the posterior distribution of $\theta_1 - \theta_2$ by integration, but it is simpler just to draw 1000 points $(\theta_1, \theta_2, \theta_3)$ from the posterior Dirichlet distribution and compute $\theta_1 - \theta_2$ for each. The result is displayed in Figure 3.3. All of the 1000 had $\theta_1 > \theta_2$; thus, the estimated posterior probability that Bush had more support than Dukakis in the survey population is over 99.9%.

In fact, the CBS survey does not use independent random sampling but rather uses a variant of a stratified sampling plan. We discuss an improved analysis of this survey, using knowledge of the sampling scheme, in Section 7.4 (see Table 7.2 on page 210).

In complicated problems—for example, analyzing the results of many survey questions simultaneously—the number of multinomial categories, and thus parameters, becomes so large that it is hard to usefully analyze a dataset of moderate size without additional structure in the model. Formally, additional information can enter the analysis through the prior distribution or the sampling model. An informative prior distribution might be used to improve inference in complicated problems, using the ideas of hierarchical modeling introduced in Chapter 5. Alternatively, loglinear models can be used to impose structure on multinomial parameters that result from cross-classifying several survey questions; Section 16.8 provides details and an example.

THE MULTIVARIATE NORMAL MODEL

3.6 The multivariate normal model

Here we give a somewhat formal account of the distributional results of Bayesian inference for the parameters of a multivariate normal distribution. In many ways, these results parallel those already given for the univariate normal model, but there are some important new aspects that play a major role in the analysis of linear models, which is the central activity of much applied statistical work (see Chapters 5, 14, and 15). This section can be viewed at this point as reference material for future chapters.

Multivariate normal likelihood

The basic model to be discussed concerns an observable vector y of d components, with the multivariate normal distribution,

$$y|\mu, \Sigma \sim N(\mu, \Sigma), \qquad (3.15)$$

where μ is a (column) vector of length d and Σ is a $d \times d$ variance matrix, which is symmetric and positive definite. The likelihood function for a single observation is

$$p(y|\mu, \Sigma) \propto |\Sigma|^{-1/2} \exp\left(-\frac{1}{2}(y-\mu)^T \Sigma^{-1}(y-\mu)\right),$$

and for a sample of n iid observations, $y_1, \ldots, y_n$, is

$$
\begin{aligned}
p(y_1, \ldots, y_n | \mu, \Sigma) &\propto |\Sigma|^{-n/2} \exp\left(-\frac{1}{2}\sum_{i=1}^{n}(y_i-\mu)^T \Sigma^{-1}(y_i-\mu)\right) \\
&= |\Sigma|^{-n/2} \exp\left(-\frac{1}{2}\text{tr}(\Sigma^{-1}S_0)\right), \qquad (3.16)
\end{aligned}
$$

where S_0 is the matrix of 'sums of squares' relative to μ,

$$S_0 = \sum_{i=1}^{n}(y_i-\mu)(y_i-\mu)^T. \qquad (3.17)$$

Multivariate normal with known variance

As with the univariate normal model, we analyze the multivariate normal model by first considering the case of known Σ.

Conjugate prior distribution for μ with known Σ. The log-likelihood is a quadratic form in μ, and therefore the conjugate prior distribution for μ is the multivariate normal distribution, which we parameterize as $\mu \sim N(\mu_0, \Lambda_0)$.

Posterior distribution for μ with known Σ. The posterior distribution of μ is

$$p(\mu|y, \Sigma) \propto \exp\left(-\frac{1}{2}\left[(\mu-\mu_0)^T \Lambda_0^{-1}(\mu-\mu_0) + \sum_{i=1}^{n}(y_i-\mu)^T \Sigma^{-1}(y_i-\mu)\right]\right),$$

which is an exponential of a quadratic form in μ. Completing the quadratic form and pulling out constant factors (see Exercise 3.13) gives

$$p(\mu|y, \Sigma) \propto \exp\left(-\frac{1}{2}(\mu - \mu_n)^T \Lambda_n^{-1}(\mu - \mu_n)\right)$$
$$= N(\mu|\mu_n, \Lambda_n),$$

where

$$\mu_n = (\Lambda_0^{-1} + n\Sigma^{-1})^{-1}(\Lambda_0^{-1}\mu_0 + n\Sigma^{-1}\bar{y})$$
$$\Lambda_n^{-1} = \Lambda_0^{-1} + n\Sigma^{-1}. \tag{3.18}$$

These are similar to the results for the univariate normal model of Section 2.6, the posterior mean being a weighted average of the data and the prior mean, with weights given by the data and prior precision matrices, $n\Sigma^{-1}$ and Λ_0^{-1}, respectively. The posterior precision is the sum of the prior and data precisions.

Posterior conditional and marginal distributions of subvectors of μ with known Σ. It follows from the properties of the multivariate normal distribution (see Appendix A) that the marginal posterior distribution of a subset of the parameters, $\mu^{(1)}$ say, is also multivariate normal, with mean vector equal to the appropriate subvector of the posterior mean vector μ_n and variance matrix equal to the appropriate submatrix of Λ_n. Also, the conditional posterior distribution of a subset $\mu^{(1)}$ given the values of a second subset $\mu^{(2)}$ is multivariate normal. If we write superscripts in parentheses to indicate appropriate subvectors and submatrices, then

$$\mu^{(1)}|\mu^{(2)}, y \sim N\left(\mu_n^{(1)} + \beta^{1|2}(\mu^{(2)} - \mu_n^{(2)}), \Lambda^{1|2}\right), \tag{3.19}$$

where the regression coefficients $\beta^{1|2}$ and conditional variance matrix $\Lambda^{1|2}$ are defined by

$$\beta^{1|2} = \Lambda_n^{(12)}\left(\Lambda_n^{(22)}\right)^{-1}$$
$$\Lambda^{1|2} = \Lambda_n^{(11)} - \Lambda_n^{(12)}\left(\Lambda_n^{(22)}\right)^{-1}\Lambda_n^{(21)}.$$

Posterior predictive distribution for new data. We can now determine the analytic form of the posterior predictive distribution for a new observation $\tilde{y} \sim N(\mu, \Sigma)$. As in the univariate normal distribution, we first note that the joint distribution, $p(\tilde{y}, \mu|y) = N(\tilde{y}|\mu, \Sigma)N(\mu|\mu_n, \Lambda_n)$, is the exponential of a quadratic form in $(\tilde{y}, \mu)$; hence $(\tilde{y}, \mu)$ have a joint normal posterior distribution, and so the marginal posterior distribution of $\tilde{y}$ is (multivariate) normal. We are still assuming the variance matrix Σ is known. As in the univariate case, we can determine the posterior mean and variance of $\tilde{y}$ using (2.7) and (2.8):

$$E(\tilde{y}|y) = E(E(\tilde{y}|\mu, y)|y)$$
$$= E(\mu|y) = \mu_n,$$

HYATT
SUMMERFIELD
SUITES™

866 XS HYATT
hyattsummerfieldsuites.com

and

$$\begin{aligned}
\text{var}(\tilde{y}|y) &= \text{E}(\text{var}(\tilde{y}|\mu, y)|y) + \text{var}(\text{E}(\tilde{y}|\mu, y)|y) \\
&= \text{E}(\Sigma|y) + \text{var}(\mu|y) = \Sigma + \Lambda_n.
\end{aligned}$$

To sample from the posterior distribution or the posterior predictive distribution, refer to Appendix A for a method of generating random draws from a multivariate normal distribution with specified mean and variance matrix.

Noninformative prior density for μ. A noninformative uniform prior density for μ is $p(\mu) \propto$ constant, obtained in the limit as the prior precision tends to zero in the sense $|\Lambda_0^{-1}| \to 0$; in the limit of infinite prior variance (zero prior precision), the prior mean is irrelevant. The posterior density is then proportional to the likelihood (3.16). This is a proper posterior distribution only if $n \geq d$, that is, if the sample size is greater than or equal to the dimension of the multivariate normal; otherwise the matrix S_0 is not full rank. If $n \geq d$, the posterior distribution for μ, given the uniform prior density, is $\mu|\Sigma, y \sim \text{N}(\overline{y}, \Sigma/n)$.

Multivariate normal with unknown mean and variance

Conjugate family of prior distributions. Recall that the conjugate distribution for the univariate normal with unknown mean and variance is the normal-inverse-χ^2 distribution (3.6). We can use the inverse-Wishart distribution, a multivariate generalization of the scaled inverse-χ^2, to describe the prior distribution of the matrix Σ. The conjugate prior distribution for (μ, Σ), the normal-inverse-Wishart, is conveniently parameterized in terms of hyperparameters $(\mu_0, \Lambda_0/\kappa_0; \nu_0, \Lambda_0)$:

$$\begin{aligned}
\Sigma &\sim \text{Inv-Wishart}_{\nu_0}(\Lambda_0^{-1}) \\
\mu|\Sigma &\sim \text{N}(\mu_0, \Sigma/\kappa_0),
\end{aligned}$$

which corresponds to the joint prior density

$$p(\mu, \Sigma) \propto |\Sigma|^{-((\nu_0+d)/2+1)} \exp\left(-\frac{1}{2}\text{tr}(\Lambda_0\Sigma^{-1}) - \frac{\kappa_0}{2}(\mu - \mu_0)^T\Sigma^{-1}(\mu - \mu_0)\right).$$

The parameters ν_0 and Λ_0 describe the degrees of freedom and the scale matrix for the inverse-Wishart distribution on Σ. The remaining parameters are the prior mean, μ_0, and the number of prior measurements, κ_0, on the Σ scale. Multiplying the prior density by the normal likelihood results in a posterior density of the same family with parameters

$$\begin{aligned}
\mu_n &= \frac{\kappa_0}{\kappa_0 + n}\mu_0 + \frac{n}{\kappa_0 + n}\overline{y} \\
\kappa_n &= \kappa_0 + n \\
\nu_n &= \nu_0 + n \\
\Lambda_n &= \Lambda_0 + S + \frac{\kappa_0 n}{\kappa_0 + n}(\overline{y} - \mu_0)(\overline{y} - \mu_0)^T,
\end{aligned}$$

where S is the sum of squares matrix about the sample mean,

$$S = \sum_{i=1}^{n}(y_i - \overline{y})(y_i - \overline{y})^T.$$

Other results from the univariate normal distribution are easily generalized to the multivariate case. The marginal posterior distribution of μ is multivariate $t_{\nu_n-d+1}(\mu_n, \Lambda_n/(\kappa_n(\nu_n - d + 1)))$. The posterior predictive distribution of a new observation $\tilde{y}$ is also multivariate Student-t with an additional factor of $\kappa_n + 1$ in the numerator of the scale matrix. Samples from the joint posterior distribution of (μ, Σ) are easily obtained using the following procedure: first, draw $\Sigma|y \sim \text{Inv-Wishart}_{\nu_n}(\Lambda_n^{-1})$, then draw $\mu|\Sigma, y \sim \text{N}(\mu_n, \Sigma/\kappa_n)$. See Appendix A for drawing from inverse-Wishart and multivariate normal distributions. To draw from the posterior predictive distribution of a new observation, draw $\tilde{y}|\mu, \Sigma, y \sim \text{N}(\mu, \Sigma)$, given the already drawn values of μ and Σ.

Noninformative prior distribution. A commonly proposed noninformative prior distribution is the multivariate Jeffreys prior density,

$$p(\mu, \Sigma) \propto |\Sigma|^{-(d+1)/2},$$

which is the limit of the conjugate prior density as $\kappa_0 \to 0, \nu_0 \to -1, |\Lambda_0| \to 0$. The corresponding posterior distribution can be written as

$$\begin{aligned}\Sigma|y &\sim \text{Inv-Wishart}_{n-1}(S) \\ \mu|\Sigma, y &\sim \text{N}(\overline{y}, \Sigma/n).\end{aligned}$$

Results for the marginal distribution of μ and the posterior predictive distribution of $\tilde{y}$, assuming that the posterior distribution is proper, follow from the previous paragraph. For example, the marginal posterior distribution of μ is multivariate $t_{n-d}(\overline{y}, S/(n(n - d)))$.

It is especially important to check that the posterior distribution is proper when using noninformative prior distributions in high dimensions.

Nonconjugate prior distributions. The multivariate normal variance matrix includes a large number of parameters, d variance parameters and $d(d - 1)/2$ covariances (or correlations). Alternatives to the conjugate Wishart family of distributions may be constructed in terms of these parameters, as we discuss in Section 19.2 in the context of multivariate models.

3.7 Example: analysis of a bioassay experiment

Beyond the normal distribution, few multiparameter sampling models allow simple explicit calculation of posterior distributions. Data analysis for such models is possible using simulation techniques. In practice, such analyses often use the Markov chain Monte Carlo simulation techniques considered in Chapter 11. Here we present an example of a nonconjugate model for a bioassay experiment, drawn from the recent literature on applied Bayesian statistics.

Dose, x_i (log g/ml)	Number of animals, n_i	Number of deaths, y_i
−0.86	5	0
−0.30	5	1
−0.05	5	3
0.73	5	5

Table 3.1 *Bioassay data from Racine et al. (1986).*

The model is a two-parameter example from the broad class of generalized linear models to be considered more thoroughly in Chapter 16. We use a particularly simple simulation approach, approximating the posterior distribution by a discrete distribution supported on a two-dimensional grid of points, that provides sufficiently accurate inferences for this two-parameter example.

The scientific problem and the data

In the development of drugs and other chemical compounds, acute toxicity tests or bioassay experiments are commonly performed on animals. Such experiments proceed by administering various dose levels of the compound to batches of animals. The animals' responses are typically characterized by a dichotomous outcome: for example, alive or dead, tumor or no tumor. An experiment of this kind gives rise to data of the form

$$(x_i, n_i, y_i); \ i = 1, \dots, k,$$

where x_i represents the ith of k dose levels (often measured on a logarithmic scale) given to n_i animals, of which y_i subsequently respond with positive outcome. An example of real data from such an experiment is shown in Table 3.1: twenty animals were tested, five at each of four dose levels.

Modeling the dose–response relation

Given what we have seen so far, we must model the outcomes of the five animals *within each group i* as exchangeable, and it seems reasonable to model them as independent with equal probabilities, which implies that the data points y_i are binomially distributed:

$$y_i|\theta_i \sim \text{Bin}(n_i, \theta_i),$$

where θ_i is the probability of death for animals given dose x_i. (An example of a situation in which independence and the binomial model would *not* be appropriate is if the deaths were caused by a contagious disease.) For this experiment, it is also reasonable to treat the outcomes in the four groups as independent of each other, given the parameters $\theta_1, \dots, \theta_4$.

The simplest analysis would treat the four parameters θ_i as exchangeable in their prior distribution, perhaps using a noninformative density such as

$p(\theta_1, \ldots, \theta_4) \propto 1$, in which case the parameters θ_i would have independent beta posterior distributions. The exchangeable prior model for the θ_i parameters has a serious flaw, however; we know the dose level x_i for each group i, and one would expect the probability of death to vary systematically as a function of dose.

The simplest model of the *dose–response relation*—that is, the relation of θ_i to x_i—is linear: $\theta_i = \alpha + \beta x_i$. Unfortunately, this model has the flaw that for very low or high doses, x_i approaches $\pm\infty$ (recall that the dose is measured on the log scale), whereas θ_i, being a probability, must be constrained to lie between 0 and 1. The standard solution is to use a transformation of the θ's, such as the logistic, in the dose–response relation:

$$\mathrm{logit}(\theta_i) = \alpha + \beta x_i, \tag{3.20}$$

where $\mathrm{logit}(\theta_i) = \log(\theta_i/(1 - \theta_i))$ as defined in (1.9). This is called a *logistic regression* model.

The likelihood

Under the model (3.20), we can write the sampling distribution, or likelihood, for each group i in terms of the parameters α and β as

$$p(y_i | \alpha, \beta, n_i, x_i) \propto [\mathrm{logit}^{-1}(\alpha + \beta x_i)]^{y_i} [1 - \mathrm{logit}^{-1}(\alpha + \beta x_i)]^{n_i - y_i}.$$

The model is characterized by the parameters α and β, whose joint posterior distribution is

$$\begin{aligned} p(\alpha, \beta | y, n, x) &\propto p(\alpha, \beta | n, x) p(y | \alpha, \beta, n, x) \tag{3.21} \\ &\propto p(\alpha, \beta) \prod_{i=1}^{k} p(y_i | \alpha, \beta, n_i, x_i). \end{aligned}$$

We consider the sample sizes n_i and dose levels x_i as fixed for this analysis and suppress the conditioning on (n, x) in subsequent notation.

The prior distribution

We present an analysis based on a prior distribution for (α, β) that is independent and locally uniform in the two parameters; that is, $p(\alpha, \beta) \propto 1$. In practice, we might use a uniform prior distribution if we really have no prior knowledge about the parameters, or if we want to present a simple analysis of this experiment alone. If the analysis using the noninformative prior distribution is insufficiently precise, we may consider using other sources of substantive information (for example, from other bioassay experiments) to construct an informative prior distribution.

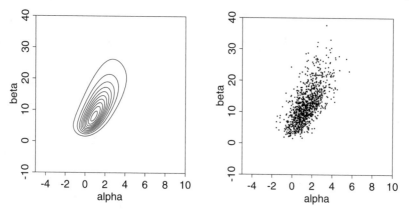

Figure 3.4 *(a) Contour plot for the posterior density of the parameters in the bioas-say example. Contour lines are at $0.05, 0.15, \ldots, 0.95$ times the density at the mode. (b) Scatterplot of 1000 draws from the posterior distribution.*

A rough estimate of the parameters

We will compute the joint posterior distribution (3.21) at a grid of points (α, β), but before doing so, it is a good idea to get a rough estimate of (α, β) so we know where to look. To obtain the rough estimate, we note that $\text{logit}(\text{E}[(y_i/n_i)|\alpha, \beta]) = \alpha + \beta x_i$. We can crudely estimate (α, β) by a linear regression of $\text{logit}(y_i/n_i)$ on x_i for the four data points in Table 3.1. The logits of 0 and 1 are not defined, so for the purposes of the approximate analysis, we temporarily change y_1 to 0.5 and y_4 to 4.5. The linear regression estimate is $(\hat{\alpha}, \hat{\beta}) = (0.1, 2.9)$, with standard errors of 0.3 and 0.5 for α and β, respectively. (Another approach, which would require even less effort in many computational environments, would be to obtain the rough estimate by performing a logistic regression; that is, finding the maximum likelihood estimate of (α, β) in (3.21) for the four data points in Table 3.1.)

Obtaining a contour plot of the joint posterior density

We are now ready to compute the posterior density at a grid of points (α, β). In our first try at a contour plot, we computed the unnormalized density (3.21), based on a uniform prior distribution, on a 200×200 grid on the range $[-1, 1] \times [1, 5]$—that is, the estimated mean plus or minus more than two standard errors in each direction—and then used a computer program to calculate contour lines of equal posterior density. The contour lines on the first plot ran off the page, indicating that the initial normal regression analysis was very crude indeed. We recomputed the posterior density in a much wider range: $(\alpha, \beta) \in [-5, 10] \times [-10, 40]$. The resulting contour plot appears in Figure 3.4a; a general justification for setting the lowest contour level at 0.05 for two-dimensional plots appears on page 103 in Section 4.1.

Sampling from the joint posterior distribution

Having computed the unnormalized posterior density at a grid of values that cover the effective range of (α, β), we can normalize by approximating the distribution as a step function over the grid and setting the total probability in the grid to 1. We sample 1000 random draws (α^l, β^l) from the posterior distribution using the following procedure.

1. Compute the marginal posterior distribution of α by numerically summing over β in the discrete distribution computed on the grid of Figure 3.4a.

2. For $l = 1, \ldots, 1000$:

   (a) Draw α^l from the discretely-computed $p(\alpha|y)$; this can be viewed as a discrete version of the inverse cdf method described in Section 1.9.

   (b) Draw β^l from the discrete conditional distribution, $p(\beta|\alpha, y)$, given the just-sampled value of α.

   (c) For each of the sampled α and β, add a uniform random jitter centered at zero with a width equal to the spacing of the sampling grid. This gives the simulation draws a continuous distribution.

The 1000 draws (α^l, β^l) are displayed on a scatterplot in Figure 3.4b. The scale of the plot, which is the same as the scale of Figure 3.4a, has been set large enough that all the 1000 draws would fit on the graph.

There are a number of practical considerations when applying this two-dimensional grid approximation. There can be difficulty finding the correct location and scale for the grid points. A grid that is defined on too small an area may miss important features of the posterior distribution that fall outside the grid. A grid defined on a large area with wide intervals between points can miss important features that fall between the grid points. It is also important to avoid overflow and underflow operations when computing the posterior distribution. It is usually a good idea to compute the logarithm of the unnormalized posterior distribution and subtract off the maximum value before exponentiating. This creates an unnormalized discrete approximation with maximum value 1, which can then be normalized (by setting the total probability in the grid to 1).

The posterior distribution of the LD50

A parameter of common interest in bioassay studies is the *LD50*—the dose level at which the probability of death is 50%. In our logistic model, a 50% survival rate means

$$\text{LD50: } \quad \mathrm{E}\left(\frac{y_i}{n_i}\right) = \text{logit}^{-1}(\alpha + \beta x_i) = 0.5;$$

thus, $\alpha + \beta x_i = \text{logit}(0.5) = 0$, and the LD50 is $x_i = -\alpha/\beta$. Computing the posterior distribution of any summaries in the Bayesian approach is straightforward, as discussed at the end of Section 1.9. Given what we have done

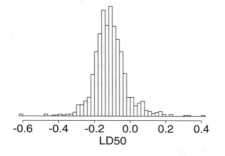

Figure 3.5 *Histogram of the draws from the posterior distribution of the LD50 (on the scale of log dose in g/ml) in the bioassay example, conditional on the parameter β being positive.*

so far, simulating the posterior distribution of the LD50 is trivial: we just compute $-\alpha/\beta$ for the 1000 draws of (α, β) pictured in Figure 3.4b.

Difficulties with the LD50 parameterization if the drug is beneficial. In the context of this example, LD50 is a meaningless concept if $\beta \leq 0$, in which case increasing the dose does not cause the probability of death to increase. If we were certain that the drug could *not* cause the tumor rate to decrease, we should constrain the parameter space to exclude values of β less than 0. However, it seems more reasonable here to allow the possibility of $\beta \leq 0$ and just note that LD50 is hard to interpret in this case.

We summarize the inference on the LD50 scale by reporting two results: (1) the posterior probability that $\beta > 0$—that is, that the drug is harmful—and (2) the posterior distribution for the LD50 conditional on $\beta > 0$. All of the 1000 simulation draws had positive values of β, so the posterior probability that $\beta > 0$ is estimated to exceed 0.999. We compute the LD50 for the simulation draws with positive values of β (which happen to be all 1000 draws for this example); a histogram is displayed in Figure 3.5. This example illustrates that the marginal posterior mean is not always a good summary of inference about a parameter. We are *not*, in general, interested in the posterior mean of the LD50, because the posterior mean includes the cases in which the dose–response relation is negative.

3.8 Summary of elementary modeling and computation

The lack of multiparameter models permitting easy calculation of posterior distributions is not a major practical handicap for three main reasons. First, when there are few parameters, posterior inference in nonconjugate multiparameter models can be obtained by simple simulation methods, as we have seen in the bioassay example. Second, sophisticated models can often be represented in a hierarchical or conditional manner, as we shall see in Chapter 5, for which effective computational strategies are available (as we discuss in

general in Part III). Finally, as we discuss in Chapter 4, we can often apply
a normal approximation to the posterior distribution, and therefore the conjugate structure of the normal model can play an important role in practice,
well beyond its application to explicitly normal sampling models.

Our successful analysis of the bioassay example suggests the following strategy for computation of simple Bayesian posterior distributions. What follows
is not truly a general approach, but it summarizes what we have done so far
and foreshadows the general methods—based on successive approximations—
presented in Part III.

1. Write the likelihood part of the model, $p(y|\theta)$, ignoring any factors that are
   free of θ.

2. Write the posterior density, $p(\theta|y) \propto p(\theta)p(y|\theta)$. If prior information is well-
   formulated, include it in $p(\theta)$. Otherwise, temporarily set $p(\theta) \propto$ constant,
   with the understanding that the prior density can be altered later to include
   additional information or structure.

3. Create a crude estimate of the parameters, θ, for use as a starting point
   and a comparison to the computation in the next step.

4. Draw simulations $\theta^1, \ldots, \theta^L$, from the posterior distribution. Use the sample draws to compute the posterior density of any functions of θ that may
   be of interest.

5. If any predictive quantities, $\tilde{y}$, are of interest, simulate $\tilde{y}^1, \ldots, \tilde{y}^L$ by drawing each $\tilde{y}^l$ from the sampling distribution conditional on the drawn value
   θ^l, $p(\tilde{y}|\theta^l)$. In Chapter 6, we discuss how to use posterior simulations of θ
   and $\tilde{y}$ to check the fit of the model to data and substantive knowledge.

For models that are not conjugate or semi-conjugate, step 4 above can be difficult. Various methods have been developed to draw posterior simulations in
complicated models, as we discuss in Part III. Occasionally, high-dimensional
problems can be solved by combining analytical and numerical simulation
methods, as in the normal model with semi-conjugate prior distribution at
the end of Section 3.4. If θ has only one or two components, it is possible to
draw simulations by computing on a grid, as we illustrated in the previous
section for the bioassay example.

3.9 Bibliographic note

Chapter 2 of Box and Tiao (1973) thoroughly treats the univariate and multivariate normal distribution problems and also some related problems such
as estimating the difference between two means and the ratio between two
variances. At the time that book was written, computer simulation methods
were much less convenient than they are now, and so Box and Tiao, and other
Bayesian authors of the period, restricted their attention to conjugate families and devoted much effort to deriving analytic forms of marginal posterior
densities.

Many textbooks on multivariate analysis discuss the unique mathematical

Survey	Bush	Dukakis	No opinion/other	Total
pre-debate	294	307	38	639
post-debate	288	332	19	639

Table 3.2 *Number of respondents in each preference category from ABC News pre- and post-debate surveys in 1988.*

features of the multivariate normal distribution, such as the property that all marginal and conditional distributions of components of a multivariate normal vector are normal; for example, see Mardia, Kent, and Bibby (1979).

Simon Newcomb's data, along with a discussion of his experiment, appear in Stigler (1977).

The multinomial model and corresponding informative and noninformative prior distributions are discussed by Good (1965) and Fienberg (1977); also see the bibliographic note on loglinear models at the end of Chapter 16.

The data and model for the bioassay example appear in Racine et al. (1986), an article that presents several examples of simple Bayesian analyses that have been useful in the pharmaceutical industry.

3.10 Exercises

1. Binomial and multinomial models: suppose data $(y_1, \ldots, y_J)$ follow a multinomial distribution with parameters $(\theta_1, \ldots, \theta_J)$. Also suppose that $\theta = (\theta_1, \ldots, \theta_J)$ has a Dirichlet prior distribution. Let $\alpha = \frac{\theta_1}{\theta_1 + \theta_2}$.

   (a) Write the marginal posterior distribution for α.

   (b) Show that this distribution is identical to the posterior distribution for α obtained by treating y_1 as an observation from the binomial distribution with probability α and sample size $y_1 + y_2$, ignoring the data $y_3, \ldots, y_J$.

   This result justifies the application of the binomial distribution to multinomial problems when we are only interested in two of the categories; for example, see the next problem.

2. Comparison of two multinomial observations: on September 25, 1988, the evening of a Presidential campaign debate, ABC News conducted a survey of registered voters in the United States; 639 persons were polled before the debate, and 639 different persons were polled after. The results are displayed in Table 3.2. Assume the surveys are independent simple random samples from the population of registered voters. Model the data with two different multinomial distributions. For $j = 1, 2$, let α_j be the proportion of voters who preferred Bush, out of those who had a preference for either Bush or Dukakis at the time of survey j. Plot a histogram of the posterior density for $\alpha_2 - \alpha_1$. What is the posterior probability that there was a shift toward Bush?

3. Estimation from two independent experiments: an experiment was performed on the effects of magnetic fields on the flow of calcium out of chicken brains. The experiment involved two groups of chickens: a control group of 32 chickens and an exposed group of 36 chickens. One measurement was taken on each chicken, and the purpose of the experiment was to measure the average flow μ_c in untreated (control) chickens and the average flow μ_t in treated chickens. The 32 measurements on the control group had a sample mean of 1.013 and a sample standard deviation of 0.24. The 36 measurements on the treatment group had a sample mean of 1.173 and a sample standard deviation of 0.20.

(a) Assuming the control measurements were taken at random from a normal distribution with mean μ_c and variance σ_c^2, what is the posterior distribution of μ_c? Similarly, use the treatment group measurements to determine the marginal posterior distribution of μ_t. Assume a uniform prior distribution on $(\mu_c, \mu_t, \log \sigma_c, \log \sigma_t)$.

(b) What is the posterior distribution for the difference, $\mu_t - \mu_c$? To get this, you may sample from the independent Student-t distributions you obtained in part (a) above. Plot a histogram of your samples and give an approximate 95% posterior interval for $\mu_t - \mu_c$.

The problem of estimating two normal means with unknown ratio of variances is called the Behrens–Fisher problem.

4. Inference for a 2×2 table: an experiment was performed to estimate the effect of beta-blockers on mortality of cardiac patients. A group of patients were randomly assigned to treatment and control groups: out of 674 patients receiving the control, 39 died, and out of 680 receiving the treatment, 22 died. Assume that the outcomes are independent and binomially distributed, with probabilities of death of p_0 and p_1 under the control and treatment, respectively.

(a) Set up a noninformative prior distribution on (p_0, p_1) and obtain posterior simulations.

(b) The *odds ratio* is defined as $(p_1/(1 - p_1))/(p_0/(1 - p_0))$. Summarize the posterior distribution for this estimand.

(c) Discuss the sensitivity of your inference to your choice of noninformative prior density.

We return to this example in Section 5.6.

5. Rounded data: it is a common problem for measurements to be observed in rounded form (for a recent review, see Heitjan, 1989). For a simple example, suppose we weigh an object five times and measure weights, rounded to the nearest pound, of 10, 10, 12, 11, 9. Assume the unrounded measurements are normally distributed with a noninformative prior distribution on the mean μ and variance σ^2.

(a) Give the posterior distribution for (μ, σ^2) obtained by pretending that the observations are exact unrounded measurements.

(b) Give the correct posterior distribution for (μ, σ^2) treating the measurements as rounded.

(c) How do the incorrect and correct posterior distributions differ? Compare means, variances, and contour plots.

(d) Let $z = (z_1, \ldots, z_5)$ be the original, unrounded measurements corresponding to the five observations above. Draw simulations from the posterior distribution of z. Compute the posterior mean of $(z_1 - z_2)^2$.

6. Binomial with unknown probability and sample size: some of the difficulties with setting prior distributions in multiparameter models can be illustrated with the simple binomial distribution. Consider data $y_1, \ldots, y_n$ modeled as iid $\text{Bin}(N, \theta)$, with both N and θ unknown. Defining a convenient family of prior distributions on (N, θ) is difficult, partly because of the discreteness of N.

Raftery (1988) considers a hierarchical approach based on assigning the parameter N a Poisson distribution with *unknown* mean μ. To define a prior distribution on (θ, N), Raftery defines $\lambda = \mu\theta$ and specifies a prior distribution on (λ, θ). The prior distribution is specified in terms of λ rather than μ because 'it would seem easier to formulate prior information about λ, the unconditional expectation of the observations, than about μ, the mean of the unobserved quantity N.'

(a) A suggested noninformative prior distribution is $p(\lambda, \theta) \propto \lambda^{-1}$. What is a motivation for this noninformative distribution? Is the distribution improper? Transform to determine $p(N, \theta)$.

(b) The Bayesian method is illustrated on counts of waterbuck obtained by remote photography on five separate days in Kruger Park in South Africa. The counts were 53, 57, 66, 67, and 72. Perform the Bayesian analysis on these data and display a scatterplot of posterior simulations of (N, θ). What is the posterior probability that $N > 100$?

(c) Why not simply use a Poisson prior distribution with fixed μ as a prior distribution for N?

7. Poisson and binomial distributions: a student sits on a street corner for an hour and records the number of bicycles b and the number of other vehicles v that go by. Two models are considered:

- The outcomes b and v have independent Poisson distributions, with unknown means θ_b and θ_v.

- The outcome b has a binomial distribution, with unknown probability p and sample size $b + v$.

Show that the two models have the same likelihood if we define $p = \frac{\theta_b}{\theta_b + \theta_v}$.

Type of street	Bike route?	Counts of bicycles/other vehicles
Residential	yes	16/58, 9/90, 10/48, 13/57, 19/103, 20/57, 18/86, 17/112, 35/273, 55/64
Residential	no	12/113, 1/18, 2/14, 4/44, 9/208, 7/67, 9/29, 8/154
Fairly busy	yes	8/29, 35/415, 31/425, 19/42, 38/180, 47/675, 44/620, 44/437, 29/47, 18/462
Fairly busy	no	10/557, 43/1258, 5/499, 14/601, 58/1163, 15/700, 0/90, 47/1093, 51/1459, 32/1086
Busy	yes	60/1545, 51/1499, 58/1598, 59/503, 53/407, 68/1494, 68/1558, 60/1706, 71/476, 63/752
Busy	no	8/1248, 9/1246, 6/1596, 9/1765, 19/1290, 61/2498, 31/2346, 75/3101, 14/1918, 25/2318

Table 3.3 *Counts of bicycles and other vehicles in one hour in each of 10 city blocks in each of six categories. (The data for two of the residential blocks were lost.) For example, the first block had 16 bicycles and 58 other vehicles, the second had 9 bicycles and 90 other vehicles, and so on. Streets were classified as 'residential,' 'fairly busy,' or 'busy' before the data were gathered.*

8. Analysis of proportions: a survey was done of bicycle and other vehicular traffic in the neighborhood of the campus of the University of California, Berkeley, in the spring of 1993. Sixty city blocks were selected at random; each block was observed for one hour, and the numbers of bicycles and other vehicles traveling along that block were recorded. The sampling was stratified into six types of city blocks: busy, fairly busy, and residential streets, with and without bike routes, with ten blocks measured in each stratum. Table 3.3 displays the number of bicycles and other vehicles recorded in the study. For this problem, restrict your attention to the first four rows of the table: the data on residential streets.

(a) Let $y_1, \ldots, y_{10}$ and $z_1, \ldots, z_8$ be the observed proportion of traffic that was on bicycles in the residential streets with bike lanes and with no bike lanes, respectively (so $y_1 = 16/(16 + 58)$ and $z_1 = 12/(12 + 113)$, for example). Set up a model so that the y_i's are iid given parameters θ_y and the z_i's are iid given parameters θ_z.

(b) Set up a prior distribution that is independent in θ_y and θ_z.

(c) Determine the posterior distribution for the parameters in your model and draw 1000 simulations from the posterior distribution. (Hint: θ_y and θ_z are independent in the posterior distribution, so they can be simulated independently.)

(d) Let $\mu_y = \mathrm{E}(y_i|\theta_y)$ be the mean of the distribution of the y_i's; μ_y will be a function of θ_y. Similarly, define μ_z. Using your posterior simulations from (c), plot a histogram of the posterior simulations of $\mu_y - \mu_z$, the expected difference in proportions in bicycle traffic on residential streets with and without bike lanes.

We return to this example in Exercise 5.11.

9. Normal likelihood with conjugate prior distribution: suppose y is an independent and identically distributed sample of size n from the distribution $N(\mu, \sigma^2)$, where (μ, σ^2) have the $N\text{-Inv-}\chi^2(\mu_0, \sigma_0^2/\kappa_0; \nu_0, \sigma_0^2)$ prior distribution, (that is, $\sigma^2 \sim \text{Inv-}\chi^2(\nu_0, \sigma_0^2)$ and $\mu|\sigma^2 \sim N(\mu_0, \sigma^2/\kappa_0)$). The posterior distribution, $p(\mu, \sigma^2|y)$, is also normal-inverse-χ^2; derive explicitly its parameters in terms of the prior parameters and the sufficient statistics of the data.

10. Comparison of normal variances: for $j = 1, 2$, suppose that

$$(y_{j1}, \ldots, y_{jn_j}|\mu_j, \sigma_j^2) \sim \text{iid } N(\mu_j, \sigma_j^2),$$
$$p(\mu_j, \sigma_j^2) \propto \sigma_j^{-2},$$

and (μ_1, σ_1^2) are independent of (μ_2, σ_2^2) in the prior distribution. Show that the posterior distribution of $(s_1^2/s_2^2)/(\sigma_1^2/\sigma_2^2)$ is F with $(n_1 - 1)$ and $(n_2 - 1)$ degrees of freedom. (Hint: to show the required form of the posterior density, you do not need to carry along all the normalizing constants.)

11. Computation: in the bioassay example, replace the uniform prior density by a joint normal prior distribution on (α, β), with $\alpha \sim N(0, 2^2)$, $\beta \sim N(10, 10^2)$, and $\text{corr}(\alpha, \beta) = 0.5$.

   (a) Repeat all the computations and plots of Section 3.7 with this new prior distribution.

   (b) Check that your contour plot and scatterplot look like a compromise between the prior distribution and the likelihood (as displayed in Figure 3.4).

   (c) Discuss the effect of this hypothetical prior information on the conclusions in the applied context.

12. Poisson regression model: expand the model of Exercise 2.13(a) by assuming that the number of fatal accidents in year t follows a Poisson distribution with mean $\alpha + \beta t$. You will estimate α and β, following the example of the analysis in Section 3.7.

   (a) Discuss various choices for a 'noninformative' prior distribution for (α, β). Choose one.

   (b) Discuss what would be a realistic informative prior distribution for (α, β). Sketch its contours and then put it aside. Do parts (c)–(h) of this problem using your noninformative prior distribution from (a).

   (c) Write the posterior density for (α, β). What are the sufficient statistics?

   (d) Check that the posterior density is proper.

   (e) Calculate crude estimates and uncertainties for (α, β) using linear regression.

(f) Plot the contours and draw 1000 random samples from the joint posterior density, of (α, β).

(g) Using your samples of (α, β), plot a histogram of the posterior density for the *expected number* of fatal accidents in 1986, $\alpha + 1986\beta$.

(h) Create simulation draws and obtain a 95% predictive interval for the *number* of fatal accidents in 1986.

(i) How does your hypothetical informative prior distribution in (b) differ from the posterior distribution in (f) and (g), obtained from the noninformative prior distribution and the data? If they disagree, discuss.

13. Multivariate normal model: derive equations (3.18) by completing the square in vector-matrix notation.

14. Improper prior and proper posterior distributions: prove that the posterior density (3.21) for the bioassay example has a finite integral over the range $(\alpha, \beta) \in (-\infty, \infty) \times (-\infty, \infty)$.

CHAPTER 4

Large-sample inference and frequency properties of Bayesian inference

We have seen that many simple Bayesian analyses based on noninformative prior distributions give similar results to standard non-Bayesian approaches (for example, the posterior t interval for the normal mean with unknown variance). The extent to which a noninformative prior distribution can be justified as an objective assumption depends on the amount of information available in the data: in the simple cases discussed in Chapters 2 and 3, it was clear that as the sample size n increases, the influence of the prior distribution on posterior inferences decreases. These ideas, sometimes referred to as asymptotic theory, because they refer to properties that hold in the limit as n becomes large, will be reviewed in the present chapter, along with some more explicit discussion of the connections between Bayesian and non-Bayesian methods. The large-sample results are not actually necessary for performing Bayesian data analysis but are often useful as approximations and as tools for understanding.

We begin this chapter with a discussion of the various uses of the normal approximation to the posterior distribution. Theorems about consistency and normality of the posterior distribution in large samples are outlined in Section 4.2, followed by several counterexamples in Section 4.3; proofs of the theorems are sketched in Appendix B. Finally, we discuss how the methods of frequentist statistics can be used to evaluate the properties of Bayesian inferences.

4.1 Normal approximations to the posterior distribution

Normal approximation to the joint posterior distribution

If the posterior distribution $p(\theta|y)$ is unimodal and roughly symmetric, it is often convenient to approximate it by a normal distribution centered at the mode; that is, the logarithm of the posterior density function is approximated by a quadratic function.

A Taylor series expansion of $\log p(\theta|y)$ centered at the posterior mode, $\hat{\theta}$ (where θ can be a vector and $\hat{\theta}$ is assumed to be in the interior of the parameter space), gives

$$\log p(\theta|y) = \log p(\hat{\theta}|y) + \frac{1}{2}(\theta - \hat{\theta})^T \left[\frac{d^2}{d\theta^2} \log p(\theta|y) \right]_{\theta=\hat{\theta}} (\theta - \hat{\theta}) + \dots, \quad (4.1)$$

where the linear term in the expansion is zero because the log-posterior density has zero derivative at its mode. As we discuss in Section 4.2, the remainder terms of higher order fade in importance relative to the quadratic term when θ is close to $\hat{\theta}$ and n is large. Considering (4.1) as a function of θ, the first term is a constant, whereas the second term is proportional to the logarithm of a normal density, yielding the approximation,

$$p(\theta|y) \approx N(\hat{\theta}, [I(\hat{\theta})]^{-1}), \tag{4.2}$$

where $I(\theta)$ is the *observed information*,

$$I(\theta) = -\frac{d^2}{d\theta^2} \log p(\theta|y).$$

If the mode, $\hat{\theta}$, is in the interior of parameter space, then $I(\hat{\theta})$ is positive; if θ is a vector parameter, then $I(\theta)$ is a matrix.

Example. Normal distribution with unknown mean and variance
We illustrate the approximate normal distribution with a simple theoretical example. Let $y_1, \ldots, y_n$ be iid observations from a $N(\mu, \sigma^2)$ distribution, and, for simplicity, we assume a uniform prior density for $(\mu, \log \sigma)$. We set up a normal approximation to the posterior distribution of $(\mu, \log \sigma)$, which has the virtue of restricting σ to positive values. To construct the approximation, we need the second derivatives of the log posterior density,

$$\log p(\mu, \log \sigma|y) = \text{constant} - n \log \sigma - \frac{1}{2\sigma^2}[(n-1)s^2 + n(\bar{y} - \mu)^2].$$

The first derivatives are

$$\frac{d}{d\mu} \log p(\mu, \log \sigma|y) = \frac{n(\bar{y} - \mu)}{\sigma^2},$$

$$\frac{d}{d(\log \sigma)} \log p(\mu, \log \sigma|y) = -n + \frac{(n-1)s^2 + n(\bar{y} - \mu)^2}{\sigma^2},$$

from which the posterior mode is readily obtained as

$$(\hat{\mu}, \log \hat{\sigma}) = \left(\bar{y}, \log \left(\sqrt{\frac{n-1}{n}} s \right) \right).$$

The second derivatives of the log posterior density are

$$\frac{d^2}{d\mu^2} \log p(\mu, \log \sigma|y) = -\frac{n}{\sigma^2}$$

$$\frac{d^2}{d\mu d(\log \sigma)} \log p(\mu, \log \sigma|y) = -2n\frac{\bar{y} - \mu}{\sigma^2}$$

$$\frac{d^2}{d(\log \sigma)^2} \log p(\mu, \log \sigma|y) = -\frac{2}{\sigma^2}((n-1)s^2 + n(\bar{y} - \mu)^2).$$

The matrix of second derivatives at the mode is then $\begin{pmatrix} -n/\hat{\sigma}^2 & 0 \\ 0 & -2n \end{pmatrix}$. From (4.2), the posterior distribution can be approximated as

$$p(\mu, \log \sigma|y) \approx N\left(\begin{pmatrix} \mu \\ \log \sigma \end{pmatrix} \middle| \begin{pmatrix} \bar{y} \\ \log \hat{\sigma} \end{pmatrix}, \begin{pmatrix} \hat{\sigma}^2/n & 0 \\ 0 & 1/(2n) \end{pmatrix} \right).$$

If we had instead constructed the normal approximation in terms of $p(\mu, \sigma^2)$, the second derivative matrix would be multiplied by the Jacobian of the transformation from $\log \sigma$ to σ^2 and the mode would change slightly, to $\tilde{\sigma}^2 = \frac{n}{n+2}\hat{\sigma}^2$. The two components, (μ, σ^2), would still be independent in their approximate posterior distribution, and $p(\sigma^2|y) \approx N(\sigma^2|\tilde{\sigma}^2, 2\tilde{\sigma}^4/(n+2))$.

Interpretation of the posterior density function relative to the density at the mode

In addition to its direct use as an approximation, the multivariate normal distribution provides a useful benchmark for interpreting the posterior density function and contour plots. In the d-dimensional normal distribution, the logarithm of the density function is a constant plus a χ_d^2 distribution divided by -2. For example, the 95th percentile of the χ_{10}^2 density is 18.31, so if a problem has $d = 10$ parameters, then approximately 95% of the posterior probability mass is associated with the values of θ for which $p(\theta|y)$ is no less than $\exp(-18.31/2) = 1.1 \times 10^{-4}$ times the density at the mode. Similarly, with $d = 2$ parameters, approximately 95% of the posterior mass corresponds to densities above $\exp(-5.99/2) = 0.05$, relative to the density at the mode. In a two-dimensional contour plot of a posterior density (for example, Figure 3.4a), the 0.05 contour line thus includes approximately 95% of the probability mass.

Summarizing posterior distributions by point estimates and standard errors

The asymptotic theory outlined in Section 4.2 shows that if n is large enough, a posterior distribution can be summarized by simple approximations based on the normal distribution. In many areas of application, a standard inferential summary is the 95% interval obtained by computing a point estimate, $\hat{\theta}$, such as the maximum likelihood estimate (which is the posterior mode under a uniform prior density), plus or minus two standard errors, with the standard error estimated from the information at the estimate, $I(\hat{\theta})$. A different asymptotic argument justifies the non-Bayesian, frequentist interpretation of this summary, but in many simple situations both interpretations hold. It is difficult to give general guidelines on when the normal approximation is likely to be adequate in practice. From the Bayesian point of view, the accuracy in any given example can be directly determined by inspecting the posterior distribution.

In many cases, convergence to normality of the posterior distribution for a parameter θ can be dramatically improved by transformation. If ϕ is a continuous transformation of θ, then both $p(\phi|y)$ and $p(\theta|y)$ approach normal distributions, but the closeness of the approximation for finite n can vary substantially with the transformation chosen.

Data reduction and summary statistics

Under the normal approximation, the posterior distribution depends on the data only through the mode, $\hat{\theta}$, and the curvature of the posterior density, $I(\hat{\theta})$; that is, asymptotically, these are sufficient statistics. In the examples at the end of the next chapter, we shall see that it is often convenient to summarize 'local-level' or 'individual-level' data from a number of sources by their normal-theory sufficient statistics. This approach using summary statistics allows the relatively easy application of hierarchical modeling techniques to improve each individual estimate. For example, in Section 5.5, each of a set of eight experiments is summarized by a point estimate and a standard error estimated from an earlier linear regression analysis. Using summary statistics is clearly most reasonable when posterior distributions are close to normal; the approach can otherwise discard important information and lead to erroneous inferences.

Lower-dimensional normal approximations

For a finite sample size n, the normal approximation is typically more accurate for conditional and marginal distributions of components of θ than for the full joint distribution. For example, if a joint distribution is multivariate normal, all its margins are normal, but the converse is not true. Determining the marginal distribution of a component of θ is equivalent to averaging over all the other components of θ, and averaging a family of distributions generally brings them closer to normality, by the same logic that underlies the central limit theorem.

The normal approximation for the posterior distribution of a low-dimensional θ is often perfectly acceptable, especially after appropriate transformation. If θ is high-dimensional, two situations commonly arise. First, the marginal distributions of many individual components of θ can be approximately normal; inference about any one of these parameters, taken individually, can then be well summarized by a point estimate and a standard error. Second, it is possible that θ can be partitioned into two subvectors, $\theta = (\theta_1, \theta_2)$, for which $p(\theta_2|y)$ is *not* necessarily close to normal, but $p(\theta_1|\theta_2, y)$ is, perhaps with mean and variance that are functions of θ_2. The approach of approximation using conditional distributions is often useful, and we consider it more systematically in Section 12.4.

Finally, approximations based on the normal distribution are often useful for debugging a computer program or checking a more elaborate method for approximating the posterior distribution.

Example. Bioassay experiment (continued)
We illustrate the use of a normal approximation by reanalyzing the model and data from the bioassay experiment of Section 3.7. The sample size in this experiment is relatively small, only twenty animals in all, and we find that the normal approximation is close to the exact posterior distribution but with important differences.

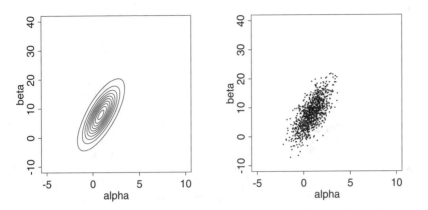

Figure 4.1 *(a) Contour plot of the normal approximation to the posterior distribution of the parameters in the bioassay example. Contour lines are at 0.05, 0.15, . . . , 0.95 times the density at the mode. Compare to Figure 3.4a. (b) Scatterplot of 1000 draws from the normal approximation to the posterior distribution. Compare to Figure 3.4b.*

The normal approximation to the joint posterior distribution of (α, β). To begin, we compute the mode of the posterior distribution (using a logistic regression program) and the normal approximation (4.2) evaluated at the mode. The posterior mode of (α, β) is the same as the maximum likelihood estimate because we have assumed a uniform prior density for (α, β). Figure 4.1 shows a contour plot of the bivariate normal approximation and a scatterplot of 1000 draws from this approximate distribution. The plots resemble the plots of the actual posterior distribution in Figure 3.4 but without the skewness in the upper right corner of the earlier plots. The effect of the skewness is apparent when comparing the mean of the normal approximation, $(\alpha, \beta) = (0.88, 7.93)$, to the mean of the actual posterior distribution, $(\alpha, \beta) = (1.37, 11.93)$, computed from the simulations displayed in Figure 3.4b.

The posterior distribution for the LD50 using the normal approximation on (α, β). Flaws of the normal approximation. The same set of 1000 draws from the normal approximation can be used to estimate the probability that β is positive and the posterior distribution of the LD50, conditional on β being positive. Out of the 1000 simulation draws, 950 had positive values of β, yielding the estimate $\Pr(\beta > 0) = 0.95$, quite a bit different from the result from the exact distribution that $\Pr(\beta > 0) > 0.999$. Continuing with the analysis based on the normal approximation, we compute the LD50 as $-\alpha/\beta$ for each of the 950 draws with $\beta > 0$; Figure 4.2a presents a histogram of the LD50 values, excluding some extreme values in both tails. (If the entire range of the simulations were included, the shape of the distribution would be nearly impossible to see.) To get a better picture of the center of the distribution, we display in Figure 4.2b a histogram of the middle 95% of the 950 simulation draws of the LD50. The histograms are centered in approximately the same place as Figure 3.5 but with substantially more variation, due to the possibility that β is very close to zero.

In summary, we have seen that posterior inferences based on the normal approxi-

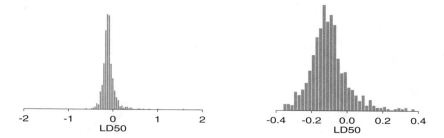

Figure 4.2 *(a) Histogram of the simulations of LD50, conditional on $\beta > 0$, in the bioassay example based on the normal approximation $p(\alpha, \beta|y)$. The wide tails of the histogram correspond to values of β very close to 0. Omitted from this histogram are five simulation draws with values of LD50 less than -2 and four draws with values greater than 2; the extreme tails are truncated to make the histogram visible. The values of LD50 for the 950 simulation draws corresponding to $\beta > 0$ had a range of $[-12.4, 5.4]$. Compare to Figure 3.5. (b) Histogram of the central 95% of the distribution.*

mation in this problem are roughly similar to the exact results, but because of the small sample, the actual joint posterior distribution is substantially more skewed than the large-sample approximation, and the posterior distribution of the LD50 actually has much shorter tails than implied by using the joint normal approximation. Whether or not these differences imply that the normal approximation is inadequate for practical use in this example depends on the ultimate aim of the analysis.

4.2 Large-sample theory

To understand why the normal approximation is often reasonable, we review some theory of how the posterior distribution behaves as the amount of data, from some fixed sampling distribution, increases.

Notation and mathematical setup

The basic tool of large sample Bayesian inference is *asymptotic normality of the posterior distribution*: as more and more data arrive from the same underlying process, the posterior distribution of the parameter vector approaches multivariate normality, even if the true distribution of the data is not within the parametric family under consideration. Mathematically, the results apply most directly to observations $y_1, \ldots, y_n$ that are independent outcomes sampled from a common distribution, $f(y)$. In many situations, the notion of a 'true' underlying distribution, $f(y)$, for the data is difficult to interpret, but it is necessary in order to develop the asymptotic theory. Suppose the data are modeled by a parametric family, $p(y|\theta)$, with a prior distribution $p(\theta)$. In general, the data points y_i and the parameter θ can be vectors. If the true data distribution is included in the parametric family—that is, if $f(y) = p(y|\theta_0)$ for some θ_0—then, in addition to asymptotic normality, the property of *consistency* holds: the posterior distribution converges to a point mass at the true parameter value, θ_0, as $n \to \infty$. When the true distribution is not included in the parametric family, there is no longer a true value θ_0, but its role in the theoretical result is replaced by a value θ_0 that makes the model distribution, $p(y|\theta)$, closest to the true distribution, $f(y)$, in a technical sense involving *Kullback-Leibler information*, as is explained in Appendix B.

In discussing the large-sample properties of posterior distributions, the concept of *Fisher information*, $J(\theta)$, introduced as (2.19) in Section 2.9 in the context of Jeffreys' prior distributions, plays an important role.

Asymptotic normality and consistency

The fundamental mathematical result given in Appendix B shows that, under some regularity conditions (notably that the likelihood is a continuous function of θ and that θ_0 is not on the boundary of the parameter space), as $n \to \infty$, the posterior distribution of θ approaches normality with mean θ_0 and variance $(nJ(\theta_0))^{-1}$. At its simplest level, this result can be understood in terms of the Taylor series expansion (4.1) of the log posterior density centered about the posterior mode. A preliminary result shows that the posterior mode is consistent for θ_0, so that as $n \to \infty$, the mass of the posterior distribution $p(\theta|y)$ becomes concentrated in smaller and smaller neighborhoods of θ_0, and the distance $|\hat{\theta} - \theta_0|$ approaches zero.

Furthermore, we can rewrite the coefficient of the quadratic term in (4.1) as

$$\left[\frac{d^2}{d\theta^2} \log p(\theta|y)\right]_{\theta=\hat{\theta}} = \left[\frac{d^2}{d\theta^2} \log p(\theta)\right]_{\theta=\hat{\theta}} + \sum_{i=1}^{n} \left[\frac{d^2}{d\theta^2} \log p(y_i|\theta)\right]_{\theta=\hat{\theta}}.$$

Considered as a function of θ, this coefficient is a constant plus the sum of n terms, each of whose expected value under the true sampling distribution of y_i, $p(y|\theta_0)$, is approximately $-J(\theta_0)$, as long as $\hat{\theta}$ is close to θ_0 (we are assuming now that $f(y) = p(y|\theta_0)$ for some θ_0). Therefore, for large n, the curvature of the log posterior density can be approximated by the Fisher information, evaluated at either $\hat{\theta}$ or θ_0 (where of course only the former is available in practice).

In summary, in the limit of large n, in the context of a specified family of models, the posterior mode, $\hat{\theta}$, approaches θ_0, and the curvature (the observed information or the negative of the coefficient of the second term in the Taylor expansion) approaches $nJ(\hat{\theta})$ or $nJ(\theta_0)$. In addition, as $n \to \infty$, the likelihood dominates the prior distribution, so we can just use the likelihood alone to obtain the mode and curvature for the normal approximation. More precise statements of the theorems and outlines of proofs appear in Appendix B.

Likelihood dominating the prior distribution

The asymptotic results formalize the notion that the importance of the prior distribution diminishes as the sample size increases. One consequence of this result is that in problems with large sample sizes we need not work especially hard to formulate a prior distribution that accurately reflects all available information. When sample sizes are small, the prior distribution is a critical part of the model specification.

4.3   Counterexamples to the theorems

A good way to understand the limitations of the large-sample results is to consider cases in which the theorems fail. The normal distribution is usually helpful as a starting approximation, but one must examine deviations, especially with unusual parameter spaces and in the extremes of the distribution. The counterexamples to the asymptotic theorems generally correspond to situations in which the prior distribution has an impact on the posterior inference, even in the limit of infinite sample sizes.

Underidentified models and nonidentified parameters.   The model is *under-identified* given data y if the likelihood, $p(\theta|y)$, is equal for a range of values of θ. This may also be called a flat likelihood (although that term is sometimes also used for likelihoods for parameters that are only very weakly identified by the data—so the likelihood function is not strictly equal for a range of values, only almost so). Under such a model, there is no single point θ_0 to which the posterior distribution can converge.

For example, consider the model,

$$\left(\begin{array}{c} u \\ v \end{array} \right) \sim \mathrm{N}\left(\left(\begin{array}{c} 0 \\ 0 \end{array} \right), \left(\begin{array}{cc} 1 & \rho \\ \rho & 1 \end{array} \right) \right),$$

in which only one of u or v is observed from each pair (u, v). Here, the parameter ρ is *nonidentified*. The data supply no information about ρ, so the posterior distribution of ρ is the same as its prior distribution, no matter how large the dataset is.

The only solution to a problem of nonidentified or underidentified parameters is to recognize that the problem exists and, if there is a desire to estimate these parameters more precisely, gather further information that can enable the parameters to be estimated (either from future data collection or from external information that can inform a prior distribution).

Number of parameters increasing with sample size.   In complicated problems, there can be large numbers of parameters, and then we need to distinguish between different types of asymptotics. If, as n increases, the model changes so that the number of parameters increases as well, then the simple results outlined in Sections 4.1 and 4.2, which assume a fixed model class $p(y_i|\theta)$, do not apply. For example, sometimes a parameter is assigned for each sampling unit in a study; for example, $y_i \sim \mathrm{N}(\theta_i, \sigma^2)$. The parameters θ_i generally cannot be estimated consistently unless the amount of data collected from each sampling unit increases along with the number of units.

As with underidentified parameters, the posterior distribution for θ_i will not converge to a point mass if new data do not bring enough information about θ_i. Here, the posterior distribution will not in general converge to a point in the expanding parameter space (reflecting the increasing dimensionality of θ), and its projection into any fixed space—for example, the marginal posterior distribution of any particular θ_i—will not necessarily converge to a point either.

Aliasing. Aliasing is a special case of underidentified parameters in which the same likelihood function repeats at a discrete set of points. For example, consider the following normal mixture model with iid data $y_1, \ldots, y_n$ and parameter vector $\theta = (\mu_1, \mu_2, \sigma_1^2, \sigma_2^2, \lambda)$:

$$p(y_i | \mu_1, \mu_2, \sigma_1^2, \sigma_2^2, \lambda) = \lambda \frac{1}{\sqrt{2\pi}\,\sigma_1} e^{-\frac{1}{2\sigma_1^2}(y_i - \mu_1)^2} + (1 - \lambda)\frac{1}{\sqrt{2\pi}\,\sigma_2} e^{-\frac{1}{2\sigma_2^2}(y_i - \mu_2)^2}.$$

If we interchange each of (μ_1, μ_2) and (σ_1^2, σ_2^2), and replace λ by $(1 - \lambda)$, the likelihood of the data remains the same. The posterior distribution of this model generally has at least two modes and consists of a $(50\%, 50\%)$ mixture of two distributions that are mirror images of each other; it does not converge to a single point no matter how large the dataset is.

In general, the problem of aliasing is eliminated by restricting the parameter space so that no duplication appears; in the above example, the aliasing can be removed by restricting μ_1 to be less than or equal to μ_2.

Unbounded likelihoods. If the likelihood function is unbounded, then there might be no posterior mode within the parameter space, invalidating both the consistency results and the normal approximation. For example, consider the previous normal mixture model; for simplicity, assume that λ is known (and not equal to 0 or 1). If we set $\mu_1 = y_i$ for any arbitrary y_i, and let $\sigma_1^2 \to 0$, then the likelihood approaches infinity. As $n \to \infty$, the number of modes of the likelihood increases. If the prior distribution is uniform on σ_1^2 and σ_2^2 in the region near zero, there will be likewise an increasing number of posterior modes, with no corresponding normal approximations. A prior distribution proportional to $\sigma_1^{-2}\sigma_2^{-2}$ just makes things worse because this puts more probability near zero, causing the posterior distribution to explode even faster at zero.

In general, this problem should arise rarely in practice, because the poles of an unbounded likelihood correspond to unrealistic conditions in a model. The problem can be solved by restricting the model to a plausible set of distributions. When the problem occurs for variance components near zero, it can be simply solved by bounding the parameters away from zero in the prior distribution.

Improper posterior distributions. If the unnormalized posterior density, obtained by multiplying the likelihood by a 'formal' prior density representing an improper prior distribution, integrates to infinity, then the asymptotic results, which rely on probabilities summing to 1, do not follow. An improper posterior distribution cannot occur except with an improper prior distribution.

A simple example arises from combining a Beta$(0,0)$ prior distribution for a binomial proportion with data consisting of n successes and 0 failures. More subtle examples, with hierarchical binomial and normal models, are discussed in Sections 5.3 and 5.4.

The solution to this problem is clear. An improper prior distribution is only a convenient approximation, and if it does not give rise to a proper posterior distribution then the sought convenience is lost. In this case a proper prior distribution is needed, or at least an improper prior density that when combined with the likelihood has a finite integral.

Prior distributions that exclude the point of convergence. If $p(\theta_0) = 0$ for a discrete parameter space, or if $p(\theta) = 0$ in a neighborhood about θ_0 for a continuous parameter space, then the convergence results, which are based on the likelihood dominating the prior distribution, do not hold. The solution is to give positive probability density in the prior distribution to all values of θ that are even remotely plausible.

Convergence to the edge of parameter space. If θ_0 is on the boundary of the parameter space, then the Taylor series expansion must be truncated in some directions, and the normal distribution will not necessarily be appropriate, even in the limit.

For example, consider the model, $y_i \sim N(\theta, 1)$, with the restriction $\theta \geq 0$. Suppose that the model is accurate, with $\theta = 0$ as the true value. The posterior distribution for θ is normal, centered at $\bar{y}$, truncated to be positive. The shape of the posterior distribution for θ, in the limit as $n \rightarrow \infty$, is half of a normal distribution, centered about 0, truncated to be positive.

For another example, consider the same assumed model, but now suppose that the true θ is -1, a value outside the assumed parameter space. The limiting posterior distribution for θ has a sharp spike at 0 with no resemblance to a normal distribution at all. The solution in practice is to recognize the difficulties of applying the normal approximation if one is interested in parameter values near the edge of parameter space. More important, one should give positive prior probability density to all values of θ that are even remotely possible, or in the neighborhood of remotely possible values.

Tails of the distribution. The normal approximation can hold for essentially all the mass of the posterior distribution but still not be accurate in the tails. For example, suppose $p(\theta|y)$ is proportional to $e^{-c|\theta|}$ as $|\theta| \rightarrow \infty$, for some constant c; by comparison, the normal density is proportional to $e^{-c\theta^2}$. The distribution function still converges to normality, but for any finite sample size n the approximation fails far out in the tail. As another example, consider any parameter that is constrained to be positive. For any finite sample size, the normal approximation will admit the possibility of the parameter being negative, because the approximation is simply not appropriate at that point in the tail of the distribution, but that point becomes farther and farther in the tail as n increases.

4.4 Frequency evaluations of Bayesian inferences

Just as the Bayesian paradigm can be seen to justify simple 'classical' techniques, the methods of frequentist statistics provide a useful approach for evaluating the properties of Bayesian inferences—their operating characteristics—when these are regarded as embedded in a sequence of repeated samples. We have already used this notion in discussing the ideas of consistency and asymptotic normality. The notion of *stable estimation*, which says that for a fixed model, the posterior distribution approaches a point as more data arrive—leading, in the limit, to inferential certainty—is based on the idea of repeated sampling. It is certainly appealing that if the hypothesized family of probability models contains the true distribution (and assigns it a nonzero prior density), then as more information about θ arrives, the posterior distribution converges to the true value of θ.

Large-sample correspondence

Suppose that the normal approximation (4.2) for the posterior distribution of θ holds; then we can transform to the standard multivariate normal:

$$[I(\hat{\theta})]^{1/2}(\theta - \hat{\theta}) \,|\, y \sim N(0, I), \qquad (4.3)$$

where $\hat{\theta}$ is the posterior mode and $[I(\hat{\theta})]^{1/2}$ is any matrix square root of $I(\hat{\theta})$. In addition, $\hat{\theta} \to \theta_0$, and so we could just as well write the approximation in terms of $I(\theta_0)$. If the true data distribution is included in the class of models, so that $f(y) \equiv p(y|\theta)$ for some θ, then in repeated sampling with fixed θ, in the limit $n \to \infty$, it can be proved that

$$[I(\hat{\theta})]^{1/2}(\theta - \hat{\theta}) \,|\, \theta \sim N(0, I), \qquad (4.4)$$

a result from classical statistical theory that is generally proved for $\hat{\theta}$ equal to the maximum likelihood estimate but is easily extended to the case with $\hat{\theta}$ equal to the posterior mode. These results mean that, for any function of $(\theta - \hat{\theta})$, the posterior distribution derived from (4.3) is asymptotically the same as the repeated sampling distribution derived from (4.4). Thus, for example, a 95% central posterior interval for θ will cover the true value 95% of the time under repeated sampling with any fixed true θ.

Point estimation, consistency, and efficiency

In the Bayesian framework, obtaining an 'estimate' of θ makes most sense in large samples when the posterior mode, $\hat{\theta}$, is the obvious center of the posterior distribution of θ and the uncertainty conveyed by $nI(\hat{\theta})$ is so small as to be practically unimportant. More generally, however, in smaller samples, it is inappropriate to summarize inference about θ by one value, especially when the posterior distribution of θ is more variable or even asymmetric. Formally, by incorporating loss functions in a decision-theoretic context (see Section 22.1 and Exercise 22.1), one can define optimal point estimates; for the purposes of Bayesian data analysis, however, we believe that representation of the full posterior distribution (as, for example, with 50% and 95% central posterior intervals) is more useful. In many problems, especially with large samples, a point estimate and its estimated standard error are adequate to summarize a posterior inference, but we interpret the estimate as an inferential summary, not as the solution to a decision problem. In any case, the large-sample frequency properties of any estimate can be evaluated, without consideration of whether the estimate was derived from a Bayesian analysis.

A point estimate is said to be *consistent* in the sampling theory sense if, as samples get larger, it converges to the true value of the parameter that it is asserted to estimate. Thus, if $f(y) \equiv p(y|\theta_0)$, then a point estimate $\hat{\theta}$ of θ is consistent if its sampling distribution converges to a point mass at θ_0 as the data sample size n increases (that is, considering $\hat{\theta}$ as a function of y, which is a random variable conditional on θ_0). A closely related concept is *asymptotic unbiasedness*, where $(\mathrm{E}(\hat{\theta}|\theta_0) - \theta_0)/\mathrm{sd}(\hat{\theta}|\theta_0)$ converges to 0 (once again, considering $\hat{\theta}(y)$ as a random variable whose distribution is determined by $p(y|\theta)$). When the truth is included in the family of models being fitted, the posterior mode $\hat{\theta}$, and also the posterior mean and median, are consistent and asymptotically unbiased under mild regularity conditions.

A point estimate $\hat{\theta}$ is said to be *efficient* if there exists no other function of y that estimates θ with lower mean squared error, that is, if the expression $\mathrm{E}[(\hat{\theta} - \theta_0)^2|\theta_0]$ is at its optimal, lowest value. More generally, the *efficiency* of $\hat{\theta}$ is the optimal mean squared error divided by the mean squared error of $\hat{\theta}$. An estimate is asymptotically efficient if its efficiency approaches 1 as the sample size $n \to \infty$. Under mild regularity conditions, the center of the posterior distribution (defined, for example, by the posterior mean, median, or mode) is asymptotically efficient.

Confidence coverage

If a region $C(y)$ includes θ_0 at least $100(1 - \alpha)\%$ of the time (no matter what the value of θ_0) in repeated samples, then $C(y)$ is called a $100(1 - \alpha)\%$ *confidence region* for the parameter θ. The word 'confidence' is carefully chosen to distinguish such intervals from probability intervals and to convey the following behavioral meaning: if one chooses α to be small enough (for example, 0.05 or 0.01), then since confidence regions cover the truth in at least $(1 - \alpha)$ of their applications, one should be confident in each application that the truth is within the region and therefore act as if it is. We saw previously that asymptotically a $100(1 - \alpha)\%$ central posterior interval for θ has the property that, in repeated samples of y, $100(1 - \alpha)\%$ of the intervals include the value θ_0.

4.5 Bibliographic note

Relatively little has been written on the practical implications of asymptotic theory for Bayesian analysis. The overview by Edwards, Lindman, and Savage (1963) remains one of the best and includes a detailed discussion of the principle of 'stable estimation' or when prior information can be satisfactorily approximated by a uniform density function. Much more has been written comparing Bayesian and non-Bayesian approaches to inference, and we have largely ignored the extensive philosophical and logical debates on this subject. Some good sources on the topic from the Bayesian point of view include Lindley (1958), Pratt (1965), and Berger and Wolpert (1984). Jaynes (1976) discusses some disadvantages of non-Bayesian methods compared to a particular Bayesian approach.

In Appendix B we provide references to the asymptotic normality theory. The counterexamples presented in Section 4.3 have arisen, in various forms, in our own applied research. Berzuini et al. (1997) discuss Bayesian inference for sequential data problems, in which the posterior distribution changes as data arrive, thus approaching the asymptotic results dynamically.

An example of the use of the normal approximation with small samples is provided by Rubin and Schenker (1987), who approximate the posterior distribution of the logit of the binomial parameter in a real application and evaluate the frequentist operating characteristics of their procedure; see also Agresti and Coull (1998). Clogg et al. (1991) provide additional discussion of this approach in a more complicated setting.

Morris (1983) and Rubin (1984) discuss, from two different standpoints, the concept of evaluating Bayesian procedures by examining long-run frequency properties (such as coverage of 95% confidence intervals). An example of frequency evaluation of Bayesian procedures in an applied problem is given by Zaslavsky (1993).

4.6 Exercises

1. Normal approximation: suppose that $y_1, \ldots, y_5$ are independent samples from a Cauchy distribution with unknown center θ and known scale 1: $p(y_i|\theta) \propto 1/(1 + (y_i - \theta)^2)$. Assume that the prior distribution for θ is uniform on $[0, 1]$. Given the observations $(y_1, \ldots, y_5) = (-2, -1, 0, 1.5, 2.5)$:

   (a) Determine the derivative and the second derivative of the log posterior density.

   (b) Find the posterior mode of θ by iteratively solving the equation determined by setting the derivative of the log-likelihood to zero.

   (c) Construct the normal approximation based on the second derivative of the log posterior density at the mode. Plot the approximate normal density and compare to the exact density computed in Exercise 2.11.

2. Normal approximation: derive the analytic form of the information matrix and the normal approximation variance for the bioassay example.

3. Normal approximation to the marginal posterior distribution of an estimand: in the bioassay example, the normal approximation to the joint posterior distribution of (α, β) is obtained. The posterior distribution of any estimand, such as the LD50, can be approximated by a normal distribution fit to its marginal posterior mode and the curvature of the marginal posterior density about the mode. This is sometimes called the 'delta method.' Expand the posterior distribution of the LD50, $-\alpha/\beta$, as a Taylor series around the posterior mode and thereby derive the asymptotic posterior median and standard deviation. Compare to the histogram in Figure 4.2.

4. Asymptotic normality: assuming the regularity conditions hold, we know that $p(\theta|y)$ approaches normality as $n \to \infty$. In addition, if $\phi = f(\theta)$ is any one-to-one continuous transformation of θ, we can express the Bayesian inference in terms of ϕ and find that $p(\phi|y)$ also approaches normality. But it is well known that a nonlinear transformation of a normal distribution is no longer normal. How can both limiting normal distributions be valid?

Part II: Fundamentals of Bayesian Data Analysis

For most problems of applied Bayesian statistics, the data analyst must go beyond the simple structure of prior distribution, likelihood, and posterior distribution. In Chapter 5, we introduce *hierarchical models*, which allow the parameters of a prior, or population, distribution themselves to be estimated from data. In Chapter 6, we discuss methods of assessing the sensitivity of posterior inferences to model assumptions and checking the fit of a probability model to data and substantive information. Model checking allows an escape from the tautological aspect of formal approaches to Bayesian inference, under which all conclusions are conditional on the truth of the posited model. Chapter 7 outlines the role of study design and methods of data collection in probability modeling, focusing on how to set up Bayesian inference for sample surveys, designed experiments, and observational studies; this chapter contains some of the most conceptually distinctive and potentially difficult material in the book. Chapter 8 presents connections between the Bayesian methods of Parts I and II and other approaches to inference. Chapter 9 serves as a review of our approach to data analysis with illustrations drawn from our experiences. These five chapters illustrate the creative choices that are required, first to set up a Bayesian model in a complex problem, and then to perform the model checking and confidence building that is typically necessary to make posterior inferences scientifically defensible.

CHAPTER 5

Hierarchical models

Many statistical applications involve multiple parameters that can be regarded as related or connected in some way by the structure of the problem, implying that a joint probability model for these parameters should reflect the dependence among them. For example, in a study of the effectiveness of cardiac treatments, with the patients in hospital j having survival probability θ_j, it might be reasonable to expect that estimates of the θ_j's, which represent a sample of hospitals, should be related to each other. We shall see that this is achieved in a natural way if we use a prior distribution in which the θ_j's are viewed as a sample from a common *population distribution*. A key feature of such applications is that the observed data, y_{ij}, with units indexed by i within groups indexed by j, can be used to estimate aspects of the population distribution of the θ_j's even though the values of θ_j are not themselves observed. It is natural to model such a problem hierarchically, with observable outcomes modeled conditionally on certain parameters, which themselves are given a probabilistic specification in terms of further parameters, known as *hyperparameters*. Such hierarchical thinking helps in understanding multiparameter problems and also plays an important role in developing computational strategies.

Perhaps even more important in practice is that nonhierarchical models are usually inappropriate for hierarchical data: with few parameters, they generally cannot fit large datasets accurately, whereas with many parameters, they tend to 'overfit' such data in the sense of producing models that fit the existing data well but lead to inferior predictions for new data. In contrast, hierarchical models can have enough parameters to fit the data well, while using a population distribution to structure some dependence into the parameters, thereby avoiding problems of overfitting. As we show in the examples in this chapter, it is often sensible to fit hierarchical models with more parameters than there are data points.

In Section 5.1, we consider the problem of constructing a prior distribution using hierarchical principles but without fitting a formal probability model for the hierarchical structure. We first consider the analysis of a single experiment, using historical data to create a prior distribution, and then we consider a plausible prior distribution for the parameters of a set of experiments. The treatment in Section 5.1 is not fully Bayesian, because, for the purpose of simplicity in exposition, we work with a point estimate, rather than a complete joint posterior distribution, for the parameters of the population distribution (the hyperparameters). In Section 5.2, we discuss how to construct a hierarchical prior distribution in the context of a fully Bayesian analysis. Sections 5.3–5.4 present a general approach to computation with hierarchical models in conjugate families by combining analytical and numerical methods. We defer details of the most general computational methods to Part III in order to explore immediately the important practical and conceptual advantages of hierarchical Bayesian models. The chapter concludes with two extended examples: a hierarchical model for an educational testing experiment and a Bayesian treatment of the method of 'meta-analysis' as used in medical research to combine the results of separate studies relating to the same research question.

5.1 Constructing a parameterized prior distribution

Analyzing a single experiment in the context of historical data

To begin our description of hierarchical models, we consider the problem of estimating a parameter θ using data from a small experiment and a prior distribution constructed from similar previous (or historical) experiments. Mathematically, we will consider the current and historical experiments to be a random sample from a common population.

Example. Estimating the risk of tumor in a group of rats

In the evaluation of drugs for possible clinical application, studies are routinely performed on rodents. For a particular study drawn from the statistical literature, suppose the immediate aim is to estimate θ, the probability of tumor in a population of female laboratory rats of type 'F344' that receive a zero dose of the drug (a control group). The data show that 4 out of 14 rats developed endometrial stromal polyps (a kind of tumor). It is natural to assume a binomial model for the number of tumors, given θ. For convenience, we select a prior distribution for θ from the conjugate family, $\theta \sim \text{Beta}(\alpha, \beta)$.

Analysis with a fixed prior distribution. From historical data, suppose we knew that the tumor probabilities θ among groups of female lab rats of type F344 follow an approximate beta distribution, with known mean and standard deviation. The tumor probabilities θ vary because of differences in rats and experimental conditions among the experiments. Referring to the expressions for the mean and variance of the beta distribution (see Appendix A), we could find values for α, β that correspond to the given values for the mean and standard deviation. Then, assuming a $\text{Beta}(\alpha, \beta)$ prior distribution for θ yields a $\text{Beta}(\alpha + 4, \beta + 10)$ posterior distribution for θ.

Approximate estimate of the population distribution using the historical data. Typically, the mean and standard deviation of underlying tumor risks are not available. Rather, historical *data* are available on previous experiments on similar groups of rats. In the rat tumor example, the historical data were in fact a set of observations of tumor incidence in 70 groups of rats (Table 5.1). In the jth historical experiment, let the number of rats with tumors be y_j and the total number of rats be n_j. We model the y_j's as independent binomial data, given sample sizes n_j and study-specific means θ_j. Assuming that the beta prior distribution with parameters (α, β) is a good description of the population distribution

Previous experiments:

0/20	0/20	0/20	0/20	0/20	0/20	0/20	0/19	0/19	0/19
0/19	0/18	0/18	0/17	1/20	1/20	1/20	1/20	1/19	1/19
1/18	1/18	2/25	2/24	2/23	2/20	2/20	2/20	2/20	2/20
2/20	1/10	5/49	2/19	5/46	3/27	2/17	7/49	7/47	3/20
3/20	2/13	9/48	10/50	4/20	4/20	4/20	4/20	4/20	4/20
4/20	10/48	4/19	4/19	4/19	5/22	11/46	12/49	5/20	5/20
6/23	5/19	6/22	6/20	6/20	6/20	16/52	15/47	15/46	9/24

Current experiment:
4/14

Table 5.1 *Tumor incidence in historical control groups and current group of rats, from Tarone (1982). The table displays the values of y_j/n_j: (number of rats with tumors)/(total number of rats).*

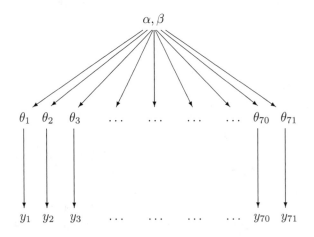

α, β

θ_1 θ_2 θ_3 $\cdots$ $\cdots$ $\cdots$ $\cdots$ θ_{70} θ_{71}

y_1 y_2 y_3 $\cdots$ $\cdots$ $\cdots$ $\cdots$ y_{70} y_{71}

Figure 5.1 *Structure of the hierarchical model for the rat tumor example.*

of the θ_j's in the historical experiments, we can display the hierarchical model schematically as in Figure 5.1, with θ_{71} and y_{71} corresponding to the current experiment.

The observed sample mean and standard deviation of the 70 values y_j/n_j are 0.136 and 0.103. If we set the mean and standard deviation of the population distribution to these values, we can solve for α and β—see (A.3) on page 582 in Appendix A. The resulting estimate for (α, β) is $(1.4, 8.6)$. This is *not* a Bayesian calculation because it is not based on any specified full probability model. We present a better, fully Bayesian approach to estimating (α, β) for this example in Section 5.3. The estimate $(1.4, 8.6)$ is simply a starting point from which we can explore the idea of estimating the parameters of the population distribution.

Using the simple estimate of the historical population distribution as a prior distribution for the current experiment yields a Beta(5.4, 18.6) posterior distribution for θ_{71}: the posterior mean is 0.223, and the standard deviation is 0.083. The prior information has resulted in a posterior mean substantially lower than the crude proportion, $4/14 = 0.286$, because the weight of experience indicates that the number of tumors in the current experiment is unusually high.

These analyses require that the current tumor risk, θ_{71}, and the 70 historical tumor risks, $\theta_1, \ldots, \theta_{70}$, be considered a random sample from a common distribution, an assumption that would be invalidated, for example, if it were known that the historical experiments were all done in laboratory A but the current data were gathered in laboratory B, or if time trends were relevant. In practice, a simple, although arbitrary, way of accounting for differences between the current and historical data is to inflate the historical variance. For the beta model, inflating the historical variance means decreasing $(\alpha + \beta)$ while holding α/β constant. Other systematic differences, such as a time trend in tumor risks, can be incorporated in a more extensive model.

Having used the 70 historical experiments to form a prior distribution for θ_{71}, we might now like also to use this same prior distribution to obtain Bayesian inferences for the tumor probabilities in the first 70 experiments, $\theta_1, \ldots, \theta_{70}$. There are several logical and practical problems with the approach of directly estimating a prior distribution from existing data:

- If we wanted to use the estimated prior distribution for inference about the first 70 experiments, then the data would be used twice: first, all the results together are used to estimate the prior distribution, and then each experiment's results are used to estimate its θ. This would seem to cause us to overestimate our precision.

- The point estimate for α and β seems arbitrary, and using any point estimate for α and β necessarily ignores some posterior uncertainty.

- We can also make the opposite point: does it make sense to 'estimate' α and β at all? They are part of the 'prior' distribution: should they be known before the data are gathered, according to the logic of Bayesian inference?

Logic of combining information

Despite these problems, it clearly makes more sense to try to estimate the population distribution from all the data, and thereby to help estimate each θ_j, than to estimate all 71 values θ_j separately. Consider the following thought experiment about inference on two of the parameters, θ_{26} and θ_{27}, each corresponding to experiments with 2 observed tumors out of 20 rats. Suppose our prior distribution for both θ_{26} and θ_{27} is centered around 0.15; now suppose that you were told after completing the data analysis that $\theta_{26} = 0.1$ exactly. This should influence your estimate of θ_{27}; in fact, it would probably make you think that θ_{27} is lower than you previously believed, since the data for the two parameters are identical, and the postulated value of 0.1 is lower than you previously expected for θ_{26} from the prior distribution. Thus, θ_{26} and θ_{27} should be dependent in the posterior distribution, and they should not be analyzed separately.

We retain the advantages of using the data to estimate prior parameters and eliminate all of the disadvantages just mentioned by putting a probability model on the entire set of parameters and experiments and then performing a Bayesian analysis on the joint distribution of all the model parameters. A complete Bayesian analysis is described in Section 5.3. The analysis using the data to estimate the prior parameters, which is sometimes called *empirical Bayes*, can be viewed as an approximation to the complete hierarchical Bayesian analysis. We prefer to avoid the term 'empirical Bayes' because it misleadingly suggests that the full Bayesian method, which we discuss here and use for the rest of the book, is not 'empirical.'

5.2 Exchangeability and setting up hierarchical models

Generalizing from the example of the previous section, consider a set of experiments $j = 1, \ldots, J$, in which experiment j has data (vector) y_j and parameter (vector) θ_j, with likelihood $p(y_j|\theta_j)$. (Throughout this chapter we use the word 'experiment' for convenience, but the methods can apply equally well to nonexperimental data.) Some of the parameters in different experiments may overlap; for example, each data vector y_j may be a sample of observations from a normal distribution with mean μ_j and common variance σ^2, in which case $\theta_j = (\mu_j, \sigma^2)$. In order to create a joint probability model for all the parameters θ, we use the crucial idea of exchangeability introduced in Chapter 1 and used repeatedly since then.

Exchangeability

If no information—other than the data y—is available to distinguish any of the θ_j's from any of the others, and no ordering or grouping of the parameters can be made, one must assume symmetry among the parameters in their prior distribution. This symmetry is represented probabilistically by exchangeability; the parameters $(\theta_1, \ldots, \theta_J)$ are *exchangeable* in their joint distribution if $p(\theta_1, \ldots, \theta_J)$ is invariant to permutations of the indexes $(1, \ldots, J)$. For example, in the rat tumor problem, suppose we have no information to distinguish the 71 experiments, other than the sample sizes n_j, which presumably are not related to the values of θ_j; we therefore use an exchangeable model for the θ_j's.

We have already encountered the concept of exchangeability in constructing iid models for unit- or individual-level data. In practice, ignorance implies exchangeability. Generally, the less we know about a problem, the more confidently we can make claims of exchangeability. (This is not, we hasten to add, a good reason to limit our knowledge of a problem before embarking on statistical analysis!) Consider the analogy to a roll of a die: we should initially assign equal probabilities to all six outcomes, but if we study the measurements of the die and weigh the die carefully, we might eventually notice imperfections, which might make us favor one outcome over the others and thus eliminate the symmetry among the six outcomes.

The simplest form of an exchangeable distribution has each of the parameters θ_j as an independent sample from a prior (or population) distribution governed by some unknown parameter vector ϕ; thus,

$$p(\theta|\phi) = \prod_{j=1}^{J} p(\theta_j|\phi). \tag{5.1}$$

In general, ϕ is unknown, so our distribution for θ must average over our uncertainty in ϕ:

$$p(\theta) = \int \left[\prod_{j=1}^{J} p(\theta_j|\phi) \right] p(\phi)d\phi, \tag{5.2}$$

This form, the mixture of iid distributions, is usually all that we need to capture exchangeability in practice.

A related theoretical result, *de Finetti's theorem*, to which we alluded in Section 1.2, states that in the limit as $J \to \infty$, any suitably well-behaved exchangeable distribution on $(\theta_1, \ldots, \theta_J)$ can be written in the iid mixture form (5.2). Formally, de Finetti's theorem does not hold when J is finite (see Exercise 5.2). Statistically, the iid mixture model characterizes parameters θ as drawn from a common 'superpopulation' that is determined by the unknown hyperparameters, ϕ. We are already familiar with exchangeable models for *data*, $y_1, \ldots, y_n$, in the form of 'iid' likelihoods, in which the n observations are independent and identically distributed, given some parameter vector θ.

Example. Exchangeability and sampling

The following thought experiment illustrates the role of exchangeability in inference from random sampling. For simplicity, we use a nonhierarchical example with exchangeability at the level of y rather than θ.

We, the authors, have selected eight states out of the United States and recorded the divorce rate per 1000 population in each state in 1981. Call these $y_1, \ldots, y_8$. What can you, the reader, say about y_8, the divorce rate in the eighth state?

Since you have no information to distinguish any of the eight states from the others, you must model them exchangeably. You might use a beta distribution for the eight y_j's, a logit normal, or some other prior distribution restricted to the range $[0, 1]$. Unless you are familiar with divorce statistics in the United States, your distribution on $(y_1, \ldots, y_8)$ should be fairly vague.

We now randomly sample seven states from these eight and tell you their divorce rates: $5.8, 6.6, 7.8, 5.6, 7.0, 7.1, 5.4$, each in numbers of divorces per 1000 population (per year). Based primarily on the data, a reasonable posterior (predictive) distribution for the remaining value, y_8, would probably be centered around 6.5 and have most of its mass between 5.0 and 8.0.

Suppose initially we had given you the further prior information that the eight states are Mountain states: Arizona, Colorado, Idaho, Montana, Nevada, New Mexico, Utah, and Wyoming, but selected in a random order; you still are not told which observed rate corresponds to which state. Now, before the seven data points were observed, the eight divorce rates should still be modeled exchangeably. However, your prior distribution (that is, *before* seeing the data), for the eight numbers should change: it seems reasonable to assume that Utah, with its large Mormon population, has a much lower divorce rate, and Nevada, with its liberal divorce laws, has a much higher divorce rate, than the remaining six states. Perhaps, given your expectation of outliers in the distribution, your prior distribution should have wide tails. Given this extra information (the names of the eight states), when you see the seven observed values and note that the numbers are so close together, it might seem a reasonable guess that the missing eighth state is Nevada or Utah. Therefore its value might be expected to be much lower or much higher than the seven values observed. This might lead to a bimodal or trimodal posterior distribution to account for the two plausible scenarios. The prior distribution on the eight values y_j is still exchangeable, however, because you have no information telling which state corresponds to which index number. (See Exercise 5.4.)

Finally, we tell you that the state not sampled (corresponding to y_8) was Nevada. Now, even before seeing the seven observed values, you cannot assign an exchangeable prior distribution to the set of eight divorce rates, since you have information that distinguishes y_8 from the other seven numbers, here suspecting it is larger than any of the others. Once $y_1, \ldots, y_7$ have been observed, a reasonable posterior distribution for y_8 plausibly should have most of its mass above the largest observed rate.

Incidentally, Nevada's divorce rate in 1981 was 13.9 per 1000 population.

Exchangeability when additional information is available on the units

In the previous example, if we knew x_j, the divorce rate in state j *last* year, for $j = 1, \ldots, 8$, but not which index corresponded to which state, then we would certainly be able to distinguish the eight values of y_j, but the joint prior distribution $p(x_j, y_j)$ would be the same for each state. In general, the usual way to model exchangeability with covariates is through conditional independence:
$p(\theta_1, \ldots, \theta_J | x_1, \ldots, x_J) = \int [\prod_{j=1}^{J} p(\theta_j | \phi, x_j)] p(\phi | x) d\phi$, with $x = (x_1, \ldots, x_J)$.

In this way, exchangeable models become almost universally applicable, because any information available to distinguish different units should be encoded in the x and y variables. For example, consider the probabilities of a given die landing on each of its six faces, *after* we have carefully measured the die and noted its physical imperfections. If we include the imperfections (such as the area of each face, the bevels of the corners, and so forth) as explanatory variables x in a realistic physical model, the probabilities $\theta_1, \ldots, \theta_6$ should become exchangeable, conditional on x. In this example, the six parameters θ_j are constrained to sum to 1 and so *cannot* be modeled with a mixture of iid distributions; nonetheless, they can be modeled exchangeably.

In the rat tumor example, we have already noted that the sample sizes n_j are the only available information to distinguish the different experiments. It does not seem likely that n_j would be a useful variable for modeling tumor rates, but if one were interested, one could create an exchangeable model for the J pairs $(n, y)_j$. A natural first step would be to plot y_j/n_j vs. n_j to see any obvious relation that could be modeled. For example, perhaps some studies

j had larger sample sizes n_j because the investigators correctly suspected rarer events; that is, smaller θ_j and thus smaller expected values of y_j/n_j. In fact, the plot of y_j/n_j versus n_j, not shown here, shows no apparent relation between the two variables.

Objections to exchangeable models

In virtually any statistical application, it is natural to object to exchangeability on the grounds that the units actually differ. For example, the 71 rat tumor experiments were performed at different times, on different rats, and presumably in different laboratories. Such information does *not*, however, invalidate exchangeability. That the experiments differ implies that the θ_j's differ, but it might be perfectly acceptable to consider them as if drawn from a common distribution. In fact, with no information available to distinguish them, we have no logical choice but to model the θ_j's exchangeably. Objecting to exchangeability for modeling ignorance is no more reasonable than objecting to an iid model for samples from a common population, objecting to regression models in general, or, for that matter, objecting to displaying points in a scatterplot without individual labels. As with regression, the valid concern is not about exchangeability, but about encoding relevant knowledge as explanatory variables where possible.

The full Bayesian treatment of the hierarchical model

Returning to the problem of inference, the key 'hierarchical' part of these models is that ϕ is not known and thus has its own prior distribution, $p(\phi)$. The appropriate Bayesian posterior distribution is of the vector (ϕ, θ). The joint prior distribution is

$$p(\phi, \theta) = p(\phi)p(\theta|\phi),$$

and the joint posterior distribution is

$$\begin{aligned} p(\phi, \theta|y) &\propto p(\phi, \theta)p(y|\phi, \theta) \\ &= p(\phi, \theta)p(y|\theta), \end{aligned} \tag{5.3}$$

with the latter simplification holding because the data distribution, $p(y|\phi, \theta)$, depends only on θ; the hyperparameters ϕ affect y only through θ. Previously, we assumed ϕ was known, which is unrealistic; now we include the uncertainty in ϕ in the model.

The hyperprior distribution

In order to create a joint probability distribution for (ϕ, θ), we must assign a prior distribution to ϕ. If little is known about ϕ, we can assign a diffuse prior distribution, but we must be careful when using an improper prior density to check that the resulting posterior distribution is proper, and we should assess whether our conclusions are sensitive to this simplifying assumption. In most real problems, one should have enough substantive knowledge about the parameters in ϕ at least to constrain the hyperparameters into a finite region, if not to assign a substantive hyperprior distribution. As in nonhierarchical models, it is often practical to start with a simple, relatively noninformative, prior distribution on ϕ and seek to add more prior information if there remains too much variation in the posterior distribution.

In the rat tumor example, the hyperparameters are (α, β), which determine the beta distribution for θ. We illustrate one approach to constructing an appropriate hyperprior distribution in the continuation of that example in the next section.

Posterior predictive distributions

Hierarchical models are characterized both by hyperparameters, ϕ, in our notation, and parameters θ. There are two posterior predictive distributions that might be of interest to the data analyst: (1) the distribution of future observations $\tilde{y}$ corresponding to an existing θ_j, or (2) the distribution of observations $\tilde{y}$ corresponding to future θ_j's drawn from the same superpopulation. We label the future θ_j's as $\tilde{\theta}$. Both kinds of replications can be used to assess model adequacy, as we discuss in Chapter 6. In the rat tumor example, future observations can be (1) additional rats from an existing experiment, or (2) results from a future experiment. In the former case, the posterior predictive draws $\tilde{y}$ are based on the posterior draws of θ_j for the existing experiment. In the latter case, one must first draw $\tilde{\theta}$ for the new experiment from the population distribution, given the posterior draws of ϕ, and then draw $\tilde{y}$ given the simulated $\tilde{\theta}$.

5.3 Computation with hierarchical models

Our computational strategy for hierarchical models follows the general approach to multiparameter problems presented in Section 3.8 but is more difficult in practice because of the large number of parameters that commonly appear in a hierarchical model. In particular, we cannot generally plot the contours or display a scatterplot of the simulations from the joint posterior distribution of (θ, ϕ). With care, however, we can follow a similar approach as before, treating θ as the vector parameter of interest and ϕ as the vector of nuisance parameters (though we recognize that both ϕ and θ will be of interest in some problems).

In this section, we present an approach that combines analytical and numerical methods to obtain simulations from the joint posterior distribution, $p(\theta, \phi|y)$, for some simple but important hierarchical models in which the population distribution, $p(\theta|\phi)$, is conjugate to the likelihood, $p(y|\theta)$. For the many nonconjugate hierarchical models that arise in practice, more advanced computational methods, presented in Part III of this book, are necessary. Even for more complicated problems, however, the approach using conjugate distributions is useful for obtaining approximate estimates and starting points for more accurate computations.

Analytic derivation of conditional and marginal distributions

We first perform the following three steps analytically.

1. Write the joint posterior density, $p(\theta, \phi|y)$, in unnormalized form as a product of the hyperprior distribution $p(\phi)$, the population distribution $p(\theta|\phi)$, and the likelihood $p(y|\theta)$.

2. Determine analytically the conditional posterior density of θ given the hyperparameters ϕ; for fixed observed y, this is a function of ϕ, $p(\theta|\phi, y)$.

3. Estimate ϕ using the Bayesian paradigm; that is, obtain its marginal posterior distribution, $p(\phi|y)$.

The first step is immediate, and the second step is easy for conjugate models because, conditional on ϕ, the population distribution for θ is just the iid model (5.1), so that the conditional posterior density is a product of conjugate posterior densities for the components θ_j.

The third step can be performed by brute force by integrating the joint posterior distribution over θ:

$$p(\phi|y) = \int p(\theta, \phi|y)d\theta. \tag{5.4}$$

For many standard models, however, including the normal distribution, the marginal posterior distribution of ϕ can be computed algebraically using the conditional probability formula,

$$p(\phi|y) = \frac{p(\theta, \phi|y)}{p(\theta|\phi, y)}, \tag{5.5}$$

which we have used already for the semi-conjugate prior distribution for the normal model in Section 3.4. This expression is useful because the numerator is just the joint posterior distribution (5.3), and the denominator is the posterior distribution for θ if ϕ were known. The difficulty in using (5.5), beyond a few standard conjugate models, is that the denominator, $p(\theta|\phi, y)$, regarded as a function of both θ and ϕ for fixed y, has a normalizing factor that depends on ϕ as well as y. One must be careful with the proportionality 'constant' in Bayes' theorem, especially when using hierarchical models, to make sure it is actually constant. Exercise 5.9 has an example of a nonconjugate model in which the integral (5.4) has no closed-form solution so that (5.5) is no help.

Drawing simulations from the posterior distribution

The following strategy is useful for simulating a draw from the joint posterior distribution, $p(\theta, \phi|y)$, for simple hierarchical models such as are considered in this chapter.

1. Draw the vector of hyperparameters, ϕ, from its marginal posterior distribution, $p(\phi|y)$. If ϕ is low-dimensional, the methods discussed in Chapter 3 can be used; for high-dimensional ϕ, more sophisticated methods such as described in Part III may be needed.

2. Draw the parameter vector θ from its conditional posterior distribution, $p(\theta|\phi, y)$, given the drawn value of ϕ. For the examples we consider in this chapter, the factorization $p(\theta|\phi, y) = \prod_j p(\theta_j|\phi, y)$ holds, and so the components θ_j can be drawn independently, one at a time.

3. If desired, draw predictive values $\tilde{y}$ from the posterior predictive distribution given the drawn θ. Depending on the problem, it might be necessary first to draw a new value $\tilde{\theta}$, given ϕ, as discussed at the end of the previous section.

As usual, the above steps are performed L times in order to obtain a set of L draws. From the joint posterior simulations of θ and $\tilde{y}$, we can compute the posterior distribution of any estimand or predictive quantity of interest.

Example. Rat tumors (continued)

We now perform a full Bayesian analysis of the rat tumor experiments described in Section 5.1. Once again, the data from experiments $j = 1, \ldots, J$, $J = 71$, are assumed to follow independent binomial distributions:

$$y_j \sim \text{Bin}(n_j, \theta_j),$$

with the number of rats, n_j, known. The parameters θ_j are assumed to be independent samples from a beta distribution:

$$\theta_j \sim \text{Beta}(\alpha, \beta),$$

and we shall assign a noninformative hyperprior distribution to reflect our ignorance about the unknown hyperparameters. As usual, the word 'noninformative' indicates our attitude toward this part of the model and is not intended to imply that this particular distribution has any special properties. If the hyperprior distribution turns out to be crucial for our inference, we should report this and if possible seek further substantive knowledge that could be used to construct a more informative prior distribution. If we wish to assign an improper prior distribution for the hyperparameters, (α, β), we must check that the posterior distribution is proper. We defer the choice of noninformative hyperprior distribution, a relatively arbitrary and unimportant part of this particular analysis, until we inspect the integrability of the posterior density.

Joint, conditional, and marginal posterior distributions. We first perform the three steps for determining the analytic form of the posterior distribution. The joint posterior distribution of all parameters is

$$p(\theta, \alpha, \beta|y) \propto p(\alpha, \beta)p(\theta|\alpha, \beta)p(y|\theta, \alpha, \beta)$$

$$\propto p(\alpha, \beta) \prod_{j=1}^{J} \frac{\Gamma(\alpha + \beta)}{\Gamma(\alpha)\Gamma(\beta)} \theta_j^{\alpha-1}(1 - \theta_j)^{\beta-1} \prod_{j=1}^{J} \theta_j^{y_j}(1 - \theta_j)^{n_j - y_j}. \quad (5.6)$$

Given (α, β), the components of θ have independent posterior densities that are of the form $\theta_j^A (1 - \theta_j)^B$—that is, beta densities—and the joint density is

$$p(\theta|\alpha, \beta, y) = \prod_{j=1}^{J} \frac{\Gamma(\alpha + \beta + n_j)}{\Gamma(\alpha + y_j)\Gamma(\beta + n_j - y_j)} \theta_j^{\alpha + y_j - 1}(1 - \theta_j)^{\beta + n_j - y_j - 1}. \quad (5.7)$$

We can determine the marginal posterior distribution of (α, β) by substituting (5.6) and (5.7) into the conditional probability formula (5.5):

$$p(\alpha, \beta|y) \propto p(\alpha, \beta) \prod_{j=1}^{J} \frac{\Gamma(\alpha + \beta)}{\Gamma(\alpha)\Gamma(\beta)} \frac{\Gamma(\alpha + y_j)\Gamma(\beta + n_j - y_j)}{\Gamma(\alpha + \beta + n_j)}. \quad (5.8)$$

The product in equation (5.8) cannot be simplified analytically but is easy to compute for any specified values of (α, β) using a standard routine to compute the gamma function.

Choosing a standard parameterization and setting up a 'noninformative' hyperprior distribution. Because we have no immediately available information about the distribution of tumor rates in populations of rats, we seek a relatively diffuse hyperprior distribution for (α, β). Before assigning a hyperprior distribution, we reparameterize in terms of logit$(\frac{\alpha}{\alpha+\beta}) = \log(\frac{\alpha}{\beta})$ and $\log(\alpha + \beta)$, which are the logit of the mean and the logarithm of the 'sample size' in the beta population distribution for θ. It would seem reasonable to assign independent hyperprior distributions to the prior mean and 'sample size,' and we use the logistic and logarithmic transformations to put each on a $(-\infty, \infty)$ scale. Unfortunately, a uniform prior density on these newly transformed parameters yields an improper *posterior* density, with an infinite integral in the limit $(\alpha + \beta) \to \infty$, and so this particular prior density cannot be used here.

In a problem such as this with a reasonably large amount of data, it is possible to set up a 'noninformative' hyperprior density that is dominated by the likelihood and yields a proper posterior distribution. One reasonable choice of diffuse hyperprior density is uniform on $(\frac{\alpha}{\alpha+\beta}, (\alpha + \beta)^{-1/2})$, which when multiplied by the appropriate Jacobian yields the following densities on the original scale,

$$p(\alpha, \beta) \propto (\alpha + \beta)^{-5/2}, \quad (5.9)$$

and on the natural transformed scale:

$$p\left(\log\left(\frac{\alpha}{\beta}\right), \log(\alpha + \beta)\right) \propto \alpha\beta(\alpha + \beta)^{-5/2}. \quad (5.10)$$

See Exercise 5.7 for a discussion of this prior density.

We could avoid the mathematical effort of checking the integrability of the posterior density if we were to use a proper hyperprior distribution. Another approach would be tentatively to use a flat hyperprior density, such as $p(\frac{\alpha}{\alpha+\beta}, \alpha + \beta) \propto 1$, or even $p(\alpha, \beta) \propto 1$, and then compute the contours and simulations from the posterior density (as detailed below). The result would clearly show the posterior contours drifting off toward infinity, indicating that the posterior density is not integrable in that limit. The prior distribution would then have to be altered to obtain an integrable posterior density.

Incidentally, setting the prior distribution for $(\log(\frac{\alpha}{\beta}), \log(\alpha + \beta))$ to uniform in a vague but finite range, such as $[-10^{10}, 10^{10}] \times [-10^{10}, 10^{10}]$, would *not* be an

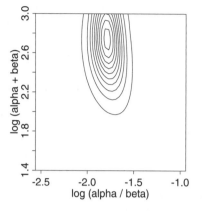

Figure 5.2 *First try at a contour plot of the marginal posterior density of* $(\log(\frac{\alpha}{\beta}), \log(\alpha + \beta))$ *for the rat tumor example. Contour lines are at* $0.05, 0.15, \ldots, 0.95$ *times the density at the mode.*

acceptable solution for this problem, as almost all the posterior mass in this case would be in the range of α and β near 'infinity,' which corresponds to a Beta(α, β) distribution with a variance of zero, meaning that all the θ_j parameters would be essentially equal in the posterior distribution. When the likelihood is not integrable, setting a faraway finite cutoff to a uniform prior density does not necessarily eliminate the problem.

Computing the marginal posterior density of the hyperparameters. Now that we have established a full probability model for data and parameters, we compute the marginal posterior distribution of the hyperparameters. Figure 5.2 shows a contour plot of the unnormalized marginal posterior density on a grid of values of $(\log(\frac{\alpha}{\beta}), \log(\alpha + \beta))$. To create the plot, we first compute the logarithm of the density function (5.8) with prior density (5.9), multiplying the Jacobian to obtain the density $p(\log(\frac{\alpha}{\beta}), \log(\alpha + \beta)|y)$. We set a grid in the range $(\log(\frac{\alpha}{\beta}), \log(\alpha + \beta)) \in [-1, -2.5] \times [1.5, 3]$, which is centered near our earlier point estimate $(-1.8, 2.3)$ (that is, $(\alpha, \beta) = (1.4, 8.6)$) and covers a factor of 4 in each parameter. Then, to avoid computational overflows, we subtract the maximum value of the log density from each point on the grid and exponentiate, yielding values of the unnormalized marginal posterior density.

The most obvious features of the contour plot are (1) the mode is not far from the point estimate (as we would expect), and (2) important parts of the marginal posterior distribution lie outside the range of the graph.

We recompute $p(\log(\frac{\alpha}{\beta}), \log(\alpha+\beta)|y)$, this time in the range $(\log(\frac{\alpha}{\beta}), \log(\alpha+\beta)) \in [-1.3, -2.3] \times [1, 5]$. The resulting grid, shown in Figure 5.3a, displays essentially all of the marginal posterior distribution. Figure 5.3b displays 1000 random draws from the numerically computed posterior distribution. The graphs show that the marginal posterior distribution of the hyperparameters, under this transformation, is approximately symmetric about the mode, roughly $(-1.75, 2.8)$. This corresponds to approximate values of $(\alpha, \beta) = (2.4, 14.0)$, which differs somewhat from the crude estimate obtained earlier.

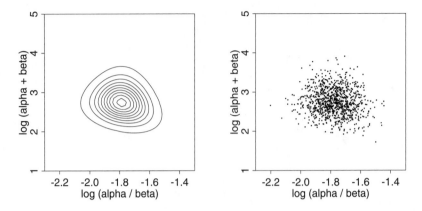

Figure 5.3 *(a) Contour plot of the marginal posterior density of* $(\log(\frac{\alpha}{\beta}), \log(\alpha+\beta))$ *for the rat tumor example. Contour lines are at* $0.05, 0.15, \ldots, 0.95$ *times the density at the mode. (b) Scatterplot of 1000 draws* $(\log(\frac{\alpha}{\beta}), \log(\alpha+\beta))$ *from the numerically computed marginal posterior density.*

Having computed the relative posterior density at a grid of values that cover the effective range of (α, β), we normalize by approximating the distribution as a step function over the grid and setting the total probability in the grid to 1.

We can then compute posterior moments based on the grid of $(\log(\frac{\alpha}{\beta}), \log(\alpha+\beta))$; for example,

$$\mathrm{E}(\alpha|y) \text{ is estimated by} \sum_{\log(\frac{\alpha}{\beta}),\log(\alpha+\beta)} \alpha\, p\left(\log\left(\frac{\alpha}{\beta}\right), \log(\alpha+\beta)\bigg| y\right).$$

From the grid in Figure 5.3, we compute $\mathrm{E}(\alpha|y) = 2.4$ and $\mathrm{E}(\beta|y) = 14.3$. This is close to the estimate based on the mode of Figure 5.3a, given above, because the posterior distribution is approximately symmetric on the scale of $(\log(\frac{\alpha}{\beta}), \log(\alpha+\beta))$. A more important consequence of averaging over the grid is to account for the posterior uncertainty in (α, β), which is not captured in the point estimate.

Sampling from the joint posterior distribution of parameters and hyperparameters. We draw 1000 random samples from the joint posterior distribution of $(\alpha, \beta, \theta_1, \ldots, \theta_J)$, as follows.

1. Simulate 1000 draws of $(\log(\frac{\alpha}{\beta}), \log(\alpha+\beta))$ from their posterior distribution displayed in Figure 5.3, using the same discrete-grid sampling procedure used to sample (α, β) for Figure 3.4b in the bioassay example of Section 3.8.

2. For $l = 1, \ldots, 1000$:

   (a) Transform the lth draw of $(\log(\frac{\alpha}{\beta}), \log(\alpha+\beta))$ to the scale (α, β) to yield a draw of the hyperparameters from their marginal posterior distribution.

   (b) For each $j = 1, \ldots, J$, sample θ_j from its conditional posterior distribution, $\theta_j|\alpha, \beta, y \sim \mathrm{Beta}(\alpha + y_j, \beta + n_j - y_j)$.

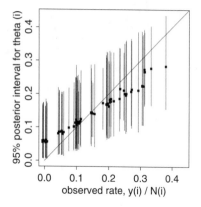

Figure 5.4 *Posterior medians and 95% intervals of rat tumor rates, θ_j (plotted vs. observed tumor rates y_j/n_j), based on simulations from the joint posterior distribution. The 45° line corresponds to the unpooled estimates, $\hat{\theta}_i = y_i/n_i$. The horizontal positions of the line have been jittered to reduce overlap.*

Displaying the results. Figure 5.4 shows posterior means and 95% intervals for the θ_j's, computed by simulation. The rates θ_j are shrunk from their sample point estimates, y_j/n_j, towards the population distribution, with approximate mean 0.14; experiments with fewer observations are shrunk more and have higher posterior variances. The results are superficially similar to what would be obtained based on a point estimate of the hyperparameters, which makes sense in this example, because of the fairly large number of experiments. But key differences remain, notably that posterior variability is higher in the full Bayesian analysis, reflecting posterior uncertainty in the hyperparameters.

5.4 Estimating an exchangeable set of parameters from a normal model

We now present a full treatment of a simple hierarchical model based on the normal distribution, in which observed data are normally distributed with a different mean for each 'group' or 'experiment,' with known observation variance, and a normal population distribution for the group means. This model is sometimes termed the one-way normal random-effects model with known data variance and is widely applicable, being an important special case of the hierarchical normal linear model, which we treat in some generality in Chapter 15. In this section, we present a general treatment following the computational approach of Section 5.3. The following section presents a detailed example; those impatient with the algebraic details may wish to look ahead at the example for motivation.

The data structure

Consider J independent experiments, with experiment j estimating the parameter θ_j from n_j independent normally distributed data points, y_{ij}, each with known error variance σ^2; that is,

$$y_{ij}|\theta_j \sim \mathrm{N}(\theta_j, \sigma^2), \text{ for } i = 1, \ldots, n_j; \ j = 1, \ldots, J. \qquad (5.11)$$

Using standard notation from the analysis of variance, we label the sample mean of each group j as

$$\overline{y}_{.j} = \frac{1}{n_j} \sum_{i=1}^{n_j} y_{ij}$$

with sampling variance

$$\sigma_j^2 = \frac{\sigma^2}{n_j}.$$

We can then write the likelihood for each θ_j in terms of the sufficient statistics, $\overline{y}_{.j}$:

$$\overline{y}_{.j}|\theta_j \sim \mathrm{N}(\theta_j, \sigma_j^2), \qquad (5.12)$$

a notation that will prove useful later because of the flexibility in allowing a separate variance σ_j^2 for the mean of each group j. For the rest of this chapter, all expressions will be implicitly conditional on the known values σ_j^2. The problem of estimating a set of means with unknown variances will require some additional computational methods, presented in Sections 11.7 and 12.5. Although rarely strictly true, the assumption of known variances at the sampling level of the model is often an adequate approximation.

The treatment of the model provided in this section is also appropriate for situations in which the variances differ for reasons other than the number of data points in the experiment. In fact, the likelihood (5.12) can appear in much more general contexts than that stated here. For example, if the group sizes n_j are large enough, then the means $\overline{y}_{.j}$ are approximately normally distributed, given θ_j, even when the data y_{ij} are not. Other applications where the actual likelihood is well approximated by (5.12) appear in the next two sections.

Constructing a prior distribution from pragmatic considerations

Rather than considering immediately the problem of specifying a prior distribution for the parameter vector $\theta = (\theta_1, \ldots, \theta_J)$, let us consider what sorts of posterior estimates might be reasonable for θ, given data (y_{ij}). A simple natural approach is to estimate θ_j by $\overline{y}_{.j}$, the average outcome in experiment j. But what if, for example, there are $J = 20$ experiments with only $n_j = 2$ observations per experimental group, and the groups are 20 pairs of assays taken from the same strain of rat, under essentially identical conditions? The two observations per group do not permit very accurate estimates. Since the 20 groups are from the same strain of rat, we might now prefer to estimate each θ_j by the pooled estimate,

$$\bar{y}_{..} = \frac{\sum_{j=1}^{J} \frac{1}{\sigma_j^2} \bar{y}_{.j}}{\sum_{j=1}^{J} \frac{1}{\sigma_j^2}}. \tag{5.13}$$

To decide which estimate to use, a traditional approach from classical statistics is to perform an analysis of variance F test for differences among means: if the J group means appear significantly variable, choose separate sample means, and if the variance between the group means is not significantly greater than what could be explained by individual variability within groups, use $\bar{y}_{..}$. The theoretical analysis of variance table is as follows, where τ^2 is the variance of $\theta_1, \ldots, \theta_J$. For simplicity, we present the analysis of variance for a balanced design in which $n_j = n$ and $\sigma_j^2 = \sigma^2$ for all j.

	df	SS	MS	$E(\mathrm{MS}|\sigma^2, \tau)$
Between groups	$J-1$	$\sum_i \sum_j (\bar{y}_{.j} - \bar{y}_{..})^2$	$\mathrm{SS}/(J-1)$	$n\tau^2 + \sigma^2$
Within groups	$J(n-1)$	$\sum_i \sum_j (y_{ij} - \bar{y}_{.j})^2$	$\mathrm{SS}/(J(n-1))$	σ^2
Total	$Jn-1$	$\sum_i \sum_j (y_{ij} - \bar{y}_{..})^2$	$\mathrm{SS}/(Jn-1)$	

In the classical random-effects analysis of variance, one computes the sum of squares (SS) and the mean square (MS) columns of the table and uses the 'between' and 'within' mean squares to estimate τ. If the ratio of between to within mean squares is significantly greater than 1, then the analysis of variance suggests separate estimates, $\hat{\theta}_j = \bar{y}_{.j}$ for each j. If the ratio of mean squares is not 'statistically significant,' then the F test cannot 'reject the hypothesis' that $\tau = 0$, and pooling is reasonable: $\hat{\theta}_j = \bar{y}_{..}$, for all j. We discuss Bayesian analysis of variance in Section 15.6 in the context of hierarchical regression models.

But we are not forced to choose between complete pooling and none at all. An alternative is to use a weighted combination:

$$\hat{\theta}_j = \lambda_j \bar{y}_{.j} + (1 - \lambda_j) \bar{y}_{..},$$

where λ_j is between 0 and 1.

What kind of prior models produce these various posterior estimates?

1. The unpooled estimate $\hat{\theta}_j = \bar{y}_{.j}$ is the posterior mean if the J values θ_j have independent uniform prior densities on $(-\infty, \infty)$.

2. The pooled estimate $\hat{\theta} = \bar{y}_{..}$ is the posterior mean if the J values θ_j are restricted to be equal, with a uniform prior density on the common θ.

3. The weighted combination is the posterior mean if the J values θ_j have iid normal prior densities.

All three of these options are exchangeable in the θ_j's, and options 1 and 2 are special cases of option 3. No pooling corresponds to $\lambda_j \equiv 1$ for all j and an infinite prior variance for the θ_j's, and complete pooling corresponds to $\lambda_j \equiv 0$ for all j and a zero prior variance for the θ_j's.

The hierarchical model

For the convenience of conjugacy (actually, semi-conjugacy in the sense of Section 3.4), we assume that the parameters θ_j are drawn from a normal distribution with hyperparameters (μ, τ):

$$p(\theta_1, \ldots, \theta_J | \mu, \tau) = \prod_{j=1}^{J} \mathrm{N}(\theta_j | \mu, \tau^2) \tag{5.14}$$

$$p(\theta_1, \ldots, \theta_J) = \int \prod_{j=1}^{J} \left[\mathrm{N}(\theta_j | \mu, \tau^2) \right] p(\mu, \tau) d(\mu, \tau).$$

That is, the θ_j's are conditionally independent given (μ, τ). The hierarchical model also permits the interpretation of the θ_j's as a random sample from a shared population distribution, as illustrated in Figure 5.1 for the rat tumor example.

We assign a noninformative uniform hyperprior distribution to μ, given τ, so that

$$p(\mu, \tau) = p(\mu | \tau) p(\tau) \propto p(\tau). \tag{5.15}$$

The uniform prior density for μ is generally reasonable for this problem; because the combined data from all J experiments are generally highly informative about μ, we can afford to be vague about its prior distribution. We defer discussion of the prior distribution of τ to later in the analysis, although relevant principles have already been discussed in the context of the rat tumor example. As usual, we first work out the answer conditional on the hyperparameters and then consider their prior and posterior distributions.

The joint posterior distribution

Combining the sampling distribution for the observable y_{ij}'s and the prior distribution yields the joint posterior distribution of all the parameters and hyperparameters, which we can express in terms of the sufficient statistics, $\bar{y}_{.j}$:

$$p(\theta, \mu, \tau | y) \propto p(\mu, \tau) p(\theta | \mu, \tau) p(y | \theta)$$

$$\propto p(\mu, \tau) \prod_{j=1}^{J} \mathrm{N}(\theta_j | \mu, \tau^2) \prod_{j=1}^{J} \mathrm{N}(\bar{y}_{.j} | \theta_j, \sigma_j^2), \tag{5.16}$$

where we can ignore factors that depend only on y and the parameters σ_j, which are assumed known for this analysis.

The conditional posterior distribution of the normal means, given the hyperparameters

As in the general hierarchical structure, the parameters θ_j are independent in the prior distribution (given μ and τ) and appear in different factors in the likelihood (5.11); thus, the conditional posterior distribution $p(\theta|\mu, \tau, y)$ factors into J components.

Conditional on the hyperparameters, we simply have J independent unknown normal means, given normal prior distributions, so we can use the methods of Section 2.6 independently on each θ_j. The conditional posterior distributions for the θ_j's are independent, and

$$\theta_j|\mu, \tau, y \sim N(\hat{\theta}_j, V_j),$$

where

$$\hat{\theta}_j = \frac{\frac{1}{\sigma_j^2}\overline{y}_{.j} + \frac{1}{\tau^2}\mu}{\frac{1}{\sigma_j^2} + \frac{1}{\tau^2}} \quad \text{and} \quad V_j = \frac{1}{\frac{1}{\sigma_j^2} + \frac{1}{\tau^2}}. \tag{5.17}$$

The posterior mean is a precision-weighted average of the prior population mean and the sample mean of the jth group; these expressions for $\hat{\theta}_j$ and V_j are functions of μ and τ as well as the data. The conditional posterior density for each θ_j given μ, τ is proper.

The marginal posterior distribution of the hyperparameters

The solution so far is only partial because it depends on the unknown μ and τ. The next step in our approach is a full Bayesian treatment for the hyperparameters. Section 5.3 mentions integration or analytic computation as two approaches for obtaining $p(\mu, \tau|y)$ from the joint posterior density $p(\theta, \mu, \tau|y)$. For the hierarchical normal model, we can simply consider the information supplied by the data about the hyperparameters directly:

$$p(\mu, \tau|y) \propto p(\mu, \tau)p(y|\mu, \tau).$$

For many problems, this decomposition is no help, because the 'marginal likelihood' factor, $p(y|\mu, \tau)$, cannot generally be written in closed form. For the normal distribution, however, the marginal likelihood has a particularly simple form. The marginal distributions of the group means $\overline{y}_{.j}$, averaging over θ, are independent (but not identically distributed) normal:

$$\overline{y}_{.j}|\mu, \tau \sim N(\mu, \sigma_j^2 + \tau^2).$$

Thus we can write the marginal posterior density as

$$p(\mu, \tau|y) \propto p(\mu, \tau) \prod_{j=1}^{J} N(\overline{y}_{.j}|\mu, \sigma_j^2 + \tau^2). \tag{5.18}$$

Posterior distribution of μ given τ. We could use (5.18) to compute directly the posterior distribution $p(\mu, \tau|y)$ as a function of two variables and proceed as in the rat tumor example. For the normal model, however, we can further simplify by integrating over μ, leaving a simple univariate numerical computation of $p(\tau|y)$. We factor the marginal posterior density of the hyperparameters as we did the prior density (5.15):

$$p(\mu, \tau|y) = p(\mu|\tau, y)p(\tau|y). \tag{5.19}$$

The first factor on the right side of (5.19) is just the posterior distribution of μ if τ were known. From inspection of (5.18) with τ assumed known, and with a uniform conditional prior density $p(\mu|\tau)$, the log posterior distribution is found to be quadratic in μ; thus, $p(\mu|\tau, y)$ must be normal. The mean and variance of this distribution can be obtained immediately by considering the group means $\overline{y}_{.j}$ as J independent estimates of μ with variances $(\sigma_j^2 + \tau^2)$. Combining the data with the uniform prior density $p(\mu|\tau)$ yields

$$\mu|\tau, y \sim \mathrm{N}(\hat{\mu}, V_\mu),$$

where $\hat{\mu}$ is the precision-weighted average of the $\overline{y}_{.j}$-values, and V_μ^{-1} is the total precision:

$$\hat{\mu} = \frac{\sum_{j=1}^{J} \frac{1}{\sigma_j^2 + \tau^2} \overline{y}_{.j}}{\sum_{j=1}^{J} \frac{1}{\sigma_j^2 + \tau^2}} \quad \text{and} \quad V_\mu^{-1} = \sum_{j=1}^{J} \frac{1}{\sigma_j^2 + \tau^2}. \tag{5.20}$$

The result is a proper posterior density for μ, given τ.

Posterior distribution of τ. We can now obtain the posterior distribution of τ analytically from (5.19) and substitution of (5.18) and (5.20) for the numerator and denominator, respectively:

$$
\begin{aligned}
p(\tau|y) &= \frac{p(\mu, \tau|y)}{p(\mu|\tau, y)} \\
&\propto \frac{p(\tau) \prod_{j=1}^{J} \mathrm{N}(\overline{y}_{.j}|\mu, \sigma_j^2 + \tau^2)}{\mathrm{N}(\mu|\hat{\mu}, V_\mu)}.
\end{aligned}
$$

This identity must hold for any value of μ (in other words, all the factors of μ must cancel when the expression is simplified); in particular, it holds if we set μ to $\hat{\mu}$, which makes evaluation of the expression quite simple:

$$
\begin{aligned}
p(\tau|y) &\propto \frac{p(\tau) \prod_{j=1}^{J} \mathrm{N}(\overline{y}_{.j}|\hat{\mu}, \sigma_j^2 + \tau^2)}{\mathrm{N}(\hat{\mu}|\hat{\mu}, V_\mu)} \\
&\propto p(\tau) V_\mu^{1/2} \prod_{j=1}^{J} (\sigma_j^2 + \tau^2)^{-1/2} \exp\left(-\frac{(\overline{y}_{.j} - \hat{\mu})^2}{2(\sigma_j^2 + \tau^2)} \right), \tag{5.21}
\end{aligned}
$$

with $\hat{\mu}$ and V_μ defined in (5.20). Both expressions are functions of τ, which means that $p(\tau|y)$ is a complicated function of τ.

Prior distribution for τ. To complete our analysis, we must assign a prior distribution to τ. For convenience, we use a diffuse noninformative prior density for τ and hence must examine the resulting posterior density to ensure it has a finite integral. For our illustrative analysis, we use the uniform prior distribution, $p(\tau) \propto 1$. We leave it as an exercise to show mathematically that the uniform prior density for τ yields a proper posterior density and that, in contrast, the seemingly reasonable 'noninformative' prior distribution for a variance component, $p(\log \tau) \propto 1$, yields an improper posterior distribution for τ. Alternatively, in applications it involves little extra effort to determine a 'best guess' and an upper bound for the population variance τ, and a reasonable prior distribution can then be constructed from the scaled inverse-χ^2 family (the natural choice for variance parameters), matching the 'best guess' to the mean of the scaled inverse-χ^2 density and the upper bound to an upper percentile such as the 99th. Once an initial analysis is performed using the noninformative 'uniform' prior density, a sensitivity analysis with a more realistic prior distribution is often desirable.

Computation

For this model, computation of the posterior distribution of θ is most conveniently performed via simulation, following the factorization used above:

$$p(\theta, \mu, \tau | y) = p(\tau | y) p(\mu | \tau, y) p(\theta | \mu, \tau, y).$$

The first step, simulating τ, is easily performed numerically using the inverse cdf method (see Section 1.9) on a grid of uniformly spaced values of τ, with $p(\tau | y)$ computed from (5.21). The second and third steps, simulating μ and then θ, can both be done easily by sampling from normal distributions, first (5.20) to obtain μ and then (5.17) to obtain the θ_j's independently.

Posterior predictive distributions

Obtaining samples from the posterior predictive distribution of new data, either from a current batch or a new batch, is straightforward given draws from the posterior distribution of the parameters. We consider two scenarios: (1) future data $\tilde{y}$ from the current set of batches, with means $\theta = (\theta_1, \ldots, \theta_J)$, and (2) future data $\tilde{y}$ from $\tilde{J}$ future batches, with means $\tilde{\theta} = (\tilde{\theta}_1, \ldots, \tilde{\theta}_{\tilde{J}})$. In the latter case, we must also specify the $\tilde{J}$ individual sample sizes $\tilde{n}_j$ for the future batches.

To obtain a draw from the posterior predictive distribution of new data $\tilde{y}$ from the current set of random effects, θ, first obtain a draw from $p(\theta, \mu, \tau | y)$ and then draw the predictive data $\tilde{y}$ from (5.11).

To obtain posterior predictive simulations of new data $\tilde{y}$ for $\tilde{J}$ new groups, perform the following three steps: first, draw (μ, τ) from their posterior distribution; second, draw $\tilde{J}$ new parameters $\tilde{\theta} = (\tilde{\theta}_1, \ldots, \tilde{\theta}_{\tilde{J}})$ from the population distribution $p(\tilde{\theta}_j | \mu, \tau)$, which is the population, or prior, distribution for θ given the hyperparameters (equation (5.14)); and third, draw $\tilde{y}$ given $\tilde{\theta}$ from the data distribution (5.11).

Difficulty with a natural non-Bayesian estimate of the hyperparameters

To see some advantages of our fully Bayesian approach, we compare it to an approximate method that is sometimes used based on a *point estimate* of μ and τ from the data. Unbiased point estimates, derived from the analysis of variance presented earlier, are

$$\hat{\mu} = \overline{y}_{..}$$
$$\hat{\tau}^2 = (\mathrm{MS}_B - \mathrm{MS}_W)/n. \tag{5.22}$$

The terms MS_B and MS_W are the 'between' and 'within' mean squares, respectively, from the analysis of variance. In this alternative approach, inference for $\theta_1, \ldots, \theta_J$ is based on the conditional posterior distribution, $p(\theta | \hat{\mu}, \hat{\tau})$, given the point estimates.

As we saw in the rat tumor example of the previous section, the main problem with substituting point estimates for the hyperparameters is that it ignores our real uncertainty about them. The resulting inference for θ cannot be interpreted as a Bayesian posterior summary. In addition, the estimate $\hat{\tau}^2$ in (5.22) has the flaw that it can be negative! The problem of a negative estimate for a variance component can be avoided by setting $\hat{\tau}^2$ to zero in the case that MS_W exceeds MS_B, but this creates new issues. Estimating $\tau^2 = 0$ whenever $\mathrm{MS}_W > \mathrm{MS}_B$ seems too strong a claim: if $\mathrm{MS}_W > \mathrm{MS}_B$, then the sample size is too small for τ^2 to be distinguished from zero, but this is not the same as saying we know that $\tau^2 = 0$. The latter claim, made implicitly by the point estimate, implies that all the group means θ_j are absolutely identical, which leads to scientifically indefensible claims, as we shall see in the example in the next section. It is possible to construct a point estimate of (μ, τ) to avoid this particular difficulty, but it would still have the problem, common to all point estimates, of ignoring uncertainty.

5.5 Example: combining information from educational testing experiments in eight schools

We illustrate the normal model with a problem in which the hierarchical Bayesian analysis gives conclusions that differ in important respects from other methods.

A study was performed for the Educational Testing Service to analyze the effects of special coaching programs on test scores. Separate randomized experiments were performed to estimate the effects of coaching programs for the SAT-V (Scholastic Aptitude Test-Verbal) in each of eight high schools. The outcome variable in each study was the score on a special administration of the SAT-V, a standardized multiple choice test administered by the Educational Testing Service and used to help colleges make admissions decisions; the scores can vary between 200 and 800, with mean about 500 and standard deviation about 100. The SAT examinations are designed to be resistant to short-term efforts directed specifically toward improving performance on the test; instead they are designed to reflect knowledge acquired and abilities developed over many years of education. Nevertheless, each of the eight schools in this study considered its short-term coaching program to be very successful at increasing SAT scores. Also, there was no prior reason to believe that any of the eight programs was more effective than any other or that some were more similar in effect to each other than to any other.

The results of the experiments are summarized in Table 5.2. All students in the experiments had already taken the PSAT (Preliminary SAT), and allowance was made for differences in the PSAT-M (Mathematics) and PSAT-V test scores between coached and uncoached students. In particular, in each school the estimated coaching effect and its standard error were obtained by an analysis of covariance adjustment (that is, a linear regression was performed of SAT-V on treatment group, using PSAT-M and PSAT-V as control variables) appropriate for a completely randomized experiment. A separate regression was estimated for each school. Although not simple sample means (because of the covariance adjustments), the estimated coaching effects, which we label y_j, and their sampling variances, σ_j^2, play the same role in our model as $\bar{y}_{\cdot j}$ and σ_j^2 in the previous section. The estimates y_j are obtained by independent experiments and have approximately normal sampling distributions with sampling variances that are known, for all practical purposes, because the sample sizes in all of the eight experiments were relatively large, over thirty students in each school (recall the discussion of data reduction in Section 4.1). Incidentally, eight more points on the SAT-V corresponds to about one more test item correct.

Inferences based on nonhierarchical models and their problems

Before applying the hierarchical Bayesian method, we first consider two simpler nonhierarchical methods—estimating the effects from the eight experiments independently, and complete pooling—and discuss why neither of these approaches is adequate for this example.

Separate estimates. A cursory examination of Table 5.2 may at first suggest that some coaching programs have moderate effects (in the range 18–28 points), most have small effects (0–12 points), and two have small negative effects; however, when we take note of the standard errors of these estimated effects, we see that it is difficult statistically to distinguish between any of the experiments. For example, treating each experiment separately and applying the simple normal analysis in each yields 95% posterior intervals that all overlap substantially.

A pooled estimate. The general overlap in the posterior intervals based on independent analyses suggests that all experiments might be estimating the same quantity. Under the hypothesis that all experiments have the same effect and produce independent estimates of this common effect, we could treat the data in Table 5.2 as eight normally distributed observations with known variances. With a noninformative prior distribution, the posterior mean for the common coaching effect in the schools is $\bar{y}_{\cdot\cdot}$, as defined in equation (5.13) with y_j in place of $\bar{y}_{\cdot j}$. This pooled estimate is 7.9, and the posterior variance is $(\sum_{j=1}^{8} \frac{1}{\sigma_j^2})^{-1} = 17.4$ because the eight experiments are independent. Thus, we

School	Estimated treatment effect, y_j	Standard error of effect estimate, σ_j
A	28	15
B	8	10
C	−3	16
D	7	11
E	−1	9
F	1	11
G	18	10
H	12	18

Table 5.2 *Observed effects of special preparation on SAT-V scores in eight random-ized experiments. Estimates are based on separate analyses for the eight experiments. From Rubin (1981).*

would estimate the common effect to be 7.9 points with standard error equal to $\sqrt{17.4} = 4.2$, which would lead to the 95% posterior interval $[-0.3, 16.0]$, or approximately $[8 \pm 8]$. Supporting this analysis, the classical test of the hypothesis that all θ_j's are estimating the same quantity yields a χ^2 statistic less than its degrees of freedom (seven, in this case): $\sum_{j=1}^{8}(y_j - \overline{y}_{..})^2/\sigma_i^2 = 4.6$. To put it another way, the estimate $\hat{\tau}^2$ from (5.22) is negative.

Would it be possible to have one school's observed effect be 28 just by chance, if the coaching effects in all eight schools were really the same? To get a feeling for the natural variation that we would expect across eight studies if this assumption were true, suppose the estimated treatment effects are eight independent draws from a normal distribution with mean 8 points and stan-dard deviation 13 points (the square root of the mean of the eight variances σ_j^2). Then, based on the expected values of normal order statistics, we would expect the largest observed value of y_j to be about 26 points and the others, in diminishing order, to be about 19, 14, 10, 6, 2, −3, and −9 points. These expected effect sizes are quite consistent with the set of observed effect sizes in Table 5.2. Thus, it would appear imprudent to believe that school A really has an effect as large as 28 points.

Difficulties with the separate and pooled estimates. To see the problems with the two extreme attitudes—the separate analyses that consider each θ_j sep-arately, and the alternative view (a single common effect) that leads to the pooled estimate—consider θ_1, the effect in school A. The effect in school A is estimated as 28.4 with a standard error of 14.9 under the separate analysis, versus a pooled estimate of 7.9 with a standard error of 4.2 under the common-effect model. The separate analyses of the eight schools imply the following posterior statement: 'the probability is $\frac{1}{2}$ that the true effect in A is more than 28.4,' a doubtful statement, considering the results for the other seven schools. On the other hand, the pooled model implies the following statement:

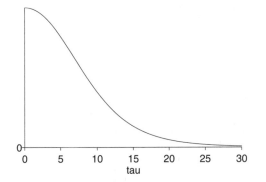

Figure 5.5 *Marginal posterior density, $p(\tau|y)$, for standard deviation of the population of school effects θ_j in the educational testing example.*

'the probability is $\frac{1}{2}$ that the true effect in A is less than 7.9,' which, despite the non-significant χ^2 test, seems an inaccurate summary of our knowledge. The pooled model also implies the statement: 'the probability is $\frac{1}{2}$ that the true effect in A is less than the true effect in C,' which also is difficult to justify given the data in Table 5.2. As in the theoretical discussion of the previous section, neither estimate is fully satisfactory, and we would like a compromise that combines information from all eight experiments without assuming all the θ_j's to be equal. The Bayesian analysis under the hierarchical model provides exactly that.

Posterior simulation under the hierarchical model

Consequently, we compute the posterior distribution of $\theta_1, \ldots, \theta_8$, based on the normal model presented in Section 5.4. (More discussion of the reasonableness of applying this model in this problem appears in Sections 6.8 and 17.4.) We draw from the posterior distribution for the Bayesian model by simulating the random variables τ, μ, and θ, in that order, from their posterior distribution, as discussed at the end of the previous section. The sampling standard deviations, σ_j, are assumed known and equal to the values in Table 5.2, and we assume independent uniform prior densities on μ and τ.

Results

The marginal posterior density function, $p(\tau|y)$ from (5.21), is plotted in Figure 5.5. Values of τ near zero are most plausible; zero is the most likely value, values of τ larger than 10 are less than half as likely as $\tau = 0$, and $\Pr(\tau > 25) \approx 0$. Inference regarding the marginal distributions of the other model parameters and the joint distribution are obtained from the simulated values. Illustrations are provided in the discussion that follows this section. In

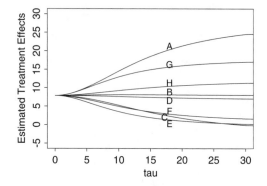

Figure 5.6 *Conditional posterior means of treatment effects,* $E(\theta_j|\tau, y)$, *as functions of the between-school standard deviation* τ, *for the educational testing example. The line for school C crosses the lines for E and F because C has a higher measurement error (see Table 5.2) and its estimate is therefore shrunk more strongly toward the overall mean in the Bayesian analysis.*

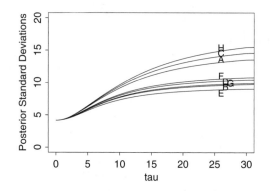

Figure 5.7 *Conditional posterior standard deviations of treatment effects,* $\text{sd}(\theta_j|\tau, y)$, *as functions of the between-school standard deviation* τ, *for the educational testing example.*

the normal hierarchical model however we learn a great deal by considering the conditional posterior distributions given τ (and averaged over μ).

The conditional posterior means $E(\theta_j|\tau, y)$ (averaging over μ) are displayed as functions of τ in Figure 5.6; the vertical axis displays the scale for the θ_j's. Comparing Figure 5.6 to Figure 5.5, which has the same scale on the horizontal axis, we see that for most of the likely values of τ, the estimated effects are relatively close together; as τ becomes larger, corresponding to more variability among schools, the estimates become more like the raw values in Table 5.2.

The lines in Figure 5.7 show the conditional standard deviations, $\text{sd}(\theta_j|\tau, y)$, as a function of τ. As τ increases, the population distribution allows the eight

| School | Posterior quantiles | | | | |
	2.5%	25%	median	75%	97.5%
A	−2	7	10	16	31
B	−5	3	8	12	23
C	−11	2	7	11	19
D	−7	4	8	11	21
E	−9	1	5	10	18
F	−7	2	6	10	28
G	−1	7	10	15	26
H	−6	3	8	13	33

Table 5.3 *Summary of 200 simulations of the treatment effects in the eight schools.*

effects to be more different from each other, and hence the posterior uncertainty in each individual θ_j increases, approaching the standard deviations in Table 5.2 in the limit of $\tau \to \infty$. (The posterior means and standard deviations for the components θ_j, given τ, are computed using the mean and variance formulas (2.7) and (2.8), averaging over μ; see Exercise 5.10.)

The general conclusion from an examination of Figures 5.5–5.7 is that an effect as large as 28.4 points in any school is unlikely. For the likely values of τ, the estimates in all schools are substantially less than 28 points. For example, even at $\tau = 10$, the probability that the effect in school A is less than 28 points is $\Phi[(28 - 14.5)/9.1] = 93\%$, where Φ is the standard normal cumulative distribution function; the corresponding probabilities for the effects being less than 28 points in the other schools are 99.5%, 99.2%, 98.5%, 99.96%, 99.8%, 97%, and 98%.

Of substantial importance, we do not obtain an accurate summary of the data if we condition on the posterior mode of τ. The technique of conditioning on a modal value (for example, the maximum likelihood estimate) of a hyperparameter such as τ is often used in practice (at least as an approximation), but it ignores the uncertainty conveyed by the posterior distribution of the hyperparameter. At $\tau = 0$, the inference is that all experiments have the same size effect, 7.9 points, and the same standard error, 4.2 points. Figures 5.5–5.7 certainly suggest that this answer represents too much pulling together of the estimates in the eight schools. The problem is especially acute in this example because the posterior mode of τ is on the boundary of its parameter space. A joint posterior modal estimate of $(\theta_1, \ldots, \theta_J, \mu, \tau)$ suffers from even worse problems in general.

Discussion

Table 5.3 summarizes the 200 simulated effect estimates for all eight schools. In one sense, these results are similar to the pooled 95% interval [8±8], in that the eight Bayesian 95% intervals largely overlap and are median-centered between 5 and 10. In a second sense, the results in the table are quite different from the

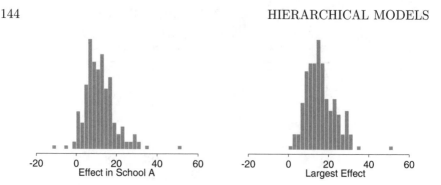

Figure 5.8 *Histograms of two quantities of interest computed from the 200 simulation draws: (a) the effect in school A, θ_1; (b) the largest effect, $\max\{\theta_j\}$. The jaggedness of the histograms is just an artifact caused by sampling variability from using only 200 random draws.*

pooled estimate in a direction toward the eight independent answers: the 95% Bayesian intervals are each almost twice as wide as the one common interval and suggest substantially greater probabilities of effects larger than 16 points, especially in school A, and greater probabilities of negative effects, especially in school C. If greater precision were required in the posterior intervals, one could simulate more simulation draws; we use only 200 draws here to illustrate that a small simulation gives adequate inference for many practical purposes.

The ordering of the effects in the eight schools as suggested by Table 5.3 are essentially the same as would be obtained by the eight separate estimates. However, there are differences in the details; for example, the Bayesian probability that the effect in school A is as large as 28 points is less than 10%, which is substantially less than the 50% probability based on the separate estimate for school A.

As an illustration of the simulation-based posterior results, 200 simulations of school A's effect are shown in Figure 5.8a. Having simulated the parameter θ, it is easy to ask more complicated questions of this model. For example, what is the posterior distribution of $\max\{\theta_j\}$, the effect of the most successful of the eight coaching programs? Figure 5.8b displays a histogram of 200 values from this posterior distribution and shows that only 22 draws are larger than 28.4; thus, $\Pr(\max\{\theta_j\} > 28.4) \approx \frac{22}{200}$. Since Figure 5.8a gives the marginal posterior distribution of the effect in school A, and Figure 5.8b gives the marginal posterior distribution of the largest effect no matter which school it is in, the latter figure has larger values. For another example, we can estimate $\Pr(\theta_1 > \theta_3 | y)$, the posterior probability that the coaching program is more effective in school A than in school C, by the proportion of simulated draws of θ for which $\theta_1 > \theta_3$; the result is $\frac{141}{200} = 0.705$.

To sum up, the Bayesian analysis of this example not only allows straightforward inferences about many parameters that may be of interest, but the hierarchical model is flexible enough to adapt to the data, thereby providing posterior inferences that account for the partial pooling as well as the uncertainty in the hyperparameters.

5.6 Hierarchical modeling applied to a meta-analysis

Meta-analysis is an increasingly popular and important process of summarizing and integrating the findings of research studies in a particular area. As a method for combining information from several parallel data sources, meta-analysis is closely connected to hierarchical modeling. In this section we consider a relatively simple application of hierarchical modeling to a meta-analysis in medicine. We return to this example in Section 19.4 and consider another meta-analysis problem in the context of decision analysis in Section 22.2.

The data in our medical example are displayed in the first three columns of Table 5.4, which summarize mortality after myocardial infarction in 22 clinical trials, each consisting of two groups of heart attack patients randomly allocated to receive or not receive beta-blockers (a family of drugs that affect the central nervous system and can relax the heart muscles). Mortality varies from 3% to 21% across the studies, most of which show a modest, though not 'statistically significant,' benefit from the use of beta-blockers. The aim of a meta-analysis is to provide a combined analysis of the studies that indicates the overall strength of the evidence for a beneficial effect of the treatment under study. Before proceeding to a formal meta-analysis, it is important to apply rigorous criteria in determining which studies are included. (This relates to concerns of ignorability in data collection for observational studies, as discussed in Chapter 7.)

Defining a parameter for each study

In the beta-blocker example, the meta-analysis involves data in the form of several 2×2 tables. If clinical trial j (in the series to be considered for meta-analysis) involves the use of n_{0j} subjects in the control group and n_{1j} in the treatment group, giving rise to y_{0j} and y_{1j} deaths in control and treatment groups, respectively, then the usual sampling model involves two independent binomial distributions with probabilities of death p_{0j} and p_{1j}, respectively. Estimands of common interest are the difference in probabilities, $p_{1j} - p_{0j}$, the probability or *risk* ratio, p_{1j}/p_{0j}, and the odds ratio, $\rho_j = (p_{1j}/(1 - p_{1j}))/(p_{0j}/(1 - p_{0j}))$. For a number of reasons, including interpretability in a range of study designs (including case-control studies as well as clinical trials and cohort studies), and the fact that its posterior distribution is close to normality even for relatively small sample sizes, we concentrate on inference for the (natural) logarithm of the odds ratio, which we label $\theta_j = \log \rho_j$.

Study, j	Raw data (deaths/total) Control	Raw data (deaths/total) Treated	Log-odds, y_j	sd, σ_j	Posterior quantiles of effect θ_j normal approx. (on log-odds scale) 2.5%	25%	median	75%	97.5%
1	3/39	3/38	0.028	0.850	−0.57	−0.33	−0.24	−0.16	0.12
2	14/116	7/114	−0.741	0.483	−0.64	−0.37	−0.28	−0.20	−0.00
3	11/93	5/69	−0.541	0.565	−0.60	−0.35	−0.26	−0.18	0.05
4	127/1520	102/1533	−0.246	0.138	−0.45	−0.31	−0.25	−0.19	−0.05
5	27/365	28/355	0.069	0.281	−0.43	−0.28	−0.21	−0.11	0.15
6	6/52	4/59	−0.584	0.676	−0.62	−0.35	−0.26	−0.18	0.05
7	152/939	98/945	−0.512	0.139	−0.61	−0.43	−0.36	−0.28	−0.17
8	48/471	60/632	−0.079	0.204	−0.43	−0.28	−0.21	−0.13	0.08
9	37/282	25/278	−0.424	0.274	−0.58	−0.36	−0.28	−0.20	−0.02
10	188/1921	138/1916	−0.335	0.117	−0.48	−0.35	−0.29	−0.23	−0.13
11	52/583	64/873	−0.213	0.195	−0.48	−0.31	−0.24	−0.17	0.01
12	47/266	45/263	−0.039	0.229	−0.43	−0.28	−0.21	−0.12	0.11
13	16/293	9/291	−0.593	0.425	−0.63	−0.36	−0.28	−0.20	0.01
14	45/883	57/858	0.282	0.205	−0.34	−0.22	−0.12	0.00	0.27
15	31/147	25/154	−0.321	0.298	−0.56	−0.34	−0.26	−0.19	0.01
16	38/213	33/207	−0.135	0.261	−0.48	−0.30	−0.23	−0.15	0.08
17	12/122	28/251	0.141	0.364	−0.47	−0.29	−0.21	−0.12	0.17
18	6/154	8/151	0.322	0.553	−0.51	−0.30	−0.23	−0.13	0.15
19	3/134	6/174	0.444	0.717	−0.53	−0.31	−0.23	−0.14	0.15
20	40/218	32/209	−0.218	0.260	−0.50	−0.32	−0.25	−0.17	0.04
21	43/364	27/391	−0.591	0.257	−0.64	−0.40	−0.31	−0.23	−0.09
22	39/674	22/680	−0.608	0.272	−0.65	−0.40	−0.31	−0.23	−0.07

Table 5.4 *Results of 22 clinical trials of beta-blockers for reducing mortality after myocardial infarction, with empirical log-odds and approximate sampling variances. Data from Yusuf et al. (1985). Posterior quantiles of treatment effects are based on 5000 simulation draws from a Bayesian hierarchical model described here. Negative effects correspond to reduced probability of death under the treatment.*

A normal approximation to the likelihood

Relatively simple Bayesian meta-analysis is possible using the normal-theory results of the previous sections if we summarize the results of each experiment j with an approximate normal likelihood for the parameter θ_j. This is possible with a number of standard analytic approaches that produce a point estimate and standard errors, which can be regarded as approximating a normal mean and standard deviation. One approach is based on *empirical logits*: for each study j, one can estimate θ_j by

$$y_j = \log\left(\frac{y_{1j}}{n_{1j} - y_{1j}}\right) - \log\left(\frac{y_{0j}}{n_{0j} - y_{0j}}\right), \tag{5.23}$$

with approximate sampling variance

$$\sigma_j^2 = \frac{1}{y_{1j}} + \frac{1}{n_{1j} - y_{1j}} + \frac{1}{y_{0j}} + \frac{1}{n_{0j} - y_{0j}}. \tag{5.24}$$

We use the notation y_j and σ_j^2 to be consistent with our expressions for the hierarchical normal model in the previous sections. There are various refinements of these estimates that improve the asymptotic normality of the sampling distributions involved (in particular, it is often recommended to add a fraction such as 0.5 to each of the four counts in the 2×2 table), but whenever study-specific sample sizes are moderately large, such details do not concern us.

The estimated log-odds ratios y_j and their estimated standard errors σ_j^2 are displayed as the fourth and fifth columns of Table 5.4. We use a hierarchical Bayesian analysis to combine information from the 22 studies and gain improved estimates of each θ_j, along with estimates of the mean and variance of the effects over all studies.

Goals of inference in meta-analysis

Discussions of meta-analysis are sometimes imprecise about the estimands of interest in the analysis, especially when the primary focus is on testing the null hypothesis of no effect in any of the studies to be combined. Our focus is on estimating meaningful parameters, and for this objective there appear to be three possibilities, accepting the overarching assumption that the studies are comparable in some broad sense. The first possibility is that we view the studies as identical replications of each other, in the sense we regard the individuals in all the studies as independent samples from a common population, with the same outcome measures and so on. A second possibility is that the studies are so different that the results of any one study provide no information about the results of any of the others. A third, more general, possibility is that we regard the studies as exchangeable but not necessarily either identical or completely unrelated; in other words we allow differences from study to study, but such that the differences are not expected *a priori* to have predictable effects favoring one study over another. As we have discussed in detail in this chapter, this third possibility represents a continuum between the two extremes, and it is this exchangeable model (with unknown hyperparameters characterizing the population distribution) that forms the basis of our Bayesian analysis.

Exchangeability does not specify the form of the joint distribution of the study effects. In what follows we adopt the convenient assumption of a normal distribution for the random effects; in practice it is important to check the validity of this assumption using some of the techniques discussed in Chapter 6.

The first potential estimand of a meta-analysis, or a hierarchically structured problem in general, is the mean of the distribution of effect sizes, since this represents the overall 'average' effect across all studies that could be regarded as exchangeable with the observed studies. Other possible estimands are the effect size in any of the observed studies and the effect size in another, comparable (exchangeable) unobserved study.

What if exchangeability is inappropriate?

When assuming exchangeability we assume there are no important covariates that might form the basis of a more complex model, and this assumption (perhaps misguidedly) is widely adopted in meta-analysis. What if other information (in addition to the data (n, y)) is available to distinguish among the J studies in a meta-analysis, so that an exchangeable model is inappropriate? In this situation, we can expand the framework of the model to be exchangeable in the observed data and covariates, for example using a hierarchical regression model, as in Chapter 15, so as to estimate how the treatment effect behaves as a function of the covariates. The real aim might in general be to estimate a *response surface* so that one could predict an effect based on known characteristics of a population and its exposure to risk.

A hierarchical normal model

A normal population distribution in conjunction with the approximate normal sampling distribution of the study-specific effect estimates allows an analysis of the same form as used for the SAT coaching example in the previous section. Let y_j represent generically the point estimate of the effect θ_j in the jth study, obtained from (5.23), where $j = 1, \ldots, J$. The first stage of the hierarchical normal model assumes that

$$y_j | \theta_j, \sigma_j^2 \sim N(\theta_j, \sigma_j^2),$$

where σ_j represents the corresponding estimated standard error from (5.24), which is assumed known without error. The simplification of known variances has little effect here because, with the large sample sizes (more than 50 persons in each treatment group in nearly all of the studies in the beta-blocker example), the binomial variances in each study are precisely estimated. At the second stage of the hierarchy, we again use an exchangeable normal prior distribution, with mean μ and standard deviation τ, which are unknown hyperparameters. Finally, a hyperprior distribution is required for μ and τ. For this problem, it is reasonable to assume a noninformative or locally uniform prior density for μ, since even with quite a small number of studies (say 5 or 10), the combined data become relatively informative about the center of the population distribution of effect sizes. As with the SAT coaching example, we also assume a locally uniform prior density for τ, essentially for convenience, although it is easy to modify the analysis to include prior information.

Estimand	Posterior quantiles				
	2.5%	25%	median	75%	97.5%
Mean, μ	-0.37	-0.29	-0.25	-0.20	-0.11
Standard deviation, τ	0.02	0.08	0.13	0.18	0.31
Predicted effect, $\tilde{\theta}_j$	-0.58	-0.34	-0.25	-0.17	0.11

Table 5.5 *Summary of posterior inference for the overall mean and standard deviation of study effects, and for the predicted effect in a hypothetical future study, from the meta-analysis of the beta-blocker trials in Table 5.4. All effects are on the log-odds scale.*

Results of the analysis and comparison to simpler methods

The analysis of our meta-analysis model now follows exactly the same methodology as in the previous sections. First, a plot (not shown here) similar to Figure 5.5 shows that the marginal posterior density of τ peaks at a nonzero value, although values near zero are clearly plausible, zero having a posterior density only about 25% lower than that at the mode. Posterior quantiles for the effects θ_j for the 22 studies on the logit scale are displayed as the last columns of Table 5.4.

Since the posterior distribution of τ is concentrated around values that are small relative to the sampling standard deviations of the data (compare the posterior median of τ, 0.13, in Table 5.5 to the values of σ_j in the fourth column of Table 5.4), considerable shrinkage is evident in the Bayes estimates, especially for studies with low internal precision (for example, studies 1, 6, and 18). The substantial degree of homogeneity between the studies is further reflected in the large reductions in posterior variance obtained when going from the study-specific estimates to the Bayesian ones, which borrow strength from each other. Using an approximate approach fixing τ would yield standard deviations that would be too small compared to the fully Bayesian ones.

Histograms (not shown) of the simulated posterior densities for each of the individual effects exhibit skewness away from the central value of the overall mean, whereas the distribution of the overall mean has greater symmetry. The imprecise studies, such as 2 and 18, exhibit longer-tailed posterior distributions than the more precise ones, such as 7 and 14.

In meta-analysis, interest often focuses on the estimate of the overall mean effect, μ. Superimposing the graphs (not shown here) of the conditional posterior mean and standard deviation of μ given τ on the posterior density of τ reveals a very small range in the plausible values of $E(\mu|\tau, y)$, from about -0.26 to just over -0.24, but $sd(\mu|\tau, y)$ varies by a factor of more than 2 across the plausible range of values of τ. The latter feature indicates the importance of averaging over τ in order to account adequately for uncertainty in its estimation. In fact, the conditional posterior standard deviation, $sd(\mu|\tau, y)$ has the value 0.060 at $\tau = 0.13$, whereas upon averaging over the posterior distribution for τ we find a value of $sd(\mu|y) = 0.071$.

Table 5.5 gives a summary of posterior inferences for the hyperparameters μ and τ and the predicted effect, $\tilde{\theta}_j$, in a hypothetical future study. The approximate 95% highest posterior density interval for μ is $[-0.37, -0.11]$, or $[0.69, 0.90]$ when converted to the odds ratio scale (that is, exponentiated). In contrast, the 95% posterior interval that results from complete pooling— that is, assuming $\tau = 0$—is considerably narrower, $[0.70, 0.85]$. In the original published discussion of these data, it was remarked that the latter seems an 'unusually narrow range of uncertainty.' The hierarchical Bayesian analysis suggests that this was due to the use of an inappropriate model that had the effect of claiming all the studies were identical. In mathematical terms, complete pooling makes the assumption that the parameter τ is exactly zero, whereas the data supply evidence that τ might be very close to zero, but might also plausibly be as high as 0.3. A related concern is that commonly used analyses tend to place undue emphasis on inference for the overall mean effect. Uncertainty about the probable treatment effect in a particular population where a study has not been performed (or indeed in a previously studied population but with a slightly modified treatment) might be more reasonably represented by inference for a new study effect, exchangeable with those for which studies have been performed, rather than for the overall mean. In this case, uncertainty is of course even greater, as exhibited in the 'Predicted effect' row of Table 5.5; uncertainty for an individual patient includes yet another component of variation. In particular, with the beta-blocker data, there is just over 10% posterior probability that the true effect, $\tilde{\theta}_j$, in a new study would be positive (corresponding to the treatment increasing the probability of death in that study).

5.7 Bibliographic note

The early non-Bayesian work on shrinkage estimation of Stein (1955) and James and Stein (1960) were influential in the development of hierarchical normal models. Efron and Morris (1971, 1972) present subsequent theoretical work on the topic. Robbins (1955, 1964) constructs and justifies hierarchical methods from a decision-theoretic perspective. De Finetti's theorem is described by de Finetti (1974); Bernardo and Smith (1994) discuss its role in Bayesian modeling. An early thorough development of the idea of Bayesian hierarchical modeling is given by Good (1965).

Mosteller and Wallace (1964) analyzed a hierarchical Bayesian model using the negative binomial distribution for counts of words in a study of authorship. Restricted to the limited computing power at the time, they used various approximations and point estimates for hyperparameters.

Recent decades have seen a multitude of papers on 'empirical Bayes' (or, in our terminology, hierarchical Bayes) methods, including very influential applications to longitudinal modeling (for example, Laird and Ware, 1982; also see Hartley and Rao, 1967) and spatial analysis of incidence rates in epidemiology (for example, Clayton and Kaldor, 1987); the review by Breslow

(1990) describes other applications in biostatistics. Morris (1983) presents a detailed review of non-Bayesian theory in this area, whereas Deely and Lindley (1981) relate 'empirical Bayes' methods to fully Bayesian analysis.

The problem of estimating several normal means using an exchangeable hierarchical model was treated in a fully Bayesian framework by Hill (1965), Tiao and Tan (1965, 1966), and Lindley (1971b). Box and Tiao (1973) present hierarchical normal models using slightly different notation from ours. They compare Bayesian and non-Bayesian methods and discuss the analysis of variance table in some detail. A generalization of the hierarchical normal model of Section 5.4, in which the parameters θ_j are allowed to be partitioned into clusters, is described by Malec and Sedransk (1992). More references on hierarchical normal models appear in the bibliographic note at the end of Chapter 15.

Recently many applied Bayesian analyses using hierarchical models have appeared. For example, an important application of hierarchical models is 'small-area estimation,' in which estimates of population characteristics for local areas are improved by combining the data from each area with information from neighboring areas (see, for example, Fay and Herriot, 1979, Dempster and Raghunathan, 1987, and Mollie and Richardson, 1991). Other recent applications have included the modeling of measurement error problems in epidemiology (see Richardson and Gilks, 1993), and the treatment of multiple comparisons in toxicology (see Meng and Dempster, 1987). Manton et al. (1989) fit a hierarchical Bayesian model to cancer mortality rates and compare maps of direct and adjusted mortality rates. Bock (1989) includes several examples of hierarchical models in educational research. We provide references to a number of other applications in later chapters dealing with specific model types.

Hierarchical models can be viewed as a subclass of 'graphical models,' and this connection has been elegantly exploited for Bayesian inference in the development of the computer package Bugs, using techniques that will be explained in Chapter 11 (see also Appendix C); see Thomas, Spiegelhalter, and Gilks (1992), and Spiegelhalter et al. (1994, 2003). Related discussion and theoretical work appears in Lauritzen and Spiegelhalter (1988), Pearl (1988), Wermuth and Lauritzen (1990), and Normand and Tritchler (1992).

The rat tumor data were analyzed hierarchically by Tarone (1982) and Dempster, Selwyn, and Weeks (1983); our approach is close in spirit to the latter paper's. Leonard (1972) and Novick, Lewis, and Jackson (1973) are early examples of hierarchical Bayesian analysis of binomial data.

Much of the material in Sections 5.4 and 5.5, along with much of Section 6.8, originally appeared in Rubin (1981), which is an early example of an applied Bayesian analysis using simulation techniques.

The material of Section 5.6 is adapted from Carlin (1992), which contains several key references on meta-analysis; the original data for the example are from Yusuf et al. (1985); a similar Bayesian analysis of these data under a slightly different model appears as an example in Spiegelhalter et al.

(1994, 2003). More general treatments of meta-analysis from a Bayesian perspective are provided by DuMouchel (1990), Rubin (1989), Skene and Wakefield (1990), and Smith, Spiegelhalter, and Thomas (1995). An example of a Bayesian meta-analysis appears in Dominici et al. (1999). DuMouchel and Harris (1983) present what is essentially a meta-analysis with covariates on the studies; this article is accompanied by some interesting discussion by prominent Bayesian and non-Bayesian statisticians.

5.8 Exercises

1. Hierarchical models and multiple comparisons:

   (a) Reproduce the computations in Section 5.5 for the educational testing example. Use the posterior simulations to estimate (i) for each school j, the probability that its coaching program is the best of the eight; and (ii) for each pair of schools, j and k, the probability that the coaching program in school j is better than that in school k.

   (b) Repeat (a), but for the simpler model with τ set to ∞ (that is, separate estimation for the eight schools). In this case, the probabilities (i) and (ii) can be computed analytically.

   (c) Discuss how the answers in (a) and (b) differ.

   (d) In the model with τ set to 0, the probabilities (i) and (ii) have degenerate values; what are they?

2. Exchangeable prior distributions: suppose it is known a priori that the $2J$ parameters $\theta_1, \ldots, \theta_{2J}$ are clustered into two groups, with exactly half being drawn from a $N(1, 1)$ distribution, and the other half being drawn from a $N(-1, 1)$ distribution, but we have not observed which parameters come from which distribution.

   (a) Are $\theta_1, \ldots, \theta_{2J}$ exchangeable under this prior distribution?

   (b) Show that this distribution cannot be written as a mixture of iid components.

   (c) Why can we not simply take the limit as $J \to \infty$ and get a counterexample to de Finetti's theorem?

   See Exercise 7.10 for a related problem.

3. Mixtures of iid distributions: suppose the distribution of $\theta = (\theta_1, \ldots, \theta_J)$ can be written as a mixture of iid components:

$$p(\theta) = \int \prod_{j=1}^{J} p(\theta_j | \phi) p(\phi) d\phi.$$

   Prove that the covariances $\text{cov}(\theta_i, \theta_j)$ are all nonnegative.

4. Exchangeable models:

(a) In the divorce rate example of Section 5.2, set up a prior distribution for the values $y_1, \ldots, y_8$ that allows for one low value (Utah) and one high value (Nevada), with an iid distribution for the other six values. This prior distribution should be *exchangeable*, because it is not known which of the eight states correspond to Utah and Nevada.

(b) Determine the posterior distribution for y_8 under this model given the observed values of $y_1, \ldots, y_7$ given in the example. This posterior distribution should probably have two or three modes, corresponding to the possibilities that the missing state is Utah, Nevada, or one of the other six.

(c) Now consider the entire set of eight data points, including the value for y_8 given at the end of the example. Are these data consistent with the prior distribution you gave in part (a) above? In particular, did your prior distribution allow for the possibility that the actual data have an outlier (Nevada) at the high end, but no outlier at the low end?

5. Continuous mixture models:

(a) If $y|\theta \sim \text{Poisson}(\theta)$, and $\theta \sim \text{Gamma}(\alpha, \beta)$, then the marginal (prior predictive) distribution of y is negative binomial with parameters α and β (or $p = \beta/(1+\beta)$). Use the formulas (2.7) and (2.8) to derive the mean and variance of the negative binomial.

(b) In the normal model with unknown location and scale (μ, σ^2), the noninformative prior density, $p(\mu, \sigma^2) \propto 1/\sigma^2$, results in a normal-inverse-χ^2 posterior distribution for (μ, σ^2). Marginally then $\sqrt{n}(\mu - \bar{y})/s$ has a posterior distribution that is t_{n-1}. Use (2.7) and (2.8) to derive the first two moments of the latter distribution, stating the appropriate condition on n for existence of both moments.

6. Discrete mixture models: if $p_m(\theta)$, for $m = 1, \ldots, M$, are conjugate prior densities for the sampling model $y|\theta$, show that the class of finite mixture prior densities given by

$$p(\theta) = \sum_{m=1}^{M} \lambda_m p_m(\theta)$$

is also a conjugate class, where the λ_m's are nonnegative weights that sum to 1. This can provide a useful extension of the natural conjugate prior family to more flexible distributional forms. As an example, use the mixture form to create a bimodal prior density for a normal mean, that is thought to be near 1, with a standard deviation of 0.5, but has a small probability of being near -1, with the same standard deviation. If the variance of each observation $y_1, \ldots, y_{10}$ is known to be 1, and their observed mean is $\bar{y} = -0.25$, derive your posterior distribution for the mean, making a sketch of both prior and posterior densities. Be careful: the prior and posterior mixture proportions are different.

7. Noninformative hyperprior distributions: consider the hierarchical binomial model in Section 5.3. Improper posterior distributions are, in fact, a general problem with hierarchical models when a uniform prior distribution is specified for the logarithm of the population standard deviation of the exchangeable parameters. In the case of the beta population distribution, the prior variance is approximately $(\alpha + \beta)^{-1}$ (see Appendix A), and so a uniform distribution on $\log(\alpha+\beta)$ is approximately uniform on the log standard deviation. The resulting unnormalized posterior density (5.8) has an infinite integral in the limit as the population standard deviation approaches 0. We encountered the problem again in Section 5.4 for the hierarchical normal model.

(a) Show that, with a uniform prior density on $(\log(\alpha/\beta), \log(\alpha + \beta))$, the unnormalized posterior density has an infinite integral.

(b) A simple way to avoid the impropriety is to assign a uniform prior distribution to the standard deviation parameter itself, rather than its logarithm. For the beta population distribution we are considering here, this is achieved approximately by assigning a uniform prior distribution to $(\alpha + \beta)^{-1/2}$. Show that combining this with an independent uniform prior distribution on $\frac{\alpha}{\alpha+\beta}$ yields the prior density (5.10).

(c) Show that the resulting posterior density (5.8) is proper as long as $0 < y_j < n_j$ for at least one experiment j.

8. Checking the integrability of the posterior distribution: consider the hierarchical normal model in Section 5.4.

(a) If the hyperprior distribution is $p(\mu, \tau) \propto \tau^{-1}$ (that is, $p(\mu, \log \tau) \propto 1$), show that the posterior density is improper.

(b) If the hyperprior distribution is $p(\mu, \tau) \propto 1$, show that the posterior density is proper if $J > 2$.

(c) How would you analyze SAT coaching data if $J = 2$ (that is, data from only two schools)?

9. Nonconjugate hierarchical models: suppose that in the rat tumor example, we wish to use a normal population distribution on the log-odds scale: $\text{logit}(\theta_j) \sim \text{N}(\mu, \tau^2)$, for $j = 1, \ldots, J$. As in Section 5.3, you will assign a noninformative prior distribution to the hyperparameters and perform a full Bayesian analysis.

(a) Write the joint posterior density, $p(\theta, \mu, \tau | y)$.

(b) Show that the integral (5.4) has no closed-form expression.

(c) Why is expression (5.5) no help for this problem?

In practice, we can solve this problem by normal approximation, importance sampling, and Markov chain simulation, as described in Part III.

10. Conditional posterior means and variances: derive analytic expressions for $\text{E}(\theta_j | \tau, y)$ and $\text{var}(\theta_j | \tau, y)$ in the hierarchical normal model (and used in Figures 5.6 and 5.7). (Hint: use (2.7) and (2.8), averaging over μ.)

11. Hierarchical binomial model: Exercise 3.8 described a survey of bicycle traffic in Berkeley, California, with data displayed in Table 3.3. For this problem, restrict your attention to the first two rows of the table: residential streets labeled as 'bike routes,' which we will use to illustrate this computational exercise.

   (a) Set up a model for the data in Table 3.3 so that, for $j = 1, \ldots, 10$, the observed number of bicycles at location j is binomial with unknown probability θ_j and sample size equal to the total number of vehicles (bicycles included) in that block. The parameter θ_j can be interpreted as the underlying or 'true' proportion of traffic at location j that is bicycles. (See Exercise 3.8.) Assign a beta population distribution for the parameters θ_j and a noninformative hyperprior distribution as in the rat tumor example of Section 5.3. Write down the joint posterior distribution.

   (b) Compute the marginal posterior density of the hyperparameters and draw simulations from the joint posterior distribution of the parameters and hyperparameters, as in Section 5.3.

   (c) Compare the posterior distributions of the parameters θ_j to the raw proportions, (number of bicycles / total number of vehicles) in location j. How do the inferences from the posterior distribution differ from the raw proportions?

   (d) Give a 95% posterior interval for the average underlying proportion of traffic that is bicycles.

   (e) A new city block is sampled at random and is a residential street with a bike route. In an hour of observation, 100 vehicles of all kinds go by. Give a 95% posterior interval for the number of those vehicles that are bicycles. Discuss how much you trust this interval in application.

   (f) Was the beta distribution for the θ_j's reasonable?

12. Hierarchical Poisson model: consider the dataset in the previous problem, but suppose only the total amount of traffic at each location is observed.

   (a) Set up a model in which the total number of vehicles observed at each location j follows a Poisson distribution with parameter θ_j, the 'true' rate of traffic per hour at that location. Assign a gamma population distribution for the parameters θ_j and a noninformative hyperprior distribution. Write down the joint posterior distribution.

   (b) Compute the marginal posterior density of the hyperparameters and plot its contours. Simulate random draws from the posterior distribution of the hyperparameters and make a scatterplot of the simulation draws.

   (c) Is the posterior density integrable? Answer analytically by examining the joint posterior density at the limits or empirically by examining the plots of the marginal posterior density above.

   (d) If the posterior density is not integrable, alter it and repeat the previous two steps.

(e) Draw samples from the joint posterior distribution of the parameters and hyperparameters, by analogy to the method used in the hierarchical binomial model.

13. Meta-analysis: perform the computations for the meta-analysis data of Table 5.4.

   (a) Plot the posterior density of τ over an appropriate range that includes essentially all of the posterior density, analogous to Figure 5.5.

   (b) Produce graphs analogous to Figures 5.6 and 5.7 to display how the posterior means and standard deviations of the θ_j's depend on τ.

   (c) Produce a scatterplot of the crude effect estimates vs. the posterior median effect estimates of the 22 studies. Verify that the studies with smallest sample sizes are 'shrunk' the most toward the mean.

   (d) Draw simulations from the posterior distribution of a new treatment effect, $\tilde{\theta}_j$. Plot a histogram of the simulations.

   (e) Given the simulations just obtained, draw simulated outcomes from replications of a hypothetical new experiment with 100 persons in each of the treated and control groups. Plot a histogram of the simulations of the crude estimated treatment effect (5.23) in the new experiment.

CHAPTER 6

Model checking and improvement

6.1 The place of model checking in applied Bayesian statistics

Once we have accomplished the first two steps of a Bayesian analysis—constructing a probability model and computing (typically using simulation) the posterior distribution of all estimands—we should not ignore the relatively easy step of assessing the fit of the model to the data and to our substantive knowledge. It is difficult to include in a probability distribution all of one's knowledge about a problem, and so it is wise to investigate what aspects of reality are *not* captured by the model.

Checking the model is crucial to statistical analysis. Bayesian prior-to-posterior inferences assume the whole structure of a probability model and can yield misleading inferences when the model is poor. A good Bayesian analysis, therefore, should include at least some check of the adequacy of the fit of the model to the data and the plausibility of the model for the purposes for which the model will be used. This is sometimes discussed as a problem of sensitivity to the prior distribution, but in practice the likelihood model is typically just as suspect; throughout, we use 'model' to encompass the sampling distribution, the prior distribution, any hierarchical structure, and issues such as which explanatory variables have been included in a regression.

Sensitivity analysis and model improvement

It is typically the case that more than one reasonable probability model can provide an adequate fit to the data in a scientific problem. The basic question of a *sensitivity analysis* is: how much do posterior inferences change when other reasonable probability models are used in place of the present model? Other reasonable models may differ substantially from the present model in the prior specification, the sampling distribution, or in what information is included (for example, predictor variables in a regression). It is possible that the present model provides an adequate fit to the data, but that posterior inferences differ under plausible alternative models.

In theory, both model checking and sensitivity analysis can be incorporated into the usual prior-to-posterior analysis. Under this perspective, model checking is done by setting up a comprehensive joint distribution, such that any data that might be observed are plausible outcomes under the joint distribution. That is, this joint distribution is a mixture of all possible 'true' models or realities, incorporating all known substantive information. The prior distribution in such a case incorporates prior beliefs about the likelihood of the competing realities and about the parameters of the constituent models. The posterior distribution of such an *exhaustive* probability model automatically incorporates all 'sensitivity analysis' but is still predicated on the truth of some member of the larger class of models.

In practice, however, setting up such a super-model to include all possibilities and all substantive knowledge is both conceptually impossible and computationally infeasible in all but the simplest problems. It is thus necessary for us to examine our models in other ways to see how they fail to fit reality and how sensitive the resulting posterior distributions are to arbitrary specifications.

Judging model flaws by their practical implications

We do not like to ask, 'Is our model true or false?', since probability models in most data analyses will not be perfectly true. Even the coin tosses and die rolls ubiquitous in probability theory texts are not truly exchangeable in reality. The more relevant question is, 'Do the model's deficiencies have a noticeable effect on the substantive inferences?'

In the examples of Chapter 5, the beta population distribution for the tumor rates and the normal distribution for the eight school effects are both chosen partly for convenience. In these examples, making convenient distributional assumptions turns out not to matter, in terms of the impact on the inferences of most interest. How to judge when assumptions of convenience can be made safely is a central task of Bayesian sensitivity analysis. Failures in the model lead to practical problems by creating clearly false inferences about estimands of interest.

6.2 Do the inferences from the model make sense?

In any applied problem, there will be knowledge that is not included formally in either the prior distribution or the likelihood, for reasons of convenience or objectivity. If the additional information suggests that posterior inferences of interest are false, then this suggests a potential for creating a more accurate probability model for the parameters and data collection process. We illustrate with an example of a hierarchical regression model.

> **Example. Evaluating election predictions by comparing to substantive political knowledge**
> Figure 6.1 displays a forecast, made in early October, 1992, of the probability that Bill Clinton would win each state in the November, 1992, U.S. Presidential election. The estimates are posterior probabilities based on a hierarchical linear regression model. For each state, the height of the shaded part of the box represents the estimated probability that Clinton would win the state. Even before the election occurred, the forecasts for some of the states looked wrong; for example, from state polls, Clinton was known in October to be much weaker in Texas and Florida than shown in the map. This does not mean that the forecast

Figure 6.1 *Summary of a forecast of the 1992 U.S. Presidential election performed one month before the election. For each state, the proportion of the box that is shaded represents the estimated probability of Clinton winning the state; the width of the box is proportional to the number of electoral votes for the state.*

is useless, but it is good to know where the weak points are. Certainly, after the election, we can do an even better job of criticizing the model and understanding its weaknesses. We return to this election forecasting example in Section 15.2 as an example of a hierarchical linear model.

More formally, we can check a model by *external validation* using the model to make predictions about future data, and then collecting those data and comparing to their predictions. Posterior means should be correct on average, 50% intervals should contain the true values half the time, and so forth. We used external validation to check the empirical probability estimates in the record-linkage example in Section 1.7, and we apply the idea again to check a toxicology model in Section 20.3. In the latter example, the external validation (see Figure 20.10 on page 514) reveals a generally reasonable fit but with some notable discrepancies between predictions and external data.

6.3 Is the model consistent with data? Posterior predictive checking

If the model fits, then replicated data generated under the model should look similar to observed data. To put it another way, the observed data should look plausible under the posterior predictive distribution. This is really a self-consistency check: an observed discrepancy can be due to model misfit or chance.

Our basic technique for checking the fit of a model to data is to draw simulated values from the posterior predictive distribution of replicated data and compare these samples to the observed data. Any systematic differences between the simulations and the data indicate potential failings of the model. We introduce posterior predictive checking with a simple example of an obviously poorly fitting model, and then in the rest of this section we lay out

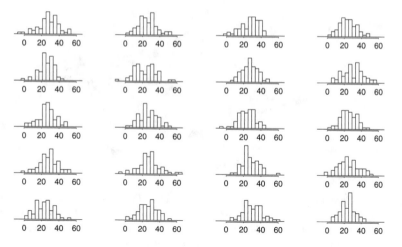

Figure 6.2 *Twenty replications, y^{rep}, of the speed of light data from the posterior predictive distribution, $p(y^{rep}|y)$; compare to observed data, y, in Figure 3.1. Each histogram displays the result of drawing 66 independent values $\tilde{y}_i$ from a common normal distribution with mean and variance (μ, σ^2) drawn from the posterior distribution, $p(\mu, \sigma^2|y)$, under the normal model.*

the key choices involved in posterior predictive checking. Sections 6.4 and 6.5 discuss graphical and numerical predictive checks in more detail.

Example. Comparing Newcomb's speed of light measurements to the posterior predictive distribution

Simon Newcomb's 66 measurements on the speed of light are presented in Section 3.2. In the absence of other information, in Section 3.2 we modeled the measurements as $N(\mu, \sigma^2)$, with a noninformative uniform prior distribution on $(\mu, \log \sigma)$. However, the lowest of Newcomb's measurements look like outliers compared to the rest of the data.

Could the extreme measurements have reasonably come from a normal distribution? We address this question by comparing the observed data to what we expect to be observed under our posterior distribution. Figure 6.2 displays twenty histograms, each of which represents a single draw from the posterior predictive distribution of the values in Newcomb's experiment, obtained by first drawing (μ, σ^2) from their joint posterior distribution, then drawing 66 values from a normal distribution with this mean and variance. All these histograms look quite different from the histogram of actual data in Figure 3.1 on page 78. One way to measure the discrepancy is to compare the smallest value in each hypothetical replicated dataset to Newcomb's smallest observation, -44. The histogram in Figure 6.3 shows the smallest observation in each of the 20 hypothetical replications; all are much larger than Newcomb's smallest observation, which is indicated by a vertical line on the graph. The normal model clearly does not capture the variation that Newcomb observed. A revised model might use an asymmetric contaminated normal distribution or a symmetric long-tailed distribution in place of the normal measurement model.

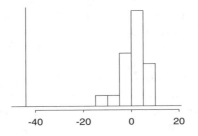

Figure 6.3 *Smallest observation of Newcomb's speed of light data (the vertical line at the left of the graph), compared to the smallest observations from each of the 20 posterior predictive simulated datasets displayed in Figure 6.2.*

Many other examples of posterior predictive checks appear throughout the book, including the educational testing example in Section 6.8, linear regressions example in Sections 14.3 and 15.2, and a hierarchical mixture model in Section 18.4.

For many problems, it is useful to examine graphical comparisons of summaries of the data to summaries from posterior predictive simulations, as in Figure 6.3. In cases with less blatant discrepancies than the outliers in the speed of light data, it is often also useful to measure the 'statistical significance' of the lack of fit, a notion we formalize here.

Notation for replications

Let y be the observed data and θ be the vector of parameters (including all the hyperparameters if the model is hierarchical). To avoid confusion with the observed data, y, we define y^{rep} as the *replicated* data that *could have been* observed, or, to think predictively, as the data we *would* see tomorrow if the experiment that produced y today were replicated with the same model and the same value of θ that produced the observed data.

We distinguish between y^{rep} and $\tilde{y}$, our general notation for predictive outcomes: $\tilde{y}$ is any future observable value or vector of observable quantities, whereas y^{rep} is specifically a replication just like y. For example, if the model has explanatory variables, x, they will be identical for y and y^{rep}, but $\tilde{y}$ may have its own explanatory variables, $\tilde{x}$.

We will work with the distribution of y^{rep} given the current state of knowledge, that is, with the posterior predictive distribution

$$p(y^{\text{rep}}|y) = \int p(y^{\text{rep}}|\theta)p(\theta|y)d\theta. \tag{6.1}$$

Test quantities

We measure the discrepancy between model and data by defining *test quantities*, the aspects of the data we wish to check. A test quantity, or *discrepancy measure*, $T(y, \theta)$, is a scalar summary of parameters and data that is used as a standard when comparing data to predictive simulations. Test quantities play the role in Bayesian model checking that test statistics play in classical testing. We use the notation $T(y)$ for a *test statistic*, which is a test quantity that depends only on data; in the Bayesian context, we can generalize test statistics to allow dependence on the model parameters under their posterior distribution. This can be useful in directly summarizing discrepancies between model and data. We discuss options for graphical test quantities in Section 6.4 and numerical summaries in Section 6.5.

Tail-area probabilities

Lack of fit of the data with respect to the posterior predictive distribution can be measured by the tail-area probability, or *p*-value, of the test quantity, and computed using posterior simulations of (θ, y^{rep}). We define the *p*-value mathematically, first for the familiar classical test and then in the Bayesian context.

Classical p-values.   The classical *p*-value for the test statistic $T(y)$ is

$$p_C = \Pr(T(y^{\text{rep}}) \geq T(y) | \theta) \tag{6.2}$$

where the probability is taken over the distribution of y^{rep} with θ fixed. (The distribution of y^{rep} given y and θ is the same as its distribution given θ alone.) Test statistics are classically derived in a variety of ways but generally represent a summary measure of discrepancy between the observed data and what would be expected under a model with a particular value of θ. This value may be a 'null' value, corresponding to a 'null hypothesis,' or a point estimate such as the maximum likelihood value. A point estimate for θ must be substituted to compute a *p*-value in classical statistics.

Posterior predictive p-values.   To evaluate the fit of the posterior distribution of a Bayesian model, we can compare the observed data to the posterior predictive distribution. In the Bayesian approach, test quantities can be functions of the unknown parameters as well as data because the test quantity is evaluated over draws from the posterior distribution of the unknown parameters. The Bayesian *p*-value is defined as the probability that the replicated data could be more extreme than the observed data, as measured by the test quantity:

$$p_B = \Pr(T(y^{\text{rep}}, \theta) \geq T(y, \theta) | y),$$

where the probability is taken over the posterior distribution of θ and the posterior predictive distribution of y^{rep} (that is, the joint distribution, $p(\theta, y^{\text{rep}} | y)$):

$$p_B = \int \int I_{T(y^{\text{rep}}, \theta) \geq T(y, \theta)} p(y^{\text{rep}} | \theta) p(\theta | y) dy^{\text{rep}} d\theta,$$

where I is the indicator function. In this formula, we have used the property of the predictive distribution that $p(y^{\text{rep}} | \theta, y) = p(y^{\text{rep}} | \theta)$.

In practice, we usually compute the posterior predictive distribution using

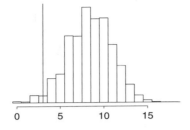

Figure 6.4 *Observed number of switches (vertical line at $T(y) = 3$), compared to 10,000 simulations from the posterior predictive distribution of the number of switches, $T(y^{rep})$.*

simulation. If we already have L simulations from the posterior density of θ, we just draw one y^{rep} from the predictive distribution for each simulated θ; we now have L draws from the joint posterior distribution, $p(y^{rep}, \theta | y)$. The posterior predictive check is the comparison between the realized test quantities, $T(y, \theta^l)$, and the predictive test quantities, $T(y^{rep\, l}, \theta^l)$. The estimated p-value is just the proportion of these L simulations for which the test quantity equals or exceeds its realized value; that is, for which $T(y^{rep\, l}, \theta^l) \geq T(y, \theta^l), l = 1, \ldots, L$.

In contrast to the classical approach, Bayesian model checking does not require special methods to handle 'nuisance parameters'; by using posterior simulations, we implicitly average over all the parameters in the model.

Example. Checking the assumption of independence in binomial trials
We illustrate posterior predictive model checking with a simple hypothetical example. Consider a sequence of binary outcomes, $y_1, \ldots, y_n$, modeled as a specified number of iid Bernoulli trials with a uniform prior distribution on the probability of success, θ. As discussed in Chapter 2, the posterior density under the model is $p(\theta | y) \propto \theta^s (1 - \theta)^{n-s}$, which depends on the data only through the sufficient statistic, $s = \sum y_i$. Now suppose the observed data are, in order, 1, 1, 0, 0, 0, 0, 0, 1, 1, 1, 1, 1, 0, 0, 0, 0, 0, 0, 0, 0. The observed autocorrelation is evidence that the model is flawed. To quantify the evidence, we can perform a posterior predictive test using the test quantity T = number of switches between 0 and 1 in the sequence. The observed value is $T(y) = 3$, and we can determine the posterior predictive distribution of $T(y^{rep})$ by simulation. To simulate y^{rep} under the model, we first draw θ from its Beta$(8, 14)$ posterior distribution, then draw $y^{rep} = (y_1^{rep}, \ldots, y_{20}^{rep})$ as independent Bernoulli variables with probability θ. Figure 6.4 displays a histogram of the values of $T(y^{rep\, l})$ for simulation draws $l = 1, \ldots, 10000$, with the observed value, $T(y) = 3$, shown by a vertical line. The observed number of switches is about one-third as many as would be expected from the model under the posterior predictive distribution, and the discrepancy cannot easily be explained by chance, as indicated by the computed p-value of $\frac{9838}{10000}$. To convert to a p-value near zero, we can change the sign of the test statistic, which amounts to computing $\Pr(T(y^{rep}, \theta) \leq T(y, \theta) | y)$, which is 0.028 in this

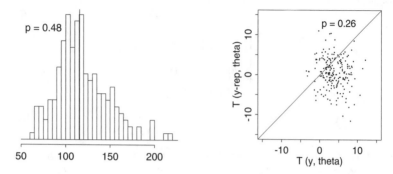

Figure 6.5 *Realized vs. posterior predictive distributions for two more test quantities in the speed of light example: (a) Sample variance (vertical line at 115.5), compared to 200 simulations from the posterior predictive distribution of the sample variance. (b) Scatterplot showing prior and posterior simulations of a test quantity: $T(y, \theta) = |y_{(61)} - \theta| - |y_{(6)} - \theta|$ (horizontal axis) vs. $T(y^{\text{rep}}, \theta) = |y_{(61)}^{\text{rep}} - \theta| - |y_{(6)}^{\text{rep}} - \theta|$ (vertical axis) based on 200 simulations from the posterior distribution of (θ, y^{rep}). The p-value is computed as the proportion of points in the upper-left half of the plot.*

case. The p-values measured from the two ends have a sum that is greater than 1 because of the discreteness of the distribution of $T(y^{\text{rep}})$.

Example. Speed of light (continued)

In Figure 6.3, we demonstrated the poor fit of the normal model to the speed of light data using $\min(y_i)$ as the test statistic. We continue this example using other test quantities to illustrate how the fit of a model depends on the aspects of the data and parameters being monitored. Figure 6.5a shows the observed sample variance and the distribution of 200 simulated variances from the posterior predictive distribution. The sample variance does not make a good test statistic because it is a sufficient statistic of the model and thus, in the absence of an informative prior distribution, the posterior distribution will automatically be centered near the observed value. We are not at all surprised to find an estimated p-value close to $\frac{1}{2}$.

The model check based on $\min(y_i)$ earlier in the chapter suggests that the normal model is inadequate. To illustrate that a model can be inadequate for some purposes but adequate for others, we assess whether the model is adequate except for the extreme tails by considering a model check based on a test quantity sensitive to asymmetry in the center of the distribution,

$$T(y, \theta) = |y_{(61)} - \theta| - |y_{(6)} - \theta|.$$

The 61st and 6th order statistics are chosen to represent approximately the 90% and 10% points of the distribution. The test quantity should be scattered about zero for a symmetric distribution. The scatterplot in Figure 6.5b shows the test quantity for the observed data and the test quantity evaluated for the simulated data for 200 simulations from the posterior distribution of (θ, σ^2). The estimated p-value is 0.26, implying that any observed asymmetry in the middle of the distribution can easily be explained by sampling variation.

Defining replications

Depending on the aspect of the model one wishes to check, one can define the reference set of replications y^{rep} by conditioning on some or all of the observed data. For example, in checking the normal model for Newcomb's speed of light data, we kept the number of observations, n, fixed at the value in Newcomb's experiment. In Section 6.8, we check the hierarchical normal model for the SAT coaching experiments using posterior predictive simulations of new data on the same eight schools. It would also be possible to examine predictive simulations on new schools drawn from the same population. In analyses of sample surveys and designed experiments, it often makes sense to consider hypothetical replications of the experiment with a new randomization of selection or treatment assignment, by analogy to classical randomization tests.

6.4 Graphical posterior predictive checks

The basic idea of graphical model checking is to display the data alongside simulated data from the fitted model, and to look for systematic discrepancies between real and simulated data. This section gives examples of three kinds of graphical display:

- Direct display of all the data (as in the comparison of the speed-of-light data in Figure 3.1 to the 20 replications in Figure 6.2).

- Display of data summaries or parameter inferences. This can be useful in settings where the dataset is large and we wish to focus on the fit of a particular aspect of the model.

- Graphs of residuals or other measures of discrepancy between model and data.

Direct data display

Figure 6.6 shows another example of model checking by displaying all the data. The left column of the figure displays a three-way array of binary data—for each of 6 persons, a possible 'yes' or 'no' to each of 15 possible reactions (displayed as rows) to 23 situations (columns)—from an experiment in psychology. The three-way array is displayed as 6 slices, one for each person. Before displaying, the reactions, situations, and persons have been ordered in increasing average response. We can thus think of the test statistic $T(y)$ as being this graphical display, complete with the ordering applied to the data y.

    The right columns of Figure 6.6 display seven independently-simulated replications y^{rep} from a fitted logistic regression model (with the rows, columns, and persons for each dataset arranged in increasing order before display, so that we are displaying $T(y^{rep})$ in each case). Here, the replicated datasets look fuzzy and 'random' compared to the observed data, which have strong rectilinear structures that are clearly not captured in the model. If the data

Figure 6.6 *Left column displays observed data y (a 15 × 23 array of binary responses from each of 6 persons); right columns display seven replicated datasets y*^{rep} *from a fitted logistic regression model. A misfit of model to data is apparent: the data show strong row and column patterns for individual persons (for example, the nearly white row near the middle of the last person's data) that do not appear in the replicates. (To make such patterns clearer, the indexes of the observed and each replicated dataset have been arranged in increasing order of average response.)*

were actually generated from the model, the observed data on the left would fit right in with the simulated datasets on the right.

Interestingly, these data have enough internal replication that the model misfit would be clear in comparison to a single simulated dataset from the model. But, to be safe, it is good to compare to several replications to see if the patterns in the observed data could be expected to occur by chance under the model.

Displaying data is not simply a matter of dumping a set of numbers on a page (or a screen). For example, we took care to align the graphs in Figure 6.6 to display the three-dimensional dataset and seven replications at once without confusion. Even more important, the arrangement of the rows, columns, and persons in increasing order is crucial to seeing the patterns in the data over and above the model. To see this, consider Figure 6.7, which presents the same information as in Figure 6.6 but without the ordering. Here, the discrepancies between data and model are not clear at all.

Figure 6.7 *Redisplay of Figure 6.6 without ordering the rows, columns, and persons in order of increasing response. Once again, the left column shows the observed data and the right columns show replicated datasets from the model. Without the ordering, it is very difficult to notice the discrepancies between data and model, which are easily apparent in Figure 6.6.*

Displaying summary statistics or inferences

A key principle of exploratory data analysis is to exploit regular structure to display data more effectively. The analogy in modeling is hierarchical or multilevel modeling, in which batches of parameters capture variation at different levels. When checking model fit, hierarchical structure can allow us to compare batches of parameters to their reference distribution. In this scenario, the replications correspond to new draws of a batch of parameters.

We illustrate with inference from a hierarchical model from psychology. This was a fairly elaborate model, whose details we do not describe here; all we need to know for this example is that the model included two vectors of parameters, $\phi_1, \ldots, \phi_{90}$, and $\psi_1, \ldots, \psi_{69}$, corresponding to patients and psychological symptoms, and that each of these 159 parameters were assigned independent Beta$(2, 2)$ prior distributions. Each of these parameters represented a probability that a given patient or symptom is associated with a particular psychological syndrome.

Data were collected (measurements of which symptoms appeared in which patients) and the full Bayesian model was fitted, yielding posterior simulations for all these parameters. If the model were true, we would expect any single simulation draw of the vectors of patient parameters ϕ and symptom

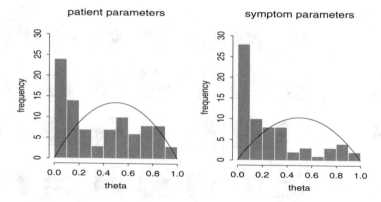

Figure 6.8 *Histograms of (a) 90 patient parameters and (b) 69 symptom parameters, from a single draw from the posterior distribution of a psychometric model. These histograms of posterior estimates contradict the assumed Beta(2, 2) prior densities (overlain on the histograms) for each batch of parameters, and motivated us to switch to mixture prior distributions. This implicit comparison to the values under the prior distribution can be viewed as a posterior predictive check in which a new set of patients and a new set of symptoms are simulated.*

parameters ψ to look like independent draws from the Beta(2, 2) distribution. We know this because of the following reasoning:

- If the model were indeed true, we could think of the observed data vector y and the vector θ of the true values of all the parameters (including ϕ and ψ) as a random draw from their joint distribution, $p(y, \theta)$. Thus, y comes from the marginal distribution, the prior predictive distribution, $p(y)$.

- A single draw θ^l from the posterior inference comes from $p(\theta^l|y)$. Since $y \sim p(y)$, this means that y, θ^l come from the model's joint distribution of y, θ, and so the marginal distribution of θ^l is the same as that of θ.

- That is, y, θ, θ^l have a combined joint distribution in which θ and θ^l have the same marginal distributions (and the same joint distributions with y).

Thus, as a model check we can plot a histogram of a single simulation of the vector of parameters ϕ or ψ and compare to the prior distribution. This corresponds to a posterior predictive check in which the inference from the observed data is compared to what would be expected if the model were applied to a new set of patients and a new set of symptoms.

Figure 6.8 shows histograms of a single simulation draw for each of ϕ and ψ as fitted to our dataset. The lines show the Beta(2, 2) prior distribution, which clearly does not fit. For both ϕ and ψ, there are too many cases near zero, corresponding to patients and symptoms that almost certainly are not associated with a particular syndrome.

Our next step was to replace the offending Beta(2, 2) prior distributions by mixtures of two beta distributions—one distribution with a spike near zero, and another that is uniform between 0 and 1—with different models for the

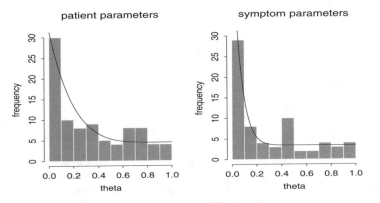

Figure 6.9 *Histograms of (a) 90 patient parameters and (b) 69 symptom parameters, as estimated from an expanded psychometric model. The mixture prior densities (overlain on the histograms) are not perfect, but they approximate the corresponding histograms much better than the Beta(2, 2) densities in Figure 6.8.*

ϕ's and the ψ's. The exact model is,

$$p(\phi_j) = 0.5\text{Beta}(\phi_j|1, 6) + 0.5\text{Beta}(\phi_j|1, 1)$$
$$p(\psi_j) = 0.5\text{Beta}(\psi_j|1, 16) + 0.5\text{Beta}(\psi_j|1, 1).$$

We set the parameters of the mixture distributions to fixed values based on our understanding of the model. It was reasonable for these data to suppose that any given symptom appeared only about half the time; however, labeling of the symptoms is subjective, so we used beta distributions peaked near zero but with some probability of taking small positive values. We assigned the Beta(1, 1) (that is, uniform) distributions for the patient and symptom parameters that were not near zero—given the estimates in Figure 6.8, these seemed to fit the data better than the original Beta(2, 2) models. (In fact, the original reason for using Beta(2, 2) rather than uniform prior distributions was so that maximum likelihood estimates would be in the interior of the interval [0, 1], a concern that disappeared when we moved to Bayesian inference; see Exercise 8.4.)

Some might object to revising the prior distribution based on the fit of the model to the data. It is, however, consistent with common statistical practice, in which a model is iteratively altered to provide a better fit to data. The natural next step would be to add a hierarchical structure, with hyperparameters for the mixture distributions for the patient and symptom parameters. This would require additional computational steps and potential new modeling difficulties (for example, instability in the estimated hyperparameters). Our main concern in this problem was to reasonably model the individual ϕ_j and ψ_j parameters without the prior distributions inappropriately interfering (which appears to be happening in Figure 6.8).

We refitted the model with the new prior distribution and repeated the

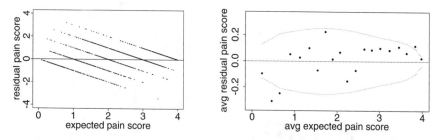

Figure 6.10 *(a) Residuals (observed − expected) vs. expected values for a model of pain relief scores (0 = no pain relief, ..., 5 = complete pain relief) (b) Average residuals vs. expected pain scores, with measurements divided into 20 equally-sized bins defined by ranges of expected pain scores. The average prediction errors are relatively small (note the scale of the y-axis), but with a consistent pattern that low predictions are too low and high predictions are too high. Dotted lines show 95% bounds under the model.*

model check, which is displayed in Figure 6.9. The fit of the prior distribution to the inferences is not perfect but is much better than before.

Residual plots and binned residual plots

Bayesian residuals. Linear and nonlinear regression models, which are the core tools of applied statistics, are characterized by a function $g(x, \theta) = E(y|x, \theta)$, where x is a vector of predictors. Then, given the unknown parameters θ and the predictors x_i for a data point y_i, the *predicted value* is $g(x_i, \theta)$ and the *residual* is $y_i - g(x_i, \theta)$. This is sometimes called a 'realized' residual in contrast to the classical or estimated residual, $y_i - g(x_i, \hat\theta)$, which is based on a point estimate $\hat\theta$ of the parameters.

A Bayesian residual graph plots a single realization of the residuals (based on a single random draw of θ). An example appears on page 513. However, classical residual plots can be thought of as approximations to the Bayesian version, ignoring posterior uncertainty in θ.

Binned residuals for discrete data. Unfortunately, for discrete data, plots of residuals can be difficult to interpret because, for any particular value of $E(y_i|X, \theta)$, the residual r_i can only take on certain discrete values; thus, even if the model is correct, the residuals will not generally be expected to be independent of predicted values or covariates in the model. Figure 6.10 illustrates with data and then residuals plotted vs. fitted values, for a model of pain relief scores, which were discretely reported as 0, 1, 2, 3, or 4. The residuals have a distracting striped pattern because predicted values plus residuals equal discrete observed data values.

A standard way to make discrete residual plots more interpretable is to work with binned or smoothed residuals, which should be closer to symmetric about zero if enough residuals are included in each bin or smoothing category (since the expectation of each residual is by definition zero, the central limit theorem ensures that the distribution of averages of many residuals will be approximately symmetric). In particular, suppose we would like to plot the vector of residuals r vs. some vector $w = (w_1, \ldots, w_n)$ that can in general be a function of X, θ, and perhaps y. We can bin the predictors and residuals by ordering the n values of w_i and sorting them into bins $k = 1, \ldots, K$, with approximately equal numbers of points n_k in each bin. For each bin, we then compute $\bar{w}_k$ and $\bar{r}_k$, the average values of w_i and r_i, respectively, for points i in bin k. The *binned residual plot* is the plot of the points $\bar{r}_k$ vs. $\bar{w}_k$, which actually must be represented by several plots (which perhaps can be overlain) representing variability due to uncertainty of θ in the posterior distribution.

Since we are viewing the plot as a test variable, it must be compared to the distribution of plots of $\bar{r}_k^{\text{rep}}$ vs. $\bar{w}_k^{\text{rep}}$, where, for each simulation draw, the values of $\bar{r}_k^{\text{rep}}$ are computed by averaging the replicated residuals $r_i^{\text{rep}} = y_i^{\text{rep}} - \text{E}(y_i | X, \theta)$ for points i in bin k. In general, the values of w_i can depend on y, and so the bins and the values of $\bar{w}_k^{\text{rep}}$ can vary among the replicated data sets.

Because we can compare to the distribution of simulated replications, the question arises: why do the binning at all? We do so because we want to understand the model misfits that we detect. Because of the discreteness of the data, the individual residuals r_i have asymmetric discrete distributions. As expected, the binned residuals are approximately symmetrically distributed. In general it is desirable for the posterior predictive reference distribution of a discrepancy variable to exhibit some simple features (in this case, independence and approximate normality of the $\bar{r}_k$'s) so that there is a clear interpretation of a misfit. This is, in fact, the same reason that one plots residuals, rather than data, vs. predicted values: it is easier to compare to an expected horizontal line than to an expected 45° line.

Under the model, the residuals are independent and, if enough are in each bin, the mean residuals $\bar{r}_k$ are approximately normally distributed. We can then display the reference distribution as 95% error bounds, as in Figure 6.10b. We never actually have to display the replicated data; the replication distribution is implicit, given our knowledge that the binned residuals are independent, approximately normally distributed, and with expected variation as shown by the error bounds.

General interpretation of graphs as model checks

More generally, we can compare any data display to replications under the model—not necessarily as an explicit model check but more to understand what the display 'should' look like if the model were true. For example, the maps and scatterplots of high and low cancer rates (Figures 2.7–2.9) show strong patterns, but these are not particularly informative if the same patterns would be expected of replications under the model. The erroneous initial interpretation of Figure 2.7—as evidence of a pattern of high cancer rates in the sparsely-populated areas in the center-west of the country—can be thought of as an erroneous model check, in which the data display was compared to a random pattern rather than to the pattern expected under a reasonable model of variation in cancer occurrences.

6.5 Numerical posterior predictive checks

Choosing test quantities

The procedure for carrying out a posterior predictive model check requires specifying a test quantity, $T(y)$ or $T(y, \theta)$, and an appropriate predictive distribution for the replications y^{rep} (which involves deciding which if any aspects of the data to condition on, as discussed at the end of Section 6.3). If $T(y)$ does not appear to be consistent with the set of values $T(y^{\text{rep}\,1}), \ldots, T(y^{\text{rep}\,L})$, then the model is making predictions that do not fit the data. The discrepancy between $T(y)$ and the distribution of $T(y^{\text{rep}})$ can be summarized by a p-value (as discussed in Section 6.3) but we prefer to look at the magnitude of the discrepancy as well as its p-value. For example, in Figure 6.4 on page 163 we see that the observed number of switches is only about one-third what would be expected from the model, *and* the p-value of 0.028 indicates that it would be highly unlikely to see such a discrepancy in replicated data.

For many problems, a function of data and parameters can directly address a particular aspect of a model in a way that would be difficult or awkward using a function of data alone. If the test quantity depends on θ as well as y, then the test quantity $T(y, \theta)$ as well as its replication $T(y^{\text{rep}}, \theta)$ are unknowns and are represented by L simulations, and the comparison can be displayed either as a scatterplot of the values $T(y, \theta^l)$ vs. $T(y^{\text{rep}\,l}, \theta^l)$ or a histogram of the differences, $T(y, \theta^l) - T(y^{\text{rep}\,l}, \theta^l)$. Under the model, the scatterplot should be symmetric about the 45° line and the histogram should include 0.

Because a probability model can fail to reflect the process that generated the data in any number of ways, posterior predictive p-values can be computed for a variety of test quantities in order to evaluate more than one possible model failure. Ideally, the test quantities T will be chosen to reflect aspects of the model that are relevant to the scientific purposes to which the inference will be applied. Test quantities are commonly chosen to measure a feature of the data not directly addressed by the probability model; for example, ranks of the sample, or correlation of model residuals with some possible explanatory variable.

> **Example. Checking the fit of hierarchical regression models for adolescent smoking**
>
> We illustrate with a model fitted to a longitudinal data set of about 2000 Australian adolescents whose smoking patterns were recorded every six months (via questionnaire) for a period of three years. Interest lay in the extent to which smoking behavior could be predicted based on parental smoking and other background variables, and the extent to which boys and girls picked up the habit of smoking during their teenage years. Figure 6.11 illustrates the overall rate of

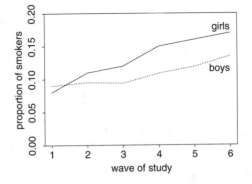

Figure 6.11 *Prevalence of regular (daily) smoking among participants responding at each wave in the study of Australian adolescents (who were on average 15 years old at wave 1).*

smoking among survey participants, who had an average age of 14.9 years at the beginning of the study.

We fit two models to these data. Our first model is a hierarchical logistic regression, in which the probability of smoking depends on sex, parental smoking, the wave of the study, and an individual parameter for the person. For person j at wave t, we model the probability of smoking as,

$$\Pr(y_{jt} = 1) = \text{logit}^{-1}(\beta_0 + \beta_1 X_{j1} + \beta_2 X_{j2} + \beta_3(1 - X_{j2})t + \beta_4 X_{j2}t + \alpha_j), \quad (6.3)$$

where X_{j1} is an indicator for parental smoking and X_{j2} is an indicator for females, so that β_3 and β_4 represent the time trends for males and females, respectively. The individual effects α_j are assigned a $N(0, \tau^2)$ distribution, with a noninformative uniform prior distribution on β, τ. (See Chapter 18 for more on hierarchical generalized linear models.)

The second model is an expansion of the first, in which each person j has an unobserved 'susceptibility' status S_j that equals 1 if the person might possibly smoke or 0 if he or she is 'immune' from smoking (that is, has no chance of becoming a smoker). This model is an oversimplification but captures the separation in the data between adolescents who often or occasionally smoke and those who never smoke at all. In this mixture model, the smoking status y_{jt} is automatically 0 at all times for non-susceptible persons. For those persons with $S_j = 1$, we use the model (6.3), understanding that these probabilities now refer to the probability of smoking, conditional on being susceptible. The model is completed with a logistic regression for susceptibility status given the individual-level predictors: $\Pr(S_j = 1) = \text{logit}^{-1}(\gamma_0 + \gamma_1 X_{j1} + \gamma_2 X_{j2})$, and a uniform prior distribution on these coefficients γ.

Table 6.1 shows the results for posterior predictive checks of the two fitted models using three different test statistics $T(y)$:

- The percentage of adolescents in the sample who never smoked.
- The percentage in the sample who smoked during all waves.
- The percentage of 'incident smokers': adolescents who began the study as non-

Test variable	$T(y)$	Model 1 95% int. for $T(y^{\mathrm{rep}})$	p-value	Model 2 95% int. for $T(y^{\mathrm{rep}})$	p-value
% never-smokers	77.3	$[75.5, 78.2]$	0.27	$[74.8, 79.9]$	0.53
% always-smokers	5.1	$[5.0, 6.5]$	0.95	$[3.8, 6.3]$	0.44
% incident smokers	8.4	$[5.3, 7.9]$	0.005	$[4.9, 7.8]$	0.004

Table 6.1 *Summary of posterior predictive checks for three test statistics for two models fit to the adolescent smoking data: (1) hierarchical logistic regression, and (2) hierarchical logistic regression with a mixture component for never-smokers. The second model better fits the percentages of never- and always-smokers, but still has a problem with the percentage of 'incident smokers,' who are defined as persons who report incidents of non-smoking followed by incidents of smoking.*

smokers, switched to smoking during the study period, and did not switch back.

From the first column of Table 6.1, we see that 77% of the sample never smoked, 5% always smoked, and 8% were incident smokers. The table then displays the posterior predictive distribution of each test statistic under each of the two fitted models. Both models accurately capture the percentage of never-smokers, but the second model better fits the percentage of always-smokers. It makes sense that the second model should fit this aspect of the data better, since its mixture form separates smokers from non-smokers. Finally, both models underpredict the proportion of incident smokers, which suggests that they are not completely fitting the variation of smoking behavior within individuals.

Posterior predictive checking is a useful direct way of assessing the fit of the model to these various aspects of the data. Our goal here is not to compare or choose among the models (a topic we discuss in Section 6.7) but rather to explore the ways in which either or both models might be lacking.

Numerical test quantities can also be constructed from patterns noticed visually (as in the test statistics chosen for the speed-of-light example in Section 6.3). This can be useful to quantify a pattern of potential interest, or to summarize a model check that will be performed repeatedly (for example, in checking the fit of a model that is applied to several different datasets).

Multiple comparisons

More generally, one might worry about interpreting the significance levels of multiple tests, or for test statistics chosen by inspection of the data. For example, we looked at three different test variables in checking the adolescent smoking models, so perhaps it is less surprising than it might seem at first that the worst-fitting test statistic had a p-value of 0.005. A 'multiple comparisons' adjustment would calculate the probability that the most extreme p-value would be as low as 0.005, which would perhaps yield an adjusted p-value somewhere near 0.015.

We do not make this adjustment, because we use the posterior predictive checks to see how particular aspects of the data would be expected to appear in replications. If we examine several test variables, we would not be surprised for some of them not to be fitted by the model—but if we are planning to apply the model, we might be interested in those aspects of the data that do not appear typical. We are not concerned with 'Type I error' rate—that is, the probability of rejecting a hypothesis conditional on it being true—because we use the checks not to accept or reject a model but rather to understand the limits of its applicability in realistic replications.

Omnibus tests such as χ^2

In addition to focused tests, it is often useful to check a model with a summary measure of fit such as the χ^2 discrepancy quantity, written here in terms of univariate responses y_i:

$$\chi^2 \text{ discrepancy:} \quad T(y, \theta) = \sum_i \frac{(y_i - E(y_i|\theta))^2}{\text{var}(y_i|\theta)}, \tag{6.4}$$

where the summation is over the sample observations. When θ is known, this test quantity resembles the classical χ^2 goodness-of-fit measure. A related option is the *deviance*, defined as $T(y, \theta) = -2 \log p(y|\theta)$.

Classical χ^2 tests are based on test statistics such as $T(y) = T(y, \theta_{\text{null}})$, or $T(y) = \min_\theta T(y, \theta)$ or perhaps $T(y) = T(y, \theta_{\text{mle}})$, where θ_{null} is a value of θ under a null hypothesis of interest and θ_{mle} is the maximum likelihood estimate. The χ^2 reference distribution in these cases is based on large-sample approximations to the posterior distribution. The same test statistics can be used in a posterior predictive model check to produce a valid p-value with no restriction on the sample size. However, in the Bayesian context it can make more sense to simply work with $T(y, \theta)$ and avoid the minimization or other additional computational effort required in computing a purely data-based summary, $T(y)$. For a Bayesian χ^2 test, as in any other posterior predictive check, the reference distribution is automatically calculated from the posterior predictive simulations.

Interpreting posterior predictive p-values

A model is suspect if a discrepancy is of practical importance and its observed value has a tail-area probability that is close to 0 or 1, thereby indicating that the observed pattern would be unlikely to be seen in replications of the data if the model were true. An extreme p-value implies that the model cannot be expected to capture this aspect of the data. A p-value is a posterior probability and can therefore be interpreted directly—although *not* as Pr(model is true|data). Major failures of the model, typically corresponding to extreme tail-area probabilities (less than 0.01 or more than 0.99), can be addressed by expanding the model in an appropriate way. Lesser failures might also suggest model improvements or might be ignored in the short term if the failure appears not to affect the main inferences. In some cases, even extreme p-values may be ignored if the misfit of the model is substantively small compared to variation within the model. We will often evaluate a model with respect to several test quantities, and we should be sensitive to the implications of this practice.

If a p-value is close to 0 or 1, it is not so important exactly how extreme it is. A p-value of 0.00001 is virtually no stronger, in practice, than 0.001; in either case, the aspect of the data measured by the test quantity is inconsistent with the model. A slight improvement in the model (or correction of a data coding error!) could bring either p-value to a reasonable range (between 0.05 and 0.95, say). The p-value measures 'statistical significance,' not 'practical significance.' The latter is determined by how different the observed data are from the reference distribution on a scale of substantive interest and depends on the goal of the study; an example in which a discrepancy is statistically but not practically significant appears at the end of Section 14.3.

The relevant goal is not to answer the question, 'Do the data come from the assumed model?' (to which the answer is almost always no), but to quantify the discrepancies between data and model, and assess whether they could have arisen by chance, under the model's own assumptions.

Limitations of posterior tests. Finding an extreme p-value and thus 'rejecting' a model is never the end of an analysis; the departures of the test quantity in question from its posterior predictive distribution will often suggest improvements of the model or places to check the data, as in the speed of light example. Conversely, even when the current model seems appropriate for drawing inferences (in that no unusual deviations between the model and the data are found), the next scientific step will often be a more rigorous experiment incorporating additional factors, thereby providing better data. For instance, in the educational testing example of Section 5.5, the data do not allow rejection of the model that all the θ_j's are equal, but that assumption is clearly unrealistic, and some of the substantive conclusions are greatly changed when the parameter τ is not restricted to be zero.

Finally, the discrepancies found by posterior predictive checking should be considered in their applied context. A model can be demonstrably wrong but can still work for some purposes, as we illustrate in a linear regression example in Section 14.3.

Relation to classical statistical tests

Bayesian posterior predictive checks are generalizations of classical tests in that they average over the posterior distribution of the unknown parameter vector θ rather than fixing it at some estimate $\hat{\theta}$. The Bayesian tests do not rely on the clever construction of pivotal quantities or on asymptotic results, and are therefore applicable to any probability model. This is not to suggest that the tests are automatic; the choice of test quantity and appropriate predictive distribution requires careful consideration of the type of inferences required for the problem being considered.

6.6 Model expansion

Sensitivity analysis

In general, the posterior distribution of the model parameters can either overestimate or underestimate different aspects of 'true' posterior uncertainty. The posterior distribution typically overestimates uncertainty in the sense that one does not, in general, include all of one's substantive knowledge in the model; hence the utility of checking the model against one's substantive knowledge. On the other hand, the posterior distribution underestimates uncertainty in two senses: first, the assumed model is almost certainly wrong—hence the need for posterior model checking against the observed data—and second, other reasonable models could have fit the observed data equally well, hence the need for sensitivity analysis. We have already addressed model checking. In this section, we consider the uncertainty in posterior inferences due to the existence of reasonable alternative models and discuss how to expand the model to account for this uncertainty. Alternative models can differ in the specification of the prior distribution, in the specification of the likelihood, or both. Model checking and sensitivity analysis go together: when conducting sensitivity analysis, it is only necessary to consider models that fit substantive knowledge and observed data in relevant ways.

The basic method of sensitivity analysis is to fit several probability models to the same problem. It is often possible to avoid surprises in sensitivity analyses by replacing improper prior distributions with proper distributions that represent substantive prior knowledge. In addition, different questions are differently affected by model changes. Naturally, posterior inferences concerning medians of posterior distributions are generally less sensitive to changes in the model than inferences about means or extreme quantiles. Similarly, predictive inferences about quantities that are most like the observed data are most reliable; for example, in a regression model, interpolation is typically less sensitive to linearity assumptions than extrapolation. It is sometimes possible to perform a sensitivity analysis by using 'robust' models, which ensure that unusual observations (or larger units of analysis in a hierarchical model) do not exert an undue influence on inferences. The typical example is the use of the Student-t distribution in place of the normal (either for the sampling or the population distribution). Such models can be quite useful but require more computational effort. We consider robust models in Chapter 17.

Adding parameters to a model

There are several possible reasons to expand a model:

1. If the model does not fit the data or prior knowledge in some important

way, it should be altered in some way, possibly by adding enough new parameters to allow a better fit.

2. If a modeling assumption is questionable or has no real justification, one can broaden the class of models (for example, replacing a normal by a Student-t, as we do in Section 17.4 for the SAT coaching example).

3. If two different models, $p_1(y, \theta)$ and $p_2(y, \theta)$, are under consideration, they can be combined into a larger model using a continuous parameterization that includes the original models as special cases. For example, the hierarchical model for SAT coaching in Chapter 5 is a continuous generalization of the complete-pooling ($\tau = 0$) and no-pooling ($\tau = \infty$) models.

4. A model can be expanded to include new data; for example, an experiment previously analyzed on its own can be inserted into a hierarchical population model. Another common example is expanding a regression model of $y|x$ to a multivariate model of (x, y) in order to model missing data in x (see Chapter 21).

All these applications of model expansion have the same mathematical structure: the old model, $p(y, \theta)$, is embedded in or replaced by a new model, $p(y, \theta, \phi)$ or, more generally, $p(y, y^*, \theta, \phi)$, where y^* represents the added data.

The joint posterior distribution of the new parameters, ϕ, and the parameters θ of the old model is,

$$p(\theta, \phi|y, y^*) \propto p(\phi)p(\theta|\phi)p(y, y^*|\theta, \phi).$$

The conditional prior distribution, $p(\theta|\phi)$, and the likelihood, $p(y, y^*|\theta, \phi)$, are determined by the expanded family. The marginal distribution of ϕ is obtained by averaging over θ:

$$p(\phi|y, y^*) \propto p(\phi) \int p(\theta|\phi)p(y, y^*|\theta, \phi)d\theta. \tag{6.5}$$

In any expansion of a Bayesian model, one must specify a set of prior distributions, $p(\theta|\phi)$, to replace the old $p(\theta)$, and also a hyperprior distribution $p(\phi)$ on the hyperparameters. Both tasks typically require thought, especially with noninformative prior distributions (see Exercises 6.12 and 6.13). For example, Section 14.7 discusses a model for unequal variances that includes unweighted and weighted linear regression as extreme cases. In Section 17.4, we illustrate the task of expanding the normal model for the SAT coaching example of Section 5.5 to a Student-t model by including the degrees of freedom of the t distribution as an additional hyperparameter. Another detailed example of model expansion appears in Section 18.4, for a hierarchical mixture model applied to data from an experiment in psychology.

Practical advice for model checking and expansion

It is difficult to give appropriate general advice for model choice; as with model building, scientific judgment is required, and approaches must vary with context.

Our recommended approach, for both model checking and sensitivity analysis, is to examine posterior distributions of substantively important parameters and predicted quantities. Then we compare posterior distributions and posterior predictions with substantive knowledge, including the observed data, and note where the predictions fail. Discrepancies should be used to suggest possible expansions of the model, perhaps as simple as putting real prior information into the prior distribution or adding a parameter such as a nonlinear term in a regression, or perhaps requiring some substantive rethinking, as for the poor prediction of the Southern states in the Presidential election model as displayed in Figure 6.1 on page 159.

Sometimes a model has stronger assumptions than are immediately apparent. For example, a regression with many predictors and a flat prior distribution on the coefficients will tend to overestimate the variation among the coefficients, just as the independent estimates for the eight schools were more spread than appropriate. If we find that the model does not fit for its intended purposes, we are obliged to search for a new model that fits; an analysis is rarely, if ever, complete with simply a rejection of some model.

If a sensitivity analysis reveals problems, the basic solution is to include the other plausible models in the prior specification, thereby forming a posterior inference that reflects uncertainty in the model specification, or simply to report sensitivity to assumptions untestable by the data at hand. Of course, one must sometimes conclude that, for practical purposes, available data cannot effectively answer some questions. In other cases, it is possible to add information to constrain the model enough to allow useful inferences; Section 9.3 presents an example in the context of a simple random sample from a nonnormal population, in which the quantity of interest is the population total.

6.7 Model comparison

There are generally many options in setting up a model for any applied problem. Our usual approach is to start with a simple model that uses only some of the available information—for example, not using some possible predictors in a regression, fitting a normal model to discrete data, or ignoring evidence of unequal variances and fitting a simple equal-variance model. Once we have successfully fitted a simple model, we can check its fit to data (as discussed in Sections 6.3–6.5) and then expand it (as discussed in Section 6.6).

There are two typical scenarios in which models are compared. First, when a model is expanded, it is natural to compare the smaller to the larger model and assess what has been gained by expanding the model (or, conversely, if a model is simplified, to assess what was lost). This generalizes into the problem of comparing a set of nested models and judging how much complexity is necessary to fit the data.

In comparing nested models, the larger model typically has the advantage of making more sense and fitting the data better but the disadvantage of being more difficult to understand and compute. The key questions of model comparison are typically: (1) is the improvement in fit large enough to justify the additional difficulty in fitting, and (2) is the prior distribution on the additional parameters reasonable?

The statistical theory of hypothesis testing is associated with methods for checking whether an improvement in fit is statistically significant—that is, whether it could be expected to occur by chance, even if the smaller model were correct. Bayes factors (see page 184) are sometimes used to make these model comparisons, but we find them generally to be irrelevant because they compute the relative probabilities of the models conditional on one of them being true. We prefer approaches that measure the distance of the data to each of the approximate models. Let θ be the vector of parameters in the smaller model and ψ be the additional parameters in the expanded model. Then we are comparing the two posterior distributions, $p(\theta|y)$ and $p(\theta, \psi|y)$, along with their predictive distributions for replicated data y^{rep}.

The second scenario of model comparison is between two or more nonnested models—neither model generalizes the other. In a regression context, one might compare regressions with completely different sets of predictors, for example, modeling political behavior using information based on past voting results or on demographics. In these settings, we are typically not interested in *choosing* one of the models—it would be better, both in substantive and predictive terms, to construct a larger model that includes both as special cases, including both sets of predictors and also potential interactions in a larger regression, possibly with an informative prior distribution if needed to control the estimation of all the extra parameters. However, it can be useful to *compare* the fit of the different models, to see how either set of predictors performs when considered alone.

Expected deviance as a measure of predictive accuracy

We first introduce some measures of prediction error and then discuss how they can be used to compare the performance of different models. We have already discussed discrepancy measures in Section 6.5 applied to checking whether data fit as well as could be expected under the model. Here, however, we are comparing the data to two (or more) different models and seeing which predicts with more accuracy. Even if none (or all) of the models fit the data, it can be informative to compare their relative fit.

Model fit can be summarized numerically by a measure such as weighted mean squared error as in (6.4): $T(y, \theta) = \frac{1}{n} \sum_{i=1}^{n} (y_i - \text{E}(y_i|\theta))^2/\text{var}(y_i|\theta)$. A more general option is to work with the 'deviance,' which is defined as -2 times the log-likelihood:

$$\text{deviance:} \quad D(y, \theta) = -2 \log p(y|\theta), \qquad (6.6)$$

and is proportional to the mean squared error if the model is normal with constant variance. The deviance has an important role in statistical model comparison because of its connection to the Kullback-Leibler information measure $H(\theta)$ defined in (B.1) on page 586. The expected deviance—computed by averaging (6.6) over the true sampling distribution $f(y)$—equals 2 times the Kullback-Leibler information, up to a fixed constant, $\int f(y) \log f(y) dy$, that does not depend on θ. As discussed in Appendix B, in the limit of large sample sizes, the model with the lowest Kullback-Leibler information—and thus, the lowest expected deviance—will have the highest posterior probability. Thus, it seems reasonable to estimate expected deviance as a measure of overall model fit. More generally, we could measure lack of fit using any discrepancy measure D, but the deviance is a standard summary.

The discrepancy between data and model depends in general on θ as well as y. To get a summary that depends only on y, one could define

$$D_{\hat{\theta}}(y) = D(y, \hat{\theta}(y)) \tag{6.7}$$

and use a point estimate for θ such as the mean of the posterior simulations. From a Bayesian perspective, it is perhaps more appealing to average the discrepancy itself over the posterior distribution:

$$D_{\text{avg}}(y) = \text{E}(D(y, \theta)|y), \tag{6.8}$$

which would be estimated using posterior simulations θ_l:

$$\widehat{D}_{\text{avg}}(y) = \frac{1}{L} \sum_{l=1}^{L} D(y, \theta^l). \tag{6.9}$$

The estimated average discrepancy (6.9) is a better summary of model error than the discrepancy (6.7) of the point estimate; the point estimate generally makes the model fit better than it really does, and $\widehat{D}_{\text{avg}}$ averages over the range of possible parameter values. This can be thought of as a special case of the numerical posterior predictive checks discussed in Section 6.5, but with a focus on comparing discrepancies between models rather than checking model fit.

Counting parameters and model complexity

The difference between the posterior mean deviance (6.9) and the deviance at $\hat{\theta}$ (6.7),

$$p_D^{(1)} = \widehat{D}_{\text{avg}}(y) - D_{\hat{\theta}}(y), \tag{6.10}$$

represents the effect of model fitting and has been used as a measure of the *effective number of parameters* of a Bayesian model. For a normal linear model with unconstrained parameters (or, equivalently, for a fixed model with large sample size; see Chapter 4), $p_D^{(1)}$ is equal to the number of parameters in the model.

A related way to measure model complexity is as half the posterior variance of the deviance, which can be estimated from the posterior simulations:

$$p_D^{(2)} = \frac{1}{2}\widehat{\mathrm{var}}(D(y,\theta)|y) = \frac{1}{2}\frac{1}{L-1}\sum_{l=1}^{L}(D(y,\theta^l) - \hat{D}_{\mathrm{avg}}(y))^2.$$

Both measures $p_D^{(1)}$ and $p_D^{(2)}$ can be derived from the asymptotic χ^2 distribution of the deviance relative to its minimum, with a mean equal to the number of parameters estimated in the model and variance equal to twice the mean.

More generally, p_D can be thought of as the number of 'unconstrained' parameters in the model, where a parameter counts as: 1 if it is estimated with no constraints or prior information; 0 if it is fully constrained or if all the information about the parameter comes from the prior distribution; or an intermediate value if both the data and prior distributions are informative.

It makes sense that both definitions of p_D depend on the data, y. For a simple example, consider the model $y \sim N(\theta, 1)$, with $\theta \sim U(0, \infty)$. That is, θ is constrained to be positive but otherwise has a noninformative uniform prior distribution. How many parameters are being estimated by this model? If the measurement y is close to zero, then the effective number of parameters p is approximately $1/2$, since roughly half the information in the posterior distribution is coming from the data and half from the prior constraint of positivity. However, if y is large, then the constraint is essentially irrelevant, and the effective number of parameters is approximately 1.

For hierarchical models, the effective number of parameters strongly depends on the variance of the group-level parameters. We shall illustrate with the example from Chapter 5 of the educational testing experiments in 8 schools. Under the hierarchical model, the effective number of parameters falls between 8 (one for each school) and 1 (for the mean of all the schools).

Estimated predictive discrepancy and the deviance information criterion

So far, we have defined the mean deviance $D_{\mathrm{avg}}(y)$ of the fitted model to the data, along with p_D, the effective number of parameters being fitted. From (6.10), p_D represents the decrease in the deviance (that is, the expected improvement in the fit) expected from estimating the parameters in the model.

A related approach is to estimate the error that would be expected when applying the fitted model to future data, for example the expected mean squared predictive error, $D_{\mathrm{avg}}^{\mathrm{pred}}(y) = \mathrm{E}\left[\frac{1}{n}\sum_{i=1}^{n}(y_i^{\mathrm{rep}} - \mathrm{E}(y_i^{\mathrm{rep}}|y))^2\right]$, where the expectation averages over the posterior predictive distribution of replicated data y^{rep}. More generally, one can compute the expected deviance for replicated data:

$$D_{\mathrm{avg}}^{\mathrm{pred}}(y) = \mathrm{E}[D(y^{\mathrm{rep}}, \hat{\theta}(y))], \qquad (6.11)$$

where $D(y^{\mathrm{rep}}, \theta) = -2\log p(y^{\mathrm{rep}}|\theta)$, the expectation averages over the distribution of y_{rep} under the unknown true sampling model, and $\hat{\theta}$ is a parameter estimate such as the posterior mean (or, more generally, the estimate that minimizes the expected deviance for replicated data). This use of a point estimate

Model	$D_{\hat{\theta}}$	$\widehat{D}_{\mathrm{avg}}$	p_D	DIC
no pooling ($\tau = \infty$)	54.9	62.6	7.7	70.3
complete pooling ($\tau = 0$)	59.5	60.5	1.0	61.5
hierarchical (τ unknown)	57.8	60.6	2.8	63.4

Table 6.2 *Point-estimate and average deviances, estimated number of parameters, and DIC for each of three models fitted to the SAT coaching experiments in Section 5.5. Lower values of deviance represent better fit to data. The no-pooling model has the best-fitting point estimate $\hat{\theta}$, the complete-pooling and hierarchical models have the best average fit to data D_{avg}, and the complete-pooling model has lowest estimated expected predictive error DIC.*

$\hat{\theta}$ departs from our usual practice of averaging over the posterior distribution. In general, the expected predictive deviance (6.11) will be higher than the expected deviance $\widehat{D}_{\mathrm{avg}}$ defined in (6.9) because the predicted data y^{rep} are being compared to a model estimated from data y.

The expected predictive deviance (6.11) has been suggested as a criterion of model fit when the goal is to pick a model with best out-of-sample predictive power. It can be approximately estimated by an expression called the *deviance information criterion* (DIC):

$$\mathrm{DIC} = \widehat{D}_{\mathrm{avg}}^{\mathrm{pred}}(y) = 2\widehat{D}_{\mathrm{avg}}(y) - D_{\hat{\theta}}(y), \qquad (6.12)$$

with $D_{\hat{\theta}}(y)$ and $\widehat{D}_{\mathrm{avg}}(y)$ as defined in (6.7) and (6.9). Expression (6.12) can be derived for normal models or in the limit of large sample sizes as in Chapter 4; see Exercise 6.8.

Example. Deviance for models of the educational testing experiments
Table 6.2 illustrates the use of deviance to compare the three models—no pooling, complete pooling, and hierarchical—fitted to the SAT coaching data in Section 5.5. For this model, the deviance is simply

$$\begin{aligned} D(y, \theta) &= -2 \sum_{j=1}^{J} \log \left[\mathrm{N}(y_j | \theta_j, \sigma_j^2) \right] \\ &= \log(2\pi J\sigma^2) + \sum_{j=1}^{J} ((y_j - \theta_j)/\sigma_j)^2. \end{aligned}$$

The first column, with $D_{\hat{\theta}}$, compares the best fits from 200 simulations for each model. The no-pooling has the best fit based on the point estimate $\hat{\theta}$, which makes sense since it allows all 8 parameters θ to be independently fitted by the data. The average discrepancy $\widehat{D}_{\mathrm{avg}}$ of the data to the parameters as summarized by the posterior distribution of the no-pooling model is 62.6. There is no easy interpretation of this level, but the difference between the average discrepancy and that of the point estimate is the expected number of parameters, p_D, which is estimated at 7.7. This differs from the exact value of 8 because of simulation variability. Finally, DIC= $62.6 + 7.7 = 70.3$ is the estimated expected discrepancy of the fitted unpooled model to replicated data.

The complete-pooling model, which assumes identical effects for all 8 schools, behaves much differently. The deviance of the point estimate is 59.5—quite a bit higher than for the no-pooling model, which makes sense, since the complete pooling model has only one parameter with which to fit the data. The average discrepancy D_{avg}, however, is lower for the complete pooling model, because there is little uncertainty in the posterior distribution. The estimated number of parameters p_D is 1.0, which makes perfect sense.

Finally, the point estimate from the hierarchical model has a deviance that is higher than the no-pooling model and lower than complete pooling, which makes sense since the hierarchical model has 8 parameters with which to fit the data, but these parameters are constrained by the prior distribution. The expected discrepancy D_{avg} is about the same as under complete pooling, and the estimated number of parameters is 2.8—closer to 1 than to 8, which makes sense, considering that for these data, τ is estimated to be low and the parameter estimates θ_j are shrunk strongly toward their common mean (see Figure 5.6). Finally, DIC for this model is 63.4, which is better than the 70.3 for the no-pooling model but worse than the 61.5 for complete pooling as applied to these data. Thus, we would expect the complete-pooling model to do the best job—in terms of log-likelihood—of predicting future data.

We would probably still prefer to use the hierarchical model, for the reasons discussed in Section 5.5—basically, we do not want to make the strong assumption that all coaching programs are identical—but for this particular dataset, the variance inherent in estimating the eight θ_j values, with a uniform hyperprior distribution on τ, does seem to add appreciable noise to predictions (of results for new schools).

In general, the predictive accuracy measures are useful in parallel with posterior predictive checks to see if there are important patterns in the data that are not captured by each model. As with predictive checking, DIC can be computed in different ways for a hierarchical model depending on whether the parameter estimates $\hat{\theta}$ and replications y^{rep} correspond to estimates and replications of new data from the existing groups (as we have performed the calculations in the above example) or new groups (additional schools from the $N(\mu, \tau^2)$ distribution in the above example).

Comparing a discrete set of models using Bayes factors

In a problem in which a discrete set of competing models is proposed, the term *Bayes factor* is sometimes used for the ratio of the marginal likelihood under one model to the marginal likelihood under a second model. If we label two competing models as H_1 and H_2, then the ratio of their posterior probabilities is

$$\frac{p(H_2|y)}{p(H_1|y)} = \frac{p(H_2)}{p(H_1)} \times \text{Bayes factor}(H_2; H_1),$$

where

$$\text{Bayes factor}(H_2; H_1) = \frac{p(y|H_2)}{p(y|H_1)} = \frac{\int p(\theta_2|H_2)p(y|\theta_2, H_2)d\theta_2}{\int p(\theta_1|H_1)p(y|\theta_1, H_1)d\theta_1}. \qquad (6.13)$$

In many cases, the competing models have a common set of parameters, but this is not necessary; hence the notation θ_i for the parameters in model H_i. As expression (6.13) makes clear, the Bayes factor is only defined when the marginal density of y under each model is proper.

The goal when using Bayes factors is to choose a single model H_i or average over a discrete set using their posterior distributions, $p(H_i|y)$. As we show by our examples in this book, we generally prefer to replace a discrete set of models with an expanded continuous family of models. To illustrate this, we consider two examples: one in which the Bayes factor is helpful and one in which it is not. The bibliographic note at the end of the chapter provides pointers to more extensive treatments of Bayes factors.

Example. An example in which Bayes factors are helpful

The Bayesian inference for the genetics example in Section 1.4 can be fruitfully reintroduced in terms of Bayes factors, with the two competing 'models' being H_1: the woman is affected, and H_2: the woman is unaffected, that is, $\theta = 1$ and $\theta = 0$ in the notation of Section 1.4. The prior odds are $p(H_2)/p(H_1) = 1$, and the Bayes factor of the data that the woman has two unaffected sons is $p(y|H_2)/p(y|H_1) = 1.0/0.25$. The posterior odds are thus $p(H_2|y)/p(H_1|y) = 4$. Computation by multiplying odds ratios makes the accumulation of evidence clear.

This example has two features that allow Bayes factors to be helpful. First, each of the discrete alternatives makes scientific sense, and there are no obvious scientific models in between. Second, the marginal distribution of the data under each model, $p(y|H_i)$, is proper.

Example. An example in which Bayes factors are a distraction

We now consider an example in which discrete model comparisons and Bayes factors are a distraction from scientific inference. Suppose we analyzed the SAT coaching experiments in Section 5.5 using Bayes factors for the discrete collection of previously proposed standard models, no pooling (H_1) and complete pooling (H_2):

$$H_1 : p(y|\theta_1, \ldots, \theta_J) = \prod_{j=1}^{J} N(y_j|\theta_j, \sigma_j^2),\ p(\theta_1, \ldots, \theta_J) \propto 1$$

$$H_2 : p(y|\theta_1, \ldots, \theta_J) = \prod_{j=1}^{J} N(y_j|\theta_j, \sigma_j^2),\ \theta_1 = \cdots = \theta_J = \theta, p(\theta) \propto 1.$$

(Recall that the standard deviations σ_j are assumed known in this example.)

If we use Bayes factors to choose or average among these models, we are immediately confronted with the fact that the Bayes factor—the ratio $p(y|H_1)/p(y|H_2)$—is not defined; because the prior distributions are improper, the ratio of density functions is $0/0$. Consequently, if we wish to continue with the approach of assigning posterior probabilities to these two discrete models, we must consider (1) proper prior distributions, or (2) improper prior distributions that are carefully constructed as limits of proper distributions. In either case, we shall see that the results are unsatisfactory.

More explicitly, suppose we replace the flat prior distributions in H_1 and H_2 by independent normal prior distributions, $N(0, A^2)$, for some large A. The resulting posterior distribution for the effect in school j is

$$p(\theta_j|y) = (1 - \lambda)p(\theta_j|y, H_1) + \lambda p(\theta_j|y, H_2),$$

where the two conditional posterior distributions are normal centered near y_j and $\bar{y}$, respectively, and λ is proportional to the prior odds times the Bayes factor, which is a function of the data and A (see Exercise 6.9). The Bayes factor for this problem is highly sensitive to the prior variance, A^2; as A increases (with fixed data and fixed prior odds, $p(H_2)/p(H_1)$) the posterior distribution becomes more and more concentrated on H_2, the complete pooling model. Therefore, the Bayes factor cannot be reasonably applied to the original models with noninformative prior densities, even if they are carefully defined as limits of proper prior distributions.

Yet another problem with the Bayes factor for this example is revealed by considering its behavior as the number of schools being fitted to the model increases. The posterior distribution for θ_j under the mixture of H_1 and H_2 turns out to be sensitive to the dimensionality of the problem, as very different inferences would be obtained if, for example, the model were applied to similar data on 80 schools (see Exercise 6.9). It makes no scientific sense for the posterior distribution to be highly sensitive to aspects of the prior distributions and problem structure that are scientifically incidental.

Thus, if we were to use a Bayes factor for this problem, we would find a problem in the model-checking stage (a discrepancy between posterior distribution and substantive knowledge), and we would be moved toward setting up a smoother, continuous family of models to bridge the gap between the two extremes. A reasonable continuous family of models is $y_j \sim N(\theta_j, \sigma_j^2)$, $\theta_j \sim N(\mu, \tau^2)$, with a flat prior distribution on μ, and τ in the range $[0, \infty)$; this, of course, is the model we used in Section 5.5. Once the continuous expanded model is fitted, there is no reason to assign discrete positive probabilities to the values $\tau = 0$ and $\tau = \infty$, considering that neither makes scientific sense.

6.8 Model checking for the educational testing example

We illustrate the ideas discussed in this chapter with the SAT coaching example introduced in Section 5.5.

Assumptions of the model

The posterior inference presented for the educational testing example is based on several model assumptions: (1) the normality of the estimates y_j given θ_j and σ_j, where the values σ_j are assumed known; (2) the exchangeability of the prior distribution of the θ_j's; (3) the normality of the prior distribution of each θ_j given μ and τ; and (4) the uniformity of the hyperprior distribution of (μ, τ).

The assumption of normality with a known variance is made routinely when a study is summarized by its estimated effect and standard error. The design (randomization, reasonably large sample sizes, adjustment for scores on earlier tests) and analysis (for example, the raw data of individual test scores were checked for outliers in an earlier analysis) were such that the assumptions seem justifiable in this case.

The second modeling assumption deserves commentary. The real-world interpretation of the mathematical assumption of exchangeability of the θ_j's is that before seeing the results of the experiments, there is no desire to include in the model features such as a belief that (a) the effect in school A is probably larger than in school B or (b) the effects in schools A and B are more similar than in schools A and C. In other words, the exchangeability assumption means that we will let the data tell us about the relative ordering and similarity of effects in the schools. Such a prior stance seems reasonable when the results of eight parallel experiments are being scientifically summarized for general presentation. Of course, generally accepted information concerning the effectiveness of the programs or differences among the schools might suggest a nonexchangeable prior distribution if, for example, schools B and C have similar students and schools A, D, E, F, G, H have similar students. Unusual types of detailed prior knowledge (for example, two schools are very similar but we do not know which schools they are) can suggest an exchangeable prior distribution that is not a mixture of iid components. In the absence of any such information, the exchangeability assumption implies that the prior distribution of the θ_j's can be considered as independent samples from a population whose distribution is indexed by some hyperparameters—in our model, (μ, τ)—that have their own hyperprior distribution.

The third and fourth modeling assumptions are harder to justify *a priori* than the first two. Why should the school effects be normally distributed rather than say, Cauchy distributed, or even asymmetrically distributed, and why should the location and scale parameters of this prior distribution be uniformly distributed? Mathematical tractability is one reason for the choice of models, but if the family of probability models is inappropriate, Bayesian answers can be quite misleading.

Comparing posterior inferences to substantive knowledge

Inference about the parameters in the model. When checking the model assumptions, our first step is to compare the posterior distribution of effects to our knowledge of educational testing. The estimated treatment effects (the posterior means) for the eight schools range from 5 to 10 points, which are plausible values. (The SAT-V is scored on a scale from 200 to 800.) The effect in school A could be as high as 31 points or as low as −2 points (a 95% posterior interval). Either of these extremes seems plausible. We could look at other summaries as well, but it seems clear that the posterior estimates of the parameters do not violate our common sense or our limited substantive knowledge about SAT preparation courses.

Inference about predicted values. Next, we simulate the posterior predictive distribution of a hypothetical replication of the experiments. Computationally, drawing from the posterior predictive distribution is nearly effortless given all that we have done so far: from each of the 200 simulations from the posterior distribution of (θ, μ, τ), we simulate a hypothetical replicated dataset, $y^{\text{rep}} = (y_1^{\text{rep}}, \ldots, y_8^{\text{rep}})$, by drawing each y_j^{rep} from a normal distribution with mean θ_j and standard deviation σ_j. The resulting set of 200 vectors y^{rep} summarizes the posterior predictive distribution. (Recall from Section 5.5 that we are treating y—the eight separate estimates—as the 'raw data' from the eight experiments.)

The model-generated values for each school in each of the 200 replications are all plausible outcomes of experiments on coaching. The smallest hypothetical observation generated was -48, and the largest was 63; because both values are possible for estimated effects from studies of coaching for the SAT-V, all estimated values generated by the model are credible.

Posterior predictive checking

But does the model fit the data? If not, we may have cause to doubt the inferences obtained under the model such as displayed in Figure 5.8 and Table 5.3. For instance, is the largest observed outcome, 28 points, consistent with the posterior predictive distribution under the model? Suppose we perform 200 posterior predictive simulations of the SAT coaching experiments and compute the largest observed outcome, $\max_j y_j^{\text{rep}}$, for each. If all 200 of these simulations lie below 28 points, then the model does not fit this important aspect of the data, and we might suspect that the normal-based inference in Section 5.5 shrinks the effect in School A too far.

In order to test the fit of the model to the observed data, we examine the posterior predictive distribution of the following test statistics: the largest of the eight observed outcomes, $\max_j y_j$, the smallest, $\min_j y_j$, the average, $\text{mean}(y_j)$, and the sample standard deviation, $\text{sd}(y_j)$. We approximate the posterior predictive distribution of each test statistic by the histogram of the values from the 200 simulations of the parameters and predictive data, and we compare each distribution to the observed value of the test statistic and our substantive knowledge of SAT coaching programs. The results are displayed in Figure 6.12.

The summaries suggest that the model generates predicted results similar to the observed data in the study; that is, the actual observations are typical of the predicted observations generated by the model.

Of course, there are many other functions of the posterior predictive distribution that could be examined, such as the differences between individual values of y_j^{rep}. Or, if we had a particular skewed nonnormal prior distribution in mind for the effects θ_j, we could construct a test quantity based on the skewness or asymmetry of the simulated predictive data as a check on whether the normal model is adequate. Often in practice we can obtain diagnostically useful displays directly from intuitively interesting quantities without having to supply a specific alternative model.

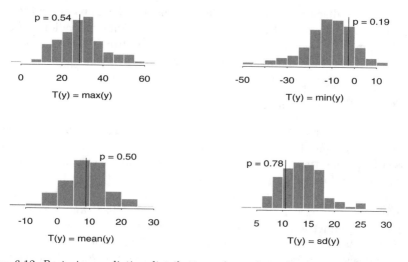

Figure 6.12 *Posterior predictive distribution, observed result, and p-value for each of four test statistics for the educational testing example.*

Sensitivity analysis

The model checks seem to support the posterior inferences for the SAT coaching example. Although we may feel confident that the data do not contradict the model, this is not enough to inspire complete confidence in our general substantive conclusions, because other reasonable models might provide just as good a fit but lead to different conclusions. Sensitivity analysis can then be used to assess the effect of alternative analyses on the posterior inferences.

The uniform prior distribution for τ. To assess the sensitivity to the prior distribution for τ we consider Figure 5.5, the graph of the marginal posterior density, $p(\tau|y)$, obtained under the assumption of a uniform prior density for τ on the positive half of the real line. One can obtain the posterior density for τ given other choices of the prior distribution by multiplying the density displayed in Figure 5.5 by the prior density. There will be little change in the posterior inferences as long as the prior density is not sharply peaked and does not put a great deal of probability mass on values of τ greater than 10.

The normal population distribution for the school effects. The normal distribution assumption on the θ_j's is made for computational convenience, as is often the case. A natural sensitivity analysis is to consider longer-tailed alternatives, such as the Student-t, as a check on robustness. We defer the details of this analysis to Section 17.4, after the required computational techniques have been presented. Any alternative model must be examined to ensure that the predictive distributions are restricted to realistic SAT improvements.

The normal likelihood. As discussed earlier, the assumption of normal data conditional on the means and standard deviations need not and cannot be seriously challenged in this example. The justification is based on the central limit theorem and the designs of the studies. Assessing the validity of this assumption would require access to the original data from the eight experiments, not just the estimates and standard errors given in Table 5.2.

6.9 Bibliographic note

The posterior predictive approach to model checking described here was presented in Rubin (1981, 1984). Gelman, Meng, and Stern (1996) discuss the use of test quantities that depend on parameters as well as data; related ideas appear in Zellner (1976) and Tsui and Weerahandi (1989). Rubin and Stern (1994) and Raghunathan (1994) provide further applied examples. The examples in Section 6.4 appear in Meulders et al. (1998) and Gelman (2003). The antisymmetric discrepancy measures discussed in Section 6.5 appear in Berkhof, Van Mechelen, and Gelman (2003a). The adolescent smoking example appears in Carlin et al. (2001). Sinharay and Stern (2003) discuss posterior predictive checks for hierarchical models, focusing on the SAT coaching example. Johnson (2002) discusses Bayesian χ^2 tests as well as the idea of using predictive checks as a debugging tool, as discussed in Section 10.3.

Model checking using simulation has a long history in statistics; for example, Bush and Mosteller (1955, p. 252) check the fit of a model by comparing observed data to a set of simulated data. Their method differs from posterior predictive checking only in that their model parameters were fixed at point estimates for the simulations rather than being drawn from a posterior distribution. Ripley (1988) applies this idea repeatedly to examine the fits of models for spatial data. Early theoretical papers featuring ideas related to Bayesian posterior predictive checks include Guttman (1967) and Dempster (1971). Bernardo and Smith (1994) discuss methods of comparing models based on predictive errors.

A related approach to model checking is *cross-validation*, in which observed data are partitioned, with each part of the data compared to its predictions conditional on the model and the rest of the data. Some references to Bayesian approaches to cross-validation include Stone (1974), Geisser and Eddy (1979), and Gelfand, Dey, and Chang (1992). Geisser (1986) discusses predictive inference and model checking in general, and Barbieri and Berger (2002) discuss Bayesian predictive model selection.

Box (1980, 1983) has contributed a wide-ranging discussion of model checking ('model criticism' in his terminology), including a consideration of why it is needed in addition to model expansion and averaging. Box proposed checking models by comparing data to the *prior predictive distribution*; in the notation of our Section 6.3, defining replications with distribution $p(y^{\mathrm{rep}}) = \int p(y^{\mathrm{rep}}|\theta)p(\theta)d\theta$. This approach has quite different implications for model checking; for example, with an improper prior distribution on θ, the prior predictive distribution is itself improper and thus the check is not generally defined, even if the posterior distribution is proper (see Exercise 6.7).

Box was also an early contributor to the literature on sensitivity analysis and robustness in standard models based on normal distributions: see Box and Tiao (1962, 1973).

Various theoretical studies have been performed on Bayesian robustness and sensitivity analysis examining the question of how posterior inferences are affected by prior assumptions; see Leamer (1978b), McCulloch (1989), Wasserman (1992), and the references at the end of Chapter 17. Kass and coworkers have developed methods based on Laplace's approximation for approximate sensitivity analysis: for example, see Kass, Tierney, and Kadane (1989) and Kass and Vaidyanathan (1992).

Nelder and Wedderburn (1972) explore the deviance as a measure of model fit, Akaike (1973) introduce the expected predictive deviance and the AIC, and Mallows (1973) derives the related C_p measure. Bayesian treatments of expected predictive errors for model comparison include Dempster (1974), Laud and Ibrahim (1995), and Gelfand and Ghosh (1998). Hansen and Yu (2001) review related ideas from an information-theoretic perspective.

The deviance information criterion (DIC) and its calculation using posterior simulations are described by Spiegelhalter et al. (2002) and is implemented in the software package Bugs; see Spiegelhalter et al. (1994, 2003). Burnham and Anderson (2002) discuss and motivate the use of the Kullback-Leibler information for model comparison, which relates to the log-likelihood deviance function used in determining DIC. The topic of counting parameters in nonlinear, constrained, and hierarchical models is discussed by Hastie and Tibshirani (1990), Gelman, Meng, and Stern (1996), Hodges and Sargent (2001), and Vaida and Blanchard (2002). The last paper discusses the different ways that information criteria can be computed in hierarchical models.

A comprehensive overview of the use of Bayes factors for comparing models and testing scientific hypotheses is given by Kass and Raftery (1995), which contains many further references in this area. Carlin and Chib (1993) discuss the problem of averaging over models that have incompatible parameterizations. Chib (1995) and Chib and Jeliazkov (2001) describe approaches for calculating the marginal likelihoods required for Bayes factors from iterative simulation output (as produced by the methods described in Chapter 11). Pauler, Wakefield, and Kass (1999) discuss Bayes factors for hierarchical models. Weiss (1996) considers the use of Bayes factors for sensitivity analysis.

Bayes factors are not defined for models with improper prior distributions, but there have been several attempts to define analogous quantities; see Spiegelhalter and Smith (1982) and Kass and Raftery (1995). A related proposal is to treat Bayes factors as posterior probabilities and then average over competing models—see Raftery (1996) for a theoretical treatment, Rosenkranz and Raftery (1994) for an application, and Hoeting et al. (1999) and Chipman, George, and McCulloch (2001) for reviews.

A variety of views on model selection and averaging appear in the articles by Draper (1995) and O'Hagan (1995) and the accompanying discussions. We refer the reader to these articles and their references for further discussion and examples of these methods. Because we emphasize continuous *families* of models rather than discrete *choices*, Bayes factors are rarely relevant in our approach to Bayesian statistics; see Raftery (1995) and Gelman and Rubin (1995) for two contrasting views on this point.

There are many examples of applied Bayesian analyses in which sensitivity to the model has been examined, for example Racine et al. (1986), Weiss (1994), and Smith, Spiegelhalter, and Thomas (1995).

Finally, many model checking methods in common practical use, including tests for outliers, plots of residuals, and normal plots, can be interpreted as Bayesian posterior predictive checks, where the practitioner is looking for discrepancies from the expected results under the assumed model (see Gelman, 2003, for an extended discussion of this point). Many non-Bayesian treatments of graphical model checking appear in the statistical literature, for example, Atkinson (1985). Tukey (1977) presents a graphical approach to data analysis that is, in our opinion, fundamentally based on model checking (see Gelman, 2003). The books by Cleveland (1985, 1993) and Tufte (1983, 1990) present many useful ideas for displaying data graphically; these ideas are fundamental to the graphical model checks described in Section 6.4.

Calvin and Sedransk (1991) provide an interesting example comparing various Bayesian and non-Bayesian methods of model checking and expansion.

6.10 Exercises

1. Posterior predictive checking:

    (a) On page 140, the data from the SAT coaching experiments were checked against the model that assumed identical effects in all eight schools: the expected order statistics of the effect sizes were (26, 19, 14, 10, 6, 2, −3, −9), compared to observed data of (28, 18, 12, 7, 3, 1, −1, −3). Express this comparison formally as a posterior predictive check comparing this model to the data. Does the model fit the aspect of the data tested here?

    (b) Explain why, even though the identical-schools model fits under this test, it is still unacceptable for some practical purposes.

2. Model checking: in Exercise 2.13, the counts of airline fatalities in 1976–1985 were fitted to four different Poisson models.

    (a) For each of the models, set up posterior predictive test quantities to check the following assumptions: (1) independent Poisson distributions, (2) no trend over time.

    (b) For each of the models, use simulations from the posterior predictive distributions to measure the discrepancies. Display the discrepancies graphically and give p-values.

    (c) Do the results of the posterior predictive checks agree with your answers in Exercise 2.13(e)?

3. Model improvement:

(a) Use the solution to the previous problem and your substantive knowledge to construct an improved model for airline fatalities.

(b) Fit the new model to the airline fatality data.

(c) Use your new model to forecast the airline fatalities in 1986. How does this differ from the forecasts from the previous models?

(d) Check the new model using the same posterior predictive checks as you used in the previous models. Does the new model fit better?

4. Model checking and sensitivity analysis: find a published Bayesian data analysis from the statistical literature.

(a) Compare the data to posterior predictive replications of the data.

(b) Perform a sensitivity analysis by computing posterior inferences under plausible alternative models.

5. Hypothesis testing: discuss the statement, 'Null hypotheses of no difference are usually known to be false before the data are collected; when they are, their rejection or acceptance simply reflects the size of the sample and the power of the test, and is not a contribution to science' (Savage, 1957, quoted in Kish, 1965). If you agree with this statement, what does this say about the model checking discussed in this chapter?

6. Variety of predictive reference sets: in the example of binary outcomes on page 163, it is assumed that the number of measurements, n, is fixed in advance, and so the hypothetical replications under the binomial model are performed with $n = 20$. Suppose instead that the protocol for measurement is to stop once 13 zeros have appeared.

(a) Explain why the posterior distribution of the parameter θ under the assumed model does not change.

(b) Perform a posterior predictive check, using the same test quantity, $T =$ number of switches, but simulating the replications y^{rep} under the new measurement protocol. Display the predictive simulations, $T(y^{\text{rep}})$, and discuss how they differ from Figure 6.4.

7. Prior vs. posterior predictive checks (from Gelman, Meng, and Stern, 1996): consider 100 observations, $y_1, \ldots, y_n$, modeled as independent samples from a $N(\theta, 1)$ distribution with a diffuse prior distribution, say, $p(\theta) = \frac{1}{2A}$ for $\theta \in [-A, A]$ with some extremely large value of A, such as 10^5. We wish to check the model using, as a test statistic, $T(y) = \max_i |y_i|$: is the maximum absolute observed value consistent with the normal model? Consider a dataset in which $\bar{y} = 5.1$ and $T(y) = 8.1$.

(a) What is the posterior predictive distribution for y^{rep}? Make a histogram for the posterior predictive distribution of $T(y^{\text{rep}})$ and give the posterior predictive p-value for the observation $T(y) = 8.1$.

(b) The prior predictive distribution is $p(y^{\text{rep}}) = \int p(y^{\text{rep}}|\theta)p(\theta)d\theta$. (Compare to equation (6.1).) What is the prior predictive distribution for y^{rep} in this example? Roughly sketch the prior predictive distribution of $T(y^{\text{rep}})$ and give the approximate prior predictive p-value for the observation $T(y) = 8.1$.

(c) Your answers for (a) and (b) should show that the data are consistent with the posterior predictive but not the prior predictive distribution. Does this make sense? Explain.

8. Deviance information criterion: show that expression (6.12) is appropriate for normal models or in the asymptotic limit of large sample sizes (see Spiegelhalter et al., 2002, p. 604).

9. Prior and posterior predictive checks when the prior distribution is improper: on page 185, we discuss Bayes factors for comparing two extreme models for the SAT coaching example.

(a) Derive the Bayes factor, $p(H_2|y)/p(H_1|y)$, as a function of $y_1, \ldots, y_J$, $\sigma_1^2, \ldots, \sigma_J^2$, and A, for the models with $N(0, A^2)$ prior distributions.

(b) Evaluate the Bayes factor in the limit $A \to \infty$.

(c) For fixed A, evaluate the Bayes factor as the number of schools, J, increases. Assume for simplicity that $\sigma_1^2 = \ldots = \sigma_J^2 = \sigma^2$, and that the sample mean and variance of the y_j's do not change.

10. Variety of posterior predictive distributions: for the educational testing example in Section 6.8, we considered a reference set for the posterior predictive simulations in which $\theta = (\theta_1, \ldots, \theta_8)$ was fixed. This corresponds to a replication of the study with the same eight coaching programs.

(a) Consider an alternative reference set, in which (μ, τ) are fixed but θ is allowed to vary. Define a posterior predictive distribution for y^{rep} under this replication, by analogy to (6.1). What is the experimental replication that corresponds to this reference set?

(b) Consider switching from the analysis of Section 6.8 to an analysis using this alternative reference set. Would you expect the posterior predictive p-values to be less extreme, more extreme, or stay about the same? Why?

(c) Reproduce the model checks of Section 6.8 based on this posterior predictive distribution. Compare to your speculations in part (b).

11. Cross-validation and posterior predictive checks:

(a) Discuss the relation of cross-validation (see page 190) to Bayesian posterior predictive checking. Is there a Bayesian version of cross-validation?

(b) Compare the two approaches with one of the examples considered so far in the book.

12. Model expansion: consider the Student-t model, $y_i|\mu, \sigma^2, \nu \sim t_\nu(\mu, \sigma^2)$, as a generalization of the normal. Suppose that, conditional on ν, you are willing to assign a noninformative uniform prior density on $(\mu, \log \sigma)$. Construct

what you consider a noninformative joint prior density on $(\mu, \log \sigma, \nu)$, for the range $\nu \in [1, \infty)$. Address the issues raised in setting up a prior distribution for the power-transformed normal model in Exercise 6.13 below.

13. Power-transformed normal models: A natural expansion of the family of normal distributions, for all-positive data, is through power transformations, which are used in various contexts, including regression models. For simplicity, consider univariate data $y = (y_1, \ldots, y_n)$, that we wish to model as iid normal after transformation.

    Box and Cox (1964) propose the model, $y_i^{(\phi)} \sim N(\mu, \sigma^2)$, where

    $$y_i^{(\phi)} = \begin{cases} (y_i^\phi - 1)/\phi & \text{for } \phi \neq 0 \\ \log y_i & \text{for } \phi = 0. \end{cases} \tag{6.14}$$

    The parameterization in terms of $y_i^{(\phi)}$ allows a continuous family of power transformations that includes the logarithm as a special case. To perform Bayesian inference, one must set up a prior distribution for the parameters, (μ, σ, ϕ).

    (a) It seems natural to apply a prior distribution of the form $p(\mu, \log \sigma, \phi) \propto p(\phi)$, where $p(\phi)$ is a prior distribution (perhaps uniform) on ϕ alone. Unfortunately, this prior distribution leads to unreasonable results. Set up a numerical example to show why. (Hint: consider what happens when all the data points y_i are multiplied by a constant factor.)

    (b) Box and Cox (1964) propose a prior distribution that has the form $p(\mu, \sigma, \phi) \propto \dot{y}^{1-\phi} p(\phi)$, where $\dot{y} = (\prod_{i=1}^n y_i)^{1/n}$. Show that this prior distribution eliminates the problem in (a).

    (c) Write the marginal posterior density, $p(\phi|y)$, for the model in (b).

    (d) Discuss the implications of the fact that the prior distribution in (b) depends on the data.

    (e) The power transformation model is used with the understanding that negative values of $y_i^{(\phi)}$ are not possible. Discuss the effect of the implicit truncation on the model.

    See Pericchi (1981) and Hinkley and Runger (1984) for further discussion of Bayesian analysis of power transformations.

14. Fitting a power-transformed normal model: Table 6.3 gives short-term radon measurements for a sample of houses in three counties in Minnesota (see Section 22.4 for more on this example). For this problem, ignore the first-floor measurements (those indicated with asterisks in the table).

    (a) Fit the power-transformed normal model from Exercise 6.13(b) to the basement measurements in Blue Earth County.

    (b) Fit the power-transformed normal model to the basement measurements in all three counties, holding the parameter ϕ equal for all three counties but allowing the mean and variance of the normal distribution to vary.

    (c) Check the fit of the model using posterior predictive simulations.

County	Radon measurements (pCi/L)
Blue Earth	5.0, 13.0, 7.2, 6.8, 12.8, 5.8*, 9.5, 6.0, 3.8, 14.3*, 1.8, 6.9, 4.7, 9.5
Clay	0.9*, 12.9, 2.6, 3.5*, 26.6, 1.5, 13.0, 8.8, 19.5, 2.5*, 9.0, 13.1, 3.6, 6.9*
Goodhue	14.3, 6.9*, 7.6, 9.8*, 2.6, 43.5, 4.9, 3.5, 4.8, 5.6, 3.5, 3.9, 6.7

Table 6.3 *Short-term measurements of radon concentration (in picoCuries/liter) in a sample of houses in three counties in Minnesota. All measurements were recorded on the basement level of the houses, except for those indicated with asterisks, which were recorded on the first floor.*

(d) Discuss whether it would be appropriate to simply fit a lognormal model to these data.

15. Model checking: check the assumed model fitted to the rat tumor data in Section 5.3. Define some test quantities that might be of scientific interest, and compare them to their posterior predictive distributions.

16. Checking the assumption of equal variance: Figures 1.1 and 1.2 display data on point spreads x and score differentials y of a set of professional football games. (The data are available at the website for this book.) In Section 1.6, a model is fit of the form, $y \sim N(x, 14^2)$. However, Figure 1.2a seems to show a pattern of decreasing variance of $y - x$ as a function of x.

(a) Simulate several replicated data sets y^{rep} under the model and, for each, create graphs like Figure 1.1 and 1.2. Display several graphs per page, and compare these to the corresponding graphs of the actual data. This is a graphical posterior predictive check as described in Section 6.4.

(b) Create a numerical summary $T(x, y)$ to capture the apparent decrease in variance of $y - x$ as a function of x. Compare this to the distribution of simulated test statistics, $T(x, y^{\text{rep}})$ and compute the p-value for this posterior predictive check.

CHAPTER 7

Modeling accounting for data collection

How does the design of a sample survey, an experiment, or an observational study affect the models that we construct for a Bayesian analysis? How does one analyze data from a survey that is not a simple random sample? How does one analyze data from hierarchical experimental designs such as randomized blocks and Latin squares, or nonrandomly generated data in an observational study? If we know that a design is randomized, how does that affect our Bayesian inference? In this chapter, we address these questions by showing how relevant features of data collection are incorporated in the process of full probability modeling.

This chapter has two general messages that are quite simple:

- The information that describes how the data were collected should be included in the analysis, typically by basing conclusions on a model (such as a regression; see Part IV) that is *conditional* on the variables that describe the data collection process.

- If partial information is available (for example, knowing that a measurement exceeds some threshold but not knowing its exact value, or having missing values for variables of interest), then a probability model should be used to relate the partially observed quantity or quantities to the other variables of interest.

Despite the simplicity of our messages, many of the examples in this chapter are mathematically elaborate, even when describing data collected under simple designs such as random sampling and randomized experiments. This chapter includes careful theoretical development of these simple examples in order to show clearly how general Bayesian modeling principles adapt themselves to particular designs. The purpose of these examples is not to reproduce or refine existing classical methods but rather to show how Bayesian methods can be adapted to deal with data collection issues such as stratification and clustering in surveys, blocking in experiments, selection in observational studies, and partial-data patterns such as censoring.

7.1 Introduction

There are a variety of settings where a data analysis must account for the design of data collection rather than simply model the observed values directly. These include classical statistical designs such as sample surveys and randomized experiments, problems of nonresponse or missing data, and studies involving censored or truncated data. Each of these, and other problems, will be considered in this chapter.

The key goal throughout is generalizing beyond existing data to a larger population representing unrecorded data, additional units not in the study (as in a sample survey), or information not recorded on the existing units (as in an experiment or observational study, in which it is not possible to apply all treatments to all units). The information used in data collection must be included in the analysis, or else inferences will not necessarily be appropriate for the general population of interest.

A naive student of Bayesian inference might claim that because all inference is conditional on the observed data, it makes no difference how those data were collected. This misplaced appeal to the likelihood principle would assert that given (1) a fixed model (including the prior distribution) for the underlying data and (2) fixed observed values of the data, Bayesian inference is determined regardless of the design for the collection of the data. Under this view there would be no formal role for randomization in either sample surveys or experiments. The essential flaw in the argument is that a complete definition of 'the observed data' should include information on how the observed values arose, and in many situations such information has a direct bearing on how these values should be interpreted. Formally then, the data analyst needs to incorporate the information describing the data collection process in the probability model used for analysis.

The notion that the method of data collection is irrelevant to Bayesian analysis can be dispelled by the simplest of examples. Suppose for instance that we, the authors, give you, the reader, a collection of the outcomes of ten rolls of a die and all are 6's. Certainly your attitude toward the nature of the die after analyzing these data would be different if we told you (i) these were the only rolls we performed, versus (ii) we rolled the die 60 times but decided to report only the 6's, versus (iii) we decided in advance that we were going to report honestly that ten 6's appeared but would conceal how many rolls it took, and we had to wait 500 rolls to attain that result. In simple situations such as these, it is easy to see that the observed data follow a different distribution from that for the underlying 'complete data.' Moreover, in such simple cases it is often easy to state immediately the marginal distribution of the observed data having properly averaged over the posterior uncertainty about the missing data. But in general such simplicity is not present.

More important than these theoretical examples, however, are the applications of the theory to analysis of data from surveys, experiments, and observational studies, and with different patterns of missing data. We shall discuss some general principles that guide the appropriate incorporation of study design and data collection in the process of data analysis:

1. The data analyst should use all relevant information; the pattern of what has been observed can be informative.

2. Ignorable designs (as defined in Section 7.3)—often based on randomiza-

Example	'Observed data'	'Complete data'
Sampling	Values from the n units in the sample	Values from all N units in the population
Experiment	Outcomes under the observed treatment for each unit treated	Outcomes under all treatments for all units
Rounded data	Rounded observations	Precise values of all observations
Unintentional missing data	Observed data values	Complete data, both observed and missing

Table 7.1 *Use of observed- and missing-data terminology for various data structures.*

tion—are likely to produce data for which inferences are less sensitive to model choice, than nonignorable designs.

3. As more explanatory variables are included in an analysis, the inferential conclusions become more valid conditionally but possibly more sensitive to the model specifications relating the outcomes to the explanatory variables.

4. Thinking about design and the data one *could have* observed helps us structure inference about models and finite-population estimands such as the population mean in a sample survey or the average causal effect of an experimental treatment. In addition, the posterior predictive checks discussed in Chapter 6 in general explicitly depend on the study design through the hypothetical replications, y^{rep}.

Generality of the observed- and missing-data paradigm

Our general framework for thinking about data collection problems is in terms of *observed data* that have been collected from a larger set of *complete data* (or *potential data*), leaving unobserved *missing data*. Inference is conditional on observed data and also on the pattern of observed and missing observations. We use the expression 'missing data' in a quite general sense to include *unintentional* missing data due to unfortunate circumstances such as survey nonresponse, censored measurements, and noncompliance in an experiment, but also *intentional* missing data such as data from units not sampled in a survey and the unobserved 'potential outcomes' under treatments not applied in an experiment (see Table 7.1).

Sections 7.2–7.3 define a general notation for data collection and introduce the concept of ignorability, which is crucial in setting up models that correctly account for data collection. The rest of this chapter discusses Bayesian analysis in different scenarios: sampling, experimentation, observational studies, and unintentional missing data. Chapter 21 goes into more detail on multivariate regression models for missing data.

To develop a basic understanding of the fundamental issues, we need a general formal structure in which we can embed the variations as special cases. As we discuss in the next section, the key idea is to expand the sample space to include, in addition to the potential (complete) data y, an indicator variable I for whether each element of y is observed or not.

7.2 Formal models for data collection

In this section, we develop a general notation for observed and potentially observed data. As noted earlier, we introduce the notation in the context of missing-data problems but apply it to sample surveys, designed experiments, and observational studies in the rest of the chapter. In a wide range of problems, it is useful to imagine what would be done if all data were completely observed—that is, if a sample survey were a census, or if all units could receive all treatments in an experiment, or if no observations were censored (see Table 7.1). We divide the modeling tasks into two parts: modeling the *complete data*, y, typically using the methods discussed in the other chapters of this book, and modeling the observation variable, I, which indexes which potential data are observed.

Notation for observed and missing data

Let $y = (y_1, \ldots, y_N)$ be the matrix of potential data (each y_i may itself be a vector with components $y_{ij}, j = 1, \ldots, J$, for example if several questions are asked of each respondent in a survey), and let $I = (I_1, \ldots, I_N)$ be a matrix of the same dimensions as y with indicators for the observation of y: $I_{ij} = 1$ means y_{ij} is observed, whereas $I_{ij} = 0$ means y_{ij} is missing. For notational convenience, let 'obs'$= \{(i, j) : I_{ij} = 1\}$ index the observed components of y and 'mis'$= \{(i, j) : I_{ij} = 0\}$ index the unobserved components of y; for simplicity we assume that I itself is always observed. (In a situation in which I is not fully observed, for example a sample survey with unknown population size, one can assign parameters to the unknown quantities so that I is fully observed, conditional on the unknown parameters.) The symbols y_{obs} and y_{mis} refer to the collection of elements of y that are observed and missing, respectively. The sample space is the product of the usual sample space for y and the sample space for I. Thus, in this chapter, and also in Chapter 21, we use the notation y_{obs} where in the other chapters we would use y.

For much of this chapter, we assume that the simple 0/1 indicator I_{ij} is adequate for summarizing the possible responses; Section 7.8 considers models in which missing data patterns are more complicated and the data points cannot simply be categorized as 'observed' or 'missing.'

Stability assumption

It is standard in statistical analysis to assume *stability*: that the process of recording or measuring does not change the values of the data. In experiments, the assumption is called the *stable unit treatment value assumption* and includes the assumption of no interference between units: the treatment applied to any particular unit should have no effect on outcomes for the other units. More generally, the assumption is that the complete-data vector (or matrix) y is not affected by the inclusion vector (or matrix) I. An example in which this assumption fails is an agricultural experiment that tests several fertilizers on closely spaced plots in which the fertilizers leach to neighboring plots.

In defining y as a fixed quantity, with I only affecting which elements of y are observed, our notation implicitly assumes stability. If instability is a possibility, then the notation must be expanded to allow all possible outcome vectors in a larger 'complete data' structure y. In order to control the computational burden, such a structure is typically created based on some specific model. We do not consider this topic further except in Exercise 7.4.

Fully observed covariates

In this chapter, we use the notation x for variables that are fully observed for all units. There are typically three reasons why we might want to include x in an analysis:

1. We may be interested in some aspect of the joint distribution of (x, y), such as the regression of y on x.

2. We may be interested in some aspect of the distribution of y, but x provides information about y: in a regression setting, if x is fully observed, then a model for $p(y|x)$ can lead to more precise inference about new values of y than would be obtained by modeling y alone.

3. Even if we are only interested in y, we must include x in the analysis if x is involved in the data collection mechanism or, equivalently, if the distribution of the inclusion indicators I depends on x. Examples of the latter kind of covariate are stratum indicators in sampling or block indicators in a randomized block experiment. We return to this topic several times in this chapter.

Data model, inclusion model, complete-data likelihood, and observed-data likelihood

It is useful when analyzing various methods of data collection to break the joint probability model into two parts: (1) the model for the underlying complete data, y—including both observed and unobserved components—and (2) the model for the inclusion vector, I. We define the *complete-data likelihood* as the product of the likelihoods of these two factors; that is, the distribution of the complete data, y, and the inclusion vector, I, given the parameters in the model:

$$p(y, I|\theta, \phi) = p(y|\theta)p(I|y, \phi). \qquad (7.1)$$

In this chapter, we use θ and ϕ to denote the parameters of the distributions of the complete data and the inclusion vectors, respectively.

In this formulation, the first factor of (7.1), $p(y|\theta)$, is a model of the underlying data without reference to the data collection process. For most problems we shall consider, the estimands of primary interest are functions of the complete data y (finite-population estimands) or of the parameters θ (superpopulation estimands). The parameters ϕ that index the missingness model are characteristic of the data collection but are not generally of scientific interest. It is possible, however, for θ and ϕ to be dependent in their prior distribution or even to be deterministically related.

Expression (7.1) is useful for setting up a probability model, but it is not actually the 'likelihood' of the data at hand unless y is completely observed. The actual information available is (y_{obs}, I), and so the appropriate likelihood for Bayesian inference is

$$p(y_{\text{obs}}, I|\theta, \phi) = \int p(y, I|\theta, \phi) dy_{\text{mis}},$$

which we call the *observed-data likelihood*. If fully observed covariates x are available, all these expressions are conditional on x.

Joint posterior distribution of parameters θ from the sampling model and ϕ from the missing-data model

The complete-data likelihood of (y, I), given parameters (θ, ϕ) and covariates x, is

$$p(y, I|x, \theta, \phi) = p(y|x, \theta)p(I|x, y, \phi),$$

where the pattern of missing data can depend on the complete data y (both observed and missing), as in the cases of censoring and truncation presented in Section 7.8. The joint posterior distribution of the model parameters θ and ϕ, given the observed information, (x, y_{obs}, I), is

$$
\begin{aligned}
p(\theta, \phi|x, y_{\text{obs}}, I) &\propto p(\theta, \phi|x)p(y_{\text{obs}}, I|x, \theta, \phi) \\
&= p(\theta, \phi|x) \int p(y, I|x, \theta, \phi) dy_{\text{mis}} \\
&= p(\theta, \phi|x) \int p(y|x, \theta)p(I|x, y, \phi) dy_{\text{mis}}.
\end{aligned}
$$

The posterior distribution of θ alone is this expression averaged over ϕ:

$$p(\theta|x, y_{\text{obs}}, I) = p(\theta|x) \int\int p(\phi|x, \theta)p(y|x, \theta)p(I|x, y, \phi) dy_{\text{mis}} d\phi. \qquad (7.2)$$

As usual, we will often avoid evaluating these integrals by simply drawing posterior simulations of the joint vector of unknowns, $(y_{\text{mis}}, \theta, \phi)$ and then focusing on the estimands of interest.

Finite-population and superpopulation inference

As indicated earlier, we distinguish between two kinds of estimands: *finite-population* quantities and *superpopulation* quantities, respectively. In the terminology of the earlier chapters, finite-population estimands are unobserved but often observable, and so sometimes may be called predictable. It is usually convenient to divide our analysis and computation into two steps: superpopulation inference—that is, analysis of $p(\theta, \phi | x, y_{\text{obs}}, I)$—and finite-population inference using $p(y_{\text{mis}} | x, y_{\text{obs}}, I, \theta, \phi)$. Posterior simulations of y_{mis} from its posterior distribution are called *multiple imputations* and are typically obtained by first drawing (θ, ϕ) from their joint posterior distribution and then drawing y_{mis} from its conditional posterior distribution given (θ, ϕ). Exercise 3.5 provides a simple example of these computations for inference from rounded data.

If all units were observed fully—for example, sampling all units in a survey or applying all treatments (counterfactually) to all units in an experiment—then finite-population quantities would be known exactly, but there would still be some uncertainty in superpopulation inferences. In situations in which a large fraction of potential observations are in fact observed, finite-population inferences about y_{mis} are more robust to model assumptions, such as additivity or linearity, than are superpopulation inferences about θ. In a model of y given x, the finite-population inferences depend only on the conditional distribution for the particular set of x's in the population, whereas the superpopulation inferences are, implicitly, statements about the infinity of unobserved values of y generated by $p(y | \theta)$.

Estimands defined from predictive distributions are of special interest because they are not tied to any particular parametric model and are therefore particularly amenable to sensitivity analysis across different models with different parameterizations.

Posterior predictive distributions

When considering prediction of future data, or replicated data for model checking, it is useful to distinguish between predicting future complete data, $\tilde{y}$, and predicting future observed data, $\tilde{y}_{\text{obs}}$. The former task is, in principle, easier because it depends only on the complete data distribution, $p(y | x, \theta)$, and the posterior distribution of θ, whereas the latter task depends also on the data collection mechanism—that is, $p(I | x, y, \phi)$.

7.3 Ignorability

If we decide to *ignore* the data collection process, we can compute the posterior distribution of θ by conditioning only on y_{obs} but not I:

$$p(\theta | x, y_{\text{obs}}) \quad \propto \quad p(\theta | x) p(y_{\text{obs}} | x, \theta)$$

$$= \quad p(\theta|x) \int p(y|x, \theta) dy_{\text{mis}}. \qquad (7.3)$$

When the missing data pattern supplies no information—that is, when the function $p(\theta|x, y_{\text{obs}})$ given by (7.3) equals $p(\theta|x, y_{\text{obs}}, I)$ given by (7.2)—the study design or data collection mechanism is called *ignorable* (with respect to the proposed model). In this case, the posterior distribution of θ and the posterior predictive distribution of y_{mis} (for example, future values of y) are entirely determined by the specification of a data model—that is, $p(y|x, \theta)p(\theta|x)$—and the observed values of y_{obs}.

'Missing at random' and 'distinct parameters'

Two general and simple conditions are sufficient to ensure ignorability of the missing data mechanism for Bayesian analysis. First, the condition of *missing at random* requires that

$$p(I|x, y, \phi) = p(I|x, y_{\text{obs}}, \phi);$$

that is, $p(I|x, y, \phi)$, evaluated at the observed value of (x, I, y_{obs}), must be free of y_{mis}—that is, given observed x and y_{obs}, it is a function of ϕ alone. 'Missing at random' is a subtle term, since the required condition is that, given ϕ, missingness depends only on x and y_{obs}. For example, a deterministic inclusion rule that depends only on x is 'missing at random' under this definition. (An example of a deterministic inclusion rule is auditing all tax returns with declared income greater than \$1 million, where 'declared income' is a fully observed covariate, and 'audited income' is y.)

Second, the condition of *distinct parameters* is satisfied when the parameters of the missing data process are independent of the parameters of the data generating process in the prior distribution:

$$p(\phi|x, \theta) = p(\phi|x).$$

Models in which the parameters are *not* distinct are considered in an extended example in Section 7.8.

The important consequence of these definitions is that, from (7.2), it follows that if data are missing at random according to a model with distinct parameters, then $p(\theta|x, y_{\text{obs}}) = p(\theta|x, y_{\text{obs}}, I)$, that is, the missing data mechanism is ignorable.

Ignorability and Bayesian inference under different data-collection schemes

The concept of ignorability supplies some justification for a relatively weak version of the claim presented at the outset of this chapter that, with fixed data and fixed models for the data, the data collection process does not influence Bayesian inference. That result is true for all *ignorable* designs when 'Bayesian inference' is interpreted strictly to refer only to the posterior distribution of the estimands with one fixed model (both prior distribution and likelihood), conditional on the data, <u>but</u> excluding both sensitivity analyses and posterior predictive checks. Our notation also highlights the incorrectness of the claim for the irrelevance of study design in general: even with a fixed likelihood function $p(y|\theta)$, prior distribution $p(\theta)$, and data y, the posterior distribution does vary with different *nonignorable* data collection mechanisms.

In addition to the ignorable/nonignorable classification for designs, we also distinguish between *known* and *unknown* mechanisms, as anticipated by the discussion in the examples of Section 7.8 contrasting inference with specified versus unspecified truncation and censoring points. The term 'known' includes data collection processes that follow a known parametric family, even if the parameters ϕ are unknown

Designs that are ignorable and known with no covariates, including simple random sampling and completely randomized experiments. The simplest data collection procedures are those that are ignorable and known, in the sense that

$$p(I|x, y, \phi) = p(I). \tag{7.4}$$

Here, there is no unknown parameter ϕ because there is a single accepted specification for $p(I)$ that does not depend on either x or y_{obs}. Only the complete-data distribution, $p(y|x, \theta)$, and the prior distribution for θ, $p(\theta)$, need be considered for inference. The obvious advantage of such an ignorable design is that the information from the data can be recovered with a relatively simple analysis. We begin Section 7.4 on surveys and Section 7.5 on experiments by considering basic examples with no covariates, x, which document this connection with standard statistical practice and also show that the potential data $y = (y_{obs}, y_{mis})$ can usefully be regarded as having many components, most of which we never expect to observe but can be used to define 'finite-population' estimands.

Designs that are ignorable and known given covariates, including stratified sampling and randomized block experiments. In practice, simple random sampling and complete randomization are less common than more complicated designs that base selection and treatment decisions on covariate values. It is not appropriate always to pretend that data $y_{obs\,1}, \ldots, y_{obs\,n}$ are collected as a simple random sample from the target population, $y_1, \ldots, y_N$. A key idea for Bayesian inference with complicated designs is to include in the model $p(y|x, \theta)$ enough explanatory variables x so that the design is ignorable. With ignorable designs, many of the models presented in other chapters of this book can be directly applied. We illustrate this approach with several examples that show that not all known ignorable data collection mechanisms are equally good for all inferential purposes.

Designs that are strongly ignorable and known. A design that is *strongly ignorable* (sometimes called *unconfounded*) satisfies

$$p(I|x, y, \phi) = p(I|x),$$

so that the only dependence is on fully-observed covariates. (For an example of a design that is ignorable but *not* strongly ignorable, consider a sequential experiment in which the unit-level probabilities of assignment depend on the observed outcomes of previously assigned units.) We discuss these designs further in the context of propensity scores below.

Designs that are ignorable and unknown, such as experiments with nonrandom treatment assignments based on fully observed covariates. When analyzing data generated by an unknown or nonrandomized design, one can still ignore the assignment mechanism if $p(I|x, y, \phi)$ depends only on fully observed covariates, and θ is distinct from ϕ. The ignorable analysis must be conditional on these covariates. We illustrate with an example in Section 7.5, with strongly ignorable (given x) designs that are unknown. Propensity scores can play a useful role in the analysis of such data.

Designs that are nonignorable and known, such as censoring. There are many settings in which the data collection mechanism is known (or assumed known) but is not ignorable. Two simple but important examples are censored data (some of the variations in Section 7.8) and rounded data (see Exercises 3.5 and 7.14).

Designs that are nonignorable and unknown. This is the most difficult case. For example, censoring at an unknown point typically implies great sensitivity to the model specification for y (discussed in Section 7.8). For another example, consider a medical study of several therapies, $j = 1, \ldots, J$, in which the treatments that are expected to have smaller effects are applied to larger numbers of patients. The sample size of the experiment corresponding to treatment j, n_j, would then be expected to be correlated with the efficacy, θ_j, and so the parameters of the data collection mechanism are *not distinct* from the parameters indexing the data distribution. In this case, we should form a parametric model for the joint distribution of (n_j, y_j). Nonignorable and unknown data collection are standard in observational studies, as we discuss in Section 7.7, where initial analyses typically treat the design as ignorable but unknown.

Propensity scores

With a strongly ignorable design, the unit-level probabilities of assignments,

$$\Pr(I_i = 1|X) = \pi_i,$$

are known as *propensity scores*. In some examples of strongly ignorable designs, the vector of propensity scores, $\pi = (\pi_1, \ldots, \pi_N)$, is an adequate summary of the covariates X in the sense that the assignment mechanism is strongly ignorable given just π rather than x. Conditioning on π rather than multivariate x may lose some precision but it can greatly simplify modeling and produce more robust inferences. Propensity scores also create another bridge to classical designs.

However, the propensity scores alone never supply enough information for posterior predictive replications (which are crucial to model checking, as discussed in Sections 6.3–6.5), because different designs can have the same propensity scores. For example, a completely randomized experiment with $\pi_i = 1/2$ for all units has the same propensity scores as an experiment with independent assignments with probabilities of $1/2$ for each treatment for each unit. Similarly, simple random sampling has the same propensity scores as some more elaborate equal-probability sampling designs involving stratified or cluster sampling.

Unintentional missing data

A ubiquitous problem in real data sets is unintentional missing data, which we discuss in more detail in Chapter 21. Classical examples include survey nonresponse, dropouts in experiments, and incomplete information in observational studies. When the amount of missing data is small, one can often perform a good analysis assuming that the missing data are ignorable (conditional on the fully observed covariates). As the fraction of missing information increases, the ignorability assumption becomes more critical. In some cases, half or more of the information is missing, and then a serious treatment of the missingness mechanism is essential. For example, in observational studies of causal effects, at most half of the potential data (corresponding to both treatments applied to all units in the study) are observed, and so it is necessary to either assume ignorability (conditional on available covariates) or explicitly model the treatment selection process.

7.4 Sample surveys

Simple random sampling of a finite population

For perhaps the simplest nontrivial example of statistical inference, consider a finite population of N persons, where y_i is the weekly amount spent on food by the ith person. Let $y = (y_1, \ldots, y_N)$, where the object of inference is average weekly spending on food in the population, $\bar{y}$. As usual, we consider the finite population as N exchangeable units; that is, we model the marginal distribution of y as an iid mixture over the prior distribution of underlying parameters θ:

$$p(y) = \int \prod_{i=1}^{N} p(y_i|\theta)p(\theta)d\theta.$$

The estimand, $\bar{y}$, will be estimated from a sample of y_i-values because a census of all N units is too expensive for practical purposes. A standard technique is to draw a simple random sample of specified size n. Let $I = (I_1, \ldots, I_N)$ be the vector of indicators for whether or not person i is included in the sample:

$$I_i = \begin{cases} 1 & \text{if } i \text{ is sampled} \\ 0 & \text{otherwise.} \end{cases}$$

Formally, simple random sampling is defined by

$$p(I|y,\phi) = p(I) = \begin{cases} \binom{N}{n}^{-1} & \text{if } \sum_{i=1}^{N} I_i = n \\ 0 & \text{otherwise.} \end{cases}$$

This method is strongly ignorable and known (compare to equation (7.4)) and therefore is straightforward to deal with inferentially. The probability of inclusion in the sample (the propensity score) is $\pi_i = n/N$ for all units.

Bayesian inference for superpopulation and finite-population estimands. We can perform Bayesian inference applying the principles of the early chapters of this book to the posterior density, $p(\theta|y_{\text{obs}}, I)$, which under an ignorable design is $p(\theta|y_{\text{obs}}) \propto p(\theta)p(y_{\text{obs}}|\theta)$. As usual, this requires setting up a model for the distribution of weekly spending on food in the population in terms of parameters θ. For this problem, however, the estimand of interest is the finite-population average, $\bar{y}$, which can be expressed as

$$\bar{y} = \frac{n}{N}\bar{y}_{\text{obs}} + \frac{N-n}{N}\bar{y}_{\text{mis}}, \tag{7.5}$$

where $\bar{y}_{\text{obs}}$ and $\bar{y}_{\text{mis}}$ are the averages of the observed and missing y_i's, respectively.

We can determine the posterior distribution of $\bar{y}$ using simulations of $\bar{y}_{\text{mis}}$ from its posterior predictive distribution. We start with simulations of θ: $\theta^l, l = 1, \ldots, L$. For each drawn θ^l we then draw a vector y_{mis} from

$$p(y_{\text{mis}}|\theta^l, y_{\text{obs}}) = p(y_{\text{mis}}|\theta^l) = \prod_{i:\, I_i=0} p(y_i|\theta^l),$$

and then average the values of the simulated vector to obtain a draw of $\bar{y}_{\text{mis}}$ from its posterior predictive distribution. Because $\bar{y}_{\text{obs}}$ is known, we can compute draws from the posterior distribution of the finite-population mean, $\bar{y}$, using (7.5) and the draws of y_{mis}. Although typically the estimand is viewed as $\bar{y}$, more generally it could be any function of y (such as the median of the y_i values, or the mean of $\log y_i$).

Large-sample equivalence of superpopulation and finite-population inference. If $N - n$ is large, then we can use the central limit theorem to approximate the sampling distribution of $\bar{y}_{\text{mis}}$:

$$p(\bar{y}_{\text{mis}}|\theta) \approx \text{N}\left(\bar{y}_{\text{mis}} \,\middle|\, \mu, \frac{1}{N-n}\sigma^2\right),$$

where $\mu = \mu(\theta) = \text{E}(y_i|\theta)$ and $\sigma^2 = \sigma^2(\theta) = \text{var}(y_i|\theta)$. If n is large as well, then the posterior distributions of θ and any of its components, such as μ and σ, are approximately normal, and so the posterior distribution of $\bar{y}_{\text{mis}}$ is approximately a normal mixture of normals and thus normal itself. More formally,

$$p(\bar{y}_{\text{mis}}|y_{\text{obs}}) \approx \int p(\bar{y}_{\text{mis}}|\mu, \sigma)p(\mu, \sigma|y_{\text{obs}})d\mu d\sigma;$$

as both N and n get large with N/n fixed, this is an approximate normal density with

$$E(\bar{y}_{\text{mis}}|y_{\text{obs}}) \approx E(\mu|y_{\text{obs}}) \approx \bar{y}_{\text{obs}},$$

and

$$\begin{aligned}
\text{var}(\bar{y}_{\text{mis}}|y_{\text{obs}}) &\approx \text{var}(\mu|y_{\text{obs}}) + E\left(\frac{1}{N-n}\sigma^2 \middle| y_{\text{obs}}\right) \\
&\approx \frac{1}{n}s^2_{\text{obs}} + \frac{1}{N-n}s^2_{\text{obs}} \\
&= \frac{N}{n(N-n)}s^2_{\text{obs}},
\end{aligned}$$

where s^2_{obs} is the sample variance of the observed values of y_{obs}. Combining this approximate posterior distribution with (7.5), it follows not only that $p(\bar{y}|y_{\text{obs}}) \to p(\mu|y_{\text{obs}})$, but, more generally, that

$$\bar{y}|y_{\text{obs}} \approx N\left(\bar{y}_{\text{obs}}, \left(\frac{1}{n}-\frac{1}{N}\right)s^2_{\text{obs}}\right). \tag{7.6}$$

This is the formal Bayesian justification of normal-theory inference for finite sample surveys.

For $p(y_i|\theta)$ normal, with the standard noninformative prior distribution, the exact result is $\bar{y}|y_{\text{obs}} \sim t_{n-1}\left(\bar{y}_{\text{obs}}, \left(\frac{1}{n}-\frac{1}{N}\right)s^2_{\text{obs}}\right)$. (See Exercise 7.7.)

Stratified sampling

In stratified random sampling, the N units are divided into J strata, and a simple random sample of size n_j is drawn using simple random sampling from each stratum $j = 1, \ldots, J$. This design is ignorable given J vectors of indicator variables, $x_1, \ldots, x_J$, with $x_j = (x_{1j}, \ldots, x_{nj})$ and

$$x_{ij} = \begin{cases} 1 & \text{if unit } i \text{ is in stratum } j \\ 0 & \text{otherwise.} \end{cases}$$

The variables x_j are effectively fully observed in the population as long as we know, for each j, the number of units N_j in the stratum, in addition to the values of x_{ij} for units in the sample. A natural analysis is to model the distributions of the measurements y_i within each stratum j in terms of parameters θ_j and then perform Bayesian inference on all the sets of parameters $\theta_1, \ldots, \theta_J$. For many applications it will be natural to assign a hierarchical model to the θ_j's, yielding a problem with structure similar to the rat tumor experiments, the educational testing experiments, and the meta-analysis of Chapter 5. We illustrate this approach with an example at the end of this section.

We obtain finite-population inferences by weighting the inferences from the separate strata in a way appropriate for the finite-population estimand. For example, we can write the population mean, $\bar{y}$, in terms of the individual stratum means, $\bar{y}_j$, as $\bar{y} = \sum_{j=1}^{J}\frac{N_j}{N}\bar{y}_j$. The finite-population quantities $\bar{y}_j$

| Stratum, j | Proportion who prefer ... | | | Sample proportion, n_j/n |
	Bush, $y_{\text{obs}\,1j}/n_j$	Dukakis, $y_{\text{obs}\,2j}/n_j$	no opinion, $y_{\text{obs}\,3j}/n_j$	
Northeast, I	0.30	0.62	0.08	0.032
Northeast, II	0.50	0.48	0.02	0.032
Northeast, III	0.47	0.41	0.12	0.115
Northeast, IV	0.46	0.52	0.02	0.048
Midwest, I	0.40	0.49	0.11	0.032
Midwest, II	0.45	0.45	0.10	0.065
Midwest, III	0.51	0.39	0.10	0.080
Midwest, IV	0.55	0.34	0.11	0.100
South, I	0.57	0.29	0.14	0.015
South, II	0.47	0.41	0.12	0.066
South, III	0.52	0.40	0.08	0.068
South, IV	0.56	0.35	0.09	0.126
West, I	0.50	0.47	0.03	0.023
West, II	0.53	0.35	0.12	0.053
West, III	0.54	0.37	0.09	0.086
West, IV	0.56	0.36	0.08	0.057

Table 7.2 *Results of a CBS News survey of 1447 adults in the United States, divided into 16 strata. The sampling is assumed to be proportional, so that the population proportions, N_j/N, are approximately equal to the sampling proportions, n_j/n.*

can be simulated given the simulated parameters θ_j for each stratum. There is no requirement that $\frac{n_j}{n} = \frac{N_j}{N}$; the finite-population Bayesian inference automatically corrects for any oversampling or undersampling of strata.

Example. Stratified sampling in pre-election polling

We illustrate the analysis of a stratified sampling design with the public opinion poll introduced in Section 3.5, in which we estimated the proportion of registered voters who supported the two major candidates in the 1988 U.S. Presidential election. In Section 3.5, we analyzed the data under the false assumption of simple random sampling. Actually, the CBS survey data were collected using a variant of stratified random sampling, in which all the primary sampling units (groups of residential phone numbers) were divided into 16 strata, cross-classified by region of the country (Northeast, Midwest, South, West) and density of residential area (as indexed by telephone exchanges). The survey results and the relative size of each stratum in the sample are given in Table 7.2. For the purposes of this example, we assume that respondents are sampled at random within each stratum and that the sampling fractions are exactly equal across strata, so that $N_j/N = n_j/n$ for each stratum j.

Complications arise from several sources, including the systematic sampling used within strata, the selection of an individual to respond from each household that is contacted, the number of times people who are not at home are called back, the use of demographic adjustments, and the general problem of nonresponse. For simplicity, we ignore many complexities in the design and restrict our attention here to the Bayesian analysis of the population of respondents answering

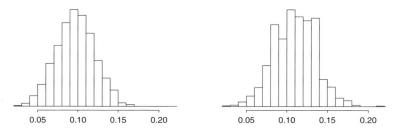

Figure 7.1 *Histograms of values of* $\sum_{j=1}^{16} \frac{N_j}{N}(\theta_{1j} - \theta_{2j})$ *for 1000 simulations from the posterior distribution for the election polling example, based on (a) the simple nonhierarchical model and (b) the hierarchical model. Compare to Figure 3.3.*

residential phone numbers, thereby assuming the nonresponse is ignorable and also avoiding the additional step of shifting to the population of registered voters. (Exercises 7.12 and 7.13 consider adjusting for the fact that the probability an individual is sampled is proportional to the number of telephone lines in his or her household and inversely proportional to the number of adults in the household.) A more complete analysis would control for covariates such as sex, age, and education that affect the probabilities of nonresponse.

Data distribution. To justify the use of an ignorable model, we must model the outcome variable conditional on the explanatory variables—region of the country and density of residential area—that determine the stratified sampling. We label the strata $j = 1, \ldots, 16$, with n_j out of N_j drawn from each stratum, and a total of $N = \sum_{j=1}^{16} N_j$ registered voters in the population. We fit the multinomial model of Section 3.5 within each stratum and a hierarchical model to link the parameters across different strata. For each stratum j, we label $y_{\text{obs}\,j} = (y_{\text{obs}\,1j}, y_{\text{obs}\,2j}, y_{\text{obs}\,3j})$, the number of supporters of Bush, Dukakis, and other/no-opinion in the sample, and we model $y_{\text{obs}\,j} \sim \text{Multin}(n_j; \theta_{1j}, \theta_{2j}, \theta_{3j})$.

A simple nonhierarchical model. The simplest analysis of these data assigns the 16 vectors of parameters $(\theta_{1j}, \theta_{2j}, \theta_{3j})$ independent prior distributions. At this point, the Dirichlet model is a convenient choice of prior distribution because it is conjugate to the multinomial likelihood (see Section 3.5). Assuming Dirichlet prior distributions with all parameters equal to 1, we can obtain posterior inferences separately for the parameters in each stratum. The resulting simulations of θ_{ij}'s constitute the 'superpopulation inference' under this model.

Finite-population inference. As discussed in the earlier presentation of this example in Section 3.5, an estimand of interest is the difference in the proportions of Bush and Dukakis supporters, that is,

$$\bar{y}_1 - \bar{y}_2 = \sum_{j=1}^{16} \frac{N_j}{N}(\bar{y}_{1j} - \bar{y}_{2j}).$$

For a national opinion poll such as this, $n/N \approx 0$, and N is in the millions, so $\bar{y}_{1j} \approx \mathrm{E}(y_{1j}|\theta)$ and $\bar{y}_{2j} \approx \mathrm{E}(y_{2j}|\theta)$, and we can use the superpopulation approximation,

$$\bar{y}_1 - \bar{y}_2 \approx \sum_{j=1}^{16} \frac{N_j}{N}(\theta_{1j} - \theta_{2j}), \tag{7.7}$$

which we can easily compute using the posterior simulations of θ and the known values of N_j/N. This is a standard case in which finite-population and superpopulation inferences are essentially identical. The results of 1000 draws from the posterior distribution of (7.7) are displayed in Figure 7.1a. The distribution is centered at a slightly smaller value than in Figure 3.3, 0.097 versus 0.098, and is slightly more concentrated about the center. The former result is likely a consequence of the choice of prior distribution. The Dirichlet prior distribution can be thought of as adding a single voter to each multinomial category within each of the 16 strata—adding a total of 16 to each of the three response categories. This differs from the nonstratified analysis in which only a single voter was added to each of the three multinomial categories. A Dirichlet prior distribution with parameters equal to 1/16 would reproduce the same median as the nonstratified analysis. The smaller spread is expected and is one of the reasons for taking account of the stratified design in the analysis.

A hierarchical model.   The simple model with prior independence across strata allows easy computation, but, as discussed in Chapter 5, when presented with a hierarchical dataset, we can improve inference by estimating the population distributions of parameters using a hierarchical model.

As an aid to constructing a reasonable model, we transform the multinomial parameters to separate the effects of partisan preference and probability of having a preference:

$$
\begin{aligned}
\alpha_{1j} &= \frac{\theta_{1j}}{\theta_{1j} + \theta_{2j}} = \begin{array}{l}\text{probability of preferring Bush, given}\\ \text{that a preference is expressed}\end{array}\\
\alpha_{2j} &= 1 - \theta_{3j} = \text{probability of expressing a preference.} \tag{7.8}
\end{aligned}
$$

We then transform these parameters to the logit scale (because they are restricted to lie between 0 and 1),

$$\beta_{1j} = \text{logit}(\alpha_{1j}) \quad \text{and} \quad \beta_{2j} = \text{logit}(\alpha_{2j}),$$

and model them as exchangeable across strata with a bivariate normal distribution indexed by hyperparameters $(\mu_1, \mu_2, \tau_1, \tau_2, \rho)$:

$$p(\beta|\mu_1, \mu_2, \tau_1, \tau_2, \rho) = \prod_{j=1}^{16} \text{N}\left(\begin{pmatrix}\beta_{1j}\\\beta_{2j}\end{pmatrix}\middle|\begin{pmatrix}\mu_1\\\mu_2\end{pmatrix}, \begin{pmatrix}\tau_1^2 & \rho\tau_1\tau_2\\\rho\tau_1\tau_2 & \tau_2^2\end{pmatrix}\right),$$

with the conditional distributions $p(\beta_{1j}, \beta_{2j}|\mu_1, \mu_2, \tau_1, \tau_2, \rho)$ independent for $j = 1, \ldots, 16$. This model, which is exchangeable in the 16 strata, does not use all the available prior information, because the strata are actually structured in a 4×4 array, but it will do for now as an improvement upon the nonhierarchical model (which is of course equivalent to the hierarchical model with the parameters τ_1 and τ_2 fixed at ∞). We complete the model by assigning a uniform prior density to the hierarchical mean and standard deviation parameters and to the correlation of the two logits.

Estimand	Posterior quantiles				
	2.5%	25%	median	75%	97.5%
Stratum 1, $\alpha_{1,1}$	0.34	0.43	0.48	0.52	0.57
Stratum 2, $\alpha_{1,2}$	0.42	0.50	0.53	0.56	0.61
Stratum 3, $\alpha_{1,3}$	0.48	0.52	0.54	0.56	0.60
Stratum 4, $\alpha_{1,4}$	0.41	0.47	0.50	0.54	0.58
Stratum 5, $\alpha_{1,5}$	0.39	0.49	0.52	0.55	0.61
Stratum 6, $\alpha_{1,6}$	0.44	0.50	0.53	0.55	0.64
Stratum 7, $\alpha_{1,7}$	0.48	0.53	0.56	0.58	0.63
Stratum 8, $\alpha_{1,8}$	0.52	0.56	0.59	0.61	0.66
Stratum 9, $\alpha_{1,9}$	0.47	0.54	0.57	0.61	0.69
Stratum 10, $\alpha_{1,10}$	0.47	0.52	0.55	0.57	0.61
Stratum 11, $\alpha_{1,11}$	0.47	0.53	0.56	0.58	0.63
Stratum 12, $\alpha_{1,12}$	0.53	0.56	0.58	0.61	0.65
Stratum 13, $\alpha_{1,13}$	0.43	0.50	0.53	0.56	0.63
Stratum 14, $\alpha_{1,14}$	0.50	0.55	0.57	0.60	0.67
Stratum 15, $\alpha_{1,15}$	0.50	0.55	0.58	0.59	0.65
Stratum 16, $\alpha_{1,16}$	0.50	0.55	0.57	0.60	0.65
Stratum 16, $\alpha_{2,16}$	0.87	0.90	0.91	0.92	0.94
$\text{logit}^{-1}(\mu_1)$	0.50	0.53	0.55	0.56	0.59
$\text{logit}^{-1}(\mu_2)$	0.89	0.91	0.91	0.92	0.93
τ_1	0.11	0.17	0.23	0.30	0.47
τ_2	0.14	0.20	0.28	0.40	0.78
ρ	−0.92	−0.71	−0.44	0.02	0.75

Table 7.3 *Summary of posterior inference for the hierarchical analysis of the CBS survey in Table 7.2. The posterior distributions for the α_{1j}'s vary from stratum to stratum much less than the raw counts do. The inference for $\alpha_{2,16}$ for stratum 16 is included above as a representative of the 16 parameters α_{2j}. The parameters μ_1 and μ_2 are transformed to the inverse-logit scale so they can be more directly interpreted.*

Results under the hierarchical model. Posterior inference for the 37-dimensional parameter vector $(\beta, \mu_1, \mu_2, \sigma_1, \sigma_2, \tau)$ is conceptually straightforward but requires computational methods beyond those developed in Parts I and II of this book. For the purposes of this example, we present the results of sampling from the joint posterior distribution obtained using the Metropolis algorithm; see Exercise 11.5.

Table 7.3 provides posterior quantities of interest for the hierarchical parameters and for the parameters α_{1j}, the proportion preferring Bush (among those who have a preference) in each stratum. The posterior medians of the α_{1j}'s vary from 0.48 to 0.59, representing considerable shrinkage from the proportions in the raw counts, $y_{\text{obs}\,1j}/(y_{\text{obs}\,1j} + y_{\text{obs}\,2j})$, which vary from 0.33 to 0.67. The posterior median of ρ, the between-stratum correlation of logit of support for Bush and logit of proportion who express a preference, is negative, but the posterior distribution has substantial variability. Posterior quantiles for the probability of expressing a preference are displayed for only one stratum, stratum 16, as an illustration.

The results of 1000 draws from the posterior distribution of (7.7) are displayed in

Figure 7.1b. Comparison to Figures 7.1a and 3.3 indicates that the hierarchical model yields a higher posterior median, 0.110, and slightly more variability than either of the other approaches. The higher median occurs because the strata in which support for Bush was lowest, strata 1 and 5, have relatively small samples (see Table 7.2) and so are pulled more toward the grand mean in the hierarchical analysis (see Table 7.3).

Model checking. The large shrinkage of the extreme values in the hierarchical analysis is a possible cause for concern, considering that the observed support for Bush in strata 1 and 5 was so much less than in the other 14 strata. Perhaps the normal model for the distribution of true stratum parameters β_{1j} is inappropriate? We check the fit of the model using a posterior predictive check, using as a test statistic $T(y) = \min_j y_{\text{obs }1j}/n_j$, which has an observed value of 0.298 (occurring in stratum 1). Using 1000 draws from the posterior predictive distribution, we find that $T(y^{\text{rep}})$ varies from 0.14 to 0.48, with $\frac{163}{1000}$ replicated values falling below 0.298. Thus, the extremely low value is plausible under the model.

Cluster sampling

In cluster sampling, the N units in the population are divided into K clusters, and sampling proceeds in two stages. First, a sample of J clusters is drawn, and second, a sample of n_j units is drawn from the N_j units within each sampled cluster $j = 1, \ldots, J$. This design is ignorable given indicator variables for the J clusters and knowledge of the number of units in each of the J clusters. Analysis of a cluster sample proceeds as for a stratified sample, except that inference must be extended to the clusters not sampled, which correspond to additional exchangeable batches in a hierarchical model (see Exercise 7.9).

Example. A survey of Australian schoolchildren
We illustrate with an example of a study estimating the proportion of children in the Melbourne area who walked to school. The survey was conducted by first sampling 72 schools from the metropolitan area and then surveying the children from two classes randomly selected within each school. The schools were selected from a list, with probability of selection proportional to the number of classes in the school. This 'probability proportional to size' sampling is a classical design for which the probabilities of selection are equal for each student in the city, and thus the average of the sample is an unbiased estimate of the population mean (see Section 8.1). The Bayesian analysis for this design is more complicated but has the advantage of being easily generalized to more elaborate inferential settings such as regressions.

Notation for cluster sampling. For this simple example, we label students, classes, and schools as i, j, k, respectively, with N_{jk} students within class j in school k, M_k classes within school k, and K schools in the city. The number of students in school k is then $N_k = \sum_{j=1}^{M_k} N_{jk}$, with a total of $N = \sum_{k=1}^{K} N_k$ students in the city.

We define y_{ijk} to equal 1 if student i (in class j and school k) walks to school and 0 otherwise, $\bar{y}_{.jk} = \frac{1}{N_{jk}} \sum_{i=1}^{N_{jk}} y_{ijk}$ to be the proportion of students in class j and school K who walk to school, and $\bar{y}_{..k} = \frac{1}{N_k} \sum_{j=1}^{M_k} N_{jk} \bar{y}_{.jk}$ to be the proportion of students in school k who walk. The estimand of interest is $\overline{Y} = \frac{1}{N} \sum_{k=1}^{K} N_k \bar{y}_{..k}$, the proportion of students in the city who walk.

The model. The general principle of modeling for cluster sampling is to include a parameter for each level of clustering. A simple model would start with independence at the student level within classes: $Pr(y_{ijk} = 1) = \theta_{jk}$. Assuming a reasonable number of students N_{jk} in the class, we approximate the distribution of the average in class j within school k as independent with distributions,

$$\bar{y}_{.jk} \sim N(\theta_{jk}, \sigma_{jk}^2), \qquad (7.9)$$

where $\sigma_{jk}^2 = \bar{y}_{.jk}(1-\bar{y}_{.jk})/N_{jk}$. We could perform the computations with the exact binomial model, but for our purposes here, the normal approximation allows us to lay out more clearly the structure of the model for cluster data.

We continue by modeling the classes within each school as independent (conditional on school-level parameters) with distributions,

$$\begin{aligned}
\theta_{jk} &\sim N(\theta_k, \tau_{\text{class}}^2) \\
\theta_k &\sim N(\alpha + \beta M_k, \tau_{\text{school}}^2).
\end{aligned} \qquad (7.10)$$

We include the number of classes M_k in the school-level model because of the principle that all information used in the design should be included in the analysis. In this survey, the sampling probabilities depend on the school sizes M_k, and so the design is ignorable only for a model that includes the M_k's. Of course, the linear form (7.10) is only one possible way to do this, but in practice it is a reasonable place to start.

Inference for the population mean. Bayesian inference for this problem proceeds in four steps:

1. Fit the model defined by (7.9) and (7.10), along with a noninformative prior distribution on the hyperparameters $\alpha, \beta, \tau_{\text{class}}, \tau_{\text{school}}$ to obtain inferences for these hyperparameters, along with the parameters θ_{jk} for the classes j in the sample and θ_k for the schools k in the sample.

2. For each posterior simulation from the model, use the model (7.10) to simulate the parameters θ_{jk} and θ_k for classes and schools that are not in the sample.

3. Use the sampling model (7.9) to obtain inferences about students in unsampled classes and unsampled schools, and then combine these to obtain inferences about the average response in each school, $\bar{y}_{..k} = \frac{1}{N_k} \sum_{j=1}^{M_k} N_{jk} \bar{y}_{.jk}$. For unsampled schools k, $\bar{y}_{..k}$ is calculated directly from the posterior simulations of its M_k classes, or more directly simulated as,

$$\bar{y}_{..k} \sim N(\alpha + \beta M_k, \tau_{\text{school}}^2 + \tau_{\text{class}}^2/M_k + \sigma^2/N_k).$$

For sampled schools k, $\bar{y}_{..k}$ is an average over sampled and unsampled classes. This can be viewed as a more elaborate version of the $\bar{y}_{\text{obs}}, \bar{y}_{\text{mis}}$ calculations described at the beginning of Section 7.4.

4. Combine the posterior simulations about the school-level averages to obtain an inference about the population mean, $\overline{Y} = \frac{1}{N} \sum_{k=1}^{K} N_k \bar{y}_{..k}$. To perform this summation (as well as the simulations in the previous step), we must know the number of children N_k within each of the schools in the population. If these

are not known, they must be estimated from the data—that is, we must jointly model N_k, θ_k—as illustrated in the Alcoholics Anonymous sampling example below.

Approximate inference when the fraction of clusters sampled is small. For inference about schools in the Melbourne area, the superpopulation quantities θ_{jk} and θ_k can be viewed as intermediate quantities, useful for the ultimate goal of estimating the finite-population average $\bar{Y}$, as detailed above. In general practice, however (and including this particular survey), the number of schools sampled is a small fraction of the total in the city, and thus we can approximate the average of any variable in the population by its mean in the superpopulation distribution. In particular, we can approximate $\bar{Y}$, the overall proportion of students in the city who walk, by the expectations in the school-level model:

$$\bar{Y} \approx \frac{1}{N} \sum_{k=1}^{K} N_k \mathrm{E}(\theta_k) = \frac{1}{N} \sum_{k=1}^{K} N_k (\alpha + \beta M_k). \qquad (7.11)$$

This expression has a characteristic feature of sample survey inference, that the quantity of interest depends both on model parameters and cluster sizes. Since we are now assuming that only a very small fraction of schools are included in the sample, the calculation of (7.11) depends only on the hyperparameters α, β and the distribution of M_k, N_k in the population, and not on the school-level parameters θ_k.

Unequal probabilities of selection

Another example of a common survey design involves random sampling with unequal sampling probabilities for different units. This design is ignorable conditional on the covariates, x, that determine the sampling probability, as long as the covariates are fully observed in the general population (or, to put it another way, as long as we know the values of the covariates in the sample and also the distribution of x in the general population—for example, if the sampling probabilities depend on sex and age, the number of young females, young males, old females, and old males in the population). The critical step then in Bayesian modeling is formulating the conditional distribution of y given x.

Example. Sampling of Alcoholics Anonymous groups
Approximately every three years, Alcoholics Anonymous performs a survey of its members in the United States, first selecting a number of meeting groups at random, then sending a packet of survey forms to each group and asking all the persons who come to the meeting on a particular day to fill out the survey. For the 1997 survey, groups were sampled with equal probability, and 400 groups throughout the nation responded, sending back an average of 20 surveys each.

For any binary survey response y_i (for example, the answer to the question, 'Have you been an AA member for more than 5 years?'), we assume independent responses within each group j with probability θ_j for a Yes response; thus,

$$y_j \sim \mathrm{Bin}(n_j, \theta_j),$$

where y_j is the number of Yes responses and n_j the total number of people who respond to the survey sent to members of group j. (This binomial model ignores the finite-population correction that is appropriate if a large proportion of the potential respondents in a group actually do receive and respond to the survey. In practice, it is acceptable to ignore this correction because, in this example, the main source of uncertainty is with respect to the unsampled clusters—a more precise within-cluster model would not make much difference.)

For any response y, interest lies in the mean of the values of θ_j across the population; that is,

$$\overline{Y} = \frac{\sum_{j=1}^{K} x_j \theta_j}{\sum_{n=1}^{K} x_j}, \tag{7.12}$$

where x_j represents the number of persons in group j, defined as the average number of persons who come to meetings of the group. To evaluate (7.12), we must perform inferences about the group sizes x_j and the probabilities θ_j, for all the groups j in the population—the 400 sampled groups and the 6000 unsampled.

For Bayesian inference, we need a joint model for (x_j, θ_j). We set this up conditionally: $p(x_j, \theta_j) = p(x_j)p(\theta_j | x_j)$. For simplicity, we set up a joint normal distribution, which we can write as a normal model for x_j and a linear regression for θ_j given x_j:

$$
\begin{aligned}
x_j &\sim \mathrm{N}(\mu_x, \tau_x^2) \\
\theta_j &\sim \mathrm{N}(\alpha + \beta x_j, \tau_\theta^2).
\end{aligned} \tag{7.13}
$$

This is not the best parameterization since it does not account for the constraint that the probabilities θ_j must lie between 0 and 1 and the group sizes x_j must be positive. The normal model is convenient, however, for illustrating the mathematical ideas beyond cluster sampling, especially when we simplify the likelihood as,

$$y_j/n_j \approx \mathrm{N}(\theta_j, \sigma_j^2), \tag{7.14}$$

where $\sigma_j^2 = y_j(n_j - y_j)/n_j^3$.

We complete the model with a likelihood on n_j, which we assume has a Poisson distribution with mean πx_j, where π represents the probability that a person will respond to the survey. For convenience, we can approximate the Poisson by a normal model:

$$n_j \approx \mathrm{N}(\pi x_j, \pi x_j), \tag{7.15}$$

The likelihood (7.14) and (7.15), along with the prior distribution (7.13), and a noninformative uniform distribution on the hyperparameters $\alpha, \beta, \tau_x, \tau_\theta, \pi$, represents a hierarchical normal model for the parameters (x_j, θ_j)—a bivariate version of the hierarchical normal model discussed in Chapter 5. We can compute posterior simulations from $p(x, \theta | y, n)$ using the methods described in Chapter 19. (It would require only a little more effort to compute with the exact posterior distribution using the binomial and Poisson likelihoods, with more realistic models for x_j and θ_j.)

In general, the population quantity (7.12) depends on the observed and unobserved groups. In this example, the number of unobserved clusters is very large, and so we can approximate (7.12) by

$$\overline{Y} = \frac{\mathrm{E}(x_j \theta_j)}{\mathrm{E}(x_j)}.$$

From the normal model (7.13), we can express these expectations as posterior means of functions of the hyperparameters:

$$\overline{Y} = \frac{\alpha\mu_x + \beta(\mu_x^2 + \tau_x^2)}{\mu_x}. \tag{7.16}$$

If the average responses θ_j and the group sizes x_j are independent, then $\beta = 0$ and (7.16) reduces to α. Otherwise, the estimate is corrected to reflect the goal of estimating the entire population, with each group counted in proportion to its size.

In general, the analysis of sample surveys with unequal sampling probabilities can be difficult because of the large number of potential categories, each with its own model for the distribution of y given x and a parameter for the frequency of that category in the population. At this point, hierarchical models may be needed to obtain good results.

7.5 Designed experiments

In statistical terminology, an *experiment* involves the assignment of treatments to units, with the assignment under the control of the experimenter. In an experiment, statistical inference is needed to generalize from the observed outcomes to the hypothetical outcomes that would have occurred if different treatments had been assigned. Typically, the analysis also should generalize to other units not studied, that is, thinking of the units included in the experiment as a sample from a larger population. We begin with a strongly ignorable design—the completely randomized experiment—and then consider more complicated experimental designs that are ignorable only conditional on information used in the treatment assignment process.

Completely randomized experiments

Notation for complete data.   Suppose, for notational simplicity, that the number of units n in an experiment is even, with half to receive the basic treatment A and half to receive the new active treatment B. We define the complete set of observables as $(y_i^A, y_i^B), i = 1, \ldots, n$, where y_i^A and y_i^B are the outcomes if the ith unit received treatment A or B, respectively. This model, which characterizes all potential outcomes as an $n \times 2$ matrix, requires the stability assumption; that is, the treatment applied to unit i is assumed to have no effect on the potential outcomes in the other $n-1$ units.

Causal effects in superpopulation and finite-population frameworks.   The A versus B causal effect for the ith unit is typically defined as $y_i^A - y_i^B$, and the overall causal estimand is typically defined to be the true average of the causal effects. In the superpopulation framework, the average causal effect is $\mathrm{E}(y_i^A - y_i^B|\theta) = \mathrm{E}(y_i^A|\theta) - \mathrm{E}(y_i^B|\theta)$, where the expectations average over the complete-data likelihood, $p(y_i^A, y_i^B|\theta)$. The average causal effect is thus a function of θ, with a posterior distribution induced by the posterior distribution of θ.

The finite population causal effect is $\overline{y}^A - \overline{y}^B$, for the finite population under study. In many experimental contexts, the superpopulation estimand is of primary interest, but it is also nice to understand the connection to finite-population inference in sample surveys.

Inclusion model. The data collection indicator for an experiment is $I = ((I_i^A, I_i^B), i = 1, \ldots, n)$, where $I_i = (1, 0)$ if the ith unit receives treatment A, $I_i = (0, 1)$ if the ith unit receives treatment B, and $I_i = (0, 0)$ if the ith unit receives neither treatment (for example, a unit whose treatment has not yet been applied). It is not possible for both I_i^A and I_i^B to equal 1. For a completely randomized experiment,

$$p(I|y, \phi) = p(I) = \begin{cases} \binom{n}{n/2}^{-1} & \text{if } \sum_{i=1}^n I_i^A = \sum_{i=1}^n I_i^B = \frac{n}{2}, \ I_i^A \neq I_i^B \text{ for all } i, \\ 0 & \text{otherwise.} \end{cases}$$

This treatment assignment is known and ignorable. The discussion of propensity scores on page 206 applied to the situation where $I_i = (0, 0)$ could not occur, and so the notation becomes $I_i = I_i^A$ and $I_i^B = 1 - I_i^A$.

Bayesian inference for superpopulation and finite-population estimands. Inference is quite simple under a completely randomized experiment, just as with simple random sampling. Because the treatment assignment is ignorable, we can perform posterior inference about the parameters θ using $p(\theta|y_{\text{obs}}) \propto p(\theta)p(y_{\text{obs}}|\theta)$. For example, under the usual iid mixture model,

$$p(y_{\text{obs}}|\theta) = \prod_{i:\, I_i=(1,0)} p(y_i^A|\theta) \prod_{i:\, I_i=(0,1)} p(y_i^B|\theta),$$

which yields to the standard Bayesian approach of the previous chapters, as applied to the parameters governing the marginal distribution of y_i^A, say θ_A, and those of the marginal distribution of y_i^B, say θ_B. (See Exercise 3.3, for example.)

Once posterior simulations have been obtained for the superpopulation parameters, θ, one can obtain inference for the finite-population quantities $(y_{\text{mis}}^A, y_{\text{mis}}^B)$ by drawing simulations from the posterior predictive distribution, $p(y_{\text{mis}}|\theta, y_{\text{obs}})$. The finite-population inference is trickier than in the sample survey example because only partial outcome information is available on the treated units. Under the iid mixture model,

$$p(y_{\text{mis}}|\theta, y_{\text{obs}}) = \prod_{i:\, I_i=(1,0)} p(y_i^B|\theta, y_i^A) \prod_{i:\, I_i=(1,0)} p(y_i^A|\theta, y_i^B),$$

which requires a model of the joint distribution, $p(y_i^A, y_i^B|\theta)$. (See Exercise 7.8.) The parameters governing the joint distribution of (y_i^A, y_i^B), say θ_{AB}, do not appear in the likelihood. The posterior distribution of θ_{AB} is found by averaging the conditional prior distribution, $p(\theta_{AB}|\theta_A, \theta_B)$ over the posterior distribution of θ_A and θ_B.

Large sample correspondence. Inferences for finite-population estimands such as $\bar{y}^A - \bar{y}^B$ are sensitive to aspects of the joint distribution of y_i^A and y_i^B, such as $\text{corr}(y_i^A, y_i^B|\theta)$, for which no data are available (in the usual experimental

B: 257	E: 230	A: 279	C: 287	D: 202
D: 245	A: 283	E: 245	B: 280	C: 260
E: 182	B: 252	C: 280	D: 246	A: 250
A: 203	C: 204	D: 227	E: 193	B: 259
C: 231	D: 271	B: 266	A: 334	E: 338

Table 7.4 *Yields of plots of millet arranged in a Latin square. Treatments A, B, C, D, E correspond to spacings of width 2, 4, 6, 8, 10 inches, respectively. Yields are in grams per inch of spacing. From Snedecor and Cochran (1989).*

setting in which each unit receives no more than one treatment). For large populations, however, the sensitivity vanishes if the causal estimand can be expressed as a comparison involving only θ_A and θ_B, the separate parameters for the outcome under each treatment. For example, suppose the n units are themselves randomly sampled from a much larger finite population of N units, and the causal estimate is the mean difference between treatments for all N units, $\bar{y}^A - \bar{y}^B$. Then it can be shown (see Exercise 7.8), using the central limit theorem, that the posterior distribution of the finite-population causal effect for large n and N/n is

$$(\bar{y}^A - \bar{y}^B)|y_{\text{obs}} \approx \text{N}\left(\bar{y}_{\text{obs}}^A - \bar{y}_{\text{obs}}^B, \frac{2}{n}(s_{\text{obs}}^{2A} + s_{\text{obs}}^{2B})\right), \tag{7.17}$$

where s_{obs}^{2A} and s_{obs}^{2B} are the sample variances of the observed outcomes under the two treatments. The practical similarity between the Bayesian results and the repeated sampling randomization-based results is striking, but not entirely unexpected considering the relation between Bayesian and sampling-theory inferences in large samples, as discussed in Section 4.4.

Randomized blocks, Latin squares, etc.

More complicated designs can be analyzed using the same principles, modeling the outcomes conditional on all factors that are used in determining treatment assignments. We present an example here of a Latin square; the exercises provide examples of other designs.

Example. Latin square experiment
Table 7.4 displays the results of an agricultural experiment carried out under a Latin square design with 25 plots (units) and five treatments labeled A, B, C, D, E. Using our general notation, the complete data y are a 25×5 matrix with one entry observed in each row, and the indicator I is a fully observed 25×5 matrix of zeros and ones. The estimands of interest are the average yields under each treatment.

Setting up a model under which the design is ignorable. The factors relevant in the design (that is, affecting the probability distribution of I) are the physical locations of the 25 plots, which can be coded as a 25×2 matrix x of horizontal and vertical coordinates. Any ignorable model must be of the form $p(y|x,\theta)$. If additional relevant information were available (for example, the location of a stream running through the field), it should of course be included in the analysis.

Inference for estimands of interest under various models. Under a model for $p(y|x,\theta)$, the design is ignorable, and so we can perform inference for θ based on the likelihood of the observed data, $p(y_{\mathrm{obs}}|x,\theta)$.

For example, the standard analysis of variance is equivalent to a linear regression, $y_{\mathrm{obs}}|X,\theta \sim \mathrm{N}(X\beta, \sigma^2 I)$, where X is a 25×13 matrix composed of five columns of indicators for the five treatments and four columns of indicators for the horizontal and vertical geographic coordinates for each plot, and the model parameters are $\theta = (\beta, \sigma^2)$. The five regression coefficients for the treatments could be grouped in a hierarchical model. (See Section 15.1 for more discussion of this sort of hierarchical linear regression model.)

A more complicated model has interactions between the treatment effects and the geographic coordinates; for example, the effect of treatment A might increase going from the left to the right side of the field. Such interactions can be included as additional terms in the regression model. The analysis is best summarized in two parts: (1) superpopulation inference for the parameters of the distribution of y given the treatments and x, and (2) finite-population inference obtained by averaging the distributions in the first step conditional on the values of x in the 25 plots.

Relevance of the Latin square design. Now suppose we are told that the data in Table 7.4 actually arose from a completely randomized experiment that just happened to be balanced in rows and columns. How would this affect our analysis? Actually, our analysis should not be affected at all, as it is still desirable to model the plot yields in terms of the plot locations as well as the assigned treatments. Nevertheless, under a completely randomized design, the treatment assignment would be ignorable under a simpler model of the form $p(y|\theta)$, and so a Bayesian analysis ignoring the plot locations would yield valid posterior inferences (without conditioning on plot location). However, the analysis conditional on plot location is more relevant given what we know, and would tend to yield more precise inference assuming the true effects of location on plot yields are modeled appropriately. This point is explored more fully in Exercise 7.6.

Sequential designs

Consider a randomized experiment in which the probabilities of treatment assignment for unit i depend on the results of the randomization or on the experimental outcomes on previously treated units. Appropriate Bayesian analysis of sequential experiments is sometimes described as essentially impossible and sometimes described as trivial, in that the data can be analyzed as if the treatments were assigned completely at random. Neither of these claims is true. A randomized sequential design is ignorable *conditional* on all the variables used in determining the treatment allocations, including time of entry in the study and the outcomes of any previous units that are used in the design. See Exercise 7.15 for a simple example. A sequential design is not strongly ignorable.

Including additional explanatory variables beyond the minimally adequate summary

From the design of a randomized study, we can determine the minimum set of explanatory variables required for ignorability; this minimal set, along with the treatment and outcome measurements, is called an *adequate summary* of the data, and the resulting inference is called a minimally adequate summary or simply a *minimal analysis*. As suggested earlier, sometimes the propensity scores can be such an adequate summary. In many examples, however, additional information is available that was not used in the design, and it is generally advisable to try to use all available information in a Bayesian analysis, thus going beyond the minimal analysis.

For example, suppose a simple random sample of size 100 is drawn from a large population that is known to be 51% female and 49% male, and the sex of each respondent is recorded along with the answer to the target question. A minimal summary of the data does not include sex of the respondents, but a better analysis models the responses conditional on sex and then obtains inferences for the general population by averaging the results for the two sexes, thus obtaining posterior inferences using the data as if they came from a stratified sample. (Posterior predictive checks for this problem could still be based on the simple random sampling design.)

On the other hand, if the population frequencies of males and females were not known, then sex would not be a fully observed covariate, and the frequencies of men and women would themselves have to be estimated in order to estimate the joint distribution of sex and the target question in the population. In that case, in principle the joint analysis could be informative for the purpose of estimating the distribution of the target variable, but in practice the adequate summary might be more appealing because it would not require the additional modeling effort involving the additional unknown parameters. See Exercise 7.6 for further discussion of this point.

Example. An experiment with treatment assignments based on observed covariates

An experimenter can sometimes influence treatment assignments even in a randomized design. For example, an experiment was conducted on 50 cows to estimate the effect of a feed additive (methionine hydroxy analog) on six outcomes related to the amount of milk fat produced by each cow. Four diets (treatments) were considered, corresponding to different levels of the additive, and three variables were recorded before treatment assignment: lactation number (seasons of lactation), age, and initial weight of cow.

Cows were initially assigned to treatments completely at random, and then the distributions of the three covariates were checked for balance across the treatment groups; several randomizations were tried, and the one that produced the 'best' balance with respect to the three covariates was chosen. The treatment assignment is ignorable (because it depends only on fully observed covariates and not on unrecorded variables such as the physical appearances of the cows or the times at which the cows entered the study) but unknown (because the decisions whether to re-randomize are not explained). In our general notation, the covariates x are a 50×3 matrix and the complete data y are a 50×24 matrix, with only one of 4 possible subvectors of dimension 6 observed for each unit (there were 6 different outcome measures relating to the cows' diet after treatment).

The minimal analysis uses a model of mean daily milk fat conditional on the treatment and the three pre-treatment variables. This analysis, based on ignorability, implicitly assumes distinct parameters; reasonable violations of this assumption should not have large effects on our inferences for this problem (see Exercise 7.3). A natural analysis of these data uses linear regression, after appropriate transformations (see Exercise 14.5). As usual, one would first compute and draw samples from the posterior distribution of the superpopulation parameters—the regression coefficients and variance. In this example, there is probably no reason to compute inferences for finite-population estimands such as the average treatment effects, since there was no particular interest in the 50 cows that happened to be in the study. Sensitivity analysis and model checking would be performed as described in Chapter 6.

If the goal is more generally to understand the effects of the treatments, it would be better to model the multivariate outcome y—the six post-treatment measurements—conditional on the treatment and the three pre-treatment variables. After appropriate transformations, a multivariate normal regression model might be reasonable (see Chapter 19).

The only issue we need to worry about is modeling y, unless ϕ is not distinct from θ (for example, if the treatment assignment rule chosen is dependent on the experimenter's belief about the treatment efficacy).

For fixed model and fixed data, the posterior distribution of θ and finite population estimands is the same for all ignorable data collection models. However, better designs are likely to yield data exhibiting less sensitivity to variations in models. In the cow experiment, a better design would have been to explicitly balance over the covariates, most simply using a randomized block design.

7.6 Sensitivity and the role of randomization

We have seen how ignorable designs facilitate Bayesian analysis by allowing us to model observed data directly, without needing to consider the data collection mechanism. To put it another way, posterior inference for θ and y_{mis} is completely insensitive to an ignorable design, given a fixed model for the data.

Complete randomization

How does randomization fit into this picture? First, consider the situation with no fully observed covariates x, in which case the *only* way to have an ignorable design—that is, a probability distribution, $p(I_1, \ldots, I_n | \phi)$, that is invariant to permutations of the indexes—is to randomize (excluding the degenerate designs in which all units get assigned one of the treatments).

However, for any given inferential goal, some ignorable designs are better than others in the sense of being more likely to yield data that provide more precise inferences about estimands of interest. For example, for estimating the average treatment effect in a group of 10 subjects with a noninformative prior distribution on the distribution of outcomes under each of two treatments, the strategy of assigning a random half of the subjects to each treatment is generally better than flipping a coin to assign a treatment to each subject independently, because the expected posterior precision of estimation is higher with equal numbers of subjects in each treatment group.

Randomization given covariates

When fully observed covariates are available, randomized designs are in competition with deterministic ignorable designs, that is, designs with propensity scores equal to 0 or 1. What are the advantages, if any, of randomization in this setting, and how does knowledge of randomization affect Bayesian data analysis? Even in situations where little is known about the units, distinguishing unit-level information is usually available in some form, for example telephone numbers in a telephone survey or physical location in an agricultural experiment. For a simple example, consider a long stretch of field divided into 12 adjacent plots, on which the relative effects of two fertilizers, A and B, are to be estimated. Compare two designs: assignment of the six plots to each treatment at random, or the systematic design ABABABBABABA. Both designs are ignorable given x, the locations of the plots, and so a usual Bayesian analysis of y given x is appropriate. The randomized design is ignorable even not given x, but in this setting it would seem advisable, for the purpose of fitting an accurate model, to include at least a linear trend for $E(y|x)$ no matter what design is used to collect these data.

So are there any potential advantages to the randomized design? Suppose the randomized design were used. Then an analysis that pretends x is unknown is still a valid Bayesian analysis. Suppose such an analysis is conducted, yielding a posterior distribution for y_{mis}, $p(y_{mis}|y_{obs})$. Now pretend x is suddenly observed; the posterior distribution can then be updated to produce $p(y_{mis}|y_{obs}, x)$. (Since the design is ignorable, the inferences are also implicitly conditional on I.) Since both analyses are correct given their respective states of knowledge, we would expect them to be consistent with each other, with $p(y_{mis}|y_{obs}, x)$ expected to be more precise as well as conditionally appropriate given x. If this is not the case, we should reconsider the modeling assumptions. This extra step of model examination is not available in the systematic design without explicitly averaging over a distribution of x.

Another potential advantage of the randomized design is the increased flexibility for carrying out posterior predictive checks in hypothetical future replications. With the randomized design, future replications give different treatment assignments to the different plots.

Finally, any particular systematic design is sensitive to associated particular model assumptions about y given x, and so repeated use of a single systematic design would cause a researcher's inferences to be systematically dependent on a particular assumption. In this sense, there is a benefit to using different patterns of treatment assignment for different experiments; if nothing else about the experiments is specified, they are exchangeable, and the global treatment assignment is necessarily randomized over the set of experiments.

Designs that 'cheat'

Another advantage of randomization is to make it more difficult for an experimenter to 'cheat,' intentionally or unintentionally, by choosing sample selections or treatment assignments in such a way as to bias the results (for example, assigning treatments A and B to sunny and shaded plots, respectively). This sort of complication can enter a Bayesian analysis in several ways:

1. If treatment assignment depends on unrecorded covariates (for example, an indicator for whether each plot is sunny or shaded), then it is not ignorable, and the resulting unknown selection effects must be modeled, at best a difficult enterprise with heightened sensitivity to model assumptions.

2. If the assignment depends on recorded covariates, then the dependence of the outcome variable the covariates, $p(y|x, \theta)$, must be modeled. Depending on the actual pattern of treatment assignments, the resulting inferences may be highly sensitive to the model; for example, if all sunny plots get A and all shaded plots get B, then the treatment indicator and the sunny/shaded indicator are identical, so the observed-data likelihood provides no information to distinguish between them.

3. Even a randomized design can be nonignorable if the parameters are not distinct. For example, consider an experimenter who uses complete randomization when he thinks that treatment effects are large but uses randomized blocks for increased precision when he suspects smaller effects. In this case, the treatment assignment mechanism depends on the prior distribution of the treatment effects; in our general notation, ϕ and θ are dependent in their prior distribution, and we no longer have ignorability. In practice, of course, one can often ignore such effects if the data dominate the prior distribution, but it is theoretically important to see how they fit into the Bayesian framework.

Bayesian analysis of nonrandomized studies

Randomization is a method of ensuring ignorability and thus making Bayesian inferences less sensitive to assumptions. Consider the following nonrandomized sampling scheme: in order to estimate the proportion of the adults in a city who hold a certain opinion, an interviewer stands on a street corner and interviews all the adults who pass by between 11 am and noon on a certain day. This design can be modeled in two ways: (1) nonignorable because the probability that adult i is included in the sample depends on that person's travel pattern, a variable that is not observed for the $N - n$ adults not in the sample; or (2) ignorable because the probability of inclusion in the sample depends on a fully observed indicator variable, x_i, which equals 1 if adult i passed by in that hour and 0 otherwise. Under the nonignorable parameterization, we must specify a model for I given y and include that factor in the likelihood. Under the ignorable model, we must perform inference for the distribution of y given x, with no data available when $x = 0$. In either case, posterior inference for the estimand of interest, $\bar{y}$, is highly sensitive to the prior distribution unless n/N is close to 1.

In contrast, if covariates x are available and a nonrandomized but ignorable and roughly balanced design is used, then inferences are typically less sensitive to prior assumptions about the design mechanism. Of course, any strongly ignorable design is *implicitly* randomized over all variables not in x, in the sense that if two units i and j have identical values for all covariates x, then their propensity scores are equal: $p(I_i|x) = p(I_j|x)$. Our usual goal in randomized or nonrandomized studies is to set up an ignorable model so we can use the standard methods of Bayesian inference developed in most of this book, working with $p(\theta|x, y_{\text{obs}}) \propto p(\theta)p(x, y_{\text{obs}}|\theta)$.

7.7 Observational studies

Comparison to experiments

In an *observational* or *nonexperimental* study, data are typically analyzed as if they came from an experiment—that is, with a treatment and an outcome recorded for each unit—but the treatments are simply *observed* and not under the experimenter's control. For example, the SAT coaching study presented in Section 5.5 involves experimental data, because the students were assigned to the two treatments by the experimenter in each school. In that case, the treatments were assigned randomly; in general, such studies are often, but not necessarily, randomized. The data would have arisen from an observational study if, for example, the students themselves had chosen whether or not to enroll in the coaching programs.

In a randomized experiment, the groups receiving each treatment will be similar, on average, or with differences that are known ahead of time by the experimenter (if a design is used that involves unequal probabilities of assignment). In an observational study, in contrast, the groups receiving each treatment can differ greatly, and by factors not measured in the study.

A well-conducted observational study can provide a good basis for drawing causal inferences, provided that it (1) controls well for background differences between the units exposed to the different treatments; (2) has enough independent units exposed to each treatment to provide reliable results (that is, narrow-enough posterior intervals); (3) the study is designed without reference to the outcome of the analysis; (4) attrition, dropout, and other forms of unintentional missing data are minimized or else appropriately modeled; and (5) the analysis takes account of the information used in the design.

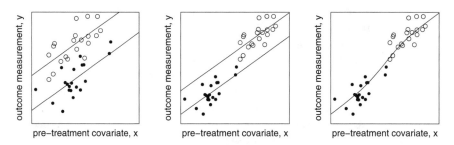

Figure 7.2 *Hypothetical-data illustrations of sensitivity analysis for observational studies. In each graph, the circles and dots represent treated and control units, respectively. (a) The first graph shows balanced data, as from a randomized experiment, and the difference between the two lines shows the estimated treatment effect from a simple linear regression. (b, c) The second and third graphs show unbalanced data, as from a poorly-conducted observational study, with two different models fit to the data. The estimated treatment effect for the unbalanced data in (b) and (c) is highly sensitive to the form of the fitted model, even when the treatment assignment is ignorable.*

Many observational studies do not satisfy these criteria. In particular, it is important that systematic pre-treatment differences between groups be included in the analysis, using available background information on the units or with a realistic nonignorable model. Minor differences between different treatment groups can be controlled, at least to some extent, using models such as regression, but for larger differences, posterior inferences will become highly sensitive to the functional form and other details used in the adjustment model. In some cases, the use of estimated propensity scores can be helpful in limiting the sensitivity of such analyses by restricting to a subset of treated and control units with similar distributions of the covariates. In this context, the propensity score is the probability (as a function of the covariates) that a unit receives the treatment.

Figure 7.2 illustrates the connection between lack of balance in data collection and sensitivity to modeling assumptions in the case with a single continuous covariates. In the first of the three graphs, which could have come from a randomized experiment, the two groups are similar with respect to the pretreatment covariate. As a result, the estimated treatment effect—the difference between the two regression lines in an additive model—is relatively insensitive to the form of the model fit to $y|x$. The second and third graphs show data that could arise from a poorly-balanced observational study (for example, an economic study in which a certain training program is taken only by people who already had relatively higher incomes). From the second graph, we see that a linear regression yields a positive estimated treatment effect. However, the third graph shows that the identical data could be fitted just as well with a mildly nonlinear relation between y and x, without any treatment effect. In this hypothetical scenario, estimated treatment effects from the observational study are extremely sensitive to the form of the fitted model.

Another way to describe this sensitivity is by using estimated propensity scores. Suppose we estimate the propensity score, $\Pr(I_i = 1 | X_i)$, assuming strongly ignorable treatment assignment, for example using a logistic regression model (see Section 3.7 and Chapter 16). In the case of Figure 7.2b,c, there will be little or no overlap in the estimated propensity scores in the two treatment conditions. This technique works even with multivariate X because the propensity score is a scalar and so can be extremely useful in observational studies with many covariates to diagnose and reveal potential sensitivity to models for causal effects.

Bayesian inference for observational studies

In Bayesian analysis of observational studies, it is typically important to gather many covariates so that the treatment assignment is close to ignorable conditional on the covariates. Once ignorability is accepted, the observational study can be analyzed as if it were an experiment with treatment assignment probabilities depending on the included covariates. We shall illustrate this approach in Section 14.3 for the example of estimating the advantage of incumbency in legislative elections in the United States. In such examples, the collection of relevant data is essential, because without enough covariates to make the design approximately ignorable, sensitivity of inferences to plausible missing data models can be so great that the observed data may provide essentially no information about the questions of interest. As illustrated in Figure 7.2, inferences can still be sensitive to the model even if ignorability is accepted, if there is little overlap in the distributions of covariates for units receiving different treatments.

Data collection and organization methods for observational studies include matched sampling, subclassification, blocking, and stratification, all of which are methods of introducing covariates x in a way to limit the sensitivity of inference to the specific form of the y given x models by limiting the range of x-space over which these models are being used for extrapolation. Specific techniques that arise naturally in Bayesian analyses include poststratification and the analysis of covariance (regression adjustment). Under either of these approaches, inference is performed conditional on covariates and then averaged over the distribution of these covariates in the population, thereby correcting for differences between treatment groups.

Two general difficulties arise when implementing this plan for analyzing observational studies:

1. Being out of the experimenter's control, the treatments can easily be unbalanced. Consider the educational testing example of Section 5.5 and suppose the data arose from observational studies rather than experiments. If, for example, good students received coaching and poor students received no coaching, then the inference in each school would be highly sensitive to

model assumptions (for example, the assumption of an additive treatment effect, as illustrated in Figure 7.2). This difficulty alone can make a dataset useless in practice for answering questions of substantive interest.

2. Typically, the actual treatment assignment in an observational study depends on several unknown and even possibly unmeasurable factors (for example, the state of mind of the student on the day of decision to enroll in a coaching program), and so inferences about θ are sensitive to assumptions about the nonignorable model for treatment assignment.

Data gathering and analysis in observational studies for causal effects is a vast area. Our purpose in raising the topic here is to connect the general statistical ideas of ignorability, sensitivity, and using available information to the specific models of applied Bayesian statistics discussed in the other chapters of this book. For example, in Sections 7.4 and 7.5, we illustrated how the information inherent in stratification in a survey and blocking in an experiment can be used in Bayesian inference via hierarchical models. In general, the use of covariate information increases expected precision under specific models and, if well done, reduces sensitivity to alternative models. However, too much blocking can make modeling more difficult and sensitive. See Exercise 7.6 for more discussion of this point.

Causal inference and principal stratification

Principal stratification refers to a method of adjusting for an outcome variable that is 'intermediate to' or 'on the causal pathway' to the final outcome y. Suppose we call this intermediate outcome C, with corresponding potential outcomes $C(1)$ and $C(0)$, respectively, if an individual is assigned to treatment condition 1 or 0. If I is an indicator variable for assignment to condition 1, then the observed value is C_{obs}, where $C_{obs,i} = I_i C_i(1) + (1 - I_i)C_i(0)$. A common mistake is to treat C_{obs} as if it were a covariate, which it is not, unless $C(1) \equiv C(0)$, and do an analysis stratified by C_{obs}. The correct procedure is to stratify on the joint values $(C(1), C(0))$, which are unaffected by assignment I and so can be treated as a vector covariate. Thus, stratifying by $C(1), C(0)$ is legitimate and is called 'principal stratification.'

There are many examples of this general principle of stratifying on intermediate outcomes, the most common being compliance with assigned treatment. This is an important topic for a few reasons. First, it is a bridge to the economists' tool of instrumental variables, as we shall discuss. Second, randomized experiments with noncompliance can be viewed as 'islands' between the 'shores' of perfect randomized experiments and purely observational studies. Third, noncompliance is an important introduction to more complex examples of principal stratification.

Example. A randomized experiment with noncompliance

A large randomized experiment assessing the effect of vitamin A supplements on infant mortality was conducted in Indonesian villages, where vitamin A was only available to those assigned to take it. In the context of this example I is an

Category	Assignment, $I_{\text{obs},i}$	Exposure, $U_{\text{obs},i}$	Survival, $Y_{\text{obs},i}$	#units in category
Complier or never-taker	0	0	0	74
Complier or never-taker	0	0	1	11514
Never-taker	1	0	0	34
Never-taker	1	0	1	2385
Complier	1	1	0	12
Complier	1	1	1	9663

Table 7.5 *Summary statistics from an experiment on vitamin A supplements, where the vitamin was available (but optional) only to those assigned the treatment. The table shows number of units in each assignment/exposure/outcome condition. From Sommer and Zeger (1991).*

indicator for assignment to the vitamin A treatment. The intermediate outcome in this case is compliance. There are two principal strata here, defined by whether or not the units would take vitamin A if assigned it: compliers and noncompliers. The strata are observed for the units assigned to take vitamin A because we known whether they comply or not, but the strata are not observed for the units assigned to control because we do not know what they would have done had they been assigned to the vitamin A group. In terms of the notation above, we know $C_i(0) = 0$ for everyone, indicating no one would take vitamin A when assigned not to take it (they have no way of getting it), and know $C_i(1)$ only for those units assigned to take vitamin A. The data are summarized in Table 7.5.

Complier average causal effects and instrumental variables

In a randomized experiment with noncompliance, such as the Indonesian vitamin A experiment just described, the objective is to estimate the causal effect of the treatment within each principal stratum (see Exercise 7.16). We shall use that setting to describe the instrumental variable approach (popular in econometrics) for estimating causal effects within each stratum and then compare it to the Bayesian likelihood approach. The average causal effects for compliers and noncompliers are called the complier average causal effect (CACE) and the noncomplier average causal effect (NACE), respectively. The overall average causal effect of being assigned to the treatment (averaged over compliance status) is

$$\overline{Y}_1 - \overline{Y}_0 = p_c \cdot \text{CACE} + (1 - p_c) \cdot \text{NACE}, \qquad (7.18)$$

where p_c is the proportion of compliers in the population. Expression (7.18) is known as the *intention-to-treat effect* because it measures the effect over the entire population that we intend to treat (including those who do not comply with the treatment assigned and therefore do not reap its potential benefits).

In the case of a randomized experiment, we can estimate the intention-to-treat effect $\overline{Y}_1 - \overline{Y}_0$ with the usual estimate, $\overline{y}_1 - \overline{y}_0$. It is also straightforward to estimate the proportion of compliers p_c in the population, with the estimate being the proportion of compliers in the random half assigned to take vitamin A. We would like to estimate CACE—the effect of the treatment for those who actually would take it if assigned.

Suppose we assume that there is no effect on mortality of being *assigned* to take vitamin A for those who would not take it even when assigned to take it. This is known as the exclusion restriction because it excludes the causal effect of treatment *assignment* for noncompliers. This assumption means that $NACE = 0$, and then a simple estimate of CACE is

$$\widehat{CACE} = (\bar{y}_1 - \bar{y}_0)/\hat{p}_c, \tag{7.19}$$

which is called the *instrumental variables estimate* in economics. The instrumental variables estimate for CACE is thus the estimated intention-to-treat effect on Y, divided by the proportion of the treatment group who are compliers, that is, who actually receive the new treatment.

Bayesian causal inference with noncompliance

The instrumental variables approach to noncompliance is very effective at revealing how simple assumptions can be used to address noncompliance. The associated estimate (7.19) is a 'method of moments' estimate, however, which is generally far from satisfactory. The Bayesian approach is to treat the unknown compliance status for each person in the control group explicitly as missing data. A particular advantage of the Bayesian approach is the freedom to relax the exclusion restriction. A more complex example with noncompliance involves a study where encouragement to receive influenza shots is randomly assigned, but many patients do not comply with their encouragements, thereby creating two kind of noncompliers (those who are encouraged to receive a shot but do not, and those who are not encouraged to receive a shot but do so anyway) in addition to compliers. The computations are easily done using iterative simulation methods of the sort discussed in Part III.

7.8 Censoring and truncation

We illustrate a variety of possible missing data mechanisms by considering a series of variations on a simple example. In all these variations, it is possible to state the appropriate model directly—although as examples become more complicated, it is useful and ultimately necessary to work within a formal structure for modeling data collection.

The example involves observation of a portion of a 'complete' data set that are N weighings y_i, $i = 1, \ldots, N$, of an object with unknown weight θ, that are assumed to follow a $N(\theta, 1)$ distribution, with a noninformative uniform prior distribution on θ. Initially N is fixed at 100. In each variation, a different data-collection rule is followed, and in each case only $n = 91$ of the original N measurements are observed. We label $\bar{y}_{obs}$ as the mean of the observed measurements.

For each case, we first describe the data collection and then present the Bayesian analysis. This example shows how a fixed data set can have different inferential implications depending on how it was collected.

1. Data missing completely at random

Suppose we weigh an object 100 times on an electronic scale with a known $N(\theta, 1)$ measurement distribution, where θ is the true weight of the object. Randomly, with probability 0.1, the scale fails to report a value, and we observe 91 values. Then the complete data y are $N(\theta, 1)$ subject to Bernoulli sampling with known probability of selection of 0.9. Even though the sample size $n = 91$ is binomially distributed under the model, the posterior distribution of θ is the same as if the sample size of 91 had been fixed in advance.

The inclusion model is $I_i \sim \text{Bernoulli}(0.9)$, independent of y, and the posterior distribution is,

$$p(\theta | y_{\text{obs}}, I) = p(\theta | y_{\text{obs}}) = N(\theta | \bar{y}_{\text{obs}}, 1/91).$$

2. Data missing completely at random with unknown probability of missingness

Consider the same situation with 91 observed values and 9 missing values, except that the probability that the scale randomly fails to report a weight is unknown. Now the complete data are $N(\theta, 1)$ subject to Bernoulli sampling with unknown probability of selection, π. The inclusion model is then $I_i | \pi \sim \text{Bernoulli}(\pi)$, independent of y.

The posterior distribution of θ is the same as in variation 1 only if θ and π are *independent* in their prior distribution, that is, are 'distinct' parameters. If θ and π are dependent, then $n = 91$, the number of reported values, provides extra information about θ beyond the 91 measured weights. For example, if it is known that $\pi = \theta/(1 + \theta)$, then $n/N = 91/100$ can be used to estimate π, and thereby $\theta = \pi/(1 - \pi)$, even if the measurements y were not recorded.

Formally, the posterior distribution is,

$$\begin{aligned} p(\theta, \pi | y_{\text{obs}}, I) &\propto p(\theta, \pi) p(y_{\text{obs}}, I | \theta, \pi) \\ &\propto p(\theta, \pi) N(\theta | \bar{y}_{\text{obs}}, 1/91) \text{Bin}(n | 100, \pi). \end{aligned}$$

This formula makes clear that if θ and π are independent in the prior distribution, then the posterior inference for θ is as above. If $\pi = \theta/(1 + \theta)$, then the posterior distribution of θ is,

$$p(\theta | y, I) \propto N(\theta | \bar{y}_{\text{obs}}, 1/91) \text{Bin}(n | 100, \theta/(1 + \theta)).$$

Given $n = 91$ and $\bar{y}_{\text{obs}}$, this density can be calculated numerically over a range of θ, and then simulations of θ can be drawn using the inverse-cdf method.

3. Censored data

Now modify the scale so that all weights produce a report, but the scale has an upper limit of 200 kg for reports: all values above 200 kg are reported as 'too heavy.' The complete data are still $N(\theta, 1)$, but the observed data are *censored*; if we observe 'too heavy,' we know that it corresponds to a weighing with a reading above 200. Now, for the same 91 observed weights and 9 'too heavy' measurements, the posterior distribution of θ differs from that of variation 2 of this example. In this case, the contributions to the likelihood from the 91 numerical measurements are normal densities, and the contributions from the 9 'too heavy' measurements are of the form $\Phi(\theta - 200)$, where Φ is the normal cumulative distribution function.

The inclusion model is $\Pr(I_i = 1 | y_i) = 1$ if $y_i \leq 200$, and 0 otherwise. The posterior distribution is,

$$
\begin{aligned}
p(\theta | y_{\text{obs}}, I) &= p(\theta) p(y_{\text{obs}}, I | \theta) \\
&= p(\theta) \int p(y_{\text{obs}}, y_{\text{mis}}, I | \theta) dy_{\text{mis}} \\
&= p(\theta) \binom{100}{91} \prod_{i=1}^{91} N(y_{\text{obs}\, i} | \theta, 1) \prod_{i=1}^{9} \Phi(\theta - 200) \\
&\propto N(\theta | \bar{y}_{\text{obs}}, 1/91)[\Phi(\theta - 200)]^9.
\end{aligned}
$$

Given $\bar{y}_{\text{obs}}$, this density can be calculated numerically over a range of θ, and then simulations of θ can be drawn using the inverse-cdf method.

4. Censored data with unknown censoring point

Now extend the experiment by allowing the censoring point to be unknown. Thus the complete data are distributed as $N(\theta, 1)$, but the observed data are censored at an unknown ϕ, rather than at 200 as in the previous variation. Now the posterior distribution of θ differs from that of the previous variation because the contributions from the 9 'too heavy' measurements are of the form $\Phi(\theta - \phi)$. Even when θ and ϕ are *a priori* independent, these 9 contributions to the likelihood create dependence between θ and ϕ in the posterior distribution, and so to find the posterior distribution of θ, we must consider the joint posterior distribution $p(\theta, \phi)$.

The posterior distribution is $p(\theta, \phi | y_{\text{obs}}, I) \propto p(\phi | \theta) N(\theta | \bar{y}_{\text{obs}}, 1/91)[\Phi(\theta - \phi)]^9$. Given $\bar{y}_{\text{obs}}$, this density can be calculated numerically over a grid of (θ, ϕ), and then simulations of (θ, ϕ) can be drawn using the grid method (as in the example in Figure 3.4 on page 91).

We can formally derive the censored-data model using the observed and missing-data notation, as follows. Label $y = (y_1, \ldots, y_N)$ as the original $N = 100$ uncensored weighings: the complete data. The observed information consists of the $n = 91$ observed values, $y_{\text{obs}} = (y_{\text{obs}\, 1}, \ldots, y_{\text{obs}\, 91})$, and the inclusion vector, $I = (I_1, \ldots, I_{100})$, which is composed of 91 ones and 9 zeros. There are no covariates, x.

The complete-data likelihood in this example is

$$p(y|\theta) = \prod_{i=1}^{100} N(y_i|\theta, 1),$$

and the likelihood of the inclusion vector, given the complete data, has a simple iid form:

$$p(I|y, \phi) = \prod_{i=1}^{100} p(I_i|y_i, \phi)$$

$$= \prod_{i=1}^{100} \begin{cases} 1 & \text{if } (I_i = 1 \text{ and } y_i \le \phi) \text{ or } (I_i = 0 \text{ and } y_i > \phi) \\ 0 & \text{otherwise.} \end{cases}$$

For valid Bayesian inference we must condition on all observed data, which means we need the joint likelihood of y_{obs} and I, which we obtain mathematically by integrating out y_{mis} from the complete-data likelihood:

$$p(y_{\text{obs}}, I|\theta, \phi) = \int p(y, I|\theta, \phi) dy_{\text{mis}}$$

$$= \int p(y|\theta, \phi) p(I|y, \theta, \phi) dy_{\text{mis}}$$

$$= \prod_{i: I_i=1} N(y_i|\theta, 1) \prod_{i: I_i=0} \int_\phi^\infty N(y_i|\theta, 1) p(I_i|y_i, \phi) dy_i$$

$$= \prod_{i: I_i=1} N(y_i|\theta, 1) \prod_{i: I_i=0} \Phi(\theta - \phi)$$

$$= \left[\prod_{i=1}^{91} N(y_{\text{obs}\,i}|\theta, 1) \right] \left[\Phi(\theta - \phi) \right]^9. \tag{7.20}$$

Since the joint posterior distribution $p(\theta, \phi|y_{\text{obs}}, I)$ is proportional to the joint prior distribution of (θ, ϕ) multiplied by the likelihood (7.20), we have provided an algebraic illustration of a case where the unknown ϕ cannot be ignored in making inferences about θ.

Thus we can see that the missing data mechanism is nonignorable. The likelihood of the observed measurements, if we (mistakenly) ignore the observation indicators, is

$$p(y_{\text{obs}}|\theta) = \prod_{i=1}^{91} N(y_{\text{obs}\,i}|\theta, 1),$$

which is wrong—crucially different from the appropriate likelihood (7.20)—because it omits the factors corresponding to the censored observations.

5. Truncated data

Now suppose the object is weighed by someone else who only provides to you the 91 observed values, but not the number of times the object was weighed. Also, suppose, as in our first censoring example above, that we know that no values over 200 are reported by the scale. The complete data can still be viewed as $N(\theta, 1)$, but the observed data are *truncated* at 200. The likelihood of each observed data point in the truncated distribution is a normal density divided by a normalizing factor of $\Phi(200 - \theta)$. We can proceed by working with this observed data likelihood, but we first demonstrate the connection between censoring and truncation. Truncated data differ from censored data in that no count of observations beyond the truncation point is available. With censoring, the *values* of observations beyond the truncation point are lost but their number is observed.

Now that N is unknown, the joint posterior distribution is,

$$p(\theta, N | y_{\text{obs}}, I) \;\propto\; p(\theta, N) p(y_{\text{obs}}, I | \theta, N)$$

$$\propto\; p(\theta, N) \binom{N}{91} N(\theta | \overline{y}_{\text{obs}}, 1/91) [\Phi(\theta - 200)]^{N-91}.$$

So the marginal posterior distribution of θ is,

$$p(\theta | y_{\text{obs}}, I) \propto p(\theta) N(\theta | \overline{y}_{\text{obs}}, 1/91) \sum_{N=91}^{\infty} p(N | \theta) \binom{N}{91} [\Phi(\theta - 200)]^{N-91}.$$

If $p(\theta, N) \propto 1/N$, then this becomes,

$$p(\theta | y_{\text{obs}}, I) \;\propto\; N(\theta | \overline{y}_{\text{obs}}, 1/91) \sum_{N=91}^{\infty} (1/N) \binom{N}{91} [\Phi(\theta - 200)]^{N-91}$$

$$\propto\; N(\theta | \overline{y}_{\text{obs}}, 1/91) \sum_{N=91}^{\infty} \binom{N-1}{90} [\Phi(\theta - 200)]^{N-91}$$

$$=\; N(\theta | \overline{y}_{\text{obs}}, 1/91)[1 - \Phi(\theta - 200)]^{-91},$$

where the last line can be derived because the summation has the form of a negative binomial density with $\theta = N - 91$, $\alpha = 91$, and $\frac{1}{\beta+1} = \Phi(\theta - 200)$ (see Appendix A). Thus there are two ways to end up with the posterior distribution for θ in this case, by using the truncated likelihood or by viewing this as a case of censoring with $p(N) \propto 1/N$.

6. Truncated data with unknown truncation point

Finally, extend the variations to allow an unknown truncation point; that is, the complete data are $N(\theta, 1)$, but the observed data are truncated at an unknown value ϕ. Here the posterior distribution of θ is a mixture of posterior distributions with known truncation points (from the previous variation), averaged over the posterior distribution of the truncation point, ϕ. This posterior distribution differs from the analogous one for censored data in variation

4. With censored data and an unknown censoring point, the proportion of values that are observed provides relatively powerful information about the censoring point ϕ (in units of standard deviations from the mean), but this source of information is absent with truncated data.

Now the joint posterior distribution is,

$$p(\theta, \phi, N | y_{\mathrm{obs}}, I) \propto p(\theta, \phi, N) \binom{N}{91} \mathrm{N}(\theta | \overline{y}_{\mathrm{obs}}, 1/91)[\Phi(\theta - \phi)]^{N-91}.$$

Once again, we can sum over N to get a marginal density of (θ, ϕ). If, as before, $p(N | \theta, \phi) \propto 1/N$, then

$$p(\theta, \phi | y_{\mathrm{obs}}, I) \propto p(\theta, \phi) \mathrm{N}(\theta | \overline{y}_{\mathrm{obs}}, 1/91)[1 - \Phi(\theta - \phi)]^{-91}.$$

With a noninformative prior density, this joint posterior density actually implies an improper posterior distribution for ϕ, because as $\phi \to \infty$, the factor $[1 - \Phi(\theta - \phi)]^{-91}$ approaches 1. In this case, the marginal posterior density for θ is just,

$$p(\theta | y_{\mathrm{obs}}) = \mathrm{N}(\theta | \overline{y}_{\mathrm{obs}}, 1/91),$$

as in the first variation of this example.

Obviously, we could continue to exhibit more and more complex variations in which the data collection mechanism influences the posterior distribution.

More complicated patterns of missing data

Incomplete data can be observed in other forms too, such as rounded or binned data (for example, heights rounded to the nearest inch, ages rounded down to the nearest integer, or income reported in discrete categories); see Exercise 3.5. With categorical data, it can happen that one knows which of a set of categories a data point belongs to, but not the exact category. For example, a survey respondent might report being Christian without specifying Protestant, Catholic, or other. Section 21.6 illustrates patterns of missing categorical data induced by nonresponse in a survey.

For more complicated missing data patterns, it is necessary to generalize the notation of Section 7.2 to allow for partial information such as censoring points, rounding, or data observed in coarse categories. The general Bayesian approach still holds, but now the observation indicators I_i are not simply 0's and 1's but more generally indicate which set in the sample space y_i can belong to.

7.9 Discussion

In general, the method of data collection dictates the minimal level of modeling required for a valid Bayesian analysis, that is, conditioning on all information used in the design—for example, conditioning on strata and clusters in a sample survey or blocks in an experiment. A Bayesian analysis that is conditional on enough information can ignore the data collection mechanism for inference although not necessarily for model checking. As long as data have been recorded on all the variables that need to be included in the model—whether for scientific modeling purposes or because they are used for data collection—one can proceed with the methods of modeling and inference discussed in the other chapters of this book, notably using regression models for $p(y_{obs}|x, \theta)$ as in the models of Parts IV and V. As usual, the greatest practical advantages of the Bayesian approach come from (1) accounting for uncertainty in a multiparameter setting, (2) hierarchical modeling, and (3) not requiring a particular point estimate (this last benefit is clearly revealed with principal stratification).

7.10 Bibliographic note

The material in this chapter on the role of study design in Bayesian data analysis develops from a sequence of contributions on Bayesian inference with missing data, where even sample surveys and studies for causal effects are viewed as problems of missing data. The general Bayesian perspective was first presented in Rubin (1976), which defined the concepts of ignorability, missing at random, and distinctness of parameters; related work appears in Dawid and Dickey (1977). The notation of potential outcomes with fixed unknown values in randomized experiments dates back to Neyman (1923) and is standard in that context (see references in Speed, 1990, and Rubin, 1990); this idea was introduced for causal inference in observational studies by Rubin (1974b). More generally, Rubin (1978a) applied the perspective to Bayesian inference for causal effects, where treatment assignment mechanisms were treated as missing-data mechanisms. Dawid (2000) and the accompanying discussions present a variety of Bayesian and related perspectives on causal inference; see also Greenland, Robins, and Pearl (1999), Robins (1998), Rotnitzky, Robins, and Scharfstein (1999), and Pearl (2000).

David et al. (1986) examined the reasonableness of the missing-at-random assumption for a problem in missing data imputation. The stability assumption in experiments was defined in Rubin (1980a), further discussed in the second chapter of Rubin (1987a), which explicates this approach to survey sampling, and extended in Rubin (1990). Smith (1983) discusses the role of randomization for Bayesian and non-Bayesian inference in survey sampling. Work on Bayesian inference before Rubin (1976) did not explicitly consider models for the data collection process, but rather developed the analysis directly from assumptions of exchangeability; see, for example, Ericson (1969) and Scott and Smith (1973) for sample surveys and Lindley and Novick (1981) for experiments. Rosenbaum and Rubin (1983a) introduced the expression 'strongly ignorable.'

The problem of Bayesian inference for data collected under sequential designs has been the subject of much theoretical study and debate, for example, Barnard (1949), Anscombe (1963), Edwards, Lindman, and Savage (1963), and Pratt (1965). Berger (1985, Chapter 7), provides an extended discussion from the perspective of decision theory. Rosenbaum and Rubin (1984a) and Rubin (1984) discuss the relation between sequential designs and robustness to model uncertainty.

There is a vast statistical literature on the general problems of missing data, surveys, experiments, observational studies, and censoring and truncation discussed in this chapter. We present a few of the recent references that apply modern Bayesian models to various data collection mechanisms. Little and Rubin (2002) present many techniques and relevant theory for handling missing data, some of which we review in Chapter 21. Heitjan and Rubin (1990, 1991) generalize missing data to coarse data, which includes rounding and heaping; previous work on rounding is reviewed by Heitjan (1989). Analysis of record-breaking data, a kind of time-varying censoring, is discussed by Carlin and Gelfand (1993).

Rosenbaum and Rubin (1983b) present a study of sensitivity to nonignorable models for observational studies. Heckman (1979) is an influential work on nonignorable models from an econometric perspective.

Hierarchical models for stratified and cluster sampling are discussed by Scott and Smith (1969), Little (1991, 1993), and Nadaram and Sedransk (1993); related material also appears in Skinner, Holt, and Smith (1989). The survey of Australian schoolchildren in Section 7.4 is described in Carlin et al. (1997). The introduction to Goldstein and Silver (1989) discusses the role of designs of surveys and experiments in gathering data for the purpose of estimating hierarchical models. Hierarchical models for experimental data are also discussed by Tiao and Box (1967) and Box and Tiao (1973).

Rubin (1978a) and Kadane and Seidenfeld (1990) discuss randomization from two different Bayesian perspectives. Rubin (1977) discusses the analysis of designs that are ignorable conditional on a covariate. Rosenbaum and Rubin (1983a) introduce the idea of propensity scores, which can be minimally adequate summaries as defined by Rubin (1985); technical applications of propensity scores to experiments and observational studies appear in Rosenbaum and Rubin (1984b, 1985). Recent work on propensity score methods includes Rubin and Thomas (1992, 2000) and Imbens (2000). There is now a relatively vast applied literature on propensity scores; see, for example, Connors et al. (1996). The example in Section 7.7 comes from Rubin (2002).

Frangakis and Rubin (2002) introduced the expression 'principal stratification' and discuss its application to sample outcomes. Rubin (1998, 2000) presents the principal stratification approach to the problem of 'censoring due to death,' and Zhang (2002) develops a Bayesian attack on the problem. The vitamin A experiment, along with a general discussion of the connection between principal stratification and instrumental variables, appears in Imbens, and Rubin (1997). Recent applications of these ideas in medicine and public policy include Dehejia and Wahba (1999), Hirano et al. (2000) and Barnard et al. (2003). Imbens and Angrist (1994) give a non-Bayesian presentation of instrumental variables for causal inference; see also McClellan, McNeil, and Newhouse (1994) and Newhouse and McClellan (1994), as well as Bloom

(1984) and Zelen (1979). Glickman and Normand (2000) connect principal stratification to continuous instrumental variables models.

7.11 Exercises

1. Definition of concepts: the concepts of *randomization, exchangeability,* and *ignorability* have often been confused in the statistical literature. For each of the following statements, explain why it is false but also explain why it has a kernel of truth. Illustrate with examples from this chapter or earlier chapters.

   (a) Randomization implies exchangeability: that is, if a randomized design is used, an exchangeable model is appropriate for the observed data, $y_{\text{obs}\,1}, \ldots, y_{\text{obs}\,n}$.

   (b) Randomization is required for exchangeability: that is, an exchangeable model for $y_{\text{obs}\,1}, \ldots, y_{\text{obs}\,n}$ is appropriate *only* for data that were collected in a randomized fashion.

   (c) Randomization implies ignorability; that is, if a randomized design is used, then it is ignorable.

   (d) Randomization is required for ignorability; that is, randomized designs are the *only* designs that are ignorable.

   (e) Ignorability implies exchangeability; that is, if an ignorable design is used, then an exchangeable model is appropriate for the observed data, $y_{\text{obs}\,1}, \ldots, y_{\text{obs}\,n}$.

   (f) Ignorability is required for exchangeability; that is, an exchangeable model for $y_{\text{obs}\,1}, \ldots, y_{\text{obs}\,n}$ is appropriate *only* for data that were collected using an ignorable design.

2. Application of design issues: choose an example from the earlier chapters of this book and discuss the relevance of the material in the current chapter to the analysis. In what way would you change the analysis, if at all, given what you have learned from the current chapter?

3. Distinct parameters and ignorability:

   (a) For the milk production experiment in Section 7.5, give an argument for why the parameters ϕ and θ may not be distinct.

   (b) If the parameters are not distinct in this example, the design is no longer ignorable. Discuss how posterior inferences would be affected by using an appropriate nonignorable model. (You need not set up the model; just discuss the direction and magnitude of the changes in the posterior inferences for the treatment effect.)

4. Interaction between units: consider a hypothetical agricultural experiment in which each of two fertilizers is assigned to 10 plots chosen completely at random from a linear array of 20 plots, and the outcome variable is the average yield of the crops in each plot. Suppose there is interference between

Block	Treatment			
	A	B	C	D
1	89	88	97	94
2	84	77	92	79
3	81	87	87	85
4	87	92	89	84
5	79	81	80	88

Table 7.6 *Yields of penicillin produced by four manufacturing processes (treatments), each applied in five different conditions (blocks). Four runs were made within each block, with the treatments assigned to the runs at random. From Box, Hunter, and Hunter (1978), who adjusted the data so that the averages are integers, a complication we ignore in our analysis.*

units, because each fertilizer leaches somewhat onto the two neighboring plots.

(a) Set up a model of potential data y, observed data y_{obs}, and inclusion indicators I. The potential data structure will have to be *larger* than a 20×4 matrix in order to account for the interference.

(b) Is the treatment assignment ignorable under this notation?

(c) Suppose the estimand of interest is the average difference in yields under the two treatments. Define the finite-population estimand mathematically in terms of y.

(d) Set up a probability model for y.

5. Analyzing a designed experiment: Table 7.6 displays the results of a randomized blocks experiment on penicillin production.

(a) Express this experiment in the general notation of this chapter, specifying x, y_{obs}, y_{mis}, N, and I. Sketch the table of units by measurements. How many observed measurements and how many unobserved measurements are there in this problem?

(b) Under the randomized blocks design, what is the distribution of I? Is it ignorable? Is it known? Is it strongly ignorable? Are the propensity scores an adequate summary?

(c) Set up a normal-based model of the data and all relevant parameters that is conditional on enough information for the design to be ignorable.

(d) Suppose one is interested in the (superpopulation) average yields of penicillin, averaging over the block conditions, under each of the four treatments. Express this estimand in terms of the parameters in your model.

We return to this example in Exercise 15.2.

6. Including additional information beyond the adequate summary:

(a) Suppose that the experiment in the previous exercise had been performed by complete randomization (with each treatment coincidentally appearing once in each block), not randomized blocks. Explain why the appropriate Bayesian modeling and posterior inference would not change.

(b) Describe how the posterior predictive check would change under the assumption of complete randomization.

(c) Why is the randomized blocks design preferable to complete randomization in this problem?

(d) Give an example illustrating why too much blocking can make modeling more difficult and sensitive to assumptions.

7. Simple random sampling:

(a) Derive the exact posterior distribution for $\bar{y}$ under simple random sampling with the normal model and noninformative prior distribution.

(b) Derive the asymptotic result (7.6).

8. Finite-population inference for completely randomized experiments:

(a) Derive the asymptotic result (7.17).

(b) Derive the (finite-population) inference for $\bar{y}_A - \bar{y}_B$ under a model in which the pairs (y_i^A, y_i^B) have iid bivariate normal distributions with mean (μ^A, μ^B), standard deviations (σ^A, σ^B), and correlation ρ.

(c) Discuss how inference in (b) depends on ρ and the implications in practice. Why does the dependence on ρ disappear in the limit of large N/n?

9. Cluster sampling:

(a) Discuss the analysis of one-stage and two-stage cluster sampling designs using the notation of this chapter. What is the role of hierarchical models in analysis of data gathered by one- and two-stage cluster sampling?

(b) Discuss the analysis of cluster sampling in which the clusters were sampled with probability proportional to some measure of size, where the measure of size is known for all clusters, sampled and unsampled. In what way do the measures of size enter into the Bayesian analysis?

See Kish (1965) and Lohr (1999) for classical methods for design and analysis of such data.

10. Cluster sampling: Suppose data have been collected using cluster sampling, but the details of the sampling have been lost, so it is not known which units in the sample came from common clusters.

(a) Explain why an exchangeable but not iid model is appropriate.

(b) Suppose the clusters are of equal size, with A clusters, each of size B, and the data came from a simple random sample of a clusters, with a simple random sample of b units within each cluster. Under what limits of a, A, b, and B can we ignore the cluster sampling in the analysis?

| Preference | Number of phone lines | | | | |
	1	2	3	4	?
Bush	557	38	4	3	7
Dukakis	427	27	1	0	3
No opinion/other	87	1	0	0	7

Table 7.7 *Respondents to the CBS telephone survey classified by opinion and number of residential telephone lines (category '?' indicates no response to the number of phone lines question).*

11. Capture-recapture (see Seber, 1992, and Barry et al., 2003): a statistician/fisherman is interested in N, the number of fish in a certain pond. He catches 100 fish, tags them, and throws them back. A few days later, he returns and catches fish until he has caught 20 tagged fish, at which point he has also caught 70 untagged fish. (That is, the second sample has 20 tagged fish out of 90 total.)

   (a) Assuming that all fish are sampled independently and with equal probability, give the posterior distribution for N based on a noninformative prior distribution. (You can give the density in unnormalized form.)

   (b) Briefly discuss your prior distribution and also make sure your posterior distribution is proper.

   (c) Give the probability that the next fish caught by the fisherman is tagged. Write the result as a sum or integral—you do not need to evaluate it, but the result should *not* be a function of N.

   (d) The statistician/fisherman checks his second catch of fish and realizes that, of the 20 'tagged' fish, 15 are definitely tagged, but the other 5 may be tagged—he is not sure. Include this aspect of missing data in your model and give the new joint posterior density for all parameters (in unnormalized form).

12. Sampling with unequal probabilities: Table 7.7 summarizes the opinion poll discussed in the examples in Sections 3.5 and 7.4, with the responses classified by Presidential preference and number of telephone lines in the household. We shall analyze these data assuming that the probability of reaching a household is proportional to the number of telephone lines. Pretend that the responding households are a simple random sample of telephone numbers; that is, ignore the stratification discussed in Section 7.4 and ignore all nonresponse issues.

   (a) Set up parametric models for (i) preference given number of telephone lines, and (ii) distribution of number of telephone lines in the population. (Hint: for (i), consider the parameterization (7.8).)

   (b) What assumptions did you make about households with no telephone lines and households that did not respond to the 'number of phone lines' question?

Number of adults	Preference	Number of phone lines				
		1	2	3	4	?
1	Bush	124	3	0	2	2
	Dukakis	134	2	0	0	0
	No opinion/other	32	0	0	0	1
2	Bush	332	21	3	0	5
	Dukakis	229	15	0	0	3
	No opinion/other	47	0	0	0	6
3	Bush	71	9	1	0	0
	Dukakis	47	7	1	0	0
	No opinion/other	4	1	0	0	0
4	Bush	23	4	0	1	0
	Dukakis	11	3	0	0	0
	No opinion/other	3	0	0	0	0
5	Bush	3	0	0	0	0
	Dukakis	4	0	0	0	0
	No opinion/other	1	0	0	0	0
6	Bush	1	0	0	0	0
	Dukakis	1	0	0	0	0
	No opinion/other	0	0	0	0	0
7	Bush	2	0	0	0	0
	Dukakis	0	0	0	0	0
	No opinion/other	0	0	0	0	0
8	Bush	1	0	0	0	0
	Dukakis	0	0	0	0	0
	No opinion/other	0	0	0	0	0
?	Bush	0	1	0	0	0
	Dukakis	1	0	0	0	0
	No opinion/other	0	0	0	0	0

Table 7.8 *Respondents to the CBS telephone survey classified by opinion, number of residential telephone lines (category '?' indicates no response to the number of phone lines question), and number of adults in the household (category '?' includes all responses greater than 8 as well as nonresponses).*

(c) Write the joint posterior distribution of all parameters in your model.

(d) Draw 1000 simulations from the joint distribution. (Use approximate computational methods.)

(e) Compute the mean preferences for Bush, Dukakis, and no opinion/other in the population of households (not phone numbers!) and display a histogram for the difference in support between Bush and Dukakis. Compare to Figure 3.3 and discuss any differences.

(f) Check the fit of your model to the data using posterior predictive checks.

(g) Explore the sensitivity of your results to your assumptions.

13. Sampling with unequal probabilities (continued): Table 7.8 summarizes the

opinion poll discussed in the examples in Sections 3.5 and 7.4, with the responses classified by vote preference, size of household, and number of telephone lines in the household. We shall analyze these data assuming that the probability of reaching an individual is proportional to the number of telephone lines and inversely proportional to the number of persons in the household. Use this additional information to obtain inferences for the mean preferences for Bush, Dukakis, and no opinion/other among individuals, rather than households, answering the analogous versions of questions (a)–(g) in the previous exercise. Compare to your results for the previous exercise and explain the differences. (A complete analysis would require the data also cross-classified by the 16 strata in Table 7.2 as well as demographic data such as sex and age that affect the probability of nonresponse.)

14. Rounded data: the last two columns of Table 2.2 on page 69 give data on passenger airline deaths and deaths per passenger mile flown. We would like to divide these to obtain the number of passenger miles flown in each year, but the 'per mile' data are rounded. (For the purposes of this exercise, ignore the column in the table labeled 'Fatal accidents.')

    (a) Using just the data from 1976 (734 deaths, 0.19 deaths per 100 million passenger miles), obtain inference for the number of passenger miles flown in 1976. Give a 95% posterior interval (you may do this by simulation). Clearly specify your model and your prior distribution.

    (b) Apply your method to obtain intervals for the number of passenger miles flown each year until 1985, analyzing the data from each year separately.

    (c) Now create a model that allows you to use data from all the years to estimate jointly the number of passenger miles flown each year. Estimate the model and give 95% intervals for each year. (Use approximate computational methods.)

    (d) Describe how you would use the results of this analysis to get a better answer for Exercise 2.13.

15. Sequential treatment assignment: consider a medical study with two treatments, in which the subjects enter the study one at a time. As the subjects enter, they must be assigned treatments. Efron (1971) evaluates the following 'biased-coin' design for assigning treatments: each subject is assigned a treatment at random with probability of receiving treatment depending on the treatment assignments of the subjects who have previously arrived. If equal numbers of previous subjects have received each treatment, then the current subject is given the probability $\frac{1}{2}$ of receiving each treatment; otherwise, he or she is given the probability p of receiving the treatment that has been assigned to *fewer* of the previous subjects, where p is a fixed value between $\frac{1}{2}$ and 1.

    (a) What covariate must be recorded on the subjects for this design to be ignorable?

    (b) Outline how you would analyze data collected under this design.

(c) To what aspects of your model is this design sensitive?

(d) Discuss in Bayesian terms the advantages and disadvantages of the biased-coin design over the following alternatives: (i) independent randomization (that is, $p = \frac{1}{2}$ in the above design), (ii) randomized blocks where the blocks consist of successive pairs of subjects (that is, $p = 1$ in the above design). Be aware of the practical complications discussed in Section 7.6.

16. Randomized experiment with noncompliance: Table 7.5 on page 230 gives data from the study of vitamin A in Indonesia described in the example on page 229 (see Imbens and Rubin, 1997).

(a) Is treatment assignment ignorable? Strongly ignorable? Known?

(b) Estimate the intention-to-treat effect: that is, the effect of assigning the treatment, irrespective of compliance, on the entire population.

(c) Give the simple instrumental variables estimate of the average effect of the treatment for the compliers.

(d) Write the likelihood, assuming compliance status is known for all units.

CHAPTER 8

Connections and challenges

8.1 Bayesian interpretations of other statistical methods

We consider three levels at which Bayesian statistical methods can be compared with other methods. First, as we have already indicated in Chapters 2–4, Bayesian methods are often similar to other statistical approaches in problems involving large samples from a fixed probability model. Second, even for small samples, many statistical methods can be considered as approximations to Bayesian inferences based on particular prior distributions; as a way of understanding a statistical procedure, it is often useful to determine the implicit underlying prior distribution. Third, some methods from classical statistics (notably hypothesis testing) can give results that differ greatly from those given by Bayesian methods. In this section, we briefly consider several statistical concepts—point and interval estimation, likelihood inference, unbiased estimation, frequency coverage of confidence intervals, hypothesis testing, multiple comparisons, nonparametric methods, and the jackknife and bootstrap—and discuss their relation to Bayesian methods.

Maximum likelihood and other point estimates

From the perspective of Bayesian data analysis, we can often interpret classical point estimates as exact or approximate posterior summaries based on some implicit full probability model. In the limit of large sample size, in fact, we can use asymptotic theory to construct a theoretical Bayesian justification for classical maximum likelihood inference. In the limit (assuming regularity conditions), the maximum likelihood estimate, $\hat{\theta}$, is a sufficient statistic—and so is the posterior mode, mean, or median. That is, for large enough n, the maximum likelihood estimate (or any of the other summaries) supplies essentially all the information about θ available from the data. The asymptotic irrelevance of the prior distribution can be taken to justify the use of convenient noninformative prior models.

In repeated sampling with $\theta = \theta_0$,

$$p(\hat{\theta}(y)|\theta\!=\!\theta_0) \approx \mathrm{N}(\hat{\theta}(y)|\theta_0, [nJ(\theta_0)]^{-1});$$

that is, the sampling distribution of $\hat{\theta}(y)$ is approximately normal with mean θ_0 and precision $nJ(\theta_0)$, where for clarity we emphasize that $\hat{\theta}$ is a function of y. Assuming that the prior distribution is locally uniform (or continuous and nonzero) near the true θ, the simple analysis of the normal mean (Section

3.6) shows that the posterior Bayesian inference is

$$p(\theta|\hat{\theta}) \approx N(\theta|\hat{\theta}, [nJ(\hat{\theta})]^{-1}).$$

This result appears directly from the asymptotic normality theorem, but deriving it indirectly through Bayesian inference given $\hat{\theta}$ gives insight into a Bayesian rationale for classical asymptotic inference based on point estimates and standard errors.

For finite n, the above approach is inefficient or wasteful of information to the extent that $\hat{\theta}$ is *not* a sufficient statistic. When the number of parameters is large, the consistency result is often not helpful, and noninformative prior distributions are hard to justify. As discussed in Chapter 5, hierarchical models are preferable when dealing with large number of parameters since then their common distribution can be estimated from data. In addition, any method of inference based on the likelihood alone can be improved if real prior information is available that is strong enough to contribute substantially to that contained in the likelihood function.

Unbiased estimates

Some non-Bayesian statistical methods place great emphasis on unbiasedness as a desirable principle of estimation, and it is intuitively appealing that, over repeated sampling, the mean (or perhaps the median) of a parameter estimate should be equal to its true value. From a Bayesian perspective, the principle of unbiasedness is reasonable in the limit of large samples, but otherwise is potentially misleading. The major difficulties arise when there are many parameters to be estimated and our knowledge or partial knowledge of some of these parameters is clearly relevant to the estimation of others. Requiring unbiased estimates will often lead to relevant information being ignored (for example, in the case of hierarchical models in Chapter 5). In sampling theory terms, minimizing bias will often lead to counterproductive increases in variance.

A general problem with unbiasedness (and point estimation in general) is that it is often not possible to estimate several parameters at once in an even approximately unbiased manner. For example, unbiased estimates of $\theta_1, \ldots, \theta_J$ yield an upwardly biased estimate of the variance of the θ_j's (except in the trivial case in which the θ_j's are known exactly).

Another simple illustration of the difficulty that can be caused by the principle of unbiasedness arises when treating a future observable value as a parameter in prediction problems.

Example. Prediction using regression
Consider the problem of estimating θ, the height of an adult daughter, given y, the height of her mother. For simplicity, we assume that the heights of mothers and daughters are jointly normally distributed, with known equal means of 160 centimeters, equal variances, and a known correlation of 0.5. Conditioning on the known value of y (in other words, using Bayesian inference), the posterior mean

of θ is

$$E(\theta|y) = 160 + 0.5(y - 160). \qquad (8.1)$$

The posterior mean is *not*, however, an unbiased estimate of θ, in the sense of repeated sampling of y given a fixed θ. Given the daughter's height, θ, the mother's height, y, has mean $E(y|\theta) = 160 + 0.5(\theta - 160)$. Thus, under repeated sampling of y given fixed θ, the posterior mean estimate (8.1) has expectation $160 + 0.25(\theta - 160)$ and is biased towards the grand mean of 160. In contrast, the estimate

$$\hat{\theta} = 160 + 2(y - 160)$$

is unbiased under repeated sampling of y, conditional on θ. Unfortunately, the estimate $\hat{\theta}$ makes no sense for values of y not equal to 160; for example, if a mother is 10 centimeters taller than average, it estimates her daughter to be 20 centimeters taller than average!

In this simple example, in which θ has an accepted population distribution, a sensible non-Bayesian statistician would not use the unbiased estimate $\hat{\theta}$; instead, this problem would be classified as 'prediction' rather than 'estimation,' and procedures would not be evaluated conditional on the random variable θ. The example illustrates, however, the limitations of unbiasedness as a general principle: it requires unknown quantities to be characterized either as 'parameters' or 'predictions,' with quite different implications for estimation but no clear substantive distinction. Chapter 5 considered similar situations in which the population distribution of θ must be estimated from data rather than conditioning on a particular value.

The important principle illustrated by the example is that of *regression to the mean*: for any given mother, the expected value of her daughter's height lies between her mother's height and the population mean. This principle was fundamental to the original use of the term 'regression' for this type of analysis by Galton in the late 19th century. In many ways, Bayesian analysis can be seen as a logical extension of the principle of regression to the mean, ensuring that proper weighting is made of information from different sources.

Confidence intervals

Even in small samples, Bayesian $(1 - \alpha)$ posterior intervals often have close to $(1 - \alpha)$ confidence coverage under repeated samples conditional on θ. But there are some confidence intervals, derived purely from sampling-theory arguments, that differ considerably from Bayesian probability intervals. From our perspective these intervals are of doubtful value. For example, many authors have shown that a general theory based on unconditional behavior can lead to clearly counterintuitive results, for example, the possibilities of confidence intervals with zero or infinite length. A simple example is the confidence interval that is empty 5% of the time and contains all of the real line 95% of the time: this always contains the true value (of any real-valued parameter) in 95% of repeated samples. Such examples do not imply that there is no value in the concept of confidence coverage but rather show that coverage alone is not a sufficient basis on which to form reasonable inferences.

Hypothesis testing

The perspective of this book has little role for the non-Bayesian concept of hypothesis tests, especially where these relate to point null hypotheses of the form $\theta = \theta_0$. In order for a Bayesian analysis to yield a nonzero probability for a point null hypothesis, it must begin with a nonzero prior probability for that hypothesis; in the case of a continuous parameter, such a prior distribution (comprising a discrete mass, of say 0.5, at θ_0 mixed with a continuous density elsewhere) usually seems contrived. In fact, most of the difficulties in interpreting hypothesis tests arise from the artificial dichotomy that is required between $\theta = \theta_0$ and $\theta \neq \theta_0$. Difficulties related to this dichotomy are widely acknowledged from all perspectives on statistical inference. In problems involving a continuous parameter θ (say the difference between two means), the hypothesis that θ is exactly zero is rarely reasonable, and it is of more interest to estimate a posterior distribution or a corresponding interval estimate of θ. For a continuous parameter θ, the question 'Does θ equal 0?' can generally be rephrased more usefully as 'What is the posterior distribution for θ?'

In various simple one-sided hypothesis tests, conventional p-values may correspond with posterior probabilities under noninformative prior distributions. For example, suppose we observe $y = 1$ from the model $y \sim \mathrm{N}(\theta, 1)$, with a uniform prior density on θ. One cannot 'reject the hypothesis' that $\theta = 0$: the one-sided p-value is 0.16 and the two-sided p-value is 0.32, both greater than the conventionally accepted cut-off value of 0.05 for 'statistical significance.' On the other hand, the posterior probability that $\theta > 0$ is 84%, which is clearly a more satisfactory and informative conclusion than the dichotomous verdict 'reject' or 'do not reject.'

Goodness-of-fit testing

In contrast to the problem of making inference about a parameter within a particular model, we do find a form of hypothesis testing to be useful when assessing the goodness of fit of a probability model. In the Bayesian framework, it is useful to *check* a model by comparing observed data to possible predictive outcomes, as discussed in Chapter 6.

Multiple comparisons

Consider a problem with independent measurements, $y_j \sim \mathrm{N}(\theta_j, 1)$, on each of J parameters, in which the goal is to detect differences among and ordering of the continuous parameters θ_j. Several competing multiple comparisons procedures have been derived in classical statistics, with rules about when various θ_j's can be declared 'significantly different.' In the Bayesian approach, the parameters have a joint posterior distribution. One can compute the posterior probability of each of the $J!$ orderings if desired. If there is posterior uncertainty in the ordering, several permutations will have substantial probabilities, which is a more reasonable conclusion than producing a list of θ_j's that can be declared different (with the false implication that other θ_j's may be exactly equal). With J large, the exact ordering is probably not important, and it might be more reasonable to give a posterior median and interval estimate of the quantile of each θ_j in the population.

We prefer to handle multiple comparisons problems using hierarchical models, as illustrated in the comparison of the eight schools in Section 5.5 (see also Exercise 5.1). Hierarchical modeling automatically shrinks estimates of different θ_j's toward each other when there is little evidence for real variation. As a result, this Bayesian procedure automatically addresses the key concern of classical multiple comparisons analysis, which is the possibility of finding large differences as a byproduct of searching through so many possibilities. For example, in the educational testing example, the eight schools give $8 \cdot 7/2 = 28$ possible comparisons, and none are close to 'statistically significant' (in the sense that zero is contained within the 95% intervals for all the differences in effects between pairs of schools), which makes sense since the between-school variation τ is estimated to be low.

Nonparametric methods, permutation tests, jackknife, bootstrap

Many non-Bayesian methods have been developed that avoid complete probability models, even at the sampling level. It is difficult to evaluate many of these from a Bayesian point of view. For instance, hypothesis tests for comparing medians based on ranks do not have direct counterparts in Bayesian inference; therefore it is hard to interpret the resulting estimates and p-values from a Bayesian point of view (for example, as posterior expectations, intervals, or probabilities for parameters or predictions of interest). In complicated problems, there is often a degree of arbitrariness in the procedures used; for example there is generally no clear method for constructing a nonparametric inference or an 'estimator' to jackknife/bootstrap in hypothetical replications. Without a specified probability model, it is difficult to see how to test the assumptions underlying a particular nonparametric method. In such problems, we find it more satisfactory to construct a joint probability distribution and check it against the data (as in Chapter 6) than to construct an estimator and evaluate its frequency properties. Nonparametric methods are useful to us as tools for data summary and description that can help us to construct models or help us evaluate inferences from a completely different perspective.

Some nonparametric methods such as permutation tests for experiments and sampling-theory inference for surveys turn out to give very similar results in simple problems to Bayesian inferences with noninformative prior distributions, if the Bayesian model is constructed to fit the data reasonably well. Such simple problems include balanced designs with no missing data and surveys based on simple random sampling. When estimating several parameters at once or including explanatory variables in the analysis (using methods such as the analysis of covariance or regression) or prior information on the parameters, the permutation/sampling theory methods give no direct answer, and this often provides considerable practical incentive to move to a model-based Bayesian approach.

Example. The Wilcoxon rank test

Another connection can be made by interpreting nonparametric methods in terms of implicit models. For example, the Wilcoxon rank test for comparing two samples $(y_1, \ldots, y_{n_y})$ and $(z_1, \ldots, z_{n_z})$ proceeds by first ranking each of the points in the combined data from 1 to $n = n_y + n_z$, then computing the difference between the average ranks of the y's and z's, and finally computing the p-value of this difference by comparing to a tabulated reference distribution calculated based on the assumption of random assignment of the n ranks. This can be formulated as a nonlinear transformation that replaces each data point by its rank in the combined data, followed by a comparison of the mean values of the two transformed samples. Even more clear would be to transform the ranks $1, 2, \ldots, n$ to quantiles $\frac{1}{2n}, \frac{3}{2n}, \ldots, \frac{2n-1}{2n}$, so that the difference between the two means can be interpreted as an average distance in the scale of the quantiles of the combined distribution. From the Central Limit Theorem, the mean difference is approximately normally distributed, and so classical normal-theory confidence intervals can be interpreted as Bayesian posterior probability statements, as discussed at the beginning of this section.

We see two major advantages of expressing rank tests as approximate Bayesian inferences. First, the Bayesian framework is more flexible than rank testing for handling the complications that arise, for example, from additional information such as regression predictors or from complications such as censored or truncated data. Second, setting up the problem in terms of a nonlinear transformation reveals the generality of the model-based approach—we are free to use any transformation that might be appropriate for the problem, perhaps now treating the combined quantiles as a convenient default choice.

8.2 Challenges in Bayesian data analysis

We try to demonstrate in this book that Bayesian methods can be used effectively and fairly routinely in a wide variety of applications. However, some difficulties remain. In addition to computational challenges, which we discuss in Chapter 13, various open problems arise in setting up and checking the fit of Bayesian models. We discuss these roughly in the order in which they occur in the book.

Noninformative prior distributions

We often want to analyze data using minimal prior information, and in practice this often seems to be possible, even with small datasets (as, for example, in the bioassay model in Section 3.7). However, it is not clear what general principles can be applied in setting up models with noninformative prior distributions, especially with many parameters. It is disturbing that seemingly reasonable 'noninformative' densities do not work for some problems—for example, the uniform prior distribution on $\log \tau$ in the hierarchical model, or a uniform prior distribution on the parameters of a mixture model (see Chapter 18). Our general practice is to consider noninformative models as expedient devices to be replaced when necessary, but it is still not always clear when this will work.

However, we should never find ourselves seeking a noninformative prior distribution for large number of parameters, since in that case it should be possible to structure the parameters into a hierarchical model.

Accounting for model choice in data analysis

We typically construct the final form of a model only after extensive data analysis, which leads to concerns that are related to the classical problems of multiple comparisons and estimation of prediction error. As discussed in Section 8.1, a Bayesian treatment of multiple comparisons uses hierarchical modeling, simultaneously estimating the joint distribution of all possible comparisons and shrinking these as appropriate (for example, in the analysis of the eight schools, the θ_j's are all shrunk toward μ, so the differences $\theta_j - \theta_k$ are automatically shrunk toward 0). Nonetheless, some potential problems arise, such as the possibility of performing many analyses on a single dataset in order to find the strongest conclusion. This is a danger with all applied statistical methods and is only partly alleviated by the Bayesian attempt to include all sources of uncertainty in a model.

Posterior predictive checks

Predictive checking is our basic method of assessing model fit, and every problem we work on potentially requires its own test summaries. Conventional summaries such as average errors, residual plots, and deviances can be helpful, but we often want to look at test statistics and discrepancy measures that address particular ways in which we intend to use the model. The graphical and numerical examples in Chapter 6 give some sense of the effectiveness of predictive checking and the challenge of adapting it to individual examples.

In addition, the model-checking phase of Bayesian data analysis shares some features of data collection and study design, in that it requires consideration of data-generating structures—in particular, the choice of whether to replicate new groups in a hierarchical model.

Deviance, model comparison, and cross-validation

The deviance is an important tool in classical model fitting, as it allows a modeler to have a sense of whether the improvement in fit resulting from a model expansion is statistically worthwhile and the extent to which it is estimated to increase (or decrease) the accuracy of predictions of new data. The deviance information criterion (DIC) plays a similar role in Bayesian data analysis (see Section 6.7). However, the justification of DIC is asymptotic, which makes interpretation awkward for Bayesian inference concerning any particular dataset.

If a model is to be evaluated by its expected predictive fit to new data, a reasonable approach would seem to be cross-validation—fitting the model to part of a dataset and evaluating it on the rest—but it is not clear how to interpret this in terms of posterior inference. Conditional on a model being correct, we can directly assess its predictive accuracy by comparing its fit to simulated replicated data sets y^{rep}—but DIC and cross-validation attempt to do better and capture predictive error including model misfit.

Selection of predictors and combining information

In regression problems there are generally many different reasonable-seeming ways to set up a model, and these different models can give dramatically different answers (as we illustrate in Section 22.2 in an analysis of the effects of incentives on survey response). Putting together existing information in the form of predictors is nearly always an issue in observational studies (see Section 7.7), and can be seen as a model specification issue. Even when only a few predictors are available, we can choose among possible transformations and interactions. Methods such as propensity scores (see the bibliographic note at the end of Chapter 7) can be useful here.

As we shall discuss in Sections 15.5 and 15.6, we prefer including as many predictors as possible in a regression and then scaling and batching them into an analysis-of-variance structure, so that they are all considered to some extent rather than being discretely 'in' or 'out.' Even so, choices must be made in selecting the variables to be included in the hierarchical model itself. Bayesian methods for discrete model averaging may be helpful here, although we have not used this approach in our own research.

A related and more fundamental issue arises when setting up regression models for causal inference in observational studies. Here, the relations among the variables in the substantive context are relevant, as in principal stratification methods (see Section 7.7), where, after the model is constructed, additional analysis is required to compute causal estimands of interest, which are not in general the same as the regression coefficients.

Alternative model formulations

We often find that adding a parameter to a model makes it much more flexible. For example, in a normal model, we prefer to estimate the variance parameter rather than set it to a pre-chosen value. At the next stage, the t model is more flexible than the normal (see Chapter 17), and this has been shown to make a practical difference in many applications. But why stop there? There is always a balance between accuracy and convenience. As discussed in Chapter 6, predictive model checks can reveal serious model misfit, but we do not yet have good general principles to justify our basic model choices. As computation of hierarchical models becomes more routine, we may begin to use more elaborate models as defaults.

We do not have a clear sense of the potential gains to be achieved using new and innovative classes of models, especially for highly structured datasets. What is standard in Bayesian modeling has changed much in the past twenty years, and with the continuing development of sophisticated highly-parameterized models such as splines, neural networks, and local regressions, further dramatic improvements may be possible.

8.3 Bibliographic note

Krantz (1999) discusses the strengths and weaknesses of p-values as used in statistical data analysis in practice. Discussions of the role of p-values in Bayesian inference appear in Bayarri and Berger (1998, 2000). Earlier work on the Bayesian analysis of hypothesis testing and the problems of interpreting conventional p-values is provided by Berger and Sellke (1987), which contains a lively discussion and many further references. A very simple and pragmatic discussion of the need to consider Bayesian ideas in hypothesis testing in a biostatistical context is given by Browner and Newman (1987), and further discussion of the role of Bayesian thinking in medical statistics appears in Goodman (1999a, b) and Sterne and Davey Smith (2001). Gelman and Tuerlinckx (2000) give a Bayesian perspective on multiple comparisons in the context of hierarchical modeling.

Stigler (1983) discusses the similarity between Bayesian inference and regression prediction that we mention in our critique of unbiasedness in Section 8.1; Stigler (1986) discusses Galton's use of regression.

Sequential monitoring and analysis of clinical trials in medical research is an important area of practical application that has been dominated by frequentist thinking but has recently seen considerable discussion of the merits of a Bayesian approach; recent reviews and examples are provided by Freedman, Spiegelhalter, and Parmar (1994), Parmar et al. (2001), and Vail et al. (2001). More references on sequential designs appear in the bibliographic note at the end of Chapter 7.

The non-Bayesian principles and methods mentioned in Section 8.1 are covered in many books, for example, Lehmann (1983, 1986), Cox and Hinkley (1974), Hastie and Tibshirani (1990), and Efron and Tibshirani (1993).

8.4 Exercises

1. Statistical decision theory: a decision-theoretic approach to the estimation of an unknown parameter θ introduces the loss function $L(\theta, a)$ which, loosely speaking, gives the cost of deciding that the parameter has the value a, when it is in fact equal to θ. The estimate a can be chosen to

minimize the *posterior expected loss*,

$$E\left[L(a|y)\right] = \int L(\theta, a)p(\theta|y)d\theta.$$

This optimal choice of a is called a *Bayes estimate* for the loss function L. Show the following:

(a) If $L(\theta, a) = (\theta - a)^2$ (squared error loss), then the posterior mean, $E(\theta|y)$, if it exists, is the unique Bayes estimate of θ.

(b) If $L(\theta, a) = |\theta - a|$, then any posterior median of θ is a Bayes estimate of θ.

(c) If k_0 and k_1 are nonnegative numbers, not both zero, and

$$L(\theta, a) = \begin{cases} k_0(\theta - a) & \text{if} \quad \theta \geq a \\ k_1(a - \theta) & \text{if} \quad \theta < a, \end{cases}$$

then any $k_0/(k_0 + k_1)$ quantile of the posterior distribution $p(\theta|y)$ is a Bayes estimate of θ.

2. Unbiasedness: prove that the Bayesian posterior mean, based on a proper prior distribution, cannot be an unbiased estimator except in degenerate problems (see Bickel and Blackwell, 1967, and Lehmann, 1983, p. 244).

3. Regression to the mean: work through the details of the example of mother's and daughter's heights on page 248, illustrating with a sketch of the joint distribution and relevant conditional distributions.

4. Point estimation: suppose a measurement y is recorded with a $N(\theta, \sigma^2)$ sampling distribution, with σ known exactly and θ known to lie in the interval $[0, 1]$. Consider two point estimates of θ: (1) the maximum likelihood estimate, restricted to the range $[0, 1]$, and (2) the posterior mean based on the assumption of a uniform prior distribution on θ. Show that if σ is large enough, estimate (1) has a higher mean squared error than (2) for any value of θ in $[0, 1]$. (The unrestricted maximum likelihood estimate has even higher mean squared error, of course.)

5. Non-Bayesian inference: replicate the analysis of the bioassay example in Section 3.7 using non-Bayesian inference. This problem does not have a unique answer, so be clear on what methods you are using.

(a) Construct an 'estimator' of (α, β); that is, a function whose input is a dataset, (x, n, y), and whose output is a point estimate $(\hat{\alpha}, \hat{\beta})$. Compute the value of the estimate for the data given in Table 5.2.

(b) The bias and variance of this estimate are functions of the true values of the parameters (α, β) and also of the sampling distribution of the data, given α, β. Assuming the binomial model, estimate the bias and variance of your estimator.

(c) Create approximate 95% confidence intervals for α, β, and the LD50 based on asymptotic theory and the estimated bias and variance.

(d) Does the inaccuracy of the normal approximation for the posterior distribution (compare Figures 3.4 and 4.1) cast doubt on the coverage properties of your confidence intervals in (c)? If so, why?

(e) Create approximate 95% confidence intervals for α, β, and the LD50 using the jackknife or bootstrap (see Efron and Tibshirani, 1993).

(f) Compare your 95% intervals for the LD50 in (c) and (e) to the posterior distribution displayed in Figure 3.5 and the posterior distribution based on the normal approximation, displayed in 4.2b. Comment on the similarities and differences among the four intervals. Which do you prefer as an inferential summary about the LD50? Why?

6. Bayesian interpretation of non-Bayesian estimates: consider the following estimation procedure, which is based on classical hypothesis testing. A matched pairs experiment is done, and the differences $y_1, \ldots, y_n$ are recorded and modeled as iid $N(\theta, \sigma^2)$. For simplicity, assume σ^2 is known. The parameter θ is estimated as the average observed difference if it is 'statistically significant' and zero otherwise:

$$\hat{\theta} = \begin{cases} \bar{y} & \text{if } \bar{y} \geq 1.96\sigma/\sqrt{n} \\ 0 & \text{otherwise.} \end{cases}$$

Can this be interpreted, in some sense, as an approximate summary (for example, a posterior mean or mode) of a Bayesian inference under some prior distribution on θ?

7. Bayesian interpretation of non-Bayesian estimates: repeat the above problem but with σ replaced by s, the sample standard deviation of $y_1, \ldots, y_n$.

8. Objections to Bayesian inference: discuss the criticism, 'Bayesianism assumes: (a) *Either* a weak or uniform prior [distribution], in which case why bother?, (b) *Or* a strong prior [distribution], in which case why collect new data?, (c) *Or* more realistically, something in between, in which case Bayesianism always seems to duck the issue' (Ehrenberg, 1986). Feel free to use any of the examples covered so far to illustrate your points.

9. Objectivity and subjectivity: discuss the statement, 'People tend to believe results that support their preconceptions and disbelieve results that surprise them. Bayesian methods encourage this undisciplined mode of thinking.'

CHAPTER 9

General advice

We conclude this part of the book with a brief review of our general approach to statistical data analysis, illustrating with some further examples from our particular experiences in applied statistics.

A pragmatic rationale for the use of Bayesian methods is the inherent flexibility introduced by their incorporation of multiple levels of randomness and the resultant ability to combine information from different sources, while incorporating all reasonable sources of uncertainty in inferential summaries. Such methods naturally lead to smoothed estimates in complicated data structures and consequently have the ability to obtain better real-world answers.

Another reason for focusing on Bayesian methods is more psychological, and involves the relationship between the statistician and the client or specialist in the subject matter area who is the consumer of the statistician's work. In most practical cases, clients will interpret interval estimates provided by statisticians as Bayesian intervals, that is, as probability statements about the likely values of unknown quantities conditional on the evidence in the data. Such direct probability statements require prior probability specifications for unknown quantities (or more generally, probability models for vectors of unknowns), and thus the kinds of answers clients will assume are being provided by statisticians, Bayesian answers, require full probability models—explicit or implicit.

Finally, Bayesian inferences are conditional on probability models that invariably contain approximations in their attempt to represent complicated real-world relationships. If the Bayesian answers vary dramatically over a range of scientifically reasonable assumptions that are unassailable by the data, then the resultant range of possible conclusions must be entertained as legitimate, and we believe that the statistician has the responsibility to make the client aware of this fact.

9.1 Setting up probability models

Our general approach is to build models hierarchically. The models described in the early chapters represent some basic building blocks. Informal methods such as scatterplots are often crucial in setting up simple, effective models, as we have seen in Figures 1.1, 1.2, and 3.1, and illustrate in later examples in Figures 14.1, 15.1, and 18.1. Several of the examples in Part V of the book show how different kinds of models can be used at different levels of a hierarchy.

Example. Setting up a hierarchical prior distribution for the parameters of a model in pharmacokinetics

An aspect of the flexibility of hierarchical models arose in our applied research in pharmacokinetics, the study of the absorption, distribution and elimination of drugs from the body. We discuss this example in detail in Section 20.3; here, we give just enough information to illustrate the relevance of a hierarchical model for this problem. The model has 15 parameters for each of six persons in a pharmacokinetic experiment; θ_{jk} is the value of the kth parameter for person j, with $j = 1, \ldots, 6$ and $k = 1, \ldots, 15$. Prior information about the parameters was available in the biological literature. For each parameter, it was important to distinguish between two sources of variation: prior uncertainty about the value of the parameter and population variation. This was represented by way of a lognormal model for each parameter, $\log \theta_{jk} \sim N(\mu_k, \tau_k^2)$, and by assigning independent prior distributions to the population geometric mean and standard deviation, μ_k and τ_k: $\mu_k \sim N(M_k, S_k^2)$ and $\tau_k^2 \sim \text{Inv-}\chi^2(\nu, \tau_{k0}^2)$. The prior distribution on μ_k, especially through the standard deviation S_k, describes our uncertainty about typical values of the parameter in the population. The prior distribution on τ_k tell us about the population variation for the parameter. Because prior knowledge about population variation was imprecise, the degrees of freedom in the prior distributions for τ_k^2 were set to the low value of $\nu = 2$.

Some parameters are better understood than others. For example, the weight of the liver, when expressed as a fraction of lean body weight, was estimated to have a population geometric mean of 0.033, with both the uncertainty on the population average and the heterogeneity in the population estimated to be of the order of 10% to 20%. The prior parameters were set to $M_k = \log(0.033)$, $S_k = \log(1.1)$, and $\tau_{k0} = \log(1.1)$. In contrast, the Michaelis–Menten coefficient (a particular parameter in the pharmacokinetic model) was poorly understood: its population geometric mean was estimated at 0.7, but with a possible uncertainty of up to a factor of 100 above or below. Despite the large uncertainty in the magnitude of this parameter, however, it was believed to vary by no more than a factor of 4 relative to the population mean, among individuals in the population. The prior parameters were set to $M_k = \log(0.7)$, $S_k = \log(10)$, and $\tau_{k0} = \log(2)$. The hierarchical model provides an essential framework for expressing the two sources of variation (or uncertainty) and combining them in the analysis.

One way to develop possible models is to examine the interpretation of crude data-analytic procedures as approximations to Bayesian inference under specific models. For example, a widely used technique in sample surveys is *ratio estimation* , in which, for example, given data from a simple random sample, one estimates $R = \bar{y}/\bar{x}$ by $\bar{y}_{\text{obs}}/\bar{x}_{\text{obs}}$, in the notation of Chapter 7. It can be shown that this estimate corresponds to a summary of a Bayesian posterior inference given independent observations $y_i|x_i \sim N(Rx_i, \sigma^2 x_i)$ and a noninformative prior distribution. Of course, ratio estimates can be useful in a wide variety of cases in which this model does not hold, but when the data deviate greatly from this model, the ratio estimate generally is not appropriate.

For another example, standard methods of selecting regression predictors, based on 'statistical significance,' correspond to Bayesian analyses under exchangeable prior distributions on the coefficients in which the prior distribution of each coefficient is a mixture of a peak at zero and a widely spread distribution, as we discuss further in Section 15.5. We believe that understanding this correspondence suggests when such models can be usefully applied and how they can be improved. Often, in fact, such procedures can be improved by including additional information, for example, in problems involving large numbers of predictors, by clustering regression coefficients that are likely to be similar into batches.

The following example is a simplified version of a problem for which accounting for uncertainty in selecting an appropriate prediction model was crucial.

Example. Importance of including substantively relevant predictors even when not statistically significant or even identifiable
From each decennial census, the U.S. Census Bureau produces public-use data files of a million or more units each, which are analyzed by many social scientists. One of the fields in these files is 'occupation,' which is used, for example, to define subgroups for studying relations between education and income across time. Thus, comparability of data from one census to the next is important. The occupational coding system used by the Census Bureau, however, changed between 1970 and 1980. In order to facilitate straightforward analyses of time trends, a common coding was needed, and the 1980 system was considered better than the 1970 system. Therefore, the objective was to have the data from 1970, which had been coded using the 1970 system, classified in terms of the 1980 occupation codes.

A randomly selected subsample of the 1970 Census data was available that included enough information on each individual to determine the 1970 occupation in terms of both the 1970 and 1980 occupation codes. We call this dataset the 'double-coded' data. For 1970 public-use data, however, only the 1970 occupation code was available, and so the double-coded data were used to impute the missing 1980 occupation code for each individual. (See Chapter 21 for more on missing-data imputation.)

A simple example makes clear the need to reflect variability due to substantively relevant variables—even when there are too few data points to estimate relationships using standard methods.

A 2 × 2 array of predictor variables. Suppose for simplicity we are examining one 1970 occupation code with 200 units equally divided into two 1980 occupation codes, OCCUP1 and OCCUP2. Further, suppose that there are only two substantively important predictors, both dichotomous: education (E=high or E=low) and income (I=high or I=low), with 100 units having 'E=high, I=high,' 100 units having 'E=low, I=low,' and no units having either 'E=high, I=low' or 'E=low, I=high.' Finally, suppose that of the 100 'E=high, I=high' units, 90 have OCCUP1 and 10 have OCCUP2, and of the 100 'E=low, I=low' units, 10 have OCCUP1 and 90 have OCCUP2. Thus, the double-coded data for this 1970 occupation are as given in Table 9.1.

The goal is to use these data to multiply impute 1980 occupation codes for single-coded 1970 Census data. This can be done using a regression model, which predicts the probability of OCCUP1 as a function of the dichotomous predictor variables, education and income, for the units with the given 1970 occupation code. Here, we concern ourselves with the predictors to be included in the pre-

Education	Income		
	high	low	
high	OCCUP1: 90 OCCUP2: 10	0	100
low	0	OCCUP1: 10 OCCUP2: 90	100
	100	100	200

Table 9.1 *Hypothetical 'double-coded' data on 200 persons with a specified 1970 occupation code, classified by income, education, and 1980 occupation code. The goal is to use these data to construct a procedure for multiply imputing 1980 occupation codes, given income and education, for 'single-coded' data: persons with known 1970 occupation codes but unknown 1980 occupation codes. From Rubin (1983c).*

diction, but not with the details of the model (logit or probit link function, and so forth), and not with issues such as pooling information from units with other 1970 occupation codes.

The usual 'noninformative' solution. The usual rule for regression prediction would suggest using either education or income as a predictor, but not both since they are perfectly correlated in the sample. Certainly, any standard regression or logistic regression package would refuse to compute an answer using both predictors, but use them we must! Here is why.

Analysis using income as the predictor. Suppose that we used only income as a predictor. Then the implication is that education is irrelevant—that is, its coefficient in a regression model is assumed to have a posterior distribution concentrated entirely at zero. How would this affect the imputations for the public-use data? Consider a set of single-coded data with perhaps 100 times as many units as the double-coded sample; in such data we might find, say, 100 'E=low, I=high' units even though there were none in the double-coded sample. The prediction model says only Income is relevant, so that at each imputation, close to 90 of these 100 'E=low, I=high' units are assigned OCCUP1 and 10 are assigned OCCUP2. Is this a reasonable answer?

Analysis using education as the predictor and the rejection of both analyses. Now suppose that education had been chosen to be the one predictor variable instead of income. By the same logic, at each imputation close to 10 rather than 90 of the 100 'E=low, I=high' units would now have been assigned to OCCUP1, and close to 90 rather than 10 would have been assigned to OCCUP2. Clearly, neither the 90/10 split nor the 10/90 is reasonable, because in the first case the researcher studying the public-use tape would conclude with relative confidence that of people with the given 1970 occupation, nearly all 'E=low, I=high' units have OCCUP1, whereas in the second case the researcher would conclude with near certainty that nearly all 'E=low, I=high' units have OCCUP2, and the truth is that we have essentially no evidence on the split for these units.

A more acceptable result. A more acceptable result is that, when using posterior simulations to express uncertainty in the missing data, the occupational split for the 'E=low, I=high' units should vary between, say, 90/10 and 10/90. The variability between the posterior simulations should be so large that the investigator knows the data cannot support firm conclusions regarding the recoded occupations of these individuals.

This concern affects only about 1% of the public-use data, so it might seem relatively unimportant for the larger purpose of modeling income and education as a function of occupation. In practice, however, the 'E=low, I=high' and 'E=high, I=low' cases may occur more frequently in some subgroups of the population— for example, female urban minorities in the Northeast—and the model for these cases can have a large effect on research findings relating to such subgroups.

The main point is that the objective of Bayesian inference is a full posterior distribution of the estimands of interest, reflecting all reasonable sources of uncertainty under a specified model. If some variable should or could be in the model on substantive grounds, then it should be included even if it is not 'statistically significant' and even if there is no information in the data to estimate it using traditional methods.

Another important principle for setting up models, but often difficult to implement in practice, is to include relevant information about a problem's structure in a model. This arises most clearly in setting up hierarchical models; for example, as discussed in Chapter 5, an analysis of SAT coaching data from several schools clustered in each of several school districts naturally follows a three-level hierarchy. For another example, information available on the individual studies of a meta-analysis can be coded as predictor variables in a larger regression model. Another way of modeling complexity is to consider multivariate outcomes in sequence in a regression context.

Moreover, when more than one proposed model is reasonable, it generally makes sense to combine them by embedding them in a larger continuous family. For instance, in Section 14.7, we consider a continuous model for unequal variances that bridges between unweighted and weighted linear regression. In fact, most of the models in this book can be interpreted as continuous families bridging extreme alternatives; for another example, the SAT coaching example in Section 5.5 averages over a continuum of possibilities between no pooling and complete pooling of information among schools.

Transformations are a simple yet important way to include substantive information and improve inferences, no less important in a Bayesian context than in other approaches to statistics. Univariate transformations often allow the fitting of simpler, well-understood models such as the normal. Multivariate transformations can give the additional benefit of reducing correlations between parameters, thereby making inferences less sensitive to assumptions about correlations, as we discuss at the end of Section 19.2.

9.2 Posterior inference

A key principle in statistical inference is to define quantities of interest in terms of parameters, unobserved data, and potential future observables. What would you do if you had *all* the potential data in a sample survey or experiment? For example, inference for the bioassay example in Section 3.7 was summarized by the LD50 rather than by either of the model parameters α and β.

A key principle when computing posterior inferences is to simulate random draws from the joint posterior distribution of all unknowns, as in Table 1.1. This principle of *multiple imputation* can be applied to all posterior inferences. Using simulation gives us the flexibility to obtain inferences immediately about any quantities of interest. Problems such as in the bioassay example in Section 3.7, where interest lies in the ratio or other nonlinear combination of parameters, arise commonly in applications, and the simulation approach provides a simple and universally applicable method of analysis.

Posterior distributions can also be useful for *data reduction*. A simple example is Table 5.2, which summarizes the results of eight experiments by eight estimates and eight standard errors, which is equivalent to summarizing by eight independent normal posterior distributions. These are then treated as 'data' in a hierarchical model. More generally, beyond using summary statistics, a simulated sample from the posterior distribution can be used to summarize the information from a study for later purposes such as decision-making and more sophisticated analysis.

The complexity of the models we can use in an effective way is of course limited to some extent by currently available experience in applied modeling and, unfortunately, also by computation. We have found *duplication* to be a useful principle in many areas of Bayesian computation: using successive approximations and a variety of computational methods, and performing multiple mode-finding and iterative simulation computations from dispersed starting points (as we discuss in detail in Part III). It is also important to remember that, in most applications, the goal is to approximate the posterior distribution—even if it is too diffuse to reach conclusions of desired precision—not to obtain a point summary such as a posterior mean or mode.

The example on double-coding census data presented in the previous section demonstrates the important distinction between inferential uncertainty, as indicated by variation among posterior simulation draws (for example, concerning the estimation of regression coefficients for income and education) and population variability (for example, approximately 90% of 100 but not all 'E=high, I=high' units are OCCUP1). Theoretically, it is straightforward to describe the steps needed to reflect both sources of uncertainty, assuming an appropriate Bayesian model has been fitted. To perform the appropriate simulations, three steps are required: (1) parameters of the regressions are drawn at random from their posterior distributions, (2) predictive probabilities of 1980 codes are calculated, and (3) imputed 1980 codes are drawn at random according to the probabilities in step (2). Each pass through these three steps generates one vector of posterior simulations of all the uncertain quantities in the model.

9.3 Model evaluation

Despite our best efforts to include information, all models are approximate. Hence, checking the fit of a model to data and prior assumptions is always important. For the purpose of model evaluation, we can think of the inferential step of Bayesian data analysis as a sophisticated way to explore all the implications of a proposed model, in such a way that these implications can be compared with observed data and other knowledge not included in the model. For example, Section 6.4 illustrates graphical predictive checks for models fitted to data for two different problems in psychological research. In each case, the fitted model captures a general pattern of the data but misses some key features. In the second example, finding the model failure leads to a model improvement—a mixture distribution for the patient and symptom parameters—that better fits the data, as seen in Figure 6.9.

Posterior inferences can often be summarized graphically. For simple problems or one or two-dimensional summaries, we can plot a histogram or scatterplot of posterior simulations, as in Figures 3.3, 3.4, and 5.8. For larger problems, summary graphs such as Figures 5.4–5.7 are useful. Plots of several independently derived inferences are useful in summarizing results so far and suggesting future model improvements. We illustrate in Figure 14.2 with a series of regression estimates of the advantage of incumbency in Congressional elections.

Graphs (or even maps, as in Figure 6.1) are also useful for model checking and sensitivity analysis, as we have shown in Chapter 6 and further illustrate in many of the examples in Sections IV and V.

When checking a model, one must keep in mind the purposes for which it will be used. For example, the normal model for football scores in Section 1.6 accurately predicts the probability of a win, but gives poor predictions for the probability that a game is exactly tied (see Figure 1.1).

It is also important to understand the limitations of automatic Bayesian inference. Even a model that fits observed data well can yield poor inferences about some quantities of interest. It is surprising and instructive to see the pitfalls that can arise when models are automatically applied and not subjected to model checks.

Example. Estimating a population total under simple random sampling using transformed normal models

We consider the problem of estimating the total population of the $N = 804$ municipalities in New York State in 1960 from a simple random sample of $n = 100$—an artificial example, but one that illustrates the role of model checking in avoiding seriously wrong inferences. Table 9.2 presents summary statistics for the population of this 'survey' along with two simple random samples (which were the first and only ones chosen). With knowledge of the population, neither sample appears particularly atypical; sample 1 is very representative of the population

	Population $(N = 804)$	Sample 1 $(n = 100)$	Sample 2 $(n = 100)$
total	13,776,663	1,966,745	3,850,502
mean	17,135	19,667	38,505
sd	139,147	142,218	228,625
lowest	19	164	162
5%	336	308	315
25%	800	891	863
median	1,668	2,081	1,740
75%	5,050	6,049	5,239
95%	30,295	25,130	41,718
highest	2,627,319	1,424,815	1,809,578

Table 9.2 *Summary statistics for populations of municipalities in New York State in 1960 (New York City was represented by its five boroughs); all 804 municipalities and two independent simple random samples of 100. From Rubin (1983a).*

according to the summary statistics provided, whereas sample 2 has a few too many large values. Consequently, it might at first glance seem straightforward to estimate the population total, perhaps overestimating the total from the second sample.

Sample 1: initial analysis. We begin the data analysis by trying to estimate the population total from sample 1 assuming that the N values in the population were drawn from a $N(\mu, \sigma^2)$ superpopulation, with a uniform prior density on $(\mu, \log \sigma)$. In the notation of Chapter 7, we wish to estimate the finite-population quantity,

$$y_{\text{total}} = N\bar{y} = n\bar{y}_{\text{obs}} + (N - n)\bar{y}_{\text{mis}}. \tag{9.1}$$

As discussed in Section 7.4, under this model, the posterior distribution of $\bar{y}$ is $t_{n-1}(\bar{y}_{\text{obs}}, (\frac{1}{n} - \frac{1}{N})s_{\text{obs}}^2)$. Using the data from the second column of Table 9.2 and the tabulated Student-t distribution, we obtain the following 95% posterior interval for y_{total}: $[-5.4 \times 10^6, 37.0 \times 10^6]$. The practical person examining this 95% interval might find the upper limit useful and simply replace the lower limit by the total in the sample, since the total in the population can be no less. This procedure gives a 95% interval estimate of $[2.0 \times 10^6, 37.0 \times 10^6]$.

Surely, modestly intelligent use of statistical models should produce a better answer because, as we can see in Table 9.2, both the population and sample 1 are very far from normal, and the standard interval is most appropriate with normal populations. Moreover, all values in the population are known to be positive. Even before seeing any data, we know that sizes of municipalities are far more likely to have a normal distribution on the logarithmic than on the untransformed scale.

We repeat the above analysis under the assumption that the $N = 804$ values in the complete data follow a lognormal distribution: $\log y_i \sim N(\mu, \sigma^2)$, with a uniform prior distribution on $(\mu, \log \sigma)$. Posterior inference for y_{total} is performed in the usual manner: drawing (μ, σ) from their posterior (normal-inverse-χ^2) distribution, then drawing $y_{\text{mis}}|\mu, \sigma$ from the predictive distribution, and finally calculating y_{total} from (9.1). Based on 100 simulation draws, the 95% interval for y_{total} is $[5.4 \times 10^6, 9.9 \times 10^6]$. This interval is narrower than the original interval and at first glance looks like an improvement.

Sample 1: checking the lognormal model. One of our major principles is to apply posterior predictive checks to models before accepting them. Because we are interested in a population total, y_{total}, we apply a posterior predictive check using, as a test statistic, the total in the sample, $T(y_{\text{obs}}) = \sum_{i=1}^{n} y_{\text{obs}\,i}$. Using our $L = 100$ sample draws of (μ, σ^2) from the posterior distribution under the lognormal model, we obtain posterior predictive simulations of L independent replicated datasets, $y_{\text{obs}}^{\text{rep}}$, and compute $T(y_{\text{obs}}^{\text{rep}}) = \sum_{i=1}^{n} y_{\text{obs}\,i}^{\text{rep}}$ for each. The result is that, for this predictive quantity, the lognormal model is *unacceptable*: all of the $L = 100$ simulated values are lower than the actual total in the sample, 1,966,745.

Sample 1: extended analysis. A natural generalization beyond the lognormal model for municipality sizes is the power-transformed normal family, which adds an additional parameter, ϕ, to the model; see (6.14) on page 195 for details. The values $\phi = 1$ and 0 correspond to the untransformed normal and lognormal models, respectively, and other values correspond to other transformations.

To fit a transformed normal family to data y_{obs}, the easiest computational approach is to fit the normal model to transformed data at several values of ϕ and then compute the marginal posterior density of ϕ. Using the data from sample 1, the marginal posterior density of ϕ is strongly peaked around the value $-1/8$ (assuming a uniform prior distribution for ϕ, which is reasonable given the relatively informative likelihood). Based on 100 simulated values under the extended model, the 95% interval for y_{total} is $[5.8 \times 10^6, 31.8 \times 10^6]$. With respect to the posterior predictive check, 15 out of 100 simulated replications of the sample total are larger than the actual sample total; the model fits adequately in this sense.

Perhaps we have learned how to apply Bayesian methods successfully to estimate a population total with this sort of data: use a power-transformed family and summarize inference by simulation draws. But we did not conduct a very rigorous test of this conjecture. We started with the log transformation and obtained an inference that initially looked respectable, but we saw that the posterior predictive check distribution indicated a lack of fit in the model with respect to predicting the sample total. We then enlarged the family of transformations and performed inference under the larger model (or, equivalently in this case, found the best-fitting transformation, since the transformation power was so precisely estimated by the data). The extended procedure seemed to work in the sense that the resultant 95% interval was plausible; moreover, the posterior predictive check on the sample total was acceptable. To check on this extended procedure, we try it on the second random sample of 100.

Sample 2. The standard normal-based inference for the population total from the second sample yields a 95% interval of $[-3.4 \times 10^6, 65.3 \times 10^6]$. Substituting the sample total for the lower limit gives the wide interval of $[3.9 \times 10^6, 65.3 \times 10^6]$.

Following the steps used on sample 1, modeling the sample 2 data as lognormal leads to a 95% interval for y_{total} of $[8.2 \times 10^6, 19.6 \times 10^6]$. The lognormal inference is quite tight. However, in the posterior predictive check for sample 2 with the lognormal model, none of 100 simulations of the sample total was as large as the observed sample total, and so once again we find this model unsuited for estimation of the population total.

Based upon our experience with sample 1, and the posterior predictive checks under the lognormal models for both samples, we should not trust the lognormal interval and instead should consider the general power family, which includes the lognormal as a special case. For sample 2, the marginal posterior distribution for the power parameter ϕ is strongly peaked at $-1/4$. The posterior predictive check generated 48 of 100 sample totals larger than the observed total—no indication of any problems, at least if we do not examine the specific values being generated.

In this example we have the luxury of knowing the correct example (corresponding to having complete data rather than a sample of 100 municipalities). Unfortunately, the inference for the population total under the power family turns out to be, from a substantive standpoint, atrocious: for example, the median of the 100 generated values of y_{total} is 57×10^7, the 97th value is 14×10^{15}, and the largest value generated is 12×10^{17}.

Need to specify crucial prior information. What is going on? How can the inferences for the population total in sample 2 be so much less realistic with a better-fitting model (that is, assuming a normal distribution for $y_i^{-1/4}$) than with a worse-fitting model (that is, assuming a normal distribution for $\log y_i$)?

The problem with the inferences in this example is not an inability of the models to fit the data, but an inherent inability of the data to distinguish between alternative models that have very different implications for estimation of the population total, y_{total}. Estimates of y_{total} depend strongly on the upper extreme of the distribution of municipality sizes, but as we fit models like the power family, the right tail of these models (especially beyond the 99.5% quantile), is being affected dramatically by changes governed by the fit of the model to the main body of the data (between the 0.5% and 99.5% quantiles). The inference for y_{total} is actually critically dependent upon tail behavior beyond the quantile corresponding to the largest observed $y_{obs\,i}$. In order to estimate the total (or the mean), not only do we need a model that reasonably fits the observed data, but we also need a model that provides realistic extrapolations beyond the region of the data. For such extrapolations, we must rely on prior assumptions, such as specification of the largest possible size of a municipality.

More explicitly, for our two samples, the three parameters of the power family are basically enough to provide a reasonable fit to the observed data. But in order to obtain realistic inferences for the population of New York State from a simple random sample of size 100, we must constrain the distribution of large municipalities. We were warned, in fact, by the specific values of the posterior simulations for the sample total from sample 2: 10 of the 100 simulations for the replicated sample total were larger than 300 million!

The substantive knowledge that is used to criticize the power-transformed normal model can also be used to improve the model. Suppose we know that no single municipality has population greater than 5×10^6. To incorporate this information as part of the model we simply draw posterior simulations in the same way as before but truncate municipality sizes to lie below that upper bound. The resulting posterior inferences for total population size are quite reasonable. For both samples, the inferences for y_{total} under the power family are tighter than with the untruncated models and are realistic. The 95% intervals under samples 1 and 2 are $[6 \times 10^6, 20 \times 10^6]$ and $[10 \times 10^6, 34 \times 10^6]$, respectively. Incidentally, the true population total is $13.7 \cdot 10^6$ (see Table 9.2), which is included in both intervals.

Why does the untransformed normal model work reasonably well for estimating the population total? Interestingly, the inferences for y_{total} based on the simple untransformed normal model for y_i are not terrible, even without supplying an upper bound for municipality size. Why? The estimate for y_{total} under the normal model is essentially based only on the assumed normal sampling distribution for $\overline{y}_{\text{obs}}$ and the corresponding χ^2 sampling distribution for s^2_{obs}. In order to believe that these sampling distributions are approximately valid, we need the central limit theorem to apply, which we achieve by *implicitly* bounding the upper tail of the distribution for y_i enough to make approximate normality work for a sample size of 100. This is not to suggest that we recommend the untransformed normal model for clearly nonnormal data; in the example considered here, the bounded power-transformed family makes more efficient use of the data. In addition, the untransformed normal model gives extremely poor inferences for estimands such as the population median. In general, a Bayesian analysis that limits large values of y_i must do so explicitly.

Well-designed samples or robust questions obviate the need for strong prior information. Of course, extensive modeling and simulation are not needed to estimate totals routinely in practice. Good survey practitioners know that a simple random sample is not a good survey design for estimating the total in a highly skewed population. If stratification variables were available, one would prefer to oversample the large municipalities (for example, sample all five boroughs of New York City, a large proportion of cities, and a smaller proportion of towns).

Inference for the population median. It should not be overlooked, however, that the simple random samples we drew, although not ideal for estimating the population total, are quite satisfactory for answering many questions *without* imposing strong prior restrictions.

For example, consider inference for the median size of the 804 municipalities. Using the data from sample 1, the simulated 95% posterior intervals for the median municipality size under the three models: (a) lognormal, (b) power-transformed normal family, and (c) power-transformed normal family truncated at 5×10^6, are [1800, 3000], [1600, 2700], and [1600, 2700], respectively. The comparable intervals based on sample 2 are [1700, 3600], [1300, 2400], and [1200, 2400]. In general, better models tend to give better answers, but for questions that are robust with respect to the data at hand, such as estimating the median from our simple random sample of size 100, the effect is rather weak. For such questions, prior constraints are not extremely critical and even relatively inflexible models can provide satisfactory answers. Moreover, the posterior predictive checks for the sample median looked fine—with the observed sample median near the middle of the distribution of simulated sample medians—for all these models (but not for the untransformed normal model).

What general lessons have we learned from considering this example? The first two messages are specific to the example and address accuracy of resultant inferences for covering the true population total.

1. The lognormal model may yield inaccurate inferences for the population total even when it appears to fit observed data fairly well.

2. Extending the lognormal family to a larger, and so better-fitting, model such as the power transformation family, may lead to less realistic inferences for the population total.

These two points are not criticisms of the lognormal distribution or power transformations. Rather, they provide warnings to be heeded when using a model that has not been subjected to posterior predictive checks (for test variables relevant to the estimands of interest) and reality checks. In this context, the naive statement, 'better fits to data mean better models which in turn mean better real-world answers,' is not necessarily true. Statistical answers rely on prior assumptions as well as data, and better real-world answers generally require models that incorporate more realistic prior assumptions (such as bounds on municipality sizes) as well as provide better fits to data. This comment naturally leads to a general message encompassing the first two points.

3. In general, inferences may be sensitive to features of the underlying distribution of values in the population that cannot be addressed by the observed data. Consequently, for good statistical answers, we not only need models that fit observed data, but we also need:

   (a) flexibility in these models to allow specification of realistic underlying features not adequately addressed by observed data, such as behavior in the extreme tails of the distribution, *or*

   (b) questions that are robust for the type of data collected, in the sense that all relevant underlying features of population values are adequately addressed by the observed values.

Finding models that satisfy (a) is a more general approach than finding questions that satisfy (b) because statisticians are often presented with hard questions that require answers of some sort, and do not have the luxury of posing easy (that is, robust) questions in their place. For example, for environmental reasons it may be important to estimate the total amount of pollutant being emitted by a manufacturing plant using samples of the soil from the surrounding geographical area, or, for purposes of budgeting a health-care insurance program, it may be necessary to estimate the total amount of medical expenses from a sample of patients. Such questions are inherently nonrobust in that their answers depend on the behavior in the extreme tails of the underlying distributions. Estimating more robust population characteristics, such as the median amount of pollutant in soil samples or the median medical expense for patients, does not address the essential questions in such examples.

Relevant inferential tools, whether Bayesian or non-Bayesian, cannot be free of assumptions. Robustness of Bayesian inference is a joint property of data, prior knowledge, and questions under consideration. For many problems, statisticians may be able to define the questions being studied so as to have robust answers. Sometimes, however, the practical, important question is inescapably nonrobust, with inferences being sensitive to assumptions that the data at hand cannot address, and then a good Bayesian analysis expresses this sensitivity.

9.4 Summary

In concluding the first two parts of the book, we draw together a brief summary of some important themes of Bayesian data analysis. We focus on the construction of models (especially hierarchical ones) to relate complicated data structures to scientific questions, checking the fit of such models, and investigating the sensitivity of conclusions to reasonable modeling assumptions. From this point of view, the strength of the Bayesian approach lies in (1) its ability to combine information from multiple sources (thereby in fact allowing greater 'objectivity' in final conclusions), and (2) its more encompassing accounting of uncertainty about the unknowns in a statistical problem.

Other important themes, many of which are common to much modern applied statistical practice, whether formally Bayesian or not, are the following:

- a willingness to use many parameters

- hierarchical structuring of models, which is the essential tool for achieving partial pooling of estimates and compromising in a scientific way between alternative sources of information

- model checking—not only by examining the internal goodness of fit of models to observed and possible future data, but also by comparing inferences about estimands and predictions of interest to substantive knowledge

- an emphasis on inference in the form of distributions or at least interval estimates rather than simple point estimates

- the use of simulation as the primary method of computation; the modern computational counterpart to a 'joint probability distribution' is a set of randomly drawn values, and a key tool for dealing with missing data is the method of multiple imputation (computation and multiple imputation are discussed in more detail in later chapters)

- the use of probability models as tools for understanding and possibly improving data-analytic techniques that may not explicitly invoke a Bayesian model

- the importance of conditioning on as much covariate information as possible in the analysis of data, with the aim of making study design ignorable

- the importance of designing studies to have the property that inferences for estimands of interest will be robust to model assumptions.

9.5 Bibliographic note

Throughout the book we refer to many papers in statistical journals illustrating applied Bayesian analysis. The three examples in this chapter are derived from Gelman, Bois, and Jiang (1996), Rubin (1983c), and Rubin (1983a), respectively. The connection between ratio estimation and modeling alluded to in Section 9.1 is discussed by Brewer (1963), Royall (1970), and, from our Bayesian approach, Rubin (1987a, p. 46).

Part III: Advanced Computation

The remainder of this book delves into more sophisticated models. Before we begin this enterprise, however, we detour to describe methods for computing posterior distributions in hierarchical models. Toward the end of Chapter 5, the algebra required for analytic derivation of posterior distributions became less and less attractive, and that was with a model based on the normal distribution! If we try to solve more complicated problems analytically, the algebra starts to overwhelm the statistical science almost entirely, making the full Bayesian analysis of realistic probability models too cumbersome for most practical applications. Fortunately, a battery of powerful methods has been developed over the past few decades for simulating from probability distributions. In the next four chapters, we survey some useful simulation methods that we apply in later chapters in the context of specific models. Some of the simpler simulation methods we present here have already been introduced in examples in earlier chapters.

Because the focus of this book is data analysis rather than computation, we move through the material of Part III briskly, with the intent that it be used as a reference when applying the models discussed in Parts IV and V. We have also attempted to place a variety of useful techniques in the context of a systematic general approach to Bayesian computation. Our general philosophy in computation, as in modeling, is pluralistic, starting with simple approximate methods and gradually working toward more precise computations.

CHAPTER 10

Overview of computation

This chapter provides a general perspective on computation for Bayesian data analysis. Our computational strategy for complex problems begins with crude initial estimates of model parameters and usually continues with Markov chain simulation, sometimes using mode-based approximations of the posterior distribution as an intermediate step. We begin with some notation and terminology that are relevant to all the chapters of this part of the book. Chapters 11, 12, and 13 cover Markov chain simulation, mode-based approximations, and more advanced methods in Bayesian computation.

Joint, conditional, and marginal target distributions

We refer to the (multivariate) distribution to be simulated as the *target distribution* and denote it as $p(\theta|y)$. At various points we consider partitioning a high-dimensional parameter vector as $\theta = (\gamma, \phi)$, where typically γ will include most of the components of θ. (This is not the same as the θ, ϕ notation in Chapter 7.) For example, in a hierarchical model, γ could be the parameters that are exchangeable given hyperparameters ϕ. Later, we will see that the same computational techniques can be used to handle unobserved indicator variables (Chapter 18) and missing data (Chapter 21). Formally, in those cases θ includes all unknown quantities in the joint distribution including unobserved or missing values. For example, in a problem with missing data, γ could be the missing values and ϕ the model parameters. Such factorizations serve as useful computational devices for partitioning high-dimensional distributions into manageable parts. When using the (γ, ϕ) notation, we work with the factorization, $p(\theta|y) = p(\gamma, \phi|y) = p(\gamma|\phi, y)p(\phi|y)$, and compute the conditional and marginal posterior densities in turn.

Normalized and unnormalized densities

Unless otherwise noted (in Section 13.4), we assume that the target density $p(\theta|y)$ can be easily computed for any value of θ, up to a proportionality constant involving only the data y; that is, we assume there is some easily computable function $q(\theta|y)$, an *unnormalized density*, for which $q(\theta|y)/p(\theta|y)$ is a constant that depends only on y. For example, in the usual use of Bayes' theorem, we work with the product $p(\theta)p(y|\theta)$, which is proportional to the posterior density.

10.1 Crude estimation by ignoring some information

Before developing more elaborate approximations or complicated methods for sampling from the target distribution, it is almost always useful to obtain a rough estimate of the location of the target distribution—that is, a point estimate of the parameters in the model—using some simple, noniterative technique. The method for creating this first estimate will vary from problem to problem but typically will involve discarding parts of the model and data to create a simple problem for which convenient parameter estimates can be found.

In a hierarchical model, one can sometimes roughly estimate the main parameters γ by first estimating the hyperparameters ϕ crudely and then treating the resulting distribution of γ given ϕ as a fixed prior distribution for γ. We applied this approach to the rat tumor example in Section 5.1, where crude estimates of the hyperparameters (α, β) were used to obtain initial estimates of the other parameters, θ_j.

For another example, in the educational testing analysis in Section 5.5, the school effects θ_j can be crudely estimated by the data y_j from the individual experiments, and the between-school standard deviation τ can then be estimated very crudely by the standard deviation of the eight y_j-values or, to be slightly more sophisticated, the estimate (5.22), restricted to be nonnegative.

When some data are missing, a good way to get started is by simplistically imputing the missing values based on available data. Or, even simpler, it may be convenient temporarily to ignore data from all experimental units that have missing observations. (Ultimately, inferences for the missing data should be included as part of the model; see Chapter 21.)

In addition to creating a starting point for a more exact analysis, crude inferences are useful for comparison with later results—if the rough estimate differs greatly from the results of the full analysis, the latter may very well have errors in programming or modeling. Crude estimates are often convenient and reliable because they can be computed using available computer programs.

10.2 Use of posterior simulations in Bayesian data analysis

Bayesian inference are usually most conveniently summarized by random draws from the posterior distribution of the model parameters. Percentiles of the posterior distribution of univariate estimands can be reported to convey the shape of the distribution. For example, reporting the 2.5%, 25%, 50%, 75%, and 97.5% points of the sampled distribution of an estimand provides a 50% and a 95% posterior interval and also conveys skewness in its marginal posterior density. Scatterplots of simulations, contour plots of density functions, or more sophisticated graphical techniques can also be used to examine the posterior distribution in two or three dimensions. Quantities of interest can be defined in terms of the parameters (for example, LD50 in the bioassay example in Section 3.7) or of parameters and data.

We also use posterior simulations to make inferences about predictive quantities. Given each simulation θ^l, we can simulate a posterior draw for any predictive quantity of interest: $\tilde{y}^l \sim p(\tilde{y}|\theta^l)$ or, for a regression model, $\tilde{y}^l \sim p(\tilde{y}|\tilde{X}, \theta^l)$. Posterior inferences and probability calculations can then be performed for each predictive quantity using the L simulations (for example, the predicted probability of Bill Clinton winning each state as displayed in Figure 6.1 on page 159).

Finally, given each simulation θ^l, we can simulate a replicated dataset $y^{\text{rep}\,l}$. As described in Chapter 6, we can then check the model by comparing the data to these posterior predictive replications.

How many simulation draws are needed?

Our goal in Bayesian computation is to obtain a set of independent draws θ^l, $l = 1, \ldots, L$, from the posterior distribution, with enough draws L so that quantities of interest can be estimated with reasonable accuracy. For most examples, $L = 100$ independent samples are enough for reasonable posterior summaries. We can see this by considering a scalar parameter θ with an approximately normal posterior distribution (see Chapter 4) with mean μ_θ and standard deviation σ_θ. We assume these cannot be calculated analytically and instead are estimated from the mean $\bar{\theta}$ and standard deviation s_θ of the L simulation draws. The posterior mean is then estimated to an accuracy of approximately $s_\theta/\sqrt{L}$. The total standard deviation of the computational parameter estimate (including *Monte Carlo error*, the uncertainty contributed by having only a finite number of simulation draws) is then $s_\theta\sqrt{1 + 1/L}$. For $L = 100$, the factor $\sqrt{1 + 1/L} = 1.005$, implying that Monte Carlo error adds almost nothing to the uncertainty coming from actual posterior variance. However, it can be convenient to have more than 100 simulations just so that the numerical summaries are more stable, even if this stability typically confers no important practical advantage.

For some posterior inferences, more simulation draws are needed to obtain desired precisions. For example, posterior probabilities are estimated to a standard deviation of $\sqrt{p(1-p)/L}$, so that $L = 100$ simulations allow estimation of a probability near 0.5 to an accuracy of 5%. $L = 2500$ simulations are needed to estimate to an accuracy of 1%. Even more simulation draws are needed to compute the posterior probability of rare events, unless analytic methods are used to assist the computations.

Example. Educational testing experiments
We illustrate with the hierarchical model fitted to the SAT coaching experiments as described in Section 5.5. First consider inference for a particular parameter, for example θ_1, the estimated effect of coaching in School A. Table 5.3 shows that from 200 simulation draws, our posterior median estimate was 10, with a 50% interval of $[7, 16]$ and a 95% interval of $[-2, 31]$. Repeating the computation, another 200 draws gave a posterior median of 9, with a 50% interval of $[6, 14]$ and a 95% interval of $[-4, 32]$. These intervals differ slightly but convey the same general information about θ_1. From $L = 10{,}000$ simulation draws, the median is

10, the 50% interval is $[6, 15]$, and the 95% interval is $[-2, 31]$. In practice, these are no different from either of the summaries obtained from 200 draws.

We now consider some posterior probability statements. Our original 200 simulations gave us an estimate of 0.73 for the posterior probability $\Pr(\theta_1 > \theta_3 | y)$, the probability that the effect is larger in school A than in school C (see the end of Section 5.5). This probability is estimated to an accuracy of $\sqrt{0.73(1 - 0.73)/200} = 0.03$, which is good enough in this example.

How about a rarer event, such as the probability that the effect in School A is greater than 50 points? None of our 200 simulations θ_1^l exceeds 50, so the simple estimate of the probability is that it is zero (or less than $1/200$). When we simulate $L = 10{,}000$ draws, we find 3 of the draws to have $\theta_1 > 50$, which yields a crude estimated probability of 0.0003.

An alternative way to compute this probability is semi-analytically. Given μ and τ, the effect in school A has a normal posterior distribution, $p(\theta_1 | \mu, \tau, y) = \mathrm{N}(\hat{\theta}_1, V_1)$, where this mean and variance depend on y_1, μ, and τ (see (5.17) on page 135). The conditional probability that θ_1 exceeds 50 is then $\Pr(\theta_1 > 50 | \mu, \tau, y) = \Phi((\hat{\theta}_1 - 50)/\sqrt{V_1})$, and we can estimate the unconditional posterior probability $\Pr(\theta_1 > 50 | y)$ as the average of these normal probabilities as computed for each simulation draw (μ^l, τ^l). Using this approach, $L = 200$ draws are sufficient for a reasonably accurate estimate.

In general, fewer simulations are needed to estimate posterior medians of parameters, probabilities near 0.5, and low-dimensional summaries than extreme quantiles, posterior means, probabilities of rare events, and higher-dimensional summaries. In most of the examples in this book, we use a moderate number of simulation draws (typically 100 to 2000) as a way of emphasizing that applied inferences do not typically require a high level of simulation accuracy.

10.3 Practical issues

Computational tools

Section 1.10 lists some of the software that is useful for Bayesian data analysis, and Appendix C illustrates with an extended example of several different ways to compute a model using the statistical packages **Bugs** and **R**. More generally, the following computational operations are useful:

- Simulation from standard distributions (see Appendix A) and the ability to sample from arbitrary discrete distributions, including discrete approximations to continuous distributions, as was done for the hyperparameter τ in the hierarchical model in Section 5.4 (see Figure 5.5 on page 141).

- Vector and matrix operations, including Cholesky factorization (matrix square roots), and linear transformations of vectors using matrix multiplication. These operations are useful for regression models, for conditioning in multivariate distributions (see (A.1) and (A.2) on page 579), and for linear transformations to make Gibbs samplers more efficient (see Section 15.4).

- Even more generally, matrices are crucial for organizing the simulations of parameters, missing data, predictions, and replicated data, as indicated in Table 1.1 on page 26.

- Linear regression computations—as discussed in Chapter 14, we like to think of the regression estimate $\hat{\beta}$ and covariance matrix V_β as basic operations on data X and y, without worrying about the intermediate matrix computations. Classical regression is also an important component of Bayesian computation for hierarchical regression and generalized linear models.

- All-purpose optimization algorithms (such as the optim function in R) can be useful in obtaining crude estimates.

- Other numerical operations that are often performed are transformations (most commonly logarithm, logit, and their inverses), and numerical derivatives (see (12.1) on page 313).

- Finally, graphical tools are needed to display data, posterior inferences, and simulated replicated data. These tools begin with histograms and scatterplots, but software should be flexible enough to display multiple plots on a single page (as in Figure 6.2 on page 160) and to allow the creation of specialized plots to capture complex data structures (as in Figure 6.6 on page 166).

Debugging using fake data

Our usual approach for building confidence in our posterior inferences is to fit different versions of the desired model, noticing when the inferences change unexpectedly. Section 10.1 discusses crude inferences from simplified models that typically ignore some structure in the data.

Within the computation of any particular model, we check convergence by running parallel simulations from different starting points, checking that they mix and converge to the same estimated posterior distribution (see Section 11.6). This can be seen as a form of debugging of the individual simulated sequences.

When a model is particularly complicated, or its inferences are unexpected enough to be not necessarily believable, one can perform more elaborate debugging using fake data. The basic approach is:

1. Pick a reasonable value for the 'true' parameter vector θ. Strictly speaking, this value should be a random draw from the prior distribution, but if the prior distribution is noninformative, then any reasonable value of θ should work.

2. If the model is hierarchical (as it generally will be), then perform the above step by picking reasonable values for the hyperparameters, then drawing the other parameters from the prior distribution conditional on the specified hyperparameters.

3. Simulate a large fake dataset y^{fake} from the data distribution $p(y|\theta)$.

4. Perform posterior inference about θ from $p(\theta|y^{\text{fake}})$.

5. Compare the posterior inferences to the 'true' θ from step 1 or 2. For instance, for any element of θ, there should be a 50% probability that its 50% posterior interval contains the truth.

Formally, this procedure requires that the model has proper prior distributions and that the frequency evaluations be averaged over many values of the 'true' θ, drawn independently from the prior distribution in step 1 above. In practice, however, the debugging procedure can be useful with just a single reasonable choice of θ in the first step. If the model does not produce reasonable inferences with θ set to a reasonable value, then there is probably something wrong, either in the computation or in the model itself.

Inference from a single fake dataset can be revealing for debugging purposes, if the true value of θ is far outside the computed posterior distribution. If the dimensionality of θ is large (as can easily happen with hierarchical models), we can go further and compute debugging checks such as the proportion of the 50% intervals that contain the true value.

To check that inferences are correct on average, one can create a 'residual plot' as follows. For each scalar parameter θ_j, define the predicted value as the average of the posterior simulations of θ_j, and the error as the true θ_j (as specified or simulated in step 1 or 2 above) minus the predicted value. If the model is computed correctly, the errors should have zero mean, and we can diagnose problems by plotting errors vs. predicted values, with one dot per parameter.

For models with only a few parameters, one can get the same effect by performing many fake-data simulations, resampling a new 'true' vector θ and a new fake dataset y^{fake} each time, and then checking that the errors have zero mean and the correct interval coverage, on average.

Model checking and convergence checking as debugging

Finally, the techniques for model checking and comparison described in Chapter 6, and the techniques for checking for poor convergence of iterative simulations, which we describe in Section 11.6, can also be interpreted as methods for debugging.

In practice, when a model grossly misfits the data, or when a histogram or scatterplot or other display of replicated data looks weird, it is often because of a computing error. These errors can be as simple as forgetting to recode discrete responses (for example, $1 = \text{Yes}$, $0 = \text{No}$, $-9 = \text{Don't Know}$) or misspelling a regression predictor, or as subtle as a miscomputed probability ratio in a Metropolis updating step (see Section 11.4), but typically they show up as predictions that do not make sense or do not fit the data. Similarly, poor convergence of an iterative simulation algorithm can sometimes occur from programming errors or even conceptual errors in the model.

Tips and tricks

Among the practical techniques for making statistical computation more reliable, probably the most important involve programming. We try wherever possible to code procedures as subroutines or functions rather than sequences of commands in the main program. A modular programming style makes it easier to expand a model by adding additional parameters to a Gibbs-Metropolis updating scheme. We illustrate in Section C.4.

When displaying statistical inferences, we recommend routinely putting several graphs on a single page (for example, using the `mfrow` option in the `par` function in R) and labeling all graphs (in R, using the `xlab`, `ylab`, and `main` options in the `plot` function). It is much easier to explore patterns in data and inferences when viewing several graphs at once.

To avoid computational overflows and underflows, one should compute with the logarithms of posterior densities whenever possible. Exponentiation should be performed only when necessary and as late as possible; for example, in the Metropolis algorithm, the required ratio of two densities (11.1) should be computed as the exponential of the difference of the log-densities.

Set up computations to be able to use the L posterior simulation draws rather than a point estimate of the parameters. Thus, any scalar parameter becomes a vector of length L, a vector parameter of length J becomes an $L \times J$ matrix, and so on. Setting everything in a vector-matrix format is convenient for displaying posterior inferences and predictive simulations.

Almost no computer program works the first time. A good way to speed the debugging process is to start with a smaller dataset, perhaps randomly sampled from the original data. This speeds computations and also makes it easier to inspect the data and inferences for possible problems. In many cases, a sample of 10 or 20 data points is enough to take us quickly through the initial stage of debugging. A related piece of advice, when running iterative simulations, is to run for only a few iterations at first, possibly 10 or 100, to check that the algorithm seems to be on the right track. There is no point in waiting an hour for the computer to get highly precise computations for the wrong model.

When the posterior inferences from a fitted model do not make sense, it is sometimes not clear whether there is a bug in the program or a fundamental error in the model itself. At this point, a useful conceptual and computational strategy is to simplify—to remove parameters from the model, or to give them fixed values or highly informative prior distributions, or to separately analyze data from different sources (that is, to un-link a hierarchical model). These computations can be performed in steps, for example first removing a parameter from the model, then setting it equal to a null value (for example, zero) just to check that adding it into the program has no effect, then fixing it at a reasonable nonzero value, then assigning it a precise prior distribution, then allowing it to be estimated more fully from the data. Model building is a gradual process, and we often find ourselves going back and forth between simpler and more complicated models, both for conceptual and computational reasons.

Finally, almost any statistical computation can be seen as part of a larger problem. As noted already in this chapter, for problems of realistic complexity we usually fit at least one preliminary model to obtain starting points for iterative simulation. To take things one step further, an existing model can be used as an approximation for a future expanded model. Another form of expansion is to add data, for example by taking an analysis that could apply individually to several datasets and linking them into a hierarchical model. This can be straightforward to program using the Gibbs sampler, alternating between updating the small model fit to each dataset and the population model that links the parameters. We illustrate with an example in Section 20.3.

10.4 Exercises

The exercises in Part III focus on computational details. Data analysis exercises using the methods described in this part of the book appear in the appropriate chapters in Parts IV and V.

1. Number of simulation draws: suppose you are interested in inference for the parameter θ_1 in a multivariate posterior distribution, $p(\theta|y)$. You draw 100 independent values θ from the posterior distribution of θ and find that the posterior density for θ_1 is approximately normal with mean of about 8 and standard deviation of about 4.

   (a) Using the average of the 100 draws of θ_1 to estimate the posterior mean, $E(\theta_1|y)$, what is the approximate standard deviation due to simulation variability?

   (b) About how many simulation draws would you need to reduce the simulation standard deviation of the posterior mean to 0.1 (thus justifying the presentation of results to one decimal place)?

   (c) A more usual summary of the posterior distribution of θ_1 is a 95% central posterior interval. Based on the data from 100 draws, what are the approximate simulation standard deviations of the estimated 2.5% and 97.5% quantiles of the posterior distribution? (Recall that the posterior density is approximately normal.)

   (d) About how many simulation draws would you need to reduce the simulation standard deviations of the 2.5% and 97.5% quantiles to 0.1?

   (e) In the SAT coaching example of Section 5.5, we simulated 200 independent draws from the posterior distribution. What are the approximate simulation standard deviations of the 2.5% and 97.5% quantiles for school A in Table 5.3?

   (f) Why was it not necessary, in practice, to simulate more than 200 draws for the SAT coaching example?

CHAPTER 11

Posterior simulation

The bulk of this chapter presents the most widely used Markov chain simulation methods—the Gibbs sampler and the Metropolis-Hastings algorithm—in the context of our general computing approach based on successive approximation. We sketch a proof of the convergence of Markov chain simulation algorithms and present a method for monitoring the convergence in practice. We illustrate these methods in Section 11.7 for a hierarchical normal model. For most of this chapter we consider simple and familiar (even trivial) examples in order to focus on the principles of iterative simulation methods as they are used for posterior simulation. Many examples of these methods appear in the recent statistical literature (see the bibliographic note at the end of this chapter) and also in Parts IV and V of this book. Appendix C shows the details of implementation in the computer languages R and Bugs for the educational testing example from Chapter 5.

11.1 Direct simulation

In simple nonhierarchical Bayesian models, it is often easy to draw from the posterior distribution directly, especially if conjugate prior distributions have been assumed. For more complicated problems, an often useful strategy is to factor the distribution analytically and simulate it in parts: for example, first obtain draws from the marginal posterior distribution of the hyperparameters, then simulate the other parameters conditional on the data and the simulated hyperparameters. It is sometimes possible to perform direct simulations and analytic integrations for parts of the larger problem, as was done in the examples of Chapter 5.

Frequently, draws from standard distributions or low-dimensional non-standard distributions are required, either as direct draws from the posterior distribution of the estimand in an easy problem, or as an intermediate step in a more complex problem. Appendix A is a relatively detailed source of advice, algorithms, and procedures specifically relating to a variety of commonly-used distributions. In this section, we describe methods of drawing a random sample of size 1, with the understanding that the methods can be repeated to draw larger samples. When obtaining more than one sample, it is often possible to reduce computation time by saving intermediate results such as the Cholesky factor for a fixed multivariate normal distribution.

Direct approximation by calculating at a grid of points

Often as a first approximation or for computational convenience, it is desirable to approximate the distribution of a continuous parameter as a discrete distribution on a grid of points. For the simplest discrete approximation, compute the target density, $p(\theta|y)$, at a set of evenly spaced values $\theta_1, \ldots, \theta_N$, that cover a broad range of the parameter space for θ, then approximate the continuous $p(\theta|y)$ by the discrete density at $\theta_1, \ldots, \theta_N$, with probabilities $p(\theta_i|y)/\sum_{j=1}^{N} p(\theta_j|y)$. Because the approximate density must be normalized anyway, this method will work just as well using an unnormalized density function, $q(\theta|y)$, in place of $p(\theta|y)$.

Once the grid of density values is computed, a random draw from $p(\theta|y)$ is obtained by drawing a random sample U from the uniform distribution on $[0, 1]$, then transforming by the inverse cdf method (see Section 1.9) to obtain a sample from the discrete approximation. When the points θ_i are spaced closely enough and miss nothing important beyond their boundaries, this method works well. The discrete approximation is more difficult to use in higher-dimensional multivariate problems, where computing at every point in a dense multidimensional grid becomes prohibitively expensive.

Rejection sampling

Many general techniques are available for simulating draws directly from the target density, $p(\theta|y)$; see the bibliographic note at the end of the chapter. Of the many techniques, we choose rejection sampling for a detailed description here because of its simplicity and generality and because it is often used as part of the more complex approaches described later in the chapter.

Suppose we want to obtain a single random draw from a density $p(\theta|y)$, or perhaps an unnormalized density $q(\theta|y)$ (with $p(\theta|y) = q(\theta|y)/\int q(\theta|y)d\theta$). In the following description we use p to represent the target distribution, but we could just as well work with the unnormalized form q instead. To perform *rejection sampling* we require a positive function $g(\theta)$ defined for all θ for which $p(\theta|y) > 0$ that has the following properties:

- We are able to draw random samples from the probability density proportional to g. It is *not* required that $g(\theta)$ integrate to 1, but $g(\theta)$ must have a finite integral.

- The *importance ratio* $p(\theta|y)/g(\theta)$ must have a known bound; that is, there must be some known constant M for which $p(\theta|y)/g(\theta) \leq M$ for all θ.

The rejection sampling algorithm proceeds in two steps:

1. Sample θ at random from the probability density proportional to $g(\theta)$.

2. With probability $p(\theta|y)/(Mg(\theta))$, *accept* θ as a draw from p. If the drawn θ is rejected, return to step 1.

Figure 11.1 illustrates rejection sampling. An accepted θ has the correct distribution, $p(\theta|y)$; that is, the distribution of drawn θ, conditional on it being

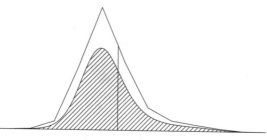

Figure 11.1 *Illustration of rejection sampling. The top curve is an approximation function, $Mg(\theta)$, and the bottom curve is the target density, $p(\theta|y)$. As required, $Mg(\theta) \geq p(\theta|y)$ for all θ. The vertical line indicates a single random draw θ from the density proportional to g. The probability that a sampled draw θ is accepted is the ratio of the height of the lower curve to the height of the higher curve at the value θ.*

accepted, is $p(\theta|y)$ (see Exercise 11.1). The boundedness condition is necessary so that the probability in step 2 is not greater than 1.

A good approximate density $g(\theta)$ for rejection sampling should be roughly proportional to $p(\theta|y)$ (considered as a function of θ). The ideal situation is $g \propto p$, in which case, with a suitable value of M, we can accept every draw with probability 1. When g is not nearly proportional to p, the bound M must be set so large that almost all samples obtained in step 1 will be rejected in step 2. A virtue of rejection sampling is that it is self-monitoring—if the method is not working efficiently, very few simulated draws will be accepted.

The function $g(\theta)$ is chosen to approximately match $p(\theta|y)$ and so in general will depend on y. We do not use the notation $g(\theta, y)$ or $g(\theta|y)$, however, because in practice we will be considering approximations to one posterior distribution at a time, and the functional dependence of g on y is not of interest.

Simulating from predictive distributions

Once simulations have been obtained from the posterior distribution, $p(\theta|y)$, it is typically easy to draw from the predictive distribution of unobserved or future data, $\tilde{y}$. For each draw of θ from the posterior distribution, just draw one value $\tilde{y}$ from the predictive distribution, $p(\tilde{y}|\theta)$. The set of simulated $\tilde{y}$'s from all the θ's characterizes the posterior predictive distribution. Posterior predictive distributions are crucial to the model-checking approach described in Chapter 6.

11.2 Markov chain simulation

Markov chain simulation (also called *Markov chain Monte Carlo*, or MCMC) is a general method based on drawing values of θ from approximate distributions and then correcting those draws to better approximate the target posterior

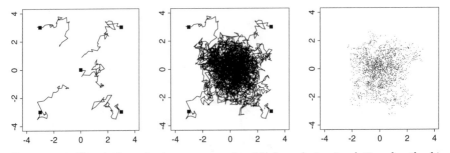

Figure 11.2 *Five independent sequences of a Markov chain simulation for the bivariate unit normal distribution, with overdispersed starting points indicated by solid squares. (a) After 50 iterations, the sequences are still far from convergence. (b) After 1000 iterations, the sequences are nearer to convergence. Figure (c) shows the iterates from the second halves of the sequences. The points in Figure (c) have been jittered so that steps in which the random walks stood still are not hidden. The simulation is a Metropolis algorithm described in the example on page 290.*

distribution, $p(\theta|y)$. The samples are drawn sequentially, with the distribution of the sampled draws depending on the last value drawn; hence, the draws form a Markov chain. (As defined in probability theory, a *Markov chain* is a sequence of random variables $\theta^1, \theta^2, \ldots$, for which, for any t, the distribution of θ^t given all previous θ's depends only on the most recent value, θ^{t-1}.) The key to the method's success, however, is not the Markov property but rather that the approximate distributions are improved at each step in the simulation, in the sense of converging to the target distribution. As we shall see in Section 11.4, the Markov property is helpful in proving this convergence.

Figure 11.2 illustrates a simple example of a Markov chain simulation—in this case, a Metropolis algorithm (see Section 11.4) in which θ is a vector with only two components, with a bivariate unit normal posterior distribution, $\theta \sim N(0, I)$. First consider Figure 11.2a, which portrays the early stages of the simulation. The space of the figure represents the range of possible values of the multivariate parameter, θ, and each of the five jagged lines represents the early path of a random walk starting near the center or the extremes of the target distribution and jumping through the distribution according to an appropriate sequence of random iterations. Figure 11.2b represents the mature stage of the same Markov chain simulation, in which the simulated random walks have each traced a path throughout the space of θ, with a common stationary distribution that is equal to the target distribution. From a simulation such as 11.2b, we can perform inferences about θ using points from the second halves of the Markov chains we have simulated, as displayed in Figure 11.2c.

In our applications of Markov chain simulation, several independent sequences of simulation draws are created; each sequence, θ^t, $t = 1, 2, 3, \ldots$, is produced by starting at some point θ^0 and then, for each t, drawing θ^t from a *transition distribution*, $T_t(\theta^t|\theta^{t-1})$ that depends on the previous draw,

θ^{t-1}. As we shall see in the discussion of combining the Gibbs sampler and Metropolis sampling in Section 11.5, it is often convenient to allow the transition distribution to depend on the iteration number t; hence the notation T_t. The transition probability distributions must be constructed so that the Markov chain converges to a unique stationary distribution that is the posterior distribution, $p(\theta|y)$.

Markov chain simulation is used when it is not possible (or not computationally efficient) to sample θ directly from $p(\theta|y)$; instead we sample *iteratively* in such a way that at each step of the process we expect to draw from a distribution that becomes closer and closer to $p(\theta|y)$. For a wide class of problems (including posterior distributions for many hierarchical models), this approach appears to be the easiest way to get reliable results, at least when used carefully. In addition, Markov chain and other iterative simulation methods have many applications outside Bayesian statistics, such as optimization, that we do not discuss here.

The key to Markov chain simulation is to create a Markov process whose stationary distribution is the specified $p(\theta|y)$ and run the simulation long enough that the distribution of the current draws is close enough to this stationary distribution. For any specific $p(\theta|y)$, or unnormalized density $q(\theta|y)$, a variety of Markov chains with the desired property can be constructed, as we demonstrate in Sections 11.3–11.5.

Once the simulation algorithm has been implemented and the simulations drawn, it is absolutely necessary to check the convergence of the simulated sequences; for example, the simulations of Figure 11.2a are far from convergence and are not close to the target distribution. We discuss how to check convergence in Section 11.6. If convergence is painfully slow, the algorithm should be altered, perhaps using the methods of Sections 11.8 and 11.9.

11.3 The Gibbs sampler

A particular Markov chain algorithm that has been found useful in many multidimensional problems is the *Gibbs sampler*, also called *alternating conditional sampling*, which is defined in terms of subvectors of θ. Suppose the parameter vector θ has been divided into d components or subvectors, $\theta = (\theta_1, \ldots, \theta_d)$. Each iteration of the Gibbs sampler cycles through the subvectors of θ, drawing each subset conditional on the value of all the others. There are thus d steps in iteration t. At each iteration t, an ordering of the d subvectors of θ is chosen and, in turn, each θ_j^t is sampled from the conditional distribution given all the other components of θ:

$$p(\theta_j|\theta_{-j}^{t-1}, y),$$

where θ_{-j}^{t-1} represents all the components of θ, except for θ_j, at their current values:

$$\theta_{-j}^{t-1} = (\theta_1^t, \ldots, \theta_{j-1}^t, \theta_{j+1}^{t-1}, \ldots, \theta_d^{t-1}).$$

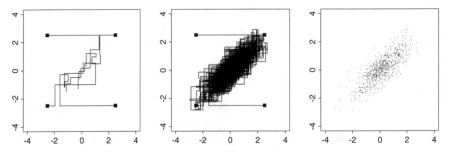

Figure 11.3 *Four independent sequences of the Gibbs sampler for a bivariate normal distribution with correlation $\rho = 0.8$, with overdispersed starting points indicated by solid squares. (a) First 10 iterations, showing the component-by-component updating of the Gibbs iterations. (b) After 500 iterations, the sequences have reached approximate convergence. Figure (c) shows the iterates from the second halves of the sequences.*

Thus, each subvector θ_j is updated conditional on the latest values of the other components of θ, which are the iteration t values for the components already updated and the iteration $t-1$ values for the others.

For many problems involving standard statistical models, it is possible to sample directly from most or all of the conditional posterior distributions of the parameters. We typically construct models using a sequence of conditional probability distributions, as in the hierarchical models of Chapter 5. It is often the case that the conditional distributions in such models are conjugate distributions that provide for easy simulation. We present an example for the hierarchical normal model at the end of this chapter and another detailed example for a normal-mixture model in Section 18.4. Here, we illustrate the workings of the Gibbs sampler with a very simple example.

Example. Bivariate normal distribution

Consider a single observation (y_1, y_2) from a bivariate normally distributed population with unknown mean $\theta = (\theta_1, \theta_2)$ and known covariance matrix $\begin{pmatrix} 1 & \rho \\ \rho & 1 \end{pmatrix}$. With a uniform prior distribution on θ, the posterior distribution is

$$\begin{pmatrix} \theta_1 \\ \theta_2 \end{pmatrix} \Bigg| y \sim \mathrm{N}\left(\begin{pmatrix} y_1 \\ y_2 \end{pmatrix}, \begin{pmatrix} 1 & \rho \\ \rho & 1 \end{pmatrix} \right).$$

Although it is simple to draw directly from the joint posterior distribution of (θ_1, θ_2), we consider the Gibbs sampler for the purpose of exposition. To apply the Gibbs sampler to (θ_1, θ_2), we need the conditional posterior distributions, which, from the properties of the multivariate normal distribution (either (A.1) or (A.2) on page 579), are

$$\begin{aligned} \theta_1 | \theta_2, y &\sim \mathrm{N}(y_1 + \rho(\theta_2 - y_2), 1 - \rho^2) \\ \theta_2 | \theta_1, y &\sim \mathrm{N}(y_2 + \rho(\theta_1 - y_1), 1 - \rho^2). \end{aligned}$$

The Gibbs sampler proceeds by alternately sampling from these two normal distributions. In general, we would say that a natural way to start the iterations would be with random draws from a normal approximation to the posterior distribution; of course, such draws would eliminate the need for iterative simulation in this trivial example. Figure 11.3 illustrates for the case $\rho = 0.8$, data $(y_1, y_2) = (0, 0)$, and four independent sequences started at $(\pm 2.5, \pm 2.5)$.

11.4 The Metropolis and Metropolis-Hastings algorithms

Many clever methods have been devised for constructing and sampling from transition distributions for arbitrary posterior distributions. The *Metropolis-Hastings algorithm* is a general term for a family of Markov chain simulation methods that are useful for drawing samples from Bayesian posterior distributions. We have already seen the Gibbs sampler in the previous section; it can be viewed as a special case of Metropolis-Hastings (as described in Section 11.5). In this section, we present the basic Metropolis algorithm and its generalization to the Metropolis-Hastings algorithm. The next section talks about combining the Gibbs sampler and the Metropolis algorithm, which is often helpful in practical problems. Subsequent sections discuss practical issues in monitoring convergence and improving the efficiency of the simulation algorithms.

The Metropolis algorithm

The Metropolis algorithm is an adaptation of a random walk that uses an acceptance/rejection rule to converge to the specified target distribution. The algorithm proceeds as follows.

1. Draw a starting point θ^0, for which $p(\theta^0|y) > 0$, from a *starting distribution* $p_0(\theta)$. The starting distribution might, for example, be based on an approximation as described in Section 12.2 in the following chapter. Or we may simply choose starting values dispersed around a crude approximate estimate of the sort discussed in Chapter 10.

2. For $t = 1, 2, \ldots$:

   (a) Sample a *proposal* θ^* from a *jumping distribution* (or *proposal distribution*) at time t, $J_t(\theta^*|\theta^{t-1})$. For the Metropolis algorithm (but not the Metropolis-Hastings algorithm, as discussed later in this section), the jumping distribution must be *symmetric*, satisfying the condition $J_t(\theta_a|\theta_b) = J_t(\theta_b|\theta_a)$ for all θ_a, θ_b, and t.

   (b) Calculate the ratio of the densities,

   $$r = \frac{p(\theta^*|y)}{p(\theta^{t-1}|y)}. \tag{11.1}$$

   (c) Set

   $$\theta^t = \begin{cases} \theta^* & \text{with probability } \min(r, 1) \\ \theta^{t-1} & \text{otherwise.} \end{cases}$$

Given the current value θ^{t-1}, the transition distribution $T_t(\theta^t|\theta^{t-1})$ of the Markov chain is thus a mixture of a point mass at $\theta^t = \theta^{t-1}$, and a weighted version of the jumping distribution, $J_t(\theta^t|\theta^{t-1})$, that adjusts for the acceptance rate.

The algorithm requires the ability to calculate the ratio r in (11.1) for all (θ, θ^*), and to draw θ from the jumping distribution $J_t(\theta^*|\theta)$ for all θ and t. In addition, step (c) above requires the generation of a uniform random number.

Note: if $\theta^t = \theta^{t-1}$—that is, the jump is not accepted—this counts as an iteration in the algorithm.

Example. Bivariate unit normal density with bivariate normal jumping kernel

For simplicity, we illustrate the Metropolis algorithm with the simple example of the bivariate unit normal distribution. The target density is the bivariate unit normal, $p(\theta|y) = N(\theta|0, I)$, where I is the 2×2 identity matrix. The jumping distribution is also bivariate normal, centered at the current iteration and scaled to $1/5$ the size: $J_t(\theta^*|\theta^{t-1}) = N(\theta^*|\theta^{t-1}, 0.2^2 I)$. At each step, it is easy to calculate the density ratio $r = N(\theta^*|0, I)/N(\theta^{t-1}|0, I)$. It is clear from the form of the normal distribution that the jumping rule is symmetric. Figure 11.2 on page 286 displays five simulation runs starting from different points. We have purposely chosen the relatively inefficient jumping rule with scale $1/5$ in order to make the random walk aspect of the algorithm obvious in the figure.

Relation to optimization

The acceptance/rejection rule of the Metropolis algorithm can be stated as follows: (a) if the jump increases the posterior density, set $\theta^t = \theta^*$; (b) if the jump decreases the posterior density, set $\theta^t = \theta^*$ with probability equal to the density ratio, r, and set $\theta^t = \theta^{t-1}$ otherwise. The Metropolis algorithm can thus be viewed as a stochastic version of a stepwise mode-finding algorithm, always accepting steps that increase the density but only sometimes accepting downward steps.

Why does the Metropolis algorithm work?

The proof that the sequence of iterations $\theta^1, \theta^2, \ldots$ converges to the target distribution has two steps: first, it is shown that the simulated sequence is a Markov chain with a unique stationary distribution, and second, it is shown that the stationary distribution equals this target distribution. The first step of the proof holds if the Markov chain is irreducible, aperiodic, and not transient. Except for trivial exceptions, the latter two conditions hold for a random walk on any proper distribution, and irreducibility holds as long as the random walk has a positive probability of eventually reaching any state from any other state; that is, the jumping distributions J_t must eventually be able to jump to all states with positive probability.

To see that the target distribution is the stationary distribution of the Markov chain generated by the Metropolis algorithm, consider starting the algorithm at time $t-1$ with a draw θ^{t-1} from the target distribution $p(\theta|y)$. Now consider any two such points θ_a and θ_b, drawn from $p(\theta|y)$ and labeled so that $p(\theta_b|y) \geq p(\theta_a|y)$. The unconditional probability density of a transition from θ_a to θ_b is

$$p(\theta^{t-1} = \theta_a, \theta^t = \theta_b) = p(\theta_a|y)J_t(\theta_b|\theta_a),$$

where the acceptance probability is 1 because of our labeling of a and b, and the unconditional probability density of a transition from θ_b to θ_a is, from (11.1),

$$
\begin{aligned}
p(\theta^t = \theta_a, \theta^{t-1} = \theta_b) &= p(\theta_b|y)J_t(\theta_a|\theta_b)\left(\frac{p(\theta_a|y)}{p(\theta_b|y)}\right) \\
&= p(\theta_a|y)J_t(\theta_a|\theta_b),
\end{aligned}
$$

which is the same as the probability of a transition from θ_a to θ_b, since we have required that $J_t(\cdot|\cdot)$ be symmetric. Since their joint distribution is symmetric, θ^t and θ^{t-1} have the same marginal distributions, and so $p(\theta|y)$ is the stationary distribution of the Markov chain of θ. For more detailed theoretical concerns, see the references at the end of this chapter.

The Metropolis-Hastings algorithm

The Metropolis-Hastings algorithm generalizes the basic Metropolis algorithm presented above in two ways. First, the jumping rules J_t need no longer be symmetric; that is, there is no requirement that $J_t(\theta_a|\theta_b) \equiv J_t(\theta_b|\theta_a)$. Second, to correct for the asymmetry in the jumping rule, the ratio r in (11.1) is replaced by a ratio of ratios:

$$r = \frac{p(\theta^*|y)/J_t(\theta^*|\theta^{t-1})}{p(\theta^{t-1}|y)/J_t(\theta^{t-1}|\theta^*)}. \tag{11.2}$$

(The ratio r is always defined, because a jump from θ^{t-1} to θ^* can only occur if both $p(\theta^{t-1}|y)$ and $J_t(\theta^*|\theta^{t-1})$ are nonzero.)

Allowing asymmetric jumping rules can be useful in increasing the speed of the random walk. Convergence to the target distribution is proved in the same way as for the Metropolis algorithm. The proof of convergence to a unique stationary distribution is identical. To prove that the stationary distribution is the target distribution, $p(\theta|y)$, consider any two points θ_a and θ_b with posterior densities labeled so that $p(\theta_b|y)J_t(\theta_a|\theta_b) \geq p(\theta_a|y)J_t(\theta_b|\theta_a)$. If θ^{t-1} follows the target distribution, then it is easy to show that the unconditional probability density of a transition from θ_a to θ_b is the same as the reverse transition.

Relation between the jumping rule and efficiency of simulations

The ideal Metropolis-Hastings jumping rule is simply to sample the proposal, θ^*, from the target distribution; that is, $J(\theta^*|\theta) \equiv p(\theta^*|y)$ for all θ. Then the ratio r in (11.2) is always exactly 1, and the iterates θ^t are a sequence of independent draws from $p(\theta|y)$. In general, however, iterative simulation is applied to problems for which direct sampling is not possible.

A good jumping distribution has the following properties:

- For any θ, it is easy to sample from $J(\theta^*|\theta)$.

- It is easy to compute the ratio r.

- Each jump goes a reasonable distance in the parameter space (otherwise the random walk moves too slowly).

- The jumps are not rejected too frequently (otherwise the random walk wastes too much time standing still).

We return to the topic of constructing efficient simulation algorithms in Sections 11.8 and 11.9.

11.5 Building Markov chain algorithms using the Gibbs sampler and Metropolis algorithm

The Gibbs sampler and the Metropolis algorithm can be used as building blocks for simulating from complicated distributions. The Gibbs sampler is the simplest of the Markov chain simulation algorithms, and it is our first choice for conditionally conjugate models, where we can directly sample from each conditional posterior distribution. For example, we could use the Gibbs sampler for the normal-normal hierarchical models in Chapter 5.

The Metropolis algorithm can be used for models that are not conditionally conjugate, for example, the two-parameter logistic regression for the bioassay experiment in Section 3.7. In this example, the Metropolis algorithm could be performed in vector form—jumping in the two-dimensional space of (α, β)— or embedded within a Gibbs sampler structure, by alternately updating α and β using one-dimensional Metropolis jumps. In either case, the Metropolis algorithm will probably have to be tuned to get a good acceptance rate, as discussed in Section 11.9.

If some of the conditional posterior distributions in a model can be sampled from directly and some cannot, then the parameters can be updated one at a time, with the Gibbs sampler used where possible and one-dimensional Metropolis updating used otherwise. More generally, the parameters can be updated in blocks, where each block is altered using the Gibbs sampler or a Metropolis jump of the parameters within the block.

A general problem with conditional sampling algorithms is that they can be slow when parameters are highly correlated in the target distribution (for example, see Figure 11.3 on page 288). This can be fixed in simple problems using reparameterization (see Section 11.8) or more generally using the more advanced algorithms mentioned in Chapter 13.

Interpretation of the Gibbs sampler as a special case of the Metropolis-Hastings algorithm

Gibbs sampling can be viewed as a special case of the Metropolis-Hastings algorithm in the following way. We first define iteration t to consist of a series of d steps, with step j of iteration t corresponding to an update of the subvector θ_j conditional on all the other elements of θ. Then the jumping distribution, $J_{j,t}(\cdot|\cdot)$, at step j of iteration t only jumps along the jth subvector, and does so with the conditional posterior density of θ_j given θ_{-j}^{t-1}:

$$J_{j,t}^{\text{Gibbs}}(\theta^*|\theta^{t-1}) = \begin{cases} p(\theta_j^*|\theta_{-j}^{t-1}, y) & \text{if } \theta_{-j}^* = \theta_{-j}^{t-1} \\ 0 & \text{otherwise.} \end{cases}$$

The only possible jumps are to parameter vectors θ^* that match θ^{t-1} on all components other than the jth. Under this jumping distribution, the ratio (11.2) at the jth step of iteration t is

$$\begin{aligned} r &= \frac{p(\theta^*|y)/J_{j,t}^{\text{Gibbs}}(\theta^*|\theta^{t-1})}{p(\theta^{t-1}|y)/J_{j,t}^{\text{Gibbs}}(\theta^{t-1}|\theta^*)} \\ &= \frac{p(\theta^*|y)/p(\theta_j^*|\theta_{-j}^{t-1}, y)}{p(\theta^{t-1}|y)/p(\theta_j^{t-1}|\theta_{-j}^{t-1}, y)} \\ &= \frac{p(\theta_{-j}^{t-1}|y)}{p(\theta_{-j}^{t-1}|y)} \\ &\equiv 1, \end{aligned}$$

and thus every jump is accepted. The second line above follows from the first because, under this jumping rule, θ^* differs from θ^{t-1} only in the jth component. The third line follows from the second by applying the rules of conditional probability to $\theta = (\theta_j, \theta_{-j})$ and noting that $\theta_{-j}^* = \theta_{-j}^{t-1}$.

Usually, one iteration of the Gibbs sampler is defined as we do, to include all d steps corresponding to the d components of θ, thereby updating all of θ at each iteration. It is possible, however, to define Gibbs sampling without the restriction that each component be updated in each iteration, as long as each component is updated periodically.

Gibbs sampler with approximations

For some problems, sampling from some, or all, of the conditional distributions $p(\theta_j|\theta_{-j}, y)$ is impossible, but one can construct approximations, which we label $g(\theta_j|\theta_{-j})$, from which sampling is possible. The general form of the Metropolis-Hastings algorithm can be used to compensate for the approximation. As in the Gibbs sampler, we choose an order for altering the d elements of θ; the jumping function at the jth Metropolis step at iteration t is then

$$J_{j,t}(\theta^*|\theta^{t-1}) = \begin{cases} g(\theta_j^*|\theta_{-j}^{t-1}) & \text{if } \theta_{-j}^* = \theta_{-j}^{t-1} \\ 0 & \text{otherwise,} \end{cases}$$

and the ratio r in (11.2) must be computed and the acceptance or rejection of θ^* decided.

11.6 Inference and assessing convergence

The basic method of inference from iterative simulation is the same as for Bayesian simulation in general: use the collection of all the simulated draws from $p(\theta|y)$ to summarize the posterior density and to compute quantiles, moments, and other summaries of interest as needed. Posterior predictive simulations of unobserved outcomes $\tilde{y}$ can be obtained by simulation conditional on the drawn values of θ. Inference using the iterative simulation draws requires some care, however, as we discuss in this section.

Difficulties of inference from iterative simulation

Iterative simulation adds two difficulties to the problem of simulation inference. First, if the iterations have not proceeded long enough, as in Figure 11.2a, the simulations may be grossly unrepresentative of the target distribution. Even when the simulations have reached approximate convergence, the early iterations still are influenced by the starting approximation rather than the target distribution; for example, consider the early iterations of Figures 11.2b and 11.3b.

The second problem with iterative simulation draws is their within-sequence correlation; aside from any convergence issues, simulation inference from correlated draws is generally less precise than from the same number of independent draws. Serial correlation in the simulations is not necessarily a problem because, at convergence, the draws are identically distributed as $p(\theta|y)$, and so when performing inferences, we ignore the order of the simulation draws in any case. But such correlation can cause inefficiencies in simulations. Consider Figure 11.2c, which displays 500 successive iterations from each of five simulated sequences of the Metropolis algorithm: the patchy appearance of the scatterplot would not be likely to appear from 2500 independent draws from the normal distribution but is rather a result of the slow movement of the simulation algorithm. In some sense, the 'effective' number of simulation draws here is far fewer than 2500. We formally define this concept in equation (11.4) on page 298.

We handle the special problems of iterative simulation in three ways. First, we attempt to design the simulation runs to allow effective monitoring of convergence, in particular by simulating multiple sequences with starting points dispersed throughout parameter space, as in Figure 11.2a. Second, we monitor the convergence of all quantities of interest by comparing variation between and within simulated sequences until 'within' variation roughly equals 'between' variation, as in Figure 11.2b. Only when the distribution of each simulated sequence is close to the distribution of all the sequences mixed together can they all be approximating the target distribution. Third, if the simulation efficiency is unacceptably low (in the sense of requiring too much real time on the computer to obtain approximate convergence of posterior inferences for quantities of interest), the algorithm can be altered, as we discuss in Sections 11.8 and 11.9.

Discarding early iterations of the simulation runs

To diminish the effect of the starting distribution, we generally discard the first half of each sequence and focus attention on the second half. Our ultimate inferences will be based on the assumption that the distributions of the simulated values θ^t, for large enough t, are close to the target distribution, $p(\theta|y)$. The practice of discarding early iterations in Markov chain simulation is referred to as *burn-in*; depending on the context, different burn-in fractions can be appropriate. For example, in the Gibbs sampler displayed in Figure 11.3, it would be necessary to discard only a few initial iterations.

We adopt the general practice of discarding the first half as a conservative choice. For example, we might run 200 iterations and discard the first half. If approximate convergence has not yet been reached, we might then run another 200 iterations, now discarding all of the initial 200 iterations.

Dependence of the iterations in each sequence

Another issue that sometimes arises, once approximate convergence has been reached, is whether to *thin* the sequences by keeping every kth simulation draw from each sequence and discarding the rest. In our applications, we have found it useful to skip iterations in problems with large numbers of parameters where computer storage is a problem, perhaps setting k so that the total number of iterations saved is no more than 1000.

Whether or not the sequences are thinned, if the sequences have reached approximate convergence, they can be directly used for inferences about the parameters θ and any other quantities of interest.

Monitoring convergence based on multiple sequences with overdispersed starting points

Our recommended approach to assessing convergence of iterative simulation is based on comparing different simulated sequences, as illustrated in Figure 11.2, which shows five parallel simulations before and after approximate convergence. In Figure 11.2a, the multiple sequences clearly have not converged; the variance within each sequence is much less than the variance between sequences. Later, in Figure 11.2b, the sequences have mixed, and the two variance components are essentially equal.

To see such disparities, we clearly need more than one independent sequence. Thus our plan is to simulate independently at least two sequences, with starting points drawn from an overdispersed distribution (either from a crude estimate such as discussed in Section 10.1 or a more elaborate approximation as discussed in the next chapter).

Monitoring scalar estimands

Our approach involves monitoring each scalar estimand or other scalar quantities of interest separately. Estimands include all the parameters of interest in the model and any other quantities of interest (for example, the ratio of two parameters or the value of a predicted future observation). It is often useful also to monitor the value of the logarithm of the posterior density, which has probably already been computed if we are using a version of the Metropolis algorithm. Since our method of assessing convergence is based on means and variances, it is best where possible to transform the scalar estimands to be approximately normal (for example, take logarithms of all-positive quantities and logits of quantities that lie between 0 and 1).

Monitoring convergence of each scalar estimand

Suppose we have simulated m parallel sequences, each of length n (after discarding the first half of the simulations). For each scalar estimand ψ, we label the simulation draws as ψ_{ij} ($i = 1, \ldots, n; j = 1, \ldots, m$), and we compute B and W, the between- and within-sequence variances:

$$B = \frac{n}{m-1} \sum_{j=1}^{m} (\overline{\psi}_{.j} - \overline{\psi}_{..})^2, \quad \text{where} \quad \overline{\psi}_{.j} = \frac{1}{n} \sum_{i=1}^{n} \psi_{ij}, \quad \overline{\psi}_{..} = \frac{1}{m} \sum_{j=1}^{m} \overline{\psi}_{.j}$$

$$W = \frac{1}{m} \sum_{j=1}^{m} s_j^2, \quad \text{where} \quad s_j^2 = \frac{1}{n-1} \sum_{i=1}^{n} (\psi_{ij} - \overline{\psi}_{.j})^2.$$

The between-sequence variance, B, contains a factor of n because it is based on the variance of the within-sequence means, $\overline{\psi}_{.j}$, each of which is an average of n values ψ_{ij}. If only one sequence is simulated (that is, if $m = 1$), then B cannot be calculated.

We can estimate var$(\psi|y)$, the marginal posterior variance of the estimand, by a weighted average of W and B, namely

$$\widehat{\text{var}}^+(\psi|y) = \frac{n-1}{n} W + \frac{1}{n} B. \tag{11.3}$$

This quantity *overestimates* the marginal posterior variance assuming the starting distribution is appropriately overdispersed, but is *unbiased* under stationarity (that is, if the starting distribution equals the target distribution), or in the limit $n \to \infty$ (see Exercise 11.4). This is analogous to the classical variance estimate with cluster sampling.

Meanwhile, for any finite n, the 'within' variance W should be an *underestimate* of var$(\psi|y)$ because the individual sequences have not had time to range over all of the target distribution and, as a result, will have less variability; in the limit as $n \to \infty$, the expectation of W approaches var$(\psi|y)$.

We monitor convergence of the iterative simulation by estimating the factor by which the scale of the current distribution for ψ might be reduced if the simulations were continued in the limit $n \to \infty$. This potential scale reduction is estimated by[*]

$$\widehat{R} = \sqrt{\frac{\widehat{\text{var}}^+(\psi|y)}{W}},$$

which declines to 1 as $n \to \infty$. If the potential scale reduction is high, then we have reason to believe that proceeding with further simulations may improve our inference about the target distribution of the associated scalar estimand.

Monitoring convergence for the entire distribution

We recommend computing the potential scale reduction for all scalar estimands of interest; if $\widehat{R}$ is not near 1 for all of them, continue the simulation runs (perhaps altering the simulation algorithm itself to make the simulations more efficient, as described in the next section). Once $\widehat{R}$ is near 1 for all scalar estimands of interest, just collect the mn simulations from the second halves of all the sequences together and treat them as a sample from the target distribution. The condition of $\widehat{R}$ being 'near' 1 depends on the problem at hand; for most examples, values below 1.1 are acceptable, but for a final analysis in a critical problem, a higher level of precision may be required.

In addition, even if an iterative simulation appears to converge and has passed all tests of convergence, it still may actually be far from convergence if important areas of the target distribution were not captured by the starting distribution and are not easily reachable by the simulation algorithm

Example. Bivariate unit normal density with bivariate normal jumping kernel (continued)
We illustrate the multiple sequence method using the Metropolis simulations of the bivariate normal distribution illustrated in Figure 11.2. Table 11.1 displays posterior inference for the two parameters of the distribution as well as the log posterior density (relative to the density at the mode). After 50 iterations, the variance between the five sequences is much greater than the variance within, for all three univariate summaries considered. However, the five simulated sequences have converged adequately after 2000 or certainly 5000 iterations for the quantities of interest. The comparison with the true target distribution shows how some variability remains in the posterior inferences even after the Markov chains have converged. (This must be so, considering that even if the simulation draws were independent, so that the Markov chains would converge in a single iteration, it would still require hundreds or thousands of draws to obtain precise estimates of extreme posterior quantiles.)

The method of monitoring convergence presented here has the key advantage of not requiring the user to examine time series graphs of simulated

---

[*] In the first edition of this book, $\widehat{R}$ was defined as $\widehat{\text{var}}^+(\psi|y)/W$. We switch to the square-root definition for notatinal convenience.

Number of	95% intervals and $\widehat{R}$ for ...		
iterations	θ_1	θ_2	$\log p(\theta_1, \theta_2|y)$
50	$[-2.14, 3.74]$, 12.3	$[-1.83, 2.70]$, 6.1	$[-8.71, -0.17]$, 6.1
500	$[-3.17, 1.74]$, 1.3	$[-2.17, 2.09]$, 1.7	$[-5.23, -0.07]$, 1.3
2000	$[-1.83, 2.24]$, 1.2	$[-1.74, 2.09]$, 1.03	$[-4.07, -0.03]$, 1.10
5000	$[-2.09, 1.98]$, 1.02	$[-1.90, 1.95]$, 1.03	$[-3.70, -0.03]$, 1.00
∞	$[-1.96, 1.96]$, 1	$[-1.96, 1.96]$, 1	$[-3.69, -0.03]$, 1

Table 11.1 *95% central intervals and estimated potential scale reduction factors for three scalar summaries of the bivariate normal distribution simulated using a Metropolis algorithm. (For demonstration purposes, the jumping scale of the Metropolis algorithm was purposely set to be inefficient; see Figure 11.2.) Displayed are inferences from the second halves of five parallel sequences, stopping after 50, 500, 2000, and 5000 iterations. The intervals for ∞ are taken from the known normal and $\chi_2^2/2$ marginal distributions for these summaries in the target distribution.*

sequences. Inspection of such plots is a notoriously unreliable method of assessing convergence (see references at the end of this chapter) and in addition is unwieldy when monitoring a large number of quantities of interest, such as can arise in complicated hierarchical models. Because it is based on means and variances, the simple method presented here is most effective for quantities whose marginal posterior distributions are approximately normal. When performing inference for extreme quantiles, or for parameters with multimodal marginal posterior distributions, one should monitor also extreme quantiles of the 'between' and 'within' sequences.

Effective number of independent draws

Once the simulated sequences have mixed, we can compute an approximate 'effective number of independent simulation draws' for any estimand of interest ψ by comparing the variances between and within the simulated sequences. We start with the observation that if the n simulation draws within each sequence were truly independent, then the between-sequence variance B would be an unbiased estimate of the posterior variance, $\text{var}(\psi|y)$, and we would have a total of mn independent simulations from the m sequences. In general, however, the simulations of ψ within each sequence will be autocorrelated, and B will be larger than $\text{var}(\psi|y)$, in expectation. We thus define the effective number of independent draws of ψ as

$$n_{\text{eff}} = mn\frac{\widehat{\text{var}}^+(\psi|y)}{B}, \tag{11.4}$$

with $\widehat{\text{var}}^+(\psi|y)$ and B as defined on page 296. If m is small, then B will have a high sampling variability, so that n_{eff} is a fairly crude estimate. We actually report $\min(n_{\text{eff}}, mn)$, to avoid claims that our simulation is more efficient than

Diet	Measurements
A	62, 60, 63, 59
B	63, 67, 71, 64, 65, 66
C	68, 66, 71, 67, 68, 68
D	56, 62, 60, 61, 63, 64, 63, 59

Table 11.2 *Coagulation time in seconds for blood drawn from 24 animals randomly allocated to four different diets. Different treatments have different numbers of observations because the randomization was unrestricted. From Box, Hunter, and Hunter (1978), who adjusted the data so that the averages are integers, a complication we ignore in our analysis.*

random sampling. (Superefficient iterative simulation is possible but in practice highly unlikely.) If desired, more precise measures of simulation accuracy and effective number of draws could be constructed based on autocorrelations within each sequence.

11.7 Example: the hierarchical normal model

We illustrate the simulation algorithms with the hierarchical normal model, extending the problem discussed in Section 5.4 by allowing the data variance, σ^2, to be unknown. The example is continued in Section 12.5 to illustrate mode-based computation. We demonstrate with the normal model because it is simple enough that the key computational ideas do not get lost in the details.

Data from a small experiment

We demonstrate the computations on a small experimental dataset, displayed in Table 11.2, that has been used previously as an example in the statistical literature. Our purpose here is solely to illustrate computational methods, not to perform a full Bayesian data analysis (which includes model construction and model checking), and so we do not discuss the applied context.

The model

Under the hierarchical normal model (restated here, for convenience), data y_{ij}, $i = 1, \ldots, n_j$, $j = 1, \ldots, J$, are independently normally distributed within each of J groups, with means θ_j and common variance σ^2. The total number of observations is $n = \sum_{j=1}^{J} n_j$. The group means are assumed to follow a normal distribution with unknown mean μ and variance τ^2, and a uniform prior distribution is assumed for $(\mu, \log \sigma, \tau)$, with $\sigma > 0$ and $\tau > 0$; equivalently, $p(\mu, \log \sigma, \log \tau) \propto \tau$. If we were to assign a uniform prior distribution to $\log \tau$, the posterior distribution would be improper, as discussed in Chapter 5.

The joint posterior density of all the parameters is

$$p(\theta, \mu, \log \sigma, \log \tau | y) \propto \tau \prod_{j=1}^{J} N(\theta_j | \mu, \tau^2) \prod_{j=1}^{J} \prod_{i=1}^{n_j} N(y_{ij} | \theta_j, \sigma^2).$$

Starting points

In this example, we can choose overdispersed starting points for each parameter θ_j by simply taking random points from the data y_{ij} from group j. We obtain 10 starting points for the simulations by drawing θ_j independently in this way for each group. We also need starting points for μ, which can be taken as the average of the starting θ_j values. No starting values are needed for τ or σ as they can be drawn as the first steps in the Gibbs sampler.

Section 12.5 presents a more elaborate procedure for constructing a starting distribution for the iterative simulations using the posterior mode and a normal approximation.

Gibbs sampler

The conditional distributions for this model all have simple conjugate forms:

1. *Conditional posterior distribution of each θ_j.* The factors in the joint posterior density that involve θ_j are the $N(\mu, \tau^2)$ prior distribution and the normal likelihood from the data in the jth group, $y_{ij}, i = 1, \ldots, n_j$. The conditional posterior distribution of each θ_j given the other parameters in the model is

$$\theta_j | \mu, \sigma, \tau, y \sim N(\hat{\theta}_j, V_{\theta_j}), \tag{11.5}$$

where the parameters of the conditional posterior distribution depend on μ, σ, and τ as well as y:

$$\hat{\theta}_j = \frac{\frac{1}{\tau^2}\mu + \frac{n_j}{\sigma^2}\overline{y}_{.j}}{\frac{1}{\tau^2} + \frac{n_j}{\sigma^2}} \tag{11.6}$$

$$V_{\theta_j} = \frac{1}{\frac{1}{\tau^2} + \frac{n_j}{\sigma^2}}. \tag{11.7}$$

These conditional distributions are independent; thus drawing the θ_j's one at a time is equivalent to drawing the vector θ all at once from its conditional posterior distribution.

2. *Conditional posterior distribution of μ.* Conditional on y and the other parameters in the model, μ has a normal distribution determined by the J values θ_j:

$$\mu | \theta, \sigma, \tau, y \sim N(\hat{\mu}, \tau^2/J), \tag{11.8}$$

where

$$\hat{\mu} = \frac{1}{J} \sum_{j=1}^{J} \theta_j. \tag{11.9}$$

3. *Conditional posterior distribution of σ^2.* The conditional posterior density for σ^2 has the form corresponding to a normal variance with known mean; there are n observations y_{ij} with means θ_j. The conditional posterior distribution is

$$\sigma^2|\theta, \mu, \tau, y \sim \text{Inv-}\chi^2(n, \hat{\sigma}^2), \tag{11.10}$$

where

$$\hat{\sigma}^2 = \frac{1}{n} \sum_{j=1}^{J} \sum_{i=1}^{n_j} (y_{ij} - \theta_j)^2. \tag{11.11}$$

4. *Conditional posterior distribution of τ^2.* Conditional on the data and the other parameters in the model, τ^2 has a scaled inverse-χ^2 distribution, with parameters depending only on μ and θ (as can be seen by examining the joint posterior density):

$$\tau^2|\theta, \mu, \sigma, y \sim \text{Inv-}\chi^2(J-1, \hat{\tau}^2), \tag{11.12}$$

with

$$\hat{\tau}^2 = \frac{1}{J-1} \sum_{j=1}^{J} (\theta_j - \mu)^2. \tag{11.13}$$

The expressions for τ^2 have $(J-1)$ degrees of freedom instead of J because $p(\tau) \propto 1$ rather than τ^{-1}.

Numerical results with the coagulation data

We illustrate the Gibbs sampler with the coagulation data of Table 11.2. Inference from ten parallel Gibbs sampler sequences appears in Table 11.3; 100 iterations were sufficient for approximate convergence.

The Metropolis algorithm

We also describe how the Metropolis algorithm can be used for this problem. It would be possible to apply the algorithm to the entire joint distribution, $p(\theta, \mu, \sigma, \tau|y)$, but we can work more efficiently in a lower-dimensional space by taking advantage of the conjugacy of the problem that allows us to compute the function $p(\mu, \log \sigma, \log \tau|y)$, as we discuss in Section 12.5. We use the Metropolis algorithm to jump through the marginal posterior distribution of $(\mu, \log \sigma, \log \tau)$ and then draw simulations of the vector θ from its normal conditional posterior distribution (11.5). Following a principle of efficient Metropolis jumping that we shall discuss in Section 11.9, we jump through the space of $(\mu, \log \sigma, \log \tau)$ using a multivariate normal jumping kernel with mean equal to the current value of the parameters and variance matrix equal to that of a normal approximation (see Section 12.5), multiplied by $2.4^2/d$, where d is the dimension of the Metropolis jumping distribution. In this case, $d = 3$.

Estimand	Posterior quantiles					$\widehat{R}$
	2.5%	25%	median	75%	97.5%	
θ_1	58.9	60.6	61.3	62.1	63.5	1.01
θ_2	63.9	65.3	65.9	66.6	67.7	1.01
θ_3	66.0	67.1	67.8	68.5	69.5	1.01
θ_4	59.5	60.6	61.1	61.7	62.8	1.01
μ	56.9	62.2	63.9	65.5	73.4	1.04
σ	1.8	2.2	2.4	2.6	3.3	1.00
τ	2.1	3.6	4.9	7.6	26.6	1.05
$\log p(\mu, \log \sigma, \log \tau \mid y)$	−67.6	−64.3	−63.4	−62.6	−62.0	1.02
$\log p(\theta, \mu, \log \sigma, \log \tau \mid y)$	−70.6	−66.5	−65.1	−64.0	−62.4	1.01

Table 11.3 *Summary of posterior inference for the individual-level parameters and hyperparameters for the coagulation example. Posterior quantiles and estimated potential scale reductions computed from the second halves of ten Gibbs sampler sequences, each of length 100. Potential scale reductions for σ and τ were computed on the log scale. The hierarchical standard deviation, τ, is estimated less precisely than the unit-level standard deviation, σ, as is typical in hierarchical modeling with a small number of batches.*

Metropolis results with the coagulation data

We ran ten parallel sequences of Metropolis algorithm simulations. In this case 500 iterations were sufficient for approximate convergence ($\widehat{R} < 1.1$ for all parameters); at that point we obtained similar results to those obtained using Gibbs sampling. The acceptance rate for the Metropolis simulations was 0.35, which is close to the expected result for the normal distribution with $d = 3$ using a jumping distribution scaled by $2.4/\sqrt{d}$ (see Section 11.8).

11.8 Efficient Gibbs samplers

Various theoretical arguments suggest methods for constructing efficient simulation algorithms and improving the efficiency of existing algorithms. This is an area of much current research (see the bibliographic notes at the end of this chapter and the next); in this section and the next we discuss two of the simplest and most general approaches: choice of parameterization and scaling of Metropolis jumping rules.

Transformations and reparameterization

The Gibbs sampler is most efficient when parameterized in terms of independent components; Figure 11.3 shows an example with highly dependent components that create slow convergence. The simplest way to reparameterize is by a linear transformation of the parameters, but posterior distributions that are not approximately normal may require special methods.

The same arguments apply to Metropolis jumps. In a normal or approximately normal setting, the jumping kernel should ideally have the same covariance structure as the target distribution, which can be approximately estimated based on the normal approximation at the mode. Markov chain simulation of a distribution with multiple modes can be greatly improved by allowing jumps between modes. Section 13.1 describes an approach for dealing with multiple modes.

Auxiliary variables

Gibbs sampler computations can often be simplified or convergence accelerated by adding auxiliary variables, for example indicator variables for mixture distributions, as described in Chapter 18. The idea of adding variables is also called *data augmentation* and is a useful conceptual and computational tool for many problems, both for the Gibbs sampler and for the EM algorithm (as discussed in Section 12.3).

Example. Student-t model

A simple but important example of auxiliary variables arises with the t distribution, which can be expressed as a mixture of normal distributions, as noted in Chapter 3 and discussed in more detail in Chapter 17. We illustrate with the example of inference for the parameters μ, σ^2 given n iid data points from the $t_\nu(\mu, \sigma^2)$ distribution, where for simplicity we assume ν is known. We also assume a uniform prior distribution on $\mu, \log \sigma$. The t likelihood for each data point is equivalent to the model,

$$
\begin{aligned}
y_i &\sim \mathrm{N}(\mu, V_i) \\
V_i &\sim \text{Inv-}\chi^2(\nu, \sigma^2),
\end{aligned}
\tag{11.14}
$$

where the V_i's are auxiliary variables that cannot be directly observed. If we perform inference using the joint posterior distribution, $p(\mu, \sigma^2, V|y)$, and then just consider the simulations for μ, σ, these will represent the posterior distribution under the original t model.

There is no direct way to sample from the parameters μ, σ^2 in the t model, but it is straightforward to perform the Gibbs sampler on V, μ, σ^2 in the augmented model:

1. *Conditional posterior distribution of each V_i.* Conditional on the data y and the other parameters of the model, each V_i is a normal variance parameter with a scaled inverse-χ^2 prior distribution, and so its posterior distribution is also inverse-χ^2 (see Section 2.7):

$$
V_i | \mu, \sigma^2, \nu, y \sim \text{Inv-}\chi^2 \left(\nu + 1, \frac{\nu\sigma^2 + (y_i - \mu)^2}{\nu + 1} \right).
$$

The n parameters V_i are independent in their conditional posterior distribution, and we can directly apply the Gibbs sampler by sampling from their scaled inverse-χ^2 distributions.

2. *Conditional posterior distribution of μ.* Conditional on the data y and the other parameters of the model, information about μ is supplied by the n data points

y_i, each with its own variance. Combining with the uniform prior distribution on μ yields,

$$\mu|\sigma^2, V, \nu, y \sim N\left(\frac{\sum_{i=1}^n \frac{1}{V_i} y_i}{\sum_{i=1}^n \frac{1}{V_i}}, \frac{1}{\sum_{i=1}^n \frac{1}{V_i}}\right).$$

3. *Conditional posterior distribution of* σ^2. Conditional on the data y and the other parameters of the model, all the information about σ comes from the variances V_i. The conditional posterior distribution is,

$$
\begin{aligned}
p(\sigma^2|\mu, V, \nu, y) \quad &\propto \quad \sigma^{-2} \prod_{i=1}^n \sigma^\nu e^{-\nu\sigma^2/(2V_i)} \\
&= \quad (\sigma^2)^{n\nu/2-1} \exp\left(-\frac{\nu}{2}\sum_{i=1}^n \frac{1}{V_i}\sigma^2\right) \\
&\propto \quad \text{Gamma}\left(\sigma^2 \left|\frac{n\nu}{2}, \frac{\nu}{2}\sum_{i=1}^n \frac{1}{V_i}\right.\right),
\end{aligned}
$$

from which we can sample directly.

Parameter expansion

For some problems, the Gibbs sampler can be slow to converge because of posterior dependence among parameters that cannot simply be resolved with a linear transformation. Paradoxically, adding an additional parameter—thus performing the random walk in a larger space—can improve the convergence of the Markov chain simulation. We illustrate with the Student-t example above.

Example. Student-t model (continued)
In the latent-parameter form (11.14) of the t model, convergence will be slow if a simulation draw of σ is close to zero, because the conditional distributions will then cause the V_i's to be sampled with values near zero, and then the conditional distribution of σ will be near zero, and so on. Eventually the simulations will get unstuck but it can be slow for some problems. We can fix things by adding a new parameter whose only role is to allow the Gibbs sampler to move in more directions and thus avoid getting stuck.

The expanded model is,

$$
\begin{aligned}
y_i &\sim \quad N(\mu, \alpha^2 U_i) \\
U_i &\sim \quad \text{Inv-}\chi^2(\nu, \tau^2),
\end{aligned}
$$

where $\alpha > 0$ can be viewed as an additional scale parameter. In this new model, $\alpha^2 U_i$ plays the role of V_i in (11.14) and $\alpha\tau$ plays the role of σ. The parameter α has no meaning on its own and we can assign it a noninformative uniform prior distribution on the logarithmic scale.

The Gibbs sampler on this expanded model now has four steps:

1. For each i, U_i is updated much as V_i was before:

$$U_i | \alpha, \mu, \tau^2, \nu, y \sim \text{Inv-}\chi^2 \left(\nu + 1, \frac{\nu \tau^2 + ((y_i - \mu)/\alpha)^2}{\nu + 1} \right).$$

2. The mean, μ, is updated as before:

$$\mu | \alpha, \tau^2, U, \nu, y \sim \text{N} \left(\frac{\sum_{i=1}^{n} \frac{1}{\alpha^2 U_i} y_i}{\sum_{i=1}^{n} \frac{1}{\alpha^2 U_i}}, \frac{1}{\sum_{i=1}^{n} \frac{1}{\alpha^2 U_i}} \right).$$

3. The variance parameter τ^2, is updated much as σ^2 was before:

$$\tau^2 | \alpha, \mu, U, \nu, y \sim \text{Gamma} \left(\frac{n\nu}{2}, \frac{\nu}{2} \sum_{i=1}^{n} \frac{1}{U_i} \right).$$

4. Finally, we must update α^2, which is easy since conditional on all the other parameters in the model it is simply a normal variance parameter:

$$\alpha^2 | \mu, \tau^2, U, \nu, y \sim \text{Inv-}\chi^2 \left(n, \frac{1}{n} \sum_{i=1}^{n} \frac{(y_i - \mu)^2}{U_i} \right).$$

The parameters α^2, U, τ in this expanded model are not identified in that the data do not supply enough information to estimate each of these quantities. However, the model as a whole is identified as long as we monitor convergence of the summaries μ, $\sigma = \alpha\tau$, and $V_i = \alpha^2 U_i$ for $i = 1, \ldots, n$. (Or, if the only goal is inference for the original t model, we can simply save μ and σ from the simulations.)

The Gibbs sampler under the expanded parameterizations converges more reliably because the new parameter α breaks the dependence between τ and the V_i's.

Parameter expansion is particularly useful for hierarchical linear models such as in Chapter 15, as we discuss in Section 15.4 and illustrate in Appendix C.

11.9 Efficient Metropolis jumping rules

For any given posterior distribution, the Metropolis-Hastings algorithm can be implemented in an infinite number of ways. Even after reparameterizing, there are still endless choices in the jumping rules, J_t. In many situations with conjugate families, the posterior simulation can be performed entirely or in part using the Gibbs sampler, which is not always efficient but generally is easy to program, as we illustrated with the hierarchical normal model in Section 11.7. For nonconjugate models we must rely on Metropolis-Hastings algorithms (either within a Gibbs sampler or directly on the multivariate posterior distribution). The choice of jumping rule then arises.

There are two main classes of simple jumping rules. The first are essentially random walks around the parameter space. These jumping rules are often normal jumping kernels with mean equal to the current value of the parameter and variance set to obtain efficient algorithms. The second approach uses proposal distributions that are constructed to closely approximate the target distribution (either the conditional distribution of a subset in a Gibbs sampler or the joint posterior distribution). In the second case the goal is to accept as many draws as possible with the Metropolis-Hastings acceptance step being used primarily to correct the approximation. There is no natural advantage to altering one parameter at a time except for potential computational savings in evaluating only part of the posterior density at each step.

It is hard to give general advice on efficient jumping rules, but some results have been obtained for normal random walk jumping distributions that seem to be useful in many problems. Suppose there are d parameters, and the posterior distribution of $\theta = (\theta_1, \ldots, \theta_d)$, after appropriate transformation, is multivariate normal with known variance matrix Σ. Further suppose that we will take draws using the Metropolis algorithm with a normal jumping kernel centered on the current point and with the same shape as the target distribution: that is, $J(\theta^*|\theta^{t-1}) = N(\theta^*|\theta^{t-1}, c^2\Sigma)$. Among this class of jumping rules, the most efficient has scale $c \approx 2.4/\sqrt{d}$, where efficiency is defined relative to independent sampling from the posterior distribution. The efficiency of this optimal Metropolis jumping rule for the d-dimensional normal distribution can be shown to be about $0.3/d$ (by comparison, if the d parameters were independent in their posterior distribution, the Gibbs sampler would have efficiency $1/d$, because after every d iterations, a new independent draw of θ would be created). Which algorithm is best for any particular problem also depends on the computation time for each iteration, which in turn depends on the conditional independence and conjugacy properties of the posterior density.

A Metropolis algorithm can also be characterized by the proportion of jumps that are accepted. For the multivariate normal random walk jumping distribution with jumping kernel the same shape as the target distribution, the optimal jumping rule has acceptance rate around 0.44 in one dimension, declining to about 0.23 in high dimensions (roughly $d > 5$). This result suggests an *adaptive* simulation algorithm:

1. Start the parallel simulations with a fixed algorithm, such as a version of the Gibbs sampler, or the Metropolis algorithm with a normal random walk jumping rule shaped like an estimate of the target distribution (using the covariance matrix computed at the joint or marginal posterior mode scaled by the factor $2.4/\sqrt{d}$).

2. After some number of simulations, update the Metropolis jumping rule as follows.

   (a) Adjust the covariance of the jumping distribution to be proportional to the posterior covariance matrix estimated from the simulations.

   (b) Increase or decrease the scale of the jumping distribution if the acceptance rate of the simulations is much too high or low, respectively. The goal is to bring the jumping rule toward the approximate optimal value of 0.44 (in one dimension) or 0.23 (when many parameters are being updated at once using vector jumping).

This algorithm can be improved in various ways, but even in its simple form, we have found it useful for drawing posterior simulations for some problems with d ranging from 1 to 50.

Adaptive algorithms

When an iterative simulation algorithm is 'tuned'—that is, modified while it is running—care must be taken to avoid converging to the wrong distribution. (If the updating rule depends on previous simulation steps, then the transition probabilities are more complicated than as stated in the Metropolis-Hastings algorithm, and the iterations will not necessarily converge to the target distribution.) To be safe, we usually run any adaptive algorithm in two phases: first, the adaptive phase, where the parameters of the algorithm can be tuned as often as desired to increase the simulation efficiency, and second, a fixed phase, where the adapted algorithm is run long enough for approximate convergence. Only simulations from the fixed phase are used in the final inferences.

11.10 Recommended strategy for posterior simulation

We summarize our recommended basic approach to Bayesian computation:

1. Start off with crude estimates and possibly a mode-based approximation to the posterior distribution (see Section 10.1 and Chapter 12).

2. If possible, simulate from the posterior distribution directly or sequentially, starting with hyperparameters and then moving to the main parameters (as in the simple hierarchical models in Chapter 5).

3. Most likely, the best approach is to set up a Markov chain simulation algorithm. The updating can be done one parameter at a time or with parameters in batches (as is often convenient in regressions and hierarchical models; see Chapter 15).

4. For parameters (or batches of parameters) whose conditional posterior distributions have standard forms, use the Gibbs sampler.

5. For parameters whose conditional distributions do not have standard forms, use Metropolis jumps. Tune the scale of each jumping distribution so that acceptance rates are near 20% (when altering a vector of parameters) or 40% (when altering one parameter at a time).

6. Construct a transformation so that the parameters are approximately independent—this will speed the convergence of the Gibbs sampler.

7. Start the Markov chain simulations with parameter values taken from the crude estimates or mode-based approximations, with noise added so they are overdispersed with respect to the target distribution. Mode-based approximations to the posterior density are discussed in Chapter 12.

8. Run multiple Markov chains and monitor the mixing of the sequences. Run until approximate convergence appears to have been reached, in the sense that the statistic $\widehat{R}$, defined in Section 11.6, is near 1 (below 1.1, say) for each scalar estimand of interest. This will typically take hundreds of iterations, at least. If approximate convergence has not been reached after a long time, consider making the simulations more efficient as discussed in Sections 11.8 and 11.9.

9. If $\widehat{R}$ is near 1 for all scalar estimands of interest, summarize inference about the posterior distribution by treating the set of all iterates from the second half of the simulated sequences as an identically distributed sample from the target distribution. These can be stored as a matrix (as in Figure 1.1 on page 26). At this point, simulations from the different sequences can be mixed.

10. Compare the posterior inferences from the Markov chain simulation to the approximate distribution used to start the simulations. If they are not close with respect to locations and approximate distributional shape, check for errors before believing that the Markov chain simulation has produced a better answer.

11.11 Bibliographic note

An excellent general book on simulation from a statistical perspective is Ripley (1987), which covers two topics that we do not address in this chapter: creating uniformly distributed (pseudo)random numbers and simulating from standard distributions (on the latter, see our Appendix A for more details). Hammersley and Handscomb (1964) is a classic reference on simulation. Thisted (1988) is a general book on statistical computation that discusses many optimization and simulation techniques. Robert and Casella (1999) covers simulation algorithms from a variety of statistical perspectives. The book by Liu (2002) reviews more advanced simulation algorithms, some of which we discuss in Chapter 13.

Kass et al. (1998) discuss many practical issues in Bayesian simulation. Gilks, Richardson, and Spiegelhalter (1996) is a book full of examples and applications of Markov chain simulation methods. Further references on Bayesian computation appear in the books by Tanner (1993) and Chen, Shao, and Ibrahim (2000). Many other applications of Markov chain simulation appear in the recent applied statistical literature.

Metropolis and Ulam (1949) and Metropolis et al. (1953) apparently were the first to describe Markov chain simulation of probability distributions (that is, the 'Metropolis algorithm'). Their algorithm was generalized by Hastings (1970); see Chib and Greenberg (1995) for an elementary introduction and Tierney (1998) for a theoretical perspective. The conditions for Markov chain convergence appear in probability texts such as Feller (1968), and more recent work such as Rosenthal (1995) has evaluated the rates of convergence of Markov chain algorithms for statistical models. The Gibbs sampler was first so named by Geman and Geman (1984) in a discussion of applications to image processing. Tanner and Wong (1987) introduced the idea of iterative simulation to many statisticians, using the special case of 'data augmentation' to emphasize the analogy to the EM algorithm (see Section 12.3). Gelfand and Smith (1990) showed how the Gibbs sampler could be used for Bayesian inference for a variety of important statistical models. The Metropolis-approximate Gibbs algorithm introduced at the end of Section 11.5 appears in Gelman (1992b) and is used by Gilks, Best, and Tan (1995).

Gelfand et al. (1990) applied Gibbs sampling to a variety of statistical problems, and many other applications of the Gibbs sampler algorithms have appeared since; for example, Clayton (1991) and Carlin and Polson (1991). Besag and Green (1993), Gilks et al. (1993), and Smith and Roberts (1993) discuss Markov simulation algorithms for Bayesian computation. Bugs (Spiegelhalter et al., 1994, 2003) is a general-purpose computer program for Bayesian inference using the Gibbs sampler; see Appendix C for details.

Inference and monitoring convergence from iterative simulation is reviewed by Gelman and Rubin (1992b), who provide a theoretical justification of the method presented in Section 11.6 and discuss a more elaborate version of the method; Brooks and Gelman (1998) and Brooks and Giudici (2000) present more recent work along these lines. Other views on assessing convergence appear in the ensuing discussion of Gelman and Rubin (1992b) and in Cowles and Carlin (1996) and Brooks and Roberts (1998). Gelman and Rubin (1992a, b) and Glickman (1993) present examples of iterative simulation in which lack of convergence is impossible to detect from single sequences but is obvious from multiple sequences.

The data on coagulation times used to illustrate the computations for the hierarchical normal model were analyzed by Box, Hunter, and Hunter (1978) using non-Bayesian methods based on the analysis of variance.

For the relatively simple ways of improving simulation algorithms mentioned in Sections 11.8 and 11.9, Tanner and Wong (1987) discuss data augmentation and auxiliary variables, and Hills and Smith (1992) and Roberts and Sahu (1997) discuss parameterization for the Gibbs sampler. Higdon (1998) discusses some more complicated auxiliary variable methods, and Liu and Wu (1999), van Dyk and Meng (2001), and Liu (2003) present different approaches to parameter expansion. The results on acceptance rates for efficient Metropolis jumping rules appear in Gelman, Roberts, and Gilks (1995); more general results for Metropolis-Hastings algorithms appear in Roberts and Rosenthal (2001) and Brooks, Giudici, and Roberts (2003).

Gelfand and Sahu (1994) discuss the difficulties of maintaining convergence to the target distribution when adapting Markov chain simulations, as discussed at the end of Section 11.9. Gilks and Berzuini (2001) and Andrieu and Robert (2001) present further work on adaptive Markov chain simulation algorithms.

11.12 Exercises

1. Rejection sampling:

   (a) Prove that rejection sampling gives draws from $p(\theta|y)$.

   (b) Why is the boundedness condition on $p(\theta|y)/q(\theta)$ necessary for rejection sampling?

2. Metropolis-Hastings algorithm: Show that the stationary distribution for the Metropolis-Hastings algorithm is, in fact, the target distribution, $p(\theta|y)$.

3. Metropolis algorithm: Replicate the computations for the bioassay example of Section 3.7 using the Metropolis algorithm. Be sure to define your starting points and your jumping rule. Run the simulations long enough for approximate convergence.

4. Monitoring convergence:

   (a) Prove that $\widehat{\text{var}}^+(\psi|y)$ as defined in (11.3) is an unbiased estimate of the marginal posterior variance of ϕ, if the starting distribution for the Markov chain simulation algorithm is the same as the target distribution, and if the m parallel sequences are computed independently. (Hint: show that $\widehat{\text{var}}^+(\psi|y)$ can be expressed as the average of the halved squared differences between simulations ϕ from different sequences, and that each of these has expectation equal to the posterior variance.)

   (b) Determine the conditions under which $\widehat{\text{var}}^+(\psi|y)$ approaches the marginal posterior variance of ϕ in the limit as the lengths n of the simulated chains approach ∞.

5. Analysis of a stratified sample survey: Section 7.4 presents an analysis of a stratified sample survey using a hierarchical model on the stratum probabilities.

   (a) Perform the computations for the simple nonhierarchical model described in the example.

   (b) Using the Metropolis algorithm, perform the computations for the hierarchical model, using the results from part (a) as a starting distribution. Check your answer by comparing your simulations to the results in Figure 7.1b.

CHAPTER 12

Approximations based on posterior modes

The early chapters of the book describe simulation approaches that work in low-dimensional problems. With complicated models, it is rare that samples from the posterior distribution can be obtained directly, and Chapter 11 describes iterative simulation algorithms that can be used with such models. In this chapter we describe an approximate approach using joint or marginal mode finding, and multivariate normal or t approximations about the modes. These approximations are useful for quick inferences and as starting points for iterative simulation algorithms such as described in Chapter 11. The approximations that we describe are relatively simple to compute, easy to understand, and can provide valuable information about the fit of the model.

We develop the following approximation strategy that is appropriate for posterior distributions that arise in many statistical problems.

1. Use *crude estimates* such as discussed in Section 10.1 as starting points for the approximation.

2. Find *posterior modes*, using methods such as conditional maximization (stepwise ascent) and Newton's method, as described in Section 12.1. For many problems, the posterior mode is more informative when applied to the *marginal posterior distribution* of a subset of the parameters, in which case the EM algorithm and its extensions can be useful computational tools, as described in Section 12.3.

3. Construct a *Normal-mixture or related approximation* to the joint posterior distribution, based on the curvature of the distribution around its modes, using a standard distribution such as the multivariate normal or t at each mode; see Section 12.2. We use logarithms or logits of parameters where appropriate to improve the approximation. If the distribution is multimodal, the relative mass of each mode can be estimated using the value of the density and the scale of the normal or t approximation at each mode.

4. Construct *separate approximations to marginal and conditional posterior densities* in high-dimensional problems such as hierarchical models in which the normal and t distributions do not fit the joint posterior density; see Section 12.4.

5. Draw *simulations from the approximate distribution*, possibly followed by *importance resampling*, as described at the end of Section 12.2. These draws can then be used as starting points for Markov chain simulation.

We conclude the chapter with an example using the hierarchical normal model. This is a relatively simple example but allows us to focus on details of the various computational algorithms without worrying about the details of a new model.

12.1 Finding posterior modes

In Bayesian computation, we search for modes not for their own sake, but as a way to begin mapping the posterior density. In particular, we have no special interest in finding the absolute maximum of the posterior density. If many modes exist, we should try to find them all, or at least all the modes with non-negligible posterior mass in their neighborhoods. In practice, we often first search for a single mode, and if it does not look reasonable in a substantive sense, we continue searching through the parameter space for other modes. To find all the local modes—or to make sure that a mode that has been found is the only important mode—sometimes one must run a mode-finding algorithm several times from different starting points.

Even better, where possible, is to find the modes of the marginal posterior density of a subset of the parameters. One then analyzes the distribution of the remaining parameters, conditional on the first subset. We return to this topic in Sections 12.3 and 12.4.

A variety of numerical methods exist for solving optimization problems and any of these, in principle, can be applied to find the modes of a posterior density. Rather than attempt to cover this vast topic comprehensively, we introduce two simple methods that are commonly used in statistical problems.

Conditional maximization

Often the simplest method of finding modes is *conditional maximization*, also called *stepwise ascent*; simply start somewhere in the target distribution—for example, setting the parameters at rough estimates—and then alter one set of components of θ at a time, leaving the other components at their previous values, at each step increasing the log posterior density. Assuming the posterior density is bounded, the steps will eventually converge to a local mode. The method of iterative proportional fitting for loglinear models (see Section 16.8) is an example of conditional maximization. To search for multiple modes, run the conditional maximization routine starting at a variety of points spread throughout the parameter space. It should be possible to find a range of reasonable starting points based on rough estimates of the parameters and problem-specific knowledge about reasonable bounds on the parameters.

For many standard statistical models, the conditional distribution of each parameter given all the others has a simple analytic form and is easily maximized. In this case, applying a conditional maximization (CM) algorithm is easy: just maximize the density with respect to one set of parameters at a time, iterating until the steps become small enough that approximate convergence has been reached. We illustrate this process in Section 12.5 for the example of the hierarchical normal model.

Newton's method

Newton's method, also called the *Newton–Raphson* algorithm, is an iterative approach based on a quadratic Taylor series approximation of the log posterior density,

$$L(\theta) = \log p(\theta|y).$$

It is also acceptable to use an unnormalized posterior density, since Newton's method uses only the derivatives of $L(\theta)$, and any multiplicative constant in p is an additive constant in L. As we have seen in Chapter 4, the quadratic approximation is generally fairly accurate when the number of data points is large relative to the number of parameters. Start by determining the functions $L'(\theta)$ and $L''(\theta)$, the vector of derivatives and matrix of second derivatives, respectively, of the logarithm of the posterior density. The derivatives can be determined analytically or numerically. The mode-finding algorithm proceeds as follows:

1. Choose a starting value, θ^0.

2. For $t = 1, 2, 3, \ldots,$

   (a) Compute $L'(\theta^{t-1})$ and $L''(\theta^{t-1})$. The Newton's method step at time t is based on the quadratic approximation to $L(\theta)$ centered at θ^{t-1}.

   (b) Set the new iterate, θ^t, to maximize the quadratic approximation; thus,

   $$\theta^t = \theta^{t-1} - [L''(\theta^{t-1})]^{-1}L'(\theta^{t-1}).$$

The starting value, θ^0, is important; the algorithm is not guaranteed to converge from all starting values, particularly in regions where $-L''$ is not positive definite. Starting values may be obtained from crude parameter estimates, or conditional maximization could be used to generate a starting value for Newton's method. The advantage of Newton's method is that convergence is extremely fast once the iterates are close to the solution, where the quadratic approximation is accurate. If the iterations do not converge, they typically move off quickly toward the edge of the parameter space, and the next step may be to try again with a new starting point.

Numerical computation of derivatives

If the first and second derivatives of the log posterior density are difficult to determine analytically, one can approximate them numerically using finite differences. Each component of L' can be estimated numerically at any specified value $\theta = (\theta_1, \ldots, \theta_d)$ by

$$L_i'(\theta) = \frac{dL}{d\theta_i} \approx \frac{L(\theta + \delta_i e_i|y) - L(\theta - \delta_i e_i|y)}{2\delta_i}, \tag{12.1}$$

where δ_i is a small value and, using linear algebra notation, e_i is the unit vector corresponding to the ith component of θ. The values of δ_i are chosen based on the scale of the problem; typically, values such as 0.0001 are low enough to approximate the derivative and high enough to avoid roundoff error on the computer. The second derivative matrix at θ is numerically estimated by applying the differencing again; for each i, j:

$$
\begin{aligned}
L''_{ij}(\theta) = \frac{d^2 L}{d\theta_i d\theta_j} \;\; &= \;\; \frac{d}{d\theta_j}\left(\frac{dL}{d\theta_i}\right) \\
&\approx \;\; \frac{L'_i(\theta + \delta_j e_j | y) - L'_i(\theta - \delta_j e_j | y)}{2\delta_j} \\
&\approx \;\; [L(\theta + \delta_i e_i + \delta_j e_j) - L(\theta - \delta_i e_i + \delta_j e_j) \\
&\qquad - L(\theta + \delta_i e_i - \delta_j e_j) + L(\theta - \delta_i e_i - \delta_j e_j)]/(4\delta_i \delta_j).
\end{aligned}
$$

$$(12.2)$$

12.2 The normal and related mixture approximations

Fitting multivariate normal densities based on the curvature at the modes

Once the mode or modes have been found, we can construct an approximation based on the (multivariate) normal distribution. For simplicity we first consider the case of a single mode at $\hat{\theta}$, where we fit a normal distribution to the first two derivatives of the log posterior density function at $\hat{\theta}$:

$$
p_{\text{normal approx}}(\theta) = \text{N}(\theta | \hat{\theta}, V_\theta).
$$

The variance matrix is the inverse of the curvature of the log posterior density at the mode,

$$
V_\theta = \left[-L''(\hat{\theta})\right]^{-1}.
$$

L'' can be calculated analytically for some problems or else approximated numerically as in (12.2). As usual, before fitting a normal density, it makes sense to transform parameters as appropriate, often using logarithms and logits, so that they are defined on the whole real line with roughly symmetric distributions (remembering to multiply the posterior density by the Jacobian of the transformation, as in Section 1.8).

Mixture approximation for multimodal densities

Now suppose we have found K modes in the posterior density. The posterior distribution can then be approximated by a mixture of K multivariate normal distributions, each with its own mode $\hat{\theta}_k$, variance matrix $V_{\theta k}$, and relative mass ω_k. That is, the target density $p(\theta | y)$ can be approximated by

$$
p_{\text{normal approx}}(\theta) \propto \sum_{k=1}^{K} \omega_k \text{N}(\theta | \hat{\theta}_k, V_{\theta k}).
$$

For each k, the mass ω_k of the kth component of the multivariate normal mixture can be estimated by equating the actual posterior density, $p(\hat{\theta}_k|y)$, or the actual unnormalized posterior density, $q(\hat{\theta}_k|y)$, to the approximate density, $p_{\text{normal approx}}(\hat{\theta}_k)$, at each of the K modes. If the modes are fairly widely separated and the normal approximation is appropriate for each mode, then we obtain

$$\omega_k = q(\hat{\theta}_k|y)|V_{\theta k}|^{1/2}, \tag{12.3}$$

which yields the normal-mixture approximation

$$p_{\text{normal approx}}(\theta) \propto \sum_{k=1}^{K} q(\hat{\theta}_k|y) \exp\left(-\frac{1}{2}(\theta - \hat{\theta}_k)^T V_{\theta k}^{-1}(\theta - \hat{\theta}_k)\right).$$

Multivariate t approximation instead of the normal

In light of the limit theorems discussed in Chapter 4, the normal distribution can often be a good approximation to the posterior distribution of a continuous parameter vector. For small samples, however, it is useful for the initial approximating distribution to be 'conservative' in the sense of covering more of the parameter space to ensure that nothing important is missed. Thus we recommend replacing each normal density by a multivariate Student-t with some small number of degrees of freedom, ν. The corresponding approximation is a mixture density function that has the functional form

$$p_{t\ approx}(\theta) \propto \sum_{k=1}^{K} q(\hat{\theta}_k|y) \left[\nu + (\theta - \hat{\theta}_k)^T V_{\theta k}^{-1}(\theta - \hat{\theta}_k)\right]^{-(d+\nu)/2},$$

where d is the dimension of θ. A Cauchy mixture, which corresponds to $\nu = 1$, is a conservative choice to ensure overdispersion, but if the parameter space is high-dimensional, most draws from a multivariate Cauchy will almost certainly be too far from the mode to generate a reasonable approximation to the target distribution. For most posterior distributions arising in practice, especially those without long-tailed underlying models, a value such as $\nu = 4$, which provides three finite moments for the approximating density, is probably dispersed enough in typical practice.

Several different strategies can be employed to improve the approximate distribution further, including (1) analytically fitting the approximation to locations other than the modes, such as saddle points or tails, of the distribution, (2) analytically or numerically integrating out some components of the target distribution to obtain a lower-dimensional approximation, or (3) bounding the range of parameter values. In addition, there are strategies based on simulation tools, like importance resampling described below, that can be used to improve the approximation. Special efforts beyond mixtures of normal or t distributions are needed for difficult problems that can arise in practice, such as 'banana-shaped' posterior distributions in many dimensions.

Sampling from the approximate posterior distributions

It is easy to draw random samples from the multivariate normal- or t-mixture approximations. To generate a single sample from the approximation, first choose one of the K mixture components using the relative probability masses of the mixture components, ω_k, as multinomial probabilities. Appendix A provides details on how to sample from a single multivariate normal or t distribution using the Cholesky factorization of the scale matrix.

Improving an approximation using importance resampling

Once we have a continuous approximation to $p(\theta|y)$, perhaps the mixture of normal or t distributions discussed above, the next step in sophistication is to improve or correct the approximation using a technique known as *importance resampling* (also called sampling-importance resampling or SIR). Let $g(\theta)$ be an unnormalized density function that we can both compute and draw samples from, and that serves as an approximation to the target distribution. Assume also that we can compute the unnormalized posterior density function, $q(\theta|y)$, for any θ. If the ratio $q(\theta|y)/g(\theta)$, known as the importance ratio or weight, is bounded, then rejection sampling (see Section 11.1) can be used to obtain draws from the target distribution. If not, or if we cannot determine whether the ratio is bounded, the importance weights can still be used to improve the approximation.

Once L draws, $\theta^1, \ldots, \theta^L$, from the approximate distribution g have been sampled, a sample of $k < L$ draws from a better approximation can be simulated as follows.

1. Sample a value θ from the set $\{\theta^1, \ldots, \theta^L\}$, where the probability of sampling each θ^l is proportional to the weight, $w(\theta^l) = q(\theta^l|y)/g(\theta^l)$.

2. Sample a second value using the same procedure, but excluding the already sampled value from the set.

3. Repeatedly sample without replacement $k - 2$ more times.

Why sample without replacement? If the importance weights are moderate, sampling with and without replacement gives similar results. Now consider a bad case, with a few very large weights and many small weights. Sampling with replacement will pick the same few values of θ repeatedly; in contrast, sampling without replacement yields a more desirable intermediate approximation somewhere between the starting and target densities. For other purposes, sampling with replacement could be superior.

Diagnosing the accuracy of the importance resampling. Unfortunately, we have no reliable rules for assessing how accurate the importance resampling draws are as an approximation of the posterior distribution. We suggest examining the histogram of the logarithms of the largest importance ratios to check that there are no extremely high values that would unduly influence the distribution. This is discussed further in the context of importance sampling in Section 13.3.

Uses of importance resampling in Bayesian computation. If importance re-sampling does not yield an accurate approximation, then it can still be helpful for obtaining starting points for an iterative simulation of the posterior distribution, as described in Chapter 11.

Importance resampling can also be useful when considering mild changes in the posterior distribution, for example replacing a normal model by a Student-t for the SAT coaching example. The idea in this case is to treat the original posterior distribution as an approximation to the modified posterior distribution. Importance resampling is used to subsample from a set of posterior distribution samples those simulated values that are most likely under the modified posterior distribution.

12.3 Finding marginal posterior modes using EM and related algorithms

In problems with many parameters, normal approximations to the joint distribution are often useless, and the joint mode is typically not helpful. It is often useful, however, to base an approximation on a marginal posterior mode of a *subset* of the parameters; we use the notation $\theta = (\gamma, \phi)$ and suppose we are interested in first approximating $p(\phi|y)$. After approximating $p(\phi|y)$ as a normal or t or a mixture of these, we may be able to approximate the conditional distribution, $p(\gamma|\phi, y)$, as normal (or t, or a mixture) with parameters depending on ϕ. In this section we address the first problem, and in the next section we address the second.

When the marginal posterior density, $p(\phi|y)$, can be determined analytically, we can maximize it using optimization methods such as those outlined in Section 12.1 and use the related normal-based approximations of Section 12.2.

The EM algorithm can be viewed as an iterative method for finding the mode of the marginal posterior density, $p(\phi|y)$, and is extremely useful for many common models for which it is hard to maximize $p(\phi|y)$ directly but easy to work with $p(\gamma|\phi, y)$ and $p(\phi|\gamma, y)$. Examples of the EM algorithm appear in the later chapters of this book, including Sections 18.4, 21.4, and 21.6; we introduce the method here.

If we think of ϕ as the parameters in our problem and γ as missing data, the EM algorithm formalizes a relatively old idea for handling missing data: start with a guess of the parameters and then (1) replace missing values by their expectations given the guessed parameters, (2) estimate parameters assuming the missing data are equal to their estimated values, (3) re-estimate the missing values assuming the new parameter estimates are correct, (4) re-estimate parameters, and so forth, iterating until convergence. In fact, the EM algorithm is more efficient than these four steps would suggest since each missing value is not estimated separately; instead, those functions of the missing data that are needed to estimate the model parameters are estimated jointly.

The name 'EM' comes from the two alternating steps: finding the *expectation* of the needed functions (the sufficient statistics) of the missing values, and *maximizing* the resulting posterior density to estimate the parameters as if these functions of the missing data were observed. For many standard models, both steps—estimating the missing values given a current estimate of the parameter and estimating the parameters given current estimates of the missing values—are straightforward. The EM algorithm is especially useful because many models, including mixture models and some hierarchical models, can be reexpressed as probability models on augmented parameter spaces, where the added parameters γ can be thought of as missing data. Rather than list the many applications of the EM algorithm here, we simply present the basic method and then apply it in later chapters where appropriate, most notably in Chapter 21 for missing data; see the bibliographic note at the end of this chapter for further references.

Derivation of the EM and generalized EM algorithms

In the notation of this chapter, the EM algorithm finds the modes of the marginal posterior distribution, $p(\phi|y)$, averaging over the parameters γ. A more conventional presentation, in terms of missing and complete data, appears in Section 21.2. We show here that each iteration of the EM algorithm *increases* the value of the log posterior density until convergence. We start with the simple identity

$$\log p(\phi|y) = \log p(\gamma, \phi|y) - \log p(\gamma|\phi, y)$$

and take expectations of both sides, treating γ as a random variable with the distribution $p(\gamma|\phi^{\text{old}}, y)$, where ϕ^{old} is the current (old) guess. The left side of the above equation does not depend on γ, so averaging over γ yields

$$\log p(\phi|y) = \text{E}_{\text{old}}(\log p(\gamma, \phi|y)) - \text{E}_{\text{old}}(\log p(\gamma|\phi, y)), \qquad (12.4)$$

where E_{old} means averaging over γ under the distribution $p(\gamma|\phi^{\text{old}}, y)$. The key result for the EM algorithm is that the last term on the right side of (12.4), $\text{E}_{\text{old}}(\log p(\gamma|\phi, y))$, is *maximized* at $\phi = \phi^{\text{old}}$ (see Exercise 12.5). The other term, the expected log joint posterior density, $\text{E}_{\text{old}}(\log p(\gamma, \phi|y))$, is repeatedly used in computation,

$$\text{E}_{\text{old}}(\log p(\gamma, \phi|y)) = \int (\log p(\gamma, \phi|y)) p(\gamma|\phi^{\text{old}}, y) d\gamma.$$

This expression is called $Q(\phi|\phi^{\text{old}})$ in the EM literature, where it is viewed as the expected complete-data log-likelihood.

Now consider any value ϕ^{new} for which $\text{E}_{\text{old}}(\log p(\gamma, \phi^{\text{new}}|y))$ is greater than $\text{E}_{\text{old}}(\log p(\gamma, \phi^{\text{old}}|y))$. If we replace ϕ^{old} by ϕ^{new}, we increase the first term on the right side of (12.4), while not increasing the second term, and so the total must increase: $\log p(\phi^{\text{new}}|y) > \log p(\phi^{\text{old}}|y)$. This idea motivates the *generalized EM* (GEM) algorithm: at each iteration, determine $\text{E}_{\text{old}}(\log p(\gamma, \phi|y))$—considered as a function of ϕ—and update ϕ to a new value that increases this function. The EM algorithm is the special case in which the new value of ϕ is chosen to *maximize* $\text{E}_{\text{old}}(\log p(\gamma, \phi|y))$, rather than merely increase it. The

EM and GEM algorithms both have the property of increasing the marginal posterior density, $p(\phi|y)$, at each iteration.

Because the marginal posterior density, $p(\phi|y)$, increases in each step of the EM algorithm, and because the Q function is maximized at each step, EM converges to a local mode of the posterior density except in some very special cases. (Because the GEM algorithm does not maximize at each step, it does not necessarily converge to a local mode.) The rate at which the EM algorithm converges to a local mode depends on the proportion of 'information' about ϕ in the joint density, $p(\gamma, \phi|y)$, that is missing from the marginal density, $p(\phi|y)$. It can be very slow to converge if the proportion of missing information is large; see the bibliographic note at the end of this chapter for many theoretical and applied references on this topic.

Implementation of the EM algorithm

The EM algorithm can be described algorithmically as follows.

1. Start with a crude parameter estimate, ϕ^0.

2. For $t = 1, 2, \ldots$:

   (a) E-step: Determine the expected log posterior density function,

   $$\mathrm{E}_{\mathrm{old}}(\log p(\gamma, \phi|y)) = \int (\log p(\gamma, \phi|y)) p(\gamma|\phi^{\mathrm{old}}, y) d\gamma,$$

   where the expectation averages over the conditional posterior distribution of γ, given the current estimate, $\phi^{\mathrm{old}} = \phi^{t-1}$.

   (b) M-step: Let ϕ^t be the value of ϕ that maximizes $E_{\mathrm{old}}(\log p(\gamma, \phi|y))$. For the GEM algorithm, it is only required that $E_{\mathrm{old}}(\log p(\gamma, \phi|y))$ be increased, not necessarily maximized.

As we have seen, the marginal posterior density, $p(\phi|y)$, increases at each step of the EM algorithm, so that, except in some very special cases, the algorithm converges to a local mode of the posterior density.

Finding multiple modes. As with any maximization algorithm, a simple way to find multiple modes with EM is to start the iterations at many points throughout the parameter space. If we find several modes, we can roughly compare their relative masses using a normal approximation, as described in the previous section.

Debugging. A useful debugging check when running an EM algorithm is to compute the logarithm of the marginal posterior density, $\log p(\phi^t|y)$, at each iteration, and check that it increases monotonically. This computation is recommended for all problems for which it is relatively easy to compute the marginal posterior density.

Example. Normal distribution with unknown mean and variance and semi-conjugate prior distribution
For a simple illustration, consider the problem of estimating the mean of a normal distribution with unknown variance. Another simple example of the EM

algorithm appears in Exercise 17.7. Suppose we weigh an object on a scale n times, and the weighings, $y_1, \ldots, y_n$, are assumed independent with a $N(\mu, \sigma^2)$ distribution, where μ is the true weight of the object. For simplicity, we assume a $N(\mu_0, \tau_0^2)$ prior distribution on μ (with μ_0 and τ_0^2 known) and the standard noninformative uniform prior distribution on $\log \sigma$; these form a 'semi-conjugate' joint prior distribution in the sense of Section 3.4.

Because the model is not fully conjugate, there is no standard form for the joint posterior distribution of (μ, σ) and no closed-form expression for the marginal posterior density of μ. We can, however, use the EM algorithm to find the marginal posterior mode of μ, averaging over σ; that is, (μ, σ) corresponds to (ϕ, γ) in the general notation.

Joint log posterior density.   The logarithm of the joint posterior density is

$$\log p(\mu, \sigma | y) = -\frac{1}{2\tau_0^2}(\mu - \mu_0)^2 - (n+1)\log \sigma - \frac{1}{2\sigma^2}\sum_{i=1}^{n}(y_i - \mu)^2 + \text{constant}, \quad (12.5)$$

ignoring terms that do not depend on μ or σ^2.

E-step.   For the E-step of the EM algorithm, we must determine the expectation of (12.5), averaging over σ and conditional on the current guess, μ^{old}, and y:

$$\begin{aligned}
\text{E}_{\text{old}} \log p(\mu, \sigma | y) &= -\frac{1}{2\tau_0^2}(\mu - \mu_0)^2 - (n+1)\text{E}_{\text{old}}(\log \sigma) \\
&\quad - \frac{1}{2}\text{E}_{\text{old}}\left(\frac{1}{\sigma^2}\right)\sum_{i=1}^{n}(y_i - \mu)^2 + \text{constant}. \quad (12.6)
\end{aligned}$$

We must now evaluate $\text{E}_{\text{old}}(\log \sigma)$ and $\text{E}_{\text{old}}(1/\sigma^2)$. Actually, we need evaluate only the latter expression, because the former expression is not linked to μ in (12.6) and thus will not affect the M-step. The expression $\text{E}_{\text{old}}(1/\sigma^2)$ can be evaluated by noting that, given μ, the posterior distribution of σ^2 is that for a normal distribution with known mean and unknown variance, which is scaled inverse-χ^2:

$$\sigma^2 | \mu, y \sim \text{Inv-}\chi^2\left(n, \frac{1}{n}\sum_{i=1}^{n}(y_i - \mu)^2\right).$$

Then the conditional posterior distribution of $1/\sigma^2$ is a scaled χ^2, and

$$\text{E}_{\text{old}}(1/\sigma^2) = \text{E}(1/\sigma^2 | \mu^{\text{old}}, y) = \left(\frac{1}{n}\sum_{i=1}^{n}(y_i - \mu^{\text{old}})^2\right)^{-1}.$$

We can then reexpress (12.6) as

$$\begin{aligned}
\text{E}_{\text{old}} \log p(\mu, \sigma | y) = {}& \\
& -\frac{1}{2\tau_0^2}(\mu - \mu_0)^2 - \frac{1}{2}\left(\frac{1}{n}\sum_{i=1}^{n}(y_i - \mu^{\text{old}})^2\right)^{-1}\sum_{i=1}^{n}(y_i - \mu)^2 + \text{constant}. \quad (12.7)
\end{aligned}$$

M-step.   For the M-step, we must find the μ that maximizes the above expression. For this problem, the task is straightforward, because (12.7) has the form of a normal log posterior density, with prior distribution $\mu \sim N(\mu_0, \tau_0^2)$ and n data points y_i, each with variance $\frac{1}{n}\sum_{i=1}^{n}(y_i - \mu^{\text{old}})^2$. The M-step is achieved by the mode of the equivalent posterior density, which is

$$\mu^{\text{new}} = \frac{\frac{1}{\tau_0^2}\mu_0 + \frac{n}{\frac{1}{n}\sum_{i=1}^{n}(y_i-\mu^{\text{old}})^2}\bar{y}}{\frac{1}{\tau_0^2} + \frac{n}{\frac{1}{n}\sum_{i=1}^{n}(y_i-\mu^{\text{old}})^2}}.$$

If we iterate this computation, μ will converge to the marginal posterior mode of $p(\mu|y)$.

Extensions of the EM algorithm

Variants and extensions of the basic EM algorithm increase the range of problems to which the algorithm can be applied, and some versions can converge much more quickly than the basic EM algorithm. In addition, the EM algorithm and its extensions can be supplemented with calculations that obtain the second derivative matrix for use as an estimate of the asymptotic variance at the mode. We describe some of these modifications here.

The ECM algorithm.   The ECM algorithm is a variant of the EM algorithm in which the M-step is replaced by a set of conditional maximizations, or CM-steps. Suppose that ϕ^t is the current iterate. The E-step is unchanged: the expected log posterior density is computed, averaging over the conditional posterior distribution of γ given the current iterate. The M-step is replaced by a set of S conditional maximizations. At the sth conditional maximization, the value of $\phi^{t+s/S}$ is found that maximizes the expected log posterior density among all ϕ such that $g_s(\phi) = g_s(\phi^{t+(s-1)/S})$ with the $g_s(\cdot)$ known as constraint functions. The output of the last CM-step, $\phi^{t+S/S} = \phi^{t+1}$, is the next iterate of the ECM algorithm. The set of constraint functions, $g_s(\cdot), s = 1, \ldots, S$, must satisfy certain conditions in order to guarantee convergence to a stationary point. The most common choice of constraint function is the indicator function for the sth subset of the parameters. The parameter vector ϕ is partitioned into S disjoint and exhaustive subsets, $(\phi_1, \ldots, \phi_S)$, and at the sth conditional maximization step, all parameters except those in ϕ_s are constrained to equal their current values, $\phi_j^{t+s/S} = \phi_j^{t+(s-1)/S}$ for $j \neq s$. An ECM algorithm based on a partitioning of the parameters is an example of a generalized EM algorithm. Moreover, if each of the CM steps maximizes by setting first derivatives equal to zero, then ECM shares with EM the property that it will converge to a local mode of the marginal posterior distribution of ϕ. Because the log posterior density, $p(\phi|y)$, increases with every iteration of the ECM algorithm, its monotone increase can still be used for debugging.

As described in the previous paragraph, the ECM algorithm performs several CM-steps after each E-step. The *multicycle ECM* algorithm performs additional E-steps during a single iteration. For example, one might perform an additional E-step before each conditional maximization. Multicycle ECM algorithms require more computation for each iteration than the ECM algorithm but can sometimes reach approximate convergence with fewer total iterations.

The ECME algorithm. The ECME algorithm is an extension of ECM that replaces some of the conditional maximization steps of the expected log joint posterior density, $E_{old}(\log p(\gamma, \phi|y))$, with conditional maximization steps of the actual log posterior density, $\log p(\phi|y)$. The last E in the acronym refers to the choice of maximizing either the actual log posterior density or the expected log posterior density. Iterations of ECME also increase the log posterior density at each iteration. Moreover, if each conditional maximization sets first derivatives equal to zero, ECME will converge to a local mode.

ECME can be especially helpful at increasing the rate of convergence relative to ECM since the actual marginal posterior density is being increased on some steps rather than the full posterior density averaged over the current estimate of the distribution of the other parameters. In fact the increase in speed of convergence can be quite dramatic when faster converging numerical methods (such as Newton's method) are applied directly to the marginal posterior density on some of the CM-steps. For example, if one CM-step requires a one-dimensional search to maximize the expected log joint posterior density then the same effort can be applied directly to the logarithm of the actual marginal posterior density of interest.

The AECM algorithm. The ECME algorithm can be further generalized by allowing different alternating definitions of γ at each conditional maximization step. This generalization is most straightforward when ϕ represents missing data, and where there are different ways of completing the data at different maximization steps. In some problems the alternation can allow much faster convergence. The AECM algorithm shares with EM the property of converging to a local mode with an increase in the posterior density at each step.

Supplemented EM and ECM algorithms

The EM algorithm is attractive because it is often easy to implement and has stable and reliable convergence properties. The basic algorithm and its extensions can be enhanced to produce an estimate of the asymptotic variance matrix at the mode, which is useful in forming approximations to the marginal posterior density. The *supplemented EM* (SEM) algorithm and the supplemented ECM (SECM) algorithm use information from the log *joint* posterior density and repeated EM- or ECM-steps to obtain the approximate asymptotic variance matrix for the parameters ϕ.

To describe the SEM algorithm we introduce the notation $M(\phi)$ for the mapping defined implicitly by the EM algorithm, $\phi^{t+1} = M(\phi^t)$. The asymptotic variance matrix V is given by

$$V = V_{joint} + V_{joint} D_M (I - D_M)^{-1}$$

where D_M is the Jacobian matrix for the EM map evaluated at the marginal mode, $\hat{\phi}$, and V_{joint} is the asymptotic variance matrix based on the logarithm of the joint posterior density averaged over γ,

$$V_{\text{joint}} = \left[\text{E} \left(-\frac{d^2 \log p(\phi, \gamma | y)}{d\theta^2} \middle| \phi, y \right) \middle|_{\phi = \hat{\phi}} \right]^{-1}.$$

Typically, V_{joint} can be computed analytically so that only D_M is required. The matrix D_M is computed numerically at each marginal mode using the E- and M-steps according to the following algorithm.

1. Run the EM algorithm to convergence, thereby obtaining the marginal mode, $\hat{\phi}$. (If multiple runs of EM lead to different modes, apply the following steps separately for each mode.)

2. Choose a starting value ϕ^0 for the SEM calculation such that ϕ^0 does not equal $\hat{\phi}$ for any component. One possibility is to use the same starting value that is used for the original EM calculation.

3. Repeat the following steps to get a sequence of matrices R^t, $t = 1, 2, 3, \ldots$, for which each element r_{ij}^t converges to the appropriate element of D_M. In the following we describe the steps used to generate R^t given the current EM iterate, ϕ^t.

(a) Run the usual E-step and M-step with input ϕ^t to obtain ϕ^{t+1}.

(b) For each element of ϕ, say ϕ_i:

   i. Define $\phi^t(i)$ equal to $\hat{\phi}$ except for the ith element, which is replaced by its current value ϕ_i^t.

   ii. Run one E-step and one M-step treating $\phi^t(i)$ as the input value of the parameter vector, ϕ. Denote the result of these E- and M-steps as $\phi^{t+1}(i)$. The ith row of R^t is computed as

$$r_{ij}^t = \frac{\phi_j^{t+1}(i) - \hat{\phi}_j}{\phi_i^t - \hat{\phi}_i}$$

for each j.

When the value of an element r_{ij} no longer changes, it represents a numerical estimate of the corresponding element of D_M. Once all of the elements in a row have converged, then we need no longer repeat the final step for that row. If some elements of ϕ are independent of γ, then EM will converge immediately to the mode for that component with the corresponding elements of D_M equal to zero. SEM can be easily modified to obtain the variance matrix in such cases.

The same approach can be used to supplement the ECM algorithm with an estimate of the asymptotic variance matrix. The SECM algorithm is based on the following result:

$$V = V_{\text{joint}} + V_{\text{joint}}(D_M^{\text{ECM}} - D_M^{\text{CM}})(I - D_M^{\text{ECM}})^{-1},$$

with D_M^{ECM} defined and computed in a manner analogous to D_M in the above discussion except with ECM in place of EM, and where D_M^{CM} is the rate of convergence for conditional maximization applied directly to $\log p(\phi|y)$. This latter matrix depends only on V_{joint} and $\nabla_s = \nabla g_s(\hat{\phi}), s = 1, \ldots, S$, the gradient of the vector of constraint functions g_s at $\hat{\phi}$:

$$D_M^{\text{CM}} = \prod_{s=1}^{S} \left[\nabla_s (\nabla_s^T V_{\text{joint}} \nabla_s)^{-1} \nabla_s^T V_{\text{joint}} \right].$$

These gradient vectors are trivial to calculate for a constraint that directly fixes components of ϕ. In general, the SECM algorithm appears to require analytic work to compute V_{joint} and $\nabla_s, s = 1, \ldots, S$, in addition to applying the numerical approach described to compute D_M^{ECM}, but some of these calculations can be performed using results from the ECM iterations (see bibliographic note).

Parameter-expanded EM (PX-EM)

The various methods discussed in Section 11.8 for improving the efficiency of Gibbs samplers have analogues for mode-finding (and in fact were originally constructed for that purpose). For example, the parameter expansion idea illustrated with the t model on page 304 was originally developed in the context of the EM algorithm. In this setting, the individual latent data variances V_i are treated as missing data, and the ECM algorithm maximizes over the parameters μ, τ, and α in the posterior distribution.

12.4 Approximating conditional and marginal posterior densities

Approximating the conditional posterior density, $p(\gamma|\phi, y)$

As stated at the beginning of Section 12.3, for many problems, the normal, multivariate t, and other analytically convenient distributions are poor approximations to the joint posterior distribution of all parameters. Often, however, we can partition the parameter vector as $\theta = (\gamma, \phi)$, in such a way that an analytic approximation works well for the conditional posterior density, $p(\gamma|\phi, y)$. In general, the approximation will depend on ϕ. We write the approximate conditional density as $p_{\text{approx}}(\gamma|\phi, y)$. For example, in the normal random-effects model in Section 5.4, we fitted a normal distribution to $p(\theta, \mu|\tau, y)$ but not to $p(\tau|y)$ (in that example, the normal conditional distribution is an exact fit).

Approximating the marginal posterior density, $p(\phi|y)$, using an analytic approximation to $p(\gamma|\phi, y)$

The mode-finding techniques and normal approximation of Sections 12.1 and 12.2 can be applied directly to the marginal posterior density if the marginal distribution can be obtained analytically. If not, then the EM algorithm (Section 12.3) may allow us to find the mode of the marginal posterior density and construct an approximation. On occasion it is not possible to construct an approximation to $p(\phi|y)$ using any of those methods. Fortunately we may still derive an approximation if we have an analytic approximation to the conditional posterior density, $p(\gamma|\phi, y)$. We can use a trick introduced with the semi-conjugate model in Section 3.4, and also used in (5.19) in Section 5.4 to generate an approximation to $p(\phi|y)$. The approximation is constructed as the ratio of the joint posterior distribution to the analytic approximation of the conditional posterior distribution:

$$p_{\text{approx}}(\phi|y) = \frac{p(\gamma, \phi|y)}{p_{\text{approx}}(\gamma|\phi, y)}. \qquad (12.8)$$

The key to this method is that if the denominator has a standard analytic form, we can compute its normalizing factor, which, in general, depends on ϕ. When using (12.8), we must also specify a value γ (possibly as a function of ϕ) since the left side does not involve γ at all. If the analytic approximation to the conditional distribution is exact, the factors of γ in the numerator and denominator cancel, as in equations (3.12) and (3.13) for the semi-conjugate prior distribution for the normal example in Section 3.4 and we obtain the marginal posterior density exactly. If the analytic approximation is not exact, a natural value to use for γ is the center of the approximate distribution (for example, $E(\gamma|\phi, y)$ under the normal or Student-t approximations).

For example, suppose we have approximated the d-dimensional conditional density function, $p(\gamma|\phi, y)$, by a multivariate normal density with mean $\hat{\gamma}$ and scale matrix V_γ, both of which depend on ϕ. We can then approximate the marginal density of ϕ by

$$p_{\text{approx}}(\phi|y) \propto p(\hat{\gamma}(\phi), \phi|y)|V_\gamma(\phi)|^{1/2}, \qquad (12.9)$$

where ϕ is included in parentheses to indicate that the mean and scale matrix must be evaluated at each value of ϕ. The same result holds if a Student-t approximation is used; in either case, the normalizing factor in the denominator of (12.8) is proportional to $|V_\gamma(\phi)|^{-1/2}$.

12.5 Example: the hierarchical normal model (continued)

We illustrate mode-based computations with the example of the hierarchical normal model that we used in Section 11.7. In that section, we illustrated the Gibbs sampler and the Metropolis algorithm as two different ways of drawing posterior samples. In this section, we describe how to get approximate inference—and thus a starting distribution for the Gibbs or Metropolis algorithm—by finding the mode of $p(\mu, \log \sigma, \log \tau|y)$, the marginal posterior density, and a normal approximation centered at the mode. Given $(\mu, \log \sigma, \log \tau)$, the individual means θ_j have independent normal conditional posterior distributions.

Parameter	Crude estimate	Stepwise ascent		
		First iteration	Second iteration	Third iteration
θ_1	61.00	61.28	61.29	61.29
θ_2	66.00	65.87	65.87	65.87
θ_3	68.00	67.74	67.73	67.73
θ_4	61.00	61.15	61.15	61.15
μ	64.00	64.01	64.01	64.01
σ	2.29	2.17	2.17	2.17
τ	3.56	3.32	3.31	3.31
$\log p(\theta, \mu, \log \sigma, \log \tau \vert y)$	-63.70	-61.42	-61.42	-61.42

Table 12.1 *Convergence of stepwise ascent to a joint posterior mode for the coagulation example. The joint posterior density increases at each conditional maximization step, as it should. The posterior mode is in terms of* $\log \sigma$ *and* $\log \tau$, *but these values are transformed back to the original scale for display in the table.*

Crude initial parameter estimates

Initial parameter estimates for the computations are easily obtained by estimating θ_j as $\bar{y}_{.j}$, the average of the observations in the jth group, for each j, and estimating σ^2 as the average of the J within-group sample variances, $s_j^2 = \sum_{i=1}^{n_j} (y_{ij} - \bar{y}_{.j})^2 / (n_j - 1)$. We then crudely estimate μ and τ^2 as the mean and variance of the J estimated values of θ_j.

For the coagulation data in Table 11.2, the crude estimates are displayed as the first column of Table 12.1.

Finding the joint posterior mode of $p(\theta, \mu, \log \sigma, \log \tau \vert y)$ using conditional maximization

Because of the conjugacy of the normal model, it is easy to perform conditional maximization on the joint posterior density, updating each parameter in turn by its conditional mode. In general, we analyze scale parameters such as σ and τ on the logarithmic scale. The conditional modes for each parameter are easy to compute, especially because we have already determined the conditional posterior density functions in computing the Gibbs sampler for this problem in Section 11.7. After obtaining a starting guess for the parameters, the conditional maximization proceeds as follows, where the parameters can be updated in any order.

1. *Conditional mode of each θ_j.* The conditional posterior distribution of θ_j, given all other parameters in the model, is normal and given by (11.6). For $j = 1, \ldots, J$, we can maximize the conditional posterior density of θ_j given (μ, σ, τ, y) (and thereby increase the joint posterior density), by replacing the current estimate of θ_j by $\hat{\theta}_j$ in (11.6).

2. *Conditional mode of μ.* The conditional posterior distribution of μ, given all other parameters in the model, is normal and given by (11.8). For conditional maximization, replace the current estimate of μ by $\hat{\mu}$ in (11.9).

3. *Conditional mode of $\log \sigma$.* The conditional posterior density for σ^2 is scaled inverse-χ^2 and given by (11.10). The mode of the conditional posterior density of $\log \sigma$ is obtained by replacing the current estimate of $\log \sigma$ with $\log \hat{\sigma}$, with $\hat{\sigma}$ defined in (11.11). (From Appendix A, the conditional mode of σ^2 is $\frac{n}{n+2} \hat{\sigma}^2$. The factor of $\frac{n}{n+2}$ does not appear in the conditional mode of $\log \sigma$ because of the Jacobian factor when transforming from $p(\sigma^2)$ to $p(\log \sigma)$; see Exercise 12.6.)

4. *Conditional mode of $\log \tau$.* The conditional posterior density for τ^2 is scaled inverse-χ^2 and given by (11.12). The mode of the conditional posterior density of $\log \tau$ is obtained by replacing the current estimate of $\log \tau$ with $\log \hat{\tau}$, with $\hat{\tau}$ defined in (11.13).

Numerical results of conditional maximization for the coagulation example are presented in Table 12.1, from which we see that the algorithm has required only three iterations to reach approximate convergence in this small example. We also see that this posterior mode is extremely close to the crude estimate, which occurs because the shrinkage factors $\frac{\sigma^2}{n_j}/(\frac{\sigma^2}{n_j} + \tau^2)$ are all near zero. Incidentally, the mode displayed in Table 12.1 is only a local mode; the joint posterior density also has another mode at the boundary of the parameter space; we are not especially concerned with that degenerate mode because the region around it includes very little of the posterior mass (see Exercise 12.7).

In a simple applied analysis, we might stop here with an approximate posterior distribution centered at this joint mode, or even just stay with the simpler crude estimates. In other hierarchical examples, however, there might be quite a bit of pooling, as in the educational testing problem of Section 5.5, in which case it is advisable to continue the analysis, as we describe below.

Factoring into conditional and marginal posterior densities

As discussed, the joint mode often does not provide a useful summary of the posterior distribution, especially when J is large relative to the n_j's. To investigate this possibility, we consider the marginal posterior distribution of a subset of the parameters. In this example, using the notation of the previous sections, we set $\gamma = (\theta_1, \ldots, \theta_J) = \theta$ and $\phi = (\mu, \log \sigma, \log \tau)$, and we consider the posterior distribution as the product of the marginal posterior distribution of ϕ and the conditional posterior distribution of θ given ϕ. The subvector $(\mu, \log \sigma, \log \tau)$ has only three components no matter how large J is, so we expect asymptotic approximations to work better for the marginal distribution of ϕ than for the joint distribution of (γ, ϕ).

From (11.5) in the Gibbs sampling analysis of the coagulation in Chapter 11, the conditional posterior density of the normal means, $p(\theta|\mu, \sigma, \tau, y)$, is a product of independent normal densities with means $\hat{\theta}_j$ and variances V_{θ_j} that are easily computable functions of (μ, σ, τ, y).

The marginal posterior density, $p(\mu, \log \sigma, \log \tau | y)$, of the remaining parameters, can be determined using formula (12.8), where the conditional distribution $p_{\mathrm{approx}}(\theta | \mu, \log \sigma, \log \tau, y)$ is actually exact. Thus,

$$
\begin{aligned}
p(\mu, \log \sigma, \log \tau | y) &= \frac{p(\theta, \mu, \log \sigma, \log \tau | y)}{p(\theta | \mu, \log \sigma, \log \tau, y)} \\
&\propto \frac{\tau \prod_{j=1}^{J} N(\theta_j | \mu, \tau^2) \prod_{j=1}^{J} \prod_{i=1}^{n_j} N(y_{ij} | \theta_j, \sigma^2)}{\prod_{j=1}^{J} N(\theta_j | \hat{\theta}_j, V_{\theta_j})}.
\end{aligned}
$$

Because the denominator is exact, this identity must hold for any θ; to simplify calculations, we set $\theta = \hat{\theta}$, to yield

$$
p(\mu, \log \sigma, \log \tau | y) \propto \tau \prod_{j=1}^{J} N(\hat{\theta}_j | \mu, \tau^2) \prod_{j=1}^{J} \prod_{i=1}^{n_j} N(y_{ij} | \hat{\theta}_j, \sigma^2) \prod_{j=1}^{J} V_{\theta_j}^{1/2}, \quad (12.10)
$$

with the final factor coming from the normalizing constant of the normal distribution in the denominator, and where $\hat{\theta}_j$ and V_{θ_j} are defined by (11.7).

Finding the marginal posterior mode of $p(\mu, \log \sigma, \log \tau | y)$ using EM

The marginal posterior mode of (μ, σ, τ)—the maximum of (12.10)—cannot be found analytically because the $\hat{\theta}_j$'s and V_{θ_j}'s are functions of (μ, σ, τ). One possible approach is Newton's method, which requires computing derivatives and second derivatives analytically or numerically. For this problem, however, it is easier to use the EM algorithm.

To obtain the mode of $p(\mu, \log \sigma, \log \tau | y)$ using EM, we average over the parameter θ in the E-step and maximize over $(\mu, \log \sigma, \log \tau)$ in the M-step. The logarithm of the joint posterior density of all the parameters is

$$
\begin{aligned}
\log p(\theta, \mu, \log \sigma, \log \tau | y) &= -n \log \sigma - (J - 1) \log \tau - \frac{1}{2\tau^2} \sum_{j=1}^{J} (\theta_j - \mu)^2 \\
&\quad - \frac{1}{2\sigma^2} \sum_{j=1}^{J} \sum_{i=1}^{n_j} (y_{ij} - \theta_j)^2 + \text{constant}. \quad (12.11)
\end{aligned}
$$

E-step. The E-step, averaging over θ in (12.11), requires determining the conditional posterior expectations $E_{\mathrm{old}}((\theta_j - \mu)^2)$ and $E_{\mathrm{old}}((y_{ij} - \theta_j)^2)$ for all j. These are both easy to compute using the conditional posterior distribution $p(\theta | \mu, \sigma, \tau, y)$, which we have already determined in (11.5).

$$
\begin{aligned}
E_{\mathrm{old}}((\theta_j - \mu)^2) &= E((\theta_j - \mu)^2 | \mu^{\mathrm{old}}, \sigma^{\mathrm{old}}, \tau^{\mathrm{old}}, y) \\
&= [E_{\mathrm{old}}(\theta_j - \mu)]^2 + \mathrm{var}_{\mathrm{old}}(\theta_j) \\
&= (\hat{\theta}_j - \mu)^2 + V_{\theta_j}.
\end{aligned}
$$

Parameter	Value at joint mode	EM algorithm		
		First iteration	Second iteration	Third iteration
μ	64.01	64.01	64.01	64.01
σ	2.17	2.33	2.36	2.36
τ	3.31	3.46	3.47	3.47
$\log p(\mu, \log \sigma, \log \tau \vert y)$	-61.99	-61.835	-61.832	-61.832

Table 12.2 *Convergence of the EM algorithm to the marginal posterior mode of* $(\mu, \log \sigma, \log \tau)$ *for the coagulation example. The marginal posterior density increases at each EM iteration, as it should. The posterior mode is in terms of* $\log \sigma$ *and* $\log \tau$, *but these values are transformed back to the original scale for display in the table.*

Using a similar calculation,

$$\mathrm{E}_{\mathrm{old}}((y_{ij} - \theta_j)^2) = (y_{ij} - \hat{\theta}_j)^2 + V_{\theta_j}.$$

For both expressions, $\hat{\theta}_j$ and V_{θ_j} are computed from equation (11.7) based on $(\mu, \log \sigma, \log \tau)^{\mathrm{old}}$.

M-step. It is now straightforward to maximize the expected log posterior density, $\mathrm{E}_{\mathrm{old}}(\log p(\theta, \mu, \log \sigma, \log \tau \vert y))$ as a function of $(\mu, \log \sigma, \log \tau)$. The maximizing values are $(\mu^{\mathrm{new}}, \log \sigma^{\mathrm{new}}, \log \tau^{\mathrm{new}})$, with $(\mu, \sigma, \tau)^{\mathrm{new}}$ obtained by maximizing (12.11):

$$\mu^{\mathrm{new}} = \frac{1}{J} \sum_{j=1}^{J} \hat{\theta}_j$$

$$\sigma^{\mathrm{new}} = \left(\frac{1}{n} \sum_{j=1}^{J} \sum_{i=1}^{n_j} [(y_{ij} - \hat{\theta}_j)^2 + V_{\theta_j}] \right)^{1/2}$$

$$\tau^{\mathrm{new}} = \left(\frac{1}{J-1} \sum_{j=1}^{J} [(\hat{\theta}_j - \mu^{\mathrm{new}})^2 + V_{\theta_j}] \right)^{1/2}. \qquad (12.12)$$

The derivation of these is straightforward (see Exercise 12.8).

Checking that the marginal posterior density increases at each step. Ideally, at each iteration of EM, we would compute the logarithm of (12.10) using the just calculated $(\mu, \log \sigma, \log \tau)^{\mathrm{new}}$. If the function does not increase, there is a mistake in the analytic calculations or the programming, or possibly there is roundoff error, which can be checked by altering the precision of the calculations.

We applied the EM algorithm to the coagulation example, using the values of (σ, μ, τ) from the joint mode as a starting point; numerical results appear in Table 12.2, where we see that the EM algorithm has approximately converged after only three steps. As typical in this sort of problem, the variance

| Parameter | Posterior quantiles | | | | |
	2.5%	25%	median	75%	97.5%
θ_1	59.15	60.63	61.38	62.18	63.87
θ_2	63.83	65.20	65.78	66.42	67.79
θ_3	65.46	66.95	67.65	68.32	69.64
θ_4	59.51	60.68	61.21	61.77	62.99
μ	60.43	62.73	64.05	65.29	67.69
σ	1.75	2.12	2.37	2.64	3.21
τ	1.44	2.62	3.43	4.65	8.19

Table 12.3 *Summary of posterior simulations for the coagulation example, based on draws from the normal approximation to $p(\mu, \log \sigma, \log \tau | y)$ and the exact conditional posterior distribution, $p(\theta | \mu, \log \sigma, \log \tau, y)$. Compare to joint and marginal modes in Tables 12.1 and 12.2.*

parameters σ and τ are larger at the marginal mode than the joint mode. The logarithm of the marginal posterior density, $\log p(\mu, \log \sigma, \log \tau | y)$, has been computed to the (generally unnecessary) precision of three decimal places for the purpose of checking that it does, indeed, monotonically increase. (Of course, if it had not, we would have debugged the program before including the example in the book!)

Constructing an approximation to the joint posterior distribution

Having found the mode, we can construct a normal approximation based on the 3×3 matrix of second derivatives of the marginal posterior density, $p(\mu, \log \sigma, \log \tau | y)$, in (12.10). To draw simulations from the approximate joint posterior density, first draw $(\mu, \log \sigma, \log \tau)$ from the approximate normal marginal posterior density, then θ from the conditional posterior distribution, $p(\theta | \mu, \log \sigma, \log \tau, y)$, which is already normal and so does not need to be approximated. Table 12.3 gives posterior intervals for the model parameters from these simulations.

Starting points for iterative simulation

If we determine that the approximate inferences are not adequate, then we may elect to continue the analysis using the Gibbs sampler or Metropolis algorithm for simulation from the joint posterior distribution of $(\theta, \mu, \sigma, \tau | y)$. The approximation that we have developed can still serve us by helping us to obtain overdispersed starting points. To obtain starting points, we can sample $L = 2000$ draws from the t_4 approximation for $(\mu, \log \sigma, \log \tau)$. We then draw a subsample of size 10 using importance resampling (see Section 12.2) and use these as starting points for the iterative simulations. (We can start the Gibbs sampler with draws of θ, so we do not need to draw θ for the starting distribution.)

In this example, one can more directly draw overdispersed starting points using the method described in Section 11.7. For more complicated problems, however, and for automated computation (as when a model will be fitted repeatedly to various different datasets), it can be useful to have a more systematic approach to obtain approximations and starting points, as described in this section.

12.6 Bibliographic note

An accessible source of general algorithms for conditional maximization (stepwise ascent), Newton's method, and other computational methods is Press et al. (1986), who give a clear treatment and include code in Fortran, C, or Pascal for implementing all the methods discussed there. Gill, Murray, and Wright (1981) is a more comprehensive book that is useful for understanding more complicated optimization problems.

As discussed in Chapter 4, the normal approximation has a long history in Bayesian computation and has thus been a natural starting point for iterative simulations. Importance resampling was introduced by Rubin (1987b), and an accessible exposition is given by Smith and Gelfand (1992).

The EM algorithm was first presented in full generality and under that name, along with many examples, by Dempster, Laird, and Rubin (1977); the formulation in that article is in terms of finding the maximum likelihood estimate, but, as the authors note, the same arguments hold for finding posterior modes. That article and the accompanying discussion contributions also give many references to earlier implementations of the EM algorithm in specific problems, emphasizing that the idea has been used in special cases on many occasions. It was first presented in a general statistical context by Orchard and Woodbury (1972) as the 'missing information principle' and first derived in mathematical generality by Baum et al. (1970). Little and Rubin (2002, Chapter 8) discuss the EM algorithm for missing data problems. The statistical literature on applications of EM is vast; for example, see the review by Meng and Pedlow (1992).

Various extensions of EM have appeared in the recent literature. In particular, SEM was introduced in Meng and Rubin (1991); ECM in Meng and Rubin (1993) and Meng (1994a); SECM in van Dyk, Meng, and Rubin (1995); and ECME in Liu and Rubin (1994). AECM appears in Meng and van Dyk (1997), and the accompanying discussion provides further connections. Many of the iterative simulation methods discussed in Chapter 11 for simulating posterior distributions can be regarded as stochastic extensions of EM; Tanner and Wong (1987) is an important paper in drawing this connection. Parameter-expanded EM was introduced by Liu, Rubin, and Wu (1998), and related ideas appear in Meng and van Dyk (1997), Liu and Wu (1999), and Liu (2003).

12.7 Exercises

1. Normal approximation and importance resampling:

    (a) Repeat Exercise 3.12 using the normal approximation to the Poisson likelihood to produce posterior simulations for (α, β).

    (b) Use importance resampling to improve on the normal approximation.

    (c) Compute the importance ratios for your simulations. Plot a histogram of the importance ratios and comment on their distribution.

2. Importance resampling with and without replacement:

    (a) Consider the bioassay example introduced in Section 3.7. Use importance resampling to approximate draws from the posterior distribution of the parameters (α, β), using the normal approximation of Section 4.1 as the starting distribution. Compare your simulations of (α, β) to Figure 3.4b and discuss any discrepancies.

    (b) Comment on the distribution of the simulated importance ratios.

    (c) Repeat part (a) using importance sampling with replacement. Discuss how the results differ.

3. Analytic approximation to a subset of the parameters: suppose that the joint posterior distribution $p(\theta_1, \theta_2 | y)$ is of interest and that it is known that the Student-t provides an adequate approximation to the conditional distribution, $p(\theta_1 | \theta_2, y)$. Show that both the normal and Student-t approaches described in the last paragraph of Section 12.4 lead to the same answer.

4. Estimating the number of unseen species (see Fisher, Corbet, and Williams, 1943, Efron and Thisted, 1976, and Seber, 1992): suppose that during an animal trapping expedition the number of times an animal from species i is caught is $x_i \sim \text{Poisson}(\lambda_i)$. For parts (a)–(d) of this problem, assume a $\text{Gamma}(\alpha, \beta)$ prior distribution for the λ_i's, with a uniform hyperprior distribution on (α, β). The only observed data are y_k, the number of species observed exactly k times during a trapping expedition, for $k = 1, 2, 3, \ldots$.

    (a) Write the distribution $p(x_i | \alpha, \beta)$.

    (b) Use the distribution of x_i to derive a multinomial distribution for y given that there are a total of N species.

    (c) Suppose that we are given $y = (118, 74, 44, 24, 29, 22, 20, 14, 20, 15, 12, 14, 6, 12, 6, 9, 9, 6, 10, 10, 11, 5, 3, 3)$, so that 118 species were observed only once, 74 species were observed twice, and so forth, with a total of 496 species observed and 3266 animals caught. Write down the likelihood for y using the multinomial distribution with 24 cells (ignoring unseen species). Use any method to find the mode of α, β and an approximate second derivative matrix.

    (d) Derive an estimate and approximate 95% posterior interval for the number of additional species that would be observed if 10,000 more animals were caught.

(e) Evaluate the fit of the model to the data using appropriate posterior predictive checks.

(f) Discuss the sensitivity of the inference in (d) to each of the model assumptions.

5. Derivation of the monotone convergence of EM algorithm: prove that the function $E_{old} \log p(\gamma|\phi, y)$ in (12.4) is maximized at $\phi = \phi^{old}$. (Hint: express the expectation as an integral and apply Jensen's inequality to the convex logarithm function.)

6. Conditional maximization for the hierarchical normal model: show that the conditional modes of σ and τ associated with (11.10) and (11.12), respectively, are correct.

7. Joint posterior modes for hierarchical models:

(a) Show that the posterior density for the coagulation example has a degenerate mode at $\tau = 0$ and $\theta_j = \mu$ for all j.

(b) The rest of this exercise demonstrates that the degenerate mode represents a very small part of the posterior distribution. First estimate an upper bound on the integral of the unnormalized posterior density in the neighborhood of the degenerate mode. (Approximate the integrand so that the integral is analytically tractable.)

(c) Now approximate the integral of the unnormalized posterior density in the neighborhood of the other mode using the density at the mode and the second derivative matrix of the log posterior density at the mode.

(d) Finally, estimate an upper bound on the posterior mass in the neighborhood of the degenerate mode.

8. The EM algorithm for the hierarchical normal model: For the model in Section 12.5 for the coagulation data, derive the expressions (12.12) for μ^{new}, σ^{new}, and τ^{new}.

CHAPTER 13

Special topics in computation

This chapter collects a number of special topics related to Bayesian computation, including ideas related to both approximation and simulation. We do not attempt to be comprehensive but rather try to give a sense of different areas of advanced Bayesian computation. The bibliographic note at the end of the chapter gives references to more thorough descriptions of these and other methods.

13.1 Advanced techniques for Markov chain simulation

The basic Gibbs sampler and Metropolis algorithm work well for a wide range of problems, especially when the user can optimize the simulation, most commonly by reparameterizing to approximate independence (for the Gibbs sampler) and tuning the jumping distribution to an acceptance rate between 20% and 45% (for the Metropolis algorithm).

Advanced Markov chain simulation algorithms are built upon the framework of the Metropolis-Hastings algorithm described in Section 11.4, generally by means of auxiliary variables, and can be useful for complex models. In general, we recommend starting with the simplest algorithms and, if convergence is slow, trying simple improvements such as discussed in Sections 11.8 and 11.9. If convergence is still a problem, the more advanced algorithms sketched here might be effective.

Hybrid Monte Carlo for moving more rapidly through the target distribution

An inherent inefficiency in the Gibbs sampler and Metropolis algorithm is their random walk behavior—as illustrated in the figures on pages 286 and 288, the simulations can take a long time zigging and zagging while moving through the target distribution. Reparameterization and efficient jumping rules can improve the situation (see Sections 11.8 and 11.9), but for complicated models this local random-walk behavior remains, especially for high-dimensional target distributions.

The algorithm of *hybrid Monte Carlo* borrows ideas from physics to add auxiliary variables that suppress the local random walk behavior in the Metropolis algorithm, thus allowing it to move much more rapidly through the target distribution. For each component θ_j in the target space, hybrid Monte Carlo adds a 'momentum' variable ϕ_j. Both θ and ϕ are then updated together in a new Metropolis algorithm, in which the jumping distribution for θ is determined largely by ϕ. Roughly, the momentum ϕ gives the expected distance and direction of the jump in θ, so that successive jumps tend to be in the same direction, allowing the simulation to move rapidly where possible through the space of θ. The Metropolis-Hastings accept/reject rule stops the movement when it reaches areas of low probability, at which point the momentum changes until the jumping can continue. Hybrid Monte Carlo is also called *Hamiltonian Monte Carlo* because it is related to the model of Hamiltonian dynamics in physics.

Langevin rule for Metropolis-Hastings jumping

The basic symmetric-jumping Metropolis algorithm is simple to apply but has the disadvantage of wasting many of its jumps by going into low-probability areas of the target distribution. For example, optimal Metropolis algorithms in high dimensions have acceptance rates below 25% (see Section 11.9), which means that, in the best case, over 3/4 of the jumps are wasted.

A potential improvement is afforded by the *Langevin algorithm*, in which each jump is associated with a small shift in the direction of the gradient of the logarithm of the target density, thus moving the jumps toward higher density regions of the distribution. This jumping rule is no longer symmetric (the probability of jumping from θ_a to θ_b depends on the density at θ_a, and the probability of the reverse jump depends on the density at θ_b), and so the Metropolis-Hastings algorithm (see page 291) must be used to decide whether to accept or reject. Langevin updating can be considered as a special case of hybrid Monte Carlo in which Hamiltonian dynamics is simulated for only one time step.

In practice, both the size of the Langevin shift and the scale of the random jump can be set to optimize simulation efficiency as estimated based on pilot runs, in a generalization of the procedure recommended in Section 11.9 for the Metropolis algorithm.

Slice sampling using the uniform distribution

A random sample of θ from the d-dimensional target distribution, $p(\theta|y)$, is equivalent to a random sample from the area under the distribution (for example, the shaded area under the curve in the illustration of rejection sampling on page 285). Formally, sampling is performed from the $d+1$-dimensional distribution of (θ, u), where, for any θ, $p(\theta, u|y) \propto 1$ for $u \in [0, p(\theta|y)]$ and 0 otherwise. *Slice sampling* refers to the application of iterative simulation algorithms on this uniform distribution. The details of implementing an effective slice sampling procedure can be complicated, but the method can be applied in great generality and can be especially useful for sampling one-dimensional conditional distributions in a Gibbs sampling structure.

Simulated tempering for moving through a multimodal target distribution

Multimodal posterior distributions can pose special problems for Markov chain simulation techniques. The goal is to sample from the entire posterior distribution and this requires sampling from each of the modes with significant posterior probability. Unfortunately it is easy for Markov chain simulations to remain in the neighborhood of a single mode for a long period of time. This occurs primarily when two (or more) modes are separated by regions of extremely low posterior density. Then it is difficult to move from one mode to the other because, for example, Metropolis jumps to the region between the two modes will be rejected.

Simulated tempering is one strategy for improving Markov chain simulation performance in this case. As usual, we take $p(\theta|y)$ to be the target density. Simulated tempering works with a set of $K+1$ Markov chain simulations, each with its own stationary distribution $p_k(\theta|y), k = 0, 1, \ldots, K$, where $p_0(\theta|y) = p(\theta|y)$, and $p_1, \ldots, p_K$ are distributions with the same basic shape but with improved opportunities for mixing across the modes. As usual with Metropolis algorithms, the distributions p_k need not be fully specified; it is only necessary that the user can compute unnormalized density functions q_k, where $q_k(\theta|y) = p_k(\theta|y)$ multiplied by a constant (which can depend on y and k but not on the parameters θ). One choice for the unnormalized densities q_k is

$$q_k(\theta|y) = p(\theta|y)^{1/T_k},$$

for a set of 'temperature' parameters $T_k > 0$. Setting $T_k = 1$ reduces to the original density, and large values of T_k produce less highly peaked modes. (That is, 'high temperatures' add 'thermal noise' to the system.) A single composite Markov chain simulation is then developed that randomly moves across the $K+1$ component Markov chains, with T_0 set to 1 so that $q_0 = p(\theta|y)$. The state of the composite Markov chain at iteration t is represented by the pair (θ^t, s^t), where s^t is an integer identifying the component Markov chain used at iteration t. Each iteration of the composite Markov chain simulation consists of two steps:

1. A new value θ^{t+1} is selected using the Markov chain simulation with stationary distribution q_{s^t}.

2. A jump from the current sampler s^t to an alternative sampler j is proposed with probability $J_{s^t,j}$. The move is then accepted with probability $\min(r, 1)$, where

$$r = \frac{c_j q_j(\theta^{t+1}) J_{j,s^t}}{c_{s^t} q_{s^t}(\theta^{t+1}) J_{s^t,j}}.$$

The constants $c_k, k = 0, 1, \ldots, K$ are set adaptively (that is, assigned initial values and then altered after the simulation has run a while) to approximate the inverses of the normalizing constants for the distributions defined by the unnormalized densities q_k. The chain will then spend an approximately equal amount of time in each sampler.

At the end of the Markov chain simulation, only those values of θ simulated from the target distribution (q_0) are used to obtain posterior inferences.

Other auxiliary variable methods have been developed that are tailored to particular structures of multivariate distributions. For example, highly correlated variables such as arise in spatial statistics can be simulated using *multigrid sampling*, in which computations are done alternately on the original scale and on coarser scales that do not capture the local details of the target distribution but allow faster movement between states.

Reversible jump sampling for moving between spaces of differing dimensions

In a number of settings it is desirable to carry out a *trans-dimensional* Markov chain simulation, in which the dimension of the parameter space can change from one iteration to the next. One example where this occurs is in model averaging where a single Markov chain simulation is constructed that includes moves among a number of plausible models (perhaps regression models with different sets of predictors). The 'parameter space' for such a Markov chain simulation includes the traditional parameters along with an indication of the current model. A second example includes finite mixture models (see Chapter 18) in which the number of mixture components is allowed to vary.

It is still possible to perform the Metropolis algorithm in such settings, using the method of *reversible jump* sampling. We use notation corresponding to the case where a Markov chain moves among a number of candidate models. Let $M_k, k = 1, \ldots, K$, denote the candidate models and θ_k the parameter vector for model k with dimension d_k. A key aspect of the reversible jump approach is the introduction of additional random variables that enable the matching of parameter space dimensions across models. Specifically if a move from k to k^* is being considered, then an auxiliary random variable u with jumping distribution $J(u|k, k^*, \theta_k)$ is generated. A series of one-to-one deterministic functions are defined that do the dimension-matching with $(\theta_{k^*}, u^*) = g_{k,k^*}(\theta_k, u)$ and $d_k + \dim(u) = d_{k^*} + \dim(u^*)$. The dimension matching ensures that the balance condition we used to prove the convergence of the Metropolis-Hastings algorithm in Chapter 11 continues to hold in this setting.

We present the reversible jump algorithm in general terms followed by an example. For the general description, let π_k denote the prior probability on model k, $p(\theta_k|M_k)$ the prior distribution for the parameters in model k, and $p(y|\theta_k, M_k)$ the sampling distribution under model k. Reversible jump Markov chain simulation generates samples from $p(k, \theta_k|y)$ using the following three steps at each iteration:

1. Starting in state (k, θ_k) (that is, model M_k with parameter vector θ_k), propose a new model M_{k^*} with probability J_{k,k^*} and generate an augmenting random variable u from proposal density $J(u|k, k^*, \theta_k)$.

2. Determine the proposed model's parameters, $(\theta_{k^*}, u^*) = g_{k,k^*}(\theta_k, u)$.

3. Define the ratio

$$r = \frac{p(y|\theta_{k^*}, M_{k^*})p(\theta_{k^*}|M_{k^*})\pi_{k^*}}{p(y|\theta_k, M_k)p(\theta_k|M_k)\pi_k} \frac{J_{k^*,k}J(u^*|k^*, k, \theta_{k^*})}{J_{k,k^*}J(u|k, k^*, \theta_k)} \left| \frac{\nabla g_{k,k^*}(\theta_k, u)}{\nabla(\theta_k, u)} \right|$$
$$(13.1)$$

and accept the new model with probability $\min(r, 1)$.

The resulting posterior draws provide inference about the posterior probability for each model as well as the parameters under that model.

Example. Testing a variance component in a logistic regression

The application of reversible jump sampling, especially the use of the auxiliary random variables u, is seen most easily through an example.

Consider a probit regression model for survival of turtles during a natural selection experiment. Let y_{ij} denote the binary response for turtle i in family j with $\Pr(y_{ij} = 1) = p_{ij}$ for $i = 1, \ldots, n_j$ and $j = 1, \ldots, J$. The weight x_{ij} of the turtle is known to affect survival probability, and it is likely that familial factors also play a role. This suggests the model, $p_{ij} = \Phi(\alpha_0 + \alpha_1 x_{ij} + b_j)$. It is natural to model the b_j's as exchangeable family effects, $b_j \sim N(0, \tau^2)$. The prior distribution $p(\alpha_0, \alpha_1, \tau)$ is not central to this discussion so we do not discuss it further here.

We suppose for the purpose of this example that it is desirable to test whether the variance component τ is needed by running a Markov chain that considers the model with and without the random effects parameters b_j. As emphasized in the model checking chapter, we much prefer to fit the model with the variance parameter and assess its importance by examining its posterior distribution. However, it might be of interest to consider the model that allows $\tau = 0$ as a discrete possibility, and we choose this example to illustrate the reversible jump algorithm.

Let M_0 denote the model with $\tau = 0$ (no variance component) and M_1 denote the model including the variance component. We use numerical integration (some techniques are described in Sections 13.2 and 13.3) to compute the marginal likelihood $p(y|\alpha_0, \alpha_1, \tau)$ for model M_1. Thus the random effects b_j are not part of the iterative simulation under model M_1. The reversible jump algorithm takes $\pi_0 = \pi_1 = 0.5$ and $J_{0,0} = J_{0,1} = J_{1,0} = J_{1,1} = 0.5$. At each step we either take a Metropolis step within the current model (with probability 0.5) or propose a jump to the other model. If we are in model 0 and are proposing a jump to model 1, then the auxiliary random variable is $u \sim J(u)$ (scaled inverse-χ^2 in this case) and we define the parameter vector for model 1 by setting $\tau^2 = u$ and leaving α_0 and α_1 as they were in the previous iteration. The ratio (13.1) is then

$$r = \frac{p(y|\alpha_0, \alpha_1, \tau^2, M_1)p(\tau^2)}{p(y|\alpha_0, \alpha_1, M_0)J(\tau^2)},$$

because the prior distributions on α and the models cancel, and the Jacobian of the transformation is 1. The candidate model is accepted with probability $\min(r, 1)$. There is no auxiliary random variable for going from model 1 to model 0. In that case we merely set $\tau = 0$, and the acceptance probability is the reciprocal of the above. In the example we chose $J(\tau^2)$ based on a pilot analysis of model M_1 (an inverse-χ^2 distribution matching the posterior mean and variance).

Regeneration and restarting distributions

Several methods have been developed for efficiently restarting Markov chain simulations. In Section 11.8 we have already considered the approach in which an initial 'pilot' run (for example, a few sequences of length 100 or 1000) is conducted and then the algorithm is altered based on that information. Other restarting approaches use more elaborate data analysis of the pilot simulation runs or special versions of the Metropolis-Hastings algorithm that can restart directly from the target distribution. Various methods are currently in the research stage and perform well for some specific models but have not yet been formulated for general problems.

Restarting methods involve analyzing the initial runs of many parallel chains before convergence in an attempt to estimate the stationary distribution of the process. That is, instead of waiting for approximate convergence after discarding initial iterations, this approach takes into account the transition from starting to target distribution. In simple situations with normal posterior distributions and few enough components in θ, standard time series analysis can provide estimates of the target distribution that can be used as improved starting distributions.

Regeneration refers to Markov chain simulation algorithms constructed to have regeneration points, or subspaces of θ, so that if a finite sequence starts and ends at a regeneration point, it can be considered as an exact (although dependent) sample from the target distribution. When a regeneration algorithm can be constructed, it eliminates the need for checking convergence; however difficulties can arise in practice since regeneration times can be long, and the convergence property only applies if the user is willing to wait until the regeneration point has been reached.

Perfect simulation refers to a class of Markov chain simulation algorithms that can be run in such a way that, after a certain number of iterations, the simulations are known to have exactly converged to the target distribution. Perfect simulation has been applied to simulation problems in statistical physics, and it is currently an active research topic to apply it to statistical models. As with restarting and regeneration, this approach shows promise but is not yet available for general implementations of Gibbs sampling and the Metropolis-Hastings algorithm.

13.2 Numerical integration

Numerical integration, also called 'quadrature,' can be applied at several points in a Bayesian analysis. For problems with only a few parameters, it may be possible to compute the posterior distribution of the parameters of interest directly by numerical integration. For example, in a problem with data from two independent binomial distributions, with independent beta prior distributions, inference about the difference between the two success probabilities can be computed directly via numerical integration (see Exercise 13.1). For more complicated problems, numerical integration can be used to obtain the marginal distribution of one or more parameters of interest (as in the reversible jump Markov chain Monte Carlo example of the previous section) or to compute the normalizing constant of a particular density. This section provides a brief summary of statistical procedures to approximately evaluate integrals. More techniques appear in Section 13.3 on importance sampling and Section 13.4 in the context of computing normalizing factors or integrals of unnormalized density functions. The bibliographic note at the end of this chapter suggests other sources.

Evaluating integrals and posterior expectations by direct simulation

We are already familiar with using direct or Markov chain simulation to summarize a posterior distribution. These simulations can directly be used to estimate integrals over this distribution. The posterior expectation of any function $h(\theta)$ is defined as an integral,

$$E(h(\theta)|y) = \int h(\theta)p(\theta|y)d\theta, \qquad (13.2)$$

where the integral has as many dimensions as θ. Conversely, we can express any integral over the space of θ as a posterior expectation by defining $h(\theta)$ appropriately. If we have posterior draws θ^l from $p(\theta|y)$, we can just estimate the integral by the sample average, $\frac{1}{L}\sum_{l=1}^{L} h(\theta^l)$. For any finite number of simulation draws, the accuracy of this estimate can be roughly gauged by the standard deviation of the $h(\theta^l)$ values. If it is not easy to draw from the posterior distribution, or if the $h(\theta^l)$ values are too variable (so that the sample average is too variable an estimate to be useful), more sophisticated methods of numerical integration are necessary.

Laplace's method for analytic approximation of integrals

Familiar numerical methods such as Simpson's rule and Gaussian quadrature (see the bibliographic note for references) can be difficult to apply in multivariate settings. Often an analytic approximation such as *Laplace's method*, based on the normal distribution, can be used to provide an adequate approximation to such integrals. In order to evaluate the integral (13.2) using Laplace's method, we first express the integrand in the form $\exp[\log(h(\theta)p(\theta|y))]$ and then expand $\log(h(\theta)p(\theta|y))$ as a function of θ in a quadratic Taylor approximation around its mode. The resulting approximation for $h(\theta)p(\theta|y)$ is proportional to a (multivariate) normal density in θ, and its integral is just

approximation of $E(h(\theta)|y)$: $\quad h(\theta_0)p(\theta_0|y)(2\pi)^{d/2}|-u''(\theta_0)|^{1/2},$

where d is the dimension of θ, $u(\theta) = \log(h(\theta)p(\theta|y))$, and θ_0 is the point at which $u(\theta)$ is maximized.

If $h(\theta)$ is a fairly smooth function, this approximation is often reasonable in practice, due to the approximate normality of the posterior distribution,

$p(\theta|y)$, for large sample sizes (recall Chapter 4). Because Laplace's method is based on normality, it is most effective for unimodal posterior densities, or when applied separately to each mode of a multimodal density. In fact, we have already used Laplace's method to compute the relative masses of the densities in a normal-mixture approximation to a multimodal density, formula (12.3) on page 315.

Laplace's method using unnormalized densities. If we are only able to compute the unnormalized density $q(\theta|y)$, we can apply Laplace's method separately to hq and q to evaluate the numerator and denominator, respectively, of the expression,

$$\mathrm{E}(h(\theta)|y) = \frac{\int h(\theta)q(\theta|y)d\theta}{\int q(\theta|y)d\theta}. \tag{13.3}$$

13.3 Importance sampling

Importance sampling is a method, related to rejection sampling (see Section 11.1) and a precursor to the Metropolis algorithm, for computing expectations using a random sample drawn from an approximation to the target distribution. Suppose we are interested in $\mathrm{E}(h(\theta)|y)$, but we cannot generate random draws from q directly and thus cannot evaluate (13.2) or (13.3) directly by simulation.

If $g(\theta)$ is a probability density from which we can generate random draws, then we can write,

$$\mathrm{E}(h(\theta|y)) = \frac{\int h(\theta)q(\theta|y)d\theta}{\int q(\theta|y)d\theta} = \frac{\int [h(\theta)q(\theta|y)/g(\theta)]\, g(\theta)d\theta}{\int [q(\theta|y)/g(\theta)]\, g(\theta)d\theta}, \tag{13.4}$$

which can be estimated using L draws $\theta^1, \ldots, \theta^L$ from $g(\theta)$ by the expression,

$$\frac{\frac{1}{L}\sum_{l=1}^{L} h(\theta^l)w(\theta^l)}{\frac{1}{L}\sum_{l=1}^{L} w(\theta^l)}, \tag{13.5}$$

where the factors

$$w(\theta^l) = \frac{q(\theta^l|y)}{g(\theta^l)}$$

are called *importance ratios* or *importance weights*. It is generally advisable to use the same set of random draws for both the numerator and denominator in order to reduce the sampling error in the estimate.

If $g(\theta)$ can be chosen such that hq/g is roughly constant, then fairly precise estimates of the integral can be obtained. Importance sampling is not a useful method if the importance ratios vary substantially. The worst possible scenario occurs when the importance ratios are small with high probability but with a low probability are very large, which happens, for example, if hq has very wide tails compared to g, as a function of θ.

Accuracy and efficiency of importance sampling estimates. In general, without some form of mathematical analysis of the exact and approximate densities, there is always the realistic possibility that we have missed some extremely large but rare importance weights. However, it may help to examine the distribution of sampled importance weights to discover possible problems. In practice, it is often useful to examine a histogram of the logarithms of the largest importance ratios: estimates will often be poor if the largest ratios are too large relative to the others. In contrast, we do not have to worry about the behavior of small importance ratios, because they have little influence on equation (13.4).

A good way to develop an understanding of importance sampling is to program simulations for simple examples, such as using a t_3 distribution as an approximation to the normal (good practice) or vice versa (bad practice); see Exercises 13.2 and 13.3. As with starting distributions in Markov chain sampling, the approximating distribution g in importance sampling should cover all the important regions of the target distribution.

Computing marginal posterior densities using importance sampling

Marginal posterior densities are often of interest in Bayesian computation. Chapter 12 describes an approach to approximating the marginal posterior distribution. Using the notation of that chapter, consider a partition of the parameter space, $\theta = (\gamma, \phi)$, with the corresponding factorization of the posterior density, $p(\theta|y) = p(\gamma|\phi, y)p(\phi|y)$. Suppose we have approximated the conditional posterior density of γ by $p_{\mathrm{approx}}(\gamma|\phi, y)$, as in Section 12.4. Equation (12.9) on page 325 gives an approximate formula for the marginal posterior density $p(\phi|y)$. We can correct the approximation with importance sampling, using draws of γ from each value of ϕ at which the approximation is computed. The problem of computing a marginal posterior density can be transformed into an application of importance sampling as follows. For any given value of ϕ, we can write the marginal posterior density as

$$
\begin{aligned}
p(\phi|y) &= \int p(\gamma, \phi|y)d\gamma \\
&= \int \frac{p(\gamma, \phi|y)}{p_{\mathrm{approx}}(\gamma|\phi, y)} p_{\mathrm{approx}}(\gamma|\phi, y)d\gamma \\
&= \mathrm{E}_{\mathrm{approx}}\left(\frac{p(\gamma, \phi|y)}{p_{\mathrm{approx}}(\gamma|\phi, y)} \right),
\end{aligned}
\tag{13.6}
$$

where $\mathrm{E}_{\mathrm{approx}}$ averages over γ using the conditional posterior distribution, $p_{\mathrm{approx}}(\gamma|\phi, y)$. The importance sampling estimate of $p(\phi|y)$ can be computed by simulating L values γ^l from the approximate conditional distribution, computing the joint density and approximate conditional density at each γ^l, and then averaging the L values of $p(\gamma^l, \phi|y)/p_{\mathrm{approx}}(\gamma^l|\phi, y)$. This procedure is then repeated for each point on the grid of ϕ. If the normalizing constant for the joint density $p(\gamma, \phi | y)$ is itself unknown, then more complicated computational procedures must be used.

Bridge and path sampling for numerical integration under a family of distributions

When computing integrals numerically, we typically want to evaluate several of them (for example, when computing the marginal posterior densities of different models) or to compute them for a range of values of a continuous parameter (as with continuous model expansion or when working with models whose normalizing factors depend on the parameters in the model and cannot be determined analytically, as discussed in Section 13.4).

In these settings with a family of normalizing factors to be computed, importance sampling can be generalized in a number of useful ways. Continuing our notation above, we let ϕ be the continuous or discrete parameter indexing the family of densities $p(\gamma | \phi, y)$. The numerical integration problem is to average over γ in this distribution, for each ϕ (or for a continuous range of values ϕ). In general, for these methods it is only necessary to compute the densities p up to arbitrary normalizing constants.

One approach is to perform importance sampling using the density at some central value, $p(\gamma | \phi_*, y)$, as the approximating distribution for the entire range of ϕ. This approach is convenient as it does not require the creation of a special p_{approx} but rather uses a distribution from a family that we already know how to handle (probably using Markov chain simulation).

If the distributions $p(\gamma | \phi, y)$ are far enough apart that no single ϕ_* can effectively cover all of them, we can move to *bridge sampling*, in which samples of γ are drawn from two distributions, $p(\gamma | \phi_0, y)$ and $p(\gamma | \phi_1, y)$. Here, ϕ_0 and ϕ_1 represent two points near the end of the space of ϕ (think of the family of distributions as a suspension bridge held up at two points). The bridge sampling estimate of the integral for any ϕ is a weighted average of the importance sampling estimates from the ϕ_0 and ϕ_1, with the weighting depending on ϕ. The weights can be computed using a simple iterative formula, and they behave in the expected way—for any ϕ, more of the weight is assigned to the closer of the bridging distributions.

Bridge sampling is a general idea that arises in many statistical contexts and can be further generalized to allow sampling from more than two points, which makes sense if the distributions vary widely over ϕ. In the limit in which samples are drawn from the entire continuous range of distributions $p(\gamma | \phi_0, y)$ indexed by ϕ, we can apply *path sampling*, a differential form of bridge sampling. In path sampling, samples (γ, ϕ) are drawn from a joint posterior distribution, and the derivative of the log posterior density, $d \log p(\gamma, \phi | y) / d\phi$, is computed at the simulated values and numerically integrated over ϕ to obtain an estimate of the log marginal density, $\log p(\phi | y)$, over a continuous range of values of ϕ. This simulation-based computation uses the identity,

$$\frac{d}{d\phi} \log p(\phi|y) = \mathrm{E}(U(\gamma, \phi, y)|\phi, y),$$

where $U(\gamma, \phi, y) = d\log p(\gamma, \phi|y)/d\phi$. Numerically integrating these values gives an estimate of $\log p(\phi|y)$ (up to an additive constant) as a function of ϕ.

Bridge and path sampling are strongly connected to simulated tempering (see page 337), which uses a similar structure of samples from an indexed family of distributions. Depending on the application, the marginal distribution of ϕ can be specified for computational efficiency or convenience (as with tempering) or estimated (as with the computations of marginal densities).

13.4 Computing normalizing factors

We next discuss the application of numerical integration to compute normalizing factors, a problem that arises in some complicated models that we largely do not discuss in this book. We include this section here to introduce the problem, which is currently an active area of research; see the bibliographic note for references on the topic.

Most of the models we present are based on combining standard classes of iid models for which the normalizing constants are known; for example, all the distributions in Appendix A have exactly known densities. Even the nonconjugate models we usually use are combinations of standard parts.

For standard models, we can compute $p(\theta)$ and $p(y|\theta)$ exactly, or up to unknown multiplicative constants, and the expression

$$p(\theta|y) \propto p(\theta)p(y|\theta)$$

has a single unknown normalizing constant—the denominator of Bayes' rule, $p(y)$. A similar result holds for a hierarchical model with data y, local parameters γ, and hyperparameters ϕ. The joint posterior density has the form

$$p(\gamma, \phi|y) \propto p(\phi)p(\gamma|\phi)p(y|\gamma, \phi),$$

which, once again, has only a single unknown normalizing constant. In each of these situations we can apply standard computational methods using the unnormalized density.

The problem of unknown normalizing factors

Unknown normalizing factors in the likelihood. A new and different problem arises when the sampling density $p(y|\theta)$ has an *unknown* normalizing factor that depends on θ. Such models often arise in problems that are specified conditionally, such as in spatial statistics. For a simple example, pretend we knew that the univariate normal density was of the form $p(y|\mu, \sigma) \propto \exp(-\frac{1}{2\sigma^2}(y - \mu)^2)$, but with the normalizing factor $1/(\sqrt{2\pi}\sigma)$ unknown. Performing our analysis as before without accounting for the factor of $1/\sigma$ would lead to an incorrect posterior distribution. (See Exercise 13.4 for a simple nontrivial example of an unnormalized density.)

In general we use the following notation:

$$p(y|\theta) = \frac{1}{z(\theta)} q(y|\theta),$$

where q is a generic notation for an unnormalized density, and

$$z(\theta) = \int q(y|\theta) dy \qquad (13.7)$$

is called the *normalizing factor* of the family of distributions—being a function of θ, we can no longer call it a 'constant'—and $q(y|\theta)$ is a family of unnormalized densities. We consider the situation in which $q(y|\theta)$ can be easily computed but $z(\theta)$ is unknown. Combining the density $p(y|\theta)$ with a prior density, $p(\theta)$, yields the posterior density

$$p(\theta|y) \propto p(\theta) \frac{1}{z(\theta)} q(y|\theta).$$

To perform posterior inference, one must determine $p(\theta|y)$, as a function of θ, up to an arbitrary multiplicative constant.

Of course, an unknown, but constant, normalizing factor in the prior density, $p(\theta)$, causes no problems because it does not depend on any model parameters.

Unknown normalizing factors in hierarchical models. An analogous situation arises in hierarchical models if the population distribution has an unknown normalizing factor that depends on the hyperparameters. Consider a model with data y, first-level parameters γ, and hyperparameters ϕ. For simplicity, assume that the likelihood, $p(y|\gamma)$, is known exactly, but the population distribution is only known up to an unnormalized density, $q(\gamma|\phi) = z(\phi)p(\gamma|\phi)$. The joint posterior density is then

$$p(\gamma, \phi|y) \propto p(\phi) \frac{1}{z(\phi)} q(\gamma|\phi) p(y|\gamma),$$

and the function $z(\phi)$ must be considered. If the likelihood, $p(y|\gamma)$, also has an unknown normalizing factor, it too must be considered in order to work with the posterior distribution.

Posterior computations involving an unknown normalizing factor

A basic computational strategy. If the integral (13.7), or the analogous expression for the hierarchical model, cannot be evaluated analytically, the numerical integration methods of Sections 13.2–13.3, especially bridge and path sampling, can be used. An additional difficulty is that one must evaluate (or estimate) the integral as a function of θ, or ϕ in the hierarchical case. The following basic strategy, combining analytic and simulation-based integration methods, can be used for computation with a posterior distribution containing unknown normalizing factors.

1. Obtain an analytic estimate of $z(\theta)$ using some approximate method, for example Laplace's method centered at a crude estimate of θ.

2. Use the analytic approximation to construct an approximate posterior distribution. Perform a preliminary analysis of the posterior distribution: finding joint modes using an approach such as conditional maximization or Newton's method, finding marginal modes, if appropriate, using a method such as EM, and constructing an approximate posterior distribution, as discussed in the previous chapter.

3. For more exact computation, evaluate $z(\theta)$ (see below) whenever the posterior density needs to be computed for a new value of θ. Computationally, this approach treats $z(\theta)$ as an approximately 'known' function that is just very expensive to compute.

Other strategies are possible in specific problems. If θ (or ϕ in the hierarchical version of the problem) is only one- or two-dimensional, it may be reasonable to compute $z(\theta)$ over a finite grid of values, and interpolate over the grid to obtain an estimate of $z(\theta)$ as a function of θ. It is still recommended to perform the approximate steps 1 and 2 above so as to get a rough idea of the location of the posterior distribution—for any given problem, $z(\theta)$ needs not be computed in regions of θ for which the posterior probability is essentially zero.

Computing the normalizing factor. The normalizing factor can be computed, for each value of θ, using any of the numerical integration approaches applied to (13.7). Applying approximation methods such as Laplace's is fairly straightforward, with the notation changed so that integration is over y, rather than θ, or changed appropriately to evaluate normalizing constants as a function of hyperparameters in a hierarchical model.

The importance sampling estimate is based on the identity

$$z(\theta) = \int \frac{q(y|\theta)}{g(y)} g(y) dy = \mathrm{E}_g \left(\frac{q(y|\theta)}{g(y)} \right),$$

where E_g averages over y under the approximate density $g(y)$. The estimate of $z(\theta)$ is $\frac{1}{L} \sum_{i=1}^{L} q(y^l|\theta)/g(y^l)$, based on simulations y^l from $g(y)$. Once again, the estimates for an unknown normalizing factor in a hierarchical model are analogous.

Some additional subtleties arise, however, when applying this method to evaluate $z(\theta)$ for many values of θ. First, we can use the same approximation function, $g(y)$, and in fact the same simulations, $y^1, \ldots, y^L$, to estimate $z(\theta)$ for different values of θ. Compared to performing a new simulation for each value of θ, using the same simulations saves computing time and increases accuracy (with the overall savings in time, we can simulate a larger number L of draws), but in general this can only be done in a local range of θ where the densities $q(y|\theta)$ are similar enough to each other that they can be approximated by the same density. Second, we have some freedom in our computations because the evaluation of $z(\theta)$ as a function of θ is required only up to a proportionality constant. Any arbitrary constant that does not depend on θ becomes part of the constant in the posterior density and does not affect posterior inference. Thus, the approximate density, $g(y)$, is *not* required to be normalized, as long as we use the same function $g(y)$ to approximate $q(y|\theta)$ for all values of θ, or if we know, or can estimate, the relative normalizing constants of the different approximation functions used in the problem.

13.5 Bibliographic note

The book by Liu (2002) covers a wide range of advanced simulation algorithms including those discussed in this chapter. The monograph by Neal (1993) also overviews many of these methods. Hybrid Monte Carlo was introduced by Duane et al. (1987) in the physics literature and Neal (1994) for statistics problems. Slice sampling is discussed by Neal (2003), and simulated tempering is discussed by Geyer and Thompson (1993) and Neal (1996b). Besag et al. (1995) and Higdon (1998) review several ideas based on auxiliary variables that have been useful in high-dimensional problems arising in genetics and spatial models. Reversible jump MCMC was introduced by Green (1995); see also Richardson and Green (1997) and Brooks, Giudici, and Roberts (2003) for more on trans-dimensional MCMC.

Restarting distributions are discussed by Liu and Rubin (2000). Mykland, Tierney, and Yu (1994) discuss regeneration for Markov chain simulation. Perfect simulation was introduced by Propp and Wilson (1996) and Fill (1998).

For further information on numerical integration techniques, see Press et al. (1986); a review of the application of these techniques to Bayesian inference is provided by Smith et al. (1985).

Laplace's method for integration was developed in a statistical context by Tierney and Kadane (1986), who demonstrated the accuracy of applying the method separately to the numerator and denominator of (13.3). Extensions and refinements were made by Kass, Tierney, and Kadane (1989) and Wong and Li (1992).

Importance sampling is a relatively old idea in numerical computation; for some early references, see Hammersley and Handscomb (1964). Geweke (1989) is a pre-Gibbs sampler discussion in the context of Bayesian computation; also see Wakefield, Gelfand, and Smith (1991). Chapters 2–4 of Liu (2002) discuss importance sampling in the context of Markov chain simulation algorithms.

Kong (1992) presents a measure of efficiency for importance sampling that may be useful for cases in which the importance ratios are fairly well behaved. Bridge sampling was introduced by Meng and Wong (1996). Gelman and Meng (1998) generalize from bridge sampling to path sampling and provide references to related work that has appeared in the statistical physics literature. Meng and Schilling (1996) provide an example in which several of these methods are applied to a problem in factor analysis. Kong et al. (2003) set up a general theoretical framework that includes importance sampling and bridge sampling as special cases.

The method of computing normalizing constants for statistical problems using importance sampling has been applied by Ott (1979) and others. Models with unknown normalizing functions arise often in spatial statistics; see, for example, Besag (1974) and Ripley (1981, 1988). Geyer (1991) and Geyer and Thompson (1992, 1993) develop the idea of estimating the normalizing function using simulations from the densities $p(y|\theta)$ and have applied these methods to problems in genetics. Pettitt, Friel, and Reeves (2003) use path sampling to estimate normalizing constants for a class of models in spatial statistics. Computing normalizing functions is an area of active current research, as more and more complicated Bayesian models are coming into use.

13.6 Exercises

1. Posterior computations for the binomial model: suppose $y_1 \sim \text{Bin}(n_1, p_1)$ is the number of successfully treated patients under an experimental new drug, and $y_2 \sim \text{Bin}(n_2, p_2)$ is the number of successfully treated patients under the standard treatment. Assume that y_1 and y_2 are independent and assume independent beta prior densities for the two probabilities of success. Let $n_1 = 10, y_1 = 6$, and $n_2 = 20, y_2 = 10$. Repeat the following for several different beta prior specifications.

   (a) Use simulation to find a 95% posterior interval for $p_1 - p_2$ and the posterior probability that $p_1 > p_2$.

   (b) Use numerical integration to estimate the posterior probability that $p_1 > p_2$.

2. Importance sampling and resampling when the importance weights are well behaved: consider a univariate posterior distribution, $p(\theta|y)$, which we wish to approximate and then calculate moments of, using importance sampling from an unnormalized density, $g(\theta)$. Suppose the posterior distribution is normal, and the approximation is t_3 with mode and curvature matched to the posterior density.

   (a) Draw $L = 100$ samples from the approximate density and compute the importance ratios. Plot a histogram of the log importance ratios.

   (b) Estimate $\text{E}(\theta|y)$ and $\text{var}(\theta|y)$ using importance sampling. Compare to the true values.

   (c) Repeat (a) and (b) for $L = 10,000$.

   (d) Using the samples obtained in (c) above, draw a subsample of $k = 100$ draws, without replacement, using importance resampling. Plot a histogram of the 100 importance-resampled draws of θ and calculate their mean and variance.

3. Importance sampling and resampling when the importance weights are too variable: repeat the previous exercise, but with a t_3 posterior distribution

and a normal approximation. Explain why the estimates of $\text{var}(\theta|y)$ are systematically too low. The importance resampling draws follow a distribution that falls between the approximate and target distributions.

4. Unknown normalizing functions: compute the normalizing factor for the following unnormalized sampling density,

$$p(y|\mu, A, B, C) \propto \exp\left[-\frac{1}{2}(A(y-\mu)^6 + B(y-\mu)^4 + C(y-\mu)^2)\right],$$

as a function of A, B, C. (Hint: it will help to integrate out analytically as many of the parameters as you can.)

Part IV: Regression Models

With modern computational tools at our disposal, we now turn to linear regression and generalized linear models, which are the statistical methods most commonly used to understand the relations between variables. Chapter 14 reviews classical regression from a Bayesian context, then Chapters 15 and 16 consider hierarchical linear regression and generalized linear models, along with the analysis of variance. Chapter 17 discusses robust alternatives to the standard normal, binomial, and Poisson distributions. We shall use regression models as building blocks for the more complex examples in Part V.

CHAPTER 14

Introduction to regression models

Linear regression is one of the most widely used statistical tools. This chapter introduces Bayesian model building and inference for normal linear models, focusing on the simple case of linear regression with uniform prior distributions. We apply the hierarchical modeling ideas of Chapter 5 in the context of linear regression in the next chapter. The analysis of the SAT coaching example in Chapter 5 is a special case of hierarchical linear modeling.

The topics of setting up and checking linear regression models are far too broad to be adequately covered in one or two chapters here. Rather than attempt a complete treatment in this book, we cover the standard forms of regression in enough detail to show how to set up the relevant Bayesian models and draw samples from posterior distributions for parameters θ and future observables $\tilde{y}$. For the simplest case of linear regression, we derive the basic results in Section 14.2 and discuss the major applied issues in Section 14.3 with an extended example of estimating the effect of incumbency in elections. In the later sections of this chapter, we discuss analytical and computational methods for more complicated models.

Throughout, we describe computations that build on the methods of standard least-squares regression where possible. In particular, we show how simple simulation methods can be used to (1) draw samples from posterior and predictive distributions, automatically incorporating uncertainty in the model parameters, and (2) draw samples for posterior predictive checks.

14.1 Introduction and notation

Many scientific studies concern relations among two or more observable quantities. A common question is: how does one quantity, y, vary as a function of another quantity or vector of quantities, x? In general, we are interested in the conditional distribution of y, given x, parameterized as $p(y|\theta, x)$, under a model in which the n observations $(x, y)_i$ are exchangeable.

Notation

The quantity of primary interest, y, is called the *response* or *outcome variable*; we assume here that it is continuous. The variables $x = (x_1, \ldots, x_k)$ are called *explanatory variables* and may be discrete or continuous. We sometimes choose a single variable x_j of primary interest and call it the *treatment* variable, labeling the other components of x as *control variables*. The distribution of

y given x is typically studied in the context of a set of *units* or experimental *subjects*, $i = 1, \ldots, n$, on which y_i and $x_{i1}, \ldots, x_{ik}$ are measured. Throughout, we use i to index units and j to index components of x. We use y to denote the vector of outcomes for the n subjects and X as the $n \times k$ matrix of explanatory variables.

The simplest and most widely used version of this model is the *normal linear model*, in which the distribution of y given X is a normal whose mean is a linear function of X:

$$\mathrm{E}(y_i | \beta, X) = \beta_1 x_{i1} + \ldots + \beta_k x_{ik},$$

for $i = 1, \ldots, n$. For many applications, the variable x_{i1} is fixed at 1, so that $\beta_1 x_{i1}$ is constant for all i.

In this chapter, we restrict our attention to the normal linear model; in Sections 14.2–14.5, we further restrict to the case of *ordinary linear regression* in which the conditional variances are equal, $\mathrm{var}(y_i | \theta, X) = \sigma^2$ for all i, and the observations y_i are conditionally independent given θ, X. The parameter vector is then $\theta = (\beta_1, \ldots, \beta_k, \sigma^2)$. We consider more complicated variance structures in Sections 14.6–14.7.

In the normal linear model framework, the key statistical modeling issues are (1) defining the variables x and y (possibly using transformations) so that the conditional expectation of y is reasonably linear as a function of the columns of X with approximately normal errors, and (2) setting up a prior distribution on the model parameters that accurately reflects substantive knowledge—a prior distribution that is sufficiently strong for the model parameters to be accurately estimated from the data at hand, yet not so strong as to dominate the data inappropriately. The statistical inference problem is to estimate the parameters θ, conditional on X and y.

Because we can choose as many variables X as we like and transform the X and y variables in any convenient way, the normal linear model is a remarkably flexible tool for quantifying relationships among variables. In Chapter 16, we discuss *generalized linear models*, which broaden the range of problems to which the linear predictor can be applied.

Formal Bayesian justification of conditional modeling

In reality, the numerical 'data' in a regression problem include both X and y. Thus, a full Bayesian model includes a distribution for X, $p(X|\psi)$, indexed by a parameter vector ψ, and thus involves a *joint* likelihood, $p(X, y|\psi, \theta)$, along with a prior distribution, $p(\psi, \theta)$. In the standard regression context, the distribution of X is assumed to provide no information about the conditional distribution of y given X; that is, the parameters θ determining $p(y|X, \theta)$ and the parameters ψ determining $p(X|\psi)$ are assumed independent in their prior distributions.

Thus, from a Bayesian perspective, the defining characteristic of a 'regression model' is that it ignores the information supplied by X about (ψ, θ). How can this be justified? Suppose that ψ and θ are independent in their prior distribution; that is, $p(\psi, \theta) = p(\psi)p(\theta)$. Then the posterior distribution factors as

$$p(\psi, \theta | X, y) = p(\psi | X)p(\theta | X, y),$$

and we can analyze the second factor by itself (that is, as a standard regression model), with no loss of information:

$$p(\theta | X, y) \propto p(\theta)p(y | X, \theta).$$

In the special case in which the explanatory variables X are chosen (for example, in a designed experiment), their probability $p(X)$ is known, and there are no parameters ψ.

The practical advantage of using such a regression model is that it is much easier to specify a realistic conditional distribution of one variable given k others than a joint distribution on all $k+1$ variables.

14.2 Bayesian analysis of the classical regression model

A large part of applied statistical analysis is based on linear regression techniques that can be thought of as Bayesian posterior inference based on a noninformative prior distribution for the parameters of the normal linear model. In Sections 14.2–14.5, we briefly outline, from a Bayesian perspective, the choices involved in setting up a regression model; these issues also apply to methods such as the analysis of variance and the analysis of covariance that can be considered special cases of linear regression. For more discussions of these issues from a non-Bayesian perspective, see any standard regression or econometrics textbook. Under a standard noninformative prior distribution, the Bayesian estimates and standard errors coincide with the classical results. However, even in the noninformative case, posterior simulations are useful for predictive inference and model checking.

Notation and basic model

In the simplest case, sometimes called *ordinary linear regression*, the observation errors are independent and have equal variance; in vector notation,

$$y | \beta, \sigma^2, X \sim \mathrm{N}(X\beta, \sigma^2 I), \tag{14.1}$$

where I is the $n \times n$ identity matrix. We discuss departures from the assumptions of the ordinary linear regression model—notably, the constant variance and zero conditional correlations in (14.1)—in Section 14.6.

The standard noninformative prior distribution

In the normal regression model, a convenient noninformative prior distribution is uniform on $(\beta, \log \sigma)$ or, equivalently,

$$p(\beta, \sigma^2 | X) \propto \sigma^{-2}. \tag{14.2}$$

When there are many data points and only a few parameters, the noninformative prior distribution is useful—it gives acceptable results (for reasons discussed in Chapter 4) and takes less effort than specifying prior knowledge in probabilistic form. For a small sample size or a large number of parameters, the likelihood is less sharply peaked, and so prior distributions and hierarchical models are more important.

We return to the issue of prior information for the normal linear model in Section 14.8 and Chapter 15.

The posterior distribution

As with the normal distribution with unknown mean and variance analyzed in Chapter 3, we determine first the posterior distribution for β, conditional on σ^2, and then the marginal posterior distribution for σ^2. That is, we factor the joint posterior distribution for β and σ^2 as $p(\beta, \sigma^2|y) = p(\beta|\sigma^2, y)p(\sigma^2|y)$. For notational convenience, we suppress the dependence on X here and in subsequent notation.

Conditional posterior distribution of β, given σ^2. The conditional posterior distribution of the (vector) parameter β, given σ^2, is the exponential of a quadratic form in β and hence is normal. We use the notation

$$\beta|\sigma^2, y \sim N(\hat{\beta}, V_\beta \sigma^2), \tag{14.3}$$

where, using the now familiar technique of completing the square (see Exercise 14.3), one finds,

$$\hat{\beta} = (X^T X)^{-1} X^T y, \tag{14.4}$$
$$V_\beta = (X^T X)^{-1}. \tag{14.5}$$

Marginal posterior distribution of σ^2. The marginal posterior distribution of σ^2 can be written as

$$p(\sigma^2|y) = \frac{p(\beta, \sigma^2|y)}{p(\beta|\sigma^2, y)},$$

which can be seen to have a scaled inverse-χ^2 form (see Exercise 14.4),

$$\sigma^2|y \sim \text{Inv-}\chi^2(n - k, s^2), \tag{14.6}$$

where

$$s^2 = \frac{1}{n-k}(y - X\hat{\beta})^T (y - X\hat{\beta}). \tag{14.7}$$

The marginal posterior distribution of $\beta|y$, averaging over σ^2, is multivariate t with $n - k$ degrees of freedom, but we rarely use this fact in practice when drawing inferences by simulation, since to characterize the joint posterior distribution we can draw simulations of σ^2 and then $\beta|\sigma^2$.

Comparison to classical regression estimates. The standard non-Bayesian estimates of β and σ^2 are $\hat{\beta}$ and s^2, respectively, as just defined. The classical standard error estimate for β is obtained by setting $\sigma^2 = s^2$ in (14.3).

Checking that the posterior distribution is proper. As for any analysis based on an improper prior distribution, it is important to check that the posterior distribution is proper (that is, has a finite integral). It turns out that $p(\beta, \sigma^2 | y)$ is proper as long as (1) $n > k$ and (2) the rank of X equals k (see Exercise 14.6). Statistically, in the absence of prior information, the first condition requires that there be at least as many data points as parameters, and the second condition requires that the columns of X be linearly independent (that is, no column can be expressed as a linear combination of the other columns) in order for all k coefficients of β to be uniquely identified by the data.

Sampling from the posterior distribution

It is easy to draw samples from the posterior distribution, $p(\beta, \sigma^2 | y)$, by (1) computing $\hat{\beta}$ from (14.4) and V_β from (14.5), (2) computing s^2 from (14.7), (3) drawing σ^2 from the scaled inverse-χ^2 distribution (14.6), and (4) drawing β from the multivariate normal distribution (14.3). In practice, $\hat{\beta}$ and V_β can be computed using standard linear regression software.

To be computationally efficient, the simulation can be set up as follows, using standard matrix computations. (See the bibliographic note at the end of the chapter for references on matrix factorization and least squares computation.) Computational efficiency is important for large datasets and also with the iterative methods required to estimate several variance parameters simultaneously, as described in Section 14.7.

1. Compute the *QR factorization*, $X = QR$, where Q is an $n \times k$ matrix of orthonormal columns and R is a $k \times k$ upper triangular matrix.

2. Compute R^{-1}—this is an easy task since R is upper triangular. R^{-1} is a Cholesky factor (that is, a matrix square root) of the covariance matrix V_β, since $R^{-1}(R^{-1})^T = (X^T X)^{-1} = V_\beta$.

3. Compute $\hat{\beta}$ by solving the linear system, $R\hat{\beta} = Q^T y$, using the fact that R is upper triangular.

Once σ^2 is simulated (using the random χ^2 draw), β can be easily simulated from the appropriate multivariate normal distribution using the Cholesky factorization and a program for generating independent standard normals (see Appendix A). The QR factorization of X is useful both for computing the mean of the posterior distribution and for simulating the random component in the posterior distribution of β.

For some large problems involving thousands of data points and hundreds of explanatory variables, even the QR decomposition can require substantial computer storage space and time, and methods such as conjugate gradient, stepwise ascent, and iterative simulation can be more effective.

The posterior predictive distribution for new data

Now suppose we apply the regression model to a new set of data, for which we have observed the matrix $\tilde{X}$ of explanatory variables, and we wish to predict the outcomes, $\tilde{y}$. If β and σ^2 were known exactly, the vector $\tilde{y}$ would have a normal distribution with mean $\tilde{X}\beta$ and variance matrix $\sigma^2 I$. Instead, our current knowledge of β and σ^2 is summarized by our posterior distribution.

Posterior predictive simulation. The posterior predictive distribution of unobserved data, $p(\tilde{y}|y)$, has two components of uncertainty: (1) the fundamental variability of the model, represented by the variance σ^2 in y not accounted for by $X\beta$, and (2) the posterior uncertainty in β and σ^2 due to the finite sample size of y. (Our notation continues to suppress the dependence on X and $\tilde{X}$.) As the sample size $n \to \infty$, the variance due to posterior uncertainty in (β, σ^2) decreases to zero, but the predictive uncertainty remains. To draw a random sample $\tilde{y}$ from its posterior predictive distribution, we first draw (β, σ^2) from their joint posterior distribution, then draw $\tilde{y} \sim \mathrm{N}(\tilde{X}\beta, \sigma^2 I)$.

Analytic form of the posterior predictive distribution. The normal linear model is simple enough that we can also determine the posterior predictive distribution analytically. Deriving the analytic form is not necessary—we can easily draw (β, σ^2) and then $\tilde{y}$, as described above—however, we can gain useful insight by studying the predictive uncertainty analytically.

We first of all consider the conditional posterior predictive distribution, $p(\tilde{y}|\sigma^2, y)$, then average over the posterior uncertainty in $\sigma^2|y$. Given σ^2, the future observation $\tilde{y}$ has a normal distribution (see Exercise 14.7), and we can derive its mean by averaging over β using (2.7):

$$
\begin{aligned}
\mathrm{E}(\tilde{y}|\sigma^2, y) &= \mathrm{E}(\mathrm{E}(\tilde{y}|\beta, \sigma^2, y)|\sigma^2, y) \\
&= \mathrm{E}(\tilde{X}\beta|\sigma^2, y) \\
&= \tilde{X}\hat{\beta},
\end{aligned}
$$

where the inner expectation averages over $\tilde{y}$, conditional on β, and the outer expectation averages over β. All expressions are conditional on σ^2 and y, and the conditioning on X and $\tilde{X}$ is implicit. Similarly, we can derive $\mathrm{var}(\tilde{y}|\sigma^2, y)$ using (2.8):

$$
\begin{aligned}
\mathrm{var}(\tilde{y}|\sigma^2, y) &= \mathrm{E}[\mathrm{var}(\tilde{y}|\beta, \sigma^2, y)|\sigma^2, y] + \mathrm{var}[\mathrm{E}(\tilde{y}|\beta, \sigma^2, y)|\sigma^2, y] \\
&= \mathrm{E}[\sigma^2 I|\sigma^2, y] + \mathrm{var}[\tilde{X}\beta|\sigma^2, y] \\
&= (I + \tilde{X}V_\beta \tilde{X}^T)\sigma^2. \qquad (14.8)
\end{aligned}
$$

This result makes sense: conditional on σ^2, the posterior predictive variance has two terms: $\sigma^2 I$, representing sampling variation, and $\tilde{X}V_\beta \tilde{X}^T \sigma^2$, due to uncertainty about β.

Given σ^2, the future observations have a normal distribution with mean $\tilde{X}\hat{\beta}$, which does not depend on σ^2, and variance (14.8) that is proportional to σ^2. To complete the determination of the posterior predictive distribution, we must average over the marginal posterior distribution of σ^2 in (14.6). The resulting posterior predictive distribution, $p(\tilde{y}|y)$, is multivariate t with center $\tilde{X}\hat{\beta}$, squared scale matrix $s^2(I + \tilde{X}V_\beta\tilde{X}^T)$, and $n - k$ degrees of freedom.

Prediction when $\tilde{X}$ is not completely observed. It is harder to predict $\tilde{y}$ if not all the explanatory variables in $\tilde{X}$ are known, because then the explanatory variables must themselves be modeled by a probability distribution. We return to the problem of multivariate missing data in Chapter 21.

Model checking and robustness

Checking the fit and robustness of a linear regression model is a well-developed topic in statistics. The standard methods such as examining plots of residuals against explanatory variables are useful and can be directly interpreted as posterior predictive checks. An advantage of the Bayesian approach is that we can compute, using simulation, the posterior predictive distribution for any data summary, so we do not need to put a lot of effort into estimating the sampling distributions of test statistics. For example, to assess the statistical and practical significance of patterns in a residual plot, we can obtain the posterior predictive distribution of an appropriate test statistic (for example, the correlation between the squared residuals and the fitted values), as we illustrate in Table 14.2 in the following example.

14.3 Example: estimating the advantage of incumbency in U.S. Congressional elections

We illustrate the Bayesian interpretation of linear regression with an example of constructing a regression model using substantive knowledge, computing its posterior distribution, interpreting the results, and checking the fit of the model to data.

Observers of legislative elections in the United States have often noted that incumbency—that is, being the current representative in a district—is an advantage for candidates. Political scientists are interested in the magnitude of the effect, formulated, for example, as 'what proportion of the vote is incumbency worth?' and 'how has incumbency advantage changed over the past few decades?' We shall use linear regression to study the advantage of incumbency in elections for the U.S. House of Representatives in the past century. In order to assess changes over time, we run a separate regression for each election year in our study. The results of each regression can be thought of as summary statistics for the effect of incumbency in each election year; these summary statistics can themselves be analyzed, formally by a hierarchical time series model or, as we do here, informally by examining graphs of the estimated effect and standard errors over time.

The electoral system for the U.S. House of Representatives is based on plurality vote in single-member districts: every two years, an election occurs in each of about 435 geographically distinct districts. Typically, about 100 to 150 of the district elections are uncontested; that is, one candidate runs unopposed.

Almost all the other district elections are contested by one candidate from each of the two major parties, the Democrats and the Republicans. In each district, one of the parties—the *incumbent party* in that district—currently holds the seat in the House, and the current officeholder—the *incumbent*—may or may not be a candidate for reelection. We are interested in the effect of the decision of the incumbent to run for reelection on the vote received by the incumbent party's candidate.

Units of analysis, outcome, and treatment variables

For each election year, the units of analysis in our study are the contested district elections. The outcome variable, y_i, is the proportion of the vote received by the incumbent party (see below) in district i, and we code the treatment variable as R_i, the decision of the incumbent to run for reelection:

$$R_i = \begin{cases} 1 & \text{if the incumbent officeholder runs for reelection} \\ 0 & \text{otherwise.} \end{cases}$$

If the incumbent does not run for reelection (that is, if $R_i = 0$), the district election is called an 'open seat.' Thus, an incumbency advantage would cause the value of y_i to increase if $R_i = 1$. We exclude from our analysis votes for third-party candidates and elections in which only one major-party candidate is running; see the bibliographic note for references discussing these and other data-preparation issues.

We analyze the data as an observational study in which we are interested in estimating the effect of the incumbency variable on the vote proportion. The estimand of primary interest in this study is the average effect of incumbency.

We define the theoretical incumbency advantage for a single legislative district election i as

$$\text{incumbency advantage}_i = y^I_{\text{complete } i} - y^O_{\text{complete } i}, \tag{14.9}$$

where

$y^I_{\text{complete } i}$ = proportion of the vote in district i received by the incumbent *legislator*, if he or she *runs for reelection* against major-party opposition in district i (thus, $y^I_{\text{complete } i}$ is unobserved in an open-seat election),

$y^O_{\text{complete } i}$ = proportion of the vote in district i received by the incumbent *party*, if the incumbent legislator *does not run* and the two major parties compete for the open seat in district i (thus, $y^O_{\text{complete } i}$ is unobserved if the incumbent runs for reelection).

The observed outcome, y_i, equals either $y^I_{\text{complete } i}$ or $y^O_{\text{complete } i}$, depending on whether the treatment variable equals 0 or 1.

We define the aggregate incumbency advantage for an entire legislature as the average of the incumbency advantages for all districts in a general election. This theoretical definition applies within a single election year and allows incumbency advantage to vary among districts. The definition (14.9) does *not* assume that the candidates under the two treatments are identical in all respects except for incumbency status.

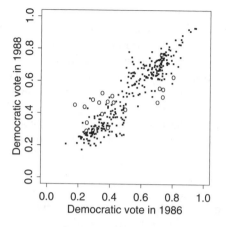

Figure 14.1 *U.S. Congressional elections: Democratic proportion of the vote in contested districts in 1986 and 1988. Dots and circles indicate district elections that in 1988 had incumbents running and open seats, respectively. Points on the left and right halves of the graph correspond to the incumbent party being Republican or Democratic, respectively.*

The incumbency advantage in a district depends on both $y^I_{\text{complete } i}$ and $y^O_{\text{complete } i}$; unfortunately, a real election in a single district will reveal only one of these. The problem can be thought of as estimating a causal effect from an observational study; as discussed in Sections 7.5 and 7.7, the average treatment effect can be estimated using a regression model, if we condition on enough control variables for the treatment assignment to be considered ignorable.

Setting up control variables so that data collection is approximately ignorable

It would be possible to estimate the incumbency advantage with only two columns in X: the treatment variable and the constant term (a column of ones). The regression would then be directly comparing the vote shares of incumbents to nonincumbents. The weakness of this simple model is that, since incumbency is not a randomly assigned experimental treatment, incumbents and nonincumbents no doubt differ in important ways other than incumbency. For example, suppose that incumbents tend to run for reelection in 'safe seats' that favor their party, but typically decline to run for reelection in 'marginal seats' that they have less chance of winning. If this were the case, then incumbents would be getting higher vote shares than non-incumbents, even in the absence of incumbency advantage. The resulting inference for incumbency advantage would be flawed because of serious nonignorability in the treatment assignment.

A partial solution is to include the vote for the incumbent party in the previous election as a control variable. Figure 14.1 shows the data for the

1988 election, using the 1986 election as a control variable. Each symbol in Figure 14.1 represents a district election; the dots represent districts in which an incumbent is running for reelection, and the open circles represent 'open seats' in which no incumbent is running. The vertical coordinate of each point is the share of the vote received by the Democratic party in the district, and the horizontal coordinate is the share of the vote received by the Democrats *in the previous election* in that district. The strong correlation confirms both the importance of using the previous election outcome as a control variable and the rough linear relation between the explanatory and outcome variables.

We include another control variable for the *incumbent party*: $P_i = 1$ if the Democrats control the seat and -1 if the Republicans control the seat before the election, whether or not the incumbent is running for reelection. This includes in the model a possible nationwide partisan swing; for example, a swing of 5% toward the Democrats would add 5% to y_i for districts i in which the Democrats are the incumbent party and -5% to y_i for districts i in which the Republicans are the incumbent party.

It might make sense to include other control variables that may affect the treatment and outcome variables, such as incumbency status in that district in the previous election, the outcome in the district two elections earlier, and so forth. At some point, of course, additional variables will add little to the ability of the regression model to predict y and will have essentially no influence on the coefficient for the treatment variable.

Implicit ignorability assumption

For our regression to estimate the actual effect of incumbency, we are implicitly assuming that the treatment assignments—the decision of an incumbent political party to run an incumbent or not in district i, and thus to set R_i equal to 0 or 1—conditional on the control variables, do not depend on any other variables that also affect the election outcome. For example, if incumbents who knew they would lose decided not to run for reelection, then the decision to run would depend on an *unobserved* outcome, the treatment assignment would be nonignorable, and the selection effect would have to be modeled.

In a separate analysis of these data, we have found that the probability an incumbent runs for reelection is approximately independent of the vote in that district in the previous election. If electoral vulnerability were a large factor in the decision to run, we would expect that incumbents with low victory margins in the previous election would be less likely to run for reelection. Since this does not occur, we believe the departures from ignorability are small. So, although the ignorability assumption is imperfect, we tentatively accept it for this analysis. The decision to accept ignorability is made based on subject-matter knowledge and additional data analysis.

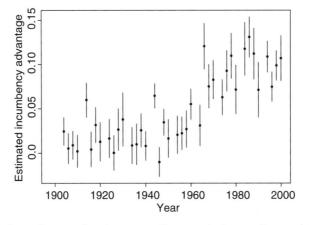

Figure 14.2 *Incumbency advantage over time: posterior median and 95% interval for each election year. The inference for each year is based on a separate regression. As an example, the results from the regression for 1988, based on the data in Figure 14.1, are displayed in Table 14.1.*

Transformations

Since the explanatory variable is restricted to lie between 0 and 1 (recall that we have excluded uncontested elections from our analysis), it would seem advisable to transform the data, perhaps using the logit transformation, before fitting a linear regression model. In practice, however, almost all the vote proportions y_i fall between 0.2 and 0.8, so the effect of such a transformation would be minor. We analyze the data on the original scale for simplicity in computation and interpretation of inferences.

Posterior inference

As an initial analysis, we estimate separate regressions for each of the election years in the twentieth century, excluding election years immediately following redrawing of the district boundaries, for it is difficult to define incumbency in those years. Posterior means and 95% posterior intervals (determined analytically from the appropriate t distributions) for the coefficient for incumbency are displayed for each election year in Figure 14.2. As usual, we can use posterior simulations of the regression coefficients to compute any quantity of interest. For example, the increase from the average incumbency advantage in the 1950s to the average advantage in the 1980s has a posterior mean of 0.050 with a central 95% posterior interval of $[0.035, 0.065]$, according to an estimate based on 1000 independent simulation draws.

The results displayed here are based on using the incumbent party and the previous election result as control variables (in addition to the constant term). Including more control variables to account for earlier incumbency and election

Variable	Posterior quantiles				
	2.5%	25%	median	75%	97.5%
Incumbency	0.084	0.103	0.114	0.124	0.144
Vote proportion in 1986	0.576	0.627	0.654	0.680	0.731
Incumbent party	−0.014	−0.009	−0.007	−0.004	0.001
Constant term	0.066	0.106	0.127	0.148	0.188
σ (residual sd)	0.061	0.064	0.066	0.068	0.071

Table 14.1 *Posterior inferences for parameters in the regression estimating the incumbency advantage in 1988. The outcome variable is the proportion of the two-party vote for the incumbent party in 1988, and only districts that were contested by both parties in both 1986 and 1988 were included. The parameter of interest is the coefficient of incumbency. Data are displayed in Figure 14.1. The posterior median and 95% interval for the coefficient of incumbency correspond to the bar for 1988 in Figure 14.2.*

results did not substantially change the inference about the coefficient of the treatment variable, and in addition made the analysis more difficult because of complications such as previous elections that were uncontested.

As an example of the results from a single regression, Table 14.1 displays posterior inferences for the coefficients β and residual standard deviation σ of the regression estimating the incumbency advantage in 1988, based on a non-informative uniform prior distribution on $(\beta, \log \sigma)$ and the data displayed in Figure 14.1. The posterior quantiles could have been computed by simulation, but for this simple case we computed them analytically from the posterior t and scaled inverse-χ^2 distributions.

Model checking and sensitivity analysis

The estimates in Figure 14.2 are plausible from a substantive standpoint and also add to our understanding of elections, in giving an estimate of the magnitude of the incumbency advantage ('what proportion of the vote is incumbency worth?') and evidence of a small positive incumbency advantage in the first half of the century.

In addition it is instructive, and crucial if we are to have any faith in our results, to check the fit of the model to our data.

Search for outliers. A careful look at Figure 14.1 suggests that the outcome variable is not normally distributed, even after controlling for its linear regression on the treatment and control variables. To examine the outliers further, we plot in Figure 14.3 the standardized residuals from the regressions from the 1980s. As in Figure 14.1, elections with incumbents and open seats are indicated by dots and circles, respectively. (We show data from just one decade because displaying the points from all the elections in our data would overwhelm the scatterplot.) For the standardized residual for the data point i,

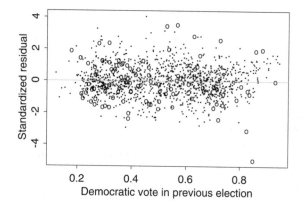

Figure 14.3 *Standardized residuals, $(y_{it} - (X_t\hat{\beta}_t)_i)/s_t$, from the incumbency advantage regressions for the 1980s, vs. Democratic vote in the previous election. (The subscript t indexes the election years.) Dots and circles indicate district elections with incumbents running and open seats, respectively.*

we just use $(y_i - X_i\hat{\beta})/s$, where s is the estimated standard deviation from equation (14.7). For simplicity, we still have a separate regression, and thus separate values of $\hat{\beta}$ and s, for each election year. If the normal linear model is correct, the standardized residuals should be approximately normally distributed, with a mean of 0 and standard deviation of about 1. Some of the standardized residuals in Figure 14.3 appear to be outliers by comparison to the normal distribution. (The residual standard deviations of the regressions are about 0.07—see Table 14.1, for example—and almost all of the vote shares lie between 0.2 and 0.8, so the fact that the vote shares are bounded between 0 and 1 is essentially irrelevant here.)

Posterior predictive checks. To perform a more formal check, we compute the proportion of district elections over a period of decades whose unstandardized residuals from the fitted regression models are greater than 0.20 in absolute value, a value that is roughly 3 estimated standard deviations away from zero. We use this unconventional measure of nonnormality partly to demonstrate the flexibility of the Bayesian approach to posterior predictive model checking and partly because the definition has an easily understood political meaning—the proportion of elections mispredicted by more than 20%. The results classified by incumbency status are displayed in the first column of Table 14.2.

As a comparison, we simulate the posterior predictive distribution of the test statistics under the model, as follows.

1. For $l = 1, \ldots, 1000$:

   (a) For each election year in the study:

      i. Draw (β, σ^2) from their posterior distribution.

      ii. Draw a hypothetical replication, y^{rep}, from the predictive distribution,

	Observed proportion of outliers	Posterior predictive dist. of proportion of outliers		
		2.5%	median	97.5%
Open seats	$41/1596 = 0.0257$	0.0013	0.0038	0.0069
Incumbent running	$84/10303 = 0.0082$	0.0028	0.0041	0.0054

Table 14.2 *Summary of district elections that are 'outliers' (defined as having absolute (unstandardized) residuals from the regression model of more than 0.2) for the incumbency advantage example. Elections are classified as open seats or incumbent running; for each category, the observed proportion of outliers is compared to the posterior predictive distribution. Both observed proportions are far higher than expected under the model.*

$y^{\text{rep}} \sim \text{N}(X\beta, \sigma^2 I)$, given the drawn values of (β, σ^2) and the existing vector X for that election year.

iii. Run a regression of y^{rep} on X and save the residuals.

(b) Combine the results from the individual election years to get the proportion of residuals that exceed 0.2 in absolute value, for elections with and without incumbents running.

2. Use the 1000 simulated values of the above test statistics to represent the posterior predictive distribution.

Quantiles from the posterior predictive distributions of the test statistics are shown as the final three columns of Table 14.2. The observed numbers of outliers in the two categories are about ten times and twice the values expected under the model and can clearly not be explained by chance.

One way to measure the seriousness of the outliers is to compute a test statistic measuring the effect on political predictions. For example, consider the number of party switches—districts in which the candidate of the incumbent party loses the election. In the actual data, $1498/11899 = 0.126$ of contested district elections result in a party switch in that district. By comparison, we can compute the posterior predictive distribution of the proportion of party switches using the same posterior predictive simulations as above; the median of the 1000 simulations is 0.136 with a central 95% posterior interval of $[0.130, 0.143]$. The posterior predictive simulations tell us that the observed proportion of party switches is lower than could be predicted under the model, but the difference is a minor one of overpredicting switches by about one percentage point.

Sensitivity of results to the normality assumption. For the purpose of estimating the average incumbency advantage, these outliers do not strongly affect our inference, so we ignore this failing of the model. We would not want to use this model for predictions of extreme outcomes, however, nor would we be surprised by occasional election outcomes far from the regression line. In political terms, the outliers may correspond to previously popular politicians who have been tainted by scandal—information that is not included in the explanatory variables of the model. It would be possible to extend our statistical modeling, for example by modeling the outliers with unequal variances for incumbents and open seats, along with a Student-t error term; but we would need a clear substantive motivation before adding in this additional complexity.

14.4 Goals of regression analysis

At least three common substantive goals are addressed by regression models: (1) understanding the behavior of y, given x (for example, 'what are the factors that aid a linear prediction of the Democratic share of the vote in a Congressional district?'); (2) predicting y, given x, for future observations ('what share of the vote might my local Congressmember receive in the next election?' or 'how many Democrats will be elected to Congress next year?'); (3) causal inference, or predicting how y would change if x were changed in a specified way ('what would be the effect on the number of Democrats elected to Congress next year, if a term limitations bill were enacted—so no incumbents would be allowed to run for reelection—compared to if no term limitations bill were enacted?').

The goal of understanding how y varies as a function of x is clear, given any particular regression model. We discuss the goals of prediction and causal inference in more detail, focusing on how the general concepts can be implemented in the form of probability models, posterior inference, and prediction.

Predicting y from x for new observations

Once its parameters have been estimated, a regression model can be used to predict future observations from units in which the explanatory variables $\tilde{X}$, but not the outcome variable $\tilde{y}$, have been observed. When using the regression model predictively, we are assuming that the old and new observations are exchangeable given the same values of x or, to put it another way, that the (vector) variable x contains all the information we have to distinguish the new observation from the old (this includes, for example, the assumption that time of observation is irrelevant if it is not encoded in x). For example, suppose we have fitted a regression model to 100 schoolchildren, with y for each child being the reading test score at the end of the second grade and x having two components: the student's test score at the beginning of the second grade and a constant term. Then we could use these predictors to construct a predictive distribution of $\tilde{y}$ for a new student for whom we have observed $\tilde{x}$. This prediction would be most trustworthy if all 101 students were randomly sampled from a common population (such as students in a particular school or students in the United States as a whole) and less reliable if the additional student differed in some known way from the first hundred—for example, if the first hundred came from a single school and the additional student attended a different school.

As with exchangeability in general, it is *not* required that the 101 students be 'identical' or even similar, just that all relevant knowledge about them be included in x. Of course, the more similar the units are, the lower the variance of the regression will be, but that is an issue of precision, not validity.

When the old and new observations are not exchangeable, the relevant information should be encoded in x. For example, if we are interested in learning about students from two different schools, we should include an indicator variable in the regression. The simplest approach is to replace the constant term in x by two indicator variables (that is, replacing the column of ones in X by two columns): x_A that equals 1 for students from school A and 0 for students from school B, and x_B that is the reverse. Now, if all the 100 students used to estimate the regression attended school A, then the data will provide no evidence about the coefficient for school B. The resulting predictive distribution of y for a new student in school B is highly dependent on our prior distribution, indicating our uncertainty in extrapolating to a new population. (With a noninformative uniform prior distribution and no data on the coefficient of school B, the improper posterior predictive distribution for a new observation in school B will have infinite variance. In a real study, it should be possible to construct some sort of weak prior distribution linking the coefficients in the two schools, which would lead to a posterior distribution with high variance.)

Causal inference

The modeling goals of describing the relationship between y and x and using the resulting model for prediction are straightforward applications of estimating $p(y|x)$. Causal inference is a far more subtle issue, however. When thinking about causal inference, as in the incumbency advantage example of Section 14.3, we think of the variable of interest as the *treatment* variable and the other explanatory variables as *control* variables or *covariates*. In epidemiology, closely related terms are *exposure* and *confounding* variables, respectively. The treatment variables represent attributes that are manipulated or at least potentially manipulable by the investigator (such as the doses of drugs applied to a patient in an experimental medical treatment), whereas the control variables measure other characteristics of the experimental unit or experimental environment, such as the patient's weight, measured *before* the treatment.

Do not control for post-treatment variables when estimating the causal effect. Some care must be taken when considering control variables for causal inference. For instance, in the incumbency advantage example, what if we were to include a control variable for campaign spending, perhaps the logarithm of the number of dollars spent by the incumbent candidate's party in the election? After all, campaign spending is generally believed to have a large effect on many election outcomes. For the purposes of predicting election outcomes, it would be a good idea to include campaign spending as an explanatory variable. For the purpose of estimating the incumbency advantage with a regression, however, total campaign spending should *not* be included, because much spending occurs after the decision of the incumbent whether to run for reelection. The causal effect of incumbency, as we have defined it, is not equivalent to the effect of the incumbent running versus not running, with total campaign spending held constant, since, if the incumbent runs, total campaign spending by the incumbent party will probably increase. Controlling for 'pre-decision' campaign spending would be legitimate, however. If we control for one of the effects of the treatment variable, our regression will probably underestimate the true causal effect. If we are interested in both predicting vote share and estimating the causal effect of incumbency, we could include both campaign spending and vote share as correlated outcome variables, using the methods described in Chapter 19.

14.5 Assembling the matrix of explanatory variables

The choice of which variables to include in a regression model depends on the purpose of the study. We discuss, from a Bayesian perspective, some issues that arise in classical regression. We have already discussed issues arising from the distinction between prediction and causal inference.

Identifiability and collinearity

The parameters in a classical regression cannot be uniquely estimated if there are more parameters than data points or, more generally, if the columns of the matrix X of explanatory variables are not linearly independent. In these cases, the data are said to be 'collinear,' and β cannot be uniquely estimated from the data alone, no matter how large the sample size. Think of a k-dimensional scatterplot of the n data points: if the n points fall in a lower-dimensional subspace (such as a two-dimensional plane sitting in a three-dimensional space), then the data are collinear ('coplanar' would be a better word). If the data are nearly collinear, falling very close to some lower-dimensional hyperplane, then they supply very little information about some linear combinations of the β's.

For example, consider the incumbency regression. If all the incumbents running for reelection had won with 70% of the vote in the previous election, and all the open seats occurred in districts in which the incumbent party won 60% of the vote in the previous election, then the three variables, incumbency, previous vote for the incumbent party, and the constant term, would be collinear (previous vote = $0.6 + 0.1 \cdot$ R), and it would be impossible to estimate the three coefficients from the data alone. To do better we need more data—or prior information—that do not fall along the plane. Now consider a hypothetical dataset that is nearly collinear: suppose all the candidates who had received more than 65% in the previous election always ran for reelection, whereas members who had won less than 65% always declined to run. The near-collinearity of the data means that the posterior variance of the regression coefficients would be high in this hypothetical case. Another problem in addition to increased uncertainty conditional on the regression model is that in practice the inferences would be highly *sensitive* to the model's assumption that $E(y|x, \theta)$ is linear in x.

Nonlinear relations

Once the variables have been selected, it often makes sense to transform them so that the relation between x and y is close to linear. Transformations such as logarithms and logits have been found useful in a variety of applications. One must take care, however: a transformation changes the interpretation of the regression coefficient to the change in transformed y per unit change in the transformed x variable. If it is thought that a variable x_j has a nonlinear effect on y, it is also possible to include more than one transformation of x_j in the regression—for example, including both x_j and x_j^2 allows an arbitrary quadratic function to be fitted.

When y is discrete, a generalized linear model can be appropriate; see Chapter 16.

Indicator variables

To include a categorical variable in a regression, a natural approach is to construct an 'indicator variable' for each category. This allows a separate effect for each level of the category, without assuming any ordering or other structure on the categories. When there are two categories, a simple 0/1 indicator works; when there are k categories, $k - 1$ indicators are required in addition to the constant term. It is often useful to incorporate the coefficients of indicator variables into hierarchical models, as we discuss in Chapter 15.

Categorical and continuous variables

If there is a natural order to the categories of a discrete variable, then it is often useful to treat the variable as if it were continuous. For example, the letter grades A, B, C, D might be coded as 4, 3, 2, 1. In epidemiology this approach is often referred to as trend analysis. It is also possible to create a categorical variable from a continuous variable by grouping the values in an appropriate way. This is sometimes helpful for examining and modeling departures from linearity in the relationship of y to a particular component of x.

Interactions

In the linear model as described, a change of one unit in x_i, with other predictors fixed, is associated with the same change in the mean response of y_i, given any fixed values of the other predictors. If the response to a unit change in x_i depends on what value another predictor x_j has been fixed at, then it is necessary to include *interaction* terms in the model. Generally the interaction can be allowed for by adding the cross-product term $(x_i - \overline{x}_i)(x_j - \overline{x}_j)$ as an additional predictor, although such terms may not be readily interpretable if both x_i and x_j are continuous (if such is the case, it is often preferable to categorize at least one of the two variables). For purposes of this exposition we treat these interactions just as we would any other explanatory variable: that is, create a new column of X and estimate a new element of β.

Controlling for irrelevant variables

In addition, we generally wish to include only variables that have some reasonable substantive connection to the problem under study. Often in regression there are a large number of potential control variables, some of which may appear to have predictive value. For one example, consider, as a possible control variable in the incumbency example, the number of letters in the last name of the incumbent party's candidate. On the face of it, this variable looks silly, but it might happen to have predictive value for our dataset. In almost all cases, length of name is determined before the decision to seek reelection, so it will not interfere with causal inference. However, if length of name has predictive value in the regression, we should try to understand what is happening, rather than blindly throwing it into the model. For example, if length of name is correlating with ethnic group, which has political implications, it would be better, if possible, to use the more substantively meaningful variable, ethnicity itself, in the final model.

Selecting the explanatory variables

Ideally, we should include all relevant information in a statistical model, which in the regression context means including in x all covariates that might possibly help predict y. The attempt to include relevant predictors is demanding in practice but is generally worth the effort. The possible loss of precision when including unimportant predictors is usually viewed as a relatively small price to pay for the general validity of predictions and inferences about estimands of interest.

In classical regression, there are direct disadvantages to increasing the number of explanatory variables. For one thing there is the restriction, when using the noninformative prior distribution, that $k < n$. In addition, using a large number of explanatory variables leaves little information available to obtain precise estimates of the variance. These problems, which are sometimes summarized by the label 'overfitting,' are of much less concern with reasonable prior distributions, such as those applied in hierarchical linear models, as we shall see in Chapter 15. We consider Bayesian approaches to handling a large number of predictor variables in Section 15.5.

14.6 Unequal variances and correlations

Departures from the ordinary linear regression model

The data distribution (14.1) makes several assumptions—linearity of the expectation, $E(y|\theta, X)$, as a function of X, normality of the error terms, and independent observations with equal variance—none of which is true in common practice. As always, the question is: does the gap between theory and practice adversely affect our inferences? Following the methods of Chapter 6, we can check posterior estimates and predictions to see how well the model fits aspects of the data of interest, and we can fit several competing models to the same dataset to see how sensitive the inferences are to various untestable model assumptions.

In the regression context, we could try to reduce *nonlinearity* of $E(y|\theta, X)$ by including explanatory variables, transformed appropriately. Nonlinearity can be diagnosed by plots of residuals against explanatory variables. If there is some concern about the proper relation between y and X, try several regressions: fitting y to various transformations of X. For example, in medicine, the degree of improvement of a patient may depend on the age of the patient. It is common for this relationship to be nonlinear, perhaps increasing for younger patients and then decreasing for older patients. Introduction of a nonlinear term such as a quadratic may improve the fit of the model.

Nonnormality is sometimes apparent for structural reasons—for example, when y only takes on discrete values—and can also be diagnosed by residual plots, as in Figure 14.3 for the incumbency example. If nonnormality is a serious concern, transformation is the first line of attack, or a generalized linear model may be appropriate; see Chapter 16 for details.

Unequal variances of the regression errors, $y_i - X_i\beta$, can sometimes be detected by plots of absolute residuals versus explanatory variables. Often the solution is just to include more explanatory variables. For example, a regression of agricultural yield in a number of geographic areas, on various factors concerning the soil and fertilizer, may appear to have unequal variances because local precipitation was not included as an explanatory variable. In other cases, unequal variances are a natural part of the data collection process. For example, if the sampling units are hospitals, and each data point is obtained as an average of patients within a hospital, then the variance is expected to be roughly proportional to the reciprocal of the sample size in each hospital. For another example, data collected by two different technicians of different proficiency will presumably exhibit unequal variances. We discuss models with more than one variance parameter in Section 14.7.

Correlations between $(y_i - X_i\beta)$ and $(y_j - X_j\beta)$ (conditional on X and the model parameters) can sometimes be detected by examining the correlation of residuals with respect to the possible cause of the problem. For example, if sampling units are collected sequentially in time, then the autocorrelation of the residual sequence should be examined and, if necessary, modeled. The usual linear model is not appropriate because the time information was not explicitly included in the model. If correlation exists in the data but is not included in the model, then the posterior inference about model parameters will typically be falsely precise, because the n sampling units will contain less information than n independent sampling units. In addition, predictions for future data will be inaccurate if they ignore correlation between relevant observed units. For example, heights of siblings remain correlated even after controlling for age and sex. But if we also control for genetically related variables such as the heights of the two parents (that is, add two more columns to the X matrix), the siblings' heights will have a lower (but in general still positive) correlation. This example suggests that it may often be possible to use more explanatory variables to reduce the complexity of the covariance matrix and thereby use more straightforward analyses. However, nonzero correlations will always be required when the values of y for particular subjects under study are related to each other through mechanisms other than systematic dependence on observed covariates. The way to proceed in problems with unknown correlations is not always clear; we return to this topic in Chapter 19.

Modeling unequal variances and correlated errors

Unequal variances and correlated errors can be included in the linear model by allowing a data covariance matrix Σ_y that is *not* necessarily proportional to the identity matrix:

$$y \sim \mathrm{N}(X\beta, \Sigma_y), \tag{14.10}$$

where Σ_y is a symmetric, positive definite $n \times n$ matrix. Modeling and estimation are, in general, more difficult than in ordinary linear regression. The data variance matrix Σ_y must be specified or given an informative prior distribution.

Bayesian regression with a known covariance matrix

We first consider the simplest case of unequal variances and correlated errors, where the variance matrix Σ_y is known. We continue to assume a noninformative uniform prior distribution for β. The computations in this section will be useful later as an intermediate step in iterative computations for more complicated models with informative prior information and hierarchical structures.

The posterior distribution. The posterior distribution is nearly identical to that for ordinary linear regression with known variance, if we apply a simple linear transformation to X and y. Let $\Sigma_y^{1/2}$ be a Cholesky factor (an upper triangular 'matrix square root') of Σ_y. Multiplying both sides of the regression equation (14.10) by $\Sigma_y^{-1/2}$ yields

$$\Sigma_y^{-1/2} y | \theta, X \sim \mathrm{N}(\Sigma_y^{-1/2} X\beta, I).$$

Drawing posterior simulations. To draw samples from the posterior distribution with data variance Σ_y known, first compute the Cholesky factor $\Sigma_y^{1/2}$ and its inverse, $\Sigma_y^{-1/2}$, then repeat the procedure of Section 14.2 with σ^2 fixed at 1, replacing y by $\Sigma_y^{-1/2}y$ and X by $\Sigma_y^{-1/2}X$. Algebraically, this means replacing (14.4) and (14.5) by

$$\hat{\beta} = (X^T\Sigma_y^{-1}X)^{-1}X^T\Sigma_y^{-1}y \tag{14.11}$$

$$V_\beta = (X^T\Sigma_y^{-1}X)^{-1}, \tag{14.12}$$

with the posterior distribution given by (14.3). As with (14.4) and (14.5), the matrix inversions never actually need to be computed since the Cholesky decomposition should be used for computation.

Prediction. Suppose we wish to sample from the posterior predictive distribution of $\tilde{n}$ new observations, $\tilde{y}$, given an $\tilde{n} \times k$ matrix of explanatory variables, $\tilde{X}$. Prediction with nonzero correlations is *more complicated* than in ordinary linear regression because we must specify the joint variance matrix for the old and new data. For example, consider a regression of children's heights in which the heights of children from the same family are correlated with a fixed known correlation. If we wish to predict the height of a new child whose brother is in the old data set, we should use that correlation in the predictions. We will use the following notation for the joint normal distribution of y and $\tilde{y}$, given the explanatory variables and the parameters of the regression model:

$$\left(\begin{array}{c} y \\ \tilde{y} \end{array} \bigg| X, \tilde{X}, \theta \right) \sim N\left(\left(\begin{array}{c} X\beta \\ \tilde{X}\beta \end{array} \right), \left(\begin{array}{cc} \Sigma_y & \Sigma_{y,\tilde{y}} \\ \Sigma_{\tilde{y},y} & \Sigma_{\tilde{y}} \end{array} \right) \right).$$

The covariance matrix for $(y, \tilde{y})$ must be symmetric and positive semidefinite.

Given (y, β, Σ_y), the heights of the new children have a joint normal posterior predictive distribution with mean and variance matrix,

$$E(\tilde{y}|y, \beta, \Sigma_y) = \tilde{X}\beta + \Sigma_{\tilde{y},y}\Sigma_y^{-1}(y - X\beta)$$

$$\text{var}(\tilde{y}|y, \beta, \Sigma_y) = \Sigma_{\tilde{y}} - \Sigma_{\tilde{y},y}\Sigma_y^{-1}\Sigma_{y,\tilde{y}},$$

which can be derived from the properties of the multivariate normal distribution; see (3.19) on page 86 and (A.1) on page 579.

Bayesian regression with unknown covariance matrix

We now derive the posterior distribution when the covariance matrix is unknown. As usual, we divide the problem of inference in two parts: posterior inference for β conditional on Σ_y—which we have just considered—and posterior inference for Σ_y. Assume that the prior distribution on β is uniform, with fixed scaling not depending on Σ_y; that is, $p(\beta|\Sigma_y) \propto 1$. Then the marginal posterior distribution of Σ_y can be written as

$$p(\Sigma_y|y) = \frac{p(\beta, \Sigma_y|y)}{p(\beta|\Sigma_y, y)}$$

$$\propto \frac{p(\Sigma_y)N(y|\beta, \Sigma_y)}{N(\beta|\hat{\beta}, V_\beta)}, \tag{14.13}$$

where $\hat{\beta}$ and V_β depend on Σ_y and are defined by (14.11) and (14.12). Expression (14.13) must hold for any β (since the left side of the equation does not depend on β at all); for convenience and computational stability we set $\beta = \hat{\beta}$:

$$p(\Sigma_y|y) \propto p(\Sigma_y)|\Sigma_y|^{-1/2}|V_\beta|^{1/2} \exp\left(-\frac{1}{2}(y - X\hat{\beta})^T\Sigma_y^{-1}(y - X\hat{\beta})\right). \quad (14.14)$$

Difficulties with the general parameterization. The density (14.14) is easy to compute but hard to draw samples from in general, because of the dependence of $\hat{\beta}$ and $|V_\beta|^{1/2}$ on Σ_y. Perhaps more important, setting up a prior distribution on Σ_y is, in general, a difficult task. In the next section we discuss several important special cases of parameterizations of the variance matrix, focusing on models with unequal variances but zero correlations. We discuss models with unequal correlations in Chapter 19.

14.7 Models for unequal variances

In this section, we consider several models for the data variance matrix Σ_y, including weighted linear regression and parametric models for variances.

Variance matrix known up to a scalar factor

We first consider the case in which we can write the data variance Σ_y as

$$\Sigma_y = Q_y\sigma^2, \quad (14.15)$$

where the matrix Q_y is known but the scale σ is unknown. As with ordinary linear regression, we start by assuming a noninformative prior distribution, $p(\beta, \sigma^2) \propto \sigma^{-2}$.

To draw samples from the posterior distribution of (β, σ^2), one must now compute $Q_y^{-1/2}$ and then repeat the procedure of Section 14.2 with σ^2 unknown, replacing y by $Q_y^{-1/2}y$ and X by $Q_y^{-1/2}X$. Algebraically, this means replacing (14.4) and (14.5) by,

$$\hat{\beta} = (X^TQ_y^{-1}X)^{-1}X^TQ_y^{-1}y \quad (14.16)$$

$$V_\beta = (X^TQ_y^{-1}X)^{-1}, \quad (14.17)$$

and (14.7) by

$$s^2 = \frac{1}{n-k}(y - X\hat{\beta})^TQ_y^{-1}(y - X\hat{\beta}), \quad (14.18)$$

with the normal and scaled inverse-χ^2 distributions (14.3) and (14.6). These formulas are just generalizations of the results for ordinary linear regression (for which $Q_y = I$). As with (14.4) and (14.5), the matrix inversions do not actually need to be computed since the Cholesky decomposition should be used for computation.

Weighted linear regression

If the data variance matrix is diagonal, then the above model is called *weighted linear regression*. We use the notation

$$\Sigma_{ii} = \sigma^2/w_i,$$

where $w_1, \ldots, w_n$ are known 'weights,' and σ^2 is an unknown variance parameter. Think of w as an additional X variable that does not affect the mean of y but does affect its variance. The procedure for weighted linear regression is the same as for the general matrix version, with the simplification that $Q_y^{-1} = \text{diag}(w_1, \ldots, w_n)$.

Parametric models for unequal variances

A more general family of models allows the variances to depend on the inverse weights in a nonlinear way:

$$\Sigma_{ii} = \sigma^2 v(w_i, \phi), \tag{14.19}$$

where ϕ is an unknown parameter and v is some function such as $v(w_i, \phi) = w_i^{-\phi}$. This parameterization has the feature of continuously changing from equal variances at $\phi = 0$ to variances proportional to $1/w_i$ when $\phi = 1$ and can thus be considered a generalization of weighted linear regression. (Another simple functional form with this feature is $v(w_i, \phi) = (1 - \phi) + \phi/w_i$.) A reasonable noninformative prior distribution for ϕ is uniform on $[0, 1]$.

Before analysis, the weights w_i are multiplied by a constant factor set so that their product is 1, so that inference for ϕ will not be affected by the scaling of the weights. If this adjustment is not done, the joint prior distribution of σ and ϕ must be set up to account for the scale of the weights (see the bibliographic note for more discussion and Exercise 6.13 for a similar example).

Drawing posterior simulations. The joint posterior distribution of all unknowns is

$$p(\beta, \sigma^2, \phi | y) \propto p(\phi) p(\beta, \sigma^2 | \phi) \prod_{i=1}^{n} \text{N}(y_i | (X\beta)_i, \sigma^2 v(w_i, \phi)). \tag{14.20}$$

Assuming the usual noninformative prior density,

$$p(\beta, \log \sigma | \phi) \propto 1,$$

we can factor the posterior distribution, and draw simulations, as follows.

First, given ϕ, the model is just weighted linear regression with

$$Q_y = \text{diag}(v(w_1, \phi), \ldots, v(w_n, \phi)).$$

To perform the computation, just replace X and y by $Q_y^{-1/2} X$ and $Q_y^{-1/2} y$, respectively, and follow the linear regression computations.

Second, the marginal posterior distribution of ϕ is

$$p(\phi|y) = \frac{p(\beta, \sigma^2, \phi|y)}{p(\beta, \sigma^2|\phi, y)}$$

$$\propto \frac{p(\phi)\sigma^{-2} \prod_{i=1}^{n} N(y_i|(X\beta)_i, \sigma^2 v(w_i, \phi))}{\text{Inv-}\chi^2(\sigma^2|n-k, s^2)N(\beta|\hat{\beta}, V_\beta \sigma^2)}.$$

This equation holds in general, so it must hold for any particular value of (β, σ^2). For analytical convenience and computational stability, we evaluate the expression at $(\hat{\beta}, s^2)$. Also, recall that the product of the weights is 1, so we now have

$$p(\phi|y) \propto p(\phi)|V_\beta|^{1/2}(s^2)^{-(n-k)/2}, \tag{14.21}$$

where $\hat{\beta}$, V_β, and s^2 depend on ϕ and are given by (14.16)–(14.18). Expression (14.21) is not a standard distribution, but for any specified set of weights w and functional form $v(w_i, \phi)$, it can be evaluated at a range of values of ϕ in $[0, 1]$ to yield a numerical posterior density, $p(\phi|y)$.

It is then easy to draw joint posterior simulations in the order ϕ, σ^2, β.

Estimating several unknown variance parameters

Thus far we have considered regression models with a single unknown variance parameter, σ, allowing us to model data with equal variances and zero correlations (in ordinary linear regression) or unequal variances with known variance ratios and correlations (in weighted linear regression). Models with several unknown variance parameters arise when different groups of observations have different variances and also, perhaps more importantly, when considering hierarchical regression models, as we discuss in Section 15.1.

When there is more than one unknown variance parameter, there is no general method to sample directly from the marginal posterior distribution of the variance parameters, and we generally must resort to iterative simulation techniques. In this section, we describe the model and computation for linear regression with unequal variances and a noninformative prior distribution; Section 14.8 and Chapter 15 explain how to modify the computation to account for prior information.

Many parametric models are possible for unequal variances; here, we discuss models in which the variance matrix Σ_y is known up to a diagonal vector of variances. If the variance matrix has unknown nondiagonal components, the computation is more difficult; we defer the problem to Chapter 19. We present the computation in the context of a specific example.

Example. Estimating the incumbency advantage (continued)
As an example, we consider the incumbency advantage problem described in Section 14.3. It is reasonable to suppose that Congressional elections with incumbents running for reelection are less variable than open-seat elections, because of the familiarity of the voters with the incumbent candidate. Love him or hate him, at least they know him, and so their votes should be predictable. Or maybe the other way around—when two unknowns are running, people vote based on

their political parties, while the incumbency advantage is a wild card that helps some politicians more than others. In any case, incumbent and open-seat elections seem quite different, and we might try modeling them with two different variance parameters.

Notation. Suppose the n observations can be divided into I batches—n_1 data points of type 1, n_2 of type 2, ..., n_I of type I—with each type of observation having its own variance parameter to be estimated, so that we must estimate I scalar variance parameters $\sigma_1, \sigma_2, \ldots, \sigma_I$, instead of just σ. This model is characterized by expression (14.10) with covariance matrix Σ_y that is diagonal with n_i instances of σ_i^2 for each $i = 1, \ldots, I$, and $\sum_{i=1}^{I} n_i = n$. In the incumbency example, $I = 2$ (incumbents and open seats).

A noninformative prior distribution. To derive the natural noninformative prior distribution for the variance components, think of the data as I separate experiments, each with its own unknown independent variance parameter. Multiplying the I separate noninformative prior distributions, along with a uniform prior distribution on the regression coefficients, yields $p(\beta, \Sigma_y) \propto \prod_{i=1}^{I} \sigma_i^{-2}$. The posterior distribution of the variance σ_i^2 is proper only if $n_i \geq 2$; if the ith batch comprises only one observation, its variance parameter σ_i^2 must have an informative prior specification.

For the incumbency example, there are enough observations in each year so that the results based on a noninformative prior distribution for $(\beta, \sigma_1^2, \sigma_2^2)$ may be acceptable. We follow our usual practice of performing a noninformative analysis and then examining the results to see where it might make sense to improve the model.

Posterior distribution. The joint posterior density of β and the variance parameters is

$$p(\beta, \sigma_1^2, \ldots, \sigma_I^2 | y) \propto \left(\prod_{i=1}^{I} \sigma_i^{-n_i - 2} \right) \exp\left(-\frac{1}{2}(y - X\beta)^T \Sigma_y^{-1}(y - X\beta) \right),$$

(14.22)

where the matrix Σ_y itself depends on the variance parameters σ_i^2. The conditional posterior distribution of β given the variance parameters is just the weighted linear regression result with known variance matrix, and the marginal posterior distribution of the variance parameters is given by

$$p(\Sigma_y | y) \propto p(\Sigma_y) |V_\beta|^{1/2} |\Sigma_y|^{-1/2} \exp\left(-\frac{1}{2}(y - X\hat{\beta})^T \Sigma_y^{-1}(y - X\hat{\beta}) \right)$$

(see (14.14)), with the understanding that Σ_y is parameterized by the vector $(\sigma_1^2, \ldots, \sigma_I^2)$, and with the prior density $p(\Sigma_y) \propto \prod_{i=1}^{I} \sigma_i^{-2}$.

A computational method based on EM and the Gibbs sampler. Here, we present one approach for drawing samples from the joint posterior distribution, $p(\beta, \sigma_1^2, \ldots, \sigma_I^2 | y)$.

1. Crudely estimate all the parameters based on a simplified regression model. For the incumbency advantage estimation, we can simply use the posterior

inferences for (β, σ^2) from the equal-variance regression in Section 14.3 and set $\sigma_1^2 = \sigma_2^2 = \sigma^2$.

2. Find the modes of the marginal posterior distribution of the variance parameters, $p(\sigma_1^2, \ldots, \sigma_I^2 | y)$, using the EM algorithm, averaging over β. Setting up the computations for the EM algorithm takes some thought, and it is tempting to skip this step and the next and go straight to the Gibbs sampler simulation, with starting points chosen from the crude parameter estimation. In simple problems for which the crude approximation is fairly close to the exact posterior density, this short-cut approach will work, but in more complicated problems the EM step is useful for finding modeling and programming mistakes and in obtaining starting points for the iterative simulation.

3. Create a t approximation to the marginal posterior distribution of the variance parameters, $p(\sigma_1^2, \ldots, \sigma_I^2 | y)$, centered at the modes. Simulate draws of the variance parameters from that approximation and draw a subset using importance resampling.

4. Simulate from the posterior distribution using the Gibbs sampler, alternately drawing β and $\sigma_1^2, \ldots, \sigma_I^2$. For starting points, use the draws obtained in the previous step.

We discuss the EM and Gibbs steps in some detail, as they are general computations that will also be useful for hierarchical linear and generalized linear models.

Finding the posterior mode of $\sigma_1^2, \ldots, \sigma_I^2$ using the EM algorithm. The marginal posterior distribution of the variance components is given by equation (14.14). In general, the mode of this distribution cannot be computed directly. It can, however, be computed using the EM algorithm, averaging over β, by iterating the following steps, which we describe below.

1. Determine the expected log posterior distribution, with β treated as a random variable, with expectations conditional on y and the last guess of Σ_y.

2. Set the new guess of Σ_y to maximize the expected joint log posterior density (to use the EM terminology).

The joint posterior density is given by (14.22). The only term in the log posterior density that depends on β, and thus the only term relevant for the E-step, is $(y - X\beta)^T \Sigma_y^{-1} (y - X\beta)$.

The two steps of the EM algorithm proceed as follows. First, compute $\hat{\beta}$ and V_β using weighted linear regression of y on X using the variance matrix Σ_y^{old}. Second, compute the following estimated variance matrix of y:

$$\widehat{S} = (y - X\hat{\beta})(y - X\hat{\beta})^T + X V_\beta X^T, \tag{14.23}$$

which is $E_{\text{old}}[(y - X\beta)(y - X\beta)^T]$, averaging over β with the conditional distribution $p(\beta | \Sigma_y, y)$. We can now write the expected joint log posterior distribution as

$$E_{\text{old}}(\log p(\beta, \Sigma_y | y)) = -\frac{1}{2} \left(\sum_{i=1}^{I} (n_i + 2) \log(\sigma_i^2) \right) - \frac{1}{2} \text{tr} \left(\Sigma_y^{-1} \widehat{S} \right) + \text{constant}.$$

The right side of this expression depends on Σ_y through $\log(\sigma_i^2)$ in the first term and Σ_y^{-1} in the second term. As a function of the variance parameters, the above is maximized by setting each σ_i^2 to the sum of the corresponding n_i elements on the diagonal of $\widehat{S}$, divided by $n_i + 2$. This new estimate of the parameters σ_i^2 is used to create Σ_y^{old} in the next iteration of the EM loop.

It is *not necessary* to compute the entire matrix $\widehat{S}$; like so many matrix expressions in regression, (14.23) is conceptually useful but should not generally be computed as written. Because we are assuming Σ_y is diagonal, we need only compute the diagonal elements of $\widehat{S}$, which entails computing the squared residuals, $(y_i - (X\hat{\beta})_i)^2$.

One way to check that the EM algorithm is programmed correctly is to compute (14.14); it should increase at each iteration. Once the posterior mode of the vector of variance parameters, $\widehat{\Sigma}_y$, has been found, it can be compared to the crude estimate as another check.

Obtaining starting points using the modal approximation. As described in Section 12.4, we can create an approximation to the marginal posterior density (14.14), centered at the posterior mode obtained from the EM algorithm. Numerically computing the second derivative matrix of (14.14) at the mode requires computing $|V_\beta|$ a few times, which is typically not too burdensome. Next we apply importance resampling using the multivariate t approximation as described in Section 12.2. Drawing 1000 samples from the multivariate t_4 and 10 resampling draws should suffice in this case.

Sampling from the posterior distribution using the Gibbs sampler. As it is naturally applied to the hierarchical normal model, the Gibbs sampler entails alternately simulating β and $\sigma_1^2, \ldots, \sigma_I^2$ from their conditional posterior distributions. Neither step is difficult.

To draw simulations from $p(\beta | \Sigma_y, y)$, the conditional posterior distribution, we just perform a weighted linear regression with known variance: compute $\hat{\beta}$ and the Cholesky factor of V_β and then sample from the multivariate normal distribution.

Simulating the variance parameters for the Gibbs sampler is also easy, because the conditional posterior distribution (just look at (14.22) and consider β to be a constant) factors into I independent scaled inverse-χ^2 densities, one for each variance parameter:

$$\sigma_i^2 \sim \text{Inv-}\chi^2(n_i, S_i/n_i), \tag{14.24}$$

where the 'sums of squares' S_i are defined (and computed) as follows. First compute

$$S = (y - X\beta)(y - X\beta)^T,$$

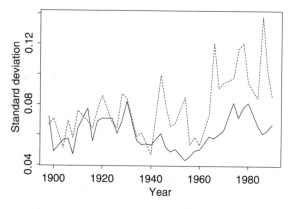

Figure 14.4 *Posterior medians of standard deviations σ_1 and σ_2 for elections with incumbents (solid line) and open-seat elections (dotted line), 1898–1990, estimated from the model with two variance components. (These years are slightly different from those in Figure 14.2 because this model was fit to a slightly different dataset.)*

which is similar to the first term of (14.23) from the EM algorithm, but now evaluated at β rather than $\hat{\beta}$. S_i is the sum of the corresponding n_i elements of the diagonal of S. To sample from the conditional posterior distribution, $p(\Sigma_y|\beta, y)$, just compute the sums S_i and simulate according to (14.24) using independent draws from $\chi^2_{n_i}$ distributions.

The convergence of parallel runs of the Gibbs sampler can be monitored using the methods described in Section 11.6, examining all scalar estimands of interest, including all components of β and $\sigma_1, \ldots, \sigma_I$.

Example. Estimating the incumbency advantage (continued)

Comparing variances by examining residuals. We can get a rough idea of what to expect by examining the average residual standard deviations for the two kinds of observations. In the post-1940 period, the residual standard deviations were, on average, higher for open seats than contested elections. As a result, a model with equal variances distorts estimates of β somewhat because it does not 'know' to treat open-seat outcomes as more variable than contested elections.

Fitting the regression model with two variance parameters. For each year, we fit the regression model, $y \sim N(X\beta, \Sigma_y)$, in which Σ_y is diagonal with Σ_{ii} equal to σ_1^2 for districts with incumbents running and σ_2^2 for open seats. We performed the computations using EM to find a marginal posterior mode of (σ_1^2, σ_2^2) and using the normal approximation about the mode as a starting point for the Gibbs sampler; in this case, the normal approximation is quite accurate (otherwise we might use a t approximation instead), and three independent sequences, each of length 100, were more than sufficient to bring the estimated potential scale reduction, $\widehat{R}$, to below 1.1 for all parameters.

The inference for the incumbency advantage coefficients over time is virtually unchanged from the equal-variance model, and so we do not bother to display the results. The inferences for the two variance components is displayed in Figure 14.4. The variance estimate for the ordinary linear regressions (not shown here) followed a pattern similar to the solid line in Figure 14.4, which makes sense considering that most of the elections have incumbents running (recall Figure 14.1). The most important difference between the two models is in the predictive distribution—the unequal-variance model realistically models the uncertainties in open-seat and incumbent elections since 1940. Further improvement could probably be made by pooling information across elections using a hierarchical model.

Even the new model is subject to criticism. For example, the spiky time series pattern of the estimates of σ_2 does not look right; more smoothness would seem appropriate, and the variability apparent in Figure 14.4 is due to the small number of open seats per year, especially in more recent years. A hierarchical time series model (which we do not cover in this book) would be an improvement on the current noninformative prior on the variances. In practice, of course, one can visually smooth the estimates in the graph, but for the purpose of estimating the size of the real changes in the variance, a hierarchical model would be preferable.

General models for unequal variances

All the models we have considered so far follow the general form,

$$\begin{aligned} \mathrm{E}(y|X,\theta) &= X\beta \\ \log(\mathrm{var}(y|X,\theta)) &= W\phi, \end{aligned}$$

where W is a specified matrix of parameters governing the variance (log weights for weighted linear regression, indicator variables for unequal variances in groups), and ϕ is the vector of variance parameters. In the general form of the model, iterative simulation methods including the Metropolis algorithm can be used to draw posterior simulations of (β, ϕ).

14.8 Including prior information

In some ways, prior information is already implicitly included in the classical regression; for example, we usually would not bother including a control variable if we thought it had no substantial predictive value. The meaning of the phrase 'substantial predictive value' depends on context, but can usually be made clear in applications; recall our discussion of the choice of variables to include in the incumbency advantage regressions.

In this section, we show how to add conjugate prior information about regression parameters to the classical regression model. This is of interest as a means of demonstrating a Bayesian approach to classical regression models and, more importantly, because the same ideas return in the hierarchical normal linear model in Chapter 15. We express all results in terms of expanded linear regressions.

Prior information about β

First, consider adding prior information about a single regression coefficient, β_j. Suppose we can express the prior information as a normal distribution:

$$\beta_j \sim N(\beta_{j0}, \sigma^2_{\beta_j}),$$

with β_{j0} and $\sigma^2_{\beta_j}$ known.

Interpreting the prior information on a regression parameter as an additional 'data point'

We can determine the posterior distribution easily by considering the prior information on β_j to be another 'data point' in the regression. The typical regression observation can be described as a normal random variable y with mean $x\beta$ and variance σ^2. The prior distribution for β_j can be seen to have the same form as a typical observation because the normal density for β_j is equivalent to a normal density for β_{j0} with mean β_j and variance $\sigma^2_{\beta_j}$:

$$p(\beta_j) \propto \frac{1}{\sigma_{\beta_j}} \exp\left(-\frac{(\beta_j - \beta_{j0})^2}{2\sigma^2_{\beta_j}}\right).$$

Thus considered as a function of β_j, the prior distribution can be viewed as an 'observation' β_{j0} with corresponding 'explanatory variables' equal to zero, except x_j, which equals 1, and a 'variance' of $\sigma^2_{\beta_j}$. To include the prior distribution in the regression, just append one data point, β_{j0}, to y, one row of all zeros except for a 1 in the jth column to X, and a diagonal element of $\sigma^2_{\beta_j}$ to the end of Σ_y (with zeros on the appended row and column of Σ_y away from the diagonal). Then apply the computational methods for a noninformative prior distribution: conditional on Σ_y, the posterior distribution for β can be obtained by weighted linear regression.

To understand this formulation, consider two extremes. In the limit of no prior information, corresponding to $\sigma^2_{\beta_j} \to \infty$, we are just adding a data point with infinite variance, which has no effect on inference. In the limit of perfect prior information, corresponding to $\sigma^2_{\beta_j} = 0$, we are adding a data point with zero variance, which has the effect of fixing β_j exactly at β_{j0}.

Interpreting prior information on several regression parameters as several additional 'data points'

Now consider prior information about the whole vector of parameters in β of the form

$$\beta \sim N(\beta_0, \Sigma_\beta).$$

We can treat the prior distribution as k prior 'data points,' and get correct posterior inference by weighted linear regression applied to 'observations' y_*, 'explanatory variables' X_*, and 'variance matrix' Σ_*, where

$$y_* = \begin{pmatrix} y \\ \beta_0 \end{pmatrix}, \quad X_* = \begin{pmatrix} X \\ I_k \end{pmatrix}, \quad \Sigma_* = \begin{pmatrix} \Sigma_y & 0 \\ 0 & \Sigma_\beta \end{pmatrix}. \tag{14.25}$$

If some of the components of β have infinite variance (that is, noninformative prior distributions), they should be excluded from these added 'prior' data points to avoid infinities in the matrix Σ_*. Or, if we are careful, we can just work with Σ_*^{-1} and its Cholesky factors and never explicitly compute Σ_*. The joint prior distribution for β is proper if all k components have proper prior distributions; that is, if Σ_β^{-1} has rank k.

This model is similar to the semi-conjugate normal model of Section 3.4. Computation conditional on Σ_* is straightforward using the methods described in Section 14.6 for regression with known covariance matrix. One can determine the marginal posterior density of Σ_* analytically and use the inverse-cdf method to draw simulations. More complicated versions of the model, such as arise in hierarchical regression, can be computed using the Gibbs sampler.

Prior information about variance parameters

In general, prior information is less important for the parameters describing the variance matrix than for the regression coefficients because σ^2 is generally of less substantive interest than β. Nonetheless, for completeness, we show how to include such prior information.

For the normal linear model (weighted or unweighted), the conjugate prior distribution for σ^2 is scaled inverse-χ^2, which we will parameterize as

$$\sigma^2 \sim \text{Inv-}\chi^2(n_0, \sigma_0^2).$$

Using the same trick as above, this prior distribution is equivalent to n_0 prior data points—n_0 need not be an integer—with sample variance σ_0^2. The corresponding posterior distribution is also scaled inverse-χ^2, and can be written as

$$\sigma^2 | y \sim \text{Inv-}\chi^2 \left(n_0 + n, \frac{n_0 \sigma_0^2 + n s^2}{n_0 + n} \right),$$

in place of (14.6). If prior information about β is also supplied, s^2 is replaced by the corresponding value from the regression of y_* on X_* and Σ_*, and n is replaced by the length of y_*. In either case, we can directly draw from the posterior distribution for σ^2. In the algebra and computations, one must replace n by $n + n_0$ everywhere and add terms $n_0 \sigma_0^2$ to every estimate of σ^2. If there are several variance parameters, then we can use independent conjugate prior distributions for them. Of course, with a number of unknown variance parameters it may be better to model them hierarchically.

Prior information in the form of inequality constraints on parameters

Another form of prior information that is easy to incorporate in our simulation framework is inequality constraints, such as $\beta_1 \geq 0$, or $\beta_2 \leq \beta_3 \leq \beta_4$. Constraints such as positivity or monotonicity occur in many problems. For example, recall the nonlinear bioassay example in Section 3.7 for which it might be sensible to constrain the slope to be nonnegative. Monotonicity can occur if the regression model includes an ordinal categorical variable that has been coded as indicator variables: we might wish to constrain the higher levels to have coefficients at least as large as the lower levels of the variable.

A simple and often effective way to handle constraints in a simulation is just to ignore them until the end. Follow all the above procedures to produce simulations of (β, σ) from the posterior distribution, then simply discard any simulations that violate the constraints. This simulation method is reasonably efficient unless the constraints eliminate a large portion of the unconstrained posterior distribution, in which case the data are tending to contradict the model.

14.9 Bibliographic note

Linear regression is described in detail in many textbooks, for example Weisberg (1985) and Neter (1996). For other presentations of Bayesian linear regression, see Zellner (1971) and Box and Tiao (1973). Fox (2002) presents linear regression using the statistical package R. The computations of linear regression, including the QR decomposition and more complicated methods that are more efficient for large problems, are described in many places; for example, Gill, Murray, and Wright (1981) and Golub and van Loan (1983).

The incumbency in elections example comes from Gelman and King (1990a); more recent work in this area includes Cox and Katz (1996) and Ansolabehere and Snyder (2002). Gelman and Huang (2003) frame the problem using a hierarchical model. The general framework used for causal inference, also discussed in Chapter 7 of this book, is presented in Rubin (1974b, 1978a). Bayesian approaches to analyzing regression residuals appear in Zellner (1975), Chaloner and Brant (1988), and Chaloner (1991).

A variety of parametric models for unequal variances have been used in Bayesian analyses; Boscardin and Gelman (1996) present some references and an example with forecasting Presidential elections (see Section 15.2). Bayesian models for correlations are discussed in Chapter 19 of this book.

14.10 Exercises

1. Analysis of radon measurements:

(a) Fit a linear regression to the logarithms of the radon measurements in Table 6.3, with indicator variables for the three counties and for whether a measurement was recorded on the first floor. Summarize your posterior inferences in nontechnical terms.

(b) Suppose another house is sampled at random from Blue Earth County. Sketch the posterior predictive distribution for its radon measurement and give a 95% predictive interval. Express the interval on the original (unlogged) scale. (Hint: you must consider the separate possibilities of basement or first-floor measurement.)

2. Causal inference using regression: discuss the difference between finite-population and superpopulation inference for the incumbency advantage example of Section 14.3.

3. Ordinary linear regression: derive the formulas for $\hat{\beta}$ and V_β in (14.4)–(14.5) for the posterior distribution of the regression parameters.

4. Ordinary linear regression: derive the formula for s^2 in (14.7) for the posterior distribution of the regression parameters.

5. Analysis of the milk production data: consider how to analyze data from the cow experiment described in Section 7.5. Specifically:

(a) Discuss the role of the treatment assignment mechanism for the appropriate analysis from a Bayesian perspective.

(b) Discuss why you would focus on finite-population inferences for these 50 cows or on superpopulation inferences for the hypothetical population of cows from which these 50 are conceptualized as a random sample. Either focus is legitimate, and a reasonable answer might be that one is easier than the other, but if this is your answer, say why it is true.

6. Ordinary linear regression: derive the conditions that the posterior distribution is proper in Section 14.2.

7. Posterior predictive distribution for ordinary linear regression: show that $p(\tilde{y}|\sigma, y)$ is a normal density. (Hint: first show that $p(\tilde{y}, \beta|\sigma, y)$ is the exponential of a quadratic form in $(\tilde{y}, \beta)$ and is thus is a normal density.)

8. Expression of prior information as additional data: give an algebraic proof of (14.25).

9. Straight-line fitting with variation in x and y: suppose we wish to model two variables, x and y, as having an underlying linear relation with added errors. That is, with data $(x, y)_i, i = 1, \ldots, n$, we model $\binom{x_i}{y_i} \sim N\left(\binom{u_i}{v_i}, \Sigma\right)$, and $v_i = a + bu_i$. The goal is to estimate the underlying regression parameters, (a, b).

(a) Assume that the values u_i follow a normal distribution with mean μ and variance τ^2. Write the likelihood of the data given the parameters; you can do this by integrating over $u_1, \ldots, u_n$ or by working with the multivariate normal distribution.

(b) Discuss reasonable noninformative prior distributions on (a, b).

Body mass (kg)	Body surface (cm^2)	Metabolic rate (kcal/day)
31.2	10750	1113
24.0	8805	982
19.8	7500	908
18.2	7662	842
9.6	5286	626
6.5	3724	430
3.2	2423	281

Table 14.3 *Data from the earliest study of metabolic rate and body surface area, measured on a set of dogs. From Schmidt-Nielsen (1984, p. 78).*

See Snedecor and Cochran (1989, p. 173) for an approximate solution, and Gull (1989b) for a Bayesian treatment of the problem of fitting a line with errors in both variables.

10. Example of straight-line fitting with variation in x and y: you will use the model developed in the previous exercise to analyze the data on body mass and metabolic rate of dogs in Table 14.3, assuming an approximate linear relation on the logarithmic scale. In this case, the errors in Σ are presumably caused primarily by failures in the model and variation among dogs rather than 'measurement error.'

 (a) Assume that log body mass and log metabolic rate have independent 'errors' of equal variance, σ^2. Assuming a noninformative prior distribution, compute posterior simulations of the parameters.

 (b) Summarize the posterior inference for b and explain the meaning of the result on the original, untransformed scale.

 (c) How does your inference for b change if you assume that the ratio of the variances is 2?

11. Example of straight-line fitting with variation in x_1, x_2, and y: adapt the model used in the previous exercise to the problem of estimating an underlying linear relation with two predictors. Estimate the relation of log metabolic rate to log body mass and log body surface area using the data in Table 14.3. How does the near-collinearity of the two predictor variables affect your inference?

CHAPTER 15

Hierarchical linear models

Hierarchical or multilevel regression modeling is an increasingly important tool in the analysis of complex data such as arise frequently in modern quantitative research. Hierarchical regression models are useful as soon as there is covariate information at different levels of variation. For example, in studying scholastic achievement we may have information about individual students (for example, family background), class-level information (characteristics of the teacher), and also information about the school (educational policy, type of neighborhood). Another situation in which hierarchical modeling arises naturally is in the analysis of data obtained by stratified or cluster sampling. A natural family of models is regression of y on indicator variables for strata or clusters, in addition to any measured covariates x. With cluster sampling, hierarchical modeling is in fact *necessary* in order to generalize to the unsampled clusters.

With covariates defined at multiple levels, the assumption of exchangeability of units or subjects at the lowest level breaks down, even after conditioning on covariate information. The simplest extension from a classical regression specification is to introduce as covariates a set of indicator variables for each of the higher-level units in the data—that is, for the classes in the educational example or for the strata or clusters in the sampling example. But this will in general dramatically increase the number of parameters in the model, and sensible estimation of these is only possible through further modeling, in the form of a population distribution. The latter may itself take a simple exchangeable or iid form, but it may also be reasonable to consider a further regression model at this second level, to allow for the effects of covariates defined at this level. In principle there is no limit to the number of levels of variation that can be handled in this way. Bayesian methods provide ready guidance on handling the estimation of unknown parameters, although computational complexities can be considerable, especially if one moves out of the realm of conjugate normal specifications. In this chapter we give a brief introduction to the very broad topic of hierarchical linear models, emphasizing the general principles used in handling normal models.

In fact, we have already considered a hierarchical linear model in Chapter 5: the problem of estimating several normal means can be considered as a special case of linear regression. In the notation of Section 5.5, the data points are y_j, $j = 1, \ldots, J$, and the regression coefficients are the school parameters θ_j. In this example, therefore, $n = J$; the number of 'data points' equals the number of explanatory variables. The X matrix is just the $J \times J$ identity matrix, and the individual observations have known variances σ_j^2. Section 5.5 discussed the flaws of no pooling and complete pooling of the data, $y_1, \ldots, y_J$, to estimate the parameters, θ_j. In the regression context, no pooling corresponds to a noninformative uniform prior distribution on the regression coefficients, and complete pooling corresponds to the J coefficients having a common prior distribution with zero variance. The favored hierarchical model corresponds to a prior distribution of the form $\beta \sim \mathrm{N}(\mu, \tau^2 I)$.

In the next section, we present notation and computations for the simple random-effects model, which constitutes the simplest version of the general hierarchical linear model (of which the SAT coaching example of Section 5.5 is in turn a very simple case). We illustrate in Section 15.2 with the example of forecasting U.S. Presidential elections, and then go on to the general form of the hierarchical linear model in Section 15.3. Throughout, we assume a normal linear regression model for the likelihood, $y \sim \mathrm{N}(X\beta, \Sigma_y)$, as in Chapter 14, and we label the regression coefficients as β_j, $j = 1, \ldots, J$.

15.1 Regression coefficients exchangeable in batches

We begin by considering hierarchical regression models in which groups of the regression coefficients are exchangeable and are modeled with normal population distributions. Each such group is called a batch of *random effects*.

Simple random-effects model

In the simplest form of the random-effects model, all of the regression coefficients contained in the vector β are exchangeable, and their population distribution can be expressed as

$$\beta \sim \mathrm{N}(1\alpha, \sigma_\beta^2 I), \tag{15.1}$$

where α and σ_β are unknown scalar parameters, and 1 is the $J \times 1$ vector of ones, $1 = (1, \ldots, 1)^T$. We use this vector-matrix notation to allow for easy generalization to regression models for the coefficients β, as we discuss in Section 15.3. Model (15.1) is equivalent to the hierarchical model we applied to the educational testing example of Section 5.5, using $(\beta, \alpha, \sigma_\beta)$ in place of (θ, μ, τ). As in that example, this general model includes, as special cases, unrelated β_j's $(\sigma_\beta = \infty)$ and all β_j's equal $(\sigma_\beta = 0)$.

It can be reasonable to start with a prior density that is uniform on α, σ_β, as we used in the educational testing example. As discussed in Section 5.4, we cannot assign a uniform prior distribution to $\log \sigma_\beta$ (the standard 'noninformative' prior distribution for variance parameters), because this leads to an improper posterior distribution with all its mass in the neighborhood of $\sigma_\beta = 0$. Another relatively noninformative prior distribution that is often used for σ_β^2 is scaled inverse-χ^2 (see Appendix A) with the degrees of freedom set to a low number such as 2. In applications one should be careful to ensure that posterior inferences are not sensitive to these choices (see the first example in

Section C.3 for an example); if they are, then greater care needs to be taken in specifying prior distributions that are defensible on substantive grounds. If there is little replication in the data at the level of variation corresponding to a particular variance parameter, then that parameter is generally not well estimated by the data and inferences may be sensitive to prior assumptions.

Intraclass correlation

There is a straightforward connection between the random-effects model just described and a simple model for within-group correlation. Suppose data $y_1, \ldots, y_n$ fall into J batches and have a multivariate normal distribution: $y \sim N(\alpha 1, \Sigma_y)$, with $var(y_i) = \eta^2$ for all i, and $cov(y_{i_1}, y_{i_2}) = \rho \eta^2$ if i_1 and i_2 are in the same batch and 0 otherwise. (We use the notation 1 for the $n \times 1$ vector of 1's.) If $\rho \geq 0$, this is equivalent to the model $y \sim N(X\beta, \sigma^2 I)$, where X is a $n \times J$ matrix of indicator variables with $X_{ij} = 1$ if unit i is in batch j and 0 otherwise, and β has the random-effects population distribution (15.1). The equivalence of the models occurs when $\eta^2 = \sigma^2 + \sigma_\beta^2$ and $\rho = \sigma_\beta^2 / (\sigma^2 + \sigma_\beta^2)$, as can be seen by deriving the marginal distribution of y, averaging over β. More generally, positive intraclass correlation in a linear regression can be subsumed into a random-effects model by augmenting the regression with J indicator variables whose coefficients have the population distribution (15.1).

Mixed-effects model

An important variation on the simple random-effects model is the 'mixed-effects model,' in which the first J_1 components of β are assigned independent improper prior distributions, and the remaining $J_2 = J - J_1$ components are exchangeable with common mean α and standard deviation σ_β. The first J_1 components, which are implicitly modeled as exchangeable with infinite prior variance, are sometimes labeled *fixed effects* in this model.*

A simple example is the hierarchical normal model considered in Chapter 5; the random-effects model with the school means normally distributed and a uniform prior density assumed for their mean α is equivalent to a mixed-effects model with a single constant 'fixed effect' and a set of random effects with mean 0.

Several sets of random effects

To generalize, allow the J components of β to be divided into K clusters of random effects, with cluster k having population mean α_k and standard

---

* The terms 'fixed' and 'random' come from the non-Bayesian statistical tradition and are somewhat confusing in a Bayesian context where all unknown parameters are treated as 'random.' The non-Bayesian view considers fixed effects to be fixed unknown quantities, but the standard procedures proposed to estimate these parameters, based on specified repeated-sampling properties, happen to be equivalent to the Bayesian posterior inference under a noninformative (uniform) prior distribution.

deviation $\sigma_{\beta k}$. A mixed-effects model is obtained by setting the variance to ∞ for one of the clusters of random effects. We return to these models in discussing the analysis of variance in Section 15.6.

Exchangeability

The essential feature of random effects models is that exchangeability of the units of analysis is achieved by conditioning on indicator variables that represent groupings in the population. The random effect parameters allow each subgroup to have a different mean outcome level, and averaging over these parameters to a marginal distribution for y induces a correlation between outcomes observed on units in the same subgroup (just as in the simple intraclass correlation model described above).

Computation

The easiest methods for computing the posterior distribution in hierarchical normal models use the Gibbs sampler, sometimes with linear transformations to speed convergence. We defer the general notation and details to Section 15.3.

15.2 Example: forecasting U.S. Presidential elections

We illustrate hierarchical linear modeling with an example in which a hierarchical model is useful for obtaining realistic forecasts. Following standard practice, we begin by fitting a nonhierarchical linear regression with a noninformative prior distribution but find that the simple model does not provide an adequate fit. Accordingly we expand the model hierarchically, including random effects to model variation at a second level in the data.

Political scientists in the U.S. have been interested in the idea that national elections are highly predictable, in the sense that one can accurately forecast election results using information publicly available several months before the election. In recent years, several different linear regression forecasts have been suggested for U.S. Presidential elections. In this chapter, we present a hierarchical linear model that was estimated from the elections through 1988 and used to forecast the 1992 election.

Unit of analysis and outcome variable

The units of analysis are results in each state from each of the 11 Presidential elections from 1948 through 1988. The outcome variable of the regression is the Democratic party candidate's share of the two-party vote for President in that state and year. For convenience and to avoid tangential issues, we discard the District of Columbia (in which the Democrats have received over 80% in every Presidential election) and states with third-party victories from our model, leaving us with 511 units from the 11 elections considered.

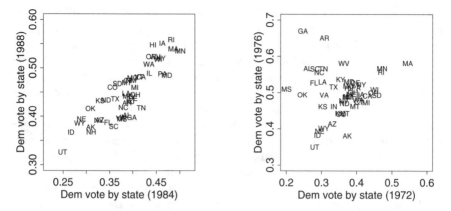

Figure 15.1 *(a) Democratic share of the two-party vote for President, for each state, in 1984 and 1988. (b) Democratic share of the two-party vote for President, for each state, in 1972 and 1976.*

Preliminary graphical analysis

Figure 15.1a suggests that the Presidential vote may be strongly predictable from one election to the next. The fifty points on the figure represent the states of the U.S. (indicated by their two-letter abbreviations); the x and y coordinates of each point show the Democratic party's share of the vote in the Presidential elections of 1984 and 1988, respectively. The points fall close to a straight line, indicating that a linear model predicting y from x is reasonable and relatively precise. This relationship is not always so strong, however; consider Figure 15.1b, which displays the votes by states in 1972 and 1976—the relation is not close to linear. Nevertheless, a careful look at the second graph reveals some patterns: the greatest outlying point, on the upper left, is Georgia ('GA'), the home state of Jimmy Carter, the Democratic candidate in 1976. The other outlying points, all on the upper left side of the 45° line, are other states in the South, Carter's home region. It appears that it may be possible to create a good linear fit by including other explanatory variables in addition to the Democratic share of the vote in the previous election, such as indicator variables for the candidates' home states and home regions. (For political analysis, the United States is typically divided into four *regions*: Northeast, South, Midwest, and West, with each region containing ten or more states.)

Fitting a preliminary, nonhierarchical, regression model

Political trends such as partisan shifts can occur nationwide, at the level of regions of the country, or in individual states; to capture these three levels of variation, we include three kinds of explanatory variables in the regression. The nationwide variables—which are the same for every state in a given election year—include national measures of the popularity of the candidates, the popularity of the incumbent President (who may or may not be running for reelection), and measures of the condition of the economy in the past two years. Regional variables include home-region indicators for the candidates and various adjustments for past elections in which regional voting had been important. Statewide variables include the Democratic party's share of the state's vote in recent Presidential elections, measures of the state's economy and political ideology, and home-state indicators. The explanatory variables used in the model are listed in Table 15.1. With 511 observations, a large number of state and regional variables can reasonably be fitted in a model of election outcome, assuming there are smooth patterns of dependence on these covariates across states and regions. Fewer relationships with national variables can be estimated, however, since for this purpose there are essentially only 11 data points—the national elections.

For a first analysis, we fit a regression model including all the variables in Table 15.1 to the data up to 1988, using ordinary linear regression (with noninformative prior distributions), as described in Chapter 14. We could then draw simulations from the posterior distribution of the regression parameters and use each of these simulations, applied to the national, regional, and state explanatory variables for 1992, to create a random simulation of the vector of election outcomes for the fifty states in 1992. These simulated results could be used to estimate the probability that each candidate would win the national election and each state election, the expected number of states each candidate would win, and other predictive quantities.

Checking the preliminary regression model

The ordinary linear regression model ignores the year-by-year structure of the data, treating them as 511 independent observations, rather than 11 sets of roughly 50 *related* observations each. Substantively, the feature of these data that such a model misses is that partisan support across the states does not vary independently: if, for example, the Democratic candidate for President receives a higher-than-expected vote share in Massachusetts in a particular year, we would expect him also to perform better than expected in Utah in that year. In other words, because of the known grouping into years, the assumption of exchangeability among the 511 observations does not make sense, *even after controlling for the explanatory variables*.

An important use of the model is to forecast the nationwide outcome of the Presidential election. One way of assessing the significance of possible failures in the model is to use the model-checking approach of Chapter 6. To check whether correlation of the observations from the same election has a substantial effect on nationwide forecasts, we create a test variable that reflects the average precision of the model in predicting the national result—the square root of the average of the squared nationwide realized residuals for the 11 general elections in the dataset. (Each nationwide realized residual is the average

Description of variable	Sample quantiles		
	min	median	max
Nationwide variables:			
Support for Dem. candidate in Sept. poll	0.37	0.46	0.69
(Presidential approval in July poll) $\times$ Inc	-0.69	-0.47	0.74
(Presidential approval in July poll) $\times$ Presinc	-0.69	0	0.74
(2nd quarter GNP growth) $\times$ Inc	-0.024	-0.005	0.018
Statewide variables:			
Dem. share of state vote in last election	-0.23	-0.02	0.41
Dem. share of state vote two elections ago	-0.48	-0.02	0.41
Home states of Presidential candidates	-1	0	1
Home states of Vice-Presidential candidates	-1	0	1
Democratic majority in the state legislature	-0.49	0.07	0.50
(State economic growth in past year) $\times$ Inc	-0.22	-0.00	0.26
Measure of state ideology	-0.78	-0.02	0.69
Ideological compatibility with candidates	-0.32	-0.05	0.32
Proportion Catholic in 1960 (compared to U.S. avg.)	-0.21	0	0.38
Regional/subregional variables:			
South	0	0	1
(South in 1964) $\times$ (-1)	-1	0	0
(Deep South in 1964) $\times$ (-1)	-1	0	0
New England in 1964	0	0	1
North Central in 1972	0	0	1
(West in 1976) $\times$ (-1)	-1	0	0

Table 15.1 *Variables used for forecasting U.S. Presidential elections. Sample minima, medians, and maxima come from the 511 data points. All variables are signed so that an increase in a variable would be expected to increase the Democratic share of the vote in a state. 'Inc' is defined to be +1 or −1 depending on whether the incumbent President is a Democrat or a Republican. 'Presinc' equals Inc if the incumbent President is running for reelection and 0 otherwise. 'Dem. share of state vote' in last election and two elections ago are coded as deviations from the corresponding national votes, to allow for a better approximation to prior independence among the regression coefficients. 'Proportion Catholic' is the deviation from the average proportion in 1960, the only year in which a Catholic ran for President. See Gelman and King (1993) and Boscardin and Gelman (1996) for details on the other variables, including a discussion of the regional/subregional variables. When fitting the hierarchical model, we also included indicators for years and regions within years.*

of $(y_i - (X\beta)_i)$ for the roughly 50 observations in that election year.) This test variable should be sensitive to positive correlations of outcomes in each year. We then compare the values of the test variable $T(y, \beta)$ from the posterior simulations of β to the hypothetical replicated values under the model, $T(y^{\text{rep}}, \beta)$. The results are displayed in Figure 15.2. As can be seen in the figure, the observed variation in national election results is larger than would

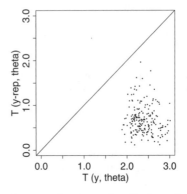

Figure 15.2 *Scatterplot showing the joint distribution of simulation draws of the realized test quantity, $T(y, \beta)$—the square root of the average of the 11 squared nationwide residuals—and its hypothetical replication, $T(y^{\text{rep}}, \beta)$, under the nonhierarchical model for the election forecasting example. The 200 simulated points are far below the $45°$ line, which means that the realized test quantity is much higher than predicted under the model.*

be expected from the model. The practical consequence of the failure of the model is that its forecasts of national election results are falsely precise.

Extending to a random-effects model

We can improve the regression model by adding an additional explanatory variable for each year to serve as an indicator for nationwide partisan shifts unaccounted for by the other national variables; this adds 11 new components of β corresponding to the 11 election years in the data. The additional columns in the X matrix are indicator vectors of zeros and ones indicating which data points correspond to which year. After controlling for the national variables already in the model, we fit an exchangeable model for the election-year variables, which in the linear model means having a common mean and variance. Since a constant term is already in the model, we can set the mean of the population distribution of the year effects to zero (recall Section 15.1). By comparison, the classical regression model we fitted earlier is a special case of the current model in which the variance of the 11 election-year coefficients is fixed at zero.

In addition to year-to-year variability not captured by the model, there are also electoral swings that follow the region of the country—Northeast, South, Midwest, and West. To capture the regional variability, we include 44 region $\times$ year indicator variables (also with the mean of the population distributions set to zero) to cover all regions in all election years. Within each region, we model these indicator variables as exchangeable. Because the South tends to act as a special region of the U.S. politically, we give the 11 Southern regional variables their own common variance, and treat the remaining 33 regional variables as exchangeable with their own variance. In total, we have added 55 new β parameters and three new variance components to the model, and we have excluded the regional and subregional corrections in Table 15.1 associated with specific years. We can write the model for data in states s, regions $r(s)$, and years t,

$$y_{st} \sim \text{N}((X\beta)_{st} + \gamma_{r(s)\,t} + \delta_t,\, \sigma^2)$$

$$\gamma_{rt} \sim \begin{cases} \text{N}(0, \tau_{\gamma 1}^2) & \text{for } r = 1, 2, 3 \quad \text{(non-South)} \\ \text{N}(0, \tau_{\gamma 2}^2) & \text{for } r = 4 \qquad \text{(South)} \end{cases}$$

$$\delta_t \sim \text{N}(0, \tau_\delta^2), \tag{15.2}$$

with a uniform hyperprior distribution on $\beta, \sigma, \tau_{\gamma 1}, \tau_{\gamma 2}, \tau_\delta$. (We also performed the analysis with a uniform prior distribution on the hierarchical variances, rather than the standard deviations, and obtained essentially identical inferences.)

To be consistent with the notation of Section 15.1, we would label β as the concatenated vector of all the random effects, (β, γ, δ) in formulation (15.2). The augmented β has a prior distribution with mean 0 and diagonal variance matrix $\Sigma_\beta = \text{diag}(\infty, \ldots, \infty, \tau_{\gamma 1}^2, \ldots, \tau_{\gamma 1}^2, \tau_{\gamma 2}^2, \ldots, \tau_{\gamma 2}^2, \tau_\delta^2, \ldots, \tau_\delta^2)$. The first 20 elements of β, corresponding to the constant term and the predictors in Table 15.1, have noninformative prior distributions—that is, $\sigma_{\beta j} = \infty$ for these elements. The next 33 values of $\sigma_{\beta j}$ are set to the parameter $\tau_{\gamma 1}$. The 11 elements corresponding to the Southern regional variables have $\sigma_{\beta j} = \tau_{\gamma 2}$. The final 11 elements correspond to the nationwide shifts and have prior standard deviation τ_δ.

Forecasting

Predictive inference is more subtle for a hierarchical model than a classical regression model, because of the possibility that new (random-effects) parameters β must be estimated for the predictive data. Consider the task of forecasting the outcome of the 1992 Presidential election, given the 50×20 matrix of explanatory variables for the linear regression corresponding to the 50 states in 1992. To form the complete matrix of explanatory variables for 1992, $\tilde{X}$, we must include 55 columns of zeros, thereby setting the indicators for previous years to zero for estimating the results in 1992. Even then, we are not quite ready to make predictions. To simulate draws from the predictive distribution of 1992 state election results using the hierarchical model, we must include another year indicator and four new region $\times$ year indicator variables for 1992—this adds five new predictors. However, we have no information on the coefficients of these predictor variables; they are not even included in the vector β that we have estimated from the data up to 1988. Since we have no data on these five new components of β, we must simulate their values from their posterior (predictive) distribution; that is, the coefficient for the year indicator is drawn as $\text{N}(0, \tau_\delta^2)$, the non-South region $\times$ year

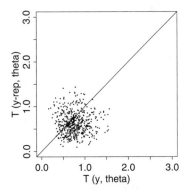

Figure 15.3 *Scatterplot showing the joint distribution of simulation draws of the realized test quantity, $T(y, \beta)$—the square root of the average of the 11 squared nationwide residuals—and its hypothetical replication, $T(y^{\text{rep}}, \beta)$, under the hierarchical model for the election forecasting example. The 200 simulated points are scattered evenly about the $45°$ line, which means that the model accurately fits this particular test quantity.*

coefficients are drawn as $N(0, \tau_{\gamma 1}^2)$, and the South $\times$ year coefficient is drawn as $N(0, \tau_{\gamma 2}^2)$, using the values $\tau_\delta, \tau_{\gamma 1}, \tau_{\gamma 2}$ drawn from the posterior simulation.

Posterior inference

We fit the hierarchical regression model using EM and the vector Gibbs sampler as described in Section 15.4, to obtain a set of draws from the posterior distribution of $(\beta, \sigma, \tau_\delta, \tau_{\gamma 1}, \tau_{\gamma 2})$. We ran ten parallel Gibbs sampler sequences; after 500 steps, the potential scale reductions, $\widehat{R}$, were below 1.1 for all parameters.

The coefficient estimates for the variables in Table 15.1 are similar to the results from the preliminary, nonhierarchical regression. The posterior medians of the coefficients all have the expected positive sign. The hierarchical standard deviations $\tau_\delta, \tau_{\gamma 1}, \tau_{\gamma 2}$ are not determined with great precision. This points out one advantage of the full Bayesian approach; if we had simply made point estimates of these variance components, we would have been ignoring a wide range of possible values for all the parameters.

When applied to data from 1992, the model yields state-by-state predictions that are summarized in Figure 6.1, with a forecasted 85% probability that the Democrats would win the national electoral vote total. The forecasts for individual states have predictive standard errors between 5% and 6%.

We tested the model in the same way as we tested the nonhierarchical model, by a posterior predictive check on the average of the squared nationwide residuals. The simulations from the hierarchical model, with their additional national and regional error terms, accurately fit the observed data, as shown in Figure 15.3.

Advantages of the hierarchical model

In summary, there are three main reasons for using a hierarchical model in this example:

- It allows the modeling of correlation within election years and regions.

- Including the year and region × year effects without a hierarchical model, or not including these effects at all, correspond to special cases of the hierarchical model with $\tau = \infty$ or 0, respectively. The more general model allows for a reasonable compromise between these extremes.

- Predictions will have additional components of variability for regions and year and should therefore be more reliable.

15.3 General notation for hierarchical linear models

More general forms of the hierarchical linear model can be created, with further levels of parameters representing additional structure in the problem. For instance, building on the brief example in the opening paragraph of this chapter, we might have a study of educational achievement in which class-level effects are not considered exchangeable but rather depend on features of the school district or state. In a similar vein, the election forecasting example might be extended to attempt some modeling of the year effects in terms of trends over time, although there is probably limited information on which to base such a model after conditioning on other observed variables. No serious conceptual or computational difficulties are added by extending the model to more levels.

A general formulation of a model with three levels of variation is:

$$\begin{aligned}
y|X, \beta, \Sigma_y &\sim N(X\beta, \Sigma_y) &\text{'likelihood'} &&n \text{ data points } y_i \\
\beta|X_\beta, \alpha, \Sigma_\beta &\sim N(X_\beta\alpha, \Sigma_\beta) &\text{'population distribution'} &&J \text{ parameters } \beta_j \\
\alpha|\alpha_0, \Sigma_\alpha &\sim N(\alpha_0, \Sigma_\alpha) &\text{'hyperprior distribution'} &&K \text{ parameters } \alpha_k
\end{aligned}$$

Interpretation as a single linear regression

The conjugacy of prior distribution and regression likelihood (see Section 14.8) allows us to express the hierarchical model as a single normal regression model using a larger 'dataset' that includes as 'observations' the information added by the population and hyperprior distributions. Specifically, for the three-level model, we can extend (14.25) to write

$$y_*|X_*, \gamma, \Sigma_* \sim N(X_*\gamma, \Sigma_*),$$

where γ is the vector (β, α) of length $J + K$, and y_*, X_*, and Σ_*^{-1} are defined by considering the likelihood, population, and hyperprior distributions as $n +$

$J + K$ 'observations' informative about γ:

$$
y_* = \begin{pmatrix} y \\ 0 \\ \alpha_0 \end{pmatrix}, \quad X_* = \begin{pmatrix} X & 0 \\ I_J & -X_\beta \\ 0 & I_K \end{pmatrix}, \quad \Sigma_*^{-1} = \begin{pmatrix} \Sigma_y^{-1} & 0 & 0 \\ 0 & \Sigma_\beta^{-1} & 0 \\ 0 & 0 & \Sigma_\alpha^{-1} \end{pmatrix}.
$$

$$(15.3)$$

If any of the components of β or α have noninformative prior distributions, the corresponding rows in y_* and X_*, as well as the corresponding rows and columns in Σ_*^{-1}, can be eliminated, because they correspond to 'observations' with infinite variance. The resulting regression then has $n + J_* + K_*$ 'observations,' where J_* and K_* are the number of components of β and α, respectively, with informative prior distributions.

For example, in the election forecasting example, β has 75 components—20 predictors in the original regression (including the constant term but excluding the five regional variables in Table 15.1 associated with specific years), 11 year effects, and 44 region × year effects—but J_* is only 55 because only the year and region × year effects have informative prior distributions. (All three groups of random effects have means fixed at zero, so in this example $K_* = 0$.)

More than one way to set up a model

A hierarchical regression model can be set up in several equivalent ways. For example, we have already noted that the hierarchical model for the 8 schools could be written as $y_j \sim N(\theta_j, \sigma_j^2)$ and $\theta_j \sim N(\mu, \tau^2)$, or as $y_j \sim N(\beta_0 + \beta_j, \sigma_j^2)$ and $\beta_j \sim N(0, \tau^2)$. The hierarchical model for the election forecasting example can be written, as described above, as a regression, $y \sim N(X\beta, \sigma^2 I)$, with 70 predictors X and normal prior distributions on the coefficients β, or as a regression with 20 predictors and three independent error terms, corresponding to year, region × year, and state × year.

In the three-level formulation described at the beginning of this section, group-level predictors can be included in either the likelihood or the population distribution, and the constant term can be included in any of the three regression levels.

15.4 Computation

There are several ways to use the Gibbs sampler to draw posterior simulations for hierarchical linear models. The different methods vary in their programming effort required, computational speed, and convergence rate, with different approaches being reasonable for different problems.

We shall discuss computation for models with independent variances at each of the hierarchical levels; computation can be adapted to structured covariance matrices as described in Section 14.6.

Crude estimation and starting points for iterative simulation

All the methods begin with crude initial estimation using nonhierarchical regressions. As always, the crude estimates are useful as starting points for iterative simulation and also to get a sense of what will be reasonable estimates. As with the educational testing example in Chapter 5, it makes sense to begin with initial pooled and unpooled estimates.

For some problems, unpooled estimates are not possible; for example, in the election forecasting example in Section 15.2, it would not be possible to estimate all the national effects as separate regression coefficients, because they would be collinear with the nationwide predictors. Instead, we can set the variance parameters to reasonable approximate values (for example, setting them all equal to 0.1 in the election example) and then perform a simple regression with this fixed prior distribution (as described in Section 14.8).

Each crude regression yields, for the regression coefficients γ, an approximate posterior distribution of the form, $N(\hat{\gamma}, V_\gamma)$. As discussed in Section 12.2, we can obtain an approximately overdispersed starting distribution for iterative simulation by drawing from the $t_4(\hat{\gamma}, V_\gamma)$ distribution, possibly following up with importance resampling of a subset of draws to use as Gibbs sampler starting points.

Gibbs sampler, one batch at a time

Perhaps the simplest computational approach for hierarchical regressions is to perform a blockwise Gibbs sampler, updating each batch of regression coefficients given all the others, and then updating the variance parameters. Given the data and all other parameters in the model, inference for a batch of coefficients corresponds to a linear regression with fixed prior distribution. We can thus update the coefficients of the entire model in batches, performing at each step an augmented linear regression, as discussed in Section 14.8. In many cases (including the statewide, regional, and national error terms in the election forecasting example), the model is simple enough that the means and variances for the Gibbs updating do not actually require a regression calculation but instead can be performed using simple averages.

The variance parameters can also be updated one at a time. For simple hierarchical models with scalar variance parameters (typically one parameter per batch of coefficients, along with one or more data-level variance parameters), the Gibbs updating distributions are scaled inverse-χ^2. For more complicated models, the variance parameters can be updated using Metropolis jumps.

The Gibbs sampler for the entire model then proceeds by cycling through all the batches of parameters in the model (including the batch, if any, of coefficients with noninformative or fixed prior distributions). One attractive feature of this algorithm is that it mimics the natural way one might try to combine information from the different levels: starting with a guess at the upper-level parameters, the lower-level regression is run, and then these simulated coefficients are used to better estimate the upper-level regression parameters, and so on.

In practice, the efficiency of the Gibbs sampler for these models can depend strongly on the placement of predictors in the different levels of the regression—for example, in the election forecasting model, we put the 11 indicators for years as predictors in the data model, but the model would be unchanged if we instead placed them as predictors in a regression model for the state × year effects.

All-at-once Gibbs sampler

As noted in Section 15.3, the different levels in a hierarchical regression context can be combined into a single regression model by appropriately augmenting the data, predictors, and variance matrix. The Gibbs sampler can then be applied, alternately updating the variance parameters (with independent inverse-χ^2 distributions given the data and regression coefficients), and the vector of regression coefficients, which can be updated at each step by running a weighted regression with weight matrix depending on the current values of the variance parameters.

In addition to its conceptual simplicity, this all-at-once Gibbs sampler has the advantages of working efficiently even if regression coefficients at different levels are correlated in their posterior distribution, as can commonly happen with hierarchical models (for example, the parameters θ_j and μ have positive posterior correlations in the 8-schools example).

The main computational disadvantage of the all-at-once Gibbs sampler is that at each step it requires a regression on a potentially large augmented dataset. For example, the election forecasting model has $n = 511$ data points and $k = 15$ predictors, but the augmented regression has $n_* = 511 + 55$ observations and $k_* = 15 + 55$ predictors. The computer time required to perform a linear regression is proportional to nk^2, and thus each step of the all-at-once Gibbs sampler can be slow when the number of parameters is large.

Scalar Gibbs sampler

Another computational approach for hierarchical linear models is to perform a pure Gibbs sampler, one parameter at a time. The variance parameters are updated as before, with independent inverse-χ^2 distributions if possible, otherwise using Metropolis jumping. For the regression coefficients, however, the scalar updating can be much faster than before. The Gibbs sampler updating of each coefficient γ_j is computationally simple: from (15.3), the relevant data and prior distribution can be combined as

$$y_* \sim N(X_{*j}\gamma_j + z_{*j}, \Sigma_*), \tag{15.4}$$

where X_{*j} is the j-th column of the augmented matrix X_*, and z_{*j} is the linear predictor corresponding to the other $J_* - 1$ predictors whose coefficients are assumed known for the purposes of this step of the Gibbs updating. We can write $z_{*j} = X_{*-j}\gamma_{-j}$, where the subscript '$-j$' refers to all the elements excluding j. Using careful storage and computation, the linear predictor $z_{*j} = X_{*-j}\gamma_{-j}$ for each j can be calculated without requiring a new matrix multiplication at each step.

Expressed as a conditional posterior distribution, (15.4) becomes a simple regression of $y_* - z_{*j}$ on the single predictor X_{*j}, which has a computation time of order n_*. Thus, updating all k_* regression coefficients in scalar Gibbs steps requires a computation time of order $n_* k_*$, which is potentially a huge improvement over the $n_* k_*^2$ time needed by the all-at-once updating.

Scalar Gibbs sampler with linear transformation

The scalar Gibbs updating algorithm described above can be slow to converge: although each complete Gibbs step (updating all the parameters in the model) is quick, the algorithm will travel only slowly through the posterior distribution if the parameters are highly correlated in the posterior distribution. (Recall the picture of the Gibbs sampler on page 288.)

A natural solution to the problem of posterior correlation is to rotate the parameters to approximate independence. If the coefficients γ have an approximate mean γ_0 and covariance matrix V_γ, then we can define,

$$\xi = A^{-1}(\gamma - \gamma_0),$$

where $A = V_\gamma^{1/2}$ is a matrix square root of V_γ such as can be obtained from the Cholesky factorization (see the discussion of the multivariate normal distribution on page 578).

We can then perform the Gibbs sampler on ξ and get rapid convergence. At the same time, the computation of each step is still fast. The scalar updating regression for each ξ_j has the same form as (15.4) but with ξ_j in place of γ_j, X_{*j} replaced by $(X_*A)_j$ (the jth column of X_*A) and z_{*j} replaced by $X_*\gamma_0 + (X_*A)_{-j}\xi_{-j}$.

The only challenge in this updating scheme is to choose the linear transformation, that is, to choose a reasonable centering point γ_0 and rotation matrix $A = V_\gamma^{1/2}$. (Actually, the Gibbs sampler is unaffected by the choice of γ_0 but for computational stability it is convenient to include it in the linear transformation rather than simply rotating about the origin.) A reasonable approach is to start by setting γ_0 and V_γ to an estimated posterior mean and variance matrix from a crude estimate of γ. The Gibbs sampler can be run for a while (perhaps 100 or 1000 iterations) using this transformation, and then a new γ_0 and V_γ can be set using the mean and variance matrix of these simulation draws.

It might seem desirable to continually update the transformation as the Gibbs sampler proceeds, but, as discussed at the end of Section 11.9, the simulations from such an adaptive algorithm will not in general converge to the target distribution. Thus, it is simpler to stop the adaptation at some point and then just save the remaining draws as posterior simulations.

Parameter expansion

Any of the above Gibbs samplers can be slow to converge when estimated hierarchical variance parameters are near zero. The problem is that, if the current draw of a hierarchical variance parameter is near 0, then in the next updating step, the corresponding batch of linear parameters γ_j will themselves be 'shrunk' to be very close to their population mean. Then, in turn, the variance parameter will be estimated to be close to 0 because it is updated based on the relevant γ_j's. Ultimately, the stochastic nature of the Gibbs sampler allows it to escape out of this trap but this may require many iterations.

The parameter-expanded Gibbs sampler and EM algorithms (see Sections 11.8 and 12.3) can be used to solve this problem. For hierarchical linear models, the basic idea is to associate with each batch of regression coefficients a multiplicative factor, which has the role of breaking the dependence between the coefficients and their variance parameter.

Example. Election forecasting (continued)

We illustrate with the Presidential election forecasting model (15.2), which in its expanded-parameter version can be written as,

$$
y_{st} \sim \begin{cases} N((X\beta)_{st} + \zeta_1^{\text{region}} \gamma_{r(s)\,t} + \zeta^{\text{year}} \delta_t, \sigma^2) & \text{if } r(s) = 1, 2, 3 \quad \text{(non-South)} \\ N((X\beta)_{st} + \zeta_2^{\text{region}} \gamma_{r(s)\,t} + \zeta^{\text{year}} \delta_t, \sigma^2) & \text{if } r(s) = 4 \quad \quad\ \ \text{(South)} \end{cases}
$$

with the same prior distributions as before. The new parameters ζ_1^{region}, ζ_2^{region}, and ζ^{year} are assigned uniform prior distributions and are not identified in the posterior distribution. The products $\zeta_1^{\text{region}} \gamma_{rt}$ (for $r = 1, 2, 3$), $\zeta_2^{\text{region}} \gamma_{rt}$ (for $r = 4$), and $\zeta^{\text{year}} \delta_t$, correspond to the parameters γ_{rt} and δ_t, respectively, in the old model. Similarly, the products $|\zeta_1^{\text{region}}| \tau_{\gamma 1}$, $|\zeta_2^{\text{region}}| \tau_{\gamma 2}$, and $|\zeta^{\text{year}}| \tau_\delta$ correspond to the variance components $\tau_{\gamma 1}$, $\tau_{\gamma 2}$, and τ_δ in the original model.

For the election forecasting, the variance parameters are estimated precisely enough that the ordinary Gibbs sampler performs reasonably efficiently. However, we can use this problem to illustrate the computational steps of parameter expansion.

The Gibbs sampler for the parameter-expanded model alternately updates the regression coefficients (β, γ, δ), the variance parameters $(\sigma, \tau_{\gamma 1}, \tau_{\gamma 2}, \tau_\delta)$, and the additional parameters $(\zeta_1^{\text{region}}, \zeta_2^{\text{region}}, \zeta^{\text{year}})$. The regression coefficients can be updated using any of the Gibbs samplers described above—by batch, all at once, or one element at a time. The ζ parameters do not change the fact that β, γ, and δ can be estimated by linear regression. Similarly, the Gibbs sampler updates for the variances are still independent inverse-χ^2.

The final step of the Gibbs sampler is to update the multiplicative parameters ζ. This step turns out to be easy: given the data and the other parameters in the model, the information about the ζ's can be expressed simply as a linear regression of the 'data' $z_{st} = y_{st} - (X\beta)_{st}$ on the 'predictors' $\gamma_{r(s)\,t}$ (for states s with regions $r(s) = 1, 2,$ or 3), $\gamma_{r(s)\,t}$ (for $r(s) = 4$), and δ_t, with variance matrix $\sigma^2 I$. The three parameters ζ are then easily updated with a linear regression.

When running the Gibbs sampler, we do not worry about the individual parameters $\zeta, \beta, \gamma, \delta$; instead, we save and monitor the convergence of the variance parameters σ and the parameters γ_{rt} and δ_t in the original parameterization (15.2). This is most easily done by just multiplying each of the parameters γ_{rt} and δ_t by the appropriate ζ parameter, and multiplying each of the variance components $\tau_{\gamma 1}, \tau_{\gamma 2}, \tau_\delta$, by the absolute value of its corresponding ζ.

More on the parameter-expanded Gibbs sampler for hierarchical models appears at the end of Section 15.6.

15.5 Hierarchical modeling as an alternative to selecting predictors

Approaches such as stepwise regression and subset selection are popular methods used in non-Bayesian statistics for choosing a set of explanatory variables to include in a regression. Mathematically, not including a variable is equivalent to setting its coefficient to exactly zero. The classical regression estimate based on a selection procedure is equivalent to obtaining the posterior mode corresponding to a prior distribution that has nonzero probability on various low-dimensional hyperplanes of β-space. The selection procedure is heavily influenced by the quantity of data available, so that important variables may be omitted because chance variation cannot be ruled out as an alternative explanation for their effect. Geometrically, if β-space is thought of as a room, the model implied by classical model selection claims that the true β has certain prior probabilities of being in the room, on the floor, on the walls, in the edge of the room, or in a corner.

In a Bayesian framework, it is appealing to include prior information in a more continuous way. If we have many explanatory variables x_j, each with a fair probability of being irrelevant to modeling the outcome variable y, we can set up a prior distribution that gives each parameter a high probability of being near zero. Suppose each variable is probably unimportant, but if it has an effect, it could be large—then one could use a t or other wide-tailed distribution for $p(\beta)$.

In practice, a commonsense Bayesian perspective indicates that the key is to use substantive knowledge, either as a formal prior distribution or more informally, in choosing which variables to include.

The largest gains in estimating regression coefficients often come from specifying structure in the model. For example, in the election forecasting problem, it is crucial that the national and regional indicator variables are clustered and modeled separately from the quantitative predictors. In general, when many predictor variables are used in a regression, they should be set up so they can be structured hierarchically, so the Bayesian analysis can do the most effective job at pooling the information about them. In Section 15.6, we discuss how to use the analysis of variance to summarize inferences about batches of regression parameters.

For example, if a set of predictors is modeled as a group of random effects, it may be important first to apply linear or other transformations to put them all on approximately a common scale. For example in the election forecasting problem, if 10 different economic measures were available on each state it would be natural to combine these in a hierarchical model after first transforming each to an approximate 0 to 100 scale. Such transformations make intuitive sense statistically, and algebraically their effect is to pull the numerical values of the true coefficients β_j closer together, reducing the random-effects variance and making the pooling more effective (consider Figure 5.6). In contrast, when a linear regression is performed with a noninformative uniform prior distribution on the coefficients β_j, linear transformations of the predictor variables have no effect on inferences or predictions.

15.6 Analysis of variance

Analysis of variance (ANOVA) represents a key idea in statistical modeling of complex data structures—the grouping of predictor variables and their coefficients into batches. In the traditional application of analysis of variance, each batch of coefficients and the associated row of the ANOVA table corresponds to a single experimental block or factor or to an interaction of two or more factors. In a hierarchical linear regression context, each row of the table corresponds to a set of regression coefficients, and we are potentially interested in the individual coefficients and also in the variance of the coefficients in each batch. We thus view the analysis of variance as a way of making sense of hierarchical regression models with many predictors or indicators that can be grouped into batches within which all the effects are exchangeable.

Notation and model

We shall work with the linear model, with the 'analysis of variance' corresponding to the batching of coefficients into 'sources of variation,' with each batch corresponding to one row of an ANOVA table. We use the notation $m = 1, \ldots, M$ for the rows of the table. Each row m represents a batch of J_m regression coefficients $\beta_j^{(m)}$, $j = 1, \ldots, J_m$. We denote the m-th subvector of coefficients as $\beta^{(m)} = (\beta_1^{(m)}, \ldots, \beta_{J_m}^{(m)})$ and the corresponding classical least-squares estimate as $\hat{\beta}^{(m)}$. These estimates are subject to c_m linear constraints, yielding $(df)_m = J_m - c_m$ degrees of freedom. We label the $c_m \times J_m$ constraint matrix as $C^{(m)}$, so that $C^{(m)}\hat{\beta}^{(m)} = 0$ for all m, and we assume that $C^{(m)}$ is of rank c_m. For notational convenience, we label the grand mean as $\beta_1^{(0)}$, corresponding to the (invisible) zeroth row of the ANOVA table and estimated with no linear constraints.

The linear model is fitted to the data points y_i, $i = 1, \ldots, n$, and can be written as

$$y_i = \sum_{m=0}^{M} \beta_{j_i^m}^{(m)}, \qquad (15.5)$$

where j_i^m indexes the appropriate coefficient j in batch m corresponding to data point i. Thus, each data point pulls one coefficient from each row in the ANOVA table. Equation (15.5) could also be expressed as a linear regression model with a design matrix composed entirely of 0's and 1's. The coefficients β_j^M of the last row of the table correspond to the residuals or error term of the model.

The analysis of variance can also be applied more generally to regression models (or to generalized linear models), in which case we can have any design matrix X, and (15.5) is replaced by

$$y_i = \sum_{m=0}^{M} \sum_{j=1}^{J_m} x_{ij}^{(m)} \beta_j^{(m)}. \tag{15.6}$$

The essence of analysis of variance is in the structuring of the coefficients into batches—hence the notation $\beta_j^{(m)}$—going beyond the usual linear model formulation that has a single indexing of coefficients β_j.

We shall use a hierarchical formulation in which the each batch of regression coefficients is modeled as a sample from a normal distribution with mean 0 and its own variance σ_m^2:

$$\beta_j^{(m)} \sim N(0, \sigma_m^2), \quad \text{for } j = 1, \ldots, J_m, \quad \text{for each batch } m = 1, \ldots, M.$$

Without loss of generality, we can center the distribution of each batch $\beta^{(m)}$ at 0—if it were appropriate for the mean to be elsewhere, we would just include the mean of the $x_j^{(m)}$'s as a separate predictor. As in classical ANOVA, we usually include interactions only if the corresponding main effects are in the model.

The conjugate hyperprior distributions for the variances are scaled inverse-χ^2 distributions:

$$\sigma_m^2 \sim \text{Inv-}\chi^2(\nu_m, \sigma_{0m}^2).$$

A natural noninformative prior density is uniform on σ_m, which corresponds to $\nu_m = -1$ and $\sigma_{0m} = 0$. For values of m in which J_m is large (that is, rows of the ANOVA table corresponding to many linear predictors), σ_m is essentially estimated from data. When J_m is small, the flat prior distribution implies that σ is allowed the possibility of taking on large values, which minimizes the amount of shrinkage in the coefficient estimates.

Computation

In this model, the posterior distribution for the parameters (β, σ) can be simulated using the Gibbs sampler, alternately updating the vector β given σ with linear regression, and updating the vector σ from the independent inverse-χ^2 conditional posterior distributions given β.

The only trouble with this Gibbs sampler is that it can get stuck with variance components σ_m near zero. A more efficient updating uses parameter expansion, as described at the end of Section 15.4. In the notation here, we reparameterize into vectors γ, ζ, and τ, which are defined as follows:

$$\beta_j^{(m)} = \zeta_m \gamma_j^{(m)}$$
$$\sigma_m = |\zeta_m| \tau_m. \tag{15.7}$$

The model can be then expressed as

$$y = X\zeta\gamma$$
$$\gamma_j^{(m)} \sim N(0, \tau_m^2) \text{ for each } m$$
$$\tau_m^2 \sim \text{Inv-}\chi^2(\nu_m, \sigma_{0m}^2).$$

The auxiliary parameters ζ are given a uniform prior distribution, and then this reduces to the original model. The Gibbs sampler then proceeds by updating γ (using linear regression with n data points and $\sum_{m=0}^{M} J_m$ predictors), ζ (linear regression with n data points and M predictors), and τ (independent inverse-χ^2 distributions). The parameters in the original parameterization, β and σ, can then be recomputed from (15.7) and stored at each step.

Finite-population and superpopulation standard deviations

One measure of the importance of each row or 'source' in the ANOVA table is the standard deviation of its constrained regression coefficients, defined as

$$s_m = \sqrt{\frac{1}{(df)_m} \beta^{(m)\,T} \left[I - C^{(m)\,T} (C^{(m)} C^{(m)\,T})^{-1} C^{(m)} \right] \beta^{(m)}}. \tag{15.8}$$

We divide by $(df)_m = J_m - c_m$ rather than $J_m - 1$ because multiplying by $C^{(m)}$ induces c_m linear constraints. We model the underlying β coefficients as unconstrained.

For each batch of coefficients $\beta^{(m)}$, there are two natural variance parameters to estimate: the *superpopulation* standard deviation σ_m and the *finite-population* standard deviation s_m as defined in (15.8). The superpopulation standard deviation characterizes the uncertainty for predicting a new coefficient from batch m, whereas the finite-population standard deviation describes the variation in the existing J_m coefficients.

Variance estimation is often presented in terms of the superpopulation standard deviations σ_m, but in our ANOVA summaries, we focus on the finite-population quantities s_m, for reasons we shall discuss here. However, for computational reasons, the parameters σ_m are useful intermediate quantities to estimate. Our general procedure is to use computational methods such as described in Section 15.4 to draw joint posterior simulations of (β, σ) and then compute the finite-sample standard deviations s_m from β using (15.8).

To see the difference between the two variances, consider the extreme case in which $J_m = 2$ (with the usual constraint that $\beta_1^{(m)} + \beta_2^{(m)} = 0$) and a large amount of data are available in both groups. Then the two parameters $\beta_1^{(m)}$ and $\beta_2^{(m)}$ will be estimated accurately and so will $s_m^2 = (\beta_1^{(m)} - \beta_2^{(m)})^2/2$. The superpopulation variance σ_m^2, on the other hand, is only being estimated by a measurement that is proportional to a χ^2 with 1 degree of freedom. We know much about the two parameters $\beta_1^{(m)}, \beta_2^{(m)}$ but can say little about others from their batch.

We believe that much of the statistical literature on fixed and random effects can be fruitfully reexpressed in terms of finite-population and superpopulation inferences. In some contexts (for example, collecting data on the 50 states of the U.S.), the finite population seems more meaningful; whereas in others (for example, subject-level effects in a psychological experiment), interest clearly lies in the superpopulation.

For example, suppose a factor has four degrees of freedom corresponding to five different medical treatments, and these are the only existing treatments and are thus considered 'fixed.' Suppose it is then discovered that these are part of a larger family of many possible treatments, and so it makes sense to model them as 'random.' In our framework, the inference about these five parameters $\beta_j^{(m)}$ and their finite-population and superpopulation standard deviations, s_m and σ_m, will not change with the news that they can actually be viewed as a random sample from a distribution of possible treatment effects. But the superpopulation variance now has an important new role in characterizing this distribution. The difference between fixed and random effects is thus not a difference in inference or computation but in the ways that these inferences will be used.

Example: a five-way factorial structure for data on Web connect times

We illustrate the analysis of variance with an example of a linear model fitted for exploratory purposes to a highly structured dataset.

Data were collected by an internet infrastructure provider on connect times for messages processed by two different companies. Messages were sent every hour for 25 consecutive hours, from each of 45 locations to 4 different destinations, and the study was repeated one week later. It was desired to quickly summarize these data to learn about the importance of different sources of variation in connect times.

Figure 15.4 shows the Bayesian ANOVA display for an analysis of logarithms of connect times on the five factors: destination ('to'), source ('from'), service provider ('company'), time of day ('hour'), and week. The data have a full factorial structure with no replication, so the full five-way interaction at the bottom represents the 'error' or lowest-level variability.

Each row of the plot shows the estimated finite-population standard deviation of the corresponding group of parameters, along with 50% and 95% uncertainty intervals. We can immediately see that the lowest-level variation is more important in variance than any of the factors except for the main effect of the destination. Company has a large effect on its own and, perhaps more interestingly, in interaction with to, from, and in the three-way interaction.

Figure 15.4 would not normally represent the final statistical analysis for

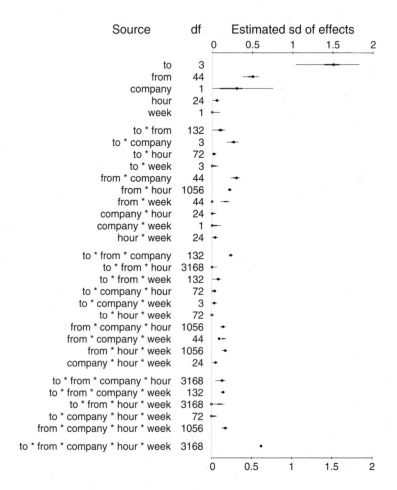

Figure 15.4 *ANOVA display for the World Wide Web data. The bars indicate 50% and 95% intervals for the finite-population standard deviations s_m. The display makes apparent the magnitudes and uncertainties of the different components of variation. Since the data are on the logarithmic scale, the standard deviation parameters can be interpreted directly. For example, $s_m = 0.20$ corresponds to a coefficient of variation of $\exp(0.2) - 1 \approx 0.2$ on the original scale, and so the exponentiated coefficients $\exp(\beta_j^{(m)})$ in this batch correspond to multiplicative increases or decreases in the range of 20%. (The dots on the bars show simple classical estimates of the variance components that can be used as starting points in a Bayesian computation.)*

this sort of problem. The ANOVA plot represents a default model and is a tool for data exploration—for learning about which factors are important in predicting the variation in the data—and can be used to construct more focused models or design future data collection.

15.7 Bibliographic note

Novick et al. (1972) describe an early application of Bayesian hierarchical regression. Lindley and Smith (1972) present the general form for the normal linear model (using a slightly different notation than ours); see also Hodges (1998). Many interesting applications of Bayesian hierarchical regression have appeared in the statistical literature since then; for example, Fearn (1975) analyzes growth curves, Hui and Berger (1983) and Strenio, Weisberg, and Bryk (1983) estimate patterns in longitudinal data, and Normand, Glickman, and Gatsonis (1997) analyze death rates in a set of hospitals. Rubin (1980b) presents a hierarchical linear regression in an educational example and goes into some detail on the advantages of the hierarchical approach. Other references on hierarchical linear models appear at the end of Chapter 5.

Random-effects regression has a long history in the non-Bayesian statistical literature; for example, see Henderson et al. (1959). Robinson (1991) provides a review, using the term 'best linear unbiased prediction'.* The Bayesian approach differs by averaging over uncertainty in the posterior distribution of the hierarchical parameters, which is important in problems such as the SAT coaching example of Section 5.5 with large posterior uncertainty in the hierarchical variance parameter.

Hierarchical linear modeling has recently gained in popularity, especially in the social sciences, where it is often called *multilevel modeling*. An excellent summary of these applications at a fairly elementary level, with some discussion of the available computer packages, is provided by Raudenbush and Bryk (2002). Other recent textbooks in this area include Kreft and DeLeeuw (1998) and Snijders and Bosker (1999). Leyland and Goldstein (2001) provide an overview of multilevel models for public health research.

Other key references on multilevel models for the social sciences are Goldstein (1995), Longford (1993), and Aitkin and Longford (1986); the latter article is an extended discussion of the practical implications of undertaking a detailed hierarchical modeling approach to controversial issues in school effectiveness studies in the United Kingdom. Sampson, Raudenbush, and Earls (1997) discuss a study of crime using a hierarchical model of city neighborhoods.

Gelman and King (1993) discuss the Presidential election forecasting prob-

---

* Posterior means of regression coefficients and 'random effects' from hierarchical models are biased 'estimates' but can be unbiased or approximately unbiased when viewed as 'predictions,' since conventional frequency evaluations condition on all unknown 'parameters' but not on unknown 'predictive quantities'; the latter distinction has no meaning within a Bayesian framework. (Recall the example on page 248 of estimating daughters' heights from mothers' heights.)

lem in more detail, with references to earlier work in the econometrics and political science literature. Boscardin and Gelman (1996) provide details on computations, inference, and model checking for the model described in Section 15.2 and some extensions.

Gelfand, Sahu, and Carlin (1995) discuss linear transformations for Gibbs samplers in hierarchical regressions, Boscardin (1996) evaluates the speed of the all-at-once and scalar Gibbs samplers, and Liu and Wu (1999) and Gelman et al. (2007) discuss the parameter-expanded Gibbs sampler for hierarchical linear and generalized linear models.

Much has been written on Bayesian methods for estimating many regression coefficients, almost all from the perspective of treating all the coefficients in a problem as exchangeable. *Ridge regression* (Hoerl and Kennard, 1970) is a procedure equivalent to an exchangeable normal prior distribution on the coefficients, as has been noted by Goldstein (1976), Wahba (1978), and others. Leamer (1978a) discusses the implicit models corresponding to stepwise regression and some other methods. George and McCulloch (1993) propose an exchangeable bimodal prior distribution for regression coefficients. Madigan and Raftery (1994) propose an approximate Bayesian approach for averaging over a distribution of potential regression models. Clyde, DeSimone, and Parmigiani (1996) and West (2003) present Bayesian methods using linear transformations for averaging over large numbers of potential predictors.

The perspective on analysis of variance given here is from Gelman (2004b); previous work along similar lines includes Plackett (1960), Yates (1967), and Nelder (1977, 1994), and Hodges and Sargent (2001).

15.8 Exercises

1. Random-effects models: express the educational testing example of Section 5.5 as a hierarchical linear model with eight observations and known observation variances. Draw simulations from the posterior distribution using the methods described in this chapter.

2. Fitting a hierarchical model for a two-way array:

   (a) Fit a standard analysis of variance model to the randomized block data discussed in Exercise 7.5, that is, a linear regression with a constant term, indicators for all but one of the blocks, and all but one of the treatments.

   (b) Summarize posterior inference for the (superpopulation) average penicillin yields, averaging over the block conditions, under each the four treatments. Under this measure, what is the probability that each of the treatments is best? Give a 95% posterior interval for the difference in yield between the best and worst treatments.

   (c) Set up a hierarchical extension of the model, in which you have indicators for all five blocks and all five treatments, and the block and treatment indicators are two sets of random effects. Explain why the means for

Reactor temperature (°C), x_1	Ratio of H_2 to n-heptane (mole ratio), x_2	Contact time (sec), x_3	Conversion of n-heptane to acetylene (%), y
1300	7.5	0.0120	49.0
1300	9.0	0.0120	50.2
1300	11.0	0.0115	50.5
1300	13.5	0.0130	48.5
1300	17.0	0.0135	47.5
1300	23.0	0.0120	44.5
1200	5.3	0.0400	28.0
1200	7.5	0.0380	31.5
1200	11.0	0.0320	34.5
1200	13.5	0.0260	35.0
1200	17.0	0.0340	38.0
1200	23.0	0.0410	38.5
1100	5.3	0.0840	15.0
1100	7.5	0.0980	17.0
1100	11.0	0.0920	20.5
1100	17.0	0.0860	19.5

Table 15.2 *Data from a chemical experiment, from Marquardt and Snee (1975). The first three variables are experimental manipulations, and the fourth is the outcome measurement.*

the block and treatment indicator groups should be fixed at zero. Write the joint distribution of all model parameters (including the hierarchical parameters).

(d) Compute the posterior mode of the three variance components of your model in (c) using EM. Construct a normal approximation about the mode and use this to obtain posterior inferences for all parameters and answer the questions in (b). (Hint: you can use the general regression framework or extend the procedure in Section 12.5.)

(e) Check the fit of your model to the data. Discuss the relevance of the randomized block design to your check; how would the posterior predictive simulations change if you were told that the treatments had been assigned by complete randomization?

(f) Obtain draws from the actual posterior distribution using the Gibbs sampler, using your results from (d) to obtain starting points. Run multiple sequences and monitor the convergence of the simulations by computing $\widehat{R}$ for all parameters in the model.

(g) Discuss how your inferences in (b), (d), and (e) differ.

3. Regression with many explanatory variables: Table 15.2 displays data from a designed experiment for a chemical process. In using these data to illustrate various approaches to selection and estimation of regression coeffi-

cients, Marquardt and Snee (1975) assume a quadratic regression form; that is, a linear relation between the expectation of the untransformed outcome, y, and the variables x_1, x_2, x_3, their two-way interactions, $x_1 x_2, x_1 x_3, x_2 x_3$, and their squares, x_1^2, x_2^2, x_3^2.

(a) Fit an ordinary linear regression model (that is, nonhierarchical with a uniform prior distribution on the coefficients), including a constant term and the nine explanatory variables above.

(b) Fit a mixed-effects linear regression model with a uniform prior distribution on the constant term and a shared normal prior distribution on the coefficients of the nine variables above. If you use iterative simulation in your computations, be sure to use multiple sequences and monitor their joint convergence.

(c) Discuss the differences between the inferences in (a) and (b). Interpret the differences in terms of the hierarchical variance parameter. Do you agree with Marquardt and Snee that the inferences from (a) are unacceptable?

(d) Repeat (a), but with a t_4 prior distribution on the nine variables.

(e) Discuss other models for the regression coefficients.

4. Analysis of variance:

(a) Create an analysis of variance plot for the educational testing example in Chapter 5, assuming that there were exactly 60 students in the study in each school, with 30 receiving the treatment and 30 receiving the control.

(b) Discuss the relevance of the finite-population and superpopulation standard deviation for each source of variation.

CHAPTER 16

Generalized linear models

This chapter reviews generalized linear models from a Bayesian perspective. We discuss noninformative prior distributions, conjugate prior distributions, and, most important, hierarchical models, focusing on the normal specification for the generalized linear model coefficients. We present a computational method based on approximating the generalized linear model by a normal linear model at the mode and then obtaining exact simulations, if necessary, using the Metropolis algorithm. Finally, we discuss the class of loglinear models, a subclass of Poisson generalized linear models that is commonly used for missing data imputation and discrete multivariate outcomes. We show how to simulate from the posterior distribution of the loglinear model with noninformative and conjugate prior distributions using a stochastic version of the iterative proportional fitting algorithm. This chapter is not intended to be exhaustive, but rather to provide enough guidance so that the reader can combine generalized linear models with the ideas of hierarchical models, posterior simulation, prediction, model checking, and sensitivity analysis that we have already presented for Bayesian methods in general, and linear models in particular. Our computational strategy is based on extending available tools of maximum likelihood for fitting generalized linear models to the Bayesian case.

16.1 Introduction

As we have seen in Chapters 14 and 15, a stochastic model based on a linear predictor $X\beta$ is easy to understand and can be appropriately flexible in a variety of problems, especially if we are careful about transformation and appropriately interpreting the regression coefficients. The purpose of the *generalized linear model* is to extend the idea of linear modeling to cases for which the linear relationship between X and $E(y|X)$ or the normal distribution is not appropriate.

In some cases, it is reasonable to apply a linear model to a suitably transformed outcome variable using suitably transformed (or untransformed) explanatory variables. For example, in the election forecasting example of Chapter 15, the outcome variable—the incumbent party candidate's share of the vote in a state election—must lie between 0 and 1, and so the linear model does not make logical sense: it is possible for a combination of explanatory variables, or the variation term, to be so extreme that y would exceed its bounds. The logit of the vote share is a more logical outcome variable. In that example, however, the boundaries are not a serious problem, since the observations are almost all between 0.2 and 0.8, and the residual standard deviation is about 0.06. Another case in which a linear model can be improved by transformation occurs when the relation between X and y is multiplicative: for example, if $y_i = x_{i1}^{b_1} x_{i2}^{b_2} \cdots x_{ik}^{b_k} \times$ variation, then $\log y_i = b_1 \log x_{i1} + \cdots + b_k \log x_{ik} +$ variation, and a linear model relating $\log y_i$ to $\log x_{ij}$ is appropriate.

However, the relation between X and $\mathrm{E}(y|X)$ cannot always be usefully modeled as normal and linear, even after transformation. For example, suppose that y cannot be negative, but might be zero. Then we cannot just analyze $\log y$, even if the relation of $\mathrm{E}(y)$ to X is generally multiplicative. If y is discrete-valued (for example, the number of occurrences of a rare disease by county) then the mean of y may be linearly related to X, but the variation term cannot be described by the normal distribution.

The class of generalized linear models unifies the approaches needed to analyze data for which either the assumption of a linear relation between x and y or the assumption of normal variation is not appropriate. A generalized linear model is specified in three stages:

1. The linear predictor, $\eta = X\beta$,

2. The *link function* $g(\cdot)$ that relates the linear predictor to the mean of the outcome variable: $\mu = g^{-1}(\eta) = g^{-1}(X\beta)$,

3. The random component specifying the distribution of the outcome variable y with mean $\mathrm{E}(y|X) = \mu$. In general, the distribution of y given x can also depend on a *dispersion parameter*, ϕ.

Thus, the mean of the distribution of y, given X, is determined by $X\beta$: $\mathrm{E}(y|X) = g^{-1}(X\beta)$. We use the same notation as in linear regression whenever possible, so that X is the $n \times p$ matrix of explanatory variables and $\eta = X\beta$ is the vector of n linear predictor values. If we denote the linear predictor for the ith case by $(X\beta)_i$ and the variance or dispersion parameter (if present) by ϕ, then the data distribution takes the form

$$p(y|X, \beta, \phi) = \prod_{i=1}^{n} p(y_i|(X\beta)_i, \phi). \qquad (16.1)$$

The most commonly used generalized linear models, for the Poisson and binomial distributions, do not require a dispersion parameter; that is, ϕ is fixed at 1. In practice, however, excess dispersion is the rule rather than the exception in most applications.

16.2 Standard generalized linear model likelihoods

Continuous data

The normal linear model is a special case of the generalized linear model, with y being normally distributed with mean μ and the identity link function,

$g(\mu) = \mu$. For continuous data that are all positive, we can use the normal model on the logarithmic scale. When this distributional family does not fit the data, the gamma and Weibull distribution are sometimes considered as alternatives.

Poisson

Counted data are often modeled using a Poisson model. The Poisson generalized linear model, often called the Poisson regression model, assumes that y is Poisson with mean μ (and therefore variance μ). The link function is typically chosen to be the logarithm, so that $\log \mu = X\beta$. The distribution for data $y = (y_1, \ldots, y_n)$ is thus

$$p(y|\beta) = \prod_{i=1}^{n} \frac{1}{y_i!} e^{-\exp(\eta_i)} (\exp(\eta_i))^{y_i}, \tag{16.2}$$

where $\eta_i = (X\beta)_i$ is the linear predictor for the i-th case. When considering the Bayesian posterior distribution, we condition on y, and so the factors of $1/y_i!$ can be absorbed into an arbitrary constant. We consider an example of a Poisson regression in Section 16.5 below.

Binomial

Perhaps the most widely used of the generalized linear models are those for binary or binomial data. Suppose that $y_i \sim \text{Bin}(n_i, \mu_i)$ with n_i known. It is common to specify the model in terms of the mean of the proportions y_i/n_i, rather than the mean of y_i. Choosing the logit transformation of the probability of success, $g(\mu_i) = \log(\mu_i/(1 - \mu_i))$, as the link function leads to the logistic regression model. The distribution for data y is

$$p(y|\beta) = \prod_{i=1}^{n} \binom{n_i}{y_i} \left(\frac{e^{\eta_i}}{1 + e^{\eta_i}} \right)^{y_i} \left(\frac{1}{1 + e^{\eta_i}} \right)^{n_i - y_i}.$$

We have already considered logistic regressions for a bioassay experiment in Section 3.7 and a public health survey in Section 6.5, and we present another example in Section 16.6 to illustrate the analysis of variance.

Other link functions are often used; for example, the probit link, $g(\mu) = \Phi^{-1}(\mu)$, is commonly used in econometrics. The data distribution for the probit model is

$$p(y|\beta) = \prod_{i=1}^{n} \binom{n_i}{y_i} (\Phi(\eta_i))^{y_i} (1 - \Phi(\eta_i))^{n_i - y_i}.$$

The probit link is obtained by retaining the normal variation process in the normal linear model while assuming that all outcomes are dichotomized. In practice, the probit and logit models are quite similar, differing mainly in the extremes of the tails. In either case, the factors of $\binom{n_i}{y_i}$ depend only on observed quantities and can be subsumed into a constant factor in the posterior density. The logit and probit models can also be generalized to model multivariate outcomes, as we discuss in Section 16.7.

The t distribution can be used as a robust alternative to the logit and probit models, as discussed in Section 17.2. Another standard link function is $g(\mu) = \log(-\log(\mu))$, the complementary log-log, which differs from the logit and probit by being asymmetrical in μ (that is, $g(\mu) \neq -g(1 - \mu)$) and is sometimes useful as an alternative for that reason.

Overdispersed models

Classical analyses of generalized linear models allow for the possibility of variation beyond that of the assumed sampling distribution, often called *overdispersion*. For an example, consider a logistic regression in which the sampling unit is a litter of mice and the proportion of the litter born alive is considered binomial with some explanatory variables (such as mother's age, drug dose, and so forth). The data might indicate more variation than expected under the binomial distribution due to systematic differences among mothers. Such variation could be incorporated in a hierarchical model using an indicator for each mother, with these indicators themselves following a distribution such as normal (which is often easy to interpret) or beta (which is conjugate to the binomial prior distribution). Section 16.5 gives an example of a Poisson regression model in which overdispersion is modeled by including a normal error term for each data point.

16.3 Setting up and interpreting generalized linear models

Canonical link functions

The description of the standard models in the previous section used what are known as the *canonical link* functions for each family. The canonical link is the function of the mean parameter that appears in the exponent of the exponential family form of the probability density (see page 42). We often use the canonical link, but nothing in our discussion is predicated on this choice; for example, the probit link for the binomial and the cumulative multinomial models (see Section 16.7) is not canonical.

Offsets

It is sometimes convenient to express a generalized linear model so that one of the explanatory variables has a known coefficient. An explanatory variable of this type is called an *offset*. The most common example of an offset occurs in models for Poisson data that arise as the number of incidents in a given exposure time T. If the rate of occurrence is μ per unit of time, then the number of incidents is Poisson with mean μT. We might like to take $\log \mu = X\beta$ as in the usual Poisson generalized linear model (16.2); however, generalized linear models are parameterized through the mean of y, which is μT, where T now represents the vector of exposure times for the units in the regression. We can apply the Poisson generalized linear model by augmenting the matrix of explanatory variables with a column containing the values $\log T$; this column of the augmented matrix corresponds to a coefficient with known value (equal to 1). An example appears in the Poisson regression of police stops in Section 16.5, where we use a measure of previous crime as an offset.

Interpreting the model parameters

The choice and parameterization of the explanatory variables x involve the same considerations as already discussed for linear models. The warnings about interpreting the regression coefficients from Chapter 14 apply here with one important addition. The linear predictor is used to predict the link function $g(\mu)$, rather than $\mu = \mathrm{E}(y)$, and therefore the effect of changing the jth explanatory variable x_j by a fixed amount depends on the current value of x. One way to translate effects onto the scale of y is to measure changes compared to a standard case with vector of predictors x_0 and standard outcome $y_0 = g^{-1}(x_0 \cdot \beta)$. Then, adding or subtracting the vector Δx leads to a change in the standard outcome from y_0 to $g^{-1}(g(y_0) \pm (\Delta x) \cdot \beta)$. This expression can also be written in differential form, but it is generally more useful to consider changes Δx that are not necessarily close to zero.

Understanding discrete-data models in terms of latent continuous data

An important idea, both in understanding discrete-data regression models and in their computation, is a reexpression in terms of unobserved (latent) continuous data. The probit model for binary data, $\Pr(y_i = 1) = \Phi((X\beta)_i)$, is equivalent to the following model on latent data u_i:

$$
\begin{aligned}
u_i &\sim \mathrm{N}((X\beta)_i, 1) \\
y_i &= \begin{cases} 1 & \text{if } u_i > 0 \\ 0 & \text{if } u_i < 0. \end{cases}
\end{aligned}
\tag{16.3}
$$

Sometimes the latent data can be given a useful interpretation. For example, in a political survey, if $y = 0$ or 1 represents support for the Democrats or the Republicans, then u can be seen as a continuous measure of partisan preference.

Another advantage of the latent parameterization for the probit model is that it allows a convenient Gibbs sampler computation. Conditional on the latent u_i's, the model is a simple linear regression (the first line of (16.3). Then, conditional on the model parameters and the data, the u_i's have truncated normal distributions, with each $u_i \sim \mathrm{N}((X\beta)_i, 1)$, truncated either to be negative or positive, depending on whether $y_i = 0$ or 1.

The latent-data idea can also be applied to logistic regression, in which case the first line of (16.3) becomes, $u_i \sim \mathrm{logistic}((X\beta)_i, 1)$, where the logistic distribution has a density of $1/(e^{x/2} + e^{-x/2})$—that is, the derivative of the inverse-logit function. This model is less convenient for computation than the probit (since the regression with logistic errors is not particularly easy to compute) but can still be useful for model understanding.

Similar interpretations can be given to ordered multinomial regressions of the sort described in Section 16.7 below. For example, if the data y can take on the values 0, 1, 2, or 3, then an ordered multinomial model can be defined in terms of cutpoints c_0, c_1, c_2, so that the response y_i equals 0 if $u_i < c_0$, 1 if $u_i \in (c_0, c_1)$, 2 if $u_i \in (c_1, c_2)$, and 3 if $u_i > c_2$. There is more than one natural way to parameterize these cutpoint models, and the choice of parameterization has implications when the model is placed in a hierarchical structure.

For example, the likelihood for a generalized linear model under this parameterization is unchanged if a constant is added to all three cutpoints c_0, c_1, c_2, and so the model is potentially nonidentifiable. The typical way to handle this is to set one of the cutpoints at a fixed value, for example setting $c_0 = 0$, so that only c_1 and c_2 need to be estimated from the data. (This generalizes the latent-data interpretation of binary data given above, for which 0 is the preset cutpoint.)

Bayesian nonhierarchical and hierarchical generalized linear models

We consider generalized linear models with noninformative prior distributions on β, informative prior distributions on β, and hierarchical models for which the prior distribution on β depends on unknown hyperparameters. We attempt to treat all generalized linear models with the same broad brush, which causes some difficulties. For example, some generalized linear models are expressed with a dispersion parameter in addition to the regression coefficients (σ in the normal case); here, we use the general notation ϕ for a dispersion parameter or parameters. A prior distribution can be placed on the dispersion parameter, and any prior information about β can be described conditional on the dispersion parameter; that is, $p(\beta, \phi) = p(\phi)p(\beta|\phi)$. In the description that follows we focus on the model for β. We defer computational details until the next section.

Noninformative prior distributions on β

The classical analysis of generalized linear models is obtained if a noninformative or flat prior distribution is assumed for β. The posterior mode corresponding to a noninformative uniform prior density is the maximum likelihood estimate for the parameter β, which can be obtained using iterative weighted linear regression (as implemented in the computer packages R or Glim, for example). Approximate posterior inference can be obtained from a normal approximation to the likelihood.

Conjugate prior distributions

A sometimes helpful approach to specifying prior information about β is in terms of hypothetical data obtained under the same model, that is, a vector, y_0, of n_0 hypothetical data points and a corresponding $n_0 \times k$ matrix of explanatory variables, X_0. As in Section 14.8, the resulting posterior distribution is identical to that from an augmented data vector $\binom{y}{y_0}$—that is, y and y_0 strung together as a vector, not the combinatorial coefficient—with matrix of explanatory variables $\binom{X}{X_0}$ and a noninformative uniform prior density on β. For computation with conjugate prior distributions, one can thus use the same iterative methods as for noninformative prior distributions.

Nonconjugate prior distributions

It is often more natural to express prior information directly in terms of the parameters β. For example, we might use the normal model, $\beta \sim N(\beta_0, \Sigma_\beta)$ with specified values of β_0 and Σ_β. A normal prior distribution on β is particularly convenient with the computational methods we describe in the next section, which are based on a normal approximation to the likelihood.

Hierarchical models

As in the normal linear model, hierarchical prior distributions for generalized linear models are a natural way to fit complex data structures and allow us to include more explanatory variables without encountering the problems of 'overfitting.'

A normal distribution for β is commonly used so that one can mimic the modeling practices and computational methods already developed for the hierarchical normal linear model, using the normal approximation to the generalized linear model likelihood described in the next section.

16.4 Computation

Posterior inference in generalized linear models typically requires the approximation and sampling tools of Part III. For example, for the hierarchical model for the rat tumor example in Section 5.1 the only reason we could compute the posterior distribution on a grid was that we assumed a conjugate beta prior distribution for the tumor rates. This trick would not be so effective, however, if we had linear predictors for the 70 experiments—at that point, it would be more appealing to set up a hierarchical logistic regression, for which exact computations are impossible.

There are different ways to apply iterative simulation to generalized linear models, but they all are based on extending algorithms that already work for linear regression and hierarchical linear models.

In general, it is useful to approximate the likelihood by a normal distribution in β and then apply the general computational methods for normal linear models described in Chapters 14 and 15. (The rat tumor example had only two regression coefficients, and so we were able to apply a less systematic treatment to compute the joint posterior distribution.) Given a method of approximating the likelihood by a normal distribution, computation can proceed as follows.

1. Find the posterior mode using an iterative algorithm. Each iteration in the mode-finding algorithm uses a quadratic approximation to the log posterior density of β and weighted linear regression.

2. Use the normal approximation as a starting point for simulations from the exact posterior distribution.

Adaptations of linear model computations

To obtain posterior simulations for generalized linear models, we build upon the basic ideas from hierarchical linear models: use linear regressions to update batches of coefficients in a block Gibbs sampler, or an augmented regression for all-at-once updating or to construct a linear transformation for one-at-a-time updating. In addition, the parameter space can be expanded to speed convergence, just as with hierarchical linear models. Each of these approaches must be altered to allow for the nonconjugacy of the generalized linear model likelihood. The most straightforward adaptation is to perform a Gibbs sampler using a normal approximation to the likelihood (described below), and then accept or reject at each step based on the Metropolis-Hastings rule (see Section 11.5). For the one-at-a-time algorithm, it can be more convenient to use symmetric Metropolis jumps at each step, with the standard deviation of the jumps set to 2.4 times the scale from the normal approximation (see Section 11.9). In any case, the basic Gibbs sampler updating structure is unchanged.

Normal approximation to the likelihood

Pseudodata and pseudovariances. For both mode-finding and the Markov chain simulation steps of the computation, we find it useful to approximate the likelihood by a normal distribution in β, conditional on the dispersion parameter ϕ, if necessary, and any hierarchical parameters. The basic method is to approximate the generalized linear model by a linear model; for each data point y_i, we construct a 'pseudodatum' z_i and a 'pseudovariance' σ_i^2 so that the generalized linear model likelihood, $p(y_i|(X\beta)_i, \phi)$, is approximated by the normal likelihood, $N(z_i|(X\beta)_i, \sigma_i^2)$. We can then combine the n pseudodata points and approximate the entire likelihood by a linear regression model of the vector $z = (z_1, \ldots, z_n)$ on the matrix of explanatory variables X, with known variance matrix, $\text{diag}(\sigma_1^2, \ldots, \sigma_n^2)$. This somewhat convoluted approach has the advantage of producing an approximate likelihood that we can analyze as if it came from normal linear regression data, thereby allowing the use of available linear regression algorithms.

Center of the normal approximation. In general, the normal approximation will depend on the value of β (and ϕ if the model has a dispersion parameter) at which it is centered. In the following development, we use the notation $(\hat{\beta}, \hat{\phi})$ for the point at which the approximation is centered and $\hat{\eta} = X\hat{\beta}$ for the corresponding vector of linear predictors. In the mode-finding stage of the computation, we iteratively alter the center of the normal approximation. Once we have approximately reached the mode, we use the normal approximation at that fixed value of $(\hat{\beta}, \hat{\phi})$.

Determining the parameters of the normal approximation. We can write the log-likelihood as

$$
\begin{aligned}
p(y_1, \ldots, y_n | \eta, \phi) &= \prod_{i=1}^{n} p(y_i | \eta_i, \phi) \\
&= \prod_{i=1}^{n} \exp(L(y_i | \eta_i, \phi)),
\end{aligned}
$$

where L is the log-likelihood function for the individual observations. We approximate each factor in the above product by a normal density in η_i, thus approximating each $L(y_i | \eta_i, \phi)$ by a quadratic function in η_i:

$$
L(y_i | \eta_i, \phi) \approx -\frac{1}{2\sigma_i^2}(z_i - \eta_i)^2 + \text{constant},
$$

where, in general, z_i, σ_i^2, and the constant depend on y, $\hat{\eta}_i = (X\hat{\beta})_i$, and $\hat{\phi}$. That is, the ith data point is approximately equivalent to an observation z_i, normally distributed with mean η_i and variance σ_i^2.

A standard way to determine z_i and σ_i^2 for the approximation is to match the first- and second-order terms of the Taylor series of $L(y_i | \eta_i, \phi)$ centered about $\hat{\eta}_i = (X\hat{\beta})_i$. Writing $dL/d\eta$ and $d^2L/d\eta^2$ as L' and L'', respectively, the result is

$$
\begin{aligned}
z_i &= \hat{\eta}_i - \frac{L'(y_i | \hat{\eta}_i, \hat{\phi})}{L''(y_i | \hat{\eta}_i, \hat{\phi})} \\
\sigma_i^2 &= -\frac{1}{L''(y_i | \hat{\eta}_i, \hat{\phi})}.
\end{aligned}
\tag{16.4}
$$

Example. The binomial-logistic model
In the binomial generalized linear model with logistic link, the log-likelihood for each observation is

$$
\begin{aligned}
L(y_i | \eta_i) &= y_i \log\left(\frac{e^{\eta_i}}{1 + e^{\eta_i}}\right) + (n_i - y_i) \log\left(\frac{1}{1 + e^{\eta_i}}\right) \\
&= y_i \eta_i - n_i \log(1 + e^{\eta_i}).
\end{aligned}
$$

(There is no dispersion parameter ϕ in the binomial model.) The derivatives of the log-likelihood are

$$
\begin{aligned}
\frac{dL}{d\eta_i} &= y_i - n_i \frac{e^{\eta_i}}{1 + e^{\eta_i}} \\
\frac{d^2L}{d\eta_i^2} &= -n_i \frac{e^{\eta_i}}{(1 + e^{\eta_i})^2}.
\end{aligned}
$$

Thus, the pseudodata z_i and variance σ_i^2 of the normal approximation for the ith sampling unit are

$$
z_i = \hat{\eta}_i + \frac{(1 + e^{\hat{\eta}_i})^2}{e^{\hat{\eta}_i}} \left(\frac{y_i}{n_i} - \frac{e^{\hat{\eta}_i}}{1 + e^{\hat{\eta}_i}} \right)
$$

$$
\sigma_i^2 = \frac{1}{n_i} \frac{(1 + e^{\hat{\eta}_i})^2}{e^{\hat{\eta}_i}}.
$$

The approximation depends on $\hat{\beta}$ through the linear predictor $\hat{\eta}$.

Finding the posterior mode

The usual first step in Bayesian computation is to find the posterior mode or modes. For noninformative or conjugate prior distributions, generalized linear model software can be used to maximize the posterior density. If no such software is available, it is still straightforward to find the modes by first obtaining a crude estimate and then applying an iterative weighted linear regression algorithm based on the normal approximation to the likelihood.

Crude estimation. Initial estimates for the parameters of a generalized linear model can be obtained using an approximate method such as fitting a normal linear model to appropriate transformations of x and y. As always, rough estimates are useful as starting points for subsequent calculations and as a check on the posterior distributions obtained from more sophisticated approaches.

Iterative weighted linear regression. Given crude estimates for the generalized linear model parameters, one can obtain the posterior mode using iterative weighted linear regression: at each step, one computes the normal approximation to the likelihood based on the current guess of (β, ϕ) and finds the mode of the resulting approximate posterior distribution by weighted linear regression. (If any prior information is available on β, it should be included as additional rows of data and explanatory variables in the regression, as described in Section 14.8 for fixed prior information and Chapter 15 for hierarchical models.) Iterating this process is equivalent to solving the system of k nonlinear equations, $dp(\beta|y)/d\beta = 0$, using Newton's method, and converges to the mode rapidly for standard generalized linear models. One possible difficulty is estimates of β tending to infinity, which can occur, for example, in logistic regression if there are some combinations of explanatory variables for which μ is nearly equal to zero or one. Substantive prior information tends to eliminate this problem.

If a dispersion parameter, ϕ, is present, one can update ϕ at each step of the iteration by maximizing its conditional posterior density (which is one-dimensional), given the current guess of β. Similarly, one can include in the iteration any hierarchical variance parameters that need to be estimated and update their values at each step.

The approximate normal posterior distribution

Once the mode $(\hat{\beta}, \hat{\phi})$ has been reached, one can approximate the conditional posterior distribution of β, given $\hat{\phi}$, by the output of the most recent weighted linear regression computation; that is,

$$p(\beta | \hat{\phi}, y) \approx \mathrm{N}(\beta | \hat{\beta}, V_\beta),$$

where V_β in this case is $(X^T \mathrm{diag}(-L''(y_i, \hat{\eta}_i, \hat{\phi})) X)^{-1}$. (In general, one need only compute the Cholesky factor of V_β, as described in Section 14.2.) If the sample size n is large, and ϕ is not part of the model (as in the binomial and Poisson distributions), we may be content to stop here and summarize the posterior distribution by the normal approximation to $p(\beta|y)$.

If a dispersion parameter, ϕ, is present, one can approximate the marginal distribution of ϕ using the method of Section 12.4 applied at the conditional mode, $\hat{\beta}(\phi)$,

$$p_{\mathrm{approx}}(\phi|y) = \frac{p(\beta, \phi|y)}{p_{\mathrm{approx}}(\beta|\phi, y)} \propto p(\hat{\beta}(\phi), \phi|y) |V_\beta(\phi)|^{1/2},$$

where $\hat{\beta}$ and V_β in the last expression are the mode and variance matrix of the normal approximation conditional on ϕ.

16.5 Example: hierarchical Poisson regression for police stops

There have been complaints in New York City and elsewhere that the police harass members of ethnic minority groups. In 1999, the New York State Attorney General's Office instigated a study of the New York City police department's 'stop and frisk' policy: the lawful practice of 'temporarily detaining, questioning, and, at times, searching civilians on the street.' The police have a policy of keeping records on stops, and we obtained all these records (about 175,000 in total) for a fifteen-month period in 1998–1999. We analyzed these data to see to what extent different ethnic groups were stopped by the police. We focused on blacks (African-Americans), hispanics (Latinos), and whites (European-Americans). The ethnic categories were as recorded by the police making the stops. We excluded members of other ethnic groups (about 4% of the stops) because of the likelihood of ambiguities in classifications. (With such a low frequency of 'other,' even a small rate of misclassifications could cause large distortions in the estimates for that group. For example, if only 4% of blacks, hispanics, and whites were mistakenly labeled as 'other,' then this would nearly double the estimates for the 'other' category while having very little effect on the three major groups.)

Aggregate data

Blacks and hispanics represented 51% and 33% of the stops, respectively, despite comprising only 26% and 24%, respectively, of the population of the city. Perhaps a more relevant comparison, however, is to the number of crimes committed by members of each ethnic group.

Data on actual crimes are not available, so as a proxy we used the number of arrests within New York City in 1997 as recorded by the Division of Criminal Justice Services (DCJS) of New York State. These were deemed to be the best available measure of local crime rates categorized by ethnicity. We used these numbers to represent the frequency of crimes that the police might suspect were committed by members of each group. When compared in that way, the ratio of stops to DCJS arrests was 1.24 for whites, 1.54 for blacks, and 1.72 for hispanics: based on this comparison, blacks are stopped 23% and hispanics 39% more often than whites.

Regression analysis to control for precincts

The analysis so far looks at average rates for the whole city. Suppose the police make more stops in high-crime areas but treat the different ethnic groups equally within any locality. Then the citywide ratios could show strong differences between ethnic groups even if stops are entirely determined by location rather than ethnicity. In order to separate these two kinds of predictors, we performed hierarchical analyses using the city's 75 precincts. Because it is possible that the patterns are systematically different in neighborhoods with different ethnic compositions, we divided the precincts into three categories in terms of their black population: precincts that were less than 10% black, 10%–40% black, and over 40% black. Each of these represented roughly 1/3 of the precincts in the city, and we performed separate analyses for each set.

For each ethnic group $e = 1, 2, 3$ and precinct p, we modeled the number of stops y_{ep} using an overdispersed Poisson regression with indicators for ethnic groups, a hierarchical model for precincts, and using n_{ep}, the number of DCJS arrests for that ethnic group in that precinct (multiplied by 15/12 to scale to a fifteen-month period), as an offset:

$$
\begin{aligned}
y_{ep} &\sim \text{Poisson}\left(n_{ep}e^{\alpha_e + \beta_p + \epsilon_{ep}}\right) \\
\beta_p &\sim \text{N}(0, \sigma_\beta^2) \\
\epsilon_{ep} &\sim \text{N}(0, \sigma_\epsilon^2),
\end{aligned}
\tag{16.5}
$$

where the coefficients α_e's control for ethnic groups, the β_p's adjust for variation among precincts, and the ϵ_{ep}'s allow for overdispersion. Of most interest are the exponentiated coefficients $\exp(\alpha_e)$, which represent relative rates of stops compared to arrests, after controlling for precinct.

By comparing to arrest rates, we can also separately analyze stops associated with different sorts of crimes. We did a separate comparison for each of four types of offenses: violent crimes, weapons offenses, property crimes, and drug crimes. For each, we modeled the number of stops y_{ep} by ethnic group e and precinct p for that crime type, using as a baseline the previous year's DCJS arrest rates n_{ep} for that crime type.

We thus estimated model (16.5) for twelve separate subsets of the data,

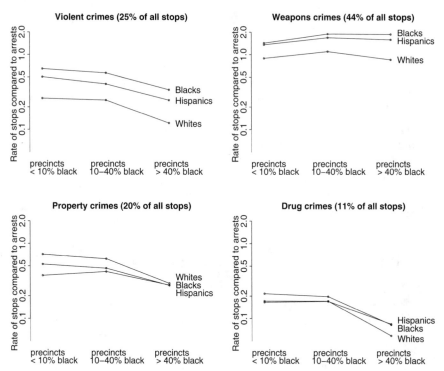

Figure 16.1 *Estimated rates* exp(α_e) *at which people of different ethnic groups were stopped for different categories of crime, as estimated from hierarchical regressions (16.5) using previous year's arrests as a baseline and controlling for differences between precincts. Separate analyses were done for the precincts that had less than 10%, 10%–40%, and more than 40% black population. For the most common stops— violent crimes and weapons offenses—blacks and hispanics were stopped about twice as often as whites. Rates are plotted on a logarithmic scale.*

corresponding to the four crime types and the three categories of precincts. We performed the computations using Bugs, linked from R (see Appendix C for general information about this approach). Figure 16.1 displays the estimated rates exp($\hat{\alpha}_e$). For each type of crime, the relative frequencies of stops for the different ethnic groups, are in the same order for each set of precincts. We also performed an analysis including the month of arrest. Rates of stops were roughly constant over the 15-month period and did not add anything informative to the comparison of ethnic groups.

Figure 16.1 shows that, for the most frequent categories of stops—those associated with violent crimes and weapons offenses—blacks and hispanics were much more likely to be stopped than whites, in all categories of precincts. For violent crimes, blacks and hispanics were stopped 2.5 times and 1.9 times as often as whites, respectively, and for weapons crimes, blacks and hispanics were stopped 1.8 times and 1.6 times as often as whites. In the less common categories of stop, whites are slightly more often stopped for property crimes and more often stopped for drug crimes, in proportion to their previous year's arrests in any given precinct.

Does the overall pattern of disproportionate stops of minorities imply that the NYPD was acting in an unfair or racist manner? Not at all. It is quite reasonable to suppose that effective policing requires many people to be stopped and questioned in order to gather information about any given crime. In the context of some difficult relations between the police and ethnic minority communities in New York City, it is useful to have some quantitative sense of the issues under dispute. Given that there have been complaints about the frequency with which the police have been stopping blacks and hispanics, it is relevant to know that this is indeed a statistical pattern. The police department then has the opportunity to explain their policies to the affected communities.

16.6 Example: hierarchical logistic regression for political opinions

We illustrate the application of the analysis of variance (see Section 15.6) to hierarchical generalized linear models with a model of public opinion.

Dozens of national opinion polls are conducted by media organizations before every election, and it is desirable to estimate opinions at the levels of individual states as well as for the entire country. These polls are generally based on national random-digit dialing with corrections for nonresponse based on demographic factors such as sex, ethnicity, age, and education. We estimated state-level opinions from these polls, while simultaneously correcting for nonresponse, in two steps. For any survey response of interest:

1. We fit a regression model for the individual response y given demographics and state. This model thus estimates an average response θ_j for each crossclassification j of demographics and state. In our example, we have sex (male or female), ethnicity (African-American or other), age (4 categories), education (4 categories), and 50 states; thus 3200 categories.

2. From the U.S. Census, we get the adult population N_j for each category j. The estimated population average of the response y in any state s is then $\theta_s = \sum_{j \in s} N_j \theta_j / \sum_{j \in s} N_j$, with each summation over the 64 demographic categories in the state.

We need a large number of categories because (a) we are interested in separating out the responses by state, and (b) nonresponse adjustments force us to include the demographics. As a result, any given survey will have few or no data in many categories. This is not a problem, however, if a multilevel model is fitted. Each factor or set of interactions in the model, corresponding to a row in the ANOVA plot, is automatically given a variance component.

As discussed in the survey sampling literature, this inferential procedure works well and outperforms standard survey estimates when estimating state-level outcomes. For this example, we choose a single outcome—the probability that a respondent prefers the Republican candidate for President—as

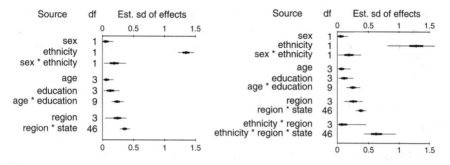

Figure 16.2 *ANOVA display for two logistic regression models of the probability that a survey respondent prefers the Republican candidate for the 1988 U.S. Presidential election, based on data from seven CBS News polls. Point estimates and error bars show posterior medians, 50% intervals, and 95% intervals of the finite-population standard deviations s_m. The demographic factors are those used by CBS to perform their nonresponse adjustments, and states and regions are included because we were interested in estimating average opinions by state. The large effects for ethnicity, region, and state suggest that it might make sense to include interactions, hence the inclusion of the ethnicity × region and ethnicity × state effects in the second model.*

estimated by a logistic regression model from a set of seven CBS News polls conducted during the week before the 1988 Presidential election. We focus here on the first stage of the estimation procedure—the inference for the logistic regression model—and use our ANOVA tools to display the relative importance of each factor in the model.

We label the survey responses y_i as 1 for supporters of the Republican candidate and 0 for supporters of the Democrat (with undecideds excluded) and model them as independent, with $\Pr(y_i = 1) = \text{logit}^{-1}((X\beta)_i)$. The design matrix X is all 0's and 1's with indicators for the demographic variables used by CBS in the survey weighting: sex, ethnicity, age, education, and the interactions of sex × ethnicity and age × education. We also include in X indicators for the 50 states and for the 4 regions of the country (northeast, midwest, south, and west). Since the states are nested within regions, no main effects for states are needed. As in our general approach for linear models, we give each batch of regression coefficients an independent normal distribution centered at zero and with standard deviation estimated hierarchically given a uniform prior density.

We fitted the model using Bugs, linked from R, which we used to compute the finite-sample standard deviations and plot the results. The left plot of Figure 16.2 displays the analysis of variance, which shows that ethnicity is by far the most important demographic factor, with state also explaining quite a bit of variation.

The natural next step is to consider interactions among the most important effects, as shown in the 'plot on the right side of Figure 16.2. The ethnicity × state × region interactions are surprisingly large: the differences between

African-Americans and others vary dramatically by state. As with the example in Section 15.6 of internet connect times, the analysis of variance is a helpful tool in understanding the importance of different components in a hierarchical model.

16.7 Models for multinomial responses

Extension of the logistic link

Appropriate models for polychotomous data can be developed as extensions of either the Poisson or binomial models. We first show how the logistic model for binary data can be extended to handle multinomial data. The notation for a multinomial random variable y_i with sample size n_i and k possible outcomes (that is, y_i is a vector with $\sum_{j=1}^{k} y_{ij} = n_i$) is $y_i \sim \text{Multin}(n_i; \alpha_{i1}, \alpha_{i2}, \ldots, \alpha_{ik})$, with α_{ij} representing the probability of the jth category, and $\sum_{j=1}^{k} \alpha_{ij} = 1$. A standard way to parameterize the multinomial generalized linear model is in terms of the logarithm of the ratio of the probability of each category relative to that of a baseline category, which we label $j = 1$, so that

$$\log(\alpha_{ij}/\alpha_{i1}) = \eta_{ij} = (X\beta_j)_i,$$

where β_j is a vector of parameters for the jth category. The data distribution is then

$$p(y|\beta) \propto \prod_{i=1}^{n} \prod_{j=1}^{k} \left(\frac{e^{\eta_{ij}}}{\sum_{l=1}^{k} e^{\eta_{il}}} \right)^{y_{ij}},$$

with β_1 set equal to zero, and hence $\eta_{i1} = 0$ for each i. The vector β_j indicates the effect of a change in X on the probability of observing outcomes in category j relative to category 1. Often the linear predictor includes a set of indicator variables for the outcome categories indicating the relative frequencies when the explanatory variables take the default value $X = 0$; in that case we can write δ_j as the coefficient of the indicator for category j and $\eta_{ij} = \delta_j + (X\beta_j)_i$, with δ_1 and β_1 typically set to 0.

Special methods for ordered categories

There is a distinction between multinomial outcomes with ordinal categories (for example, grades A, B, C, D) and those with nominal categories (for example, diseases). For ordinal categories the generalized linear model is often expressed in terms of cumulative probabilities ($\pi_{ij} = \sum_{l \leq j} \alpha_{il}$) rather than category probabilities, with $\log(\pi_{ij}/(1 - \pi_{ij})) = \delta_j + (X\beta_j)_i$, where once again we typically take $\delta_1 = \beta_1 = 0$. Due to the ordering of the categories, it may be reasonable to consider a model with a common set of regression parameters $\beta_j = \beta$ for each j. Another common choice for ordered categories is the multinomial probit model. Either of these models can also be expressed in latent variable form as described at the end of Section 16.2

| White | | | | Black pieces | | | | |
pieces	Kar	Kas	Kor	Lju	Sei	Sho	Spa	Tal
Karpov		1-0-1	1-0-0	0-1-1	1-0-0	0-0-2	0-0-0	0-0-0
Kasparov	0-0-0		1-0-0	0-0-0	0-0-1	1-0-0	1-0-0	0-0-2
Korchnoi	0-0-1	0-2-0		0-0-0	0-1-0	0-0-2	0-0-1	0-0-0
Ljubojevic	0-1-0	0-1-1	0-0-2		0-1-0	0-0-1	0-0-2	0-0-1
Seirawan	0-1-1	0-0-1	1-1-0	0-2-0		0-0-0	0-0-0	1-0-0
Short	0-0-1	0-2-0	0-0-0	1-0-1	2-0-1		0-0-1	1-0-0
Spassky	0-1-0	0-0-2	0-0-1	0-0-0	0-0-1	0-0-1		0-0-0
Tal	0-0-2	0-0-0	0-0-3	0-0-0	0-0-1	0-0-0	0-0-1	

Table 16.1 *Subset of the data from the 1988-1989 World Cup of chess: results of games between eight of the 29 players. Results are given as wins, losses, and draws; for example, when playing with the white pieces against Kasparov, Karpov had one win, no losses, and one draw. For simplicity, this table aggregates data from all six tournaments.*

Using the Poisson model for multinomial responses

Multinomial response data can also be analyzed using Poisson models by conditioning on appropriate totals. As this method is useful in performing computations, we describe it briefly and illustrate its use with an example. Suppose that $y = (y_1, \ldots, y_k)$ are independent Poisson random variables with means $\lambda = (\lambda_1, \ldots, \lambda_k)$. Then the conditional distribution of y, given $n = \sum_{j=1}^{k} y_j$, is multinomial:

$$p(y|n, \alpha) = \text{Multin}(y|n; \alpha_1, \ldots, \alpha_k), \tag{16.6}$$

with $\alpha_j = \lambda_j / \sum_{i=1}^{k} \lambda_i$. This relation can also be used to allow data with multinomial response variables to be fitted using Poisson generalized linear models. The constraint on the sum of the multinomial probabilities is imposed by incorporating additional covariates in the Poisson regression whose coefficients are assigned uniform prior distributions. We illustrate with an example.

Example. World Cup chess

The 1988–1989 World Cup of chess consisted of six tournaments involving 29 of the world's top chess players. Each of the six tournaments was a single round-robin (that is, each player played every other player exactly once) with 16 to 18 players. In total, 789 games were played; for each game in each tournament, the players, the outcome of the game (win, lose, or draw), and the identity of the player making the first move (thought to provide an advantage) are recorded. A subset of the data is displayed in Table 16.1.

Multinomial model for paired comparisons with ties. A standard model for analyzing paired comparisons data, such as the results of a chess competition, is the Bradley–Terry model. In its most basic form the model assumes that the probability that player i defeats player j is $p_{ij} = \exp(\alpha_i - \alpha_j)/(1 + \exp(\alpha_i - \alpha_j))$, where α_i, α_j are parameters representing player abilities. The parameterization using α_i rather than $\gamma_i = \exp(\alpha_i)$ anticipates the generalized linear model approach. This basic model does not address the possibility of a draw nor the advantage of moving first; an extension of the model follows for the case when i moves first:

$$p_{ij1} = \Pr(i \text{ defeats } j|\theta) = \frac{e^{\alpha_i}}{e^{\alpha_i} + e^{\alpha_j+\gamma} + e^{\delta+\frac{1}{2}(\alpha_i+\alpha_j+\gamma)}}$$

$$p_{ij2} = \Pr(j \text{ defeats } i|\theta) = \frac{e^{\alpha_j+\gamma}}{e^{\alpha_i} + e^{\alpha_j+\gamma} + e^{\delta+\frac{1}{2}(\alpha_i+\alpha_j+\gamma)}}$$

$$p_{ij3} = \Pr(i \text{ draws with } j|\theta) = \frac{e^{\delta+\frac{1}{2}(\alpha_i+\alpha_j+\gamma)}}{e^{\alpha_i} + e^{\alpha_j+\gamma} + e^{\delta+\frac{1}{2}(\alpha_i+\alpha_j+\gamma)}}, \quad (16.7)$$

where γ determines the relative advantage or disadvantage of moving first, and δ determines the likelihood of a draw.

Parameterization as a Poisson regression model with logarithmic link. Let $y_{ijk}, k = 1, 2, 3$, represent the number of wins, losses, and draws for player i in games with player j for which i had the first move. We create a Poisson generalized linear model that is equivalent to the desired multinomial model. The y_{ijk} are assumed to be independent Poisson random variables given the parameters. The mean of y_{ijk} is $\mu_{ijk} = n_{ij}p_{ijk}$, where n_{ij} is the number of games between players i and j for which i had the first move. The Poisson generalized linear model equates the logarithms of the means of the y_{ijk}'s to a linear predictor. The logarithms of the means for the components of y are

$$\log \mu_{ij1} = \log n_{ij} + \alpha_i - A_{ij}$$

$$\log \mu_{ij2} = \log n_{ij} + \alpha_j + \gamma - A_{ij}$$

$$\log \mu_{ij3} = \log n_{ij} + \delta + \frac{1}{2}\alpha_i + \frac{1}{2}\alpha_j + \frac{1}{2}\gamma - A_{ij}, \quad (16.8)$$

with A_{ij} the logarithm of the denominator of the model probabilities in (16.7). The A_{ij} terms allow us to impose the constraint on the sum of the three outcomes so that the three Poisson random variables describe the multinomial distribution; this is explained further below.

Setting up the vectors of data and explanatory variables. To describe completely the generalized linear model that is suggested by the previous paragraph, we explain the various components in some detail. The outcome variable y is a vector of length $3 \times 29 \times 28$ containing the frequency of the three outcomes for each of the 29×28 ordered pairs (i, j). The mean vector is of the same length consisting of triples $(\mu_{ij1}, \mu_{ij2}, \mu_{ij3})$ as described above. The logarithmic link expresses the logarithm of the mean vector as the linear model $X\beta$. The parameter vector β consists of the 29 player ability parameters ($\alpha_i, i = 1, \ldots, 29$), the first-move advantage parameter γ, the draw parameter δ, and the 29×28 nuisance parameters A_{ij} that were introduced to create the Poisson model. The columns of the model matrix X can be obtained by examining the expressions (16.8). For example, the first column of X, corresponding to α_1, is 1 in any row corresponding to a win for player 1 and 0.5 in any row corresponding to a draw for player 1. Similarly the column of the model matrix X corresponding to δ is 1 for each row corresponding to a draw and 0 elsewhere. The final 29×28 columns of X correspond to the parameters A_{ij} which are not of direct interest. Each column is 1 for the 3 rows that correspond to games between i and j for which i has the first move and zero elsewhere. In simulating from the posterior distribution, the parameters A_{ij} will not be sampled but instead are used to ensure that $y_{ij1} + y_{ij2} + y_{ij3} = n_{ij}$ as required by the multinomial distribution.

Using an offset to make the Poisson model correspond to the multinomial distribution. The sample size n_{ij} is a slight complication since the means of the Poisson counts clearly depend on this sample size. According to the model (16.8), $\log n_{ij}$ should be included as a column of the model matrix X with known coefficient (equal to 1). A predictor with known coefficient is known as an *offset* in the terminology of generalized linear models (see page 418). Assuming a noninformative prior distribution for all the model parameters, this Poisson generalized linear model for the chess data is overparameterized, in the sense that the probabilities specified by the model are unchanged if a constant is added to each of the α_i. It is common to require $\alpha_1 = 0$ to resolve this problem. Similarly, one of the A_{ij} must be set to zero. A natural extension would be to treat the abilities as random effects, in which case the restriction $\alpha_1 = 0$ is no longer required.

16.8 Loglinear models for multivariate discrete data

A standard approach to describe association among several categorical variables uses the family of *loglinear models*. In a loglinear model, the response or outcome variable is multivariate and discrete: the contingency table of counts cross-classified according to several categorical variables. The counts are modeled as Poisson random variables, and the logarithms of the Poisson means are described by a linear model incorporating indicator variables for the various categorical levels. Alternatively, the counts in each of several margins of the table may be modeled as multinomial random variables if the total sample size or some marginal totals are fixed by design. Loglinear models can be fitted as a special case of the generalized linear model. Why then do we include a separate section concerned with loglinear models? Basically because loglinear models are commonly used in applications with multivariate discrete data analysis—especially for multiple imputation (see Chapter 21)—and because there is an alternative computing strategy that is useful when interest focuses on the expected counts and a conjugate prior distribution is used.

The Poisson or multinomial likelihood

Consider a table of counts $y = (y_1, \ldots, y_n)$, where $i = 1, \ldots, n$ indexes the cells of the possibly multi-way table. Let $\mu = (\mu_1, \ldots, \mu_n)$ be the vector of expected counts. The Poisson model for the counts y has the distribution,

$$p(y|\mu) = \prod_{i=1}^{n} \frac{1}{y_i!} \mu_i^{y_i} e^{-\mu_i}.$$

If the total of the counts is fixed by the design of the study, then a multinomial distribution for y is appropriate, as in (16.6). If other features of the data are fixed by design—perhaps row or column sums in a two-way table—then the likelihood is the product of several independent multinomial distributions. (For example, the data could arise from a stratified sample survey that was constrained to include exactly 500 respondents from each of four geographical regions.) In the remainder of this section we discuss the Poisson model, with additional discussion where necessary to describe the modifications required for alternative models.

Setting up the matrix of explanatory variables

The loglinear model constrains the all-positive expected cell counts μ to fall on a regression surface, $\log \mu = X\beta$. The *incidence matrix* X is assumed known, and its elements are all zeros and ones; that is, all the variables in x are indicator variables. We assume that there are no 'structural' zeros—cells i for which the expected count μ_i is zero by definition, and thus $\log \mu_i = -\infty$. (An example of a structural zero would be the category of 'women with prostate cancer' in a two-way classification of persons by sex and medical condition.)

The choice of indicator variables to include depends on the important relations among the categorical variables. As usual, when assigning a noninformative prior distribution for the effect of a categorical variable with k categories, one should include only $k-1$ indicator variables. Interactions of two or more effects, represented by products of main effect columns, are used to model lack of independence. Typically a range of models is possible.

The *saturated model* includes all variables and interactions; with the noninformative prior distribution, the saturated model has as many parameters as cells in the table. The saturated model has more practical use when combined with an informative or hierarchical prior distribution, in which case there are actually *more* parameters than cells because all k categories will be included for each factor. At the other extreme, the *null model* assigns equal probabilities to each cell, which is equivalent to fitting only a constant term in the hierarchical model. A commonly used simple model is *independence*, in which parameters are fitted to all one-way categories but no two-way or higher categories. With three categorical variables z_1, z_2, z_3, the joint independence model has no interactions; the saturated model has all main (one-way) effects, two-way interactions, and the three-way interaction; and models in between are used to describe different degrees of association among variables. For example, in the loglinear model that includes $z_1 z_3$ and $z_2 z_3$ interactions but no others, z_1 and z_2 are conditionally independent given z_3.

Prior distributions

Conjugate Dirichlet family.   The conjugate prior density for the expected counts μ resembles the Dirichlet density:

$$p(\mu) \propto \prod_{i=1}^{n} \mu_i^{k_i - 1}, \qquad (16.9)$$

with the constraint that the cell counts fit the loglinear model; that is, $p(\mu) = 0$ unless some β exists for which $\log \mu = X\beta$. For a Poisson sampling model, the usual Dirichlet constraint, $\sum_{i=1}^{n} \mu_i = 1$, is not present. Densities from the family (16.9) with values of k_i set between 0 and 1 are commonly used for noninformative prior specifications.

Nonconjugate distributions. Nonconjugate prior distributions arise, for example, from hierarchical models in which parameters corresponding to high-order interactions are treated as exchangeable. Unfortunately, such models are not amenable to the special computational methods for loglinear models described below.

Computation

Finding the mode. As a first step to computing posterior inferences, we always recommend obtaining initial estimates of β using some simple procedure. In loglinear models, these crude estimates will often be obtained using standard loglinear model software with the original data supplemented by the 'prior cell counts' k_i if a conjugate prior distribution is used as in (16.9). For some special loglinear models, the maximum likelihood estimate, and hence the expected counts, can be obtained in closed form. For the saturated model, the expected counts equal the observed counts, and for the null model, the expected counts are all equal to $\sum_{i=1}^{n} y_i/n$. In the independence model, the estimates for the loglinear model parameters β are obtained directly from marginal totals in the contingency table.

For more complicated models, however, the posterior modes cannot generally be obtained in closed form. In these cases, an iterative approach, *iterative proportional fitting* (IPF), can be used to obtain the estimates. In IPF, an initial estimate of the vector of expected cell counts, μ, chosen to fit the model, is iteratively refined by multiplication by appropriate scale factors. For most problems, a convenient starting point is $\beta = 0$, so that $\mu = 1$ for all cells (assuming the loglinear model contains a constant term).

At each step of IPF, the table counts are adjusted to match the model's sufficient statistics (marginal tables). The iterative proportional fitting algorithm is generally expressed in terms of γ, the factors in the multiplicative model, which are exponentials of the loglinear model coefficients:

$$\gamma_j = \exp(\beta_j).$$

The prior distribution is assumed to be the conjugate Dirichlet-like distribution (16.9). Let

$$y_{j+} = \sum_i x_{ij}(y_i + k_i)$$

represent the margin of the table corresponding to the jth column of X. At each step of the IPF algorithm, a single parameter is altered. The basic step, updating γ_j, assigns

$$\gamma_j^{\text{new}} = \frac{y_{j+}}{\sum_{i=1}^{n} x_{ij} \mu_i^{\text{old}}} \gamma_j^{\text{old}}.$$

Then the expected cell counts are modified accordingly,

$$\mu_i^{\text{new}} = \mu_i^{\text{old}} \left(\frac{\gamma_j^{\text{new}}}{\gamma_j^{\text{old}}} \right)^{x_{ij}}, \qquad (16.10)$$

rescaling the expected count in each cell i for which $x_{ij} = 1$. These two steps are repeated indefinitely, cycling through all of the parameters j. The resulting series of tables converges to the mode of the posterior distribution of μ given the data and prior information. Cells with values equal to zero that are assumed to have occurred by chance (random zeros as opposed to structural zeros) are generally not a problem unless a saturated model is fitted or all of the cells needed to estimate a model parameter are zero. The iteration is continued until the expected counts do not change appreciably.

Bayesian IPF.   Bayesian computations for loglinear models apply a stochastic version of the IPF algorithm, a Gibbs sampler applied to the vector γ. As in IPF, an initial estimate is required, the simplest choice being $\gamma_j = 1$ for all j, which corresponds to all expected cell counts equal to 1 (recall that the expected cell counts are just products of the γ's). The only danger in choosing the initial estimates, as with iterative proportional fitting, is that the initial choices cannot incorporate structure corresponding to interactions that are not in the model. The simple choice recommended above is always safe. At each step of the Bayesian IPF algorithm, a single parameter is altered. To update γ_j, we assign

$$\gamma_j^{\text{new}} = \frac{A}{2y_{j+}} \frac{y_{j+}}{\sum_{i=1}^{n} x_{ij} \mu_i^{\text{old}}} \gamma_j^{\text{old}},$$

where A is a random draw from a χ^2 distribution with $2y_{j+}$ degrees of freedom. This step is identical to the usual IPF step except for the random first factor. Then the expected cell counts are modified using the formula (16.10). The two steps are repeated indefinitely, cycling through all of the parameters j. The resulting series of tables converges to a series of draws from the posterior distribution of μ given the data and prior information.

Bayesian IPF can be modified for non-Poisson sampling schemes. For example, for multinomial data, after each step of the Bayesian IPF, when expected cell counts are modified, we just rescale the vector μ to have the correct total. This amounts to using the β_j (or equivalently the γ_j) corresponding to the intercept of the loglinear model to impose the multinomial constraint. In fact, we need not sample this γ_j during the algorithm because it is used at each step to satisfy the multinomial constraint. Similarly, a product multinomial sample would have several parameters determined by the fixed marginal totals.

The computational approach presented here can be used only for models of categorical measurements. If there are both categorical and continuous measurements then more appropriate analyses are obtained using the normal linear model (Chapter 14) if the outcome is continuous, or a generalized linear model from earlier in this chapter if the outcome is categorical.

Example. Analysis of a survey on modes of information

We illustrate the computational methods of Bayesian loglinear models with an example from the statistical literature. A sample of 1729 persons was interviewed concerning their knowledge about cancer and its sources. Each person was scored dichotomously with respect to whether information was obtained from (1) the news media, (2) light reading, (3) solid reading, and (4) lectures. Each person was also scored as having either good or poor knowledge about cancer; the first column of numbers in Table 16.2 displays the number of persons in each category. We treat the data as a multinomial sample of size 1729 rather than as Poisson cell counts. We fit a model including all two-way interactions, so that β has 16 parameters: five main effects, ten two-way interactions, and the intercept (which is used to impose the multinomial constraint as described on page 433). We used the noninformative prior distribution with $k_i = 0.5$ for each cell i. The second column of numbers in the table shows the posterior mode of the expected cell counts obtained by IPF.

Ten independent sequences of the Bayesian IPF algorithm were run for 500 steps each (that is, 500 draws of each γ_j), at which point the potential scale reductions $\widehat{R}$ were below 1.05 for all parameters in the model. The final three columns of Table 16.2 give quantiles from the posterior simulations of the cell counts. The posterior medians are close to the mode, indicating that the posterior distribution is roughly symmetric, a consequence of having a large number of counts and relatively few parameters. The differences between the posterior modes and the observed cell counts arise from the constraints in the fitted model, which sets all three-way and higher-order interactions to zero.

Posterior inferences for the model parameters γ_j are displayed in Table 16.3. The 'Baseline' parameter corresponds to the constant term in the loglinear model; all the other parameters can be interpreted as the ratio of 'yes' to 'no' responses for a question or pair of questions. The first five parameters are estimated to be less than 1, which fits the aspect of the data that 'yes' responses are less common than 'no' for all five questions (see Table 16.2). The second five parameters are all estimated to be certainly or probably more than 1, which fits the aspect of the data that respondents who use one mode of information tend to use other modes of information, and respondents who are more informed tend to be more knowledgeable about cancer. For example, under the model, people who have attended lectures about cancer are one to two times as likely to be knowledgeable about cancer, compared to non-attenders.

Deficiencies of the model could be studied by examining the residuals between observed counts and expected counts under the model, using simulations from the posterior predictive distribution to assess whether the observed discrepancies are plausible under the model.

Sources of information				Know-		Estimates of expected counts			
							Post.	Posterior quantiles	
(1)	(2)	(3)	(4)	ledge	Count	mode	2.5%	med.	97.5%
Y	Y	Y	Y	Good	23	26.9	19.3	26.6	35.5
Y	Y	Y	Y	Poor	8	10.9	7.2	10.7	15.4
Y	Y	Y	N	Good	102	104.8	86.8	103.3	122.2
Y	Y	Y	N	Poor	67	65.1	52.9	64.2	77.9
Y	Y	N	Y	Good	8	5.2	3.2	5.0	7.7
Y	Y	N	Y	Poor	4	5.6	3.4	5.4	8.6
Y	Y	N	N	Good	35	34.6	27.0	33.9	42.9
Y	Y	N	N	Poor	59	56.8	45.4	56.5	69.1
Y	N	Y	Y	Good	27	27.1	19.8	26.7	35.1
Y	N	Y	Y	Poor	18	15.1	10.2	14.7	20.4
Y	N	Y	N	Good	201	204.2	179.1	202.3	226.7
Y	N	Y	N	Poor	177	172.9	149.5	170.3	193.8
Y	N	N	Y	Good	7	5.7	3.5	5.6	8.6
Y	N	N	Y	Poor	6	8.4	5.1	8.2	12.3
Y	N	N	N	Good	75	73.4	60.4	72.4	86.3
Y	N	N	N	Poor	156	164.2	142.9	163.4	186.4
N	Y	Y	Y	Good	1	2.2	1.3	2.1	3.7
N	Y	Y	Y	Poor	3	1.7	0.9	1.6	2.8
N	Y	Y	N	Good	16	13.0	9.1	12.7	17.1
N	Y	Y	N	Poor	16	15.3	11.2	15.0	20.0
N	Y	N	Y	Good	4	1.8	1.0	1.7	2.9
N	Y	N	Y	Poor	3	3.6	2.1	3.5	5.6
N	Y	N	N	Good	13	17.5	12.6	17.0	22.8
N	Y	N	N	Poor	50	54.8	43.3	54.0	67.3
N	N	Y	Y	Good	3	5.4	3.4	5.2	8.2
N	N	Y	Y	Poor	8	5.7	3.4	5.4	8.5
N	N	Y	N	Good	67	60.4	48.7	59.9	72.3
N	N	Y	N	Poor	83	97.3	82.3	95.5	111.3
N	N	N	Y	Good	2	4.7	2.7	4.5	7.1
N	N	N	Y	Poor	10	13.0	8.1	12.4	18.5
N	N	N	N	Good	84	89.0	73.7	87.6	103.4
N	N	N	N	Poor	393	378.6	344.3	375.2	410.3

Table 16.2 *Contingency table describing a survey of sources and quality of information about cancer for 1729 people. Sources of information are: (1) news media, (2) light reading, (3) solid reading, (4) lectures. For each category, the observed count* y_i, *the posterior mode of expected count* μ_i, *and summaries of the posterior distribution of* μ_i *are given, based on the model fitting one- and two-way interactions. From Fienberg (1977).*

Factor	Posterior quantiles				
	2.5%	25%	med.	75%	97.5%
(1) News media	0.37	0.41	0.43	0.46	0.51
(2) Light reading	0.11	0.13	0.14	0.16	0.18
(3) Solid reading	0.21	0.24	0.25	0.27	0.30
(4) Lectures	0.02	0.03	0.03	0.04	0.05
(5) Knowledge	0.19	0.22	0.23	0.25	0.28
(1) and (2)	1.80	2.19	2.40	2.64	3.12
(1) and (3)	3.32	3.83	4.10	4.39	5.08
(1) and (4)	0.98	1.30	1.52	1.76	2.29
(1) and (5)	1.51	1.75	1.91	2.06	2.36
(2) and (3)	0.84	1.01	1.09	1.19	1.37
(2) and (4)	1.35	1.71	1.94	2.20	2.81
(2) and (5)	1.07	1.25	1.36	1.47	1.72
(3) and (4)	1.16	1.49	1.72	1.98	2.56
(3) and (5)	2.17	2.49	2.68	2.89	3.30
(4) and (5)	1.08	1.36	1.54	1.74	2.26
Baseline	344.3	364.1	375.2	387.6	410.3

Table 16.3 *Posterior inferences for the loglinear model parameters γ_j for the modes of information example. The 'baseline' parameter corresponds to the constant term in the loglinear model; it is the expected count for the cell with all 'no' responses and thus corresponds to the last line in Table 16.2.*

16.9 Bibliographic note

The term 'generalized linear model' was coined by Nelder and Wedderburn (1972), who modified Fisher's scoring algorithm for maximum likelihood estimation. An excellent (non-Bayesian) reference is McCullagh and Nelder (1989). Hinde (1982) and Liang and McCullagh (1993) discuss various models of overdispersion in generalized linear models and examine how they fit actual data.

Albert and Chib (1995) and Gelman, Goegebeur, et al. (2000) discuss Bayesian residual analysis and other model checks for discrete-data regressions; see also Landwehr, Pregibon, and Shoemaker (1984).

Knuiman and Speed (1988) and Albert (1988) present Bayesian analyses of contingency tables based on analytic approximations. Bedrick, Christensen, and Johnson (1996) discuss prior distributions for generalized linear models.

Dempster, Selwyn, and Weeks (1983) is an early example of fully-Bayesian inference for logistic regression (using a normal approximation corrected by importance sampling to compute the posterior distribution for the hyperparameters). Zeger and Karim (1991), Karim and Zeger (1992), and Albert (1992) use Gibbs sampling to incorporate random effects in generalized linear models. Dellaportas and Smith (1993) describe Gibbs sampling using the rejection method of Gilks and Wild (1992) to sample each component of β conditional on the others; they show that this approach works well if the canonical link function is used. Albert and Chib (1993) perform Gibbs sampling for binary and polychotomous response data by introducing continuous latent scores. Gelfand and Sahu (1999) discuss Gibbs sampling for generalized linear models.

The police stop-and-frisk study is described in Spitzer (1999) and Gelman, Fagan, and Kiss (2007). The pre-election polling example comes from Park, Gelman, and Bafumi (2003); see also Gelman and Little (1997).

Belin et al. (1993) fit hierarchical logistic regression models to missing data in a census adjustment problem, performing approximate Bayesian computations using the ECM algorithm. This article also includes extensive discussion of the choices involved in setting up the model and the sensitivity to assumptions. Imai and van Dyk (2003) discuss Bayesian computation for unordered multinomial probit models. The basic paired comparisons model is due to Bradley and Terry (1952); the extension for ties and order effects is due to Davidson and Beaver (1977). Other references on models for paired comparisons include Stern (1990) and David (1988). The World Cup chess data are analyzed by Glickman (1993). Johnson (1996) presents a Bayesian analysis of categorical data that was ordered by multiple raters, and Bradlow and Fader (2001) use hierarchical Bayesian methods to model parallel time series of rank data. Jackman (2001) and Martin and Quinn (2002a) apply hierarchical Bayesian models to estimate ideal points from political data. Johnson (1997) presents a detailed example of a hierarchical discrete-data regression model for university grading; this article is accompanied by several interesting discussions.

Books on loglinear models include Fienberg (1977) and Agresti (2002). Goodman (1991) provides a review of models and methods for contingency table analysis. See Fienberg (2000) for a recent overview. Good (1965) discusses a variety of Bayesian models, including hierarchical models, for contingency tables based on the multinomial distribution. The example in Section 16.8 on modes of information is analyzed by Fienberg (1977).

Iterative proportional fitting was first presented by Deming and Stephan (1940). The Bayesian iterative proportional fitting algorithm was proposed by Gelman and Rubin (1991); a related algorithm using ECM to find the mode for a loglinear model with missing data appears in Meng and Rubin (1993). Dobra, Tebaldi, and West (2003) present recent work on Bayesian inference for contingency tables.

16.10 Exercises

1. Normal approximation for generalized linear models: derive equations (16.4).

2. Computation for a simple generalized linear model:

   (a) Express the bioassay example of Section 3.7 as a generalized linear model and obtain posterior simulations using the computational techniques presented in Section 16.4.

Length of roll	# faults	Length of roll	# faults	Length of roll	# faults
551	6	441	8	657	9
651	4	895	28	170	4
832	17	458	4	738	9
375	9	642	10	371	14
715	14	492	4	735	17
868	8	543	8	749	10
271	5	842	9	495	7
630	7	905	23	716	3
491	7	542	9	952	9
372	7	522	6	417	2
645	6	122	1		

Table 16.4 *Numbers of faults found in each of 32 rolls of fabric produced in a particular factory. Also shown is the length of each roll. From Hinde (1982).*

(b) Fit a probit regression model instead of the logit (you should be able to use essentially the same steps after altering the likelihood appropriately). Discuss any changes in the posterior inferences.

3. Overdispersed models:

(a) Express the bioassay example of Section 3.7 as a generalized linear model, but replacing (3.20) by

$$\text{logit}(\theta_i) \sim N(\alpha + \beta x_i, \sigma^2),$$

so that the logistic regression holds approximately but not exactly. Set up a noninformative prior distribution and obtain posterior simulations of (α, β, σ) under this model. Discuss the effect that this model expansion has on scientific inferences for this problem.

(b) Repeat (a) with the following hypothetical data: $n = (5000, 5000, 5000, 5000)$, $y = (500, 1000, 3000, 4500)$, and x unchanged from the first column of Table 3.1.

4. Computation for a hierarchical generalized linear model:

(a) Express the rat tumor example of Section 5.1 as a generalized linear model and obtain posterior simulations using the computational techniques presented in Section 16.4.

(b) Use the posterior simulations to check the fit of the model using the realized χ^2 discrepancy (6.4).

5. Poisson model with overdispersion: Table 16.4 contains data on the incidence of faults in the manufacturing of rolls of fabric.

(a) Fit a standard Poisson regression model relating the log of the expected number of faults linearly to the length of roll. Does the number of faults appear to be proportional to the length of roll?

(b) Perform some model checking on the simple model proposed in (a), and show that there is evidence of overdispersion.

(c) Fit a hierarchical model assuming a normal prior distribution of log fault rates across rolls, with a 'fixed effect' for the linear dependence on length of roll.

(d) What evidence is there that this model provides a better fit to the data?

(e) Experiment with other forms of hierarchical model, in particular a mixture model that assumes a discrete prior distribution on two or three points for the log fault rates, and perhaps also a Student-t prior distribution. Explore the fit of the various models to the data and examine the sensitivity of inferences about the dependence of fault rate on length of roll.

6. Paired comparisons: consider the subset of the chess data in Table 16.1.

(a) Perform a simple exploratory analysis of the data to estimate the relative abilities of the players.

(b) Using some relatively simple (but reasonable) model, estimate the probability that player i wins if he plays White against player j, for each pair of players, (i, j).

(c) Fit the model described in Section 16.7 and use it to estimate these probabilities.

7. Iterative proportional fitting:

(a) Prove that the IPF algorithm increases the posterior density for γ at each step.

(b) Prove that the Bayesian IPF algorithm is in fact a Gibbs sampler for the parameters γ_j.

8. Loglinear models:

(a) Reproduce the analysis for the example in Section 16.8, but fitting the model with only one-way interactions (that is, independence of the five factors).

(b) Reproduce the analysis for the example in Section 16.8, but fitting the model with all one-, two-, and three-way interactions.

(c) Discuss how the inferences under these models differ from the model with one- and two-way interactions that was fitted in Section 16.8.

CHAPTER 17

Models for robust inference

17.1 Introduction

So far, we have relied primarily upon the normal, binomial, and Poisson distributions, and hierarchical combinations of these, for modeling data and parameters. The use of a limited class of distributions results, however, in a limited and potentially inappropriate class of inferences. Many problems fall outside the range of convenient models, and models should be chosen to fit the underlying science and data, not simply for their analytical or computational convenience. As illustrated in Chapter 5, often the most useful approach for creating realistic models is to work hierarchically, combining simple univariate models. If, for convenience, we use simplistic models, it is important to answer the following question: in what ways does the posterior inference depend on extreme data points and on unassessable model assumptions? We have already discussed, in Chapter 6, the latter part of this question, which is essentially the subject of sensitivity analysis; here we return to the topic in greater detail, using more advanced computational methods.

Robustness of inferences to outliers

Models based on the normal distribution are notoriously 'nonrobust' to outliers, in the sense that a single aberrant data point can strongly affect the inference for all the parameters in the model, even those with little substantive connection to the outlying observation.

For example, in the educational testing example of Section 5.5, our estimates for the eight treatment effects were obtained by shifting the individual school means toward the grand mean (or, in other words, shifting toward the prior information that the true effects came from a common normal distribution), with the proportionate shifting for each school j determined only by its sampling error, σ_j, and the variation τ between school effects. Suppose that the observation for the eighth school in the study, y_8 in Table 5.2 on page 140, had been 100 instead of 12, so that the eight observations were 28, 8, -3, 7, -1, 1, 18, and 100, with the same standard errors as reported in Table 5.2. If we were to apply the hierarchical normal model to this dataset, our posterior distribution would tell us that τ has a high value, and thus each estimate $\hat{\theta}_j$ would be essentially equal to its observed effect y_j; see equation (5.17) and Figure 5.6. But does this make sense in practice? After all, given these hypothetical observations, the eighth school would seem to have an extremely effective SAT coaching program, or maybe the 100 is just the result of a data recording error. In either case, it would not seem right for the single observation y_8 to have such a strong influence on how we estimate $\theta_1, \ldots, \theta_7$.

In the Bayesian framework, we can reduce the influence of the aberrant eighth observation by replacing the normal population model for the θ_j's by a longer-tailed family of distributions, which allows for the possibility of extreme observations. By *long-tailed*, we mean a distribution with relatively high probability content far away from the center, where the scale of 'far away' is determined, for example, relative to the diameter of a region containing 50% of the probability in the distribution. Examples of long-tailed distributions include the family of t distributions, of which the most extreme case is the Cauchy or t_1, and also (finite) *mixture* models, which generally use a simple distribution such as the normal for the bulk of values but allow a discrete probability of observations or parameter values from an alternative distribution that can have a different center and generally has a much larger spread. In the hypothetical modification of the SAT coaching example, performing an analysis using a long-tailed distribution for the θ_j's would result in the observation 100 being interpreted as arising from an extreme draw from the long-tailed distribution rather than as evidence that the normal distribution of effects has a high variance. The resulting analysis would shrink the eighth observation somewhat toward the others, but not nearly as much (relative to its distance from the overall mean) as the first seven are shrunk toward each other. (Given this hypothetical dataset, the posterior probability $\Pr(\theta_8 > 100|y)$ should presumably be somewhat less than 0.5, and this justifies some shrinkage.)

As our hypothetical example indicates, we do not have to abandon Bayesian principles to handle outliers. For example, a long-tailed model such as a Cauchy distribution or even a two-component mixture (see Exercise 17.1) is still an exchangeable prior distribution for $(\theta_1, \ldots, \theta_8)$, as is appropriate when there is no *prior* information distinguishing among the eight schools. The choice of exchangeable prior model affects the manner in which the estimates of the θ_j's are shrunk, and we can thereby reduce the effect of an outlying observation without having to treat it in a fundamentally different way in the analysis. (This should not of course replace careful examination of the data and checking for possible recording errors in outlying values.) A distinction is sometimes made between methods that search for outliers—possibly to remove them from the analysis—and robust procedures that are invulnerable to outliers. In the Bayesian framework, the two approaches should not be distinguished. For instance, using mixture models (either finite mixture models as in Chapter 18 or overdispersed versions of standard models) not only results in categorizing extreme observations as arising from high-variance mixture components (rather than simply surprising 'outliers') but also implies that these points have less influence on inferences for estimands such as population means and medians.

Sensitivity analysis

In addition to compensating for outliers, robust models can be used to assess the sensitivity of posterior inferences to model assumptions. For example, one can use a robust model that applies the Student-t in place of a normal distribution to assess sensitivity to the normal assumption by varying the degrees of freedom from large to small. As discussed in Chapter 6, the basic idea of sensitivity analysis is to try a variety of different distributions (for likelihood and prior models) and see how posterior inferences vary for estimands and predictive quantities of interest. Once samples have already been drawn from the posterior distribution under one model, it is often straightforward to draw from alternative models using importance resampling with enough accuracy to detect major differences in inferences between the models (see Section 17.3). If the posterior distribution of estimands of interest is highly sensitive to the model assumptions, iterative simulation methods might be required for more accurate computation.

In a sense, much of the analysis of the SAT coaching experiments in Section 5.5, especially Figures 5.6 and 5.7, is a sensitivity analysis, in which the parameter τ is allowed to vary from 0 to ∞. As discussed in Section 5.5, the observed data are actually consistent with the model of all equal effects (that is, $\tau = 0$), but that model makes no substantive sense, so we fit the model allowing τ to be any positive value. The result is summarized in the marginal posterior distribution for τ (shown in Figure 5.5), which describes a range of values of τ that are supported by the data.

17.2 Overdispersed versions of standard probability models

Sometimes it will appear natural to use one of the standard models—binomial, normal, Poisson, exponential—except that the data are too dispersed. For example, the normal distribution should not be used to fit a large sample in which 10% of the points lie a distance more than 1.5 times the interquartile range away from the median. In the hypothetical example of the previous section we suggested that the prior or population model for the $\theta'_j s$ should have longer tails than the normal. For each of the standard models, there is in fact a natural extension in which a single parameter is added to allow for overdispersion. Each of the extended models has an interpretation as a mixture distribution.

A feature of all these distributions is that they can never be *underdispersed*. This makes sense in light of formulas (2.7) and (2.8) and the mixture interpretations: the mean of the generalized distribution is equal to that of the underlying family, but the variance is higher. If the data are believed to be underdispersed relative to the standard distribution, different models should be used.

The t distribution in place of the normal

The t distribution has a longer tail than the normal and can be used for accommodating (1) occasional unusual observations in a data distribution or (2) occasional extreme parameters in a prior distribution or hierarchical model. The t family of distributions—$t_\nu(\mu, \sigma^2)$—is characterized by three parameters: center μ, scale σ, and a 'degrees of freedom' parameter ν that determines the shape of the distribution. The t densities are symmetric, and ν must fall in the range $(0, \infty)$. At $\nu = 1$, the t is equivalent to the Cauchy distribution, which is so long-tailed it has infinite mean and variance, and as $\nu \to \infty$, the t approaches the normal distribution. If the t distribution is part of a probability model attempting accurately to fit a long-tailed distribution, based on a reasonably large quantity of data, then it is generally appropriate to include the degrees of freedom as an unknown parameter. In applications for which the t is chosen simply as a robust alternative to the normal, the degrees of freedom can be fixed at a small value to allow for outliers, but no smaller than prior understanding dictates. For example, t's with one or two degrees of freedom have infinite variance and are not usually realistic in the far tails.

Mixture interpretation.   Recall from Sections 3.2 and 11.8 that the $t_\nu(\mu, \sigma^2)$ distribution can be interpreted as a mixture of normal distributions with a common mean and variances distributed as scaled inverse-χ^2. For example, the model $y_i \sim t_\nu(\mu, \sigma^2)$ is equivalent to

$$
\begin{aligned}
y_i | V_i &\sim \mathrm{N}(\mu, V_i) \\
V_i &\sim \text{Inv-}\chi^2(\nu, \sigma^2),
\end{aligned}
\tag{17.1}
$$

an expression we have already introduced as (11.14) on page 303 to illustrate the computational methods of auxiliary variables and parameter expansion. Statistically, the observations with high variance can be considered the outliers in the distribution. A similar interpretation holds when modeling exchangeable parameters θ_j.

Negative binomial alternative to Poisson

A common difficulty in applying the Poisson model to data is that the Poisson model requires that the variance equal the mean; in practice, distributions of counts often are *overdispersed*, with variance greater than the mean. We have already discussed overdispersion in the context of generalized linear models (see Section 16.2), and Section 16.5 gives an example of a hierarchical normal model for overdispersed Poisson regression.

Another way to model overdispersed count data is using the negative binomial distribution, a two-parameter family that allows the mean and variance to be fitted separately, with variance at least as great as the mean. Data $y_1, \ldots, y_n$ that follow a Neg-bin(α, β) distribution can be thought of as Poisson observations with means $\lambda_1, \ldots, \lambda_n$, which follow a Gamma$(\alpha, \beta)$ distribution. The variance of the negative binomial distribution is $\frac{\beta+1}{\beta} \frac{\alpha}{\beta}$, which is always greater than the mean, $\frac{\alpha}{\beta}$, in contrast to the Poisson, whose variance is always equal to its mean. In the limit as $\beta \to \infty$ with $\frac{\alpha}{\beta}$ remaining constant, the underlying gamma distribution approaches a spike, and the negative binomial distribution approaches the Poisson.

Beta-binomial alternative to binomial

Similarly, the binomial model for discrete data has the practical limitation of having only one free parameter, which means the variance is determined by the mean. A standard robust alternative is the beta-binomial distribution, which, as the name suggests, is a beta mixture of binomials. The beta-binomial is used, for example, to model educational testing data, where a 'success' is a correct response, and individuals vary greatly in their probabilities of getting a correct response. Here, the data y_i—the number of correct responses for each individual $i = 1, \ldots, n$—are modeled with a Beta-bin(m, α, β) distribution and are thought of as binomial observations with a common number of trials m and unequal probabilities $\pi_1, \ldots, \pi_n$ that follow a Beta(α, β) distribution. The variance of the beta-binomial with mean probability $\frac{\alpha}{\alpha+\beta}$ is greater by a factor of $\frac{\alpha+\beta+m}{\alpha+\beta+1}$ than that of the binomial with the same probability; see Table A.2 in Appendix A. When $m = 1$, no information is available to distinguish between the beta and binomial variation, and the two models have equal variances.

The t distribution alternative to logistic and probit regression

Logistic and probit regressions can be nonrobust in the sense that for large absolute values of the linear predictor $X\beta$, the inverse logit or probit transformations give probabilities very close to 0 or 1. Such models could be made more robust by allowing the occasional misprediction for large values of $X\beta$. This form of robustness is defined not in terms of the data y—which equal 0 or 1 in binary regression—but with respect to the predictors X. A more robust model allows the discrete regression model to fit most of the data while occasionally making isolated errors.

A robust model, *robit regression*, can be implemented using the latent-variable formulation of discrete-data regression models (see page 419), replacing the logistic or normal distribution of the latent continuous data u with the model, $u_i \sim t_\nu((X\beta)_i, 1)$. In realistic settings it is impractical to estimate ν from the data—since the latent data u_i are never directly observed, it is essentially impossible to form inference about the shape of their continuous underlying distribution—so it is set at a low value to ensure robustness. Setting $\nu = 4$ yields a distribution that is very close to the logistic, and as $\nu \to \infty$, the model approaches the probit. Computation for the binary t regression can be performed using the EM algorithm and Gibbs sampler with the normal-mixture formulation (17.1) for the t distribution of the latent data u. In that approach, u_i and the variance of each u_i are treated as missing data.

Why ever use a nonrobust model?

The t family includes the normal as a special case, so why do we ever use the normal at all, or the binomial, Poisson, or other standard models? To start with, each of the standard models has a logical status that makes it plausible for many applied problems. The binomial and multinomial distributions apply to discrete counts for independent, identically distributed outcomes with a fixed total number of counts. The Poisson and exponential distributions fit the number of events and the waiting time for a Poisson process, which is a natural model for independent discrete events indexed by time. Finally, the central limit theorem tells us that the normal distribution is an appropriate model for data that are formed as the sum of a large number of independent components. In the educational testing example in Section 5.5, each of the observed effects, y_j, is an average of adjusted test scores with $n_j \approx 60$ (that is, the estimated treatment effect is based on about 60 students in school j). We can thus accurately approximate the sampling distribution of y_j by normality: $y_j | \theta_j, \sigma_j^2 \sim N(\theta_j, \sigma_j^2)$.

Even when they are not naturally implied by the structure of a problem, the standard models are computationally convenient, since conjugate prior distributions often allow direct calculation of posterior means and variances and easy simulation. That is why it is easy to fit a normal population model to the θ_j's in the educational testing example and why it is common to fit a normal model to the logarithm of all-positive data or the logit of data that are constrained to lie between 0 and 1. When a model is assigned in this more or less arbitrary manner, it is advisable to check the fit of the data using the posterior predictive distribution, as discussed in Chapter 6. But if we are worried that an assumed model is not robust, then it makes sense to perform a sensitivity analysis and see how much the posterior inference changes if we switch to a larger family of distributions, such as the t distributions in place of the normal.

17.3 Posterior inference and computation

As always, we can draw samples from the posterior distribution (or distributions, in the case of sensitivity analysis) using the methods described in Part III. In this section, we briefly describe the use Gibbs sampling under the mixture formulation of a robust model. The approach is illustrated for a hierarchical normal-t model in Section 17.4. When expanding a model, however, we have the possibility of a less time-consuming approximation as an alternative: we can use the draws from the original posterior distribution as a starting point for simulations from the new models. In this section, we also describe two techniques that can be useful for robust models and sensitivity analysis: importance weighting for computing the marginal posterior density (Section 13.3) in a sensitivity analysis, and importance resampling (Section 12.2) for approximating a robust analysis.

Notation for robust model as expansion of a simpler model

We use the notation $p_0(\theta|y)$ for the posterior distribution from the original model, which we assume has already been fitted to the data, and ϕ for the hyperparameter(s) characterizing the expanded model used for robustness or sensitivity analysis. Our goal is to sample from

$$p(\theta|\phi, y) \propto p(\theta|\phi)p(y|\theta, \phi), \qquad (17.2)$$

using either a pre-specified value of ϕ (such as $\nu = 4$ for a Student-t robust model) or for a range of values of ϕ. In the latter case, we also wish to compute the marginal posterior distribution of the sensitivity analysis parameter, $p(\phi|y)$.

The robust family of distributions can enter the model (17.2) through the distribution of the parameters, $p(\theta|\phi)$, or the data distribution, $p(y|\theta, \phi)$. For example, Section 17.2 focuses on robust data distributions, and our reanalysis of the SAT coaching experiments in Section 17.4 uses a robust distribution for model parameters. We must then set up a joint prior distribution, $p(\theta, \phi)$, which can require some care because it captures the prior dependence between θ and ϕ.

Gibbs sampling using the mixture formulation

Markov chain simulation can be used to draw from the posterior distributions, $p(\theta|\phi, y)$. This can be done using the mixture formulation, by sampling from the joint posterior distribution of θ and the extra unobserved scale parameters (V_i's in the Student-t model, λ_i's in the negative binomial, and π_i's in the beta-binomial).

For a simple example, consider the $t_\nu(\mu, \sigma^2)$ distribution fitted to data $y_1, \ldots, y_n$, with μ and σ^2 unknown. Given ν, we have already discussed in Section 11.8 how to program the Gibbs sampler in terms of the parameterization (17.1) involving $\mu, \sigma^2, V_1, \ldots, V_n$. If ν is itself unknown, the Gibbs sampler must be expanded to include a step for sampling from the conditional posterior distribution of ν. No simple method exists for this step, but a Metropolis step can be used instead. Another complication is that such models commonly have multimodal posterior densities, with different modes corresponding to different observations in the tails of the t distributions, meaning that additional work is required to search for modes initially and jump between modes in the simulation, for example using simulated tempering (see Section 13.1).

Sampling from the posterior predictive distribution for new data

To perform sensitivity analysis and robust inference for predictions $\tilde{y}$, follow the usual procedure of first drawing θ from the posterior distribution, $p(\theta|\phi, y)$, and then drawing $\tilde{y}$ from the predictive distribution, $p(\tilde{y}|\phi, \theta)$. To simulate data from a mixture model, first draw the mixture indicators for each future observation, then draw $\tilde{y}$, given the mixture parameters. For example, to draw

$\tilde{y}$ from a $t_\nu(\mu, \sigma^2)$ distribution, first draw $V \sim \text{Inv-}\chi^2(\nu, \sigma^2)$, then draw $\tilde{y} \sim N(\mu, V)$.

Computing the marginal posterior distribution of the hyperparameters by importance weighting

During a check for model robustness or sensitivity to assumptions, we might like to avoid the additional programming effort required to apply Markov chain simulation to a robust model. If we have simulated draws from $p_0(\theta|y)$, then it is possible to obtain approximate inference under the robust model using importance weighting and importance resampling. We assume in the remainder of this section that simulation draws $\theta^l, l = 1, \ldots, L$, have already been obtained from $p_0(\theta|y)$. We can use importance weighting to evaluate the marginal posterior distribution, $p(\phi|y)$, using identity (13.6) on page 343, which in our current notation becomes

$$p(\phi|y) \quad \propto \quad p(\phi) \int \frac{p(\theta|\phi)p(y|\theta, \phi)}{p_0(\theta)p_0(y|\theta)} p_0(\theta)p_0(y|\theta)d\theta$$

$$\propto \quad p(\phi) \int \frac{p(\theta|\phi)p(y|\theta, \phi)}{p_0(\theta)p_0(y|\theta)} p_0(\theta|y)d\theta.$$

In the first line above, the constant of proportionality is $1/p(y)$, whereas in the second it is $p_0(y)/p(y)$. For any ϕ, the value of $p(\phi|y)$, up to a proportionality constant, can be estimated by the average importance ratio for the simulations θ^l,

$$p(\phi)\frac{1}{L}\sum_{l=1}^{L} \frac{p(\theta^l|\phi)p(y|\theta^l, \phi)}{p_0(\theta^l)p_0(y|\theta^l)}, \tag{17.3}$$

which can be evaluated, using a fixed set of L simulations, at each of a range of values of ϕ, and then graphed as a function of ϕ.

Drawing approximately from the robust posterior distributions by importance resampling

To perform importance resampling, it is best to start with a large number of draws, say $L = 5000$, from the original posterior distribution, $p_0(\theta|y)$. Now, for each distribution in the expanded family indexed by ϕ, draw a smaller subsample, say $n = 500$, from the L draws, without replacement, using importance resampling, in which each of the n samples is drawn with probability proportional to its importance ratio,

$$\frac{p(\theta|\phi, y)}{p_0(\theta|y)} = \frac{p(\theta|\phi)p(y|\theta, \phi)}{p_0(\theta)p_0(y|\theta)}.$$

A new set of subsamples must be drawn for each value of ϕ, but the same set of L original draws may be used. Details are given in Section 12.2. This procedure is effective as long as the largest importance ratios are plentiful and not too variable; if they do vary greatly, this is an indication of potential sensitivity because $p(\theta|\phi, y)/p_0(\theta|y)$ is sensitive to the drawn values of θ. If the importance weights are too variable for importance resampling to be considered accurate, and accurate inferences under the robust alternatives are desired, then we must rely on Markov chain simulation.

17.4 Robust inference and sensitivity analysis for the educational testing example

Consider the hierarchical model for SAT coaching effects based on the data in Table 5.2 in Section 5.5. Given the large sample sizes in the eight original experiments, there should be little concern about assuming the data model that has $y_j \sim \text{N}(\theta_j, \sigma_j^2)$, with the variances σ_j^2 known. The population model, $\theta_j \sim \text{N}(\mu, \tau^2)$, is more difficult to justify, although the model checks in Section 6.8 suggest that it is adequate for the purposes of obtaining posterior intervals for the school effects. In general, however, posterior inferences can be highly sensitive to the assumed model, even when the model provides a good fit to the observed data. To illustrate methods for robust inference and sensitivity analysis, we explore an alternative family of models that fit t distributions to the population of school effects:

$$\theta_j|\nu, \mu, \tau \sim t_\nu(\mu, \tau^2), \quad \text{for } j = 1, \dots, 8. \tag{17.4}$$

We use the notation $p(\theta, \mu, \tau|\nu, y) \propto p(\theta, \mu, \tau|\nu)p(y|\theta, \mu, \tau, \nu)$ for the posterior distribution under the t_ν model and $p_0(\theta, \mu, \tau|y) \equiv p(\theta, \mu, \tau|\nu = \infty, y)$ for the posterior distribution under the normal model evaluated in Section 5.5.

Robust inference based on a t_4 population distribution

As discussed at the beginning of this chapter, one might be concerned that the normal population model causes the most extreme estimated school effects to be pulled too much toward the grand mean. Perhaps the coaching program in school A, for example, is different enough from the others that its estimate should not be shrunk so much to the average. A related concern would be that the largest observed effect, in school A, may be exerting undue influence on estimation of the population variance, τ^2, and thereby also on the Bayesian estimates of the other effects. From a modeling standpoint, there is a great variety of different SAT coaching programs, and the population of their effects might be better fitted by a long-tailed distribution. To assess the importance of these concerns, we perform a robust analysis, replacing the normal population distribution by the Student-t model (17.4) with $\nu = 4$ and leaving the rest of the model unchanged; that is, the likelihood is still $p(y|\theta, \nu) = \prod_j \text{N}(y_j|\theta_j, \sigma_j^2)$, and the hyperprior distribution is still $p(\mu, \tau|\nu) \propto 1$.

Gibbs sampling. We carry out Gibbs sampling using the approach described in Section 11.8 with $\nu = 4$. (See Appendix C for details on performing the Gibbs sampler using Bugs or R.) The resulting inferences for the eight schools,

School	Posterior quantiles				
	2.5%	25%	median	75%	97.5%
A	-2	6	11	16	34
B	-5	4	8	12	21
C	-14	2	7	11	21
D	-6	4	8	12	21
E	-9	1	6	9	17
F	-9	3	7	10	19
G	-1	6	10	15	26
H	-8	4	8	13	26

Table 17.1 *Summary of 2500 simulations of the treatment effects in the eight schools, using the t_4 population distribution in place of the normal. Results are similar to those obtained under the normal model and displayed in Table 5.3.*

based on 2500 draws from the posterior distribution (the last halves of five chains, each of length 1000), are provided in Table 17.1. The results are essentially identical, for practical purposes, to the inferences under the normal model displayed in Table 5.3 on page 143, with just slightly less shrinkage for the more extreme schools such as school A.

Computation using importance resampling. Though we have already done the Markov chain simulation, we discuss briefly how to apply importance resampling to approximate the posterior distribution with $\nu = 4$. First, we sample 5000 draws of (θ, μ, τ) from $p_0(\theta, \mu, \tau | y)$, the posterior distribution under the normal model, as described in Section 5.4. Next, we compute the importance ratio for each draw:

$$\frac{p(\theta, \mu, \tau | \nu, y)}{p_0(\theta, \mu, \tau | y)} \propto \frac{p(\mu, \tau | \nu)p(\theta | \mu, \tau, \nu)p(y | \theta, \mu, \tau, \nu)}{p_0(\mu, \tau)p_0(\theta | \mu, \tau)p_0(y | \theta, \mu, \tau)} = \prod_{j=1}^{8} \frac{t_\nu(\theta_j | \mu, \tau^2)}{N(\theta_j | \mu, \tau^2)}. \quad (17.5)$$

The factors for the likelihood and hyperprior density cancel in the importance ratio, leaving only the ratio of the population densities. We sample 500 draws of (θ, μ, τ), without replacement, from the sample of 5000, using importance resampling. In this case the approximation is probably sufficient for assessing robustness, but the long tail of the distribution of the logarithms of the importance ratios (not shown) does indicate serious problems for obtaining accurate inferences using importance resampling.

Sensitivity analysis based on t_ν population distributions with varying values of ν

A slightly different concern from robustness is the sensitivity of the posterior inference to the prior assumption of a normal population distribution. To study the sensitivity, we now fit a range of t distributions, with 1, 2, 3, 5, 10,

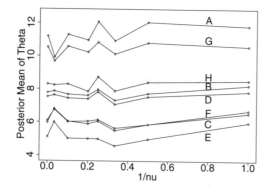

Figure 17.1 *Posterior means of treatment effects as functions of ν, on the scale of $1/\nu$, for the sensitivity analysis of the educational testing example. The values at $1/\nu=0$ come from the simulations under the normal distribution in Section 5.5. Much of the scatter in the graphs is due to simulation variability.*

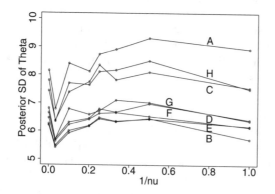

Figure 17.2 *Posterior standard deviations of treatment effects as functions of ν, on the scale of $1/\nu$, for the sensitivity analysis of the educational testing example. The values at $1/\nu=0$ come from the simulations under the normal distribution in Section 5.5. Much of the scatter in the graphs is due to simulation variability.*

and 30 degrees of freedom. We have already fitted infinite degrees of freedom (the normal model) and 4 degrees of freedom (the robust model above).

For each value of ν, we perform a Markov chain simulation to obtain draws from $p(\theta, \mu, \tau | \nu, y)$. Instead of displaying a table of posterior summaries such as Table 17.1 for each value of ν, we summarize the results by the posterior mean and standard deviation of each of the eight school effects θ_j. Figures 17.1 and 17.2 display the results as a function of $1/\nu$. The parameterization in terms of $1/\nu$ rather than ν has the advantage of including the normal distribution at $1/\nu = 0$ and encompassing the entire range from normal to Cauchy distributions in the finite interval $[0, 1]$. There is some variation in the figures but no apparent systematic sensitivity of inferences to the hyperparameter, ν.

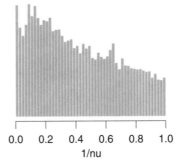

Figure 17.3 *Posterior simulations of $1/\nu$ from the Gibbs-Metropolis computation of the robust model for the educational testing example, with ν treated as unknown.*

Treating ν as an unknown parameter

Finally, we consider the sensitivity analysis parameter, ν, as an unknown quantity and average over it in the posterior distribution. In general, this computation is a key step, because we are typically only concerned with sensitivity to models that are supported by the data. In this particular example, inferences are so insensitive to ν that computing the marginal posterior distribution is unnecessary; we include it here as an illustration of the general method.

Prior distribution. Before computing the posterior distribution for ν, we must assign it a prior distribution. We try a uniform density on $1/\nu$ for the range $[0, 1]$ (that is, from the normal to the Cauchy distributions). This prior distribution favors long-tailed models, with half of the prior probability falling between the t_1 (Cauchy) and t_2 distributions.

In addition, the conditional prior distributions, $p(\mu, \tau | \nu) \propto 1$, are improper, so we must specify their dependence on ν; we use the notation $p(\mu, \tau | \nu) \propto g(\nu)$. In the t family, the parameters μ and τ characterize the median and the second derivative of the density function at the median, not the mean and variance, of the distribution of the θ_j's. The parameter μ seems to have a reasonable invariant meaning (and in fact is equal to the mean except in the limiting case of the Cauchy where the mean does not exist), but the interquartile range would perhaps be a more reasonable parameter than the curvature for setting up a prior distribution. We cannot parameterize the t_ν distributions in terms of their variance, because the variance is infinite for $\nu \leq 2$. The interquartile range varies quite mildly as a function of ν, and so for simplicity we use the convenient parameterization in terms of (μ, τ) and set $g(\nu) \propto 1$. Combining this with our prior distribution on ν yields an improper joint uniform prior density on $(\mu, \tau, 1/\nu)$. If our posterior inferences under this model turn out to depend strongly on ν, we should consider refining this prior distribution.

Posterior inference. To treat ν as an unknown parameter, we modify the Gibbs sampling simulation used in the robust analyses to include a Metropolis step for sampling from the conditional distribution of $1/\nu$. An example of the implementation of such an approach can be found in Appendix C. Figure 17.3 displays a histogram of the simulations of $1/\nu$. An alternative to extending the model is to approximate the marginal posterior density using importance sampling and (17.3).

The sensitivity analysis showed that ν has only minor effects on the posterior inference; the results in Section 5.5 are thus not strongly dependent on the normal assumption for the population distribution of the parameters θ_j. If Figures 17.1 and 17.2 had shown a strong dependence on ν—as Figures 5.5– 5.7 showed dependence on τ—then it might make sense to include ν as a hyperparameter, after thinking more seriously about a joint prior distribution for the parameters with noninformative prior distributions—(μ, τ, ν).

Discussion

Robustness and sensitivity to modeling assumptions depend on the estimands being studied. In the SAT coaching example, posterior medians, 50%, and 95% intervals for the eight school effects are insensitive to the assumption of a normal population distribution (at least as compared to the t family). In contrast, it may be that 99.9% intervals are strongly dependent on the tails of the distributions and sensitive to the degrees of freedom in the t distribution— fortunately, these extreme tails are unlikely to be of substantive interest in this example.

17.5 Robust regression using Student-t errors

As with other models based on the normal distribution, inferences under the normal linear regression model of Chapter 14 are sensitive to unusual or out- lying values. Robust regression analyses are obtained by considering robust alternatives to the normal distribution for regression errors. Robust error dis- tributions, such as the Student-t with small number of degrees of freedom, treat observations far from the regression line as high-variance observations, yielding results similar to those obtained by downweighting outliers. (Recall that the 'weights' in weighted linear regression are inverse variances.)

Iterative weighted linear regression and the EM algorithm

To illustrate robust regression calculations, we consider the t_ν regression model with fixed degrees of freedom as an alternative to the normal linear regres- sion model. The conditional distribution of the individual response variable y_i given the vector of explanatory variables X_i is $p(y_i|X_i\beta, \sigma^2) = t_\nu(y_i|X_i\beta, \sigma^2)$. The t_ν distribution can be expressed as a mixture as in equation (17.1) with $X_i\beta$ as the mean. As a first step in the robust analysis, we find the mode of the posterior distribution $p(\beta, \sigma^2|\nu, y)$ given the vector y consist- ing of n observations. Here we assume that a noninformative prior distribution is used, $p(\mu, \log \sigma | \nu) \propto 1$; more substantial information about the regression parameters can be incorporated exactly as in Sections 14.8 and Chapter 15. The posterior mode of $p(\beta, \log \sigma | \nu, y)$ under the Student-t model can be obtained directly using Newton's method (Section 12.1) or any other mode-finding technique. Alternatively, we can take advantage of the mixture form of the t model and use the EM algorithm with the variances V_i treated as 'missing data' (that is, parameters to be averaged over); in the notation of Section 12.3, $\gamma = (V_1, \ldots, V_n)$. The E-step of the EM algorithm computes the expected value of the sufficient statistics for the normal model $(\sum_{i=1}^{n} y_i^2 / V_i, \sum_{i=1}^{n} y_i / V_i, \sum_{i=1}^{n} 1 / V_i)$, given the current parameter estimates $(\beta^{\text{old}}, \sigma^{\text{old}})$ and averaging over $(V_1, \ldots, V_n)$. It is sufficient to note that

$$p(V_i | y_i, \beta^{\text{old}}, \sigma^{\text{old}}, \nu) \sim \text{Inv-}\chi^2 \left(\nu + 1, \frac{\nu(\sigma^{\text{old}})^2 + (y_i - X_i \beta^{\text{old}})^2}{\nu + 1} \right) \qquad (17.6)$$

and that

$$\text{E} \left(\frac{1}{V_i} \middle| y_i, \beta^{\text{old}}, \sigma^{\text{old}}, \nu \right) = \frac{\nu + 1}{\nu(\sigma^{\text{old}})^2 + (y_i - X_i \beta^{\text{old}})^2}.$$

The M-step of the EM algorithm is a weighted linear regression with diagonal weight matrix W containing the conditional expectations of $1/V_i$ on the diagonal. The updated parameter estimates are

$$\hat{\beta}^{\text{new}} = (X^T W X)^{-1} X^T W y \quad \text{and} \quad (\hat{\sigma}^{\text{new}})^2 = \frac{1}{n} (y - X \hat{\beta}^{\text{new}})^T W (y - X \hat{\beta}^{\text{new}}),$$

where X is the $n \times p$ matrix of explanatory variables. The iterations of the EM algorithm are equivalent to those performed in an iterative weighted least squares algorithm. Given initial estimates of the regression parameters, weights are computed for each case, with those cases having large residuals given less weight. Improved estimates of the regression parameters are then obtained by weighted linear regression.

When the degrees of freedom parameter, ν, is treated as unknown, the ECME algorithm can be applied, with an additional step added to the iteration for updating the degrees of freedom.

Other robust models. Iterative weighted linear regression, or equivalently the EM algorithm, can be used to obtain the posterior mode for a number of robust alternative models. Changing the probability model used for the observation variances, V_i, creates alternative robust models. For example, a two-point distribution can be used to model a regression with contaminated errors. The computations for robust models of this form are as described above, except that the E-step is modified to reflect the appropriate posterior conditional mean.

Gibbs sampler and Metropolis algorithm

Posterior draws from robust regression models can be obtained using Gibbs sampling and the Metropolis algorithm, as with the linear and generalized linear models discussed in Chapters 14–16. Using the mixture parameterization of the t_ν distribution, we can obtain draws from the posterior distribution $p(\beta, \sigma^2, V_1, \ldots, V_n | \nu, y)$ by alternately sampling from $p(\beta, \sigma^2 | V_1, \ldots, V_n, \nu, y)$ using the usual posterior distribution from weighted linear regression, and sampling from $p(V_1, \ldots, V_n | \beta, \sigma^2, \nu, y)$, a set of independent scaled inverse-χ^2 distributions as in equation (17.6). It can be even more effective to use parameter expansion as explained in Section 11.8.

If the degrees of freedom parameter, ν, is included as an unknown parameter in the model, then an additional Metropolis step is required in each iteration. In practice, these computations can be difficult to implement, because with low degrees of freedom ν, the posterior distribution can have many modes, and the Gibbs sampler and Metropolis algorithms can get stuck. It is important to run many simulations with overdispersed starting points for complicated models of this form.

17.6 Bibliographic note

Mosteller and Wallace (1964) use the negative binomial distribution, instead of the Poisson, for count data, and extensively study the sensitivity of their conclusions to model assumptions. Box and Tiao (1968) provide another early discussion of Bayesian robustness, in the context of outliers in normal models. Smith (1983) extends Box's approach and also discusses the t family using the same parameterization (inverse degrees of freedom) as we have. A review of models for overdispersion in binomial data, from a non-Bayesian point of view, is given by Anderson (1988), who cites many further references. Gaver and O'Muircheartaigh (1987) discuss the use of hierarchical Poisson models for robust Bayesian inference. O'Hagan (1979) and Gelman (1992a) discuss the connection between the tails of the population distribution of a hierarchical model and the shrinkage in the associated Bayesian posterior distribution.

In a series of papers, Berger and coworkers have explored theoretical aspects of Bayesian robustness, examining, for example, families of prior distributions that provide maximum robustness against the influence of aberrant observations; see for instance Berger (1984, 1990) and Berger and Berliner (1986). Related work appears in Wasserman (1992). An earlier overview from a pragmatic point of view close to ours was provided by Dempster (1975). Rubin (1983a) provides an illustration of the limitations of data in assessing model fit and the resulting inevitable sensitivity of some conclusions to untestable assumptions.

With the recent advances in computation, modeling with the t distribution has become increasingly common in statistics. Dempster, Laird, and Rubin (1977) show how to apply the EM algorithm to t models, and Liu and Rubin (1995) and Meng and van Dyk (1997) discuss faster computational methods using extensions of EM. Lange, Little, and Taylor (1989) discuss the use of the t distribution in a variety of statistical contexts. Raghunathan and Rubin

Number of occurrences in a block	0	1	2	3	4	5	6	> 6
Number of blocks (Hamilton)	128	67	32	14	4	1	1	0
Number of blocks (Madison)	156	63	29	8	4	1	1	0

Table 17.2 *Observed distribution of the word 'may' in papers of Hamilton and Madison, from Mosteller and Wallace (1964). Out of the 247 blocks of Hamilton's text studied, 128 had no instances of 'may,' 67 had one instance of 'may,' and so forth, and similarly for Madison.*

(1990) present an example using importance resampling. Liu (2002) presents the 'robit' model as an alternative to logistic and probit regression.

Rubin (1983b) and Lange and Sinsheimer (1993) review the connections between robust regression, the t and related distributions, and iterative regression computations.

Taplin and Raftery (1994) present an example of an application of a finite mixture model for robust Bayesian analysis of agricultural experiments.

17.7 Exercises

1. Prior distributions and shrinkage: in the educational testing experiments, suppose we think that most coaching programs are almost useless, but some are strongly effective; a corresponding population distribution for the school effects is a mixture, with most of the mass near zero but some mass extending far in the positive direction; for example,

$$p(\theta_1, \ldots, \theta_8) = \prod_{j=1}^{8} [\lambda_1 N(\theta_j | \mu_1, \tau_1^2) + \lambda_2 N(\theta_j | \mu_2, \tau_2^2)].$$

All these parameters could be estimated from the data (as long as we restrict the parameter space, for example by setting $\mu_1 > \mu_2$), but to fix ideas, suppose that $\mu_1 = 0$, $\tau_1 = 10$, $\mu_2 = 15$, $\tau_2 = 25$, $\lambda_1 = 0.9$, and $\lambda_2 = 0.1$.

   (a) Compute the posterior distribution of $(\theta_1, \ldots, \theta_8)$ under this model for the data in Table 5.2.

   (b) Graph the posterior distribution for θ_8 under this model for $y_8 = 0$, 25, 50, and 100, with the same standard deviation σ_8 as given in Table 5.2. Describe qualitatively the effect of the two-component mixture prior distribution.

2. Poisson and negative binomial distributions: as part of their analysis of the Federalist papers, Mosteller and Wallace (1964) recorded the frequency of use of various words in selected articles by Alexander Hamilton and James Madison. The articles were divided into 247 blocks of about 200 words each, and the number of instances of various words in each block were recorded. Table 17.2 displays the results for the word 'may.'

(a) Fit the Poisson model to these data, with different parameters for each author and a noninformative prior distribution. Plot the posterior density of the Poisson mean parameter for each author.

(b) Fit the negative binomial model to these data with different parameters for each author. What is a reasonable noninformative prior distribution to use? For each author, make a contour plot of the posterior density of the two parameters and a scatterplot of the posterior simulations.

3. Model checking with the Poisson and binomial distributions: we examine the fit of the models in the previous exercise using posterior predictive checks.

(a) Considering the nature of the data and of likely departures from the model, what would be appropriate test statistics?

(b) Compare the observed test statistics to their posterior predictive distribution (see Section 6.3) to test the fit of the Poisson model.

(c) Perform the same test for the negative binomial model.

4. Robust models and model checking: fit a robust model to Newcomb's speed of light data (Figure 3.1). Check the fit of the model using appropriate techniques from Chapter 6.

5. Contamination models: construct and fit a normal mixture model to the dataset used in the previous exercise.

6. Robust models:

(a) Choose a dataset from one of the examples or exercises earlier in the book and analyze it using a robust model.

(b) Check the fit of the model using the posterior predictive distribution and appropriate test variables.

(c) Discuss how inferences changed under the robust model.

7. Computation for the Student-t model: consider the model $y_1, \ldots, y_n \sim$ iid $t_\nu(\mu, \sigma^2)$, with ν fixed and a uniform prior distribution on $(\mu, \log \sigma)$.

(a) Work out the steps of the EM algorithm for finding posterior modes of $(\mu, \log \sigma)$, using the specification (17.1) and averaging over $V_1, \ldots, V_n$. Clearly specify the joint posterior density, its logarithm, the function $E_{\text{old}} \log p(\mu, \log \sigma, V_1, \ldots, V_n | y)$, and the updating equations for the M-step.

(b) Work out the steps of the Gibbs sampler for drawing posterior simulations of $(\mu, \log \sigma, V_1, \ldots, V_n)$.

(c) Illustrate the analysis with the speed of light data of Figure 3.1, using a t_2 model.

8. Robustness and sensitivity analysis: repeat the computations of Section 17.4 with the dataset altered as described on page 443 so that the observation y_8 is replaced by 100. Verify that, in this case, inferences are quite sensitive to ν. Which values of ν have highest marginal posterior density?

Part V: Specific Models and Problems

This final part of our book discusses some families of models that are often useful in applications. Rather than attempt to do the impossible and be exhaustive, we focus on particular examples of each kind of model. In these examples, we discuss various aspects of modeling, computation, model checking, and model improvement. We conclude with two chapters on specific problems that arise in statistical inference. Chapter 21 discusses models for the imputation of missing data, and Chapter 22 considers several examples of the use of Bayesian inference for decision analysis.

To some extent, the reader could replace this part of the book by reading across the wide range of applied Bayesian research that has recently appeared in the statistical literature. We feel, however, that the book would be incomplete without attempting to draw together some fully Bayesian treatments of classes of models that are used commonly in applications. By necessity, some of these treatments are brief and lacking in realistic detail, but we hope that they give a good sense of our principles and methods as applied to genuine complicated problems.

CHAPTER 18

Mixture models

18.1 Introduction

Mixture distributions arise in practical problems when the measurements of a random variable are taken under two different conditions. For example, the distribution of heights in a population of adults reflects the mixture of males and females in the population, and the reaction times of schizophrenics on an attentional task might be a mixture of trials in which they are or are not affected by an attentional delay (an example discussed later in this chapter). For the greatest flexibility, and consistent with our general hierarchical modeling strategy, we construct such distributions as mixtures of simpler forms. For example, it is best to model male and female heights as separate univariate, perhaps normal, distributions, rather than a single bimodal distribution. This follows our general principle of using conditioning to construct realistic probability models. The schizophrenic reaction times cannot be handled in the same way because it is not possible to identify which trials are affected by the attentional delay. *Mixture models* can be used in problems of this type, where the population of sampling units consists of a number of subpopulations within each of which a relatively simple model applies. In this chapter we discuss methods for analyzing data using mixture models.

The basic principle for setting up and computing with mixture models is to introduce unobserved *indicators*—random variables, which we usually label as a vector or matrix ζ, that specify the mixture component from which each particular observation is drawn. Thus the mixture model is viewed hierarchically; the observed variables y are modeled conditionally on the vector ζ, and the vector ζ is itself given a probabilistic specification. Often it is useful to think of the mixture indicators as missing data. Inferences about quantities of interest, such as parameters within the probability model for y, are obtained by averaging over the distribution of the indicator variables. In the simulation framework, this means drawing (θ, ζ) from their joint posterior distribution.

18.2 Setting up mixture models

Finite mixtures

Suppose that, based on scientific considerations, it is considered desirable to model the distribution of $y = (y_1, \ldots, y_n)$, or the distribution of $y|x$, as a mixture of M components. It is assumed that it is not known which mixture component underlies each particular observation. Any information that makes it possible to specify a nonmixture model for some or all of the observations, such as sex in our discussion of the distribution of adult heights, should be used to simplify the model. For $m = 1, \ldots, M$, the mth component distribution, $f_m(y_i|\theta_m)$, is assumed to depend on a parameter vector θ_m; the parameter denoting the proportion of the population from component m is λ_m, with $\sum_{m=1}^{M} \lambda_m = 1$. It is common to assume that the mixture components are all from the same parametric family, such as the normal, with different parameter vectors. The sampling distribution of y in that case is

$$p(y_i|\theta, \lambda) = \lambda_1 f(y_i|\theta_1) + \lambda_2 f(y_i|\theta_2) + \ldots + \lambda_M f(y_i|\theta_M). \qquad (18.1)$$

The form of the sampling distribution invites comparison with the standard Bayesian setup. The mixture distribution $\lambda = (\lambda_1, \ldots, \lambda_M)$ might be thought of as a discrete prior distribution on the parameters θ_m; however, it seems more appropriate to think of this prior, or mixing, distribution as a description of the variation in θ across the population of interest. In this respect the mixture model more closely resembles a hierarchical model. This resemblance is enhanced with the introduction of unobserved (or missing) indicator variables ζ_{im}, with

$$\zeta_{im} = \begin{cases} 1 & \text{if the } i\text{th unit is drawn from the } m\text{th mixture component} \\ 0 & \text{otherwise.} \end{cases}$$

Given λ, the distribution of each unobserved vector $\zeta_i = (\zeta_{i1}, \ldots, \zeta_{iM})$ is Multin$(1; \lambda_1, \ldots, \lambda_M)$. In this case the mixture parameters λ are thought of as hyperparameters determining the distribution of ζ. The joint distribution of the observed data y and the unobserved indicators ζ conditional on the model parameters can be written

$$p(y, \zeta|\theta, \lambda) = p(\zeta|\lambda)p(y|\zeta, \theta) = \prod_{i=1}^{n} \prod_{m=1}^{M} (\lambda_m f(y_i|\theta_m))^{\zeta_{im}}, \qquad (18.2)$$

with exactly one of ζ_{im} equaling 1 for each i. At this point, M, the number of mixture components, is assumed to be known and fixed. We consider this issue further when discussing model checking. If observations y are available for which their mixture components are known (for example, the heights of a group of adults whose sexes are recorded), the mixture model (18.2) is easily modified; each such observation adds a single factor to the product with a known value of the indicator vector ζ_i.

Continuous mixtures

The finite mixture is a special case of the more general specification, $p(y_i) = \int p(y_i|\theta)\lambda(\theta)d\theta$. The hierarchical models of Chapter 5 can be thought of as continuous mixtures in the sense that each observable y_i is a random variable with distribution depending on parameters θ_i; the prior distribution or population distribution of the parameters θ_i is given by the mixing distribution $\lambda(\theta)$. Continuous mixtures were used in the discussion of robust alternatives to standard models; for example, the Student-t distribution, a mixture on the scale parameter of normal distributions, yields robust alternatives to normal models, as discussed in Chapter 17. The negative binomial and beta-binomial distributions of Chapter 17 are discrete distributions that are obtained as continuous mixtures of Poisson and binomial distributions, respectively.

The computational approach for continuous mixtures follows that for finite mixtures quite closely (see also the discussion of computational methods for hierarchical models in Chapter 5). We briefly discuss the setup of a probability model based on the Student-t distribution with ν degrees of freedom in order to illustrate how the notation of this chapter is applied to continuous mixtures. The observable y_i given the location parameter μ, variance parameter σ^2, and scale parameter ζ_i (similar to the indicator variables in the finite mixture) is $N(\mu, \sigma^2 \zeta_i)$. The location and variance parameters can be thought of as the mixture component parameters θ. The ζ_i are viewed as a random sample from the mixture distribution, in this case a scaled Inv-$\chi^2(\nu, 1)$ distribution. The marginal distribution of y_i, after averaging over ζ_i, is $t_\nu(\mu, \sigma^2)$. The degrees of freedom parameter, ν, which may be fixed or unknown, describes the mixing distribution for this continuous mixture in the same way that the multinomial parameters λ do for finite mixtures. The posterior distribution of the ζ_i's may also be of interest in this case for assessing which observations are possible outliers. In the remainder of this chapter we focus on finite mixtures; the modifications required for continuous mixtures are typically minor.

Identifiability of the mixture likelihood

Parameters in a model are not identified if the same likelihood function is obtained for more than one choice of the model parameters. All finite mixture models are nonidentifiable in one sense; the distribution is unchanged if the group labels are permuted. For example, there is ambiguity in a two-component mixture model concerning which component should be designated as component 1 (see the discussion of aliasing in Section 4.3). When possible the parameter space should be defined to clear up any ambiguity, for example by specifying the means of the mixture components to be in nondecreasing order or specifying the mixture proportions λ_m to be nondecreasing. For many problems, an informative prior distribution has the effect of identifying specific components with specific subpopulations.

Prior distribution

The prior distribution for the finite mixture model parameters (θ, λ) is taken in most applications to be a product of independent prior distributions on θ and λ. If the vector of mixture indicators $\zeta_i = (\zeta_{i1}, \ldots, \zeta_{iM})$ is thought of as a multinomial random variable with parameter λ, then the natural conjugate prior distribution is the Dirichlet distribution, $\lambda \sim \text{Dirichlet}(\alpha_1, \ldots, \alpha_M)$. The relative sizes of the Dirichlet parameters α_m describe the mean of the prior distribution for λ, and the sum of the α_m's is a measure of the strength of the prior distribution, the 'prior sample size.' We use θ to represent the vector consisting of all of the parameters in the mixture components, $\theta = (\theta_1, \ldots, \theta_M)$. Some parameters may be common to all components and other parameters specific to a single component. For example, in a mixture of normals, we might assume that the variance is the same for each component but that the means differ. For now we do not make any assumptions about the prior distribution, $p(\theta)$. In continuous mixtures, the parameters of the mixture distribution (for example, the degrees of freedom in the Student-t model) require a hyperprior distribution.

Ensuring a proper posterior distribution

As has been emphasized throughout, it is critical to check before applying an improper prior distribution that the resulting model is well specified. An improper noninformative prior distribution for λ (corresponding to $\alpha_i = 0$) may cause a problem if the data do not indicate that all M components are present in the data. It is more common for problems to arise if improper prior distributions are used for the component parameters. In Section 4.3, we mention the difficulty in assuming an improper prior distribution for the separate variances of a mixture of two normal distributions. There are a number of 'uninteresting' modes that correspond to a mixture component consisting of a single observation with no variance. The posterior distribution of the parameters of a mixture of two normals is proper if the ratio of the two unknown variances is fixed or assigned a proper prior distribution, but *not* if the parameters $(\log \sigma_1, \log \sigma_2)$ are assigned a joint uniform prior density.

Number of mixture components

For finite mixture models there is often uncertainty concerning the number of mixture components M to include in the model. The computational cost of models with large values of M is sufficiently large that it is desirable to begin with a small mixture and assess the adequacy of the fit. It is often appropriate to begin with a small model for scientific reasons as well, and then determine whether some features of the data are not reflected in the current model. The posterior predictive distribution of a suitably chosen test quantity can be used to determine whether the current number of components describes the range of observed data. The test quantity must be chosen to measure aspects of the data that are not sufficient statistics for model parameters. An alternative approach is to view M as a hierarchical parameter that can attain the values $1, 2, 3, \ldots$ and average inferences about y over the posterior distribution of mixture models.

18.3 Computation

The computational approach to mixture models that have been specified using indicator variables proceeds along the same lines as our analysis of hierarchical models. Posterior inferences for elements of the parameter vector θ are obtained by averaging over the mixture indicators which are thought of as nuisance parameters or alternatively as 'missing data.' An application to an experiment in psychology is analyzed in detail later in this chapter.

Crude estimates

Initial estimates of the mixture component parameters and the relative proportion in each component can be obtained using various simple techniques. Graphical or clustering methods can be used to identify tentatively the observations drawn from the various mixture components. Ordinarily, once the observations have been classified, it is straightforward to obtain estimates of the parameters of the different mixture components. This type of analysis completely ignores the uncertainty in the indicators and thus can overestimate the differences between mixture components. Crude estimates of the indicators—that is, estimates of the mixture component to which each observation belongs—can also be obtained by clustering techniques. However, crude estimates of the indicators are not usually required because they can be averaged over in EM or drawn as the first step in a Gibbs sampler.

Finding the modes of the posterior distribution using EM

As discussed in Chapter 12, posterior modes are most useful in low-dimensional problems. With mixture models, it is not useful to find the joint mode of parameters and indicators. Instead, the EM algorithm can easily be used to estimate the parameters of a finite mixture model, averaging over the indicator variables. This approach also works for continuous mixtures, averaging over the continuous mixture variables. In either case, the E-step requires the computation of the expected value of the sufficient statistics of the joint model of (y, ζ), using the log of the complete-data likelihood (18.2), conditional on the last guess of the value of the mixture component parameters θ and the mixture proportions λ. In finite mixtures this is often equivalent to computing the conditional expectation of the indicator variables by Bayes' rule. For some problems, including the schizophrenic reaction times example discussed later in this chapter, the ECM algorithm or some other EM alternative is useful. It is important to find all of the modes of the posterior distribution and assess the relative posterior masses near each mode. We suggest choosing a fairly large number (perhaps 50 or 100) starting points by simplifying the model or random sampling. To obtain multiple starting points, a single crude estimate as in the previous paragraph is not enough. Instead, various starting points might be obtained by adding randomness to the crude estimate or from a simplification of the mixture model, for example eliminating random effects and other hierarchical parameters (as in the example in Section 18.4 below).

Posterior simulation using the Gibbs sampler

Starting values for the Gibbs sampler can be obtained via importance resampling from a suitable approximation to the posterior (a mixture of t_4 distributions located at the modes). For mixture models, the Gibbs sampler alternates two major steps: obtaining draws from the distribution of the indicators given the model parameters and obtaining draws from the model parameters given the indicators. The second step may itself incorporate several steps to update all the model parameters. Given the indicators, the mixture model reduces to an ordinary, (possibly hierarchical) model, such as we have already studied. Thus the use of conjugate families as prior distributions can be helpful. Obtaining draws from the distribution of the indicators is usually straightforward: these are multinomial draws in finite mixture models. Modeling errors hinted at earlier, such as incorrect application of an improper prior density, are often found during the iterative simulation stage of computations. For example, a Gibbs sequence started near zero variance may never leave the area. Identifiability problems may also become apparent if the Gibbs sequences appear not to converge because of aliasing in permutations of the components.

Posterior inference

When the Gibbs sampler has reached approximate convergence, posterior inferences about model parameters are obtained by ignoring the drawn indicators. The posterior distribution of the indicator variables contains information about the likely components from which each observation is drawn. The fit of the model can be assessed by a variety of posterior predictive checks, as we illustrate in Section 18.4. If robustness is a concern, the sensitivity of inferences to the assumed parametric family can be evaluated using the methods of Chapter 17.

18.4 Example: reaction times and schizophrenia

We conclude this chapter with a data analysis that incorporates a two-component mixture model. In turn, we describe the scientific questions motivating the analysis, the probability model, the computational approach, and some model evaluation and refinement.

The scientific question and the data

We illustrate the methods in this chapter with an application to an experiment in psychology. Each of 17 subjects—11 non-schizophrenics and 6 schizophrenics—had their reaction times measured 30 times. We present the data in Figure 18.1 and briefly review the basic statistical approach here.

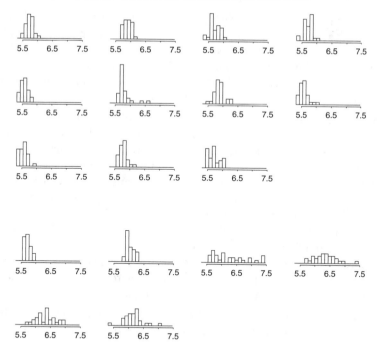

Figure 18.1 *Log response times (in milliseconds) for 11 non-schizophrenic individu-*
als (above) and 6 schizophrenic individuals (below). All histograms are on a common
scale, and there are 30 measurements for each individual. Data from Belin and Rubin
(1990).

It is clear from Figure 18.1 that the response times are higher on average
for schizophrenics. In addition, the response times for at least some of the
schizophrenic individuals are considerably more variable than the response
times for the non-schizophrenic individuals. Psychological theory from the last
half century and before suggests a model in which schizophrenics suffer from an
attentional deficit on some trials, as well as a general motor reflex retardation;
both aspects lead to relatively slower responses for the schizophrenics, with
motor retardation affecting all trials and attentional deficiency only some.

Initial statistical model

To address the questions of scientific interest, we fit the following basic model,
basic in the sense of minimally addressing the scientific knowledge underlying
the data. Response times for non-schizophrenics are described by a normal
random-effects model, in which the responses of person $j = 1, \ldots, 11$ are
normally distributed with distinct person mean α_j and common variance σ_y^2.

Finite mixture likelihood model.   To reflect the attentional deficiency, the response times for each schizophrenic individual $j = 12, \ldots, 17$ are modeled

as a two-component mixture: with probability $(1 - \lambda)$ there is no delay, and the response is normally distributed with mean α_j and variance σ_y^2, and with probability λ responses are delayed, with observations having a mean of $\alpha_j + \tau$ and the same variance, σ_y^2. Because reaction times are all positive and their distributions are positively skewed, even for non-schizophrenics, the above model was fitted to the logarithms of the reaction time measurements.

Hierarchical population model. The comparison of the typical components of $\alpha = (\alpha_1, \ldots, \alpha_{17})$ for schizophrenics versus non-schizophrenics addresses the magnitude of schizophrenics' motor reflex retardation. We include a hierarchical parameter β measuring this motor retardation. Specifically, variation among individuals is modeled by having the means α_j follow a normal distribution with mean μ for non-schizophrenics and $\mu + \beta$ for schizophrenics, with each distribution having a variance of σ_α^2. That is, the mean of α_j in the population distribution is $\mu + \beta S_j$, where S_j is an observed indicator variable that is 1 if person j is schizophrenic and 0 otherwise.

The three parameters of primary interest are: β, which measures motor reflex retardation; λ, the proportion of schizophrenic responses that are delayed; and τ, the size of the delay when an attentional lapse occurs.

Mixture model expressed in terms of indicator variables. Letting y_{ij} be the ith response of individual j, the model can be written in the following hierarchical form.

$$y_{ij}|\alpha_j, \zeta_{ij}, \phi \;\sim\; \mathrm{N}(\alpha_j + \tau\zeta_{ij}, \sigma_y^2),$$
$$\alpha_j|\zeta, \phi \;\sim\; \mathrm{N}(\mu + \beta S_j, \sigma_\alpha^2),$$
$$\zeta_{ij}|\phi \;\sim\; \mathrm{Bernoulli}(\lambda S_j),$$

where $\phi = (\sigma_\alpha^2, \beta, \lambda, \tau, \mu, \sigma_y^2)$, and ζ_{ij} is an unobserved indicator variable that is 1 if measurement i on person j arose from the delayed component and 0 if it arose from the undelayed component. In the following description we occasionally use $\theta = (\alpha, \phi)$ to represent all of the parameters except the indicator variables.

The indicators ζ_{ij} are not necessary to formulate the model but simplify the conditional distributions in the model, allowing us to use the ECM algorithm and the Gibbs sampler for easy computation. Because there are only two mixture components, we only require a single indicator, ζ_{ij}, for each observation, y_{ij}. In our general notation, $M = 2$, the mixture probabilities are $\lambda_1 = \lambda$ and $\lambda_2 = 1 - \lambda$, and the corresponding mixture indicators are $\zeta_{ij1} = \zeta_{ij}$ and $\zeta_{ij2} = 1 - \zeta_{ij}$.

Hyperprior distribution. We start by assigning a noninformative uniform joint prior density on ϕ. In this case the model is not identified, because the trials unaffected by a positive attentional delay could instead be thought of as being affected by a negative attentional delay. We restrict τ to be positive to identify the model. The variance components σ_α^2 and σ_y^2 are of course restricted to be positive as well. The mixture component λ is actually taken to be uniform on $[0.001, 0.999]$ as values of zero or one would not correspond to mixture distributions. Science and previous analysis of the data suggest that a simple model without the mixture is inadequate for this dataset.

Crude estimate of the parameters

The first step in the computation is to obtain crude estimates of the model parameters. For this example, each α_j can be roughly estimated by the sample mean of the observations on subject j, and σ_y^2 can be estimated by the average sample variance within non-schizophrenic subjects. Given the estimates of α_j, we can obtain a quick estimate of the hyperparameters by dividing the α_j's into two groups, non-schizophrenics and schizophrenics. We estimate μ by the average of the estimated α_j's for non-schizophrenics, β by the average difference between the two groups, and σ_α^2 by the variance of the estimated α_j's within groups. We crudely estimate $\hat{\lambda} = 1/3$ and $\hat{\tau} = 1.0$ based on a visual inspection of the lower 6 histograms in 18.1, which display the schizophrenics' response times. It is not necessary to create a preliminary estimate of the indicator variables, ζ_{ij}, because we update ζ_{ij} as the first step in the ECM and Gibbs sampler computations.

Finding the modes of the posterior distribution using ECM

We devote the next few pages to implementation of ECM and Gibbs sampler computations for the finite mixture model. The procedure looks difficult but in practice is straightforward, with the advantage of being easily extended to more complicated models, as we shall see later in this section.

We draw 100 points at random from a simplified distribution for ϕ and use each as a starting point for the ECM maximization algorithm to search for modes. The simplified distribution is obtained by adding some randomness to the crude parameter estimates. Specifically, to obtain a sample from the simplified distribution, we start by setting all the parameters at the crude point estimates above and then divide each parameter by an independent χ_1^2 random variable in an attempt to ensure that the 100 draws were sufficiently spread out so as to cover the modes of the parameter space.

The ECM algorithm has two steps. In the E-step, we determine the expected joint log posterior density, averaging ζ over its posterior distribution, given the last guessed value of θ^{old}; that is, the expression E_{old} refers to averaging ζ over the distribution $p(\zeta|\theta^{\text{old}}, y)$. For our hierarchical mixture model, the expected 'complete-data' log posterior density is

$$\text{E}_{\text{old}}(\log p(\zeta, \theta|y)) = \text{constant} + \sum_{j=1}^{17} \log(\text{N}(\alpha_j|\mu + \beta S_j, \sigma_\alpha^2)) +$$

$$+ \sum_{j=1}^{17}\sum_{i=1}^{30} \left[\log(\text{N}(y_{ij}|\alpha_j, \sigma_y^2))(1 - \text{E}_{\text{old}}(\zeta_{ij})) + \log(\text{N}(y_{ij}|\alpha_j + \tau, \sigma_y^2))\text{E}_{\text{old}}(\zeta_{ij})\right]$$

$$+ \sum_{j=1}^{17} \sum_{i=1}^{30} S_j \left[\log(1 - \lambda)(1 - \mathrm{E}_{\mathrm{old}}(\zeta_{ij})) + \log(\lambda) \mathrm{E}_{\mathrm{old}}(\zeta_{ij}) \right].$$

For the E-step, we must compute $\mathrm{E}_{\mathrm{old}}(\zeta_{ij})$ for each observation (i,j). Given θ^{old} and y, the indicators ζ_{ij} are independent, with conditional posterior densities,

$$\Pr(\zeta_{ij} = 0 | \theta^{\mathrm{old}}, y) = 1 - z_{ij}$$
$$\Pr(\zeta_{ij} = 1 | \theta^{\mathrm{old}}, y) = z_{ij},$$

where

$$z_{ij} = \frac{\lambda^{\mathrm{old}} \mathrm{N}(y_{ij} | \alpha_j^{\mathrm{old}} + \tau^{\mathrm{old}}, (\sigma_y^{\mathrm{old}})^2)}{(1 - \lambda^{\mathrm{old}}) \mathrm{N}(y_{ij} | \alpha_j^{\mathrm{old}}, (\sigma_y^{\mathrm{old}})^2) + \lambda^{\mathrm{old}} \mathrm{N}(y_{ij} | \alpha_j^{\mathrm{old}} + \tau^{\mathrm{old}}, (\sigma_y^{\mathrm{old}})^2)}. \quad (18.3)$$

For each i, j, the above expression is a function of (y, θ) and can be computed based on the data y and the current guess, θ^{old}.

In the M-step, we must alter θ to increase $\mathrm{E}_{\mathrm{old}}(\log p(\zeta, \theta | y))$. Using ECM, we alter one set of components at a time, for each set finding the conditional maximum given the other components. The conditional maximizing steps are easy:

1. Update λ by computing the proportion of trials by schizophrenics exhibiting delayed reaction times. We actually add up the possible fractional contributions of the 30 trials for each schizophrenic subject:

$$\lambda^{\mathrm{new}} = \frac{1}{(6)(30)} \sum_{j=12}^{17} \sum_{i=1}^{30} z_{ij}.$$

2. For each j, update α_j given the current values of the other parameters in θ by combining the normal population distribution of α_j with the normal-mixture distribution for the 30 data points on subject j:

$$\alpha_j^{\mathrm{new}} = \frac{\frac{1}{\sigma_\alpha^2}(\mu + \beta S_j) + \sum_{i=1}^{30} \frac{1}{\sigma_y^2}(y_{ij} - z_{ij}\tau)}{\frac{1}{\sigma_\alpha^2} + \sum_{i=1}^{30} \frac{1}{\sigma_y^2}}. \quad (18.4)$$

3. Given the vector α, the updated estimates for τ and σ_y^2 are obtained from the delayed components of the schizophrenics' reaction times:

$$\tau^{\mathrm{new}} = \frac{\sum_{j=12}^{17} \sum_{i=1}^{30} z_{ij}(y_{ij} - \alpha_j)}{\sum_{j=12}^{17} \sum_{i=1}^{30} z_{ij}}$$

$$(\sigma_y^{\mathrm{new}})^2 = \frac{1}{(17)(30)} \sum_{j=1}^{17} \sum_{i=1}^{30} (y_{ij} - \alpha_j - z_{ij}\tau^{\mathrm{new}})^2.$$

4. Given the vector α, the updated estimates for the population parameters μ, β, and σ_α^2 follow immediately from the normal population distribution

(with uniform hyperprior density); the conditional modes for ECM satisfy

$$\mu^{\text{new}} = \frac{1}{17} \sum_{j=1}^{17} (\alpha_j - \beta^{\text{new}} S_j)$$

$$\beta^{\text{new}} = \frac{1}{6} \sum_{j=12}^{17} (\alpha_j - \mu^{\text{new}})$$

$$(\sigma_\alpha^{\text{new}})^2 = \frac{1}{17} \sum_{j=1}^{17} (\alpha_j - \mu^{\text{new}} - \beta^{\text{new}} S_j)^2, \quad (18.5)$$

which is equivalent to

$$\mu^{\text{new}} = \frac{1}{11} \sum_{j=1}^{11} \alpha_j$$

$$\beta^{\text{new}} = \frac{1}{6} \sum_{j=12}^{17} \alpha_j - \frac{1}{11} \sum_{j=1}^{11} \alpha_j,$$

and $(\sigma_\alpha^{\text{new}})^2$ in (18.5).

After 100 iterations of ECM from each of 100 starting points, we find three local maxima of (α, ϕ): a major mode and two minor modes. The minor modes are substantively uninteresting, corresponding to near-degenerate models with the mixture parameter λ near zero, and have little support in the data, with posterior density ratios less than e^{-20} with respect to the major mode. We conclude that the minor modes can be ignored and, to the best of our knowledge, the target distribution can be considered unimodal for practical purposes.

Normal and t approximations at the major mode

The marginal posterior distribution of the model parameters θ, averaging over the indicators ζ_{ij}, is an easily computed product of mixture forms:

$$p(\theta|y) \propto \prod_{j=1}^{17} N(\alpha|\mu + \beta S_j, \sigma_\alpha^2) \times$$

$$\times \prod_{j=1}^{17} \prod_{i=1}^{30} ((1 - \lambda S_j) N(y_{ij}|\alpha_j, \sigma_y^2) + \lambda S_j N(y_{ij}|\alpha_j + \tau, \sigma_y^2)). \quad (18.6)$$

We compute this function while running the ECM algorithm to check that the marginal posterior density indeed increases at each step. Once the modes have been found, we construct a multivariate t_4 approximation for θ, centered at the major mode with scale determined by the numerically computed second derivative matrix at the mode.

We use the t_4 approximation as a starting distribution for importance resampling of the parameter vector θ. We draw $L = 2000$ independent samples of θ from the t_4 distribution.

Had we included samples from the neighborhoods of the minor modes up to this point, we would have found them to have minuscule importance weights.

Simulation using the Gibbs sampler

We drew a set of ten starting points by importance resampling from the t_4 approximation centered at the major mode to create the starting distribution for the Gibbs sampler. This distribution is intended to approximate our ideal starting conditions: for each scalar estimand of interest, the mean is close to the target mean and the variance is greater than the target variance.

The Gibbs sampler is easy to apply for our model because the full conditional posterior distributions—$p(\phi|\alpha, \zeta, y)$, $p(\alpha|\phi, \zeta, y)$, and $p(\zeta|\alpha, \phi, y)$—have standard forms and can be easily sampled from. The required steps are analogous to the ECM steps used to find the modes of the posterior distribution. Specifically, one complete cycle of the Gibbs sampler requires the following sequence of simulations:

1. For ζ_{ij}, $i = 1, \ldots, 30$, $j = 12, \ldots, 17$, independently draw ζ_{ij} as independent Bernoulli(z_{ij}), with probabilities z_{ij} defined in (18.3). The indicators ζ_{ij} are fixed at 0 for the non-schizophrenic subjects ($j < 12$).

2. For each individual j, draw α_j from a normal distribution with mean α_j^{new}, as defined in (18.4), but with the factor z_{ij} in that expression replaced by ζ_{ij}, because we are now conditional on ζ rather than averaging over it. The variance of the normal conditional distribution for α_j is just the reciprocal of the denominator of (18.4).

3. Draw the mixture parameter λ from a Beta($h+1, 180-h+1$) distribution, where $h = \sum_{j=12}^{17} \sum_{i=1}^{30} \zeta_{ij}$, the number of trials with attentional lapses out of 180 trials for schizophrenics. The simulations are subject to the constraint that λ is restricted to the interval $[0.001, 0.999]$.

4. For the remaining parameters, we proceed as in the normal distribution with unknown mean and variance. Draws from the posterior distribution of $(\beta, \mu, \tau, \sigma_y^2, \sigma_\alpha^2)$ given (α, λ, ζ) are obtained by first sampling from the marginal posterior distribution of the variance parameters and then sampling from the conditional posterior distribution of the others. First,

$$\sigma_y^2 | \alpha, \lambda, \zeta \sim \text{Inv-}\chi^2 \left(508, \frac{1}{508} \sum_{j=1}^{17} \sum_{i=1}^{30} (y_{ij} - \alpha_j - \zeta_{ij}\tau)^2 \right),$$

and

$$\sigma_\alpha^2 | \alpha, \lambda, \zeta \sim \text{Inv-}\chi^2 \left(15, \frac{1}{15} \sum_{j=1}^{17} (\alpha_j - \mu - \beta S_j)^2 \right).$$

Then, conditional on the variances, τ can be simulated from a normal distribution,

$$\tau | \alpha, \lambda, \zeta, \sigma_y^2, \sigma_\alpha^2 \sim N \left(\frac{\sum_{j=12}^{17} \sum_{i=1}^{30} \zeta_{ij} (y_{ij} - \alpha_j)}{\sum_{j=12}^{17} \sum_{i=1}^{30} \zeta_{ij}}, \frac{\sigma_y^2}{\sum_{j=12}^{17} \sum_{i=1}^{30} \zeta_{ij}} \right).$$

The conditional distribution of μ given all other parameters is normal and depends only on α, β, and σ_α^2:

$$\mu | \alpha, \lambda, \zeta, \beta, \sigma_y^2, \sigma_\alpha^2 \sim N \left(\frac{1}{17} \sum_{j=1}^{17} (\alpha_j - \beta S_j), \frac{1}{17} \sigma_\alpha^2 \right).$$

Finally, β also has a normal conditional distribution:

$$\beta | \alpha, \lambda, \zeta, \mu, \sigma_y^2, \sigma_\alpha^2 \sim N \left(\frac{1}{6} \sum_{j=12}^{17} (\alpha_j - \mu), \frac{1}{6} \sigma_\alpha^2 \right). \tag{18.7}$$

As is common in conjugate models, the steps of the Gibbs sampler are simply stochastic versions of the ECM steps; for example, variances are drawn from the relevant scaled inverse-χ^2 distributions rather than being set to the posterior mode, and the centers of most of the distributions use the ECM formulas for conditional modes with z_{ij}'s replaced by ζ_{ij}'s.

Possible difficulties at a degenerate point

If all the ζ_{ij}'s are zero, then the mean and variance of the conditional distribution (18.7) are undefined, because τ has an improper prior distribution and, conditional on $\sum_{ij} \zeta_{ij} = 0$, there are no delayed reactions and thus no information about τ. Strictly speaking, this means that our posterior distribution is improper. For the data at hand, however, this degenerate point has extremely low posterior probability and is not reached by any of our simulations. If the data were such that $\sum_{ij} \zeta_{ij} = 0$ were a realistic possibility, it would be necessary to assign an informative prior distribution for τ.

Inference from the iterative simulations

In the mixture model example, we computed several univariate estimands: the 17 random effects α_j and their standard deviation σ_α, the shift parameters τ and β, the standard deviation of observations σ_y, the mixture parameter λ, the ratio of standard deviations σ_α / σ_y, and the log posterior density. After an initial run of ten sequences for 200 simulations, we computed the estimated potential scale reductions, $\hat{R}$, for all scalar estimands, and found them all to be below 1.1; we were thus satisfied with the simulations (see the discussion in Section 11.6). The potential scale reductions were estimated on the logarithmic scale for the variance parameters and the logit scale for λ. We obtain posterior intervals for all quantities of interest from the quantiles of the 1000 simulations from the second halves of the sequences.

	Old model			New model		
Parameter	2.5%	median	97.5%	2.5%	median	97.5%
λ	0.07	0.12	0.18	0.46	0.64	0.88
τ	0.74	0.85	0.96	0.21	0.42	0.60
β	0.17	0.32	0.48	0.07	0.24	0.43
ω		fixed at 1		0.24	0.56	0.84

Table 18.1 *Posterior quantiles for parameters of interest under the old and new mixture models for the reaction time experiment. Introducing the new mixture parameter* ω, *which represents the proportion of schizophrenics with attentional delays, changes the interpretation of the other parameters in the model.*

The first three columns of Table 18.1 display posterior medians and 95% intervals from the Gibbs sampler simulations for the parameters of most interest to the psychologists:

- λ, the probability that an observation will be delayed for a schizophrenic subject to attentional delays;

- τ, the attentional delay (on the log scale);

- β, the average log response time for the undelayed observations of schizophrenics minus the average log response time for nonschizophrenics.

For now, ignore the final row and the final three columns of Table 18.1. Under this model, there is strong evidence that the average reaction times are slower for schizophrenics (the factor of $\exp(\beta)$, which has a 95% posterior interval of $[1.18, 1.62]$), with a fairly infrequent (probability in the range $[0.07, 0.18]$) but large attentional delay ($\exp(\tau)$ in the range $[2.10, 2.61]$).

Posterior predictive distributions

We obtain draws from the posterior predictive distribution by using the draws from the posterior distribution of the model parameters θ and mixture components ζ. Two kinds of predictions are possible:

1. For additional measurements on a person j in the experiment, posterior simulations $\tilde{y}_{ij}$ should be made conditional on the individual parameter α_j from the posterior simulations.

2. For measurements on an entirely new person j, one should first draw a new parameter $\tilde{\alpha}_j$ conditional on the hyperparameters, $\mu, \beta, \sigma_\alpha^2$. Simulations of measurements $\tilde{y}_{ij}$ can then be performed conditional on $\tilde{\alpha}$. That is, for each simulated parameter vector θ^l, draw $\tilde{\alpha}^l$, then $\tilde{y}^l$.

The different predictions are useful for different purposes. For checking the fit of the model, we use the first kind of prediction so as to compare the observed data with the posterior predictive distribution of the measurements on those particular individuals.

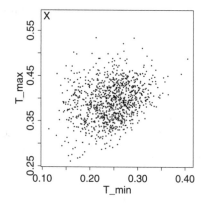

Figure 18.2 *Scatterplot of the posterior predictive distribution of two test quantities: the smallest and largest observed within-schizophrenic variances. The × represents the observed value of the test quantity in the dataset.*

Checking the model

Defining test quantities to assess aspects of poor fit. The model was chosen to fit accurately the unequal means and variances in the two groups of subjects in the study, but there was still some question about the fit to individuals. In particular, the histograms of schizophrenics' reaction times (Figure 18.1) indicate that there is substantial variation in the within-person response time variance. To investigate whether the model can explain this feature of the data, we compute s_j, the standard deviation of the 30 log reaction times y_{ij}, for each schizophrenic individual $j = 12, \ldots, 17$. We then define two test quantities: $T_{\min}$ and $T_{\max}$, the smallest and largest of the six values s_j. To obtain the posterior predictive distribution of the two test quantities, we simulate predictive datasets from the normal-mixture model for each of the 1000 simulation draws of the parameters from the posterior distribution. For each of those 1000 simulated datasets, y^{rep}, we compute the two test quantities, $T_{\min}(y^{\text{rep}}), T_{\max}(y^{\text{rep}})$.

Graphical display of realized test quantities in comparison to their posterior predictive distribution. In general, we can look at test quantities individually by plotting histograms of their posterior predictive distributions, with the observed value marked. In this case, however, with exactly two test quantities, it is natural to look at a scatterplot of the joint distribution. Figure 18.2 displays a scatterplot of the 1000 simulated values of the test quantities, with the observed values indicated by an ×. With regard to these test quantities, the observed data y are atypical of the posterior predictive distribution—$T_{\min}$ is too low and $T_{\max}$ is too high—with estimated p-values of 0.000 and 1.000 (to three decimal places).

Example of a poor test quantity. In contrast, a test quantity such as the average value of s_j is not useful for model checking since this is essentially the sufficient statistic for the model parameter σ_y^2, and thus the model will automatically fit it well.

Expanding the model

An attempt was made to fit the data more accurately by adding two further parameters to the model, one parameter to allow some schizophrenics to be unaffected by attentional delays, and a second parameter that allows the delayed observations to be more variable than undelayed observations. We add the parameter ω as the probability that a schizophrenic individual has attentional delays and the parameter σ_{y2}^2 as the variance of attention-delayed measurements. In the expanded model, we give both these parameters uniform prior distributions. The model we have previously fitted can be viewed as a special case of the new model, with $\omega = 1$ and $\sigma_{y2}^2 = \sigma_y^2$.

For computational purposes, we also introduce another indicator variable, W_j, that is 1 if individual j is prone to attention delays and 0 otherwise. The indicator W_j is automatically 0 for non-schizophrenics and is 1 with probability ω for each schizophrenic. Both of these parameters are appended to the parameter vector θ to yield the model,

$$
\begin{aligned}
y_{ij}|\zeta_{ij}, \theta &\sim \mathrm{N}(\alpha_j + \tau\zeta_{ij}, (1 - \zeta_{ij})\sigma_y^2 + \zeta_{ij}\sigma_{y2}^2) \\
\alpha_j|\zeta, S, \mu, \beta, \sigma_\alpha^2 &\sim \mathrm{N}(\mu + \beta S_j, \sigma_\alpha^2) \\
\zeta_{ij}|S, W, \theta &\sim \mathrm{Bernoulli}(\lambda S_j W_j) \\
W_j|S, \theta &\sim \mathrm{Bernoulli}(\omega S_j).
\end{aligned}
$$

It is straightforward to fit the new model by just adding three new steps in the Gibbs sampler to update ω, σ_{y2}^2, and W. In addition, the Gibbs sampler steps for the old parameters must be altered somewhat to be conditional on the new parameters. We do not give the details here but just present the results. We use ten randomly selected draws from the previous posterior simulation as starting points for ten parallel runs of the Gibbs sampler (values of the new parameters are drawn as the first Gibbs steps). Ten simulated sequences each of length 500 were sufficient for approximate convergence, with estimated potential scale reductions less than 1.1 for all model parameters. As usual, we discarded the first half of each sequence, leaving a set of 2500 draws from the posterior distribution of the larger model.

Before performing posterior predictive checks, it makes sense to compare the old and new models in their posterior distributions for the parameters. The last three columns of Table 18.1 display inferences for the parameters of applied interest under the new model and show significant differences from the old model. Under the new model, a greater proportion of schizophrenic observations is delayed, but the average delay is shorter. We have also included a row in the table for ω, the probability that a schizophrenic will be subject to attentional delays, which was fixed at 1 in the old model. Since the old

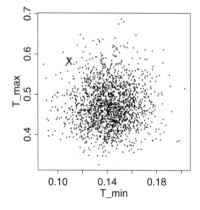

Figure 18.3 *Scatterplot of the posterior predictive distribution, under the expanded model, of two test quantities: the smallest and largest within-schizophrenic variance. The × represents the observed value of the test quantity in the dataset.*

model is nested within the new model, the differences between the inferences suggest a real improvement in fit.

Checking the new model

The expanded model is an improvement, but how well does it fit the data? We expect that the new model should show an improved fit with respect to the test quantities considered in Figure 18.2, since the new parameters describe an additional source of person-to-person variation. (The new parameters have substantive interpretations in psychology and are not merely 'curve fitting.') We check the fit of the expanded model using posterior predictive simulation of the same test quantities under the new posterior distribution. The results are displayed in Figure 18.3, based on posterior predictive simulations from the new model.

Once again, the × indicates the observed test quantity. Compared to Figure 18.2, the × is in the same place, but the posterior predictive distribution has moved closer to it. (The two figures are on different scales.) The fit of the new model, however, is by no means perfect: the × is still in the periphery, and the estimated p-values of the two test quantities are 0.97 and 0.05. Perhaps most important, the lack of fit has greatly diminished in magnitude, as can be seen by examining the scales of Figures 18.2 and 18.3. We are left with an improved model that still shows some lack of fit, suggesting possible directions for improved modeling and data collection.

18.5 Bibliographic note

Application of EM to mixture models is described in Dempster, Laird, and Rubin (1977). Gelman and King (1990b) fit a hierarchical mixture model using the Gibbs sampler in an analysis of elections, using an informative prior distribution to identify the mixture components separately. Other Bayesian applications of mixture models include Box and Tiao (1968), Turner and West (1993), and Belin and Rubin (1995b). Richardson and Green (1997) and Stephens (2000) discuss Bayesian analysis of mixtures with unknown numbers of components

A comprehensive text emphasizing non-Bayesian approaches to finite mixture models is Titterington, Smith, and Makov (1985). West (1992) provides a brief review from a Bayesian perspective. Muller, and Rosner (1997) present an application of a Bayesian hierarchical mixture model.

The schizophrenia example is discussed more completely in Belin and Rubin (1990, 1995a) and Gelman and Rubin (1992b). The posterior predictive checks for this example are presented in a slightly different graphical form in Gelman and Meng (1996). An expanded model applied to more complex data from schizophrenics appears in Rubin and Wu (1997). Rubin and Stern (1994) and Gelman, Meng, and Stern (1996) demonstrate the use of posterior predictive checks to determine the number of mixture components required for an accurate model fit in a different example in psychology.

CHAPTER 19

Multivariate models

It is common for data to be collected with a multivariate structure, for example, from a sample survey in which each individual is asked several questions or an experiment in which several outcomes are measured on each unit. In Sections 3.5 and 3.6, we introduced the most commonly used and useful models for multivariate data: the multinomial and multivariate normal models for categorical and continuous data, respectively. Why, then, have a separate chapter for multivariate models? Because, when combined with hierarchical models, or regression models, the standard distributions require additional work in specifying prior distributions and in computation. In this chapter, we discuss how to graft multivariate distributions onto the hierarchical and regression models that have been our focus in most of this book. Our basic approach, in modeling and computation, is to use normal distributions at the hierarchical level, while paying attention to the additional difficulties required in specifying distributions and computing with mean vectors and covariance matrices. We illustrate the resulting procedures with several examples.

19.1 Linear regression with multiple outcomes

Consider a study of n units, with which we wish to study how d outcome variables, y, vary as a function of k explanatory variables, x. For each unit i, we label the *vectors* of d outcome measurements and k explanatory variables as y_i and x_i, respectively. The simplest normal linear model for this relation, sometimes called *multivariate regression*, is expressed in terms of n independent multivariate observations: $y_i|B, \Lambda \sim N(x_i^T B, \Lambda)$, or equivalently,

$$p(y|x, B, \Lambda) = \prod_{i=1}^{n} N(y_i|x_i^T B, \Lambda), \qquad (19.1)$$

where B is a $k \times d$ matrix of regression coefficients and Λ is a $d \times d$ covariance matrix.

Noninformative prior distributions

The unknown parameters of the multivariate regression model are B and Λ. For an initial analysis, if $n > k$, it is often natural to consider a noninformative uniform prior distribution on the components of B and an inverse-Wishart with -1 degrees of freedom for Λ, as discussed in Section 3.6.

Modeling the regression coefficients and covariance parameters

Useful information can be included in the model in a variety of ways. In many problems, the regression coefficients will follow a natural hierarchical structure of some sort, which can be modeled as discussed in Chapters 5 and 15.

Prior distributions and models for variance and covariance parameters are discussed more fully in the next section. The conjugate prior distribution is the inverse-Wishart distribution but this is somewhat restrictive. Unfortunately the alternative approaches are more complicated to implement. We restrict attention here to short descriptions of some modeling approaches that may be useful; we elaborate in the next section. In many cases, it is natural to model the diagonal elements of Λ and the correlations separately and possibly apply a hierarchical model to each set. Also, our usual approach of successive approximations can be useful. For example, consider an educational study in which the outcomes y are the scores of each student on a battery of d tests, where some tests are known to have larger variances than others, and some pairs of tests—dealing with the same areas of knowledge—are known to be more highly correlated than other pairs. It may make sense to consider a model in which the correlations in Λ are equal to a constant ρ (perhaps with a uniform prior distribution on ρ). Having obtained posterior simulations from that model, one can then generalize to allow different correlations from a more realistic distribution.

Modeling as a nested sequence of regression models

Another often useful approach for modeling multivariate data is to place the variables y in a natural order and then apply successive normal linear regression models to each of the factors, $p(y_1|x), p(y_2|y_1, x), \ldots, p(y_d|y_1, \ldots, y_{d-1}, x)$. This sort of nested model can be equivalent to the full multivariate regression model (19.1), when only linear functions of y are conditioned upon. However, the nested-regression and multivariate-normal models can differ substantially in practice for two reasons. First, the natural prior distributions for the nested regressions do not combine to a simple form in the multivariate parameterization, so that simply-expressed models in one form become complex in the other. Second, the flexibility to condition on nonlinear functions of y can be important in fitting realistic models and allows the nested regressions to go beyond the multivariate normal model.

The equivalent univariate regression model

The multivariate regression model (19.1) can be expressed equivalently as a univariate regression of a $dn \times 1$ matrix y (a concatenation of the data vectors $y_1, \ldots, y_n$) on a $dn \times dk$ matrix X with a $dn \times dn$ covariance matrix Σ_y. This univariate parameterization can be useful conceptually because it allows us to use the ideas and methods developed for univariate regression models.

19.2 Prior distributions for covariance matrices

The initial discussion of modeling multivariate normal data in Section 3.6 of Part I discusses only the conjugate inverse-Wishart distribution for the covariance matrix. In practice the inverse-Wishart is quite restrictive. There is but a single degrees of freedom parameter, which implies the same amount of prior information about each of the variance parameters in the covariance matrix. There is no reason to expect this to generally be the case. The inverse-Wishart distribution is invariant under rotations, which can be helpful in some applications but inappropriate in others. It has been difficult however to come up with a alternative prior distribution that can be applied in general situations and is easy to work with. The most promising suggestions are based on factoring the variance matrix in some way. Matrix factorizations that have been considered include the spectral decomposition ($\Lambda = VDV^T$, where columns of V are eigenvectors and D is diagonal with elements equal to the eigenvalues), the spectral decomposition with V further factored as the product of Givens rotations, a separation of the covariance matrix into variance parameters and correlation parameters, and a diagonalization ($A\Lambda A^T = D$ with A a unit lower triangular matrix and D a diagonal matrix with positive entries). The bibliographic note at the end of the chapter gives references on all these approaches, and we describe the latter two in more detail here. In each case the goal is to build models on the factorization components. The various methods vary in terms of ease of interpretation, range of problems to which they apply, and computational convenience.

Separation strategy

An intuitive approach is to write the covariance matrix in terms of the standard deviations and correlations

$$\Lambda = SRS, \tag{19.2}$$

where $S = \text{Diag}(\sigma)$ is the diagonal matrix of residual standard deviations of the d outcomes, $\sigma = (\sigma_1, \ldots, \sigma_d)$, and R is the $d \times d$ correlation matrix. With this separation one can use the factorization $p(\sigma, R) = p(\sigma)p(R|\sigma)$ to specify the prior distribution. For example, we may want to incorporate prior information about σ but use a diffuse prior distribution for the correlation matrix R.

The vector of standard deviations σ is non-negative but otherwise unconstrained. Two natural choices of model are independent inverse-χ^2 distributions for the elements of σ^2 (with possibly different degrees of freedom) or a multivariate normal distribution for $(\log \sigma_1, \ldots, \log \sigma_d)$.

Modeling R gets more complicated because of the restrictions that the correlation matrix be positive definite. (For example, if $d = 3$, it is mathematically impossible for all three correlations r_{12}, r_{13}, r_{23} to equal -0.7.) One possible noninformative model for R is the uniform distribution, $p(R) \propto 1$, under the constraint that the matrix R be a positive definite. The marginal distribution on individual correlations is not uniform in this case; the constraints rule out extreme combinations of correlations, and as a result the marginal distribution on each correlation favors values closer to zero. However, in low dimensions (for example, $d < 10$), the marginal distributions are close to uniform over a wide range, which is probably reasonable in most cases.

An alternative model that has been proposed is,

$$p(R|\nu) \propto |R|^{\frac{1}{2}(\nu-1)(d-1)-1} \left(\prod_{i=1}^{d} |R_{(ii)}|^{-\nu/2} \right), \qquad \nu \geq d$$

where $R_{(ii)}$ is the ith principal submatrix of R. This is the marginal distribution of R when Λ has an inverse-Wishart distribution with identity scale matrix and ν degrees of freedom. The choice $\nu = d + 1$ provides a uniform marginal distribution for each pairwise correlation; other values of ν yield lighter ($\nu > d + 1$) or heavier ($d \leq \nu < d + 1$) than uniform tails for the marginal distribution.

Both choices of $p(R|\sigma)$ that we have discussed here are independent of σ. The separation strategy makes it easy to have the prior for R depend on σ if that is appropriate. Both the jointly uniform prior distribution and the marginal uniform prior distribution treat all pairwise correlations the same; again, alternatives to allow for known structure, such as blocks of correlations to be treated similarly, are possible. The separation strategy also makes it easy to set up a hierarchical model for a group of covariance matrices. If we take $\Lambda_j = S_j R_j S_j$, then it is possible to restrict the S_j's to be constant across groups, the R_j's to be constant across groups, or to model those components as draws from population distributions.

The separation strategy allows us to work with statistically relevant parameters for which it may be possible to specify our prior beliefs or model relationships to other variables. It also overcomes some difficulties of the standard inverse-Wishart prior distribution. A disadvantage, however, is that computation becomes more difficult.

For example, consider a Gibbs sampling strategy in which each element of σ and R is drawn conditional on all other elements (and other parameters in the model, such as the regression coefficients). The drawing of a given r_{ij} is complicated by the restriction that the correlation matrix R be positive definite. The range of allowable r_{ij} values can be computed as follows. Let $R(r)$ denote the matrix obtained by changing r_{ij} to r with all other components of the matrix unchanged. If the determinant $|R(r)|$ is positive, then the matrix will be positive definite. It can be easily shown from the definition of the determinant (see Exercise 19.1) that $|R(r)|$ is a quadratic function of r, and so we can write it as $ar^2 + br + c$, with coefficients that can be determined by evaluating $|R(r)|$ at three values of r, for example taking $c = |R(0)|$, $b = (|R(1)| - |R(-1)|)/2$, and $a = (|R(1)| + |R(-1)| - 2|R(0)|)/2$. From the quadratic formula, we can determine the range of r_{ij} values for which R is positive definite (conditional on the rest of the correlation matrix). Then a suitable Gibbs sampling approximation or a Metropolis step can be constructed. Generating draws from each σ_i given all other parameters is less complex because the only restriction is that σ_i be non-negative.

A strategy for longitudinal data

In the special case where the variables whose covariance is being modeled have a natural ordering, for example as a series of measurements for an individual over time, a different parameterization offers interpretable parameters and computational convenience. The key idea is that the covariance matrix Λ can be diagonalized,

$$A\Lambda A^T = D$$

where A is a unique unit lower triangular matrix and D is diagonal with positive entries. By analogy to the separation strategy, we can then specify prior distributions as $p(\Lambda) = p(D)p(A|D)$. If the covariance matrix corresponds to a vector of longitudinal measurements $y = \{y_t\}$, then the elements of A can be interpreted as autoregressive parameters, expressing the dependence of y_t on earlier elements of the response vector, and the elements of D are innovation variances. This approach is closely related to the idea mentioned in the previous section of modeling the multivariate y as a nested sequence of regression models.

The triangular matrix parameterization offers the potential for convenient computation using Gibbs sampling. If the elements of A are given a multivariate normal prior distribution and the elements of D are given inverse-χ^2 distributions, then the full conditional distributions are in the same families. Within these families that produce this conditional conjugacy, it is possible to model the elements of A and D in terms of covariates or to shrink towards values consistent with a hypothesized value for Λ, for example the identity matrix.

A limitation of this approach is that it requires an ordering for the items being considered. If the items are a set of correlated measurements then there is no obvious order. The reparameterization can still be carried out (there is a unique A, D pair), but the factorization no longer yields interpretable elements for which we might expect to be able to elicit prior information.

Prior distributions for inverse covariance matrices

A slightly different approach is to apply the separation strategy (19.2) to the inverse covariance matrix Λ^{-1}. In this case, the resulting parameters are inverse partial variances (that is, the inverse of the variance of a given element conditional on all the others) and negatives of the partial correlation coefficients. In many problems, the inverse covariance matrix has many zero elements corresponding to conditional independence relationships among the variables. Then prior distributions can be used to reflect this preference, perhaps shrinking the partial correlations to zero or using a prior distribution that attaches positive probability to the event that a partial correlation is zero.

19.3 Hierarchical multivariate models

Perhaps the most important use of multivariate models in Bayesian statistics is as population distributions in hierarchical models when the parameters have a matrix structure. For example, consider the SAT coaching example of Section 5.5. For each of the eight schools, a preliminary regression analysis was performed on students' SAT-V scores with four explanatory variables: the treatment indicator, a constant term, and two previous test scores. The analysis in Section 5.5 fitted a hierarchical normal model to the estimated coefficients of the treatment indicator displayed in Table 5.2. A more complete analysis, however, would fit a hierarchical model to the eight vectors of four regression coefficients, with a natural model being the multivariate normal:

$$p(\beta) \sim \prod_{j=1}^{8} N(\beta_j | \alpha, \Lambda_\beta),$$

where β_j is the vector of four regression coefficients for school j (in the notation of Section 5.5, $\theta_j = \beta_{j1}$), α is an unknown vector of length 4, and Λ_β is an unknown 4×4 covariance matrix. In this example, the parameters of interest would still be the treatment effects, $\theta_j = \beta_{j1}$, but the analysis based on the multivariate hierarchical model would pool the other regression coefficients somewhat, thus affecting the estimates of the parameters of interest. Of course, such an analysis is not possible given only the data in Table 5.2, and considering the sample sizes within each school, we would not expect much effect on our final inferences in any case.

The problems arising in setting up and computing with multivariate hierarchical models are similar to those for multivariate data models described in Section 19.1. We illustrate with an example of regression prediction in which the data are so sparse that a hierarchical model is *necessary* for inference about some of the estimands of interest.

Example. Predicting business school grades for different groups of students

It is common for schools of business management in the United States to use regression equations to predict the first-year grade point average of prospective students from their scores on the verbal and quantitative portions of the Graduate Management Admission Test (GMAT-V and GMAT-Q) as well as their undergraduate grade point average (UGPA). This equation is important because the predicted score derived from it may play a central role in the decision to admit the student. The coefficients of the regression equation are typically estimated from the data collected from the most recently completed first-year class.

A concern was raised with this regression model about possible biased predictions for identifiable subgroups of students, particularly black students. A study was performed based on data from 59 business schools over a two-year period, involving about 8500 students of whom approximately 4% were black. For each school, a separate regression was performed of first-year grades on four explanatory variables: a constant term, GMAT-V, GMAT-Q, and UGPA. By looking at the residuals for all schools and years, it was found that the regressions tended to *overpredict* the first-year grades of blacks.

At this point, it might seem natural to add another term to the regression model corresponding to an indicator variable that is 1 if a student is black and 0 otherwise. However, such a model was considered too restrictive; once blacks and non-blacks were treated separately in the model, it was desired to allow different regression models for the two groups. For each school, the expanded model then has eight explanatory variables: the four mentioned above, and then the same four variables multiplied by the indicator for black students. For student $i = 1, \ldots, n_j$ in school $j = 1, \ldots, 59$ we model the first-year grade point average y_{ij}, given the vector of eight covariates x_{ij}, as a linear regression with coefficient vector β_j and residual variance σ_j^2. Then the model for the entire vector of responses y is

$$p(y|\beta, \sigma^2) \sim \prod_{j=1}^{59} \prod_{i=1}^{n_j} \mathrm{N}(y_{ij}|X_{ij}^T \beta_j, \sigma_j^2).$$

Geometrically, the model is equivalent to requiring two different regression planes: one for blacks and one for non-blacks. For each school, nine parameters must be estimated: $\beta_{1j}, \ldots, \beta_{8j}, \sigma_j$. Algebraically, all eight terms of the regression are used to predict the scores of blacks but only the first four terms for non-blacks.

At this point, the procedure of estimating separate regressions for each school becomes impossible using standard least-squares methods, which are implicitly based on noninformative prior distributions. Blacks comprise only 4% of the students in the dataset, and many of the schools are all non-black or have so few blacks that the regression parameters cannot be estimated under classical regression (that is, based on a noninformative prior distribution on the nine parameters in each regression). Fortunately, it is possible to estimate all 8×59 regression parameters simultaneously using a hierarchical model. To use the most straightforward approach, the 59 vectors β_j are modeled as independent samples from a multivariate $\mathrm{N}(\alpha, \Lambda_\beta)$ distribution, with unknown vector α and 8×8 matrix Λ_β.

The unknown parameters of the model are then β, α, Λ_β, and $\sigma_1, \ldots, \sigma_{59}$. We first assumed a uniform prior distribution on $\alpha, \Lambda_\beta, \log \sigma_1, \ldots, \log \sigma_{59}$. This noninformative approach of course is not ideal (at the least, one would want to embed the 59 σ_j parameters in a hierarchical model) but is a reasonable start.

A crude approximation to α and the parameters σ_j^2 was obtained by running a regression of the combined data vector y on the eight explanatory variables, pooling the data from all 59 schools. Using the crude estimates as a starting point, the posterior mode of $(\alpha, \Lambda_\beta, \sigma_1^2, \ldots, \sigma_{59}^2)$ was found using EM. Here, we describe the conclusions of the study, which were based on this modal approximation.

One conclusion from the analysis was that the multivariate hierarchical model is a substantial improvement over the standard model, because the predictions for both black and non-black students are relatively accurate. Moreover, the analysis revealed systematic differences between predictions for black and non-black students. In particular, conditioning the test scores at the mean scores for the black students, in about 85% of schools, non-blacks were predicted to have higher first-year grade-point averages, with over 60% of the differences being more than one posterior standard deviation above zero, and about 20% being more than two posterior standard deviations above zero. This sort of comparison, conditional on school and test scores, could not be reasonably estimated with a nonhierarchical model in this dataset, in which the number of black students per school was so low.

19.4 Multivariate models for nonnormal data

When modeling multivariate data with nonnormal distributions, it is often useful to apply a normal distribution to some transformation of the parameters, as in our approach to generalized linear models in Chapter 16. As an example, consider the analysis of the stratified sample survey discussed on page 209. Here, the trivariate outcome—numbers of Bush supporters, Dukakis supporters, and those with no-opinion—is naturally modeled with a multinomial distribution in each of 16 strata. Within each stratum j, we transform the probabilities $(\theta_{1j}, \theta_{2j}, \theta_{3j})$, which are constrained to sum to 1, into a two-dimensional parameter of logits, (β_{1j}, β_{2j}). We fit the 16 vectors β_j with a bivariate normal distribution, resulting in a hierarchical multivariate model.

For another example, we reanalyze the meta-analysis example of Section 5.6 using a binomial model with a hierarchical normal model for the parameters describing the individual studies.

Example. Meta-analysis with binomial outcomes

In this example, the results of each of 22 clinical trials are summarized by a 2×2 table of death and survival under each of two treatments, and we are interested in the distribution of the effects of the treatment on the probability of death. The analysis of Section 5.6 was based on a normal approximation to the empirical log-odds ratio in each study. Because of the large sample sizes, the normal approximation is fairly accurate in this case, but it is desirable to have a more exact procedure for the general problem.

In addition, the univariate analysis in Section 5.6 used the ratio, but not the average, of the death rates in each trial; ignoring this information can have an effect, even with large samples, if the average death rates are correlated with the treatment effects.

Data model. Continuing the notation of Section 5.6, let y_{ij} be the number of deaths out of n_{ij} patients for treatment $i = 0, 1$ and study $j = 1, \ldots, 22$. As in the earlier discussion we take $i = 1$ to represent the treated groups, so that negative values of the log-odds ratio represent reduced frequency of death under the treatment. Our data model is binomial:

$$y_{ij}|n_{ij}, p_{ij} \sim \text{Bin}(n_{ij}, p_{ij}),$$

where p_{ij} is the probability of death under treatment i in study j. We must now model the 44 parameters p_{ij}, which naturally follow a multivariate model, since they fall into 22 groups of two. We first transform the p_{ij}'s to the logit scale, so they are defined on the range $(-\infty, \infty)$ and can plausibly be fitted by the normal distribution.

Hierarchical model in terms of transformed parameters. Rather than fitting a normal model directly to the parameters $\text{logit}(p_{ij})$, we transform to the average and difference effects for each experiment:

$$\beta_{1j} = (\text{logit}(p_{0j}) + \text{logit}(p_{1j}))/2$$
$$\beta_{2j} = \text{logit}(p_{1j}) - \text{logit}(p_{0j}). \tag{19.3}$$

The parameters β_{2j} correspond to the θ_j's of Section 5.6. We model the 22 exchangeable *pairs* (β_{1j}, β_{2j}) as following a bivariate normal distribution with unknown parameters:

$$p(\beta|\alpha, \Lambda) = \prod_{j=1}^{22} \text{N}\left(\begin{pmatrix}\beta_{1j} \\ \beta_{2j}\end{pmatrix}\middle|\begin{pmatrix}\alpha_1 \\ \alpha_2\end{pmatrix}, \Lambda\right).$$

This is equivalent to a normal model on the parameter pairs $(\text{logit}(p_{0j}), \text{logit}(p_{1j}))$; however, the linear transformation should leave the β's roughly independent in their population distribution, making our inference less sensitive to the prior distribution for their correlation.

Hyperprior distribution. We use the usual noninformative uniform prior distribution for the parameters α_1 and α_2. For the hierarchical variance matrix Λ, there is no standard noninformative choice; for this problem, we assign independent uniform prior distributions to the variances Λ_{11} and Λ_{22} and the correlation, $\rho_{12} = \Lambda_{12}/(\Lambda_{11}\Lambda_{22})^{1/2}$. The resulting posterior distribution is proper (see Exercise 19.2).

Posterior computations. We drew samples from the posterior distribution in the usual way based on successive approximations, following the general strategy described in Part III. (This model could be fit easily using Bugs, but we use this example to illustrate how such computations can be constructed directly.) The computational method we used here is almost certainly not the most efficient in terms of computer time, but it was relatively easy to program in a general way and yielded believable inferences. The model was parameterized in terms of β, α, $\log(\Lambda_{11})$, $\log(\Lambda_{22})$, and Fisher's z-transform of the correlation, $\frac{1}{2}\log(\frac{1+\rho_{12}}{1-\rho_{12}})$, to transform the ranges of the parameters to the whole real line. We sampled random draws from an approximation based on conditional modes, followed by importance resampling, to obtain starting points for ten parallel runs of the Metropolis algorithm. We used a normal jumping kernel with covariance from the curvature of the posterior density at the mode, scaled by a factor of $2.4/\sqrt{49}$ (because the jumping is in 49-dimensional space; see page 306). The simulations were run for 40,000 iterations, at which point the estimated scale reductions, $\hat{R}$, for all parameters were below 1.2 and most were below 1.1. We use the resulting 200,000 simulations of (β, α, Λ) from the second halves of the simulated sequences to summarize the posterior distribution in Table 19.1.

Results from the posterior simulations. The posterior distribution for ρ_{12} is centered near 0.21 with considerable variability. Consequently, the multivariate model would have only a small effect on the posterior inferences obtained from the univariate analysis concerning the log-odds ratios for the individual studies or the relevant hierarchical parameters. Comparing the results in Table 19.1 to those in Tables 5.4 and 5.5 shows that the inferences are quite similar. The multivariate analysis based on the exact posterior distribution fixes any deficiencies in the

Estimand	Posterior quantiles				
	2.5%	25%	median	75%	97.5%
Study 1 avg logit, $\beta_{1,1}$	-3.16	-2.67	-2.42	-2.21	-1.79
Study 1 effect, $\beta_{2,1}$	-0.61	-0.33	-0.23	-0.13	0.14
Study 2 effect, $\beta_{2,2}$	-0.63	-0.37	-0.28	-0.19	0.06
Study 3 effect, $\beta_{2,3}$	-0.58	-0.35	-0.26	-0.16	0.08
Study 4 effect, $\beta_{2,4}$	-0.44	-0.30	-0.24	-0.17	-0.03
Study 5 effect, $\beta_{2,5}$	-0.43	-0.27	-0.18	-0.08	0.16
Study 6 effect, $\beta_{2,6}$	-0.68	-0.37	-0.27	-0.18	0.04
Study 7 effect, $\beta_{2,7}$	-0.64	-0.47	-0.38	-0.31	-0.20
Study 8 effect, $\beta_{2,8}$	-0.41	-0.27	-0.20	-0.11	0.10
Study 9 effect, $\beta_{2,9}$	-0.61	-0.37	-0.29	-0.21	-0.01
Study 10 effect, $\beta_{2,10}$	-0.49	-0.36	-0.29	-0.23	-0.12
Study 11 effect, $\beta_{2,11}$	-0.50	-0.31	-0.24	-0.16	0.01
Study 12 effect, $\beta_{2,12}$	-0.49	-0.32	-0.22	-0.11	0.13
Study 13 effect, $\beta_{2,13}$	-0.70	-0.37	-0.24	-0.14	0.08
Study 14 effect, $\beta_{2,14}$	-0.33	-0.18	-0.08	0.04	0.30
Study 15 effect, $\beta_{2,15}$	-0.58	-0.38	-0.28	-0.18	0.05
Study 16 effect, $\beta_{2,16}$	-0.52	-0.34	-0.25	-0.15	0.08
Study 17 effect, $\beta_{2,17}$	-0.49	-0.29	-0.20	-0.10	0.17
Study 18 effect, $\beta_{2,18}$	-0.54	-0.27	-0.17	-0.06	0.21
Study 19 effect, $\beta_{2,19}$	-0.56	-0.30	-0.18	-0.05	0.25
Study 20 effect, $\beta_{2,20}$	-0.57	-0.36	-0.26	-0.17	0.04
Study 21 effect, $\beta_{2,21}$	-0.65	-0.41	-0.32	-0.24	-0.08
Study 22 effect, $\beta_{2,22}$	-0.66	-0.39	-0.27	-0.18	-0.02
mean of avg logits, α_1	-2.59	-2.42	-2.34	-2.26	-2.09
sd of avg logits, $\sqrt{\Lambda_{11}}$	0.39	0.48	0.55	0.63	0.83
mean of effects, α_2	-0.38	-0.29	-0.24	-0.20	-0.11
sd of effects, $\sqrt{\Lambda_{22}}$	0.04	0.11	0.16	0.21	0.34
correlation, ρ_{12}	-0.61	-0.13	0.21	0.53	0.91

Table 19.1 *Summary of posterior inference for the bivariate analysis of the meta-analysis of the beta-blocker trials in Table 5.4. All effects are on the log-odds scale. Inferences are similar to the results of the univariate analysis of logit differences in Section 5.6: compare the individual study effects to Table 5.4 and the mean and standard deviation of average logits to Table 5.5. 'Study 1 avg logit' is included above as a representative of the 22 parameters β_{1j}. (We would generally prefer to display all these inferences graphically but use tables here to give a more detailed view of the posterior inferences.)*

normal approximation required in the previous analysis but does not markedly change the posterior inferences for the quantities of essential interest.

19.5 Time series and spatial models

It is common for observations or variables to be ordered in time or related spatially. Formally, one can include time and spatial information as explanatory variables in a model and proceed using the principles of Bayesian inference for the joint distribution of all observables, including time and location. We will use the general notation t_i for the time or spatial coordinates of the ith observational unit; in our standard notation, the data thus consist of n exchangeable vectors $(y, x, t)_i$. In many cases, one is interested only in variables conditional on time and space, in which case a regression-type model is appropriate. All distributions we consider in this chapter are implicitly conditional on t and X.

There is an important new complication that arises, however. Almost all the models we have considered up to this point have treated exchangeable variables as iid given parameters (or hyperparameters). The only exception so far has been the weighted linear regression model of Section 14.6, and then only in the restrictive case that we know the variance matrix up to a constant. In modeling processes over time and space, however, it is natural to assume correlated outcomes, in which the correlations are unknown and are themselves parameters in the model.

We do not even remotely approach a systematic survey of the vast literature on models and methods for time series and spatial statistics but rather provide a very brief overview of how these problems fit into the multivariate regression framework.

Trend models

The simplest sort of time series model is the *trend*, or regression model on time. For example, if data y_i are observed at times t_i, then a linear trend corresponds to a regression of y on t, a quadratic trend to a regression of y on t and t^2, and so forth. These can be combined with other explanatory variables x into a larger linear or generalized linear model, but no new difficulties appear in the analysis. Trend models are rarely used by themselves to model time series but are often useful as components in more complicated models, as we shall see shortly.

Modeling correlations

A different kind of model does not account for any systematic trend but instead allows correlations between units, with the correlation depending on the time separation between units. (More generally, one can go beyond the normal model and consider other forms of dependence, but this is rarely done in practice.) It is generally useful to model the correlations as a function of time separation, using a model of the form $\text{corr}(y_i, y_j) = \rho(|t_i - t_j|)$, with a parametric form such as $\rho(t) = \exp(-\alpha t)$ or $(1 + t)^{-\alpha}$. The prior distribution must then be set up jointly on all the parameters in the model.

Combining trend, correlation, and regression models

It is often useful to combine the features of trend and correlation into a larger time series model, by fitting correlations to trend-corrected data. Regression on additional explanatory variables can also be incorporated in such a model, with the correlation model applied to the conditional distribution of y given X, the matrix of all explanatory variables, including the relevant functions of time. The normally distributed version of this model has likelihood $p(y|\beta, \alpha, \sigma^2) \propto \text{N}(y|X\beta, \Sigma)$, where the variance matrix Σ has components

$$\Sigma_{ij} = \sigma^2 \rho(|t_i - t_j|),$$

and parameters α index the correlation function ρ.

Time domain models: one-way representations

Correlation models are often usefully parameterized in terms of the conditional distribution of each observation y_i on past observations. To put it another way, one can specify the joint distribution of a time series model conditionally, in temporal order:

$$p(y|\alpha) = p(y_1|t_1, \alpha)p(y_2|y_1, t_2, \alpha) \cdots p(y_n|y_1, y_2, \ldots, y_{n-1}, t_n, \alpha), \quad (19.4)$$

where, for simplicity, we have indexed the observations in temporal order: $t_1 < t_2 < \ldots < t_n$, in abuse of our usual convention not to convey information in the indexes (see page 6).

The simplest nontrivial conditional models are *autoregressions*, for which the distribution of any y_i, given all earlier observations and the model parameters α, depends only on α and the previous k observations, $y_{i(-k)} = (y_{i-k}, \ldots, y_{i-1})$, where this expression is understood to refer only as far back as the first observation in the series if $i \leq k$. For a kth order autoregression, (19.4) becomes $p(y|\alpha) = \prod_{i=1}^n p(y_i|y_{i(-k)}, t_i, \alpha)$, which is a much simpler expression, as the parameter α is only required to characterize a single conditional distribution. (The observation times, t_i, are kept in the model to allow for the possibility of modeling trends.)

Frequency domain models

Another standard way to model correlation in time is through a mixture of sinusoids at different frequencies. As usual, we consider the normal model for simplicity. The simplest version has a single frequency:

$$y_i|a, b, \omega, \beta, \sigma \sim \text{N}((X\beta)_i + a\sin(\omega t_i) + b\cos(\omega t_i), \sigma^2).$$

More generally, one can have a mixture of frequencies $\omega_1, \ldots, \omega_k$:

$$y_i | a, b, \beta, \sigma \sim \mathrm{N}\left((X\beta)_i + \sum_{j=1}^{k}(a_j \sin(\omega_j t_i) + b_j \cos(\omega_j t_i)),\, \sigma^2\right),$$

with the observations y_i *independent* given the model parameters (and implicitly the covariates X_i and t_i). In this model, all the correlation is included in the sinusoids. The range of frequencies ω_j can be considered fixed (for example, if n observations are equally spaced in time with gaps T, then it is common to consider $\omega_j = 2\pi j/(nT)$, for $j = 0, \ldots, n/2$), and the model parameters are thus the vectors a and b, as well as the regression parameters β and residual variance σ^2. For equally spaced time series, the frequency domain, or *spectral*, model can be considered just another parameterization of the correlation model, with $(n-1)$ frequency parameters instead of the same number of correlation parameters. As with the time domain models previously considered, the frequency domain model can be restricted—for example, by setting some of the coefficients a_j and b_j to zero—but it seems generally more reasonable to include them in some sort of hierarchical model.

The spectral model, with frequencies ω_j fixed, is just a regression on the fixed vectors $\sin(\omega_j t_i)$ and $\cos(\omega_j t_i)$ and can thus be thought of as a special case of trend or regression models. If a full set of frequencies ω_j is specified, it is acceptable to treat their values as fixed, since we incorporate uncertainty about their importance in the model through the a_j's and b_j's.

Spatial models

As with time series, the simplest spatial models are 'trends'; that is, regressions of the outcome variable on spatial coordinates. For example, in the Latin square experiment described on page 220, an analysis must take account of the spatial information for the design to be ignorable, and the simplest models of this form include regressions on the continuous spatial coordinates and perhaps indicator variables for the discrete locations. More complicated spatial models present all the problems of time series models with the additional difficulty that there is no natural ordering as in time, and thus there are no natural one-way models. A related difficulty is that the normalizing factors of spatial models generally have no closed-form expression (recall Section 13.4), which can make computations for hierarchical models much more difficult.

19.6 Bibliographic note

Chapter 8 of Box and Tiao (1973) presents the multivariate regression model with noninformative prior distributions. Prior distributions and Bayesian inference for the covariance matrix of a multivariate normal distribution are discussed in Leonard and Hsu (1992), Yang and Berger (1994), Daniels and Kass (1999, 2001), and Barnard, McCulloch, and Meng (2000). Each of the above works on a different parameterization of the covariance matrix. Wong,

Carter, and Kohn (2002) discuss prior distributions for the inverse covariance matrix. Verbeke and Molenberghs (2000) and Daniels and Pourahmadi (2002) discuss hierarchical linear models for longitudinal data.

The business school prediction example of Section 19.3 is taken from Braun et al. (1983), who perform the approximate Bayesian inference described in the text. Dempster, Rubin, and Tsutakawa (1981) analyze educational data using a hierarchical regression model in which the regression coefficients are divided into 81 ordered pairs, thus requiring a 2×2 covariance matrix to be estimated at the hierarchical level; they describe how to use EM to find the posterior mode of the covariance matrix and perform an approximate Bayesian analysis by conditioning on the modal estimate. Rubin (1980b) analyzes the same data using a slightly different model.

An example of time series analysis from our own research appears in Carlin Dempster (1989). Books on Bayesian time series analysis using time-domain models include West and Harrison (1989) and Pole, West, and Harrison (1994), which includes computer programs. Jaynes (1982, 1987) and Bretthorst (1988) discuss Bayesian analysis of frequency-domain models and the relation to maximum entropy methods. The non-Bayesian texts by Box and Jenkins (1976) and Brillinger (1981) are also useful for Bayesian analysis, because they discuss a number of probability models for time series.

Recent overviews of spatial statistics and many references can be found in Besag (1974, 1986), Besag, York, and Mollie (1991), Besag et al. (1995), Cressie (1993), Grenander (1983), and Ripley (1981, 1988), all of which feature probability modeling prominently.

Recent applied Bayesian analyses of spatial and spatio-temporal and data include Besag and Higdon (1999), Waller et al. (1997), Waller, Louis, and Carlin (1997), Mugglin, Carlin, and Gelfand (2000), and Wikle et al. (2001). Chipman, Kolaczyk, and McCulloch (1997), and Simoncelli (1999) present Bayesian implementations of wavelet models, which are a generalization of frequency-domain models for time series and spatial data.

19.7 Exercises

1. Modeling correlation matrices:

   (a) Show that the determinant of a correlation matrix R is a quadratic function of any of its elements. (This fact can be used in setting up a Gibbs sampler for multivariate models; see page 484.)

   (b) Suppose that the off-diagonal elements of a 3×3 correlation matrix are 0.4, 0.8, and r. Determine the range of possible values of r.

   (c) Suppose all the off-diagonal elements of a d-dimensional correlation matrix R are equal to the same value, r. Prove that R is positive definite if and only if $-1/(d-1) < r < 1$.

2. Improper prior distributions and proper posterior distributions: consider the hierarchical model for the meta-analysis example in Section 19.4.

(a) Show that, for any value of ρ_{12}, the posterior distribution of all the remaining parameters is proper, conditional on ρ_{12}.

(b) Show that the posterior distribution of all the parameters, including ρ_{12}, is proper.

3. Analysis of a two-way stratified sample survey: Section 7.4 and Exercise 11.5 present an analysis of a stratified sample survey using a hierarchical model on the stratum probabilities. That analysis is not fully appropriate because it ignores the two-way structure of the stratification, treating the 16 strata as exchangeable.

(a) Set up a linear model for logit(ϕ) with three groups of random effect parameters, for the four regions, the four place sizes, and the 16 strata.

(b) Simplify the model by assuming that the ϕ_{1j}'s are independent of the ϕ_{2j}'s. This separates the problem into two generalized linear models, one estimating Bush vs. Dukakis preferences, the other estimating 'no opinion' preferences. Perform the computations for this model to yield posterior simulations for all parameters.

(c) Expand to a multivariate model by allowing the ϕ_{1j}'s and ϕ_{2j}'s to be correlated. Perform the computations under this model, using the results from Exercise 11.5 and part (b) above to construct starting distributions.

(d) Compare your results to those from the simpler model treating the 16 strata as exchangeable.

CHAPTER 20

Nonlinear models

20.1 Introduction

The linear regression approach of Part IV suggests a presentation of statistical models in menu form, with a set of possible distributions for the response variable, a set of transformations to facilitate the use of those distributions, and the ability to include information in the form of linear predictors. In a linear model, the expected value of the data y is a linear function of parameters β and predictors X: $E(y|X, \beta) = X\beta$. In a generalized linear model, the expected value of y is a nonlinear function of the linear predictor: $E(y|X, \beta) = g^{-1}(X\beta)$. Robust (Chapter 17) and mixture models (Chapter 18) generalize these by adding a latent (unobserved) mixture parameter for each data point.

The generalized linear modeling approach is flexible and powerful, with the advantage that linear parameters β are relatively easy to interpret, especially when comparing to each other (since they all act on the same linear predictor). However, not all phenomena behave linearly, even under transformation, and we should not be restricted to that assumption. This chapter considers the more general case where parameters and predictors do not combine linearly. Simple examples include a ratio of two linear predictors such as $E(y) = \frac{a_1 + b_1 x_1}{a_2 + b_2 x_2}$, or a sum of nonlinear functions such as $E(y) = A_1 \exp(-\alpha_1 x) + A_2 \exp(-\alpha_2 x)$; see Exercise 20.3.

The general principles of inference and computation can be directly applied to nonlinear models. We briefly consider the three steps of Bayesian data analysis: model building, computation, and model checking. The most flexible approaches to nonlinear modeling typically involve complicated relations between predictors and outcomes, but generally without the necessity for unusual probability distributions. Computation can present challenges, because we cannot simply adapt linear regression computations, as was done in Chapter 16 for generalized linear models. Model checking can sometimes be performed using residual plots, χ^2 tests, and other existing summaries, but sometimes new graphs need to be created in the context of a particular model. In addition, new difficulties arise in interpretation of parameters that cannot simply be understood in terms of a linear predictor. A key step in any inference for a nonlinear model is to display the fitted nonlinear relation graphically.

Because nonlinear models come in many flavors, there is no systematic menu of options to present. Generally each new modeling problem must be tackled afresh. However, we hope that presenting in some detail two examples from our

Std	Std	Unk1	Unk2	Unk3	Unk4	Unk5	Unk6	Unk7	Unk8	Unk9	Unk10
1	1	1	1	1	1	1	1	1	1	1	1
1/2	1/2	1/3	1/3	1/3	1/3	1/3	1/3	1/3	1/3	1/3	1/3
1/4	1/4	1/9	1/9	1/9	1/9	1/9	1/9	1/9	1/9	1/9	1/9
1/8	1/8	1/27	1/27	1/27	1/27	1/27	1/27	1/27	1/27	1/27	1/27
1/16	1/16	1	1	1	1	1	1	1	1	1	1
1/32	1/32	1/3	1/3	1/3	1/3	1/3	1/3	1/3	1/3	1/3	1/3
1/64	1/64	1/9	1/9	1/9	1/9	1/9	1/9	1/9	1/9	1/9	1/9
0	0	1/27	1/27	1/27	1/27	1/27	1/27	1/27	1/27	1/27	1/27

Figure 20.1 *Typical setup of a plate with 96 wells for a serial dilution assay. The first two columns are dilutions of 'standards' with known concentrations, and the other columns are ten different 'unknowns.' The goal of the assay is to estimate the concentrations of the unknowns, using the standards as calibration.*

own applied research will illustrate some of the unique features of nonlinear modeling and provide some useful pointers for practitioners.

20.2 Example: serial dilution assay

A common design for estimating the concentrations of compounds in biological samples is the serial dilution assay, in which measurements are taken at several different dilutions of a sample. The reason for the serial dilutions of each sample is that the concentration in an assay is quantified by an automated optical reading of a color change, and there is a limited range of concentrations for which the color change is informative: at low values, the color change is imperceptible, and at high values, the color saturates. Thus, several dilutions give several opportunities for an accurate measurement.

More precisely, the dilutions give several measurements of differing accuracy, and a likelihood or Bayesian approach should allow us to combine the information in these measurements appropriately. An assay comes on a plate with a number of wells each containing a sample or a specified dilution of a sample. There are two sorts of samples: *unknowns*, which are the samples to be measured and their dilutions, and *standards*, which are dilutions of a known compound, used to calibrate the measurements. Figure 20.1 shows a typical plate with 96 wells; the first two columns of the plate contain the standard and its dilutions and the remaining ten columns are for unknown quantities. The dilution values for the unknown samples are more widely spaced than for the standards in order to cover a wider range of concentrations with a given number of assays.

Laboratory data

Figure 20.2 shows data from a single plate in a study of cockroach allergen concentrations in the homes of asthma sufferers. Each graph shows the optical measurements y versus the dilutions for a single compound along with an estimated curve describing the relationship between dilution and measurement.

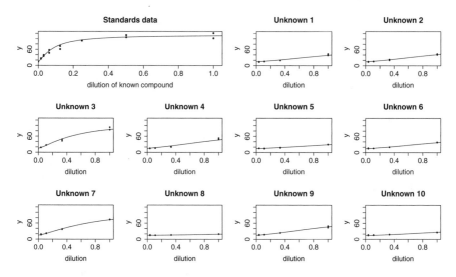

Figure 20.2 *Data from a single plate of a serial dilution assay. The large graph shows the calibration data, and the ten small graphs show the data for the unknown compounds. The goal of the analysis is to figure out how to scale the x-axes of the unknowns so they will line up with the curve estimated from the standards. (The curves shown on these graphs are estimated from the model as described in Section 20.2.)*

The estimation of the curves relating dilutions to measurements is described below.

Figure 20.3 illustrates certain difficulties with a currently standard approach to estimating unknown concentrations. The left part of the figure shows the standards data (corresponding to the first graph in Figure 20.2): the two initial samples have known concentrations of 0.64, with each followed by several dilutions and a zero measurement. The optical color measurements y start above 100 for the samples with concentration 0.64 and decrease to around 14 for the zero concentration (all inert compound) samples. The right part of Figure 20.3 shows, for two of the ten unknowns on the plate, the color measurements y and corresponding concentration estimates from a standard method.

All the estimates for Unknown 8 are shown by asterisks, indicating that they were recorded as 'below detection limit,' and the standard computer program for analyzing these data gives no estimate at all. A casual glance at the data (see the plot for Unknown 8 in Figure 20.2) might suggest that these data are indeed all noise, but a careful look at the numbers reveals that the measurements decline consistently from concentrations of 1 to 1/3 to 1/9, with only the final dilutions apparently lost in the noise (in that the measurements at 1/27 are no lower than at 1/9). A clear signal is present for the first six measurements of this unknown sample.

Standards data			Data from two of the unknown samples			
Conc.	Dilution	y	Sample	Dilution	y	Est. conc.
0.64	1	101.8	Unknown 8	1	19.2	*
0.64	1	121.4		1	19.5	*
0.32	1/2	105.2		1/3	16.1	*
0.32	1/2	114.1		1/3	15.8	*
0.16	1/4	92.7		1/9	14.9	*
0.16	1/4	93.3		1/9	14.8	*
0.08	1/8	72.4		1/27	14.3	*
0.08	1/8	61.1		1/27	16.0	*
0.04	1/16	57.6	Unknown 9	1	49.6	0.040
0.04	1/16	50.0		1	43.8	0.031
0.02	1/32	38.5		1/3	24.0	0.005
0.02	1/32	35.1		1/3	24.1	0.005
0.01	1/64	26.6		1/9	17.3	*
0.01	1/64	25.0		1/9	17.6	*
0	0	14.7		1/27	15.6	*
0	0	14.2		1/27	17.1	*
				...		

Figure 20.3 *Example of some measurements y from a plate as analyzed by a standard software package used for dilution assays. The standards data are used to estimate the calibration curve, which is then used to estimate the unknown concentrations. The concentrations indicated by asterisks are labeled as 'below detection limit.' However, information is present in these low observations, as can be seen by noting the decreasing pattern of the measurements from dilutions 1 to 1/3 to 1/9 in each sample.*

Unknown 9 shows a better outcome, in which four of the eight measurements are above detection limit. The four more diluted measurements yield readings that are below the detection limit. Once again, however, information seems to be present in the lower measurements, which decline consistently with dilution.

As can be seen in Figure 20.2, Unknowns 8 and 9 are not extreme cases but rather are somewhat typical of the data from this plate. In measurements of allergens, even low concentrations can be important, and we need to be able to distinguish between zero concentrations and values that are merely low. The Bayesian inference described here makes this distinction far more precisely than the previous method by which such data were analyzed.

The model

Notation. The parameters of interest in a study such as illustrated in Figures 20.1–20.3 are the concentrations of the unknown samples; we label these as $\theta_1, \ldots, \theta_{10}$ for the plate layout shown in Figure 20.1. The known concentration of the standard is denoted by θ_0. We use the notation x_i for the concentration in well i and y_i for the corresponding color intensity measurement, with $i = 1, \ldots, 96$ for our plate. Each x_i is a specified dilution of the corresponding sample. We lay out the model for the color intensity observations y in stages: a parametric model for the expected color intensity for a given concentration, measurement errors for the optical readings, errors introduced during the dilution preparation process, and, finally, prior distributions for all the parameters.

Curve of expected measurements given concentration. We follow the usual practice in this field and fit the following four-parameter nonlinear model for the expected optical reading given concentration x:

$$\text{E}(y|x, \beta) = g(x, \beta) = \beta_1 + \frac{\beta_2}{1 + (x/\beta_3)^{-\beta_4}} \qquad (20.1)$$

where β_1 is the color intensity at zero concentration, β_2 is the increase to saturation, β_3 is the concentration at which the gradient of the curve turns, and β_4 is the rate at which saturation occurs. All parameters are restricted to nonnegative values. This model is equivalent to a scaled and shifted logistic function of $\log(x)$. The model fits the data fairly well, as can be seen in Figure 20.2 on page 499.

Measurement error. The measurement errors are modeled as normally distributed with unequal variances:

$$y_i \sim \text{N}\left(g(x_i, \beta), \left(\frac{g(x_i, \beta)}{A}\right)^{2\alpha} \sigma_y^2 \right), \qquad (20.2)$$

where the parameter α, which is restricted to lie between 0 and 1, models the pattern that variances are higher for larger measurements (for example, see Figure 20.2). The constant A in (20.2) is arbitrary; we set it to the value 30, which is in the middle of the range of the data. It is included in the model so that the parameter σ_y has a more direct interpretation as the error standard deviation for a 'typical' measurement.

The model (20.2) reduces to an equal-variance normal model if $\alpha = 0$ and approximately corresponds to the equal-variance model on the log scale if $\alpha = 1$. Getting the variance relation correct is important here because many of our data are at very low concentrations, and we do not want our model to use these measurements but not to overstate their precision.

Dilution errors. The dilution process introduces errors in two places: the *initial dilution,* in which a measured amount of the standard is mixed with a measured amount of an inert liquid; and *serial dilutions,* in which a sample is diluted by a fixed factor such as 2 or 3. For the cockroach allergen data, serial dilution errors were very low, so we include only the initial dilution error in our model.

We use a normal model on the (natural) log scale for the initial dilution error associated with preparing the standard sample. The known concentration of the standard solution is θ_0, and d_0^{init} is the (known) initial dilution of the standard that is called for. Without dilution error, the concentration of the initial dilution would thus be $d_0^{\text{init}}\theta_0$. Let x_0^{init} be the actual (unknown) concentration of the initial dilution, with

$$\log(x_0^{\text{init}}) \sim \text{N}(\log(d_0^{\text{init}} \cdot \theta_0), (\sigma^{\text{init}})^2). \qquad (20.3)$$

For the unknowns, there is no initial dilution, and so the unknown initial concentration for sample j is $x_j^{\text{init}} = \theta_j$ for $j = 1, \ldots, 10$. For the further dilutions of standards and unknowns, we simply set

$$x_i = d_i \cdot x_{j(i)}^{\text{init}}, \tag{20.4}$$

where $j(i)$ is the sample (0, 1, 2, ..., or 10) corresponding to observation i, and d_i is the dilution of observation i relative to the initial dilution. (The d_i's are the numbers displayed in Figure 20.1.) The relation (20.4) reflects the assumption that serial dilution errors are low enough to be ignored.

Prior distributions. We assign noninformative uniform prior distributions to the parameters of the calibration curve: $\log(\beta_k) \sim U(-\infty, \infty)$ for $k = 1, \ldots, 4$; $\sigma_y \sim U(0, \infty)$; $\alpha \sim U(0, 1)$. A design such as displayed in Figure 20.1 with lots of standards data allows us to estimate all these parameters fairly accurately. We also assign noninformative prior distributions for the unknown concentrations: $p(\log \theta_j) \propto 1$ for each unknown $j = 1, \ldots, 10$. Another option would be to fit a hierarchical model of the form, $\log \theta_j \sim \text{N}(\mu_\theta, \sigma_\theta^2)$, but for simplicity we use a no-pooling model (corresponding to $\sigma_\theta = \infty$) in this analysis.

There is one parameter in the model—σ^{init}, the scale of the initial dilution error—that cannot be estimated from a single plate. For our analysis, we fix it at the value 0.02 (that is, an initial dilution error with standard deviation 2%), which was obtained from a previous analysis of data from plates with several different initial dilutions of the standard.

Inference

We fitted the model using the Bugs package (use of this software is introduced in Appendix C). We obtained approximate convergence (the potential scale reduction factors $\widehat{R}$ were below 1.1 for all parameters) after 50,000 iterations of two parallel chains of the Gibbs sampler. To save memory and computation time in processing the simulations, we save every 20th iteration of each chain. When fitting the model in Bugs, it is helpful to use reasonable starting points (which can be obtained using crude estimates from the data) and to parameterize in terms of the logarithms of the parameters β_j and the unknown concentrations θ_j. To speed convergence, we actually work with the parameters $\log \beta$, α, σ_y and $\log \gamma$, where $\log(\gamma_j) = \log(\theta_j/\beta_3)$. The use of γ_j in place of θ_j gets around the problem of the strong posterior correlation between the unknown concentrations and the parameter β_3, which indexes the x-position of the calibration curve (see (20.1)).

The posterior median estimates (and posterior 50% intervals) for the parameters of the calibration curve are $\hat{\beta}_1 = 14.7 \ [14.5, 14.9]$, $\hat{\beta}_2 = 99.7 \ [96.8, 102.9]$, $\hat{\beta}_3 = 0.054 \ [0.051, 0.058]$, and $\hat{\beta}_4 = 1.34 \ [1.30, 1.38]$. The posterior median estimate of β defines a curve $g(x, \beta)$ which is displayed in the upper-left plot of Figure 20.2. As expected, the curve goes through the data used to estimate it. The variance parameters σ_y and α are estimated as 2.2 and 0.97 (with 50%

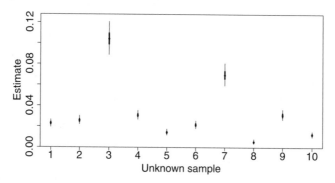

Figure 20.4 *Posterior medians, 50% intervals, and 95% intervals for the concentrations of the 10 unknowns for the data displayed in Figure 20.2. Estimates are obtained for all the samples, even Unknown 8, all of whose data were 'below detection limit' (see Figure 20.3).*

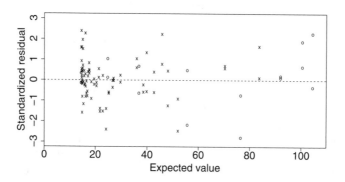

Figure 20.5 *Standardized residuals $(y_i - E(y_i|x_i))/sd(y_i|x_i))$ vs. expected values $E(y_i|x_i)$, for the model fit to standards and unknown data from a single plate. Circles and crosses indicate measurements from standards and unknowns, respectively. No major problems appear with the model fit.*

intervals of $[2.1, 2.3]$ and $[0.94, 0.99]$, respectively). The high precision of the measurements (as can be seen from the replicates in Figure 20.2) allowed the parameters to be accurately estimated from a relatively small dataset.

Figure 20.4 displays the inferences for the concentrations of the 10 unknown samples. We used these estimates, along with the estimated calibration curve, to draw scaled curves for each of the 10 unknowns displayed in Figure 20.2. Finally, Figure 20.5 displays the residuals, which seem generally reasonable.

Comparison to existing estimates

The method that is standard practice in the field involves first estimating the calibration curve and then transforming each measurement from the unknown samples directly to an estimated concentration, by inverting the fitted calibration curve. For each unknown sample, the estimated concentrations are then divided by their dilutions and averaged to obtain a single estimate. (For example, using this approach, the estimated concentration for Unknown 9 from the data displayed in Figure 20.3 is $\frac{1}{4}(0.040 + 0.031 + 3 \cdot 0.005 + 3 \cdot 0.005) = 0.025$.)

The estimates from the Bayesian analysis are generally similar to those of the standard method but with higher accuracy. An advantage of the Bayesian approach is that it yields a concentration estimate for all unknowns, even Unknown 8 for which there is no standard estimate because all its measurements are 'below detection limit'. We also created concentration estimates for each unknown based on each of the two halves of the data (in the setup of Figure 20.1, using only the top four wells or the bottom four wells for each unknown). For the standard and Bayesian approaches the two estimates are similar, but the reliability (that is, the agreement between the two estimates) is much stronger for the Bayesian estimate. Of course, we would not want to make too strong a claim based on data from a single plate. We performed a more thorough study (not shown here) to compare the old and new methods under a range of experimental conditions.

20.3 Example: population toxicokinetics

In this section we discuss a much more complicated nonlinear model used in toxicokinetics (the study of the flow and metabolism of toxins in the body) for the ultimate goal of assessing the risk in the general population associated with a particular air pollutant. This model is hierarchical and multivariate, with a vector of parameters to be estimated on each of several experimental subjects. As discussed briefly in Section 9.1, the prior distributions for this model are informative and hierarchical, with separate variance components corresponding to uncertainty about the average level in the population and variation around that average.

Background

Perchloroethylene (PERC) is one of many industrial products that cause cancer in animals and is believed to do so in humans as well. PERC is breathed in, and the general understanding is that it is metabolized in the liver and that its metabolites are carcinogenic. Thus, a relevant 'dose' to study when calibrating the effects of PERC is the amount metabolized in the liver. Not all the PERC that a person breathes will be metabolized. We focus here on estimating the fraction metabolized as a function of the concentration of the compound in the breathed air, and how this function varies across the population. To give an idea of our inferential goals, we skip ahead to show some output from our analysis. Figure 20.6 displays the estimated fraction of inhaled PERC that is metabolized as a function of concentration in air, for 10 randomly selected draws from the estimated population of young adult white

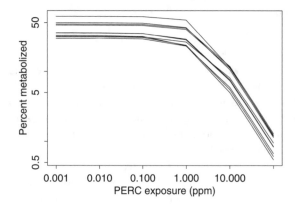

Figure 20.6 *Estimated fraction of PERC metabolized, as a function of steady-state concentration in inhaled air, for 10 hypothetical individuals randomly selected from the estimated population of young adult white males.*

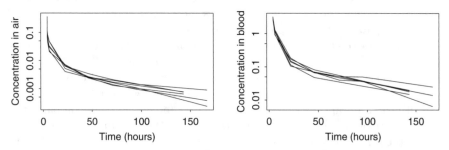

Figure 20.7 *Concentration of PERC (in milligrams per liter) in exhaled air and in blood, over time, for one of two replications in each of six experimental subjects. The measurements are displayed on logarithmic scales.*

males (the group on which we had data). The shape of the curve is discussed below after the statistical modeling is described.

It is not possible to estimate curves of this type with reasonable confidence using simple procedures such as direct measurement of metabolite concentrations (difficult even at high exposures and not feasible at low exposures) or extrapolation from animal results. Instead a mathematical model of the flow of the toxin through the bloodstream and body organs, and of its metabolism in the liver is used to estimate the fraction of PERC metabolized.

A sample of the experimental data we used to fit the model of toxin flow is shown in Figure 20.7. Each of six volunteers was exposed to PERC at a high level for four hours (believed long enough for the PERC concentrations in most of their bodily organs to come to equilibrium) and then PERC concentrations in exhaled air and in blood were measured over a period of a week (168 hours). In addition, the data on each subject were repeated at a second PERC exposure level (data not shown).

Toxicokinetic model

Our analysis is based on a standard physiological model, according to which the toxin enters and leaves through the breath, is distributed by blood flow to four 'compartments'—well-perfused tissues, poorly perfused tissues, fat, and the liver—and is metabolized in the liver. This model has a long history in toxicology modeling and has been showed to reproduce most features of such data. A simpler one or two-compartment model might be easier to estimate, but such models provide a poor fit to our data and, more importantly, do not have the complexity to accurately fit varying exposure conditions.

We briefly describe the nature of the toxicokinetic model, omitting details not needed for understanding our analysis. Given a known concentration of the compound in the air, the concentration of the compound in each compartment over time is governed by a first-order differential equation, with parameters for the volume, blood flow, and partition coefficient (equilibrium concentration relative to the blood) of each compartment. The liver compartment where metabolism occurs has a slightly different equation than the other compartments and is governed by the parameters mentioned above and a couple of additional parameters. The four differential equations give rise to a total of 15 parameters for each individual. We use the notation $\theta_k = (\theta_{k1}, \ldots, \theta_{kL})$ for the vector of $L = 15$ parameters associated with person k.

Given the values of the physiological parameters and initial exposure conditions, the differential equations can be solved using specialized numerical algorithms to obtain concentrations of the compound in each compartment and the rate of metabolism as a function of time. We can combine predictions about the PERC concentration in exhaled air and blood based on the numerical solution of the differential equations with our observed concentration measurements to estimate the model parameters for each individual.

Difficulties in estimation and the role of prior information

A characteristic difficulty of estimating toxicokinetic models (and pharmacokinetic models in general) is that they predict a pattern of concentration over time that is close to a mixture of declining exponential functions, with the amplitudes and decay times of the different components corresponding to functions of the model parameters. It is well known that the estimation of the decay times of a mixture of exponentials is an ill-conditioned problem (see Exercise 20.3); that is, the parameters in such a model are hard to estimate simultaneously.

Solving the problem of estimating metabolism from indirect data is facilitated by using a *physiological* pharmacokinetic model; that is, one in which the individual and population parameters have direct physical interpretations (for example, blood flow through the fatty tissue, or tissue/blood partition coefficients). These models permit the identification of many of their parameter values through prior (for example, published) physiological data. Since the parameters of these models are essentially impossible to estimate from the data alone, it is crucial that they have physical meaning and can be assigned informative prior distributions.

Measurement model

We first describe how the toxicological model is used as a component of the nonlinear model for blood and air concentration measurements. Following that is a description of the population model which allows us to infer the distribution of population characteristics related to PERC metabolism. At the individual level, for each subject, a series of measurements of exhaled air and blood concentrations are taken. We label these as y_{jkmt}, with j indexing replications ($j = 1, 2$ for the two exposure levels in our data), k indexing individuals, m indexing measurements ($m = 1$ for blood concentration and $m = 2$ for air concentration), and t indexing time. The expected values of the exhaled air and blood concentrations are nonlinear functions $g_m(\theta_k, E_j, t)$ of the individual's parameters θ_k, the exposure level E_j, and time t. The functions $g_m(\cdot)$ are our shorthand notation for the solution of the system of differential equations relating the physiological parameters to the expected concentration. Given the input conditions for replication j (that is, E_j) and the parameters θ_k (as well as a number of additional quantities measured on each individual but suppressed in our notation here), one can numerically evaluate the pharmacokinetic differential equations over time and compute g_1 and g_2 for all values at which measurements have been taken, thus obtaining the expected values of all the measurements.

The concentrations actually observed in expired air and blood are also affected by measurement errors, which are assumed, as usual, to be independent and lognormally distributed, with a mean of zero and a standard deviation of σ_m (on the log scale) for $m = 1, 2$. These measurement error distributions also implicitly account for errors in the model. We allow the two components of σ to differ, because the measurements in blood and exhaled air have different experimental protocols and therefore are likely to have different precisions. We have no particular reason to believe that modeling or measurement errors for air and blood measurements will be correlated, so we assign independent uniform prior distributions to $\log \sigma_1$ and $\log \sigma_2$. (After fitting the model, we examined the residuals and did not find any evidence of high correlations in the errors.)

Population model for parameters

One of the goals of this project is to estimate the distribution of the individual pharmacokinetic parameters, and in particular the distribution of predicted values such as fraction metabolized (which are complex functions of the individual parameters), in the general population. In an experiment with K individuals, we set up a hierarchical model on the K vectors of parameters to allow us to draw inferences about the general population from which the individuals are drawn.

A skewed, lognormal-like distribution is generally observed for biological parameters. Most, if not all, of the biological parameters also have physiological bounds. Based on this information the individual pharmacokinetic parameters after log-transformation and appropriate scaling (see below), are modeled with normal distributions having population mean truncated at ±3 standard deviations from the mean, where k indexes individuals and $l = 1, \ldots, L$ indexes the pharmacokinetic parameters in the model. The distributions are truncated to restrict the model parameters to scientifically reasonable values. In addition, the truncations serve a useful role when we monitor the simulations of the parameters from their posterior distribution: if the simulations for a parameter are stuck near truncation points, this indicates that the data and the pharmacokinetic model strongly contradict the prior distribution, and some part of the model should be re-examined.

The vector of parameters for individual k is $\theta_k = (\theta_{k1}, \ldots, \theta_{kL})$, with $L = 15$. Some of the parameters are constrained by definition: in the model under discussion, the parameters $\theta_{k2}, \theta_{k3}, \theta_{k4}, \theta_{k5}$ represent the fractions of blood flow to each compartment, and so are constrained to sum to 1. Also, the parameters $\theta_{k6}, \theta_{k7}, \theta_{k8}$ correspond to the scaling coefficients of the organ volumes, and are constrained to sum to 0.873 (the standard fraction of lean body mass not including bones), for each individual. Of these three parameters, θ_{k8}, the volume of the liver, is much smaller than the others and there is considerable prior information about this quantity. For the purposes of modeling and computation, we transform the model in terms of a new set of parameters ψ_{kl} defined as follows:

$$\theta_{kl} = \frac{e^{\psi_{kl}}}{e^{\psi_{k2}} + e^{\psi_{k3}} + e^{\psi_{k4}} + e^{\psi_{k5}}}, \quad \text{for } l = 2, 3, 4, 5$$

$$\theta_{kl} = (0.873 - e^{\psi_{k8}}) \frac{e^{\psi_{kl}}}{e^{\psi_{k6}} + e^{\psi_{k7}}}, \quad \text{for } l = 6, 7$$

$$\theta_{kl} = e^{\psi_{kl}}, \quad \text{for } l = 1 \text{ and } 8\text{--}15. \tag{20.5}$$

The parameters $\psi_{k2}, \ldots, \psi_{k5}$ and ψ_{k6}, ψ_{k7} are not identified (for example, adding any constant to $\psi_{k2}, \ldots, \psi_{k5}$ does not alter the values of the physiological parameters, $\theta_{k2}, \ldots, \theta_{k5}$), but they are assigned proper prior distributions, so we can formally manipulate their posterior distributions.

Each set of ψ_{kl} parameters is assumed to follow a normal distribution with mean μ_l and standard deviation τ_l, truncated at three standard deviations. Modeling on the scale of ψ respects the constraints on θ while retaining the truncated lognormal distributions for the unconstrained components. All computations are performed with the ψ's, which are then transformed back to θ's at the end to interpret the results on the natural scales.

In the model, the population distributions for the $L = 15$ physiological parameters are assumed independent. After fitting the model, we checked the $15 \cdot 14/2$ correlations among the parameter pairs across the six subjects and found no evidence that they differed from zero. If we did find large correlations, we would either want to add correlations to the model (as described in Section 19.2) or reparameterize to make the correlations smaller. In fact, the parameters in our model were already transformed to reduce correlations (for example, by working with proportional rather than absolute blood flows and organ volumes, as described above equation (20.5)).

Prior information

In order to fit the population model, we assign prior distributions to the means and variances, μ_l and τ_l^2, of the L (transformed) physiological parameters. We specify a prior distribution for each μ_l (normal with parameters M_l and S_l^2 based on substantive knowledge) and τ_l^2 (inverse-χ^2, centered at an estimate τ_{0l}^2 of the true population variance and with a low number of degrees of freedom ν_l—typically set to 2—to indicate large uncertainties).

The hyperparameters M_l, S_l, and τ_{0l} are based on estimates available in the biological literature. Sources include studies on humans and allometric scaling from animal measurements. We set independent prior distributions for the μ_l's and τ_l's because our prior information about the parameters is essentially independent, to the best of our knowledge, given the parameterization and scaling used (for example, blood flows as a proportion of the total rather than on absolute scales). In general, setting up independent population distributions means that between-subject information about one parameter will not be used to help estimate other parameters in the model. In setting uncertainties, we try to be conservative and set the prior variances higher rather than lower when there is ambiguity in the biological literature.

Joint posterior distribution for the hierarchical model

For Bayesian inference, we obtain the posterior distribution (up to a multiplicative constant) for all the parameters of interest, given the data and the prior information, by multiplying all the factors in the hierarchical model: the data distribution, $p(y|\psi, E, t, \sigma)$, the population model, $p(\psi|\mu, \tau)$, and the prior distribution, $p(\mu, \tau|M, S, \tau_0)$,

$$p(\psi, \mu, \tau^2, \sigma^2|y, E, t, \phi, M, S, \tau_0^2, \nu)$$

$$\propto p(y|\psi, \phi, E, t, \sigma^2)p(\psi|\mu, \tau^2)p(\mu, \tau^2|M, S, \tau_0^2)p(\sigma^2)$$

$$\propto \left[\prod_{j=1}^{J}\prod_{k=1}^{K}\prod_{m=1}^{2}\prod_{t} N(\log y_{jkmt}|\log g_m(\theta_k, E_j, t), \sigma_m^2)\right]\sigma_1^{-2}\sigma_2^{-2} \times$$

$$\left[\prod_{k=1}^{K}\prod_{l=1}^{L} N_{\text{trunc}}(\psi_{kl}|\mu_l, \tau_l^2)\right]\left[\prod_{l=1}^{L} N(\mu_l|M_l, S_l^2)\text{Inv-}\chi^2(\tau_l^2|\nu_l, \tau_{0l}^2)\right], \quad (20.6)$$

where ψ is the set of vectors of individual-level parameters, μ and τ are the vectors of population means and standard deviations, σ is the pair of measurement variances, y is the vector of concentration measurements, E and t

are the exposure concentrations and times, and M, S, τ, and ν are the hyper-parameters. We use the notation N_{trunc} for the normal distribution truncated at the specified number of standard deviations from the mean. The indexes j, k, l, m, and t refer to replication, subject, parameter, type of measurement (blood or air), and time of measurement. To compute (20.6) as a function of the parameters, data, and experimental conditions, the functions g_m must be computed numerically over the range of time corresponding to the experimental measurements.

Computation

Our goals are first to fit a pharmacokinetic model to experimental data, and then to use the model to perform inferences about quantities of interest, such as the population distribution of the fraction of the compound metabolized at a given dose. We attain these goals using random draws of the parameters from the posterior distribution, $p(\psi, \mu, \tau, \sigma | y, E, t, M, S, \tau_0, \nu)$. We use a Gibbs sampling approach, iteratively updating the parameters in the following sequence: σ, τ, μ, $\psi_1, \ldots, \psi_K$. Each of these is actually a vector parameter. The conditional distributions for the components of σ^2, τ^2, and μ are inverse-χ^2, inverse-χ^2, and normal. The conditional distributions for the parameters ψ have no closed form, so we sample from them using steps of the Metropolis algorithm, which requires only the ability to compute the posterior density up to a multiplicative constant, as in (20.6).

Our implementation of the Metropolis algorithm alters the parameters one person at a time (thus, K jumps in each iteration, with each jump affecting an L-dimensional vector ψ_k). The parameter vectors are altered using a normal proposal distribution centered at the current value, with covariance matrix proportional to one obtained from some initial runs and scaled so that the acceptance rate is approximately 0.23. Updating the parameters of one person at a time means that the only factors of the posterior density that need to be computed for the Metropolis step are those corresponding to that person. This is an important concern, because evaluating the functions g_m to obtain expected values of measurements is the costliest part of the computation. An alternative approach would be to alter a single component ψ_{kl} at a time; this would require KL jumps in each iteration.

We performed five independent simulation runs, each of 50,000 iterations, with starting points obtained by sampling each ψ_{kl} at random from its prior distribution and then setting the population averages μ_l at their prior means, M_l. We then began the simulations by drawing σ^2 and τ^2. Because of storage limitations, we saved only every tenth iteration of the parameter vector. We monitored the convergence of the simulations by comparing within and between-simulation variances, as described in Section 11.6. In practice, the model was gradually implemented and debugged over a period of months, and one reason for our trust in the results is their general consistency with earlier simulations of different variants of the model.

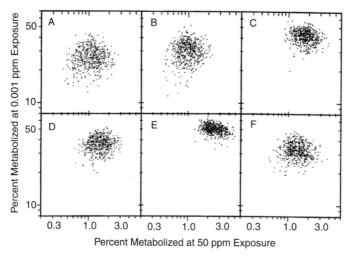

Figure 20.8 *Posterior inferences for the quantities of interest—the fraction metab-olized at high and low exposures—for each of the six subjects in the PERC experi-ment. The scatter within each plot represents posterior uncertainty about each per-son's metabolism. The variation among these six persons represents variation in the population studied of young adult white males.*

Inference for quantities of interest

First off, we examined the inferences about the model parameters and their population variability, and checked that these made sense and were consis-tent with the prior distribution. After this, the main quantities of interest were the fraction of PERC metabolized under different exposure scenarios, as computed by evaluating the differential equation model numerically under the appropriate input conditions. For each individual k, we can compute the fraction metabolized for each simulated parameter vector ψ_k; using the set of simulations yields a distribution of the fraction metabolized for that individ-ual. The variance in the distribution for each individual is due to uncertainty in the posterior distribution of the physiological parameters, ψ_k.

Figure 20.8 shows the posterior distributions of the PERC fraction metab-olized, by each subject of the experiment under analysis, at a high exposure level of 50 parts per million (ppm) and a low level of 0.001 ppm. The six sub-jects were not actually exposed to these levels; the inferences were obtained by running the differential equation model with these two hypothesized input conditions along with the estimated parameters for each person. We selected these two levels to illustrate the inferences from the model; the high level corresponds to occupational exposures and the low level to environmental ex-posures of PERC. We can and did consider other exposure scenarios too. The figure shows the correlation between the high-dose and the low-dose estimates of the fraction metabolized in the six subjects. Large variations exist between individuals; for example, a factor of two difference is seen between subjects A and E.

Similar simulations were performed for an additional person from the population (that is, a person exchangeable with the subjects in the study) by simulating random vectors of the physiological parameters from their population distributions. The variance in the resulting population distribution of fraction of PERC metabolized includes posterior uncertainty in the parameter estimates and real inter-individual variation in the population. Interval estimates for the fraction metabolized can be obtained as percentiles of the simulated distributions. At high exposure (50 ppm) the 95% interval for the fraction metabolized in the population is $[0.5\%, 4.1\%]$; at low exposure (0.001 ppm) it is $[15\%, 58\%]$.

We also studied the fraction of PERC metabolized in one day (after three weeks continuous inhalation exposure) as a function of exposure level. This is the relation we showed in introducing the PERC example; it is shown in Figure 20.6. At low exposures the fraction metabolized remains constant, since metabolism is linear. Saturation starts occurring above 1 ppm and is about complete at 10 ppm. At higher levels the fraction metabolized decreases linearly with exposure since the quantity metabolized per unit time is at its maximum.

When interpreting these results, one must remember that they are based on a single experiment. This study appears to be one of the best available; however, it included only six subjects from a homogeneous population, measured at only two exposure conditions. Much of the uncertainty associated with the results is due to these experimental limitations. Posterior uncertainty about the parameters for individual persons in the study could be reduced by collecting and analyzing additional data on these individuals. To learn more about the population we would need to include additional individuals. Population variability, which in this study is approximately as large as posterior uncertainty, might increase if a more heterogeneous group of subjects were included.

Evaluating the fit of the model

In addition to their role in inference *given* the model, the posterior simulations can be used in several ways to check the fit of the model.

Most directly, we can examine the errors of measurement and modeling by comparing observed data, y_{jkmt}, to their expectations, $g_m(\theta_k, E_j, t)$, for all the measurements, based on the posterior simulations of θ. Figure 20.9 shows a scatterplot of the relative prediction errors of all our observed data (that is, observed data divided by their predictions from the model) versus the model predictions. (Since the analysis was Bayesian, we have many simulation draws of the parameter vector, each of which yields slightly different predicted data. Figure 20.9, for simplicity, shows the predictions from just one of these simu-

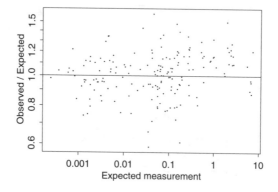

Figure 20.9 *Observed PERC concentrations (for all individuals in the study) divided by expected concentrations, plotted vs. expected concentrations. The x and y-axes are on different (logarithmic) scales: observations vary by a factor of 10,000, but the relative errors are mostly between 0.8 and 1.25. Because the expected concentrations are computed based on a random draw of the parameters from their posterior distribution, the figure shows the actual misfit estimated by the model, without the need to adjust for fitting.*

lation draws, selected at random.) The magnitude of these errors is reasonably low compared to other fits of this kind of data.

We can also check the model by comparing its predictions to additional data not used in the original fit. The additional data that we use are from a second inhalation experiment on human volunteers, in which 6 subjects were exposed to constant levels of PERC ranging from 0.5 to 9 ppm (much lower than the concentrations in our study) and the concentration in exhaled air and blood was measured during exposure for up to 50 minutes (a much shorter time period than in our study). Since these are new individuals we created posterior simulations of the blood/exhaled air concentration ratio (this is the quantity studied by the investigators in the second experiment) by using posterior draws from the population model $p(\theta|\mu, \tau^2)$ to run the nonlinear model. Figure 20.10 presents the observed data and the model prediction (with 95% and 99% simulation bounds). The model fit is good overall, even though exposure levels were 5 to 100 times lower than those used in our data. However, short-term kinetics (less than 15 minutes after the onset of exposure) are not well described by the model, which includes only a simple description of pulmonary exchanges. This is unlikely to seriously affect the quality of predictions for long-term, constant exposures.

Use of a complex model with an informative prior distribution

Our analysis has five key features, all of which work in combination: (1) a physiological model, (2) a population model, (3) prior information on the population physiological parameters, (4) experimental data, and (5) Bayesian

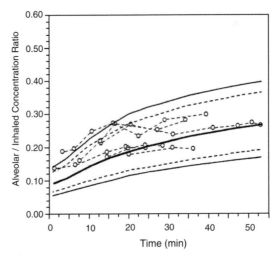

Figure 20.10 *External validation data and 95% predictive intervals from the model fit to the PERC data. The model predictions fit the data reasonably well but not in the first 15 minutes of exposure, a problem we attribute to the fact that the model assumes that all compartments are in instantaneous equilibrium, whereas this actually takes about 15 minutes to approximately hold.*

inference. If any of these five features are missing, the model will not work: (1) without a physiological model, there is no good way to obtain prior information on the parameters, (2) without a population model, there is not generally enough data to estimate the model independently on each individual, (3 and 4) the parameters of a multi-compartment physiological model cannot be determined accurately by data or prior information alone, and (5) Bayesian inference yields a distribution of parameters consistent with both prior information and data, if such agreement is possible. Because it automatically includes both inferential uncertainty and population variability, the hierarchical Bayesian approach yields a posterior distribution that can be directly used for an uncertainty analysis of the risk assessment process.

20.4 Bibliographic note

Carroll, Ruppert, and Stefanski (1995) is a recent treatment of nonlinear statistical models. Giltinan and Davidian (1995) discuss hierarchical nonlinear models. An early example of a serial dilution assay, from Fisher (1922), is discussed by McCullagh and Nelder (1989, p. 11). Assays of the form described in Section 20.2 are discussed by Racine-Poon, Weihs, and Smith (1991) and Higgins et al. (1998). The analysis described in Section 20.2 appears in Gelman, Chew, and Shnaidman (2003).

The toxicology example is described in Gelman, Bois, and Jiang (1996). Hierarchical pharmacokinetic models have a long history; see, for example,

Distance (feet)	Number of tries	Number of successes
2	1443	1346
3	694	577
4	455	337
5	353	208
6	272	149
7	256	136
8	240	111
9	217	69
10	200	67
11	237	75
12	202	52
13	192	46
14	174	54
15	167	28
16	201	27
17	195	31
18	191	33
19	147	20
20	152	24

Table 20.1 *Number of attempts and successes of golf putts, by distance from the hole, for a sample of professional golfers. From Berry (1996).*

Sheiner, Rosenberg, and Melmon (1972), Sheiner and Beal (1982), Wakefield (1996), and the discussion of Wakefield, Aarons, and Racine-Poon (1999). Other biomedical applications in which Bayesian analysis has been used for nonlinear models include magnetic resonance imaging (Genovese, 2001).

Nonlinear models with large numbers of parameters are a bridge to classical nonparametric statistical methods and to methods such as neural networks that are popular in computer science. Neal (1996a) and Denison et al. (2002) discuss these from a Bayesian perspective. Chipman , George, and McCulloch (1998, 2002) give a Bayesian presentation of nonlinear regression-tree models. Bayesian discussions of spline models, which can be viewed as nonparametric generalizations of linear regression models, include Wahba (1978), DiMatteo (2001), and Denison et al. (2002) among others. Zhao (2000) gives a theoretical discussion of Bayesian nonparameteric models.

20.5 Exercises

1. Nonlinear modeling:

2. Table 20.1 presents dataon the success rate of putts by professional golfers (see Berry, 1996, and Gelman and Nolan, 2002c).

(a) Fit a nonlinear model for the probability of success as a function of distance. Does your fitted model make sense in the range of the data and over potential extrapolations?

(b) Use posterior predictive checks to assess the fit of the model to these data.

3. Ill-posed systems: Generate n independent observations y_i from the model, $y_i \sim N(Ae^{-\alpha_1 x_i} + Be^{-\alpha_2 x_i}, \sigma^2)$, where the predictors $x_1, \ldots, x_n$ are uniformly distributed on $[0, 10]$, and you have chosen some particular true values for the parameters.

(a) Fit the model using a uniform prior distribution for the logarithms of the four parameters.

(b) Do simulations with different values of n. How large does n have to be until the Bayesian inferences match the true parameters with reasonable accuracy?

CHAPTER 21

Models for missing data

Our discussions of probability models in previous chapters, with few exceptions, assume that the desired dataset is completely observed. In this chapter we consider probability models and Bayesian methods for problems with missing data. This chapter applies some of the terminology and notation of Chapter 7, which describes aspects of data collection that affect Bayesian data analysis, including mechanisms that lead to missing data.

We show how the analysis of problems involving missing data can often be separated into two main tasks: (1) multiple imputation—that is, simulating draws from the posterior predictive distribution of unobserved y_{mis} conditional on observed values y_{obs}—and (2) drawing from the posterior distribution of model parameters θ. The general idea is to extend the model specification to incorporate the missing observations and then to perform inference by averaging over the distribution of the missing values.

Bayesian inference draws no distinction between missing data and parameters; both are uncertain, and they have a joint posterior distribution, conditional on observed data. The practical distinction arises when setting up the joint model for observed data, unobserved data, and parameters. As discussed in Chapter 7, it is most natural to set this up in three parts: a prior distribution for the parameters (along with a hyperprior distribution if the model is hierarchical), a joint model for all the data (missing and observed), and an inclusion model for the missingness process. If the missing data mechanism is ignorable (see Section 7.3), then inference about the parameters and missing data can proceed without modeling the inclusion process. However, this process does need to be modeled for the purpose of simulating replicated datasets for model checking.

21.1 Notation

We begin by reviewing some notation from Chapter 7, focusing on the problem of unintentional missing data. As in Chapter 7, let y represent the 'complete data' that would be observed in the absence of missing values. The notation is intended to be quite general; y may be a vector of univariate measures or a matrix with each row containing the multivariate response variables of a single unit. Furthermore, it may be convenient to think of the complete data y as incorporating covariates, for example using a multivariate normal model for the vector of predictors and outcomes jointly in a regression context. We write $y = (y_{\text{obs}}, y_{\text{mis}})$, where y_{obs} denotes the observed values and y_{mis} denotes the missing values. We also include in the model a random variable indicating whether each component of y is observed or missing. The *inclusion indicator* I is a data structure of the same size as y with each element of I equal to 1 if the corresponding component of y is observed and 0 if it is missing; we assume that I is completely observed. In a sample survey, item nonresponse corresponds to $I_{ij} = 0$ for unit i and item j, and unit nonresponse corresponds to $I_{ij} = 0$ for unit i and all items j.

The joint distribution of (y, I), given parameters (θ, ϕ), can be written as

$$p(y, I|\theta, \phi) = p(y|\theta)p(I|y, \phi).$$

The conditional distribution of I given the complete dataset y, indexed by the unknown parameter ϕ, describes the missing-data mechanism. The observed information is (y_{obs}, I); the distribution of the observed data is obtained by integrating over the distribution of y_{mis}:

$$p(y_{\text{obs}}, I|\theta, \phi) = \int p(y_{\text{obs}}, y_{\text{mis}}|\theta)p(I|y_{\text{obs}}, y_{\text{mis}}, \phi)dy_{\text{mis}}. \tag{21.1}$$

Missing data are said to be *missing at random* (MAR) if the distribution of the missing-data mechanism does not depend on the missing values,

$$p(I|y_{\text{obs}}, y_{\text{mis}}, \phi) = p(I|y_{\text{obs}}, \phi),$$

so that the distribution of the missing-data mechanism is permitted to depend on other observed values (including fully observed covariates) and parameters ϕ. Formally, missing at random only requires the evaluation of $p(I|y, \phi)$ at the observed values of y_{obs}, not all possible values of y_{obs}. Under MAR, the joint distribution (21.1) of y_{obs}, I can be written as

$$
\begin{aligned}
p(y_{\text{obs}}, I|\theta, \phi) &= p(I|y_{\text{obs}}, \phi) \int p(y_{\text{obs}}, y_{\text{mis}}|\theta)dy_{\text{mis}} \tag{21.2}\\
&= p(I|y_{\text{obs}}, \phi)p(y_{\text{obs}}|\theta).
\end{aligned}
$$

If, in addition, the parameters governing the distribution of the missing data mechanism, ϕ, and the parameters of the probability model, θ, are distinct, in the sense of being independent in the prior distribution, then Bayesian inferences for the model parameters θ can be obtained by considering only the observed-data likelihood, $p(y_{\text{obs}}|\theta)$. In this case, the missing-data mechanism is said to be *ignorable*.

In addition to the terminology of the previous paragraph, we speak of data that are *observed at random* as well as missing at random if the distribution of the missing-data mechanism is completely independent of y:

$$p(I|y_{\text{obs}}, y_{\text{mis}}, \phi) = p(I|\phi). \tag{21.3}$$

In such cases, we say the missing data are *missing completely at random* (MCAR). The preceding paragraph shows that the weaker pair of assumptions of MAR and distinct parameters is sufficient for obtaining Bayesian inferences without requiring further modeling of the missing-data mechanism. Since it is relatively rare in practical problems for MCAR to be plausible, we focus in this chapter on methods suitable for the more general case of MAR.

The plausibility of MAR (but not MCAR) is enhanced by including as many observed characteristics of each individual or object as possible when defining the dataset y. Increasing the pool of observed variables (with relevant variables) decreases the degree to which missingness depends on unobservables given the observed variables.

We conclude this section with a discussion of several examples that illustrate the terminology and principles described above. Suppose that the measurements consist of two variables $y = $ (age, income) with age recorded for all individuals but income missing for some individuals. For simplicity of discussion, we model the joint distribution of the outcomes as bivariate normal. If the probability that income is recorded is the same for all individuals, independent of age and income, then the data are missing at random and observed at random, and therefore missing completely at random. If the probability that income is missing depends on the age of the respondent but not on the income of the respondent given age, then the data are missing at random but not observed at random. The missing-data mechanism is ignorable when, in addition to MAR, the parameters governing the missing-data process are distinct from those of the bivariate normal distribution (as is typically the case with standard models). If, as seems likely, the probability that income is missing depends on age group and moreover on the value of income within each age group, then the data are neither missing nor observed at random. The missing data mechanism in this last case is said to be nonignorable.

The relevance of the missing-data mechanism depends on the goals of the data analysis. If we are only interested in the mean and variance of the age variable, then we can discard all recorded income data and construct a model in which the missing-data mechanism is ignorable. On the other hand, if we are interested in the marginal distribution of income, then the missing-data mechanism is of paramount importance and must be carefully considered.

If information about the missing-data mechanism is available, then it may be possible to perform an appropriate analysis even if the missing-data mechanism is nonignorable, as discussed in Section 7.

21.2 Multiple imputation

Any single imputation provides a complete dataset that can be used by a variety of researchers to address a variety of questions. Assuming the imputation model is reasonable, the results from an analysis of the imputed dataset are likely to provide more accurate estimates than would be obtained by discarding data with missing values.

The key idea of *multiple imputation* is to create more than one set of replacements for the missing values in a data set. This addresses one of the difficulties of single imputation in that the uncertainty due to nonresponse under a particular missing-data model can be properly reflected. The data augmentation algorithm that is used in this chapter to obtain posterior inference can be viewed as iterative multiple imputation.

The paradigmatic setting for missing data imputation is regression, where we are interested in the model $p(y|X, \theta)$ but have missing values in the matrix X. The full Bayesian approach would yield joint inference on $X_{\mathrm{mis}}, \theta | X_{\mathrm{obs}}, y$. However, this involves the difficulty of constructing a joint probability model for the matrix X along with the (relatively simple) model for $y|X$. More fully, if X is modeled given parameters ψ, we would need to perform inference on $X_{\mathrm{mis}}, \psi, \theta | X_{\mathrm{obs}}, y$. From a Bayesian perspective, there is no way around this problem, but it requires serious modeling effort that could take resources away from the primary goal, which is the model $p(y|X, \theta)$.

Bayesian computation in a missing-data problem is based on the joint posterior distribution of parameters and missing data, given modeling assumptions and observed data. The result of the computation is a set of vectors of simulations of all unknowns, $(y_{\mathrm{mis}}^l, \theta^l), l = 1, \ldots, L$. At this point, there are two possible courses of action:

- Obtain inferences for any parameters, missing data, and predictive quantities of interest.

- Report the results in the form of the observed data and the simulated vectors y_{mis}^l, which are called *multiple imputations*. Other users of the data can then use these multiply imputed complete datasets and perform analysis without needing to model the missing-data mechanism.

In the context of this book, the first option seems most natural, but in practice, especially when most of the data values are *not* missing, it is often useful to divide a data analysis in two parts: first, cleaning the data and multiply imputing missing values, and second, performing inference about quantities of interest using the imputed datasets.

In the multiple imputation paradigm, the inferences for X_{mis} and θ are separated into the following steps:

1. First model X, y together and, as described in the previous paragraph, obtain joint inferences for the missing data and parameters of the model. At this point, the imputer takes the surprising step of discarding the inferences about the parameters, keeping only the completed data sets $X^l = (X_{\mathrm{obs}}, X_{\mathrm{mis}}^l)$ for a few random simulation draws l.

2. For each imputed X^l, perform the desired inference for θ based on the model $p(y|X^l, \theta)$, treating the imputed data as if they were known. These are the sorts of models discussed throughout Part IV of this book.

3. Combine the inferences from the separate imputations. With Bayesian simulation, this is simple—just mix together the simulations from the separate inferences. At the end of Section 21.2, we briefly discuss some analytic methods for combining imputations.

In this chapter, we give a quick overview of theory and methods for multiple imputation—that is, step 1 above—and missing-data analysis, illustrating with two examples from survey sampling.

Computation using EM and data augmentation

The process of generating missing data imputations usually begins with crude methods of imputation based on approximate models such as MCAR. The initial imputations are used as starting points for iterative mode-finding and simulation algorithms.

Chapters 11 and 12 describe the Gibbs sampler and the EM algorithm in some detail as approaches for drawing simulations and obtaining the posterior mode in complex problems. As was mentioned there, the Gibbs sampler and EM algorithms formalize a fairly old approach to handling missing data: replace missing data by estimated values, estimate model parameters, and perhaps, repeat these two steps several times. Often, a problem with no missing data can be easier to analyze if the dataset is augmented by some unobserved values, which may be thought of as missing data.

Here, we briefly review the EM algorithm and its extensions using the notation of this chapter. Similar ideas apply for the Gibbs sampler, except that the goal is simulation rather than point estimation of the parameters. The algorithms can be applied whether the missing data are ignorable or not by including the missing-data model in the likelihood, as discussed in Chapter 7. For ease of exposition, we assume the missing-data mechanism is ignorable and therefore omit the inclusion indicator I in the following explanation. The generalization to specified nonignorable models is relatively straightforward. We assume that any augmented data, for example, mixture component indicators, are included as part of y_{mis}. Converting to the notation of Sections 12.3 and 12.4:

Notation in Section 12.3	Notation for missing data
Data, y	Observed data including inclusion information, (y_{obs}, I)
Marginal mode of parameters ϕ	Posterior mode of parameters θ (if the missingness mechanism is being estimated, (θ, ϕ))
Averaging over parameters γ	Averaging over missing data y_{mis}.

The EM algorithm is best known as it is applied to exponential families. In that case, the expected complete-data log posterior density is linear in the expected complete-data sufficient statistics so that only the latter need be evaluated or imputed. Examples are provided in the next section.

EM for nonignorable models. For a nonignorable missingness mechanism, the EM algorithm can also be applied as long as a model for the missing data is specified (for example, censored or rounded data with known censoring point or rounding rule). The only change in the EM algorithm is that all calculations explicitly condition on the inclusion indicator I. Specifically, the expected complete-data log posterior density is a function of model parameters θ and missing-data-mechanism parameters ϕ, conditional on the observed data y_{obs}

and the inclusion indicator I, averaged over the distribution of y_{mis} at the current values of the parameters $(\theta^{\mathrm{old}}, \phi^{\mathrm{old}})$.

Computational shortcut with monotone missing-data patterns. A dataset is said to have a *monotone pattern of missing data* if the variables can be ordered in blocks such that the first block of variables is more observed than the second block of variables (that is, values in the first block are present whenever values in the second are present but the converse does not necessarily hold), the second block of variables is more observed than the third block, and so forth. Many datasets have this pattern or nearly so. Obtaining posterior modes can be especially easy when the data have a monotone pattern. For instance, with normal data, rather than compute a separate regression estimate conditioning on y_{obs} in the E-step for each observation, the monotone pattern implies that there are only as many patterns of missing data as there are blocks of variables. Thus, all of the observations with the same pattern of missing data can be handled in a single step. For data that are close to the monotone pattern, the EM algorithm can be applied as a combination of two approaches: first, the E-step can be carried out for those values of y_{mis} that are outside the monotone pattern; then, the more efficient calculations can be carried out for the missing data that are consistent with the monotone pattern. An example of a monotone data pattern appears in Figure 21.1 on page 530.

Inference with multiple imputations

In some application areas such as sample surveys, a relatively small number of multiple imputations can typically be used to investigate the variability of the missing-data model, and some simple approximate inferences are widely applicable. To be specific, if there are K sets of imputed values under a single model, let $\hat{\theta}_k$ and $\widehat{W}_k$, $k = 1, \ldots, K$, be the K complete-data parameter estimates and associated variance estimates for the scalar parameter θ. The K complete-data analyses can be combined to form the combined estimate of θ,

$$\bar{\theta}_K = \frac{1}{K} \sum_{k=1}^{K} \hat{\theta}_k.$$

The variability associated with this estimate has two components: the average of the complete-data variances (the within-imputation component),

$$\overline{W}_K = \frac{1}{K} \sum_{k=1}^{K} \widehat{W}_k,$$

and the variance across the different imputations (the between-imputation component),

$$B_K = \frac{1}{K-1} \sum_{k=1}^{K} (\hat{\theta}_k - \bar{\theta}_K)^2.$$

The total variance associated with $\bar{\theta}_K$ is

$$T_K = \overline{W}_K + \frac{K+1}{K} B_K.$$

The standard approximate formula for creating interval estimates for θ uses a Student-t distribution with degrees of freedom,

$$\text{d.f.} = (K-1)\left(1 + \frac{1}{K+1}\frac{\overline{W}_K}{B_K}\right)^2,$$

an approximation computed by matching the moments of the variance estimate.

If the fraction of missing information is not too high, then posterior inference will likely not be sensitive to modeling assumptions about the missing-data mechanism. One approach is to create a 'reasonable' missing-data model, and then check the sensitivity of the posterior inferences to other missing-data models. In particular, it often seems helpful to begin with an ignorable model and explore the sensitivity of posterior inferences to plausible nonignorable models.

21.3 Missing data in the multivariate normal and t models

We consider the basic continuous-data model in which y represents a sample of size n from a d-dimensional multivariate normal distribution $N_d(\mu, \Sigma)$ with y_{obs} the set of observed values and y_{mis} the set of missing values. We present the methods here for the uniform prior distribution and then in the next section give an example with a hierarchical multivariate model.

Finding posterior modes using EM

The multivariate normal is an exponential family with sufficient statistics equal to

$$\sum_{i=1}^{n} y_{ij}, \quad j = 1, \ldots, d,$$

and

$$\sum_{i=1}^{n} y_{ij} y_{ik}, \quad j, k = 1, \ldots, d.$$

Let $y_{\text{obs}\,i}$ denote the components of $y_i = (y_{i1}, \ldots, y_{id})$ that are observed and $y_{\text{mis}\,i}$ denote the missing components. Let $\theta^{\text{old}} = (\mu^{\text{old}}, \Sigma^{\text{old}})$ denote the current estimates of the model parameters. The E step of the EM algorithm computes the expected value of these sufficient statistics conditional on the observed values and the current parameter estimates. Specifically,

$$E\left(\sum_{i=1}^{n} y_{ij} \,|\, y_{\text{obs}}, \theta^{\text{old}}\right) = \sum_{i=1}^{n} y_{ij}^{\text{old}}$$

$$E\left(\sum_{i=1}^{n} y_{ij}y_{ik}|y_{\text{obs}},\theta^{\text{old}}\right) = \sum_{i=1}^{n}(y_{ij}^{\text{old}}y_{ik}^{\text{old}} + c_{ijk}^{\text{old}}),$$

where

$$y_{ij}^{\text{old}} = \begin{cases} y_{ij} & \text{if } y_{ij} \text{ is observed} \\ E(y_{ij}|y_{\text{obs}},\theta^{\text{old}}) & \text{if } y_{ij} \text{ is missing,} \end{cases}$$

and

$$c_{ijk}^{\text{old}} = \begin{cases} 0 & \text{if } y_{ij} \text{ or } y_{ik} \text{ are observed} \\ \text{cov}(y_{ij},y_{ik}|y_{\text{obs}},\theta^{\text{old}}), & \text{if } y_{ij} \text{ and } y_{ik} \text{ are missing.} \end{cases}$$

The conditional expectation and covariance are easy to compute: the conditional posterior distribution of the missing elements of y_i, $y_{\text{mis}\,i}$, given y_{obs} and θ, is multivariate normal with mean vector and variance matrix obtained from the full mean vector and variance matrix θ^{old} as in Appendix A.

The M-step of the EM algorithm uses the expected complete-data sufficient statistics to compute the next iterate, $\theta^{\text{new}} = (\mu^{\text{new}}, \Sigma^{\text{new}})$. Specifically,

$$\mu_j^{\text{new}} = \frac{1}{n}\sum_{i=1}^{n} y_{ij}^{\text{old}}, \text{ for } j = 1,\ldots,d,$$

and

$$\sigma_{jk}^{\text{new}} = \frac{1}{n}\sum_{i=1}^{n}(y_{ij}^{\text{old}}y_{ik}^{\text{old}} + c_{ijk}^{\text{old}}) - \mu_j^{\text{new}}\mu_k^{\text{new}}, \text{ for } j,k = 1,\ldots,d.$$

Starting values for the EM algorithm can be obtained using crude methods. As always when finding posterior modes, it is wise to use several starting values in case of multiple modes. It is crucial that the initial estimate of the variance matrix be positive definite; thus various estimates based on complete cases (that is, units with all outcomes observed), if available, can be useful.

Drawing samples from the posterior distribution of the model parameters

One can draw imputations for the missing values from the normal model using the modal estimates as starting points for data augmentation (the Gibbs sampler) on the joint posterior distribution of missing values and parameters, alternately drawing y_{mis}, μ, and Σ from their conditional posterior distributions. For more complicated models, some of the steps of the Gibbs sampler must be replaced by Metropolis steps.

As with the EM algorithm, considerable gains in efficiency are possible if the missing data have a monotone pattern. In fact, for monotone missing data, it is possible under an appropriate parameterization to draw directly from the incomplete-data posterior distribution, $p(\theta|y_{\text{obs}})$. Suppose that y_1 is more observed than y_2, y_2 is more observed than y_3, and so forth. To be specific, let $\psi = \psi(\theta) = (\psi_1,\ldots,\psi_k)$, where ψ_1 denotes the parameters of the marginal distribution of the first block of variables in the monotone pattern y_1, ψ_2 denotes the parameters of the conditional distribution of y_2 given y_1, and so on (the normal distribution is d-dimensional, but in general the monotone pattern is defined by $k \leq d$ blocks of variables). For multivariate normal data, ψ_j contains the parameters of the linear regression of y_j on $y_1, \ldots, y_{j-1}$—the regression coefficients and the residual variance matrix. The parameter ψ is a one-to-one function of the parameter θ, and the complete parameter space of ψ is the product of the parameter spaces of $\psi_1, \ldots, \psi_k$. The likelihood factors into k distinct pieces:

$$
\begin{aligned}
\log p(y_{\text{obs}}|\psi) = {} & \log p(y_{\text{obs}\,1}|\psi_1) + \log p(y_{\text{obs}\,2}|y_{\text{obs}\,1}, \psi_2) + \\
& \cdots + \log p(y_{\text{obs}\,k}|y_{\text{obs}\,1}, y_{\text{obs}\,2}, \ldots, y_{\text{obs}\,k-1}, \psi_k),
\end{aligned}
$$

with the jth piece depending only on the parameters ψ_j. If the prior distribution $p(\psi)$ can be written in closed form in the factorization,

$$
p(\psi) = p(\psi_1)p(\psi_2|\psi_1) \cdots p(\psi_k|\psi_1, \ldots, \psi_{k-1}),
$$

then it is possible to draw directly from the posterior distribution in sequence: first draw ψ_1 conditional on y_{obs}, then ψ_2 conditional on ψ_1 and y_{obs}, and so forth.

For a missing-data pattern that is not precisely monotone, we can define a monotone data augmentation algorithm that imputes only enough data to obtain a monotone pattern. The imputation step draws a sample from the conditional distribution of the elements in y_{mis} that are needed to create a monotone pattern. The posterior step then draws directly from the posterior distribution taking advantage of the monotone pattern. Typically, the monotone data augmentation algorithm will be more efficient than ordinary data augmentation if the departure from monotonicity is not substantial, because fewer imputations are being done and analytic calculations are being used to replace the other simulation steps. There may be several ways to order the variables that each lead to nearly monotone patterns. Determining the best such choice is complicated since 'best' is defined by providing the fastest convergence of an iterative simulation method. One simple approach is to choose y_1 to be the variable with the fewest missing values, y_2 to be the variable with the second fewest, and so on.

Student-t extensions of the normal model

Chapter 17 described robust alternatives to the normal model based on the Student-t distribution. Such models can be useful for accommodating data prone to outliers, or as a means of performing a sensitivity analysis on a normal model. Suppose now that the intended data consist of multivariate observations,

$$
y_i|\theta, V_i \sim N_d(\mu, V_i \Sigma),
$$

where V_i are unobserved iid random variables with an Inv-$\chi^2(\nu, 1)$ distribution. For simplicity, we consider ν to be specified; if unknown, it is another parameter to be estimated.

Data augmentation can be applied to the t model with missing values in y_i by adding a step that imputes values for the V_i, which are thought of as additional missing data. The imputation step of data augmentation consists of two parts. First, a value is imputed for each V_i from its posterior distribution given $y_{\text{obs}}, \theta, \nu$. This posterior distribution is a product of the normal distribution for $y_{\text{obs}\,i}$ given V_i and the scaled inverse-χ^2 prior distribution for V_i,

$$p(V_i | y_{\text{obs}}, \theta, \nu) \propto \text{N}(y_{\text{obs}\,i} | \mu_{\text{obs}\,i}, \Sigma_{\text{obs}\,i} V_i) \text{Inv-}\chi^2(V_i | \nu, 1), \qquad (21.4)$$

where $\mu_{\text{obs}\,i}, \Sigma_{\text{obs}\,i}$ refer to the elements of the mean vector and variance matrix corresponding to components of y_i that are observed. The conditional posterior distribution (21.4) is easily recognized as scaled inverse-χ^2, so obtaining imputed values for V_i is straightforward. The second part of each iteration step is to impute the missing values $y_{\text{mis}\,i}$ given $(y_{\text{obs}}, \theta, V_i)$, which is identical to the imputation step for the ordinary normal model since given V_i, the value of $y_{\text{mis}\,i}$ is obtained as a draw from the conditional normal distribution. The posterior step of the data augmentation algorithm treats the imputed values as if they were observed and is, therefore, a complete-data weighted multivariate normal problem. The complexity of this step depends on the prior distribution for θ.

The E-step of the EM algorithm for the t extensions of the normal model is obtained by replacing the imputation steps above with expectation steps. Thus the conditional expectation of V_i from its scaled inverse-χ^2 posterior distribution and conditional means and variances of $y_{\text{mis}\,i}$ would be used in place of random draws. The M-step finds the conditional posterior mode rather than sampling from the posterior distribution. When the degrees of freedom parameter for the t distribution is allowed to vary, the ECM and ECME algorithms of Section 12.3 can be used.

Nonignorable models

The principles for performing Bayesian inference in nonignorable models based on the normal or t distributions are analogous to those presented in the ignorable case. At each stage, θ is supplemented by any parameters of the missing-data mechanism, ϕ, and inference is conditional on observed data y_{obs} and the inclusion indicator I.

21.4 Example: multiple imputation for a series of polls

In this section we present an extended example illustrating many of the practical issues involved in a Bayesian analysis of a survey data problem. The example concerns a series of polls with relatively large amounts of missing data.

Background

The U.S. Presidential election of 1988 was unusual in that the Democratic candidate, Michael Dukakis, was far ahead in the polls four months before the election, but he eventually lost by a large margin to the Republican candidate, George Bush. We studied public opinion during this campaign using data from 51 national polls conducted by nine different major polling organizations during the six months before the election. One of the major purposes of the study was to examine changes in voting intentions over time for different subgroups of the population (for example: men and women; self-declared Democrats, Republicans, and independents; low-income and high-income; and so forth).

Performing analyses of the 51 surveys required some care in handling the missing data, because not all questions of interest were asked in all surveys. For example, respondent's self-reported ideology (liberal, moderate, or conservative), a key variable, was missing in 10 of the 51 surveys, including our only available surveys during the Democratic nominating convention. Questions about the respondent's views of the national economy and of the perceived ideologies of Bush and Dukakis were asked in fewer than half of the surveys, making them difficult to study. Imputing the responses to these questions would simplify our analysis by allowing us to analyze all questions in the same way.

When imputing missing data from several sample surveys, there are two obvious ways to use existing single-survey methods: (1) separately imputing the missing data from each survey, or (2) combining the data from all the surveys and imputing the missing data in the combined 'data matrix.' Both of these methods have problems. The first approach is difficult if there is a large amount of missingness in each individual survey. For example, if a particular question is not asked in one survey, then there is no general way to impute it without using information from other surveys or some additional knowledge about the relation between responses to that question and to other questions asked in the survey. The second method does not account for differences between the surveys—for example, if they are conducted at different times, use different sampling methodologies, or are conducted by different survey organizations.

Our approach is to compromise by fitting a separate imputation model for each survey, but with the parameters in the different survey models linked with a hierarchical model. This method should have the effect that imputations of item nonresponse in a survey will be determined largely by the data from that survey, whereas imputations for questions that are not asked in a survey will be determined by data from the other surveys in the population as well as by available responses to other questions in that survey.

Multivariate missing-data framework

Creating the imputations requires a hierarchical multivariate model of $Q = 15$ questions asked in the $S = 51$ surveys, with not all questions asked in all surveys. The 15 questions included the outcome variable of interest (Presidential vote preference), the variables that were believed to have the strongest relation to vote preference, and several demographic variables that were fully observed or nearly so, which would have the effect of explanatory variables in the imputation. We also include in our analysis the date at which each survey was conducted.

To handle both sources of missingness—'not asked' and 'not answered'—we augment the data in such a way that the complete data consist of the same Q questions in all the S surveys. We denote by $\mathbf{y}_{si} = (\mathbf{y}_{si1}, \ldots, \mathbf{y}_{siQ})$ the responses of individual i in survey s to all the Q questions. Some of the elements of $\mathbf{y}_{si}$ may be missing. Letting N_s be the number of respondents in survey s, the (partially unobserved) complete data have the form,

$$((\mathbf{y}_{si1}, \ldots, \mathbf{y}_{siQ}): \quad i = 1, \ldots, N_s, \ s = 1, \ldots, S). \qquad (21.5)$$

We assume that the missingness is ignorable (see Section 7.3), which is plausible here because almost all the missingness is due to unasked questions. If clear violations of ignorability occur (for example, a question about defense policy may be more likely to be asked when the country is at war), then we would add survey-level variables to the model until missing at random is once again a reasonable assumption (for example, including a variable for the level of international tension).

A hierarchical model for multiple surveys

The simplest model for imputing the missing values in (21.5) that makes use of the data structure of the multiple surveys is a hierarchical model. We assume a multivariate normal distribution at the individual level with mean vector μ_s for survey s and a variance matrix Ψ assumed common to all surveys,

$$y_{si}|\mu_s, \Psi, \theta, \Sigma \stackrel{\text{ind}}{\sim} N_Q(\mu_s, \Psi), \quad i = 1, \ldots, N_s, \ s = 1, \ldots, S. \qquad (21.6)$$

The assumption that the variance matrix Ψ is the same for all surveys could be tested by, for example, dividing the surveys nonrandomly into two groups (for example, early surveys and late surveys) and estimating separate matrices Φ for the two groups.

We might continue by modeling the survey means μ_s as exchangeable but our background discussion suggests that factors at the survey level, such as organization effects and time trend, be included in the model. Suppose we have data on P covariates of interest at the survey level. Let $x_s = (x_{s1}, \ldots, x_{sP})$ be the vector of these covariates for survey s. We assume that x_s is fully observed for each of the S surveys. We use the notation $\mathbf{X}$ for the $S \times P$ matrix of the fully observed covariates (with row s containing the covariates for survey s). Then our model for the survey means is a multivariate multiple regression

$$\mu_s|X, \beta, \Sigma \stackrel{\text{ind}}{\sim} N_Q(\beta x_s, \Sigma), \quad s = 1, \ldots, S, \qquad (21.7)$$

where β is the $Q \times P$ matrix of the regression coefficients of the mean vector on the survey covariate vector and Σ is a diagonal matrix, $\text{Diag}(\sigma_1^2, \ldots, \sigma_Q^2)$. Since Σ is diagonal, equation (21.7) represents Q linear regression models with normal errors:

$$\mu_{sj}|X, \beta, \Sigma \sim N(\beta_j x_s, \sigma_j^2), \quad s = 1, \ldots, S, \; j = 1, \ldots, Q, \tag{21.8}$$

where μ_{sj} is the jth component of μ_s and β_j is the jth row of β. The model in equations (21.6) and (21.7) allows for pooling information from all the S surveys and imputing all the missing values, including those to the questions that were not asked in some of the surveys.

For mathematical convenience we use the following noninformative prior distribution for (Ψ, β, Σ):

$$p(\Psi, \beta, \Sigma) = p(\Psi)p(\beta) \prod_{q=1}^{Q} p(\sigma_Q^2) \propto |\Psi|^{-(Q+1)/2}. \tag{21.9}$$

If there are fewer than Q individuals with responses on all the questions, then it is necessary to use a proper prior distribution for Ψ. (Prior distributions for variance matrices are discussed in Sections 19.1 and 19.2.)

Use of the continuous model for discrete responses

There is a natural concern when using a continuous imputation model based on the normal distribution for survey responses, which are coded at varying levels of discretization. Some variables in our analysis (sex, ethnicity) are coded as unordered and discrete, others (voting intention, education) are ordered and discrete, whereas others (age, income, and the opinion questions on 1–5 and 1–7 scales) are potentially continuous, but are coded as ordered and discrete. We recode the responses from different survey organizations as appropriate so that the responses from each question fall on a common scale. (For example, for the surveys in which the 'perceived ideology' questions are framed as too-liberal/just-right/too-conservative, the responses are recoded based on the respondent's stated ideology.)

There are several possible ways to adapt a continuous model to impute discrete responses. From the most elaborate to the simplest, these include (1) modeling the discrete responses conditional on a latent continuous variable as described in Chapter 16 in the context of generalized linear models, (2) modeling the data as continuous and then using some approximate procedure to impute discrete values for the missing responses, and (3) modeling the data as continuous and imputing continuous values. We follow the third, simplest approach. In our example, little is lost by this simplification, because the variables that are the most 'discrete' (sex, ethnicity, vote intention) are fully observed or nearly so, whereas the variables with the most nonresponse (the opinion questions) are essentially continuous. When it is necessary to have discrete values, we round off the continuous imputations, essentially using approach (2).

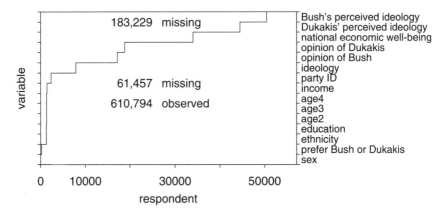

Figure 21.1 *Approximate monotone data pattern for an analysis of 51 polls conducted during the 1988 U.S. Presidential election campaign. Not all questions were asked in all surveys.*

Computation

The model is complicated enough and the dataset large enough that it was necessary to write a special program to perform posterior simulations. A simple Gibbs or Metropolis algorithm would run too slowly to work here. Our method uses two basic steps: data augmentation to form a monotone missing data pattern (as described in Section 21.3), and the Gibbs sampler to draw simulations from the joint posterior distribution of the missing data and parameters under the monotone pattern.

Monotone data augmentation is effective for multiple imputation for multiple surveys because the data can be sorted so that a large portion of the missing values fall into a monotone pattern due to the fact that some questions are not asked in some surveys. Figure 21.1 illustrates a constructed monotone data pattern for the pre-election surveys, with the variables arranged in decreasing order of proportion of missing data. The most observed variable is sex, and the least observed concern the candidates' perceived ideologies. We let k index the Q possible observed data patterns, n_k be the number of respondents with that pattern, and s_i the survey containing the ith respondent in the kth observed data pattern. The resulting data matrix can be written as

$$y_{\mathrm{mp}} = ((\mathbf{y}^{(k)}_{s_i i 1}, \ldots, \mathbf{y}^{(k)}_{s_i i k}) : i = 1, \ldots, n_k, \ k = 1, \ldots, Q), \qquad (21.10)$$

where the subscript 'mp' denotes that this is the data matrix for the monotone pattern. The monotone pattern data matrix in (21.10) still contains missing values; that is, some of the values in the bottom right part of Figure 21.1 are not observed even though the question was asked of the respondent. We denote by $y_{\mathrm{mp,mis}}$ the set of all the missing values in (21.10) and by y_{obs} the set of all the observed values. Thus, we have $y_{\mathrm{mp}} = (y_{\mathrm{obs}}, y_{\mathrm{mp,mis}})$.

Using the data in the form (21.10) and the model defined by equations

(21.6), (21.7), and (21.9), we have the observed data y_{obs} and the unknowns, $(y_{\text{mp,mis}}, \Psi, \mu_1, \ldots, \mu_S, \beta, \Sigma)$. To take draws from the posterior distribution of $(y_{\text{mp,mis}}, \Psi, \mu_1, \ldots, \mu_S, \beta, \Sigma)$ given the observed data y_{obs}, we use a version of the monotone Gibbs sampler where each iteration consists of the following three steps:

1. Impute $y_{\text{mp,mis}}$ given $\Psi, \mu_1, \ldots, \mu_S, \beta, \Sigma, y_{\text{obs}}$. Under the normal model these imputations require a series of draws from the joint normal distribution for survey responses.

2. Draw (Ψ, β, Σ) given $\mu_1, \ldots, \mu_S, y_{\text{obs}}, y_{\text{mp,mis}}$. These are parameters in a hierarchical linear model with 'complete' (including the imputations) monotone pattern.

3. Draw $(\mu_1, \ldots, \mu_S)$ given $\Psi, \beta, \Sigma, y_{\text{obs}}, y_{\text{mp,mis}}$. The parameters $\mu_1, \ldots, \mu_S$ are mutually independent and normally distributed in this conditional posterior distribution.

Accounting for survey design and weights

The respondents for each survey were assigned weights based on sampling and poststratification. These were not used in the imputation procedure because the variables on which the weights were based were, by and large, included in the imputation model already. Thus, the weights do not provide additional information about the missing responses. (Though they might impact some of the model parameters.)

We do, however, use the survey weights when computing averages based on the imputed data, in order to obtain unbiased estimates of population averages unconditional on the demographic variables.

Results

We examine the results of the imputation for the 51 surveys with a separate graph for each variable; we illustrate in Figure 21.2 with two of the variables: 'income' (chosen because it should remain stable during the four-month period under study) and 'perceived ideology of Dukakis' (which we expect to change during the campaign).

First consider Figure 21.2a, 'income.' Each symbol on the graph represents a different survey, plotting the estimated average income for the survey vs. the date of the survey, with the symbol itself indicating the survey organization. The size of the symbol is proportional to the fraction of survey respondents who responded to the particular question, with the convention that when the question is not asked (indicated by circled symbols on the graph), the symbol is tiny but not of zero size. The vertical bars show ±1 standard error in the posterior mean of the average income, where the standard error includes within-imputation sampling variance and between-imputation variance. Finally, the inner brackets on the vertical bars show the within-imputation standard deviation alone. All of the complete-data means and standard deviations displayed

The posterior means of "income" in the 51 surveys

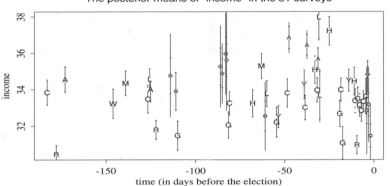

H: Harris, A: ABC; C: CBS; Y: Yankelovich; G: Gallup; L: LATimes; M: MediaGeneral/AP; R: Roper; W: WashPost

The posterior means of "Dukakis's perceived ideology" in the 51 surveys

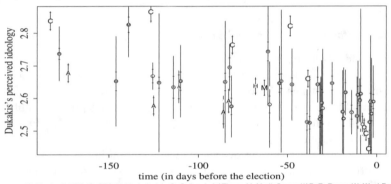

H: Harris, A: ABC; C: CBS; Y: Yankelovich; G: Gallup; L: LATimes; M: MediaGeneral/AP; R: Roper; W: WashPost

Figure 21.2 *Estimates and standard error bars for the population mean response for two questions: (a) income (in thousands of dollars), and (b) perceived ideology of Dukakis (on a 1–7 scale from liberal to conservative), over time. Each symbol represents a different survey, with different letters indicating different survey organization. The size of the letter indicates the number of responses to the question, with large-sized letters for surveys with nearly complete response and small-sized letters for surveys with few responses. Circled letters indicate surveys for which the question was not asked; the estimates for these surveys have much larger standard errors. The inner brackets on the vertical bars show the within-imputation standard deviation for the estimated mean from each survey.*

in the figure are weighted as described above. Even those surveys in which the question was asked (and answered by most respondents) have nonzero standard errors for the estimated population average income because of the finite sample sizes of the surveys. For surveys in which the income question was asked, the within-imputation variance almost equals the total variance, which makes sense since, when a question was asked, most respondents answered. The multiple imputation procedure makes very weak statements (that is, there are large standard errors) about missing income responses for those surveys missing the income question. This makes sense because income is not highly correlated with the other questions.

Figure 21.2a also shows some between-survey variability in average income, from \$31,000 to \$37,000—more than can be explained by sampling variability, as is indicated by the error bars on the surveys for which the question was asked. Since we do not believe that the average income among the population of registered or likely voters is changing that much, the explanation must lie in the surveys. In fact, different survey organizations use different codings for incomes (for example, 0–10K, 10–20K, 20–30K, ..., or 0–7.5K, 7.5–15K, 15–25K, ...). Since the point of our method is to produce imputations close to what the surveys would look like if all the questions had been asked and answered, rather than to adjust all the observed and unobserved data to estimate population quantities, this variability is reasonable. The large error bars for average income for the surveys in which the question was not asked reflect the large between-survey variation in average income, which is captured by our hierarchical model. For this study, we are interested in income as a predictor variable rather than for its own sake, and we are willing to accept this level of uncertainty.

Figure 21.2b shows a similar plot for Dukakis's perceived ideology. Both the observed and imputed survey responses show a time trend in this variable—Dukakis was perceived as more liberal near the end of the campaign. In an earlier version of the model (not shown here) that did not include time as a predictor, the plot of Dukakis's perceived ideology showed a problem: the observed data showed the time trend, but the imputations did not. This plot was useful in revealing the model flaw: the imputed survey responses looked systematically different from the data.

21.5 Missing values with counted data

The analysis of fully observed counted data is discussed in Section 3.5 for saturated multinomial models and in Chapter 16 for loglinear models. Here we consider how those techniques can be applied to missing discrete data problems.

Multinomial samples. Suppose the hypothetical complete data are a multinomial sample of size n with cells $c_1, \ldots, c_J$, probabilities $\theta = (\pi_1, \ldots, \pi_J)$, and cell counts $n_1, \ldots, n_J$. Conjugate prior distributions for θ are in the Dirichlet family (see Appendix A and Section 3.5). The observed data are m completely classified observations with m_j in the jth cell, and $n-m$ partially classified observations (the missing data). The partially classified observations are known to fall in subsets of the J cells. For example, in a $2 \times 2 \times 2$ table, $J = 8$, and an observation with known classification for the first two dimensions but with missing classification for the third dimension is known to fall in one of two possible cells.

Partially classified observations. It is convenient to organize each of the $n-m$ partially classified observations according to the subset of cells to which it can belong. Thus suppose there are K types of partially classified observation, and the r_k observations of the kth type are known to fall in one of the cells in subset S_k.

The iterative procedure used for normal data in the previous subsection, data augmentation, can be used here to iterate between imputing cells for the partially classified observations and obtaining draws from the posterior distribution of the parameters θ. The imputation step draws from the conditional distribution of the partially classified cells given the observed data and the current set of parameters θ. For each $k = 1, \ldots, K$, the r_k partially classified observations known to fall in the subset of cells S_k are assigned randomly to each of the cells in S_k with probability

$$\frac{\pi_j I_{j \in S_k}}{\sum_{l=1}^{J} \pi_l I_{l \in S_k}},$$

where $I_{j \in S_k}$ is the indicator function equal to 1 if cell j is part of the subset S_k and 0 otherwise. When the prior distribution is Dirichlet, then the parameter updating step requires drawing from the conjugate Dirichlet posterior distribution, treating the imputed data and the observed data as a complete dataset. As usual, it is possible to use other, nonconjugate, prior distributions, although this makes the posterior computation more difficult. The EM algorithm for exploring modes is a nonstochastic version of data augmentation: the E-step computes the number of the r_k partially classified observations that are expected to fall in each cell (the mean of the multinomial distribution), and the M-step computes updated cell probability estimates by combining the observed cell counts with the results of the E-step.

As usual, the analysis is simplified when the missing-data pattern is monotone or nearly monotone, so that the likelihood can be written as a product of the marginal distribution of the most observed set of variables and a set of conditional distributions for each subsequent set of variables conditional on all of the preceding, more observed variables. If the prior density is also factored, for example, as a product of Dirichlet densities for the parameters of each factor in the likelihood, then the posterior distribution can be drawn from directly. The analysis of nearly monotone data requires iterating two steps: imputing values for those partially classified observations required to complete the monotone pattern, and drawing from the posterior distribution, which can be done directly for the monotone pattern.

Further complications arise when the cell probabilities θ are modeled, as in loglinear models; see the references at the end of this chapter.

21.6 Example: an opinion poll in Slovenia

We illustrate the methods described in the previous section with the analysis of an opinion poll concerning independence in Slovenia, formerly a province

| Secession | Attendance | Independence | | |
		Yes	No	Don't know
Yes	Yes	1191	8	21
	No	8	0	4
	Don't Know	107	3	9
No	Yes	158	68	29
	No	7	14	3
	Don't Know	18	43	31
Don't Know	Yes	90	2	109
	No	1	2	25
	Don't Know	19	8	96

Table 21.1 $3 \times 3 \times 3$ table of results of 1990 preplebiscite survey in Slovenia, from Rubin, Stern, and Vehovar (1995). We treat 'don't know' responses as missing data. Of most interest is the proportion of the electorate whose 'true' answers are 'yes' on both 'independence' and 'attendance.'

of Yugoslavia and now a nation. In 1990, a plebiscite was held in Slovenia at which the adult citizens voted on the question of independence. The rules of the plebiscite were such that nonattendance, as determined by an independent and accurate census, was equivalent to voting 'no'; only those attending and voting 'yes' would be counted as being in favor of independence. In anticipation of the plebiscite, a Slovenian public opinion survey had been conducted that included several questions concerning likely plebiscite attendance and voting. In that survey, 2074 Slovenians were asked three questions: (1) Are you in favor of independence?, (2) Are you in favor of secession?, and (3) Will you attend the plebiscite? The results of the survey are displayed in Table 21.1. Let α represent the estimand of interest from the sample survey, the proportion of the population planning to attend and vote in favor of independence. It follows from the rules of the plebiscite that 'don't know' (DK) can be viewed as missing data (at least accepting that 'yes' and 'no' responses to the survey are accurate for the plebiscite). Every survey participant will vote yes or no— perhaps directly or perhaps indirectly by not attending.

Why ask three questions when we only care about two of them? The response to question 2 is not directly relevant but helps us more accurately impute the missing data. The survey participants may provide some information about their intentions by their answers to question 2; for example, a 'yes' response to question 2 might be indicative of likely support for independence for a person who did not answer question 1.

Crude estimates

As an initial estimate of α, the proportion planning to attend and vote 'yes,' we ignore the DK responses for these two questions; considering only the 'available cases' (those answering the attendance and independence questions) yields a crude estimate $\hat{\alpha} = 1439/1549 = 0.93$, which seems to suggest that the outcome of the plebiscite is not in doubt. However, given that only 1439/2074, or 69%, of the survey participants definitely plan to attend and vote 'yes,' and given the importance of the outcome, improved inference is desirable, especially considering that if we were to assume that the DK responses correspond to 'no,' we would obtain a very different estimate.

The 2074 responses include those of substitutes for original survey participants who could not be contacted after several attempts. Although the substitutes were chosen from the same clusters as the original participants to minimize differences between substitutes and nonsubstitutes, there may be some concern about differences between the two groups. We indicate the most pessimistic estimate for α by noting that only 1251/2074 of the original intended survey sample (using information not included in the table) plans to attend and vote 'yes.' For simplicity, we treat substitutes as original respondents for the remainder of this section and ignore the effects of clustering.

The likelihood and prior distribution

The complete data can be viewed as a sample of 2074 observations from a multinomial distribution on the eight cells of the $2 \times 2 \times 2$ contingency table, with corresponding vector of probabilities θ; the DK responses are treated as missing data. We use θ_{ijk} to indicate the probability of the multinomial cell in which the respondent gave answer i to question 1, answer j to question 2, and answer k to question 3, with $i, j, k = 0$ for 'no' and 1 for 'yes.' The estimand of most interest, α, is the sum of the appropriate elements of θ, $\alpha = \theta_{101} + \theta_{111}$. In our general notation, y_{obs} are the observed 'yes' and 'no' responses, and y_{mis} are the 'true' 'yes' and 'no' responses corresponding to the DK responses. The 'complete data' form a 2074×3 matrix of 0's and 1's that can be recoded as a contingency table of 2074 counts in eight cells.

The Dirichlet prior distribution for θ with parameters all equal to zero is noninformative in the sense that the posterior mode is the maximum likelihood estimate. Since one of the observed cell counts is 0 ('yes' on secession, 'no' on attendance, 'no' on independence), the improper prior distribution does not lead to a proper posterior distribution. It would be possible to proceed with the improper prior density if we thought of this cell as being a structural zero—a cell for which a nonzero count is impossible. The assumption of a structural zero does not seem particularly plausible here, and we choose to use a Dirichlet distribution with parameters all equal to 0.1 as a convenient (though arbitrary) way of obtaining a proper posterior distribution while retaining a diffuse prior distribution. A thorough analysis should explore the sensitivity of conclusions to this choice of prior distribution.

The model for the 'missing data'

We treat the DK responses as *missing values*, each known only to belong to some subset of the eight cells. Let $n = (n_{ijk})$ represent the hypothetical complete data and let $m = (m_{ijk})$ represent the number of completely classified respondents in each cell. There are 18 types of partially classified observations; for example, those answering 'yes' to questions 1 and 2 and DK to question 3, those answering 'no' to question 1 and DK to questions 2 and 3, and so on. Let r_p denote the number of partially classified observations of type p; for example, let r_1 represent the number of those answering 'yes' to questions 1 and 2 and DK to question 3. Let S_p denote the set of cells to which the partially classified observations of the pth type might belong; for example, S_1 includes the 111 and 110 cells. We assume that the DK responses are missing at random, which implies that the probability of a DK response may depend on the answers to other questions but not on the unobserved response to the question at hand.

The complete-data likelihood is

$$p(n|\theta) \propto \prod_{i=0}^{1} \prod_{j=0}^{1} \prod_{k=0}^{1} \theta_{ijk}^{n_{ijk}},$$

with complete-data sufficient statistics $n = (n_{ijk})$. If we let $\pi_{ijk\,p}$ represent the probability that a partially classified observation of the pth type belongs in cell ijk, then the MAR model implies that, given a set of parameter values θ, the distribution of the r_p observations with the pth missing-data pattern is multinomial with probabilities

$$\pi_{ijk\,p} = \frac{\theta_{ijk} I_{ijk \in S_p}}{\sum_{i'j'k'} \theta_{i'j'k'} I_{i'j'k' \in S_p}},$$

where the indicator $I_{ijk \in S_p}$ is 1 if cell ijk is in subset S_p and 0 otherwise.

Using the EM algorithm to find the posterior mode of θ

The EM algorithm in this case finds the mode of the posterior distribution of the multinomial probability vector θ by averaging over the missing data (the DK responses) and is especially easy here with the assumption of distinct parameters. For the multinomial distribution, the E-step, computing the expected complete-data log posterior density, is equivalent to computing the expected counts in each cell of the contingency table given the current parameter estimates. The expected count in each cell consists of the fully observed cases in the cell and the expected number of the partially observed cases that fall in the cell. Under the missing at random assumption and the resulting multinomial distribution, the DK responses in each category (that is, the pattern of missing data) are allocated to the possible cells in proportion to the current estimate of the model parameters. Mathematically, given current parameter estimate θ^{old}, the E-step computes

$$n_{ijk}^{\text{old}} = \text{E}(n_{ijk}|m, r, \theta^{\text{old}}) = m_{ijk} + \sum_p r_p \pi_{ijk\,p}.$$

The M-step computes new parameter estimates based on the latest estimates of the expected counts in each cell; for the saturated multinomial model here (with distinct parameters), $\theta^{\text{new}} = (n_{ijk}^{\text{old}} + 0.1)/(n + 0.8)$. The EM algorithm is considered to converge when none of the parameter estimates changes by more than a tolerance criterion, which we set here to the unnecessarily low value of 10^{-16}. The posterior mode of α is 0.882, which turned out to be quite close to the eventual outcome in the plebiscite.

Using SEM to estimate the posterior variance matrix and obtain a normal approximation

To complete the normal approximation, estimates of the posterior variance matrix are required. The SEM algorithm numerically computes estimates of the variance matrix using the EM program and the complete-data variance matrix (which is available since the complete data are modeled as multinomial).

The SEM algorithm is applied to the logit transformation of the components of θ, since the normal approximation is generally more accurate on this scale. Posterior central 95% intervals for $\text{logit}(\alpha)$ are transformed back to yield a 95% interval for α, $[0.857, 0.903]$. The standard error was inflated to account for the design effect of the clustered sampling design using approximate methods based on the normal distribution.

Multiple imputation using data augmentation

Even though the sample size is large in this problem, it seems prudent, given the missing data, to perform posterior inference that does not rely on the asymptotic normality of the maximum likelihood estimates. The data augmentation algorithm, a special case of the Gibbs sampler, can be used to obtain draws from the posterior distribution of the cell probabilities θ under a noninformative Dirichlet prior distribution. As described earlier, for count data, the data augmentation algorithm iterates between imputations and posterior draws. At each imputation step, the r_p cases with the pth missing-data pattern are allocated among the possible cells as a draw from a multinomial distribution. Conditional on these imputations, a draw from the posterior distribution of θ is obtained from the Dirichlet posterior distribution. A total of 1000 draws from the posterior distribution of θ were obtained—the second half of 20 data augmentation series, each run for 100 iterations, at which point the potential scale reductions, $\widehat{R}$, were below 1.1 for all parameters.

Posterior inference for the estimand of interest

The posterior median of α, the population proportion planning to attend and vote yes, is 0.883. We construct an approximate posterior central 95% interval for α by inflating the variance of the 95% interval from the posterior simulations to account for the clustering in the design (to avoid the complications but approximate the results of a full Bayesian analysis of this sampling design); the resulting interval is [0.859, 0.904]. It is not surprising, given the large sample size, that this interval matches the interval obtained from the asymptotic normal distribution.

By comparison, neither of the initial crude calculations is very close to the actual plebiscite outcome, in which 88.5% of the eligible population attended and voted 'yes.'

21.7 Bibliographic note

The jargon 'missing at random,' 'observed at random,' and 'ignorability' originated with Rubin (1976). Multiple imputation was proposed in Rubin (1978b) and is discussed in detail in Rubin (1987a) with a focus on sample surveys; Rubin (1996) is a recent review of the topic. Kish (1965) and Madow et al. (1983) discuss less formal ways of handling missing data in sample surveys. The book edited by Groves et al. (2002) is a recent overview of work on survey nonresponse.

Little and Rubin (1992) is a comprehensive text on statistical analysis with missing data. Tanner and Wong (1987) describe the use of data augmentation to calculate posterior distributions. Schafer (1997) applies data augmentation for multivariate exchangeable models, including the normal and loglinear models discussed briefly in this chapter; Liu (1995) extends these methods to t models. The approximate variance estimate in Section 21.2 is derived from the Satterthwaite (1946) approximation; see Rubin (1987a), Meng, Raghunathan, and Rubin (1991), and Meng and Rubin (1992).

Meng (1994b) discusses the theory of multiple imputation when different models are used for imputation and analysis. Raghunathan et al. (2001) and Gelman and Raghunathan (2001) discuss the 'inconsistent Gibbs' method for multiple imputation. Van Buuren, Boshuizen, and Knook (1999) present a related approach, which has been implemented in S; see Van Buuren and Oudshoom (2000).

Clogg et al. (1991) and Belin et al. (1993) describe hierarchical logistic regression models used for imputation for the U.S. Census. There has been growing use of multiple imputation using nonignorable models for missing data; for example, Heitjan and Landis (1994) set up a model for unobserved medical outcomes and multiply impute using matching to appropriate observed cases. David et al. (1986) present a thorough discussion and comparison of a variety of imputation methods for a missing data problem in survey imputation.

The monotone method for estimating multivariate models with missing data dates from Anderson (1957) and was extended by Rubin (1974a, 1976, 1987a);

see the rejoinder to Gelman, King, and Liu (1998) for more discussion of computational details. The example of missing data in opinion polls comes from Gelman, King, and Liu (1998). The Slovenia survey is described in more detail in Rubin, Stern, and Vehovar (1995).

21.8 Exercises

1. Computation for discrete missing data: reproduce the results of Section 21.6 for the 2×2 table involving independence and attendance. You can ignore the clustering in the survey and pretend it was obtained from a simple random sample. Specifically:

   (a) Use EM to obtain the posterior mode of α, the proportion who will attend and will vote 'yes.'
   (b) Use SEM to obtain the asymptotic posterior variance of logit(α), and thereby obtain an approximate 95% interval for α.
   (c) Use Markov chain simulation of the parameters and missing data to obtain the approximate posterior distribution of θ. Clearly say what your starting distribution is for your simulations. Be sure to simulate more than one sequence and to include some diagnostics on the convergence of the sequences.

2. Monotone missing data: create a monotone pattern of missing data for the opinion poll data of Section 21.6 by discarding some observations. Compare the results of analyzing these data with the results given in that section.

CHAPTER 22

Decision analysis

What happens after the data analysis, after the model has been built and the inferences computed, and after the model has been checked and expanded as necessary so that its predictions are consistent with observed data? What use is made of the inferences once the data analysis is done?

One form of answer to this question came in Section 7, which discussed the necessity of extending the data model to encompass the larger population of interest, including missing observations, unobserved outcomes of alternative treatments, and units not included in the study. For a Bayesian model to generalize, it must account for potential differences between observed data and the population.

This concluding chapter considers a slightly different aspect of the problem: how can inferences be used in decision making? In a general sense, we expect to be using predictive distributions, but the details depend on the particular problem. In Section 22.1 we outline the theory of decision making under uncertainty, and the rest of the chapter presents examples of the application of Bayesian inference to decisions in social science, medicine, and public health.

The first example, in Section 22.2, is the simplest: we fit a hierarchical regression on the effects of incentives on survey response rates, and then we use the predicted values to estimate costs. The result of the analysis is a graph estimating expected increase in response rate vs. the additional cost required, which allows us to apply general inferences from the regression model to making decisions for a particular survey of interest. From a decision-making point of view, this example is interesting because regression coefficients that are not 'statistically significant' (that is, that have high posterior probabilities of being positive or negative) are still highly relevant for the decision problem, and we cannot simply set them to zero.

Section 22.3 presents a more complicated decision problem, on the option of performing a diagnostic test before deciding on a treatment for cancer. This is a classic problem of the 'value of information,' balancing the risks of the screening test against the information that might lead to a better treatment decision. The example presented here is typical of the medical decision-making literature in applying a relatively sophisticated Bayesian decision analysis using point estimates of probabilities and risks taken from simple summaries of published studies.

We conclude in Section 22.4 with an example that combines full Bayesian hierarchical modeling, probabilistic decision analysis, and utility analysis, balancing the risks of exposure to radon gas against the costs of measuring the level of radon in a house and potentially remediating it. We see this analysis as a prototype of full integration of inference with decision analysis, beyond what is practical or feasible for most applications but indicating the connections between Bayesian hierarchical regression modeling and individually focused decision making.

22.1 Bayesian decision theory in different contexts

Many if not most statistical analyses are performed for the ultimate goal of decision making. In most of this book we have left the decision-making step implicit: we perform a Bayesian analysis, from which we can summarize posterior inference for quantities of interest such as the probability of dying of cancer, or the effectiveness of a medical treatment, or the vote by state in the next Presidential election.

When explicitly balancing the costs and benefits of decision options under uncertainty, we use Bayesian inference in two ways. First, a decision will typically depend on predictive quantities (for example, the probability of recovery under a given medical treatment, or the expected value of a continuous outcome such as cost or efficacy under some specified intervention) which in turn depend on unknown parameters such as regression coefficients and population frequencies. We use posterior inferences to summarize our uncertainties about these parameters, and hence about the predictions that enter into the decision calculations. We give examples in Sections 22.2 and 22.4, in both cases using inferences from hierarchical regressions.

The second way we use Bayesian inference is *within* a decision analysis, to determine the conditional distribution of relevant parameters and outcomes, given information observed as a result of an earlier decision. This sort of calculation arises in multistage decision trees, in particular when evaluating the expected value of information. We illustrate with a simple case in Section 22.3 and a more elaborate example in Section 22.4.

Bayesian inference and decision trees

Decision analysis is inherently more complicated than statistical inference because it involves optimization over decisions as well as averaging over uncertainties. We briefly lay out the elements of Bayesian decision analysis here. The implications of these general principles should become clear in the examples that follow.

Bayesian *decision analysis* can be defined mathematically by the following steps:

1. Enumerate the space of all possible decisions d and outcomes x. In a business context, x might be dollars; in a medical context, lives or life-years. More generally, outcomes can have multiple attributes and would be expressed as vectors. Section 22.2 presents an example in which outcomes are in dollars and survey response rates, and in the example of Section 22.4,

outcomes are summarized as dollars and lives. The vector of outcomes x can include observables (that is, predicted values $\tilde{y}$ in our usual notation) as well as unknown parameters θ. For example, x could include the total future cost of an intervention (which would ultimately be observed) as well as its effectiveness in the population (which might never be measured).

2. Determine the probability distribution of x for each decision option d. In Bayesian terms, this is the conditional posterior distribution, $p(x|d)$. In the decision-analytic framework, the decision d does not have a probability distribution, and so we cannot speak of $p(d)$ or $p(x)$; all probabilities must be conditional on d.

3. Define a *utility function* $U(x)$ mapping outcomes onto the real numbers. In simple problems, utility might be identified with a single continuous outcome of interest x, such as years of life, or net profit. If the outcome x has multiple attributes, the utility function must trade off different goods, for example quality-adjusted life years (in Section 22.3).

4. Compute the expected utility $E(U(x)|d)$ as a function of the decision d, and choose the decision with highest expected utility. In a *decision tree*—in which a sequence of two or more decisions might be taken—the expected utility must be calculated at each decision point, conditional on all information available up to that point.

A full decision analysis includes all four of these steps, but in many applications, we simply perform the first two, leaving it to decision makers to balance the expected gains of different decision options.

This chapter includes three case studies of the use of Bayesian inference for decision analysis. In Section 22.2, we present an example in which decision making is carried halfway: we consider various decision options and evaluate their expected consequences, but we do not create a combined utility function or attempt to choose an optimal decision. Section 22.3 analyzes a more complicated decision problem that involves conditional probability and the value of information. We conclude in Section 22.4 with a full decision analysis including utility maximization.

Distinction between decision analysis and 'statistical decision theory'

Before getting to the interesting applications, we first discuss what might be called traditional 'statistical decision theory,' a mathematical framework that is formally Bayesian but which we find too abstract to be directly useful for real decision problems. We need to discuss this theoretical approach because it is historically connected in the statistical literature with Bayesian inference.

Traditional statistical decision theory is most often applied in the estimation context where there is some unknown parameter or quantity of interest θ, the 'decision' is the choice of point estimate, $\hat{\theta}(y)$, and the utility function is some measure of closeness between θ and $\hat{\theta}$. The optimal estimate maximizes the expected utility and is determined by the form of the utility function.

For example, suppose $U(\theta, \hat{\theta}) = 1$ if $\theta = \hat{\theta}$ and 0 otherwise. Then expected utility is maximized by setting $\hat{\theta}$ to the posterior mode. If $U(\theta, \hat{\theta}) = |\theta - \hat{\theta}|^\alpha$, penalizing the absolute error to the power α, then the optimal estimate is the posterior median if $\alpha = 1$ or the posterior mean if $\alpha = 2$ (see Exercise 8.1).

These mathematical results are interesting but we do not see their relevance in practice. If the goal is inference, why choose an 'optimal' estimate or any point estimate at all? We prefer summarizing with the entire posterior distribution, or at least a posterior interval (see Section 4.4). Decision analysis comes into its own when we go beyond abstract rules such as squared error loss and look at actual outcomes, as we illustrate in Sections 22.2 and 22.4.

22.2 Using regression predictions: incentives for telephone surveys

Our first example shows the use of a meta-analysis—fitted from historical data using hierarchical linear regression—to estimate predicted costs and benefits for a new situation. The decision analysis for this problem is implicit, but the decision-making framework makes it clear why it can be important to include predictors in a regression model even when they are not statistically significant.

Background on survey incentives

Common sense and evidence (in the form of randomized experiments within surveys) both suggest that giving incentives to survey participants tends to increase response rates. From a survey designer's point of view, the relevant questions are:

- Do the benefits of incentives outweigh the costs?
- If an incentive is given, how and when should it be offered, whom should it be offered to, what form should it take, and how large should its value be?

We consider these questions in the context of the New York City Social Indicators Survey, a telephone study conducted every two years that has had a response rate below 50%. Our decision analysis proceeds in two steps: first, we perform a meta-analysis to estimate the effects of incentives on response rate, as a function of the amount of the incentive and the way it is implemented. Second, we use this inference to estimate the costs and benefits of incentives in our particular survey.

We consider the following factors that can affect the efficacy of an incentive:

- The *value* of the incentive (in tens of 1999 dollars),
- The *timing* of the incentive payment (given before the survey or after),
- The *form* of the incentive (cash or gift),
- The *mode* of the survey (face-to-face or telephone),
- The *burden*, or effort, required of the survey respondents (a survey is characterized as high burden if it is over one hour long and has sensitive or difficult questions, and low burden otherwise).

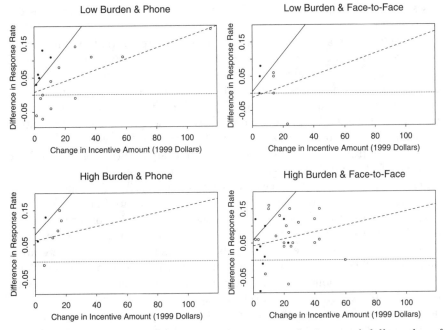

Figure 22.1 *Observed increase z_i in response rate vs. the increased dollar value of incentive compared to the control condition, for experimental data within 39 surveys. Prepaid and postpaid incentives are indicated by closed and open circles, respectively. (The graphs show more than 39 points because many surveys had multiple treatment conditions.) The lines show the expected increases for prepaid (solid lines) and postpaid (dashed lines) cash incentives as estimated from the hierarchical regression model.*

Data from 39 experiments

Data were collected on 39 surveys that had embedded experiments testing different incentive conditions. For example, a survey could, for each person contacted, give a $5 prepaid incentive with probability 1/3, a $10 prepaid incentive with probability 1/3, or no incentive with probability 1/3. The surveys in the meta-analysis were conducted on different populations and at different times, and between them they covered a range of different interactions of the five factors mentioned above (value, timing, form, mode, and burden). In total, the 39 surveys include 101 experimental conditions. We use the notation y_i to indicate the observed response rate for observation $i = 1, \ldots, 101$.

It is challenging to model the response rates y_i directly, since the surveys differ quite a bit in response rate. A reasonable starting point uses the differences, $z_i = y_i - y_i^0$, where y_i^0 corresponds to the lowest-valued incentive condition in the survey that includes condition i (in most surveys, this is simply the control case of no incentive). Working with z_i reduces the number of cases in the analysis from 101 conditions to 62 differences and eliminates the between-survey variation in baseline response rates.

Figure 22.1 displays the difference in response rates z_i vs. the difference in incentive amounts, for each of the 62 differences i. The points are partitioned into subgraphs corresponding to the mode and burden of their surveys. Within each graph, solid and open circles indicate prepaid and postpaid incentives, respectively. We complete the graphs by including a dotted line at zero, to represent the comparison case of no incentive. (The graphs also include fitted regression lines from the hierarchical model described below.)

It is clear from the graphs in Figure 22.1 that incentives generally have positive effects, and that prepaid incentives tend to be smaller in dollar value. Some of the observed differences are negative, but this can be expected from sampling variability, given that some of the 39 surveys are fairly small.

A natural way to use these data to support the planning of future surveys is to fit a classical regression model relating z_i to the value, timing, and form of incentive as well as the mode and burden of survey. However there are a number of difficulties with this approach. From the sparse data, it is difficult to estimate interactions which might well be important. For example, it seems reasonable to expect that the effect of prepaid versus postpaid incentive may depend on the amount of the incentive. In addition, a traditional regression would not reflect the hierarchical structure of the data: the 62 differences are clustered in 39 surveys. It is also not so easy in a regression model to account for the unequal sample sizes for the experimental conditions, which range from below 100 to above 2000. A simple weighting proportional to sample size is not appropriate since the regression residuals include model error as well as binomial sampling error.

We shall set up a slightly more elaborate hierarchical model because, for the purpose of estimating the costs and benefits in a particular survey, we need to estimate interactions in the model (for example, the interaction between timing and value of incentive), even if these are not statistically significant.

Setting up a Bayesian meta-analysis

We set up a hierarchical model with 101 data points i, nested within 39 surveys j. We start with a binomial model relating the number of respondents, n_i, to the number of persons contacted, N_i (thus, $y_i = n_i/N_i$), and the population response probabilities π_i:

$$n_i \sim \text{Bin}(N_i, \pi_i). \qquad (22.1)$$

The next stage is to model the probabilities π_i in terms of predictors X, including an indicator for survey incentives, the five incentive factors listed above, and various interactions. In general it would be advisable to use a transformation before modeling these probabilities since they are constrained to lie between 0 and 1. However, in our particular application area, response probabilities in telephone and face-to-face surveys are far enough from 0 and

1 that a linear model is acceptable:

$$\pi_i \sim \mathrm{N}((X\beta)_i + \alpha_{j(i)}, \sigma^2). \tag{22.2}$$

Here, $(X\beta)_i$ is the linear predictor for the condition corresponding to data point i, $\alpha_{j(i)}$ is a random effect for the survey $j = 1, \ldots, 39$ (necessary in the model because underlying response rates vary greatly), and σ represents the lack of fit of the linear model. We use the notation $j(i)$ because the conditions i are nested within surveys j. The use of the survey random effects allows us to incorporate the 101 conditions in the analysis rather than working with the 62 differences as was done earlier. The α_j's also address the hierarchical structure of the data.

Modeling (22.2) on the untransformed scale is not simply an approximation but rather a choice to set up a more interpretable model. Switching to the logistic, for example, would have no practical effect on our conclusions, but it would make all the regression coefficients much more difficult to interpret.

We next specify prior distributions for the parameters in the model. We model the survey-level random effects α_j using a normal distribution:

$$\alpha_j \sim \mathrm{N}(0, \tau^2). \tag{22.3}$$

There is no loss of generality in assuming a zero mean for the α_j's if a constant term is included in the set of predictors X. Finally, we assign uniform prior densities to the standard deviation parameters σ and τ and to the regression coefficients β. The parameters σ and τ are estimated precisely enough that the inferences are not sensitive to the particular choice of noninformative prior distribution.

Computation

Equations (22.1)–(22.3) can be combined to form a posterior distribution. For moderate and large values of N (such as are present in the meta-analysis data), we can simplify the computation by replacing the binomial distribution (22.1) by a normal distribution for the observed response rate $y_i = n_i/N_i$:

$$y_i \approx \mathrm{N}(\pi_i, V_i), \tag{22.4}$$

where $V_i = y_i(1 - y_i)/N_i$. It would be possible to use the exact binomial likelihood but in our example this would just add complexity to the computations without having any effect on the inferences.

We then combine the linear model (22.2) and the normal approximation to the binomial likelihood (22.4) to yield,

$$y_i \approx \mathrm{N}((X\beta)_i + \alpha_{j(i)}, \sigma^2 + V_i). \tag{22.5}$$

We obtain estimates and uncertainties for the parameters $\alpha, \beta, \sigma, \tau$ in our models using Bayesian posterior simulation, working with the approximate model (22.5) and the prior distribution (22.3). For any particular choice of predictive variables, this is a hierarchical linear regression with two variance parameters, σ and τ. As discussed in Chapter 15, computations from this model can be performed in several different ways.

Given σ and τ, we can easily compute the conditional posterior distribution of the linear parameters, which we write as the vector $\gamma = (\beta, \alpha)$. The computation is a simple linear regression of y_* on X_* with variance matrix Σ_*, where:

W_α is the 101×39 indicator matrix mapping conditions i to surveys $j(i)$,

y is the vector of response rates, $(y_1, \ldots, y_{101})$,

y_* the vector of length 140 represented by y followed by 39 zeroes,

$$X_* = \begin{pmatrix} X & W_\alpha \\ 0 & I_{39} \end{pmatrix},$$

$$\Sigma_y = \mathrm{Diag}(\sigma^2 + V_i),$$

$$\Sigma_* = \begin{pmatrix} \Sigma_y & 0 \\ 0 & \tau^2 I_{39} \end{pmatrix}.$$

The linear regression computation gives an estimate $\hat{\gamma}$ and variance matrix V_γ, and the conditional posterior distribution is $\gamma | \sigma, \tau, X, y \sim \mathrm{N}(\hat{\gamma}, V_\gamma)$. The variance parameters σ and τ are not known, however. They can be incorporated in a Markov chain simulation algorithm; the conditional distribution of τ^2 will be inverse-χ^2 but the conditional distribution of σ^2 will require a Metropolis algorithm because of the V_i terms in the data variance.

An alternative in this case is to compute the joint marginal posterior density for σ and τ numerically on a two-dimensional grid using the formula,

$$
\begin{aligned}
p(\sigma, \tau | X, y) &= \frac{p(\gamma, \sigma, \tau | X, y)}{p(\gamma | \sigma, \tau, X, y)} \\[2mm]
&\propto \frac{\prod_{j=1}^{39} \mathrm{N}(\alpha_j | 0, \tau^2) \prod_{i=1}^{101} \mathrm{N}(y_i | (X\beta)_i + \alpha_{j(i)}, \sigma^2 + V_i)}{\mathrm{N}(\gamma | \hat{\gamma}, V_\gamma)} \\[2mm]
&\propto \frac{|\Sigma_*|^{-1/2} \exp\left(-\tfrac{1}{2}(y_* - X_*\gamma)^t \Sigma_*^{-1}(y_* - X_*\gamma)\right)}{|V_\gamma|^{-1/2} \exp\left(-\tfrac{1}{2}(\gamma - \hat{\gamma})^t V_\gamma^{-1}(\gamma - \hat{\gamma})\right)} \\[2mm]
&\propto |V_\gamma|^{1/2} |\Sigma_*|^{-1/2} \exp\left(-\frac{1}{2}(y_* - X_*\hat{\gamma})^t \Sigma_*^{-1}(y_* - X_*\hat{\gamma})\right),
\end{aligned}
$$

where the last line follows by taking γ (which does not appear on the left hand side) equal to $\hat{\gamma}$. Then we obtain posterior simulations by drawing 1000 values of σ, τ from the two-dimensional grid and then drawing $\gamma = (\beta, \alpha)$ from the normal distribution with mean $\hat{\gamma}$ and variance matrix V_γ for each draw of the variance parameters.

Alternative computational strategies

We used the grid computation in this example because it is easy to implement using linear regressions. However, other approaches could work as well or better, depending on the computational tools directly available. We have already mentioned the possibility of using Markov chain simulation. Another possibility would be to simply put the model in the Bayesian software program

Bugs (see Appendix C) with reasonable starting values. Since the observations y_i are response rates, which vary between 0 and 1, one could simply draw the regression coefficients from independent $N(0,1)$ distributions to obtain overdispersed starting points.

Inferences from the model

Thus far we have not addressed the choice of predictors to include in the covariate matrix X. The main factors are those described earlier (which we denote as Value, Timing, Form, Mode, and Burden) along with Incentive, an indicator for whether a given condition includes an incentive (not required when we were working with the differences). Because there are restrictions (Value, Timing, and Form are only defined if Incentive $= 1$), there are 36 possible regression predictors, including the constant term and working up to the interaction of all five factors with incentive. The number of predictors would of course increase if we allowed for nonlinear functions of incentive value.

Of the predictors, we are particularly interested in those that include interactions with the Incentive indicator, since these indicate effects of the various factors. The two-way interactions in the model that include Incentive can thus be viewed as main effects of the factors included in the interactions, the three-way interactions can be viewed as two-way interactions of the included factors, and so forth.

We fit a series of models, starting with the simplest, then adding interactions until we pass the point where the existing data could estimate them effectively, then finally choosing a model that includes the key interactions needed for our decision analysis. Our chosen model includes the main effects for Mode, Burden, and the Mode × Burden interaction, which all have the anticipated large impacts on the response rate of a survey. It also includes Incentive (on average, the use of an incentive increases the response rate by around 3 percentage points), all two-way interactions of Incentive with the other factors, and the three-way interactions that include Incentive × Value interacting with Timing and Burden. We do not provide detailed results here, but some of the findings are that an extra \$10 in incentive is expected to increase the response rate by 3–4 percentage points, cash incentives increase the response rate by about 1 percentage point relative to noncash, prepaid incentives increase the response rate by 1–2 percentage points relative to postpaid, and incentives have a bigger impact (by about 5 percentage points) on high burden surveys compared to low burden surveys.

The within-study standard deviation σ is around 3 or 4 percentage points, indicating the accuracy with which differential response rates can be predicted within any survey. The between-study standard deviation τ is about 18 percentage points, indicating that the overall response rates vary greatly, even after accounting for the survey-level predictors (Mode, Burden, and their interaction).

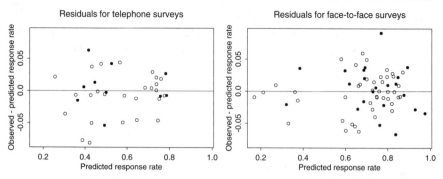

Figure 22.2 *Residuals of response rate meta-analysis data plotted vs. predicted values. Residuals for telephone and face-to-face surveys are shown separately. As in Figure 22.1, solid and open circles indicate surveys with prepaid and postpaid incentives, respectively.*

Figure 22.1 on page 545 displays the model fit as four graphs corresponding to the two possible values of the Burden and Mode variables. Within each graph, we display solid lines for the prepaid condition and dotted lines for postpaid incentives, in both cases showing only the results with cash incentives, since these were estimated to be better than gifts of the same value.

To check the fit, we display in Figure 22.2 residual plots of prediction errors for the individual data points y_i, showing telephone and face-to-face surveys separately and, as with the previous plots, using symbols to distinguish pre- and post-incentives. There are no apparent problems with the basic fit of the model, although other models could also fit these data equally well.

Inferences about costs and response rates for the Social Indicators Survey

The Social Indicators survey is a low-burden telephone survey. If we use incentives at all, we would use cash, since this appears to be more effective than gifts of the same value. We then have the choice of value and timing of incentives. Regarding timing, prepaid incentives are more effective than postpaid incentives per dollar of incentive (compare the slopes of the solid and dashed lines in Figure 22.1). But this does not directly address our decision problem. Are prepaid incentives still more effective than postpaid incentives when we look at total dollars spent? This is not immediately clear, since prepaid incentives must be sent to all potential respondents, whereas postpaid are given only to the people who actually respond. It can be expensive to send the prepaid incentives to the potential respondents who cannot be reached, refuse to respond, or are eliminated in the screening process.

We next describe how the model inferences are used to inform decisions in the context of the Social Indicators Survey. This survey is conducted by random digit dialing in two parts. 750 respondents come from an 'individual survey,' in which an attempt is made to survey an adult from every residen-

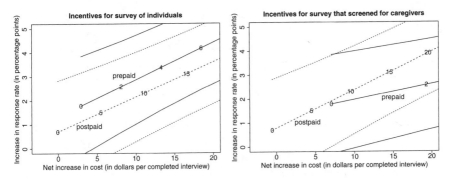

Figure 22.3 *Expected increase in response rate vs. net cost of incentive per respondent, for prepaid (solid lines) and postpaid (dotted lines) incentives, for the individual and caregiver surveys. On each plot, the heavy lines correspond to the estimated effects, with light lines showing ±1 standard error bounds. The numbers on the lines indicate incentive payments. At zero incentive payments, estimated effects and costs are nonzero because the models have nonzero intercepts (corresponding to the effect of making any contact at all) and also we are assuming a $1.25 processing cost per incentive.*

tial phone number that is called. 1500 respondents come from a 'caregiver survey,' which includes only adults who are taking care of children. The caregiver survey begins with a screening question to eliminate households without children.

For each of the two surveys, we use our model to estimate the expected increase in response rate for any hypothesized incentive and the net increase in cost to obtain that increase in response rate. It is straightforward to use the fitted hierarchical regression model to estimate the expected increase in response rate. Then we work backward and estimate the number of telephone calls required to reach the same number of respondents with this higher response rate. The net cost of the hypothesized incentive is the dollar value of the incentive (plus $1.25 to account for the cost of processing and mailing) times the number of people to whom the incentive is sent less the savings that result because fewer contacts are required.

For example, consider a $5 postpaid incentive for the caregiver survey. From the fitted model, this would lead to an expected increase of 1.5% in response rate, which would increase it from the 38.9% in the actual survey to a hypothesized 40.4%. The cost of the postpaid incentives for 1500 respondents at $6.25 each ($5 incentive plus $1.25 processing and mailing cost) is $9375. With the number of responses fixed, the increased response rate implies that only $1500/0.404 = 3715$ eligible households would have to be reached, instead of the 3658 households contacted in the actual survey. Propagating back to the screening stage leads to an estimated number of telephone numbers that would need to be contacted and an estimated number of calls to reach those numbers. In this case we estimate that 3377 fewer calls would be required, yielding an estimated savings of $2634 (based on the cost of interviewers and the average length per non-interview call). The net cost of the incentive is then $9375 - $2634 = $6741, which when divided by the 1500 completed interviews yields a cost of $4.49 per interview for this 1.5% increase in response rate. We perform similar calculations for other hypothesized incentive conditions.

Figure 22.3 summarizes the results for a range of prepaid and postpaid incentive values, assuming we are willing to spend up to $20 per respondent in incentives. For either survey, incentives are expected to raise response rates by only a few percentage points. Prepaid incentives are expected to be slightly better for the individual survey, and postpaid are preferred for the (larger) caregiver survey. For logistical reasons, we would use the same form of incentive for both, so we would recommend postpaid. In any case, we leave the final step of the decision analysis—picking the level of the incentive—to the operators of the survey, who must balance the desire to increase response rate with the cost of the incentive itself.

Loose ends

Our study of incentives is far from perfect; we use it primarily to demonstrate how a relatively routine regression analysis can be used to make inferences about the potential consequences of decision options. The most notable weaknesses are the high level of uncertainty about individual coefficients (not shown here) and the arbitrariness of the decision as to which interactions should be included/excluded. These two problems go together: when we tried including more interactions, the standard errors became even larger and the inferences became less believable. The problem is with the noninformative uniform prior distribution on the regression coefficients. It would make more sense to include all interactions and make use of prior information that might shrink the higher-order interactions without fixing them at zero. It would also be reasonable to allow the effects of incentives to vary among surveys. We did not expand the model in these ways because we felt we were at the limit of our knowledge about this problem, and we thought it better to stop and summarize our inference and uncertainties about the costs and benefits of incentives.

Another weakness of the model is its linearity, which implies undiminishing effects as incentives rise. It would be possible to add an asymptote to the model to fix this, but we do not do so, since in practice we would not attempt to extrapolate our inferences beyond the range of the data in the meta-analysis (prepaid incentives up to $20 and postpaid up to $60 or $100; see Figure 22.1).

22.3 Multistage decision making: medical screening

Decision analysis becomes more complicated when there are two or more decision points, with later decisions depending on data gathered after the first decision has been made. Such decision problems can be expressed as *trees*, alternating between decision and uncertainty nodes. In these multistage problems, Bayesian inference is particularly useful in updating the state of knowledge with the information gained at each step.

Example with a single decision point

We illustrate with a simplified example from the medical decision making literature. A 95-year-old man with an apparently malignant tumor in the lung must decide between the three options of radiotherapy, surgery, or no treatment. The following assumptions are made about his condition and life expectancy (in practice, these probabilities and life expectancies are based on extrapolations from the medical literature):

- There is a 90% chance that the tumor is malignant.
- If the man does not have lung cancer, his life expectancy is 34.8 months.
- If the man does have lung cancer,

  1. With radiotherapy, his life expectancy is 16.7 months.
  2. With surgery, there is a 35% chance he will die immediately, but if he survives, his life expectancy is 20.3 months.
  3. With no treatment, his life expectancy is 5.6 months.

Aside from mortality risk, the treatments themselves cause considerable discomfort for slightly more than a month. We shall determine the decision that maximizes the patient's quality-adjusted life expectancy, which is defined as the expected length of time the patient survives, minus a month if he goes through one of the treatments. The subtraction of a month addresses the loss in 'quality of life' due to treatment-caused discomfort.

Quality-adjusted life expectancy under each treatment is then

1. With radiotherapy: $0.9 \cdot 16.7 + 0.1 \cdot 34.8 - 1 = 17.5$ months.
2. With surgery: $0.35 \cdot 0 + 0.65 \cdot (0.9 \cdot 20.3 + 0.1 \cdot 34.8 - 1) = 13.5$ months.
3. With no treatment: $0.9 \cdot 5.6 + 0.1 \cdot 34.8 = 8.5$ months.

These simple calculations show radiotherapy to be the preferred treatment for this 95-year-old man.

Adding a second decision point

The decision problem becomes more complicated when we consider a fourth decision option, which is to perform a test to see if the cancer is truly malignant. The test, called bronchoscopy, is estimated to have a 70% chance of detecting the lung cancer if the tumor is indeed malignant, and a 2% chance of falsely finding cancer if the tumor is actually benign. In addition, there is an estimated 5% chance that complications from the test itself will kill the patient.

Should the patient choose bronchoscopy? To make this decision, we must first determine what he would do *after* the test. Bayesian inference with discrete probabilities gives the probability of cancer given the test result T as

$$\text{Pr(cancer}|T) = \frac{\text{Pr(cancer)}p(T|\text{cancer})}{\text{Pr(cancer)}p(T|\text{cancer}) + \text{Pr(no cancer)}p(T|\text{no cancer})},$$

and we can use this probability in place of the prior probability $\text{Pr(cancer)} = 0.9$ in the single decision point decision-making calculations above.

- If the test is *positive* for cancer, then the updated probability he has cancer is $\frac{0.9 \cdot 0.7}{0.9 \cdot 0.7 + 0.1 \cdot 0.02} = 0.997$, and his quality-adjusted life expectancy under each of the three treatments becomes

  1. With radiotherapy: $0.997 \cdot 16.7 + 0.003 \cdot 34.8 - 1 = 15.8$ months.
  2. With surgery: $0.35 \cdot 0 + 0.65(0.997 \cdot 20.3 + 0.003 \cdot 34.8 - 1) = 12.6$ months.
  3. With no treatment: $0.997 \cdot 5.6 + 0.003 \cdot 34.8 = 5.7$ months.

  So, if the test is positive, radiotherapy would be the best treatment, with a quality-adjusted life expectancy of 15.8 months.

- If the test is *negative* for cancer, then the updated probability he has cancer is $\frac{0.9 \cdot 0.3}{0.9 \cdot 0.3 + 0.1 \cdot 0.98} = 0.734$, and his quality-adjusted life expectancy under each of the three treatments becomes

  1. With radiotherapy: $0.734 \cdot 16.7 + 0.266 \cdot 34.8 - 1 = 20.5$ months.
  2. With surgery: $0.35 \cdot 0 + 0.65(0.734 \cdot 20.3 + 0.266 \cdot 34.8 - 1) = 15.1$ months.
  3. With no treatment: $0.734 \cdot 5.6 + 0.266 \cdot 34.8 = 13.4$ months.

  If the test is negative, radiotherapy would still be the best treatment, this time with a quality-adjusted life expectancy of 20.5 months.

At this point, it is clear that bronchoscopy is not a good idea, since whichever way the treatment goes, it will not affect the decision that is made. To complete the analysis, however, we work out the quality-adjusted life expectancy for this decision option. The bronchoscopy can yield two possible results:

- Test is positive for cancer. The probability that this outcome will occur is $0.9 \cdot 0.7 + 0.1 \cdot 0.02 = 0.632$, and the quality-adjusted life expectancy (accounting for the 5% chance that the test can be fatal) is $0.95 \cdot 15.8 = 15.0$ months.

- Test is negative for cancer. The probability that this outcome will occur is $0.9 \cdot 0.3 + 0.1 \cdot 0.98 = 0.368$, and the quality-adjusted life expectancy (accounting for the 5% chance that the test can be fatal) is $0.95 \cdot 20.5 = 19.5$ months.

The total quality-adjusted life expectancy for the bronchoscopy decision is then $0.632 \cdot 15.0 + 0.368 \cdot 19.5 = 16.6$ months. Since radiotherapy without a bronchoscopy yields an expected quality-adjusted survival of 17.5 months, it is clear that the patient should not choose bronchoscopy.

The decision analysis reveals the perhaps surprising result that, in this

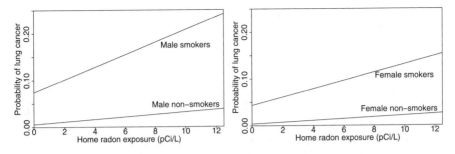

Figure 22.4 *Lifetime added risk of lung cancer, as a function of average radon exposure in picoCuries per liter (pCi/L). The median and mean radon levels in ground-contact houses in the U.S. are 0.67 and 1.3 pCi/L, respectively, and over 50,000 homes have levels above 20 pCi/L.*

scenario, bronchoscopy is pointless, since it would not affect the decision that is to be made. Any other option (for example, bronchoscopy, followed by a decision to do radiotherapy if the test is positive or do no treatment if the test is negative) would be even worse in expected value.

22.4 Decision analysis using a hierarchical model: home radon measurement and remediation

Associated with many household environmental hazards is a decision problem: whether to (1) perform an expensive remediation to reduce the risk from the hazard, (2) do nothing, or (3) take a relatively inexpensive measurement to assess the risk and use this information to decide whether to (a) remediate or (b) do nothing. This decision can often be made at the individual, household, or community level. Performing this decision analysis requires estimates for the risks. Given the hierarchical nature of the decision-making units—individuals are grouped within households which are grouped within counties, and so forth—it is natural to apply a hierarchical modeling approach to estimate the risks.

We illustrate with the example of risks and remediation for home radon exposure. We provide a fair amount of background detail to make sure that the context of the decision analysis is clear.

Background

Radon is a carcinogen—a naturally occurring radioactive gas whose decay products are also radioactive—known to cause lung cancer in high concentration, and estimated to cause several thousand lung cancer deaths per year in the U.S. Figure 22.4 shows the estimated additional lifetime risk of lung cancer death for male and female smokers and nonsmokers, as a function of average radon exposure. At high levels of exposure, the risks are large, and even the risks at low exposures are not trivial when multiplied by the millions of people affected.

The distribution of annual-average living area home radon concentrations in U.S. houses, as measured by a national survey (described in more detail below), is approximately lognormal with geometric mean 0.67 pCi/L and geometric standard deviation 3.1 (the median of this distribution is 0.67 pCi/L and the mean is 1.3 pCi/L). The vast majority of houses in the U.S. do not have high radon levels: about 84% have concentrations under 2 pCi/L, and about 90% have concentrations below 3 pCi/L. However, the survey data suggest that between 50,000 and 100,000 homes have radon concentrations in primary living space in excess of 20 pCi/L. This level causes an annual radiation exposure roughly equal to the occupational exposure limit for uranium miners.

Our decision problem includes as one option measuring the radon concentration and using this information to help decide whether to take steps to reduce the risk from radon. The most frequently used measurement protocol in the U.S. has been the 'screening' measurement: a short-term (2–7 day) charcoal-canister measurement made on the lowest level of the home (often an unoccupied basement), at a cost of about $15 to $20. Because they are usually made on the lowest level of the home (where radon levels are highest), short-term measurements are upwardly biased measures of annual living area average radon level. The magnitude of this bias varies by season and by region of the country and depends on whether the basement (if any) is used as living space. After correcting for biases, short-term measurements in a house have approximate lognormal distributions with geometric standard deviation of roughly 1.8.

A radon measure that is far less common than the screening measurement, but is much better for evaluating radon risk, is a 12-month integrated measurement of the radon concentration. These long-term observations directly measure the annual living-area average radon concentration, with a geometric standard deviation of about 1.2, at a cost of about $50. In the discussion below we find that long-term measurements are more effective, in a cost-benefit sense, than short-term measurements.

If the radon level in a home is sufficiently high, then an individual may take action to control the risk due to radon. Several radon control or remediation techniques have been developed, tested, and implemented. The currently preferred remediation method for most homes, 'sub-slab depressurization,' seals the floors and increases ventilation, at a cost of about $2000, including additional heating and cooling costs. Studies suggest that almost all homes can be remediated to below 4 pCi/L, while reductions under 1 pCi/L are rarely attained with conventional methods. For simplicity, we make the assumption that remediation will reduce radon concentration to 2 pCi/L. For obvious reasons, little is known about effects of remediation on houses that already have low radon levels; we assume that if the initial annual living area average level is less than 2 pCi/L, then remediation will have no effect.

The individual decision problem

We consider the individual homeowner to have three options:

1. Remediate without monitoring: spend the $2000 to remediate the home and reduce radon exposure to 2 pCi/L.

2. Do nothing and accept the current radon exposure.

3. Take a long-term measurement of your home at a cost of $50. Based on the result of the measurement, decide whether to remediate or do nothing.

As described above, a short-term measurement is another possibility, but in our analysis we find this not to be cost-effective.

The measurement/remediation decision must generally be made under uncertainty, because most houses have not been measured for radon. Even after measurement, the radon level is not known exactly—just as in the cancer treatment example in Section 22.3, the cancer status is not perfectly known even after the test. The decision analysis thus presents two challenges: first, deciding whether to remediate if the radon exposure were known; and second, deciding whether it is worth it to measure radon exposure given the current state of knowledge about home radon—that is, given the homeowner's prior distribution. This prior distribution is not a subjective quantity; rather, we determine it by a hierarchical analysis of a national sample of radon measurements, as we discuss below.

Decision-making under certainty

Before performing the statistical analysis, we investigate the optimal decision for the homeowner with a known radon exposure. The problem is difficult because it trades off dollars and lives.

We express decisions under certainty in terms of three quantities, equivalent under a linear no-threshold dose-response relationship:

1. D_d, the dollar value associated with a reduction of 10^{-6} in probability of death from lung cancer (essentially the value of a 'microlife');

2. D_r, the dollar value associated with a reduction of 1 pCi/L in home radon level for a 30-year period;

3. R_{action}, the home radon level above which you should remediate if your radon level is known.

The dollar value of radon reduction, D_r, depends on the number of lives (or microlives) saved by a drop in the radon level. This in turn depends on a variety of factors including the number, gender and smoking status of household occupants as well as the decrease in cancer risk due to the decrease in radon exposure. We do not discuss the details of such a calculation here but only report that for a 'typical' U.S. household (one with an average number of male and female smokers and nonsmokers) $D_r = \$4800\,D_d$. The appropriate radon level to act upon, R_{action}, depends on the dollar value of radon reduction and the benefits of remediation. We assume that remediation takes a house's annual-average living-area radon level down to a level $R_{\text{remed}} = 2$ pCi/L if it was above that, but leaves it unchanged if it was below that. Then the action level is determined as the value at which the benefit of remediation ($\$D_r(R_{\text{action}} - R_{\text{remed}})$) is equal to the cost ($\$2000$),

$$R_{\text{action}} = \frac{\$2000}{D_r} + R_{\text{remed}}. \qquad (22.6)$$

The U.S., English, Swedish, and Canadian recommended remediation levels are $R_{\text{action}} = 4$, 5, 10, and 20 pCi/L, which, with $R_{\text{remed}} = 2$ pCi/L, correspond to equivalent costs per pCi/L of $D_r = \$1000$, $\$670$, $\$250$, and $\$111$, respectively. For an average U.S. household this implies dollar values per microlife of $D_d = \$0.21$, $\$0.14$, $\$0.05$, and $\$0.02$, respectively.

From the risk assessment literature, typical values of D_d for medical interventions are in the range $\$0.10$ to $\$0.50$. Higher values are often attached to life in other contexts (for example, jury awards for deaths due to negligence). The lower values seem reasonable in this case because radon remediation, like medical intervention, is voluntary and addresses reduction of future risk rather than compensation for current loss.

With these as a comparison, the U.S. and English recommendations for radon action levels correspond to the low end of the range of acceptable risk-reduction expenditures. The Canadian and Swedish recommendations are relatively cavalier about the radon risk, in the sense that the implied dollar value per microlife is lower than ordinarily assumed for other risks.

Our calculation (which assumes an average U.S. household) obscures dramatic differences among individual households. For example, a household of one male nonsmoker and one female nonsmoker that is willing to spend $\$0.21$ per person to reduce the probability of lung cancer by 10^{-6} (so that $D_d = \$0.21$) should spend $\$370$ per pCi/L of radon reduction because their risk of lung cancer is less than for the average U.S. household. As a result, a suitable action level for such a household is $R_{\text{action}} = 7.4$ pCi/L, which can be compared to $R_{\text{action}} = 4$ for the average household. In contrast, if the male and female are both smokers, they should be willing to spend the much higher value of $\$1900$ per pCi/L, because of their higher risk of lung cancer, and thus should have an action level of $R_{\text{action}} = 3.1$ pCi/L.

Other sources of variation in R_{action} across households, in addition to household composition, are (a) variation in risk preferences, (b) variation in individual beliefs about the risks of radon and the effects of remediation, and (c) variation in the perceived dollar value associated with a given risk reduction. Through the rest of our analysis we use $R_{\text{action}} = 4$ pCi/L as an exemplary value, but rational informed individuals might plausibly choose quite different values of R_{action}, depending on financial resources, general risk tolerance, attitude towards radon risk, as well as the number of people in the household and their smoking habits.

Bayesian inference for county radon levels

The previous discussion concerns decision making under certainty. Individual homeowners are likely to have limited information about the radon exposure level for their home. A goal of some researchers has been to identify locations and predictive variables associated with high-radon homes so that monitoring and remediation programs can be focused efficiently.

Two data sets are readily available for such a study:

- Long-term measurements from approximately 5000 houses, selected as a cluster sample from 125 randomly selected counties.

- Short-term measurements from about 80,000 houses, sampled at random from all the counties in the U.S.

This is a pattern we sometimes see: a relatively small amount of accurate data, along with a large amount of biased and imprecise data. The challenge is to use the good data to calibrate the bad data, so that inference can be made about the entire country, not merely the 125 counties in the sample of long-term measurements.

Hierarchical model. We simultaneously calibrate the data and predict radon levels by fitting a hierarchical model to both sets of measurements, using predictors at the house and county level, with a separate model fit to each of the 10 regions of the U.S. Let y_i denote the logarithm of the radon measurement of house i within county $j(i)$ and X denote a matrix of household-level predictors including indicators for whether house has a basement and whether the basement is a living area, along with an indicator variable that equals 1 if measurement i is a short-term screening measurement. Including the indicator corrects for the biases in the screening measurements. We assume a normal linear regression model,

$$y_i \sim N((X\beta)_i + \alpha_{j(i)}, \sigma_i^2), \text{ for houses } i = 1, \ldots, n,$$

where $\alpha_{j(i)}$ is a county effect, and the data-level variance parameter σ_i can take on two possible values depending on whether measurement i is long- or short-term.

The county parameters α_j are also assumed normally distributed,

$$\alpha_j \sim N((W\gamma)_j + \delta_{k(j)}, \tau^2), \text{ for counties } j = 1, \ldots, J,$$

with county-level predictors W including climate data and a measure of the uranium level in the soil, and the indicator $\delta_{k(j)}$ characterizing the geology of county j as one of $K = 19$ types. Finally, the coefficients δ_k for the 19 geological types are themselves estimated from the data,

$$\delta_k \sim N(0, \kappa^2), \text{ for geologic types } k = 1, \ldots, K,$$

as are the hierarchical variance components τ and κ. Finally, we divide the country into ten regions and fit the model separately within each region.

Combining long- and short-term measurements allows us to estimate the distribution of radon levels in nearly every county in the U.S., albeit with widely varying uncertainties depending primarily on the number of houses in the sample within the county.

Computation using the Gibbs sampler. The model can easily be computed using the Gibbs sampler, alternately updating $\alpha, \beta, \gamma, \delta$, and the variance components. The conditional posterior distribution for each set of linear parameters is a simple regression, and the variances each have inverse-χ^2 conditional posterior distributions. We use the scalar updating algorithm with transformations as described in Section 15.4.

Inferences. Unfortunately (from the standpoint of radon mitigation programs), indoor radon concentrations are highly variable even within small areas. Given the predictors included in the model, the radon level of an individual house in a specified county can be predicted only to within a factor of at best about 1.9 (that is to say, the posterior geometric standard deviation is about 1.9), with a factor of 2.3 being more typical, a disappointingly large predictive uncertainty considering the factor of 3.1 that would hold given no information on the home other than that it is in the U.S. On the other hand, this seemingly modest reduction in uncertainty is still enough to identify some areas where high-radon homes are very rare or very common. For instance, in the mid-Atlantic states, more than half the houses in some counties have long-term living area concentrations over the EPA's recommended action level of 4 pCi/L, whereas in other counties fewer than one-half of one percent exceed that level.

Bayesian inference for the radon level in an individual house

We apply the fitted hierarchical regression model to perform inferences and decision analyses for previously unmeasured houses i, using the following notation:

$$
\begin{aligned}
R_i &= \text{radon concentration in house } i \\
\theta_i &= \log(R_i).
\end{aligned}
$$

For the decision analysis for house i, we need the posterior predictive distribution for given θ_i, averaging over the posterior uncertainties in regression coefficients, county effects, and variance components; it will be approximately normal (because the variance components are so well estimated), and we label it as

$$
\theta_i \sim N(M_i, S_i^2), \tag{22.7}
$$

where M_i and S_i are computed from the posterior simulations of the model estimation. The mean is $M_i = (X_i \hat{\beta}) + \hat{\alpha}_{j(i)}$, where X_i is a row vector containing the house-level predictors (indicators for whether the house has a basement and whether the basement is a living area) for house i, and $(\hat{\beta}, \hat{\alpha})$ are the posterior means from the analysis in the appropriate region of the country. The variance S_i^2 includes the posterior uncertainty in the coefficients α, β and also the hierarchical variance components τ^2 and κ^2. (We are predicting actual radon levels, not measurements, and so σ^2 does not play a role here.) It turns out that the geometric standard deviations e^S of the predictive distributions for home radon levels vary from 2.1 to 3.0, and they are in the range $(2.1, 2.5)$ for most U.S. houses. (The houses with $e^S > 2.5$ lie in small-population counties for which little information was available in the radon surveys, resulting in relatively high predictive uncertainty within these counties.) The geometric means of the house predictive distributions, e^M, vary from 0.1 to 14.6 pCi/L, with 95% in the range $[0.3, 3.7]$ and 50% in the range $[0.6, 1.6]$. The houses with the highest predictive geometric means are houses with basement living areas in high-radon counties; the houses with lowest predictive geometric means have no basements and lie in low-radon counties.

The distribution (22.7) summarizes the state of knowledge about the radon level in a house given only its county and basement information. In this respect it serves as a prior distribution for the homeowner. Now suppose a measurement $y \sim N(\theta, \sigma^2)$ is taken in the house. (We are assuming an unbiased measurement. If a short-term measurement is being used, it will have to be corrected for the biases which were estimated in the regression models.) In our notation, y and θ are the logarithms of the measurement and the true home radon level, respectively. The posterior distribution for θ is

$$\theta | M, y \sim N(\Lambda, V), \tag{22.8}$$

where

$$\Lambda = \frac{\frac{M}{S^2} + \frac{y}{\sigma^2}}{\frac{1}{S^2} + \frac{1}{\sigma^2}} \qquad V = \frac{1}{\frac{1}{S^2} + \frac{1}{\sigma^2}}. \tag{22.9}$$

We base our decision analysis of when to measure and when to remediate on the distributions (22.7) and (22.8).

Decision analysis for individual homeowners

We now work out the optimal decisions of measurement and remediation conditional on the predicted radon level in a home, the additional risk of lung cancer death from radon, the effects of remediation, and individual attitude toward risk.

Given an action level under certainty, R_{action}, we address the question of whether to pay for a home radon measurement and whether to remediate. The decision of whether to measure depends on the prior distribution (22.7) of radon level for your house, given your predictors X. We use the term 'prior distribution' to refer to the predictive distribution based on our hierarchical model; the predictive distribution conditions on the survey data but is prior to any specific measurements for the house being considered. The decision of whether to remediate depends on the posterior distribution (22.8) if a measurement has been taken or the prior distribution (22.7) otherwise. In our computations, we use the following results from the normal distribution: if $z \sim N(\mu, s^2)$, then $E(e^z) = e^{\mu + \frac{1}{2}s^2}$ and $E(e^z | z > a) \Pr(z > a) =$

$e^{\mu + \frac{1}{2}s^2}(1 - \Phi(\frac{\mu + s^2 - a}{s}))$, where Φ is the standard normal cumulative distribution function.

The decision tree is set up with three branches. In each branch, we evaluate the expected loss in dollar terms, converting radon exposure (over a 30-year period) to dollars using $D_r = \$2000/(R_{\text{action}} - R_{\text{remed}})$ as the equivalent cost per pCi/L for additional home radon exposure. In the expressions below we let $R = e^\theta$ be the unknown radon exposure level in the home being considered; the prior and posterior distributions are normal distributions for $\theta = \log R$.

1. **Remediate without monitoring.** Expected loss is remediation cost + equivalent dollar cost of radon exposure after remediation:

$$L_1 = \$2000 + D_r E(\min(R, R_{\text{remed}}))$$
$$= \$2000 + D_r \left[R_{\text{remed}} \Pr(R \geq R_{\text{remed}}) + E(R|R < R_{\text{remed}}) \Pr(R < R_{\text{remed}}) \right]$$
$$= \$2000 + D_r \left[R_{\text{remed}} \Phi(\frac{M - \log(R_{\text{remed}})}{S}) + \right.$$
$$\left. + e^{M + \frac{1}{2}S^2} \left(1 - \Phi \left(\frac{M + S^2 - \log(R_{\text{remed}})}{S} \right) \right) \right]. \quad (22.10)$$

2. **Do not monitor or remediate.** Expected loss is the equivalent dollar cost of radon exposure:

$$L_2 = D_r E(R) = D_r e^{M + \frac{1}{2}S^2}. \quad (22.11)$$

3. **Take a measurement y (measured in log pCi/L).** The immediate loss is the measurement cost (assumed to be $50) and, in addition, the radon exposure during the year that you are taking the measurement (which is $\frac{1}{30}$ of the 30-year exposure (22.11)). The inner decision has two branches:

(a) **Remediate.** Expected loss is the immediate loss due to measurement plus the remediation loss which is computed as for decision 1, but using the posterior rather than the prior distribution:

$$L_{3a} = \$50 + D_r \frac{1}{30} e^{M + \frac{1}{2}S^2} + \$2000 + D_r \left[R_{\text{remed}} \Phi \left(\frac{\Lambda - \log(R_{\text{remed}})}{\sqrt{V}} \right) \right.$$
$$\left. + e^{\Lambda + \frac{1}{2}V} \left(1 - \Phi \left(\frac{\Lambda + V - \log(R_{\text{remed}})}{\sqrt{V}} \right) \right) \right], \quad (22.12)$$

where Λ and V are the posterior mean and variance from (22.9).

(b) **Do not remediate.** Expected loss is:

$$L_{3b} = \$50 + D_r \frac{1}{30} e^{M + \frac{1}{2}S^2} + D_r e^{\Lambda + \frac{1}{2}V}. \quad (22.13)$$

Deciding whether to remediate given a measurement. To evaluate the decision tree, we must first consider the inner decision between 3(a) and 3(b), conditional on the measurement y. Let y_0 be the point (on the logarithmic scale) at which you will choose to remediate if $y > y_0$, or do nothing if $y < y_0$. (Because of measurement error, $y \neq \theta$, and consequently $y_0 \neq \log(R_{\text{action}})$.)

We determine y_0, which depends on the prior mean M, the prior standard deviation S, and the measurement standard deviation σ, by numerically solving the implicit equation

$$L_{3a} = L_{3b} \text{ at } y = y_0. \tag{22.14}$$

Details of our approach for solving the equation are not provided here.

Deciding among the three branches. The expected loss for immediate remediation (22.10) and the expected loss for no action (22.11) can be determined directly for a given prior mean M, prior standard deviation S, and specified dollar value D_r for radon reduction. We determine the expected loss for branch 3 of the decision tree,

$$L_3 = \mathrm{E}(\min(L_{3a}, L_{3b})), \tag{22.15}$$

by averaging over the prior uncertainty in the measurement y (given a value for the measurement variability σ) as follows.

1. Simulate 5000 draws of $y \sim \mathrm{N}(M, S^2 + \sigma^2)$.

2. For each draw of y, compute $\min(L_{3a}, L_{3b})$ from (22.12) and (22.13).

3. Estimate L_3 as the average of these 5000 values.

Of course, this expected loss is valid only if we assume that you will make the recommended optimal decision once the measurement is taken.

We can now compare the expected losses L_1, L_2, L_3, and choose among the three decisions. The recommended decision is the one with the lowest expected loss. An individual homeowner can apply this approach simply by specifying R_{action} (the decision threshold under certainty), looking up the prior mean and standard deviation for the home's radon level as estimated by the hierarchical model, and determining the optimal decision. In addition, our approach makes it possible for a homeowner to take account any additional information that is available. For example, if a measurement is available for a neighbor's house, then one can update the prior mean and standard deviation to include this information.

If we are willing to make the simplifying assumption that $\sigma = \log(1.2)$ and $S = \log(2.3)$ for all counties, then we can summarize the decision recommendations by giving threshold levels M_{low} and M_{high} for which decision 1 (remediate immediately) is preferred if $M > M_{\text{high}}$, decision 2 (do not monitor or remediate) is preferred if $M < M_{\text{low}}$, and decision 3 (take a measurement) is preferred if $M \in [M_{\text{low}}, M_{\text{high}}]$. Figure 22.5 displays these cutoffs as a function of R_{action}, and thus displays the recommended decision as a function of (R_{action}, e^M). For example, setting $R_{\text{action}} = 4$ pCi/L leads to the following recommendation based on e^M, the prior GM of your home radon based on your county and house type:

- If e^M is less than 1.0 pCi/L (which corresponds to 68% of U.S. houses), do nothing.

- If e^M is between 1.0 and 3.5 pCi/L (27% of U.S. houses), perform a long-term measurement (and then decide whether to remediate).

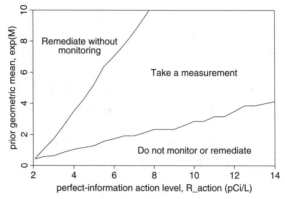

Figure 22.5 *Recommended radon remediation/measurement decision as a function of the perfect-information action level R_{action} and the prior geometric mean radon level e^M, under the simplifying assumption that $e^S = 2.3$. You can read off your recommended decision from this graph and, if the recommendation is 'take a measurement,' you can do so and then perform the calculations to determine whether to remediate, given your measurement. The horizontal axis of this figure begins at 2 pCi/L because remediation is assumed to reduce home radon level to 2 pCi/L, so it makes no sense for R_{action} to be lower than that value. Wiggles in the lines are due to simulation variability.*

- If e^M is greater than 3.5 pCi/L (5% of U.S. houses), remediate immediately without measuring. Actually, in this circumstance, short-term monitoring can turn out to be (barely) cost-effective if we include it as an option. We ignore this additional complexity to the decision tree, since it occurs rarely and has very little impact on the overall cost-benefit analysis.

Aggregate consequences of individual decisions

Now that we have made idealized recommendations for individual homeowners, we consider the aggregate effects if the recommendations are followed by all homeowners in the U.S. In particular, we compare the consequences of individuals following our recommendations to the consequences of other policies such that implicitly recommended by the EPA, of taking a short-term measurement as a condition of a home sale and performing remediation if the measurement exceeds 4 pCi/L.

Applying the recommended decision strategy to the entire country. Figure 22.6 displays the geographic pattern of recommended measurements (and, after one year, recommended remediations), based on an action level R_{action} of 4 pCi/L. Each county is shaded according to the proportion of houses for which measurement (and then remediation) is recommended. These recommendations incorporate the effects of parameter uncertainties in the models that predict radon distributions within counties, so these maps would be expected to change somewhat as better predictions become available.

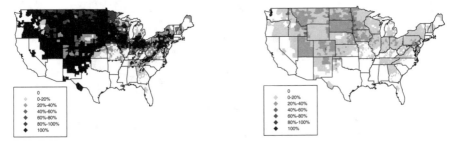

Figure 22.6 *Maps showing (a) fraction of houses in each county for which measurement is recommended, given the perfect-information action level of $R_{action} = 4$ pCi/L; (b) expected fraction of houses in each county for which remediation will be recommended, once the measurement y has been taken. For the present radon model, within any county the recommendations on whether to measure and whether to remediate depend only on the house type: whether the house has a basement and whether the basement is used as living space. Apparent discontinuities across the boundaries of Utah and South Carolina arise from irregularities in the radon measurements from the radon surveys conducted by those states, an issue we ignore here.*

From a policy standpoint, perhaps the most significant feature of the maps is that even if the EPA's recommended action level of 4 pCi/L is assumed to be correct—and, as we have discussed, it does lead to a reasonable value of D_d, under standard dose-response assumptions—monitoring is still not recommended in most U.S. homes. Indeed, only 28% of U.S. homes would perform radon monitoring. A higher action level of 8 pCi/L, a reasonable value for nonsmokers under the standard assumptions, would lead to even more restricted monitoring and remediation: only about 5% of homes would perform monitoring.

Evaluation of different decision strategies. We estimate the total monetary cost and lives saved if each of the following decision strategies were to be applied nationally:

1. The recommended strategy from the decision analysis (that is, monitor homes with prior mean estimates above a given level, and remediate those with high measurements).

2. Performing long-term measurements on all houses and then remediating those for which the measurement exceeds the specified radon action level R_{action}.

3. Performing short-term measurements on all houses and then remediating those for which the bias-corrected measurement exceeds the specified radon action level R_{action} (with the bias estimated from the hierarchical regression model).

4. Performing short-term measurements on all houses and then remediating those for which the uncorrected measurement exceeds the specified radon action level R_{action}.

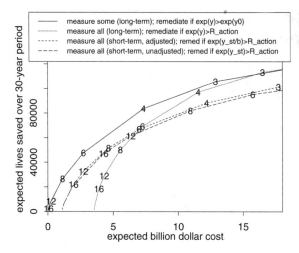

Figure 22.7 *Expected lives saved vs. expected cost for various radon measurement/ remediation strategies. Numbers indicate values of R_{action}. The solid line is for the recommended strategy of measuring only certain homes; the others assume that all homes are measured. All results are estimated totals for the U.S. over a 30-year period.*

We evaluate each of the above strategies in terms of aggregate lives saved and dollars cost, with these outcomes parameterized by the radon action level R_{action}. Both lives saved and costs are considered for a 30-year period. For each strategy, we assume that the level R_{action} is the same for all houses (this would correspond to a uniform national recommendation). To compute the lives saved, we assume that the household composition for each house is the same as the average in the U.S. We evaluate the expected cost and the expected number of lives saved by aggregating over the decisions for individual homes in the country. In practice for our model we need only consider three house types defined by our predictors (no basement, basement is not living space, basement is living space) for each of the 3078 counties.

We describe the results of our expected cost and expected lives saved calculation in some detail only for the decision strategy based on our hierarchical model. If strategy were followed everywhere with $R_{action} = 4$ pCi/L (as pictured in the maps in Figure 22.6), about 26% of the 70 million ground-contact houses in the U.S. would monitor and about 4.5% would remediate. The houses being remediated include 2.8 million homes with radon levels above 4 pCi/L (74% of all such homes), and 840,000 of the homes above 8 pCi/L (91% of all such homes). The total monetary cost is estimated at $7.3 billion—$1 billion for measurement and $6.3 billion for remediation—and would be expected to save the lives of 49,000 smokers and 35,000 nonsmokers over a 30-year period. Total cost and total lives saved for other action levels and other decision strategies are calculated in the same way.

Figure 22.7 displays the tradeoff between expected cost and expected lives saved over a thirty-year period for the four strategies. The numbers on the curves are action levels, R_{action}. This figure allows us to compare the effectiveness of alternative strategies of equal expected cost or equal expected lives saved. For example, the recommended strategy (the solid line on the graph) at $R_{action} = 4$ pCi/L would result in an expected 83,000 lives saved at an expected cost of \$7.3 billion. Let us compare this to the EPA's implicitly recommended strategy based on uncorrected short-term measurements (the dashed line on the figure). For the same cost of \$7.3 billion, the uncorrected short-term strategy is expected to save only 64,000 lives; to achieve the same expected savings of 83,000 lives, the uncorrected short-term strategy would cost about \$12 billion.

22.5 Personal vs. institutional decision analysis

Statistical inference has an ambiguous role in decision making. Under a 'subjective' view of probability (which we do not generally find useful; see Sections 1.5–1.7), posterior inferences represent the personal beliefs of the analyst, given his or her prior information and data. These can then be combined with a subjective utility function and input into a decision tree to determine the optimal decision, or sequence of decisions, so as to maximize subjective expected utility. This approach has serious drawbacks as a procedure for personal decision making, however. It can be more difficult to define a utility function and subjective probabilities than to simply choose the most appealing decision. The formal decision-making procedure has an element of circular reasoning, in that one can typically come to any desired decision by appropriately setting the subjective inputs to the analysis.

In practice, then, personal decision analysis is most useful when the inputs (utilities and probabilities) are well defined. For example, in the cancer screening example discussed in Section 22.3, the utility function is noncontroversial— years of life, with a slight adjustment for quality of life—and the relevant probabilities are estimated from the medical literature. Bayesian decision analysis then serves as a mathematical tool for calculating the expected value of the information that would come from the screening.

In institutional settings—for example, businesses, governments, or research organizations—decisions need to be justified, and formal decision analysis has a role to play in clarifying the relation between the assumptions required to build and apply a relevant probability model and the resulting estimates of costs and benefits. We introduce the term *institutional decision analysis* to refer to the process of transparently setting up a probability model, utility function, and an inferential framework leading to cost estimates and decision recommendations. Depending on the institutional setting, the decision analysis can be formalized to different extents. For example, the meta-analysis in Section 22.2 leads to fairly open-ended recommendations about incentives for sample surveys—given the high levels of posterior uncertainties, it would not make sense to give a single recommendation, since it would be so sensitive to the assumptions about the relative utility of dollars and response rate. For the cancer-screening example in Section 22.3, the decision analysis is potentially useful both for its direct recommendation (not to perform bronchoscopy for this sort of patient) and also because it can be taken apart to reveal the sensitivity of the conclusion to the different assumptions taken from the medical literature on probabilities and expected years of life.

In contrast, the key assumptions in the hierarchical decision analysis for radon exposure in Section 22.4 have to do with cost-benefit tradeoffs. By making a particular assumption about the relative importance of dollars and cancer risk (corresponding to $R_{\text{action}} = 4$ pCi/L), we can make specific recommendations by county (see the maps on page 565). It would be silly to believe that all households in the United States have utility functions equivalent to this constant level of R_{action}, but the analysis resulting in the maps is useful to give a sense of a uniform recommendation that could be made by the government.

In general, there are many ways in which statistical inferences can be used to inform decision-making. The essence of the 'objective' or 'institutional' Bayesian approach is to clearly identify the model assumptions and data used to form the inferences, evaluate the reasonableness and the fit of the model's predictions (which include decision recommendations as a special case), and then expand the model as appropriate to be more realistic. The most useful model expansions are typically those that allow more information to be incorporated into the inferences.

22.6 Bibliographic note

Berger (1985) and DeGroot (1970) both give clear presentations of the theoretical issues in decision theory and the connection to Bayesian inference. Many introductory books have been written on the topic; Luce and Raiffa (1957) is particularly interesting for its wide-ranging discussions. Savage (1954) is an influential early work that justifies Bayesian statistical methods in terms of decision theory.

Clemen (1996) provides a thorough introduction to applied decision analysis. Parmigiani (2002) is a textbook on medical decision making from a Bayesian perspective. The articles in Kahneman, Slovic, and Tversky (1982) and Gilovich, Griffin, and Kahneman (2002) address many of the component problems in decision analysis from a psychological perspective.

The decision analysis for incentives in telephone surveys appears in Gelman, Stevens, and Chan (2003). The meta-analysis data were collected by Singer et al. (1999), and Groves (1989) discusses many practical issues in sampling, including the effects of incentives in mail surveys. More generally, Dehejia (2003) discusses the connection between decision analysis and causal inference in models with interactions.

Parmigiani (2003) discusses the value of information in medical diagnostics.

Parmigiani et al. (1999) and the accompanying discussions include several perspectives on Bayesian inference in medical decision making for breast cancer screening. The cancer screening example in Section 22.3 is adapted and simplified from Moroff and Pauker (1983). The journal *Medical Decision Making*, where this article appears, contains many interesting examples and discussions of applied decision analysis. Heitjan, Moskowitz, and Whang (1999) discuss Bayesian inference for cost-effectiveness of medical treatments.

The radon problem is described by Lin et al. (1999), and the individual decision analysis based on the estimated hierarchical model is accessible at the website `http://www.stat.columbia.edu/radon/`. Boscardin (1996) describes the computations for the hierarchical model for the radon example in more detail. Ford et al. (1999) presents a cost-benefit analysis of the radon problem without using a hierarchical model.

22.7 Exercises

1. Conditional probability and elementary decision theory: Oscar has lost his dog; there is a 70% probability it is in forest A and a 30% chance it is in forest B. If the dog is in forest A and Oscar looks there for a day, he has a 50% chance of finding the dog. If the dog is in forest B and Oscar looks there for a day, he has an 80% chance of finding the dog.

   (a) If Oscar can search only one forest for a day, where should he look to maximize his probability of finding the dog? What is the probability that the dog is still lost after the search?

   (b) Assume Oscar made the rational decision and the dog is still lost (and is still in the same forest as yesterday). Where should he search for the dog on the second day? What is the probability that the dog is still lost at the end of the second day?

   (c) Again assume Oscar makes the rational decision on the second day and the dog is still lost (and is still in the same forest). Where should he search on the third day? What is the probability that the dog is still lost at the end of the third day?

   (d) (Expected value of additional information.) You will now figure out the expected value of knowing, at the beginning, which forest the dog is in. Suppose Oscar will search for at most three days, with the following payoffs: -1 if the dog is found in one day, -2 if the dog is found on the second day, and -3 if the dog is found on the third day, and -10 otherwise.

      i. What is Oscar's expected payoff without the additional information?
      ii. What is Oscar's expected payoff if he knows the dog is in forest A?
      iii. What is Oscar's expected payoff if he knows the dog is in forest B?
      iv. Before the search begins, how much should Oscar be willing to pay to be told which forest his dog is in?

2. Decision analysis:

   (a) Formulate an example from earlier in this book as a decision problem. (For example, in the bioassay example of Section 3.7, there can be a cost of setting up a new experiment, a cost per rat in the experiment, and a benefit to estimating the dose-response curve more accurately. Similarly, in the meta-analysis example in Section 5.6, there can be a cost per study, a cost per patient in the study, and a benefit to accurately estimating the efficacy of beta-blockers.)

   (b) Set up a utility function and determine the expected utility for each decision option within the framework you have set up.

   (c) Explore the sensitivity of the results of your decision analysis to the assumptions you have made in setting up the decision problem.

Appendixes

Standard probability distributions

A.1 Introduction

Tables A.1 and A.2 present standard notation, probability density functions, parameter descriptions, means, modes, and standard deviations for standard probability distributions. The rest of this appendix provides additional information including typical areas of application and methods for simulation.

We use the standard notation θ for the random variable (or random vector), except in the case of the Wishart and inverse-Wishart, for which we use W for the random matrix. The parameters are given conventional labels; all probability distributions are implicitly conditional on the parameters. Most of the distributions here are simple univariate distributions. The multivariate normal and related Wishart and multivariate t, and the multinomial and related Dirichlet distributions, are the principal exceptions. Realistic distributions for complicated multivariate models, including hierarchical and mixture models, can usually be constructed from these building blocks.

For simulating random variables from these distributions, we assume that a computer subroutine or command is available that generates pseudorandom samples from the uniform distribution on the unit interval. Some care must be taken to ensure that the pseudorandom samples from the uniform distribution are appropriate for the task at hand. For example, a sequence may appear uniform in one dimension while m-tuples are not randomly scattered in m dimensions. Many statistical software packages are available for simulating random deviates from the distributions presented here.

A.2 Continuous distributions

Uniform

The uniform distribution is used to represent a variable that is known to lie in an interval and equally likely to be found anywhere in the interval. A noninformative distribution is obtained in the limit as $a \to -\infty$, $b \to \infty$. If u is drawn from a standard uniform distribution $U(0, 1)$, then $\theta = a + (b - a)u$ is a draw from $U(a, b)$.

Table A.1   **Continuous distributions**

Distribution	Notation	Parameters
Uniform	$\theta \sim U(\alpha, \beta)$ $p(\theta) = U(\theta \mid \alpha, \beta)$	boundaries α, β with $\beta > \alpha$
Normal	$\theta \sim N(\mu, \sigma^2)$ $p(\theta) = N(\theta \mid \mu, \sigma^2)$	location μ scale $\sigma > 0$
Multivariate normal	$\theta \sim N(\mu, \Sigma)$ $p(\theta) = N(\theta \mid \mu, \Sigma)$ (implicit dimension d)	symmetric, pos. definite, $d \times d$ variance matrix Σ
Gamma	$\theta \sim \text{Gamma}(\alpha, \beta)$ $p(\theta) = \text{Gamma}(\theta \mid \alpha, \beta)$	shape $\alpha > 0$ inverse scale $\beta > 0$
Inverse-gamma	$\theta \sim \text{Inv-gamma}(\alpha, \beta)$ $p(\theta) = \text{Inv-gamma}(\theta \mid \alpha, \beta)$	shape $\alpha > 0$ scale $\beta > 0$
Chi-square	$\theta \sim \chi^2_\nu$ $p(\theta) = \chi^2_\nu(\theta)$	degrees of freedom $\nu > 0$
Inverse-chi-square	$\theta \sim \text{Inv-}\chi^2_\nu$ $p(\theta) = \text{Inv-}\chi^2_\nu(\theta)$	degrees of freedom $\nu > 0$
Scaled inverse-chi-square	$\theta \sim \text{Inv-}\chi^2(\nu, s^2)$ $p(\theta) = \text{Inv-}\chi^2(\theta \mid \nu, s^2)$	degrees of freedom $\nu > 0$ scale $s > 0$
Exponential	$\theta \sim \text{Expon}(\beta)$ $p(\theta) = \text{Expon}(\theta \mid \beta)$	inverse scale $\beta > 0$
Wishart	$W \sim \text{Wishart}_\nu(S)$ $p(W) = \text{Wishart}_\nu(W \mid S)$ (implicit dimension $k \times k$)	degrees of freedom ν symmetric, pos. definite $k \times k$ scale matrix S
Inverse-Wishart	$W \sim \text{Inv-Wishart}_\nu(S^{-1})$ $p(W) = \text{Inv-Wishart}_\nu(W \mid S^{-1})$ (implicit dimension $k \times k$)	degrees of freedom ν symmetric, pos. definite $k \times k$ scale matrix S

Density function	Mean, variance, and mode
$p(\theta) = \frac{1}{\beta-\alpha}, \ \theta \in [\alpha, \beta]$	$\mathrm{E}(\theta) = \frac{\alpha+\beta}{2}, \ \mathrm{var}(\theta) = \frac{(\beta-\alpha)^2}{12}$ no mode
$p(\theta) = \frac{1}{\sqrt{2\pi}\sigma} \exp\left(-\frac{1}{2\sigma^2}(\theta-\mu)^2\right)$	$\mathrm{E}(\theta) = \mu, \ \mathrm{var}(\theta) = \sigma^2$ $\mathrm{mode}(\theta) = \mu$
$p(\theta) = (2\pi)^{-d/2}|\Sigma|^{-1/2}$ $\times \exp\left(-\frac{1}{2}(\theta-\mu)^T\Sigma^{-1}(\theta-\mu)\right)$	$\mathrm{E}(\theta) = \mu, \ \mathrm{var}(\theta) = \Sigma$ $\mathrm{mode}(\theta) = \mu$
$p(\theta) = \frac{\beta^\alpha}{\Gamma(\alpha)}\theta^{\alpha-1}e^{-\beta\theta}, \ \theta > 0$	$\mathrm{E}(\theta) = \frac{\alpha}{\beta}$ $\mathrm{var}(\theta) = \frac{\alpha}{\beta^2}$ $\mathrm{mode}(\theta) = \frac{\alpha-1}{\beta}$, for $\alpha \geq 1$
$p(\theta) = \frac{\beta^\alpha}{\Gamma(\alpha)}\theta^{-(\alpha+1)}e^{-\beta/\theta}, \ \theta > 0$	$\mathrm{E}(\theta) = \frac{\beta}{\alpha-1}$, for $\alpha > 1$ $\mathrm{var}(\theta) = \frac{\beta^2}{(\alpha-1)^2(\alpha-2)}, \alpha > 2$ $\mathrm{mode}(\theta) = \frac{\beta}{\alpha+1}$
$p(\theta) = \frac{2^{-\nu/2}}{\Gamma(\nu/2)}\theta^{\nu/2-1}e^{-\theta/2}, \ \theta > 0$ same as Gamma$(\alpha = \frac{\nu}{2}, \beta = \frac{1}{2})$	$\mathrm{E}(\theta) = \nu, \ \mathrm{var}(\theta) = 2\nu$ $\mathrm{mode}(\theta) = \nu - 2$, for $\nu \geq 2$
$p(\theta) = \frac{2^{-\nu/2}}{\Gamma(\nu/2)}\theta^{-(\nu/2+1)}e^{-1/(2\theta)}, \ \theta > 0$ same as Inv-gamma$(\alpha = \frac{\nu}{2}, \beta = \frac{1}{2})$	$\mathrm{E}(\theta) = \frac{1}{\nu-2}$, for $\nu > 2$ $\mathrm{var}(\theta) = \frac{2}{(\nu-2)^2(\nu-4)}, \nu > 4$ $\mathrm{mode}(\theta) = \frac{1}{\nu+2}$
$p(\theta) = \frac{(\nu/2)^{\nu/2}}{\Gamma(\nu/2)}s^\nu\theta^{-(\nu/2+1)}e^{-\nu s^2/(2\theta)}, \ \theta > 0$ same as Inv-gamma$(\alpha = \frac{\nu}{2}, \beta = \frac{\nu}{2}s^2)$	$\mathrm{E}(\theta) = \frac{\nu}{\nu-2}s^2$ $\mathrm{var}(\theta) = \frac{2\nu^2}{(\nu-2)^2(\nu-4)}s^4$ $\mathrm{mode}(\theta) = \frac{\nu}{\nu+2}s^2$
$p(\theta) = \beta e^{-\beta\theta}, \ \theta > 0$ same as Gamma$(\alpha = 1, \beta)$	$\mathrm{E}(\theta) = \frac{1}{\beta}, \ \mathrm{var}(\theta) = \frac{1}{\beta^2}$ $\mathrm{mode}(\theta) = 0$
$p(W) = \left(2^{\nu k/2}\pi^{k(k-1)/4}\prod_{i=1}^k \Gamma\left(\frac{\nu+1-i}{2}\right)\right)^{-1}$ $\times |S|^{-\nu/2}|W|^{(\nu-k-1)/2}$ $\times \exp\left(-\frac{1}{2}\mathrm{tr}(S^{-1}W)\right)$, W pos. definite	$\mathrm{E}(W) = \nu S$
$p(W) = \left(2^{\nu k/2}\pi^{k(k-1)/4}\prod_{i=1}^k \Gamma\left(\frac{\nu+1-i}{2}\right)\right)^{-1}$ $\times |S|^{\nu/2}|W|^{-(\nu+k+1)/2}$ $\times \exp\left(-\frac{1}{2}\mathrm{tr}(SW^{-1})\right)$, W pos. definite	$\mathrm{E}(W) = (\nu-k-1)^{-1}S$

Table A.1   **Continuous distributions** *continued*

Distribution	Notation	Parameters
Student-t	$\theta \sim t_\nu(\mu, \sigma^2)$ $p(\theta) = t_\nu(\theta\|\mu, \sigma^2)$ t_ν is short for $t_\nu(0, 1)$	degrees of freedom $\nu > 0$ location μ scale $\sigma > 0$
Multivariate Student-t	$\theta \sim t_\nu(\mu, \Sigma)$ $p(\theta) = t_\nu(\theta\|\mu, \Sigma)$ (implicit dimension d)	degrees of freedom $\nu > 0$ location $\mu = (\mu_1, .., \mu_d)$ symmetric, pos. definite $d \times d$ scale matrix Σ
Beta	$\theta \sim \text{Beta}(\alpha, \beta)$ $p(\theta) = \text{Beta}(\theta\|\alpha, \beta)$	'prior sample sizes' $\alpha > 0, \beta > 0$
Dirichlet	$\theta \sim \text{Dirichlet}(\alpha_1, .., \alpha_k)$ $p(\theta) = \text{Dirichlet}(\theta\|\alpha_1, .., \alpha_k)$	'prior sample sizes' $\alpha_j > 0;\ \alpha_0 \equiv \sum_{j=1}^{k} \alpha_j$

Table A.2   **Discrete distributions**

Distribution	Notation	Parameters
Poisson	$\theta \sim \text{Poisson}(\lambda)$ $p(\theta) = \text{Poisson}(\theta\|\lambda)$	'rate' $\lambda > 0$
Binomial	$\theta \sim \text{Bin}(n, p)$ $p(\theta) = \text{Bin}(\theta\|n, p)$	'sample size' n (positive integer) 'probability' $p \in [0, 1]$
Multinomial	$\theta \sim \text{Multin}(n; p_1, .., p_k)$ $p(\theta) = \text{Multin}(\theta\|n; p_1, .., p_k)$	'sample size' n (positive integer) 'probabilities' $p_j \in [0, 1]$; $\sum_{j=1}^{k} p_j = 1$
Negative binomial	$\theta \sim \text{Neg-bin}(\alpha, \beta)$ $p(\theta) = \text{Neg-bin}(\theta\|\alpha, \beta)$	shape $\alpha > 0$ inverse scale $\beta > 0$
Beta-binomial	$\theta \sim \text{Beta-bin}(n, \alpha, \beta)$ $p(\theta) = \text{Beta-bin}(\theta\|n, \alpha, \beta)$	'sample size' n (positive integer) 'prior sample sizes' $\alpha > 0, \beta > 0$

Density function	Mean, variance, and mode
$p(\theta) = \frac{\Gamma((\nu+1)/2)}{\Gamma(\nu/2)\sqrt{\nu\pi}\sigma}(1 + \frac{1}{\nu}(\frac{\theta-\mu}{\sigma})^2)^{-(\nu+1)/2}$	$E(\theta) = \mu$, for $\nu > 1$ $var(\theta) = \frac{\nu}{\nu-2}\sigma^2$, for $\nu > 2$ $mode(\theta) = \mu$
$p(\theta) = \frac{\Gamma((\nu+d)/2)}{\Gamma(\nu/2)\nu^{d/2}\pi^{d/2}}|\Sigma|^{-1/2}$ $\times(1 + \frac{1}{\nu}(\theta-\mu)^T\Sigma^{-1}(\theta-\mu))^{-(\nu+d)/2}$	$E(\theta) = \mu$, for $\nu > 1$ $var(\theta) = \frac{\nu}{\nu-2}\Sigma$, for $\nu > 2$ $mode(\theta) = \mu$
$p(\theta) = \frac{\Gamma(\alpha+\beta)}{\Gamma(\alpha)\Gamma(\beta)}\theta^{\alpha-1}(1-\theta)^{\beta-1}$ $\theta \in [0,1]$	$E(\theta) = \frac{\alpha}{\alpha+\beta}$ $var(\theta) = \frac{\alpha\beta}{(\alpha+\beta)^2(\alpha+\beta+1)}$ $mode(\theta) = \frac{\alpha-1}{\alpha+\beta-2}$
$p(\theta) = \frac{\Gamma(\alpha_1+\cdots+\alpha_k)}{\Gamma(\alpha_1)\cdots\Gamma(\alpha_k)}\theta_1^{\alpha_1-1}\cdots\theta_k^{\alpha_k-1}$ $\theta_1,\ldots,\theta_k \geq 0; \sum_{j=1}^k \theta_j = 1$	$E(\theta_j) = \frac{\alpha_j}{\alpha_0}$ $var(\theta_j) = \frac{\alpha_j(\alpha_0-\alpha_j)}{\alpha_0^2(\alpha_0+1)}$ $cov(\theta_i,\theta_j) = -\frac{\alpha_i\alpha_j}{\alpha_0^2(\alpha_0+1)}$ $mode(\theta_j) = \frac{\alpha_j-1}{\alpha_0-k}$

Density function	Mean, variance, and mode
$p(\theta) = \frac{1}{\theta!}\lambda^\theta \exp(-\lambda)$ $\theta = 0,1,2,\ldots$	$E(\theta) = \lambda$, $var(\theta) = \lambda$ $mode(\theta) = \lfloor\lambda\rfloor$
$p(\theta) = \binom{n}{\theta}p^\theta(1-p)^{n-\theta}$ $\theta = 0,1,2,\ldots,n$	$E(\theta) = np$ $var(\theta) = np(1-p)$ $mode(\theta) = \lfloor(n+1)p\rfloor$
$p(\theta) = \binom{n}{\theta_1\ \theta_2\cdots\theta_k}p_1^{\theta_1}\cdots p_k^{\theta_k}$ $\theta_j = 0,1,2,\ldots,n; \sum_{j=1}^k \theta_j = n$	$E(\theta_j) = np_j$ $var(\theta_j) = np_j(1-p_j)$ $cov(\theta_i,\theta_j) = -np_ip_j$
$p(\theta) = \binom{\theta+\alpha-1}{\alpha-1}\left(\frac{\beta}{\beta+1}\right)^\alpha\left(\frac{1}{\beta+1}\right)^\theta$ $\theta = 0,1,2,\ldots$	$E(\theta) = \frac{\alpha}{\beta}$ $var(\theta) = \frac{\alpha}{\beta^2}(\beta+1)$
$p(\theta) = \frac{\Gamma(n+1)}{\Gamma(\theta+1)\Gamma(n-\theta+1)}\frac{\Gamma(\alpha+\theta)\Gamma(n+\beta-\theta)}{\Gamma(\alpha+\beta+n)}$ $\times\frac{\Gamma(\alpha+\beta)}{\Gamma(\alpha)\Gamma(\beta)},\quad \theta = 0,1,2,\ldots,n$	$E(\theta) = n\frac{\alpha}{\alpha+\beta}$ $var(\theta) = n\frac{\alpha\beta(\alpha+\beta+n)}{(\alpha+\beta)^2(\alpha+\beta+1)}$

Univariate normal

The normal distribution is ubiquitous in statistical work. Sample averages are approximately normally distributed by the central limit theorem. A noninformative or flat distribution is obtained in the limit as the variance $\sigma^2 \to \infty$. The variance is usually restricted to be positive; $\sigma^2 = 0$ corresponds to a point mass at θ. There are no restrictions on θ. The density function is always finite, the integral is finite as long as σ^2 is finite. A subroutine for generating random draws from the standard normal distribution ($\mu = 0, \sigma = 1$) is available in many computer packages. If not, a subroutine to generate standard normal deviates from a stream of uniform deviates can be obtained from a variety of simulation texts; see Section A.4 for some references. If z is a random deviate from the standard normal distribution, then $\theta = \mu + \sigma z$ is a draw from $N(\mu, \sigma^2)$.

Two properties of the normal distribution that play a large role in model building and Bayesian computation are the addition and mixture properties. The sum of two independent normal random variables is normally distributed. If θ_1 and θ_2 are independent with $N(\mu_1, \sigma_1^2)$ and $N(\mu_2, \sigma_2^2)$ distributions, then $\theta_1 + \theta_2 \sim N(\mu_1 + \mu_2, \sigma_1^2 + \sigma_2^2)$. The mixture property states that if $(\theta_1|\theta_2) \sim N(\theta_2, \sigma_1^2)$ and $\theta_2 \sim N(\mu_2, \sigma_2^2)$, then $\theta_1 \sim N(\mu_2, \sigma_1^2 + \sigma_2^2)$. This is useful in the analysis of hierarchical normal models.

Lognormal

If θ is a random variable that is restricted to be positive, and $\log \theta \sim N(\mu, \sigma^2)$, then θ is said to have a *lognormal* distribution. Using the Jacobian of the log transformation, one can directly determine that the density is $p(\theta) = (\sqrt{2\pi}\sigma\theta)^{-1} \exp(-\frac{1}{2\sigma^2}(\log\theta - \mu)^2)$, the mean is $\exp(\mu + \frac{1}{2}\sigma^2)$, the variance is $\exp(2\mu)\exp(\sigma^2)(\exp(\sigma^2) - 1)$, and the mode is $\exp(\mu - \sigma^2)$. The geometric mean and geometric standard deviation of a lognormally-distributed random variable θ are simply e^μ and e^σ.

Multivariate normal

The multivariate normal density is always finite; the integral is finite as long as $\det(\Sigma^{-1}) > 0$. A noninformative distribution is obtained in the limit as $\det(\Sigma^{-1}) \to 0$; this limit is not uniquely defined. A random draw from a multivariate normal distribution can be obtained using the Cholesky decomposition of Σ and a vector of univariate normal draws. The Cholesky decomposition of Σ produces a lower-triangular matrix A (the 'Cholesky factor') for which $AA^T = \Sigma$. If $z = (z_1, \ldots, z_d)$ are d independent standard normal random variables, then $\theta = \mu + Az$ is a random draw from the multivariate normal distribution with covariance matrix Σ.

The marginal distribution of any subset of components (for example, θ_i or (θ_i, θ_j)) is also normal. Any linear transformation of θ, such as the projection of θ onto a linear subspace, is also normal, with dimension equal to the rank of the transformation. The conditional distribution of θ, constrained to lie on any linear subspace, is also normal. The addition property holds: if θ_1 and θ_2 are independent with $N(\mu_1, \Sigma_1)$ and $N(\mu_2, \Sigma_2)$ distributions, then $\theta_1 + \theta_2 \sim N(\mu_1 + \mu_2, \Sigma_1 + \Sigma_2)$ as long as θ_1 and θ_2 have the same dimension. We discuss the generalization of the mixture property shortly.

The conditional distribution of any subvector of θ given the remaining elements is once again multivariate normal. If we partition θ into subvectors $\theta = (U, V)$, then $p(U|V)$ is (multivariate) normal:

$$
\begin{aligned}
E(U|V) &= E(U) + \text{cov}(U, V)\text{var}(V)^{-1}(V - E(V)), \\
\text{var}(U|V) &= \text{var}(U) - \text{cov}(U, V)\text{var}(V)^{-1}\text{cov}(V, U), \quad \text{(A.1)}
\end{aligned}
$$

where $\text{cov}(V, U)$ is a rectangular matrix (submatrix of Σ) of the appropriate dimensions, and $\text{cov}(U, V) = \text{cov}(V, U)^T$. In particular, if we define the matrix of conditional coefficients,

$$
C = I - [\text{diag}(\Sigma^{-1})]^{-1}\Sigma^{-1},
$$

then

$$
(\theta_i \mid \theta_j, \text{ all } j \neq i) \sim N(\mu_i + \sum_{j \neq i} c_{ij}(\theta_j - \mu_j), [(\Sigma^{-1})_{ii}]^{-1}). \quad \text{(A.2)}
$$

Conversely, if we parameterize the distribution of U and V hierarchically:

$$
U|V \sim N(XV, \Sigma_{U|V}), \quad V \sim N(\mu_V, \Sigma_V),
$$

then the joint distribution of θ is the multivariate normal,

$$
\theta = \begin{pmatrix} U \\ V \end{pmatrix} \sim N\left(\begin{pmatrix} X\mu_V \\ \mu_V \end{pmatrix}, \begin{pmatrix} X\Sigma_V X^T + \Sigma_{U|V} & X\Sigma_V \\ \Sigma_V X^T & \Sigma_V \end{pmatrix} \right).
$$

This generalizes the mixture property of univariate normals.

The 'weighted sum of squares,' $SS = (\theta - \mu)^T \Sigma^{-1}(\theta - \mu)$, has a χ_d^2 distribution. For any matrix A for which $AA^T = \Sigma$, the conditional distribution of $A^{-1}(\theta - \mu)$, given SS, is uniform on a $(d-1)$-dimensional sphere.

Gamma

The gamma distribution is the conjugate prior distribution for the inverse of the normal variance and for the mean parameter of the Poisson distribution. The gamma integral is finite if $\alpha > 0$; the density function is finite if $\alpha \geq 1$. A noninformative distribution is obtained in the limit as $\alpha \to 0$, $\beta \to 0$. Many computer packages generate gamma random variables directly; otherwise, it is possible to obtain draws from a gamma random variable using draws from a uniform as input. The most effective method depends on the parameter α; see the references for details.

There is an addition property for independent gamma random variables with the same inverse scale parameter. If θ_1 and θ_2 are independent with $\text{Gamma}(\alpha_1, \beta)$ and $\text{Gamma}(\alpha_2, \beta)$ distributions, then $\theta_1 + \theta_2 \sim \text{Gamma}(\alpha_1 + \alpha_2, \beta)$. The logarithm of a gamma random variable is approximately normal;

raising a gamma random variable to the one-third power provides an even better normal approximation.

Inverse-gamma

If θ^{-1} has a gamma distribution with parameters α, β, then θ has the inverse-gamma distribution. The density is finite always; its integral is finite if $\alpha > 0$. The inverse-gamma is the conjugate prior distribution for the normal variance. A noninformative distribution is obtained as $\alpha, \beta \to 0$.

Chi-square

The χ^2 distribution is a special case of the gamma distribution, with $\alpha = \nu/2$ and $\beta = \frac{1}{2}$. The addition property holds since the inverse scale parameter is fixed: if θ_1 and θ_2 are independent with $\chi^2_{\nu_1}$ and $\chi^2_{\nu_2}$ distributions, then $\theta_1 + \theta_2 \sim \chi^2_{\nu_1 + \nu_2}$.

Inverse chi-square

The inverse-χ^2 is a special case of the inverse-gamma distribution, with $\alpha = \nu/2$ and $\beta = \frac{1}{2}$. We also define the *scaled* inverse chi-square distribution, which is useful for variance parameters in normal models. To obtain a simulation draw θ from the Inv-$\chi^2(\nu, s^2)$ distribution, first draw X from the χ^2_ν distribution and then let $\theta = \nu s^2 / X$.

Exponential

The exponential distribution is the distribution of waiting times for the next event in a Poisson process and is a special case of the gamma distribution with $\alpha = 1$. Simulation of draws from the exponential distribution is straightforward. If U is a draw from the uniform distribution on $[0, 1]$, then $-\log(U)/\beta$ is a draw from the exponential distribution with parameter β.

Weibull

If θ is a random variable that is restricted to be positive, and $(\theta/\beta)^\alpha$ has an Expon(1) distribution, then θ is said to have a *Weibull* distribution with shape parameter $\alpha > 0$ and scale parameter $\beta > 0$. The Weibull is often used to model failure times in reliability analysis. Using the Jacobian of the log transformation, one can directly determine that the density is $p(\theta) = \frac{\alpha}{\beta^\alpha}\theta^{\alpha-1}\exp(-(\theta/\beta)^\alpha)$, the mean is $\beta\Gamma(1 + \frac{1}{\alpha})$, the variance is $\beta^2[\Gamma(1 + \frac{2}{\alpha}) - (\Gamma(1 + \frac{1}{\alpha}))^2]$, and the mode is $\beta(1 - \frac{1}{\alpha})^{1/\alpha}$.

Wishart

The Wishart is the conjugate prior distribution for the inverse covariance matrix in a multivariate normal distribution. It is a multivariate generalization of the gamma distribution. The integral is finite if the degrees of freedom parameter, ν, is greater than or equal to the dimension, k. The density is finite if $\nu \geq k + 1$. A noninformative distribution is obtained as $\nu \to 0$. The sample covariance matrix for iid multivariate normal data has a Wishart distribution. In fact, multivariate normal simulations can be used to simulate a draw from the Wishart distribution, as follows. Simulate $\alpha_1, \ldots, \alpha_\nu$, ν independent samples from a k-dimensional multivariate $N(0, S)$ distribution, then let $\theta = \sum_{i=1}^{\nu} \alpha_i \alpha_i^T$. This only works when the distribution is proper; that is, $\nu \geq k$.

Inverse-Wishart

If $W^{-1} \sim \text{Wishart}_\nu(S)$ then W has the inverse-Wishart distribution. The inverse-Wishart is the conjugate prior distribution for the multivariate normal covariance matrix. The inverse-Wishart density is always finite, and the integral is always finite. A degenerate form occurs when $\nu < k$.

Student-t

The t is the marginal posterior distribution for the normal mean with unknown variance and conjugate prior distribution and can be interpreted as a mixture of normals with common mean and variances that follow an inverse-gamma distribution. The t is also the ratio of a normal random variable and the square root of an independent gamma random variable. To simulate t, simulate z from a standard normal and x from a χ_ν^2, then let $\theta = \mu + \sigma z \sqrt{\nu/x}$. The t density is always finite; the integral is finite if $\nu > 0$ and σ is finite. In the limit $\nu \to \infty$, the t distribution approaches $N(\mu, \sigma^2)$. The case of $\nu = 1$ is called the *Cauchy distribution*. The t distribution can be used in place of a normal distribution in a robust analysis.

To draw from the multivariate $t_\nu(\mu, \Sigma)$ distribution, generate a vector $z \sim N(0, I)$ and a scalar $x \sim \chi_\nu^2$, then compute $\mu + Az\sqrt{\nu/x}$, where A satisfies $AA^T = \Sigma$.

Beta

The beta is the conjugate prior distribution for the binomial probability. The density is finite if $\alpha, \beta \geq 1$, and the integral is finite if $\alpha, \beta > 0$. The choice $\alpha = \beta = 1$ gives the standard uniform distribution; $\alpha = \beta = 0.5$ and $\alpha = \beta = 0$ are also sometimes used as noninformative densities. To simulate θ from the beta distribution, first simulate x_α and x_β from $\chi_{2\alpha}^2$ and $\chi_{2\beta}^2$ distributions, respectively, then let $\theta = \frac{x_\alpha}{x_\alpha + x_\beta}$.

It is sometimes useful to estimate quickly the parameters of the beta distribution using the method of moments:

$$\alpha + \beta = \frac{E(\theta)(1 - E(\theta))}{\text{var}(\theta)} - 1$$

$$\alpha = (\alpha + \beta)E(\theta), \qquad \beta = (\alpha + \beta)(1 - E(\theta)). \qquad (A.3)$$

The beta distribution is also of interest because the kth order statistic from a sample of n iid $U(0, 1)$ variables has the $\text{Beta}(k, n - k + 1)$ distribution.

Dirichlet

The Dirichlet is the conjugate prior distribution for the parameters of the multinomial distribution. The Dirichlet is a multivariate generalization of the beta distribution. As with the beta, the integral is finite if all of the α's are positive, and the density is finite if all are greater than or equal to one. A noninformative prior is obtained as $\alpha_j \to 0$ for all j.

The marginal distribution of a single θ_j is $\text{Beta}(\alpha_j, \alpha_0 - \alpha_j)$. The marginal distribution of a subvector of θ is Dirichlet; for example $(\theta_i, \theta_j, 1 - \theta_i - \theta_j) \sim$ $\text{Dirichlet}(\alpha_i, \alpha_j, \alpha_0 - \alpha_i - \alpha_j)$. The conditional distribution of a subvector given the remaining elements is Dirichlet under the condition $\sum_{j=1}^{k} \theta_j = 1$.

There are two standard approaches to sampling from a Dirichlet distribution. The fastest method generalizes the method used to sample from the beta distribution: draw $x_1, \ldots, x_k$ from independent gamma distributions with common scale and shape parameters $\alpha_1, \ldots, \alpha_k$, and for each j, let $\theta_j = x_j / \sum_{i=1}^{k} x_i$. A less efficient algorithm relies on the univariate marginal and conditional distributions being beta and proceeds as follows. Simulate θ_1 from a $\text{Beta}(\alpha_1, \sum_{i=2}^{k} \alpha_i)$ distribution. Then simulate $\theta_2, \ldots, \theta_{k-1}$ in order, as follows. For $j = 2, \ldots, k-1$, simulate ϕ_j from a $\text{Beta}(\alpha_j, \sum_{i=j+1}^{k} \alpha_i)$ distribution, and let $\theta_j = (1 - \sum_{i=1}^{j-1} \theta_i)\phi_j$. Finally, set $\theta_k = 1 - \sum_{i=1}^{k-1} \theta_i$.

A.3 Discrete distributions

Poisson

The Poisson distribution is commonly used to represent count data, such as the number of arrivals in a fixed time period. The Poisson distribution has an addition property: if θ_1 and θ_2 are independent with $\text{Poisson}(\lambda_1)$ and $\text{Poisson}(\lambda_2)$ distributions, then $\theta_1 + \theta_2 \sim \text{Poisson}(\lambda_1 + \lambda_2)$. Simulation for the Poisson distribution (and most discrete distributions) can be cumbersome. Table lookup can be used to invert the cumulative distribution function. Simulation texts describe other approaches.

Binomial

The binomial distribution is commonly used to represent the number of 'successes' in a sequence of n iid Bernoulli trials, with probability of success p in each trial. A binomial random variable with large n is approximately normal. If θ_1 and θ_2 are independent with $\text{Bin}(n_1, p)$ and $\text{Bin}(n_2, p)$ distributions, then $\theta_1 + \theta_2 \sim \text{Bin}(n_1 + n_2, p)$. For small n, a binomial random variable can be simulated by obtaining n independent standard uniforms and setting θ equal to the number of uniform deviates less than or equal to p. For larger n, more efficient algorithms are often available in computer packages. When $n = 1$, the binomial is called the *Bernoulli* distribution.

Multinomial

The multinomial distribution is a multivariate generalization of the binomial distribution. The marginal distribution of a single θ_i is binomial. The conditional distribution of a subvector of θ is multinomial with 'sample size' parameter reduced by the fixed components of θ and 'probability' parameters rescaled to have sum equal to one. We can simulate a multivariate draw using a sequence of binomial draws. Draw θ_1 from a $\text{Bin}(n, p_1)$ distribution. Then draw $\theta_2, \ldots, \theta_{k-1}$ in order, as follows. For $j = 2, \ldots, k - 1$, draw θ_j from a $\text{Bin}(n - \sum_{i=1}^{j-1} \theta_i, p_j / \sum_{i=j}^{k} p_i)$ distribution. Finally, set $\theta_k = n - \sum_{i=1}^{k-1} \theta_i$. If at any time in the simulation the binomial sample size parameter equals zero, use the convention that a $\text{Bin}(0, p)$ variable is identically zero.

Negative binomial

The negative binomial distribution is the marginal distribution for a Poisson random variable when the rate parameter has a $\text{Gamma}(\alpha, \beta)$ prior distribution. The negative binomial can also be used as a robust alternative to the Poisson distribution, because it has the same sample space, but has an additional parameter. To simulate a negative binomial random variable, draw $\lambda \sim \text{Gamma}(\alpha, \beta)$ and then draw $\theta \sim \text{Poisson}(\lambda)$. In the limit $\alpha \to \infty$, and $\alpha/\beta \to$ constant, the distribution approaches a Poisson with parameter α/β. Under the alternative parametrization, $p = \frac{\beta}{\beta+1}$, the random variable θ can be interpreted as the number of Bernoulli failures obtained before the α successes, where the probability of success is p.

Beta-binomial

The beta-binomial arises as the marginal distribution of a binomial random variable when the probability of success has a $\text{Beta}(\alpha, \beta)$ prior distribution. It can also be used as a robust alternative to the binomial distribution. The mixture definition gives an algorithm for simulating from the beta-binomial: draw $\phi \sim \text{Beta}(\alpha, \beta)$ and then draw $\theta \sim \text{Bin}(n, \phi)$.

A.4 Bibliographic note

Many software packages contain subroutines to simulate draws from these distributions. Texts on simulation typically include information about many of these distributions; for example, Ripley (1987) discusses simulation of all of these in detail, except for the Dirichlet and multinomial. Johnson and Kotz (1972) give more detail, such as the characteristic functions, for the distributions. Fortran and C programs for uniform, normal, gamma, Poisson, and binomial distributions are available in Press et al. (1986).

Outline of proofs of asymptotic theorems

The basic result of large-sample Bayesian inference is that as more and more data arrive, the posterior distribution of the parameter vector approaches a multivariate normal distribution. If the likelihood model happens to be correct, then we can also prove that the limiting posterior distribution is centered at the true value of the parameter vector. In this appendix, we outline a proof of the main results; we give references at the end for more thorough and rigorous treatments. The practical relevance of the theorems is discussed in Chapter 4.

We derive the limiting posterior distribution in three steps. The first step is the convergence of the posterior distribution to a point, for a discrete parameter space. If the data truly come from the hypothesized family of probability models, the point of convergence will be the true value of the parameter. The second step applies the discrete result to regions in continuous parameter space, to show that the mass of the continuous posterior distribution becomes concentrated in smaller and smaller neighborhoods of a particular value of parameter space. Finally, the third step of the proof shows the accuracy of the normal approximation to the posterior distribution in the vicinity of the posterior mode.

Mathematical framework

The key assumption for the results presented here is that data are independent and identically distributed: we label the data as $y = (y_1, \ldots, y_n)$, with probability density $\prod_{i=1}^{n} f(y_i)$. We use the notation $f(\cdot)$ for the *true distribution* of the data, in contrast to $p(\cdot|\theta)$, the distribution of our probability model. The data y may be discrete or continuous.

We are interested in a (possibly vector) parameter θ, defined on a space Θ, for which we have a prior distribution, $p(\theta)$, and a likelihood, $p(y|\theta) = \prod_{i=1}^{n} p(y_i|\theta)$, which assumes the data are independent and identically distributed. As illustrated in the counterexamples discussed in Section 4.3, some conditions are required on the prior distribution and the likelihood, as well as on the space Θ, for the theorems to hold.

It is necessary to assume a true distribution for y, because the theorems only hold in probability; for almost every problem, it is possible to construct data sequences y for which the posterior distribution of θ will not have the desired limit. The theorems are of the form, 'The posterior distribution of θ converges in probability (as $n \to \infty$) to ...'; the 'probability' in 'converges in probability' is with respect to $f(y)$, the true distribution of y.

We define θ_0 as the value of θ that minimizes the *Kullback-Leibler information*, $H(\theta)$, of the distribution $p(\cdot|\theta)$ in the model relative to the true distribution, $f(\cdot)$. The Kullback-Leibler information is defined at any value θ by

$$H(\theta) = E\left(\log\left(\frac{f(y_i)}{p(y_i|\theta)}\right)\right)$$
$$= \int \log\left(\frac{f(y_i)}{p(y_i|\theta)}\right) f(y_i)dy_i. \tag{B.1}$$

This is a measure of 'discrepancy' between the model distribution $p(y_i|\theta)$ and the true distribution $f(y)$, and θ_0 may be thought of as the value of θ that minimizes this distance. We assume that θ_0 is the unique minimizer of $H(\theta)$. It turns out that as n increases, the posterior distribution $p(\theta|y)$ becomes concentrated about θ_0.

Suppose that the likelihood model is correct; that is, there is some *true parameter value* θ for which $f(y_i) = p(y_i|\theta)$. In this case, it is easily shown via Jensen's inequality that (B.1) is minimized at the true parameter value, which we can then label as θ_0 without risk of confusion.

Convergence of the posterior distribution for a discrete parameter space

Theorem. If the parameter space Θ is finite and $\Pr(\theta = \theta_0) > 0$, then $\Pr(\theta = \theta_0|y) \to 1$ as $n \to \infty$, where θ_0 is the value of θ that minimizes the Kullback-Leibler information (B.1).

Proof. We will show that $p(\theta|y) \to 0$ as $n \to \infty$ for all $\theta \neq \theta_0$. Consider the log posterior odds relative to θ_0:

$$\log\left(\frac{p(\theta|y)}{p(\theta_0|y)}\right) = \log\left(\frac{p(\theta)}{p(\theta_0)}\right) + \sum_{i=1}^{n} \log\left(\frac{p(y_i|\theta)}{p(y_i|\theta_0)}\right). \tag{B.2}$$

The second term on the right can be considered a sum of n independent, identically distributed random variables, in which θ and θ_0 are considered fixed and the y_i's are random, with distributions f. Each term in the summation has a mean of

$$E\left(\log\left(\frac{p(y_i|\theta)}{p(y_i|\theta_0)}\right)\right) = H(\theta_0) - H(\theta),$$

which is zero if $\theta = \theta_0$ and negative otherwise, as long as θ_0 is the unique minimizer of $H(\theta)$.

Thus, if $\theta \neq \theta_0$, the second term on the right of (B.2) is the sum of n iid random variables with negative mean. By the law of large numbers, the sum approaches $-\infty$ as $n \to \infty$. As long as the first term on the right of (B.2) is finite (that is, as long as $p(\theta_0) > 0$), the whole expression approaches $-\infty$ in the limit. Then, $p(\theta|y)/p(\theta_0|y) \to 0$, and so $p(\theta|y) \to 0$. Since all probabilities sum to 1, $p(\theta_0|y) \to 1$.

Convergence of the posterior distribution for a continuous parameter space

If θ has a continuous distribution, then $p(\theta_0|y)$ is always zero for any finite sample, and so the above theorem cannot apply. We can, however, show that the posterior probability distribution of θ becomes more and more concentrated about θ_0 as $n \to \infty$. Define a neighborhood of θ_0 as the open set of all points in Θ within a fixed nonzero distance of θ_0.

Theorem. If θ is defined on a compact set and A is a neighborhood of θ_0 with nonzero prior probability, then $\Pr(\theta \in A|y) \to 1$ as $n \to \infty$, where θ_0 is the value of θ that minimizes (B.1).

Proof. The theorem can be proved by placing a small neighborhood about each point in Θ, with A being the only neighborhood that includes θ_0, and then cover Θ with a finite subset of these neighborhoods. If Θ is compact, such a finite subcovering can always be obtained. The proof of the convergence of the posterior distribution to a point is then adapted to show that the posterior probability for any neighborhood except A approaches zero as $n \to \infty$, and thus $\Pr(\theta \in A|y) \to 1$.

Convergence of the posterior distribution to normality

We have just shown that by increasing n, we can put as much of the mass of the posterior distribution as we like in any arbitrary neighborhood of θ_0. Obtaining the limiting posterior distribution requires two more steps. The first is to show that the posterior mode is consistent; that is, that the mode of the posterior distribution falls within the neighborhood where almost all the mass lies. The second step is a normal approximation centered at the posterior mode.

Theorem. Under some regularity conditions (notably that θ_0 not be on the boundary of Θ), as $n \to \infty$, the posterior distribution of θ approaches normality with mean θ_0 and variance $(nJ(\theta_0))^{-1}$, where θ_0 is the value that minimizes the Kullback-Leibler information (B.1) and J is the Fisher information (2.19).

Proof. For convenience in exposition, we first derive the result for a scalar θ.
   Define $\hat{\theta}$ as the posterior mode. The proof of the consistency of the maximum likelihood estimate (see the bibliographic note at the end of the chapter) can be mimicked to show that $\hat{\theta}$ is also consistent; that is $\hat{\theta} \to \theta_0$ as $n \to \infty$.
   Given the consistency of the posterior mode, we approximate the log posterior density by a Taylor expansion centered about $\hat{\theta}$, confident that (for large n) the neighborhood near $\hat{\theta}$ has almost all the mass in the posterior distribution. The normal approximation for θ is a quadratic approximation for the log posterior distribution of θ, a form that we derive via a Taylor series expansion

of $\log p(\theta|y)$ centered at $\hat{\theta}$:

$$
\begin{aligned}
\log p(\theta|y) = {}& \log p(\hat{\theta}|y) + \frac{1}{2}(\theta - \hat{\theta})^2 \frac{d^2}{d\theta^2} \left[\log p(\theta|y)\right]_{\theta=\hat{\theta}} + \\
& + \frac{1}{6}(\theta - \hat{\theta})^3 \frac{d^3}{d\theta^3} \left[\log p(\theta|y)\right]_{\theta=\hat{\theta}} + \cdots.
\end{aligned}
$$

(The linear term in the expansion is zero because the log posterior density has zero derivative at its interior mode.)

Consider the above equation as a function of θ. The first term is a constant. The coefficient for the second term can be written as

$$
\frac{d^2}{d\theta^2} \left[\log p(\theta|y)\right]_{\theta=\hat{\theta}} = \frac{d^2}{d\theta^2} \log p(\hat{\theta}) + \sum_{i=1}^{n} \frac{d^2}{d\theta^2} \left[\log p(y_i|\theta)\right]_{\theta=\hat{\theta}},
$$

which is a constant plus the sum of n independent, identically distributed random variables with negative mean (once again, it is the y_i's that are considered random here). If $f(y) \equiv p(y|\theta_0)$ for some θ_0, then the terms each have mean $-J(\theta_0)$. If the true data distribution $f(y)$ is not in the model class, then the mean is $\mathrm{E}_f \left(\frac{d^2}{d\theta^2} \log p(y|\theta) \right)$ evaluated at $\theta = \theta_0$, which is the negative second derivative of the Kullback-Leibler information, $H(\theta_0)$, and is thus negative, because θ_0 is defined as the point at which $H(\theta)$ is minimized. Thus, the coefficient for the second term in the Taylor expansion increases with order n. A similar argument shows that coefficients for the third- and higher-order terms increase no faster than order n.

We can now prove that the posterior distribution approaches normality. As $n \to \infty$, the mass of the posterior distribution $p(\theta|y)$ becomes concentrated in smaller and smaller neighborhoods of θ_0, and the distance $|\hat{\theta} - \theta_0|$ also approaches zero. Thus, in considering the Taylor expansion about the posterior mode, we can focus on smaller and smaller neighborhoods about $\hat{\theta}$. As $|\theta - \hat{\theta}| \to 0$, the third-order and succeeding terms of the Taylor expansion fade in importance, relative to the quadratic term, so that the distance between the quadratic approximation and the log posterior distribution approaches 0, and the normal approximation becomes increasingly accurate.

Multivariate form

If θ is a vector, the Taylor expansion becomes

$$
\log p(\theta|y) = \log p(\hat{\theta}|y) + \frac{1}{2}(\theta - \hat{\theta})^T \frac{d^2}{d\theta^2} \left[\log p(\theta|y)\right]_{\theta=\hat{\theta}} (\theta - \hat{\theta}) + \cdots,
$$

where the second derivative of the log posterior distribution is now a matrix whose expectation is the negative of a positive definite matrix which is the Fisher information matrix (2.19) if $f(y) \equiv p(y|\theta_0)$ for some θ_0.

B.1 Bibliographic note

The asymptotic normality of the posterior distribution was known by Laplace (1810) but first proved rigorously by Le Cam (1953); a general survey of previous and subsequent theoretical results in this area is given by Le Cam and Yang (1990). Like the central limit theorem for sums of random variables, the consistency and asymptotic normality of the posterior distribution also hold in far more general conditions than independent and identically distributed data. The key condition is that there be 'replication' at some level, as, for example, if the data come in a time series whose correlations decay to zero.

The Kullback-Leibler information comes from Kullback and Leibler (1951). Chernoff (1972, Sections 6 and 9.4), has a clear presentation of consistency and limiting normality results for the maximum likelihood estimate. Both proofs can be adapted to the posterior distribution. DeGroot (1970, Chapter 10) derives the asymptotic distribution for the posterior distribution in more detail; Shen and Wasserman (2001) provide more recent results in this area.

APPENDIX C

Example of computation in R and Bugs

Weillustrate some of the practical issues of simulation by fitting a single example—the hierarchical normal model for the eight educational testing experiments described in Section 5.5. After some background in Section C.1, we show in Section C.2 how to fit the model using the Bayesian inference package Bugs, operating from within the general statistical package R. We proceed from data input, through statistical inference, to posterior predictive simulation and graphics. Section C.3 describes some alternative ways of parameterizing the model in Bugs. Section C.4 presents several different ways of programming the model directly in R. These algorithms require programming efforts that are unnecessary for the Bugs user but are useful knowledge for programming more advanced models for which Bugs might not work. We conclude in Section C.5 with some comments on practical issues of programming and debugging. It may also be helpful to read the computational tips at the end of Section 10.3.

C.1 Getting started with R and Bugs

Follow the instructions at www.stat.columbia.edu/~gelman/bugsR/ to install the latest versions of R and Bugs (both of which are free of charge at the time of the writing of this book), as well as to configure the two programs appropriately and set up a working directory for R. Also download the functions at that website for running Bugs from R and postprocessing regression output. Once you have followed the instructions, your working directory will automatically be set every time you run R, and the functions for running Bugs will be automatically be loaded. You can call Bugs within R as illustrated in the example on the webpage.

R (the open-source version of S and S-Plus) is a general-purpose statistical package that is fully programmable and also has available a large range of statistical tools, including flexible graphics, simulation from probability distributions, numerical optimization, and automatic fitting of many standard probability models including linear regression and generalized linear models. For Bayesian computation, one can directly program Gibbs sampler and Metropolis algorithms, as we illustrate in Section C.4. Computationally intensive tasks can be programmed in Fortran or C and linked from R.

Bugs is a high-level language in which the user specifies a model and starting values, and then a Markov chain simulation is automatically implemented for the resulting posterior distribution. It is possible to set up models and run

them entirely within Bugs, but in practice it is almost always necessary to
process data before entering them into a model, and to process the inferences
after the model is fitted, and so we run Bugs by calling it from R, using the
bugs() function in R, as we illustrate in Section C.2. As of the writing of this
book, the current version of Bugs is called WinBUGS and must be run in the
environment of the Microsoft Windows operating system. When we mention
the Bugs package, we are referring to the latest version of WinBUGS.

R and Bugs have online help, and further information is available at the
webpages www.r-project.org and www.mrc-bsu.cam.ac.uk/bugs/. We an-
ticipate continuing improvements in both packages in the years after this book
is released (and we will update our webpage accordingly), but the general computational strategies presented here should remain relevant.

When working in Bugs and R, it is helpful to set up the computer to simultaneously display three windows: an R console and graphics window, and
text editors with the Bugs model file and the R script. Rather than typing
commands directly into R, we prefer to enter them into the text editor and
then use cut-and-paste to transfer them to the R console. Using the text editor
is convenient because it allows more flexibility in writing functions and loops.

C.2 Fitting a hierarchical model in Bugs

In this section, we describe all the steps by which we would use Bugs to fit
the hierarchical normal model to the SAT coaching data in Section 5.5. These
steps include writing the model in Bugs and using R to set up the data and
starting values, call Bugs, create predictive simulations, and graph the results.
Section C.3 gives some alternative parameterizations of the model in Bugs.

Bugs model file

The hierarchical model can be written in Bugs in the following form, which
we save in the file schools.bug in our R working directory. (The file must be
given a .bug or .txt extension.)

```
model {
  for (j in 1:J){
    y[j] ~ dnorm (theta[j], tau.y[j])
    theta[j] ~ dnorm (mu.theta, tau.theta)
    tau.y[j] <- pow(sigma.y[j], -2)
  }
  mu.theta ~ dnorm (0, 1.0E-6)
  tau.theta <- pow(sigma.theta, -2)
  sigma.theta ~ dunif (0, 1000)
}
```

This model specification looks similar to how it would be written in this
book, with two differences. First, in Bugs, the normal distribution is parameterized by its precision (1/variance), rather than its standard deviation. When

working in Bugs, we use the notation τ for precisions and σ for standard deviations (departing from our notation in the book, where we use σ and τ for data and prior standard deviations, respectively).

The second difference from our model in Chapter 5 is that Bugs requires proper prior distributions. Hence, we express noninformative prior information by proper distributions with large uncertainties: μ_θ is given a normal distribution with mean 0 and standard deviation 1000, and σ_θ has a uniform distribution from 0 to 1000. These are certainly noninformative, given that the data y all fall well below 100 in absolute value.

R script for data input, starting values, and running Bugs

We put the data into a file, `schools.dat`, in the R working directory, with headers describing the data:

```
school estimate sd
A    28   15
B     8   10
C    -3   16
D     7   11
E    -1    9
F     1   11
G    18   10
H    12   18
```

From R, we then execute the following script, which reads in the dataset, puts it in list format to be read by Bugs, sets up a function to compute initial values for the Bugs run, and identifies parameters to be saved.

```
schools <- read.table ("schools.dat", header=T)
J <- nrow (schools)
y <- schools$estimate
sigma.y <- schools$sd
data <- list ("J", "y", "sigma.y")
inits <- function()
  list (theta=rnorm(J,0,100), mu.theta=rnorm(1,0,100),
        sigma.theta=runif(1,0,100))
parameters <- c("theta", "mu.theta", "sigma.theta")
```

In general, initial values can be set from crude estimates, as discussed in Section 10.1. In this particular example, we use the range of the data to construct roughly overdispersed distributions for the θ_j's, μ_θ, and σ_θ. Bugs does not require all parameters to be initialized, but it is a good idea to do so. We prefer to explicitly construct the starting points randomly as above, but it would also be possible to start the simulations with simple initial values; for example,

```
inits1 <- list (theta=y, mu.theta=0, sigma.theta=1)
inits2 <- list (theta=y, mu.theta=0, sigma.theta=10)
inits3 <- list (theta=y, mu.theta=0, sigma.theta=100)
inits <- list (inits1, inits2, inits3)
```

```
Inference for Bugs model at "c:/bugsR/schools.bug"
 3 chains, each with 1000 iterations (first 500 discarded)
 n.sims = 1500 iterations saved

                mean  sd  2.5%  25%  50%  75% 97.5% Rhat n.eff
theta[1]        11.0 8.5  -2.1  5.4  9.9 15.2  30.5  1.0    76
theta[2]         7.4 6.2  -5.0  3.7  7.1 11.4  20.1  1.0   290
theta[3]         5.5 7.8 -12.3  1.5  6.1 10.1  19.9  1.0  1400
theta[4]         7.4 6.6  -5.7  3.7  7.2 11.3  21.2  1.0   170
theta[5]         4.6 6.4  -9.6  0.5  5.4  8.8  16.0  1.0   390
theta[6]         5.6 6.6  -8.0  1.4  6.0 10.1  18.1  1.0   430
theta[7]        10.0 6.7  -2.1  5.6  9.4 14.0  25.0  1.0    73
theta[8]         8.0 7.8  -6.5  3.8  7.4 12.1  24.9  1.0   140
mu.theta         7.5 5.4  -2.8  4.3  7.3 10.7  19.0  1.0   160
sigma.theta      6.5 5.7   0.2  2.7  4.9  9.1  21.4  1.1    23
deviance        60.3 2.2  56.9 58.9 59.8 61.1  66.0  1.0    92
 pD = 2.4 and DIC = 62.7 (using the rule, pD = var(deviance)/2)
```

Figure C.1 *Numerical output from the* bugs() *function after being fitted to the hierarchical model for the educational testing example. For each parameter, n_{eff} is a rough measure of effective sample size, and $\widehat{R}$ is the potential scale reduction factor defined in Section 11.6 (at convergence, $\widehat{R} = 1$). The effective number of parameters p_D and the expected predictive error DIC are defined in Section 6.7.*

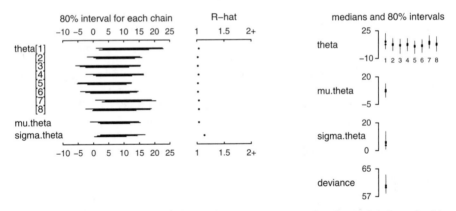

Figure C.2 *Graphical output from the* bugs() *function after being fitted to the hierarchical model for the educational testing example. The left side of the display shows the overlap of the three parallel chains, and the right side shows the posterior inference for each parameter and the deviance (−2 times the log likelihood). The on-screen display uses colors to distinguish the different chains.*

In any case, we can now run Bugs with 3 chains for 1000 iterations:

```
schools.sim <- bugs (data, inits, parameters, "schools.bug",
    n.chains=3, n.iter=1000)
```

While Bugs is running, it opens a new window and freezes R. When the computations are finished, summaries of the inferences and convergence are displayed as a table in the R console (see Figure C.1) and in an R graphics window (see Figure C.2). In this example, the sequences have mixed well—the estimated potential scale reduction factors $\widehat{R}$ are close to 1 for all parameters—and so we stop. (If some of the $\widehat{R}$ factors were still large, we would try running the simulation a bit longer—perhaps 2000 or 5000 iterations per chain—and if convergence were still a problem, we would consider reparameterizing the model. We give some examples of reparameterizing this model in Section C.3.) Running three chains for 1000 iterations and saving the second half of each chain (the default option) yields 1500 simulation draws of the parameter vector.

The output in Figure C.1 also gives information on the effective sample size n_{eff} for each parameter (as defined at the end of Section 11.6) and the deviance and DIC (see Section 6.7).

Accessing the posterior simulations in R

The output of the R function bugs() is a list which includes several components, most notably the summary of the inferences and convergence and a list containing the simulation draws of all the saved parameters. In the example above, the bugs() call is assigned to the R object schools.sim, and so typing schools.sim$summary from the R console will display the summary shown in Figure C.1.

The posterior simulations are saved in the list, schools.sim$sims.list. We we can directly access them by typing attach.all(schools.sim$sims.list) from the R console, at which point mu.theta is a vector of 1500 simulations of μ_θ, sigma.theta is a vector of 1500 simulations of σ_θ, and theta is a 1500×8 matrix of simulations of θ. Other output from the bugs() function is explained in comment lines within the function itself and can be viewed by typing bugs from the R console or examining the file bugs.R.

Posterior predictive simulations and graphs in R

Replicated data in the existing schools. Having run Bugs to successful convergence, we can work directly in R with the saved parameters, $\theta, \mu_\theta, \sigma_\theta$. For example, we can simulate posterior predictive replicated data in the original 8 schools:

```
y.rep <- array (NA, c(n.sims, J))
for (sim in 1:n.sims)
  y.rep[sim,] <- rnorm (J, theta[sim,], sigma.y)
```

We now illustrate a graphical posterior predictive check. There are not many ways to display a set of eight numbers. One possibility is as a histogram; the possible values of y^{rep} are then represented by an array of histograms as in Figure 6.2 on page 160. In R, this could be programmed as

```
par (mfrow=c(5,4), mar=c(4,4,2,2))
hist (y, xlab="", main="y")
for (sim in 1:19)
  hist (y.rep[sim,], xlab="", main=paste("y.rep",sim))
```

The upper-right histogram displays the observed data, and the other 19 histograms are posterior predictive replications, which in this example look similar to the data.

We could also compute a numerical test statistic such as the difference between the best and second-best of the 8 coaching programs:

```
test <- function (y){
  y.sort <- rev(sort(y))
  return (y.sort[1] - y.sort[2])
}
t.y <- test(y)
t.rep <- rep (NA, n.sims)
for (sim in 1:n.sims)
  t.rep[sim] <- test(y.rep[sim,])
```

We then can summarize the posterior predictive check. The following R code gives a numerical comparison of the test statistic to its replication distribution, a p-value, and a graph like those on pages 161 and 163:

```
cat ("T(y) =", round(t.y,1), "and T(y.rep) has mean",
       round(mean(t.rep),1), "and sd", round(sd(t.rep),1),
       "\nPr (T(y.rep) > T(y)) =", round(mean(t.rep>t.y),2), "\n")
hist0 <- hist (t.rep, xlim=range(t.y,t.rep), xlab="T(y.rep)")
lines (rep(t.y,2), c(0,1e6))
text (t.y, .9*max(hist0$count), "T(y)", adj=0)
```

Replicated data in new schools. As discussed in Section 6.8, another form of replication would simulate new parameter values and new data for eight *new* schools. To simulate data $y_j \sim N(\theta_j, \sigma_j^2)$ from new schools, it is necessary to make some assumption or model for the data variances σ_j^2. For the purpose of illustration, we assume these are repeated from the original 8 schools.

```
theta.rep <- array (NA, c(n.sims, J))
y.rep <- array (NA, c(n.sims, J))
for (sim in 1:n.sims){
  theta.rep[sim,] <- rnorm (J, mu.theta[sim], sigma.theta[sim])
  y.rep[sim,] <- rnorm (J, theta.rep[sim,], sigma.y)
}
```

Numerical and graphical comparisons can be performed as before.

C.3 Options in the Bugs implementation

We next explore alternative ways that the model can be expressed in Bugs.

Alternative prior distributions

The model as programmed above has a uniform prior distribution on the hyperparameters μ_θ and σ_θ. An alternative is to parameterize in terms of τ_θ, the precision parameter, for which a gamma distribution is conditionally conjugate in this model. For example, the last two lines of the Bugs model in Section C.3 can be replaced by,

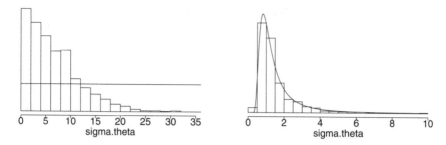

Figure C.3 *Histograms of posterior simulations of the between-school standard deviation, σ_θ, from models with two different prior distributions: (a) uniform prior distribution on σ_θ, and (b) conjugate Gamma$(1, 1)$ prior distribution on the precision parameter $1/\sigma_\theta^2$. The two histograms are on different scales. Overlain on each is the corresponding prior density function for σ_θ. (For model (b), the density for σ_θ is calculated using the gamma density function from Appendix A multiplied by the Jacobian of the $1/\sigma_\theta^2$ transformation.) In model (b), posterior inferences are highly constrained by the prior distribution.*

```
tau.theta ~ dgamma (1, 1)
sigma.theta <- 1/sqrt(tau.theta)
```

The Gamma$(1, 1)$ prior distribution is an attempt at noninformativeness within the conjugate family. (Recall that Bugs does not allow improper prior distributions.)

The initial values in the call to Bugs from R must now be redefined in terms of τ_θ rather than σ_θ; for example,

```
inits <- function ()
    list (theta=rnorm(J,0,100), mu.theta=rnorm(1,0,100),
          tau.theta=runif(1,0,100))
```

Otherwise, we fit the model as before. This new prior distribution leads to changed inferences. In particular, the posterior mean and median of σ_θ are lower and shrinkage of the θ_j's is greater than in the previously-fitted model with a uniform prior distribution on σ_θ. To understand this, it helps to graph the prior distribution in the range for which the posterior distribution is substantial. Figure C.3b shows that the prior distribution is concentrated in the range $[0.5, 5]$, a narrow zone in which the likelihood is close to flat (we can see this because the distribution of the posterior simulations of σ_θ closely matches the prior distribution, $p(\sigma_\theta)$). By comparison, in Figure C.3a, the uniform prior distribution on σ_θ seems closer to 'noninformative' for this problem, in the sense that it does not appear to be constraining the posterior inference.

Parameter expansion

A different direction to go in the Bugs programming is toward more efficient simulation by parameterizing the model so that the Gibbs sampler runs more effectively. Such tools are discussed in Section 11.8; here we illustrate how to set them up in Bugs. Introducing the new multiplicative parameter α, the model becomes

$$
\begin{aligned}
y_j &\sim N(\mu + \alpha\gamma_j, \sigma_j^2) \\
\gamma_j &\sim N(0, \sigma_\gamma^2) \\
p(\mu, \alpha, \sigma_\gamma) &\propto 1,
\end{aligned}
\tag{C.1}
$$

which we code in Bugs as follows:

```
model {
  for (j in 1:J){
    y[j] ~ dnorm (theta[j], tau.y[j])
    theta[j] <- mu.theta + alpha*gamma[j]
    gamma[j] ~ dnorm (0, tau.gamma)
    tau.y[j] <- pow(sigma.y[j], -2)
  }
  mu.theta ~ dnorm (0.0, 1.0E-6)
  tau.gamma <- pow(sigma.gamma, -2)
  sigma.gamma ~ dunif (0, 1000)
  alpha ~ dunif(0,1000)
  sigma.theta <- abs(alpha)*sigma.gamma
}
```

The parameters of interest, θ and σ_θ, are defined in the Bugs model in terms of the expanded parameter set, $\gamma, \alpha, \mu_\theta, \sigma_\gamma$. The call from R must then specify initial values in the new parameterization; for example

```
inits <- function ()
  list (gamma=rnorm(J,0,100), alpha=runif(1,0,100),
        mu.theta=rnorm(1,0,100), sigma.gamma=runif(1,0,100))
```

and then the rest of the R program is unchanged.

Posterior predictive simulations

At the end of Section C.2 we illustrated the computation of posterior predictive simulations in R. It is also possible to perform these simulations in Bugs. For example, to simulate new data from the existing schools, we can add the following line in the Bugs model, inside the 'for (j in 1:J)' loop (see the model on page 592, for example):

```
    y.rep[j] ~ dnorm (theta[j], tau.y[j])
```

Or, to simulate new data from new schools:

```
    theta.rep[j] ~ dnorm (mu.theta, tau.theta)
    y.rep[j] ~ dnorm (theta.rep[j], tau.y[j])
```

We can also include in the Bugs model the computation of a test statistic, for example, the difference between the two best coaching programs:

```
    J.1 <- J-1
    t.rep <- ranked(y.rep[],J) - ranked(y.rep[],J.1)
```

We must then add these derived quantities to the list of parameters saved in R:

```
parameters <- c("theta", "mu.theta", "sigma.theta", "y.rep", "t.rep")
```

We can then call the Bugs program from R as before.

Using the t model

It is straightforward to expand the hierarchical normal distribution for the coaching effects to a t distribution as discussed in Section 17.4. For simplicity, we return to the original form of the model without parameter expansion.

```
model {
    for (j in 1:J){
        y[j] ~ dnorm (theta[j], tau.y[j])
        theta[j] ~ dt (mu.theta, tau.theta, nu.theta)
        tau.y[j] <- pow(sigma.y[j], -2)
    }
    mu.theta ~ dnorm (0.0, 1.0E-6)
    tau.theta <- pow(sigma.theta, -2)
    sigma.theta ~ dunif (0, 1000)
    nu.theta <- 1/nu.inv.theta
    nu.inv.theta ~ dunif (0, .5)
}
```

Here, we are assigning a uniform distribution to the inverse of the shape parameter ν_θ, as in Section 17.4. In addition, Bugs requires the t degrees of freedom to be at least 2, so we have implemented that restriction on ν_θ. We run this model from R as before, adding nu.inv.theta to the set of initial values and nu.theta to the parameters saved:

```
inits <- function()
    list (theta=rnorm(J,0,100), mu.theta=rnorm(1,0,100),
          sigma.theta=runif(1,0,100), nu.inv.theta=runif(1,0,.5))
parameters <- c("theta", "mu.theta", "sigma.theta", "nu.theta")
```

Alternatively, the t model can be expressed as a normal model with a scaled inverse-χ^2 distribution for the variances, which corresponds to a gamma distributions for the precisions $\tau_{\theta j}$:

```
model {
    for (j in 1:J){
        y[j] ~ dnorm (theta[j], tau.y[j])
        theta[j] ~ dnorm (mu.theta, tau.theta[j])
        tau.y[j] <- pow(sigma.y[j], -2)
        tau.theta[j] ~ dgamma (a, b)
    }
    mu.theta ~ dnorm (0.0, 1.0E-6)
    a <- nu.theta/2
    b <- (nu.theta/2)*pow(sigma.theta, 2)
    sigma.theta ~ dunif (0, 1000)
    nu.theta <- 1/nu.inv.theta
    nu.inv.theta ~ dunif (0, 1)
}
```

In this indirect parameterization, Bugs allows the parameter ν_θ to be any positive value, and so we can apply a $U(0, 1)$ prior distribution to $1/\nu_\theta$.

It would be possible to further elaborate the model in Bugs by applying parameter expansion to the t model, but we do not consider any further extensions here.

C.4 Fitting a hierarchical model in R

In this section we demonstrate several different computational approaches for analyzing the SAT coaching data by directly programming the computations in R. Compared to using Bugs, computation in R requires more programming effort but gives us direct control of the simulation algorithm, which is helpful in settings where the Gibbs sampler or Metropolis algorithm is slow to converge.

Marginal and conditional simulation for the normal model

We begin by programming the calculations in Section 5.4. The programs provided here return to the notation of Chapter 5 (for example, τ is the population standard deviation of the θ's) as this allows for easy identification of some of the variables in the programs (for example, mu.hat and V.mu are the quantities denoted by the corresponding symbols in (5.20)).

We assume that the dataset has been read into R as in Section C.2, with J the number of schools, y the vector of data values, and sigma.y the vector of standard deviations. Then the first step of our programming is to set up a grid for τ, evaluate the marginal posterior distribution (5.21) for τ at each grid point, and sample 1000 draws from the grid. The grid here is n.grid=2000 points equally spread from 0 to 40. Here we use the grid as a discrete approximation to the posterior distribution of τ. We first define $\hat{\mu}$ and V_μ of (5.20) as functions of τ and the data, as these quantities are needed here and in later steps, and then compute the log density for τ.

```
mu.hat <- function (tau, y, sigma.y){
  sum(y/(sigma.y^2 + tau^2))/sum(1/(sigma.y^2 + tau^2))
}
V.mu <- function (tau, y, sigma.y){
  1/sum(1/(tau^2 + sigma.y^2))
}
n.grid <- 2000
tau.grid <- seq (.01, 40, length=n.grid)
log.p.tau <- rep (NA, n.grid)
for (i in 1:n.grid){
  mu <- mu.hat (tau.grid[i], y, sigma.y)
  V <- V.mu (tau.grid[i], y, sigma.y)
  log.p.tau[i] <- .5*log(V) -
    .5*sum(log(sigma.y^2 + tau.grid[i]^2)) -
    .5*sum((y-mu)^2/(sigma.y^2 + tau.grid[i]^2))
}
```

We compute the posterior density for τ on the log scale and rescale it to eliminate the possibility of computational overflow or underflow that can occur when multiplying many factors.

```
log.p.tau <- log.p.tau - max(log.p.tau)
p.tau <- exp(log.p.tau)
p.tau <- p.tau/sum(p.tau)
n.sims <- 1000
tau <- sample (tau.grid, n.sims, replace=T, prob=p.tau)
```

The last step draws the simulations of τ from the approximate discrete distribution. The remaining steps are sampling from normal conditional distributions for μ and the θ_j's as in Section 5.4. The sampled values of the eight θ_j's are collected in an array.

```
mu <- rep (NA, n.sims)
theta <- array (NA, c(n.sims,J))
for (i in 1:n.sims){
  mu[i] <- rnorm (1, mu.hat(tau[i],y,sigma.y),
    sqrt(V.mu(tau[i],y,sigma.y)))
  theta.mean <- (mu[i]/tau[i]^2 + y/sigma.y^2)/
    (1/tau[i]^2 + 1/sigma.y^2)
  theta.sd <- sqrt(1/(1/tau[i]^2 + 1/sigma.y^2))
  theta[i,] <- rnorm (J, theta.mean, theta.sd)
}
```

We now have created 1000 draws from the joint posterior distribution of τ, μ, θ. Posterior predictive distributions are easily generated using the random number generation capabilities of R as described above in the Bugs context.

Gibbs sampler for the normal model

Another approach, actually simpler to program, is to use the Gibbs sampler. This computational approach follows the outline of Section 11.7 with the simplification that the observation variances σ_j^2 are known.

```
theta.update <- function (){
  theta.hat <- (mu/tau^2 + y/sigma.y^2)/(1/tau^2 + 1/sigma.y^2)
  V.theta <- 1/(1/tau^2 + 1/sigma.y^2)
  rnorm (J, theta.hat, sqrt(V.theta))
}
mu.update <- function (){
  rnorm (1, mean(theta), tau/sqrt(J))
}
tau.update <- function (){
  sqrt(sum((theta-mu)^2)/rchisq(1,J-1))
}
```

We now generate five independent Gibbs sampling sequences of length 1000. We initialize μ and τ with overdispersed values based on the range of the data y and then run the Gibbs sampler, saving the output in a large array, sims, that contains posterior simulation draws for θ, μ, τ.

```
n.chains <- 5
n.iter <- 1000
sims <- array (NA, c(n.iter, n.chains, J+2))
dimnames (sims) <- list (NULL, NULL,
  c (paste ("theta[", 1:8, "]", sep=""), "mu", "tau"))
for (m in 1:n.chains){
  mu <- rnorm (1, mean(y), sd(y))
  tau <- runif (1, 0, sd(y))
  for (t in 1:n.iter){
    theta <- theta.update ()
    mu <- mu.update ()
    tau <- tau.update ()
    sims[t,m,] <- c (theta, mu, tau)
  }
}
```

We then check the mixing of the sequences using the R function `monitor` that carries out the multiple-chain diagnostic described in Section 11.6. We round to one decimal place to make the results more readable:

```
round (monitor (sims), 1)
```

The `monitor` function has automatically been loaded if you have followed the instructions for setting up R at the beginning of this chapter. The function takes as input an array of posterior simulations from multiple chains, and it returns an estimate of the potential scale reduction $\widehat{R}$, effective sample size n_{eff}, and summary statistics for the posterior distribution (based on the last half of the simulated Markov chains).

The model can also be computed using the alternative parameterizations and prior distributions that we implemented in Bugs in Section C.3. For example, in the parameter-expanded model (C.1), the Gibbs sampler steps can be programmed as

```
gamma.update <- function (){
  gamma.hat <- (alpha*(y-mu)/sigma.y^2)/(1/tau^2 + alpha^2/sigma.y^2)
  V.gamma <- 1/(1/tau^2 + alpha^2/sigma.y^2)
  rnorm (J, gamma.hat, sqrt(V.gamma))
}
alpha.update <- function (){
  alpha.hat <- sum(gamma*(y-mu)/sigma.y^2)/sum(gamma^2/sigma.y^2)
  V.alpha <- 1/sum(gamma^2/sigma.y^2)
  rnorm (1, alpha.hat, sqrt(V.alpha))
}
mu.update <- function (){
  mu.hat <- sum((y-alpha*gamma)/sigma.y^2)/sum(1/sigma.y^2)
  V.mu <- 1/sum(1/sigma.y^2)
  rnorm (1, mu.hat, sqrt(V.mu))
}
tau.update <- function (){
  sqrt(sum(gamma^2)/rchisq(1,J-1))
}
```

The Gibbs sampler can then be implemented as

```
sims <- array (NA, c(n.iter, n.chains, J+2))
dimnames (sims) <- list (NULL, NULL,
  c (paste ("theta[", 1:8, "]", sep=""), "mu", "tau"))
for (m in 1:n.chains){
  alpha <- 1
  mu <- rnorm (1, mean(y), sd(y))
  tau <- runif (1, 0, sd(y))
  for (t in 1:n.iter){
    gamma <- gamma.update ()
    alpha <- alpha.update ()
    mu <- mu.update ()
    tau <- tau.update ()
    sims[t,m,] <- c (mu+alpha*gamma, mu, abs(alpha)*tau)
  }
}
round (monitor (sims), 1)
```

Gibbs sampling for the t model with fixed degrees of freedom

As described in Chapter 17, the t model can be implemented using the Gibbs sampler using the normal/inverse-χ^2 parameterization for the θ_j's and their variances. Following the notation of that chapter, we take V_j to be the variance for θ_j and model the V_j's as draws from an inverse-χ^2 distribution with degrees of freedom ν and scale τ. As with the normal model, we use a uniform prior distribution on (μ, τ).

As before, we first create the separate updating functions, including a new function to update the individual-school variances V_j.

```
theta.update <- function (){
  theta.hat <- (mu/V + y/sigma.y^2)/(1/V + 1/sigma.y^2)
  V.theta <- 1/(1/V + 1/sigma.y^2)
  rnorm (J, theta.hat, sqrt(V.theta))
}
mu.update <- function (){
  mu.hat <- sum(theta/V)/sum(1/V)
  V.mu <- 1/sum(1/V)
  rnorm (1, mu.hat, sqrt(V.mu))
}
tau.update <- function (){
  sqrt (rgamma (1, J*nu/2+1, (nu/2)*sum(1/V)))
}
V.update <- function (){
  (nu*tau^2 + (theta-mu)^2)/rchisq(J,nu+1)
}
```

Initially we fix the degrees of freedom at 4 to provide a robust analysis of the data.

```
sims <- array (NA, c(n.iter, n.chains, J+2))
dimnames (sims) <- list (NULL, NULL,
  c (paste ("theta[", 1:8, "]", sep=""), "mu", "tau"))
nu <- 4
for (m in 1:n.chains){
  mu <- rnorm (1, mean(y), sd(y))
  tau <- runif (1, 0, sd(y))
  V <- runif (J, 0, sd(y))^2
  for (t in 1:n.iter){
    theta <- theta.update ()
    V <- V.update ()
    mu <- mu.update ()
    tau <- tau.update ()
    sims[t,m,] <- c (theta, mu, tau)
  }
}
round (monitor (sims), 1)
```

Gibbs-Metropolis sampling for the t model with unknown degrees of freedom

We can also include ν, the degrees of freedom in the above analysis, as an unknown parameter and update it conditional on all the others using the Metropolis algorithm. We follow the discussion in Chapter 17 and use a uniform prior distribution on $(\mu, \tau, 1/\nu)$.

The Metropolis updating function calls a function log.post to calculate the logarithm of the conditional posterior distribution of $1/\nu$ given all of the other parameters. (We work on the logarithmic scale to avoid computational overflows, as mentioned in Section 10.3.) The log posterior density function for this model has three terms—the logarithm of a normal density for the data points y_j, the logarithm of a normal density for the school effects θ_j, and the logarithm of an inverse-χ^2 density for the variances V_j. Actually, only the last term involves ν, but for generality we compute the entire log-posterior density in the log.post function.

```
log.post <- function (theta, V, mu, tau, nu, y, sigma.y){
  sum(dnorm(y,theta,sigma.y,log=T)) +
    sum(dnorm(theta,mu,sqrt(V),log=T)) +
    sum (.5*nu*log(nu/2) + nu*log(tau) -
      lgamma(nu/2) - (nu/2+1)*log(V) - .5*nu*tau^2/V)
}
```

We introduce the function that performs the Metropolis step and then describe how to alter the R code given earlier to incorporate the Metropolis step. The following function performs the Metropolis step for the degrees of freedom (recall that we work with the reciprocal of the degrees of freedom). The jumping distribution is normal with mean at the current value and standard deviation sigma.jump.nu (which is set as described below). We compute the jumping probability as described on page 289, setting it to zero if the proposed value of $1/\nu$ is outside the interval $(0, 1]$ to ensure that such proposals are rejected.

```
nu.update <- function (sigma.jump.nu){
  nu.inv.star <- rnorm(1, 1/nu, sigma.jump.nu)
  if (nu.inv.star<=0 | nu.inv.star>1)
    p.jump <- 0
  else {
    nu.star <- 1/nu.inv.star
    log.post.old <- log.post (theta, V, mu, tau, nu, y, sigma.y)
    log.post.star <- log.post (theta, V, mu, tau, nu.star,y,sigma.y)
    r <- exp (log.post.star - log.post.old)
    nu <- ifelse (runif(1) < r,  nu.star, nu)
    p.jump <- min(r,1)
  }
  return (nu=nu, p.jump=p.jump)
}
```

This updating function stores the acceptance probability p.jump.nu which is used in adaptively setting the jumping scale sigma.jump.nu, as we discuss when describing the Gibbs-Metropolis loop.

Given these functions, it is relatively easy to modify the R code that we have already written for the t model with fixed degrees of freedom. When computing the Metropolis updates, we store the acceptance probabilities in an array, p.jump.nu, to monitor the efficiency of the jumping. Theoretical results given in Chapter 11 suggest that for a single parameter the optimal acceptance rate—that is, the average probability of successfully jumping—is approximately 44%. Thus we can vary sigma.jump.nu and run a pilot study to determine an acceptable value. In this case we can settle on a value such as sigma.jump.nu=1, which has an average jumping probability of about 0.4 for these data.

```
sigma.jump.nu <- 1
p.jump.nu <- array (NA, c(n.iter, n.chains))
sims <- array (NA, c(n.iter, n.chains, J+3))
dimnames (sims) <- list (NULL, NULL,
  c (paste ("theta[", 1:8, "]", sep=""), "mu", "tau", "nu"))
for (m in 1:n.chains){
  mu <- rnorm (1, mean(y), sd(y))
  tau <- runif (1, 0, sd(y))
  V <- runif (J, 0, sd(y))^2
  nu <- 1/runif(1, 0, 1)
  for (t in 1:n.iter){
    theta <- theta.update ()
    V <- V.update ()
    mu <- mu.update ()
    tau <- tau.update ()
    temp <- nu.update (sigma.jump.nu)
    nu <- temp$nu
    p.jump.nu[t,m] <- temp$p.jump
```

```
    sims[t,m,] <- c (theta, mu, tau, nu)
  }
}
print (mean (p.jump.nu))
round (monitor (sims), 1)
```

Parameter expansion for the t model

Finally, we can make the computations for the t model more efficient by applying parameter expansion. In the expanded parameterization, the new Gibbs sampler steps can be programmed in R as

```
gamma.update <- function (){
  gamma.hat <- (alpha*(y-mu)/sigma.y^2)/(1/V + alpha^2/sigma.y^2)
  V.gamma <- 1/(1/V + alpha^2/sigma.y^2)
  rnorm (J, gamma.hat, sqrt(V.gamma))
}
alpha.update <- function (){
  alpha.hat <- sum(gamma*(y-mu)/sigma.y^2)/sum(gamma^2/sigma.y^2)
  V.alpha <- 1/sum(gamma^2/sigma.y^2)
  rnorm (1, alpha.hat, sqrt(V.alpha))
}
mu.update <- function (){
  mu.hat <- sum((y-alpha*gamma)/sigma.y^2)/sum(1/sigma.y^2)
  V.mu <- 1/sum(1/sigma.y^2)
  rnorm (1, mu.hat, sqrt(V.mu))
}
tau.update <- function (){
  sqrt (rgamma (1, J*nu/2+1, (nu/2)*sum(1/V)))
}
V.update <- function (){
  (nu*tau^2 + gamma^2)/rchisq(J,nu+1)
}
nu.update <- function (sigma.jump){
  nu.inv.star <- rnorm(1, 1/nu, sigma.jump)
  if (nu.inv.star<=0 | nu.inv.star>1)
    p.jump <- 0
  else {
    nu.star <- 1/nu.inv.star
    log.post.old <- log.post (mu+alpha*gamma, alpha^2*V, mu,
                              abs(alpha)*tau, nu, y, sigma.y)
    log.post.star <- log.post (mu+alpha*gamma, alpha^2*V, mu,
                               abs(alpha)*tau, nu.star, y, sigma.y)
    r <- exp (log.post.star - log.post.old)
    nu <- ifelse (runif(1) < r, nu.star, nu)
    p.jump <- min(r,1)
  }
  return (nu=nu, p.jump=p.jump)
}
```

The posterior density can conveniently be calculated in terms of the original parameterization, as shown in the function `nu.update()` above. We can then run the Gibbs-Metropolis algorithm as before (see the program on the bottom part of page 605 and the very top of page 606), adding an initialization step for α just before the 'for (t in 1:n.iter)' loop:

```
alpha <- rnorm (1, 0, 1)
```

adding an updating step for α inside the loop,

```
alpha <- alpha.update ()
```

and replacing the last line inside the loop with simulations transformed to the original θ, μ, τ parameterization:

```
sims[t,m,] <- c (mu+alpha*gamma, mu, abs(alpha)*tau, nu)
```

We must once again tune the scale of the Metropolis jumps. We started for convenience at `sigma.jump.nu`= 1, and this time the average jumping probability for the Metropolis step is 17%. This is quite a bit lower than the optimal rate of 44% for one-dimensional jumping, and so we would expect to get a more efficient algorithm by decreasing the scale of the jumps (see Section 11.9). Reducing `sigma.jump.nu` to 0.5 yields an average acceptance probability `p.jump.nu` of 32%, and `sigma.jump.nu`= 0.3 yields an average jumping probability of 46% and somewhat more efficient simulations—that is, the draws of ν from the Gibbs-Metropolis algorithm are less correlated and yield a more accurate estimate of the posterior distribution. Decreasing `sigma.jump.nu` any further would make the acceptance rate too high and reduce the efficiency of the algorithm.

C.5 Further comments on computation

We have already given general computational tips at the end of Section 10.3: start by computing with simple models and compare to previous inferences when adding complexity. We also recommend getting started with smaller or simplified datasets, but this strategy was not really relevant to the current example with only 8 data points. Other practical issues that arise, and which we have discussed in Part III, include starting values, the choice of simulation algorithm, and methods for increasing simulation efficiency.

There are various ways in which the programs in this appendix could be made more computationally efficient. For example, in the Metropolis updating function `nu.update` for the t degrees of freedom in Section C.4, the log posterior density can be saved so that it does not need to be calculated twice at each step. It would also probably be good to use a more structured programming style in our R code (for example, in our updating functions `mu.update()`, `tau.update()`, and so forth) and perhaps to store the parameters and data as lists and pass them directly to the functions. We expect that there are many other ways in which our programs could be improved. Our general approach is to start with transparent (and possibly inefficient) code and then reprogram more efficiently once we know it is working.

We made several mistakes in the process of implementing the computations described in this appendix. Simplest were syntax errors in Bugs and related problems such as feeding in the wrong inputs when calling the bugs() function from R. We discovered and fixed these problems by using the debug option in bugs(); for example,

```
schools.sim <- bugs (data, inits, parameters, "schools.bug",
    n.chains=3, n.iter=10, debug=T)
```

and then inspecting the log file within the Bugs window.

We fixed syntax errors and other minor problems in the R code by cutting and pasting to run the scripts one line at a time, and by inserting print statements inside the R functions to display intermediate values.

We debugged the Bugs and R programs in this appendix by comparing them against each other, and by comparing each model to previously-fitted simpler models. We found many errors, including treating variances as standard deviations (for example, the command rnorm(1,alpha.hat,V.alpha) instead of rnorm(1,alpha.hat,sqrt(V.alpha)) when simulating from a normal distribution in R), confusion between ν and $1/\nu$, forgetting a term in the log-posterior density, miscalculating the Metropolis updating condition, and saving the wrong output in the sims array in the Gibbs sampling loop.

More serious conceptual errors included the poor choice of conjugate prior distribution for τ_θ in the Bugs model at the beginning of Section C.3, which we realized was a problem by comparing to the posterior simulations as shown in Figure C.3b. We also originally had an error in the programming of the reparameterized model (C.1), both in R and Bugs (we included the parameter μ in the model for γ_j rather than for y_j). We discovered this mistake because the inferences differed dramatically from the simpler parameterization.

As the examples in this appendix illustrate, Bayesian computation is not always easy, even for relatively simple models. However, once a model has been debugged, it can be applied and then generalized to work for a range of problems. Ultimately, we find Bayesian simulation to be a flexible tool for fitting realistic models to simple and complex data structures, and the steps required for debugging are often parallel to the steps required to build confidence in a model. We can use R to graphically display posterior inferences and predictive checks.

C.6 Bibliographic note

R is available at R Project (2002), and its parent software package S is described by Becker, Chambers, and Wilks (1988). Two statistics texts that use R extensively are Fox (2002) and Venables and Ripley (2002). Information about Bugs appears at Spiegelhalter et al. (1994, 2003), and many examples appear in the textbooks of Congdon (2001, 2003). R and Bugs have online documentation, and their websites have pointers to various help files and examples. Several efforts are currently underway to develop Bayesian inference tools using R, for example Martin and Quinn (2002b), Plummer (2003), and Warnes (2003).

References

The literature of Bayesian statistics is vast, especially in recent years. Instead of trying to be exhaustive, we supply here a selective list of references that may be useful for applied Bayesian statistics. Many of these sources have extensive reference lists of their own, which may be useful for an in-depth exploration of a topic. We also include references from non-Bayesian statistics and numerical analysis that present probability models or calculations relevant to Bayesian methods.

Agresti, A. (2002). *Categorical Data Analysis*, second edition. New York: Wiley.

Agresti, A., and Coull, B. A. (1998). Approximate is better than exact for interval estimation of binomial proportions. *American Statistician* **52**, 119–126.

Aitchison, J., and Dunsmore, I. R. (1975). *Statistical Prediction Analysis*. Cambridge University Press.

Aitkin, M., and Longford, N. (1986). Statistical modelling issues in school effectiveness studies (with discussion). *Journal of the Royal Statistical Society A* **149**, 1–43.

Akaike, H. (1973). Information theory and an extension of the maximum likelihood principle. In *Proceedings of the Second International Symposium on Information Theory*, ed. B. N. Petrov and F. Csaki, 267–281. Budapest: Akademiai Kiado. Reprinted in *Breakthroughs in Statistics*, ed. S. Kotz, 610–624. New York: Springer-Verlag (1992).

Albert, J. H. (1988). Bayesian estimation of Poisson means using a hierarchical log-linear model. In *Bayesian Statistics 3*, ed. J. M. Bernardo, M. H. DeGroot, D. V. Lindley, and A. F. M. Smith, 519–531. Oxford University Press.

Albert, J. H. (1992). Bayesian estimation of normal ogive item response curves using Gibbs sampling. *Journal of Educational Statistics* **17**, 251–269.

Albert, J. H., and Chib, S. (1993). Bayesian analysis of binary and polychotomous response data. *Journal of the American Statistical Association* **88**, 669–679.

Albert, J. H., and Chib, S. (1995). Bayesian residual analysis for binary response regression models. *Biometrika* **82**, 747–759.

Alpert, M., and Raiffa, H. (1984). A progress report on the training of probability assessors. In *Judgment Under Uncertainty: Heuristics and Biases*, ed.

Kahneman, D., Slovic, P., and Tversky, A., 294–305. Cambridge University Press.

Anderson, D. A. (1988). Some models for overdispersed binomial data. *Australian Journal of Statistics* **30**, 125–148.

Anderson, T. W. (1957). Maximum likelihood estimates for a multivariate normal distribution when some observations are missing. *Journal of the American Statistical Association* **52**, 200–203.

Andrieu, C., and Robert, C. (2001). Controlled MCMC for optimal sampling. Technical report, Department of Mathematics, University of Bristol.

Angrist, J., Imbens, G., and Rubin, D. B. (1996). Identification of causal effects using instrumental variables. *Journal of the American Statistical Association* **91**, 444–455.

Anscombe, F. J. (1963). Sequential medical trials. *Journal of the American Statistical Association* **58**, 365–383.

Ansolabehere, S., and Snyder, J. M. (2002). The incumbency advantage in U.S. elections: an analysis of state and federal offices, 1942–2000. *Election Law Journal*.

Atkinson, A. C. (1985). *Plots, Transformations, and Regression*. Oxford University Press.

Barbieri, M. M., and Berger, J. O. (2002). Optimal predictive model selection. Discussion Paper 02-02, Institute of Statistics and Decision Sciences, Duke University.

Barnard, G. A. (1949). Statistical inference (with discussion). *Journal of the Royal Statistical Society B* **11**, 115–139.

Barnard, G. A. (1985). Pivotal inference. In *Encyclopedia of Statistical Sciences*, Vol. 6, ed. S. Kotz, N. L. Johnson, and C. B. Read, 743–747. New York: Wiley.

Barnard, J., Frangakis, C., Hill, J., and Rubin, D. B. (2003). A principal stratification approach to broken randomized experiments: a case study of vouchers in New York City (with discussion). *Journal of the American Statistical Association*.

Barnard, J., McCulloch, R. E., and Meng, X. L. (2000). Modeling covariance matrices in terms of standard deviations and correlations, with application to shrinkage. *Statistica Sinica* **10**, 1281–1311.

Barry, S. C., Brooks, S. P., Catchpole, E. A., and Morgan, B. J. T. (2003). The analysis of ring-recovery data using random effects. *Biometrics* **59**, 54–65.

Baum, L. E., Petrie, T., Soules, G., and Weiss, N. (1970). A maximization technique occurring in the statistical analysis of probabilistic functions of Markov chains. *Annals of Mathematical Statistics* **41**, 164–171.

Bayarri, M. J. and Berger, J. (1998). Quantifying surprise in the data and model verification (with discussion). In *Bayesian Statistics 6*, ed. J. M. Bernardo, J. O. Berger, A. P. Dawid, and A. F. M. Smith, 53–82. Oxford University Press.

Bayarri, M. J. and Berger, J. (2000). P-values for composite null models (with discussion). *Journal of the American Statistical Association* **95**, 1127–1142.

Bayes, T. (1763). An essay towards solving a problem in the doctrine of chances. *Philosophical Transactions of the Royal Society*, 330–418. Reprinted, with biographical note by G. A. Barnard, in *Biometrika* **45**, 293–315 (1958).

Becker, R. A., Chambers, J. M., and Wilks, A. R. (1988). *The New S Language: A Programming Environment for Data Analysis and Graphics.* Pacific Grove, Calif.: Wadsworth.

Bedrick, E. J., Christensen, R., and Johnson, W. (1996). A new perspective on priors for generalized linear models. *Journal of the American Statistical Association* **91**, 1450–1460.

Belin, T. R., Diffendal, G. J., Mack, S., Rubin, D. B., Schafer, J. L., and Zaslavsky, A. M. (1993). Hierarchical logistic regression models for imputation of unresolved enumeration status in undercount estimation (with discussion). *Journal of the American Statistical Association* **88**, 1149–1166.

Belin, T. R., and Rubin, D. B. (1990). Analysis of a finite mixture model with variance components. *Proceedings of the American Statistical Association, Social Statistics Section*, 211–215.

Belin, T. R., and Rubin, D. B. (1995a). The analysis of repeated-measures data on schizophrenic reaction times using mixture models. *Statistics in Medicine* **14**, 747–768.

Belin, T. R., and Rubin, D. B. (1995b). A method for calibrating false-match rates in record linkage. *Journal of the American Statistical Association* **90**, 694–707.

Berger, J. O. (1984). The robust Bayesian viewpoint (with discussion). In *Robustness in Bayesian Statistics*, ed. J. Kadane. Amsterdam: North-Holland.

Berger, J. O. (1985). *Statistical Decision Theory and Bayesian Analysis*, second edition. New York: Springer-Verlag.

Berger, J. O. (1990). Robust Bayesian analysis: sensitivity to the prior. *Journal of Statistical Planning and Inference* **25**, 303–328.

Berger, J. O., and Berliner, L. M. (1986). Robust Bayes and empirical Bayes analysis with epsilon-contaminated priors. *Annals of Statistics* **14**, 461–486.

Berger, J. O., and Sellke, T. (1987). Testing a point null hypothesis: the irreconcilability of *P* values and evidence (with discussion). *Journal of the American Statistical Association* **82**, 112–139.

Berger, J. O., and Wolpert, R. (1984). *The Likelihood Principle.* Hayward, Calif.: Institute of Mathematical Statistics.

Berkhof, J., Van Mechelen, I., and Gelman, A. (2003). A Bayesian approach to the selection and testing of latent class models. *Statistica Sinica.*

Bernardinelli, L., Clayton, D. G., and Montomoli, C. (1995). Bayesian estimates of disease maps: how important are priors? *Statistics in Medicine* **14**, 2411–2431.

Bernardo, J. M. (1979). Reference posterior distributions for Bayesian inference (with discussion). *Journal of the Royal Statistical Society B* **41**, 113–147.

Bernardo, J. M., and Smith, A. F. M. (1994). *Bayesian Theory*. New York: Wiley.

Berry, D. A. (1996). *Statistics: A Bayesian Perspective*. Belmont, Calif.: Wadsworth.

Berzuini, C., Best, N. G., Gilks, W. R., and Larizza, C. (1997). Dynamic conditional independence models and Markov chain Monte Carlo methods. *Journal of the American Statistical Association* **92**, 1403–1412.

Besag, J. (1974). Spatial interaction and the statistical analysis of lattice systems (with discussion). *Journal of the Royal Statistical Society B* **36**, 192–236.

Besag, J. (1986). On the statistical analysis of dirty pictures (with discussion). *Journal of the Royal Statistical Society B* **48**, 259–302.

Besag, J., and Green, P. J. (1993). Spatial statistics and Bayesian computation. *Journal of the Royal Statistical Society B* **55**, 25–102.

Besag, J., Green, P., Higdon, D., and Mengersen, K. (1995). Bayesian computation and stochastic systems (with discussion). *Statistical Science* **10**, 3–66.

Besag, J., and Higdon, D. (1999). Bayesian analysis of agricultural field experiments (with discussion). *Journal of the Royal Statistical Society B* **61**, 691–746.

Besag, J., York, J., and Mollie, A. (1991). Bayesian image restoration, with two applications in spatial statistics (with discussion). *Annals of the Institute of Statistical Mathematics* **43**, 1–59.

Bickel, P., and Blackwell, D. (1967). A note on Bayes estimates. *Annals of Mathematical Statistics* **38**, 1907–1911.

Bloom, H. (1984). Accounting for no-shows in experimental evaluation designs. *Evaluation Review* **8**, 225–246.

Bock, R. D., ed. (1989). *Multilevel Analysis of Educational Data*. New York: Academic Press.

Boscardin, W. J. (1996). Bayesian computation for some hierarchical linear models. Ph.D. thesis, Department of Statistics, University of California, Berkeley.

Boscardin, W. J., and Gelman, A. (1996). Bayesian regression with parametric models for heteroscedasticity. *Advances in Econometrics* **11A**, 87–109.

Box, G. E. P. (1980). Sampling and Bayes inference in scientific modelling and robustness. *Journal of the Royal Statistical Society A* **143**, 383–430.

Box, G. E. P. (1983). An apology for ecumenism in statistics. In *Scientific Inference, Data Analysis, and Robustness*, ed. G. E. P. Box, T. Leonard, T., and C. F. Wu, 51–84. New York: Academic Press.

Box, G. E. P., and Cox, D. R. (1964). An analysis of transformations (with discussion). *Journal of the Royal Statistical Society B*, **26**, 211–252.

Box, G. E. P., Hunter, W. G., and Hunter, J. S. (1978). *Statistics for Experimenters*. New York: Wiley.

Box, G. E. P., and Jenkins, G. M. (1976). *Time Series Analysis: Forecasting and Control*, second edition. San Francisco: Holden-Day.

Box, G. E. P., and Tiao, G. C. (1962). A further look at robustness via Bayes's theorem. *Biometrika* **49**, 419–432.

Box, G. E. P., and Tiao, G. C. (1968). A Bayesian approach to some outlier problems. *Biometrika* **55**, 119–129.

Box, G. E. P., and Tiao, G. C. (1973). *Bayesian Inference in Statistical Analysis*. New York: Wiley Classics.

Bradley, R. A., and Terry, M. E. (1952). The rank analysis of incomplete block designs. 1. The method of paired comparisons. *Biometrika* **39**, 324–345.

Bradlow, E. T., and Fader, P. S. (2001). A Bayesian lifetime model for the "Hot 100" Billboard songs. *Journal of the American Statistical Association* **96**, 368–381.

Braun, H. I., Jones, D. H., Rubin, D. B., and Thayer, D. T. (1983). Empirical Bayes estimation of coefficients in the general linear model from data of deficient rank. *Psychometrika* **48**, 171–181.

Breslow, N. (1990). Biostatistics and Bayes (with discussion). *Statistical Science* **5**, 269–298.

Bretthorst, G. L. (1988). *Bayesian Spectrum Analysis and Parameter Estimation*. New York: Springer-Verlag.

Brewer, K. W. R. (1963). Ratio estimation in finite populations: some results deducible from the assumption of an underlying stochastic process. *Australian Journal of Statistics* **5**, 93–105.

Brillinger, D. R. (1981). *Time Series: Data Analysis and Theory*, expanded edition. San Francisco: Holden-Day.

Brooks, S. P., and Gelman, A. (1998). General methods for monitoring convergence of iterative simulations. *Journal of Computational and Graphical Statistics* **7**, 434–455.

Brooks, S. P., and Guidici, P. (2000). MCMC convergence assessment via two-way ANOVA. *Journal of Computational and Graphical Statistics* **9**, 266–285.

Brooks, S. P., Giudici, P., and Roberts, G. O. (2003). Efficient construction of reversible jump MCMC proposal distributions (with discussion). *Journal of the Royal Statistical Society B* **65**, 3–55.

Brooks, S. P., and Roberts, G. O. (1998). Assessing convergence of Markov chain Monte Carlo algorithms. *Statistics and Computing* **8**, 319–335.

Browner, W. S., and Newman, T. B. (1987). Are all significant P values created equal? *Journal of the American Medical Association* **257**, 2459–2463.

Burnham, K. P., and Anderson, D. R. (2002). *Model Selection and Multi-*

model Inference: A Practical Information Theoretic Approach. New York: Springer-Verlag.

Bush, R. R., and Mosteller, F. (1955). *Stochastic Models for Learning.* New York: Wiley.

Calvin, J. A., and Sedransk, J. (1991). Bayesian and frequentist predictive inference for the patterns of care studies. *Journal of the American Statistical Association* **86**, 36–48.

Carlin, B. P., and Chib, S. (1993). Bayesian model choice via Markov chain Monte Carlo. *Journal of the Royal Statistical Society B* **57**, 473–484.

Carlin, B. P., and Gelfand, A. E. (1993). Parametric likelihood inference for record breaking problems. *Biometrika* **80**, 507–515.

Carlin, B. P., and Louis, T. A. (2001). *Bayes and Empirical Bayes Methods for Data Analysis*, second edition. New York: Chapman & Hall.

Carlin, B. P., and Polson, N. G. (1991). Inference for nonconjugate Bayesian models using the Gibbs sampler. *Canadian Journal of Statistics* **19**, 399–405.

Carlin, J. B. (1992). Meta-analysis for 2×2 tables: a Bayesian approach. *Statistics in Medicine* **11**, 141–158.

Carlin, J. B., and Dempster, A. P. (1989). Sensitivity analysis of seasonal adjustments: empirical case studies (with discussion). *Journal of the American Statistical Association* **84**, 6–32.

Carlin, J. B., Stevenson, M. R., Roberts, I., Bennett, C. M., Gelman, A., and Nolan, T. (1997). Walking to school and traffic exposure in Australian children. *Australian and New Zealand Journal of Public Health* **21**, 286–292.

Carlin, J. B., Wolfe, R., Brown, C. H., and Gelman, A. (2001). A case study on the choice, interpretation, and checking of multilevel models for longitudinal binary outcomes. *Biostatistics* **2**, 397–416.

Carroll, R. J., Ruppert, D., and Stefanski, L. A. (1995). *Measurement Error in Nonlinear Models.* New York: Chapman & Hall.

Chaloner, K. (1991). Bayesian residual analysis in the presence of censoring. *Biometrika* **78**, 637–644.

Chaloner, K., and Brant, R. (1988). A Bayesian approach to outlier detection and residual analysis. *Biometrika* **75**, 651–659.

Chambers, J. M., Cleveland, W. S., Kleiner, B., and Tukey, P. A. (1983). *Graphical Methods for Data Analysis.* Pacific Grove, Calif.: Wadsworth.

Chen, M. H., Shao, Q. M., and Ibrahim, J. G. (2000). *Monte Carlo Methods in Bayesian Computation.* New York: Springer-Verlag.

Chernoff, H. (1972). *Sequential Analysis and Optimal Design.* Philadelphia: Society for Industrial and Applied Mathematics.

Chib, S. (1995). Marginal likelihood from the Gibbs output. *Journal of the American Statistical Association* **90**, 1313–1321.

Chib, S., and Greenberg, E. (1995). Understanding the Metropolis-Hastings algorithm. *American Statistician* **49**, 327–335.

Chib, S., and Jeliazkov, I. (2001). Marginal likelihood from the Metropolis-Hastings output. *Journal of the American Statistical Association* **96**, 270–281.

Chipman, H., George, E. I., and McCulloch, R. E. (1998). Bayesian CART model search (with discussion). *Journal of the American Statistical Association* **93**, 935–960.

Chipman, H., George, E. I., and McCulloch, R. E. (2001). The practical implementation of Bayesian model selection (with discussion). In *Model Selection* (Institute of Mathematical Statistics Lecture Notes 38), ed. P. Lahiri, 67–116.

Chipman, H., George, E. I., and McCulloch, R. E. (2002). Bayesian treed models. *Machine Learning* **48**, 299–320.

Chipman, H., Kolaczyk, E., and McCulloch, R. E. (1997). Adaptive Bayesian wavelet shrinkage. *Journal of the American Statistical Association* **92**, 1413–1421.

Clayton, D. G. (1991). A Monte Carlo method for Bayesian inference in frailty models. *Biometrics* **47**, 467–485.

Clayton, D. G., and Bernardinelli, L. (1992). Bayesian methods for mapping disease risk. In *Geographical and Environmental Epidemiology: Methods for Small-Area Studies*, ed. P. Elliott, J. Cusick, D. English, and R. Stern, 205–220. Oxford University Press.

Clayton, D. G., and Kaldor, J. M. (1987). Empirical Bayes estimates of age-standardized relative risks for use in disease mapping. *Biometrics* **43**, 671–682.

Clemen, R. T. (1996). *Making Hard Decisions*, second edition. Belmont, Calif.: Duxbury Press.

Cleveland, W. S. (1985). *The Elements of Graphing Data*. Monterey, Calif.: Wadsworth.

Cleveland, W. S. (1993). *Envisioning Information*. Summit, N.J.: Hobart.

Clogg, C. C., Rubin, D. B., Schenker, N., Schultz, B., and Wideman, L. (1991). Multiple imputation of industry and occupation codes in Census public-use samples using Bayesian logistic regression. *Journal of the American Statistical Association* **86**, 68–78.

Clyde, M., DeSimone, H. and Parmigiani, G. (1996). Prediction via orthogonalized model mixing. *Journal of the American Statistical Association* **91**, 1197–1208.

Congdon, P. (2001). *Bayesian Statistical Modelling*. New York: Wiley.

Congdon, P. (2003). *Applied Bayesian Modelling*. New York: Wiley.

Connors, A. F., Speroff, T., Dawson, N. V., Thomas, C., Harrell, F. E., Wagner, D., Desbiens, N., Goldman, L., Wu, A. W., Califf, R. M., Fulkerson, W. J., Vidaillet, H., Broste, S., Bellamy, P., Lynn, J., and Knauss, W. A. (1996). The effectiveness of right heart catheterization in the initial care of critically ill patients. *Journal of the American Medical Association* **276**, 889–997.

Cowles, M. K., and Carlin, B. P. (1996). Markov chain Monte Carlo convergence diagnostics: a comparative review. *Journal of the American Statistical Association* **91**, 833–904.

Cox, D. R., and Hinkley, D. V. (1974). *Theoretical Statistics*. New York: Chapman & Hall.

Cox, D. R., and Snell, E. J. (1981). *Applied Statistics*. New York: Chapman & Hall.

Cox, G. W., and Katz, J. (1996). Why did the incumbency advantage grow? *American Journal of Political Science* **40**, 478–497.

Cressie, N. A. C. (1993). *Statistics for Spatial Data*, second edition. New York: Wiley.

Dalal, S. R., Fowlkes, E. B., and Hoadley, B. (1989). Risk analysis of the space shuttle: pre-Challenger prediction of failure. *Journal of the American Statistical Association* **84**, 945–957.

Daniels, M. J., and Kass, R. E. (1999). Nonconjugate Bayesian estimation of covariance matrices and its use in hierarchical models. *Journal of the American Statistical Association* **94**, 1254–1263.

Daniels, M. J., and Kass, R. E. (2001). Shrinkage estimators for covariance matrices. *Biometrics* **57**, 1173–1184.

Daniels, M. J., and Pourahmadi, M. (2002). Bayesian analysis of covariance matrices and dynamic models for longitudinal data. *Biometrika* **89**, 553–566.

David, H. A. (1988). *The Method of Paired Comparisons*, second edition. Oxford University Press.

David, M. H., Little, R. J. A., Samuhel, M. E., and Triest, R. K. (1986). Alternative methods for CPS income imputation. *Journal of the American Statistical Association* **81**, 29–41.

Datta, G. S., Lahiri, P., Maiti, T., and Lu, K. L. (1999). Hierarchical Bayes estimation of unemployment rates for the states of the U.S. *Journal of the American Statistical Association* **94**, 1074–1082.

Davidson, R. R., and Beaver, R. J. (1977). On extending the Bradley–Terry model to incorporate within-pair order effects. *Biometrics* **33**, 693–702.

Dawid, A. P. (1982). The well-calibrated Bayesian (with discussion). *Journal of the American Statistical Association* **77**, 605–610.

Dawid, A. P. (1986). Probability forecasting. In *Encyclopedia of Statistical Sciences*, Vol. 7, ed. S. Kotz, N. L. Johnson, and C. B. Read, 210–218. New York: Wiley.

Dawid, A. P. (2000). Causal inference without counterfactuals (with discussion). *Journal of the American Statistical Association* **95**, 407–448.

Dawid, A. P., and Dickey, J. M. (1977). Likelihood and Bayesian inference from selectively reported data. *Journal of the American Statistical Association* **72**, 845–850.

Dawid, A. P., Stone, M., and Zidek, J. V. (1973). Marginalization paradoxes in Bayesian and structural inferences (with discussion). *Journal of the Royal Statistical Society B*, **35**, 189–233.

Deely, J. J., and Lindley, D. V. (1981). Bayes empirical Bayes. *Journal of the American Statistical Association* **76**, 833–841.

de Finetti, B. (1974). *Theory of Probability.* New York: Wiley.

DeGroot, M. H. (1970). *Optimal Statistical Decisions.* New York: McGraw-Hill.

Dehejia, R. (2003). Program evaluation as a decision problem. *Journal of Econometrics.*

Dehejia, R., and Wahba, S. (1999). Causal effects in non-experimental studies: re-evaluating the evaluation of training programs. *Journal of the American Statistical Association* **94**, 1053–1062.

Dellaportas, P., and Smith, A. F. M. (1993). Bayesian inference for generalized linear and proportional hazards models via Gibbs sampling. *Applied Statistics* **42**, 443–459.

Deming, W. E., and Stephan, F. F. (1940). On a least squares adjustment of a sampled frequency table when the expected marginal totals are known. *Annals of Mathematical Statistics* **11**, 427–444.

Dempster, A. P. (1971). Model searching and estimation in the logic of inference. In *Proceedings of the Symposium on the Foundations of Statistical Inference*, ed. V. P. Godambe and D. A. Sprott, 56–81. Toronto: Holt, Rinehart, and Winston.

Dempster, A. P. (1974). The direct use of likelihood for significance testing. In *Proceedings of the Conference on the Foundational Questions of Statistical Inference*, ed. O. Barndorff-Nielsen, P. Blaesild, and G. Schou, 335–352. University of Aaarhus, Denmark. Reprinted in *Statistics and Computing* **7**, 247–252 (1997).

Dempster, A. P. (1975). A subjectivist look at robustness. *Bulletin of the International Statistical Institute* **46**, 349–374.

Dempster, A. P., Laird, N. M., and Rubin, D. B. (1977). Maximum likelihood from incomplete data via the EM algorithm (with discussion). *Journal of the Royal Statistical Society B* **39**, 1–38.

Dempster, A. P., and Raghunathan, T. E. (1987). Using a covariate for small area estimation: a common sense Bayesian approach. In *Small Area Statistics: An International Symposium*, ed. R. Platek, J. N. K. Rao, C. E. Sarndal, and M. P. Singh, 77–90. New York: Wiley.

Dempster, A. P., Rubin, D. B., and Tsutakawa, R. K. (1981). Estimation in covariance components models. *Journal of the American Statistical Association* **76**, 341–353.

Dempster, A. P., Selwyn, M. R., and Weeks, B. J. (1983). Combining historical and randomized controls for assessing trends in proportions. *Journal of the American Statistical Association* **78**, 221–227.

Denison, D. G. T., Holmes, C. C., Malick, B. K., and Smith, A. F. M. (2002). *Bayesian Methods for Nonlinear Classification and Regression.* New York: Wiley.

DiMatteo, I., Genovese, C. R., and Kass, R. E. (2001). Bayesian curve-fitting with free-knot splines. *Biometrika* **88**, 1055–1071.

Dobra, A., Tebaldi, C., and West, M. (2003). Bayesian inference for incomplete multi-way tables. Technical report, Institute of Statistics and Decision Sciences, Duke University.

Dominici, F., Parmigiani, G., Wolpert, R. L., and Hasselblad, V. (1999). Meta-analysis of migraine headache treatments: combining information from heterogeneous designs. *Journal of the American Statistical Association* **94**, 16–28.

Donoho, D. L., Johnstone, I. M., Hoch, J. C., and Stern, A. S. (1992). Maximum entropy and the nearly black object (with discussion). *Journal of the Royal Statistical Society B* **54**, 41–81.

Draper, D. (1995). Assessment and propagation of model uncertainty (with discussion). *Journal of the Royal Statistical Society B* **57**, 45–97.

Draper, D., Hodges, J. S., Mallows, C. L., and Pregibon, D. (1993). Exchangeability and data analysis. *Journal of the Royal Statistical Society A* **156**, 9–37.

Duane, S., Kennedy, A. D., Pendleton, B. J., and Roweth, D. (1987). Hybrid Monte Carlo. *Physics Letters B* **195**, 216–222.

DuMouchel, W. M. (1990). Bayesian meta-analysis. In *Statistical Methodology in the Pharmaceutical Sciences*, ed. D. A. Berry, 509–529. New York: Marcel Dekker.

DuMouchel, W. M., and Harris, J. E. (1983). Bayes methods for combining the results of cancer studies in humans and other species (with discussion). *Journal of the American Statistical Association* **78**, 293–315.

Edwards, W., Lindman, H., and Savage, L. J. (1963). Bayesian statistical inference for psychological research. *Psychological Review* **70**, 193–242.

Efron, B. (1971). Forcing a sequential experiment to be balanced. *Biometrika* **58**, 403–417.

Efron, B., and Morris, C. (1971). Limiting the risk of Bayes and empirical Bayes estimators—Part I: The Bayes case. *Journal of the American Statistical Association* **66**, 807–815.

Efron, B., and Morris, C. (1972). Limiting the risk of Bayes and empirical Bayes estimators—Part II: The empirical Bayes case. *Journal of the American Statistical Association* **67**, 130–139.

Efron, B., and Morris, C. (1975). Data analysis using Stein's estimator and its generalizations. *Journal of the American Statistical Association* **70**, 311–319.

Efron, B., and Thisted, R. (1976). Estimating the number of unseen species: How many words did Shakespeare know? *Biometrika* **63**, 435–448.

Efron, B., and Tibshirani, R. (1993). *An Introduction to the Bootstrap.* New York: Chapman & Hall.

Ehrenberg, A. S. C. (1986). Discussion of Racine et al. (1986). *Applied Statistics* **35**, 135–136.

Ericson, W. A. (1969). Subjective Bayesian models in sampling finite populations, I. *Journal of the Royal Statistical Society B* **31**, 195–234.

Fay, R. E., and Herriot, R. A. (1979). Estimates of income for small places: an application of James-Stein procedures to census data. *Journal of the American Statistical Association* **74**, 269–277.

Fearn, T. (1975). A Bayesian approach to growth curves. *Biometrika* **62**, 89–100.

Feller, W. (1968). *An Introduction to Probability Theory and its Applications,* Vol. 1, third edition. New York: Wiley.

Fienberg, S. E. (1977). *The Analysis of Cross-Classified Categorical Data.* Cambridge, Mass.: M.I.T. Press.

Fienberg, S. E. (2000). Contingency tables and log-linear models: basic results and new developments. *Journal of the American Statistical Association* **95**, 643–647.

Fill, J. A. (1998). An interruptible algorithm for perfect sampling. *Annals of Applied Probability* **8**, 131–162.

Fisher, R. A. (1922). On the mathematical foundations of theoretical statistics. *Philosophical Transactions of the Royal Society* **222**, 309–368.

Fisher, R. A., Corbet, A. S., and Williams, C. B. (1943). The relation between the number of species and the number of individuals in a random sample of an animal population. *Journal of Animal Ecology* **12**, 42–58.

Ford, E. S., Kelly, A. E., Teutsch, S. M., Thacker, S. B., and Garbe, P. L. (1999). Radon and lung cancer: a cost-effectiveness analysis. *American Journal of Public Health* **89**, 351–357.

Fox, J. (2002). *An R and S-Plus Companion to Applied Regression.* Sage.

Frangakis, C. and Rubin, D. B. (2002). Principal stratification in causal inference. *Biometrics* **58**, 21–29.

Freedman, L. S., Spiegelhalter, D. J., and Parmar, M. K. B. (1994). The what, why and how of Bayesian clinical trials monitoring. *Statistics in Medicine* **13**, 1371–1383.

Gatsonis, C., Hodges, J. S., Kass, R. E., Singpurwalla, N. D., West, M., Carlin, B. P., Carriquiry, A., Gelman, A., Pauler, D., Verdinelli, I., and Wakefield, J., eds. (1993–2002). *Case Studies in Bayesian Statistics,* volumes 1–7. New York: Springer-Verlag.

Gaver, D. P., and O'Muircheartaigh, I. G. (1987). Robust empirical Bayes analyses of event rates. *Technometrics* **29**, 1–15.

Geisser, S. (1986). Predictive analysis. In *Encyclopedia of Statistical Sciences,*

Vol. 7, ed. S. Kotz, N. L. Johnson, and C. B. Read, 158–170. New York: Wiley.

Geisser, S., and Eddy, W. F. (1979). A predictive approach to model selection. *Journal of the American Statistical Association* **74**, 153–160.

Gelfand, A. E., Dey, D. K., and Chang, H. (1992). Model determination using predictive distributions with implementation via sampling-based methods (with discussion). In *Bayesian Statistics 4*, ed. J. M. Bernardo, J. O. Berger, A. P. Dawid, and A. F. M. Smith, 147–167. Oxford University Press.

Gelfand, A. E., and Ghosh, S. (1998). Model choice: a minimum posterior predictive loss approach. *Biometrika* **85**, 1–11.

Gelfand, A. E., Hills, S. E., Racine-Poon, A., and Smith, A. F. M. (1990). Illustration of Bayesian inference in normal data models using Gibbs sampling. *Journal of the American Statistical Association* **85**, 972–985.

Gelfand, A. E., and Sahu, S. K. (1994). On Markov chain Monte Carlo acceleration. *Journal of Compuational and Graphical Statistics* **3**, 261–276.

Gelfand, A. E., and Sahu, S. K. (1999). Identifiability, improper priors, and Gibbs sampling for generalized linear models. *Journal of the American Statistical Association* **94**, 247–253.

Gelfand, A. E., Sahu, S. K., and Carlin, B. P. (1995). Efficient parametrizations for normal linear mixed models. *Biometrika* **82**, 479–488.

Gelfand, A. E., and Smith, A. F. M. (1990). Sampling-based approaches to calculating marginal densities. *Journal of the American Statistical Association* **85**, 398–409.

Gelman, A. (1992a). Discussion of 'Maximum entropy and the nearly black object,' by Donoho et al. *Journal of the Royal Statistical Society B* **54**, 72.

Gelman, A. (1992b). Iterative and non-iterative simulation algorithms. *Computing Science and Statistics* **24**, 433–438.

Gelman, A. (2003). A Bayesian formulation of exploratory data analysis and goodness-of-fit testing. *International Statistical Review*.

Gelman, A. (2004a). Exploratory data analysis for complex models. *Journal of Computational and Graphical Statistics*.

Gelman, A. (2004b). Analysis of variance: why it is more important than ever (with discussion). *Annals of Statistics*.

Gelman, A., Bois, F. Y., and Jiang, J. (1996). Physiological pharmacokinetic analysis using population modeling and informative prior distributions. *Journal of the American Statistical Association* **91**, 1400–1412.

Gelman, A., and Carlin, J. B. (2001). Poststratification and weighting adjustments. In *Survey Nonresponse*, ed. R. M. Groves, D. A. Dillman, J. L. Eltinge, and R. J. A. Little. New York: Wiley.

Gelman, A., Chew, G. L., and Shnaidman, M. (2004). Bayesian analysis of serial dilution assays. *Biometrics*.

Gelman, A., Fagan, J., and Kiss, A. (2007). An analysis of the NYPD's stop-

and-frisk policy in the context of claims of racial bias. *Journal of the American Statistical Association*.

Gelman, A., Goegebeur, Y., Tuerlinckx, F., and Van Mechelen, I. (2000). Diagnostic checks for discrete-data regression models using posterior predictive simulations. *Applied Statistics* **49**, 247–268.

Gelman, A., and Huang, Z. (2003). Estimating incumbency advantage and its variation, as an example of a before-after study. Technical report, Department of Statistics, Columbia University.

Gelman, A., Katz, J. N., and Tuerlinckx, F. (2002). The mathematics and statistics of voting power. *Statistical Science* **17**, 420–435.

Gelman, A., and King, G. (1990a). Estimating incumbency advantage without bias. *American Journal of Political Science* **34**, 1142–1164.

Gelman, A., and King, G. (1990b). Estimating the electoral consequences of legislative redistricting. *Journal of the American Statistical Association* **85**, 274–282.

Gelman, A., and King, G. (1993). Why are American Presidential election campaign polls so variable when votes are so predictable? *British Journal of Political Science* **23**, 409–451.

Gelman, A., King, G., and Boscardin, W. J. (1998). Estimating the probability of events that have never occurred: when does your vote matter? *Journal of the American Statistical Association* **93**, 1–9.

Gelman, A., King, G., and Liu, C. (1998). Multiple imputation for multiple surveys (with discussion). *Journal of the American Statistical Association* **93**, 846–874.

Gelman, A., and Little, T. C. (1997). Poststratification into many categories using hierarchical logistic regression. *Survey Methodology* **23**, 127–135.

Gelman, A., and Meng, X. L. (1998). Simulating normalizing constants: from importance sampling to bridge sampling to path sampling. *Statistical Science* **13**, 163–185.

Gelman, A., Meng, X. L., and Stern, H. S. (1996). Posterior predictive assessment of model fitness via realized discrepancies (with discussion). *Statistica Sinica* **6**, 733–807.

Gelman, A., and Nolan, D. (2002a). *Teaching Statistics: A Bag of Tricks*. Oxford University Press.

Gelman, A., and Nolan, D. (2002b). You can load a die but you can't bias a coin. *American Statistician* **56**, 308–311.

Gelman, A., and Nolan, D. (2002c). A probability model for golf putting. *Teaching Statistics* **24**, 93–95.

Gelman, A., and Price, P. N. (1999). All maps of parameter estimates are misleading. *Statistics in Medicine* **18**, 3221–3234.

Gelman, A., and Raghunathan, T. E. (2001). Using conditional distributions for missing-data imputation. Discussion of 'Conditionally specified distributions,' by Arnold et al. *Statistical Science* **3**, 268–269.

Gelman, A., Roberts, G., and Gilks, W. (1995). Efficient Metropolis jumping rules. In *Bayesian Statistics 5*, ed. J. M. Bernardo, J. O. Berger, A. P. Dawid, and A. F. M. Smith. Oxford University Press.

Gelman, A., and Rubin, D. B. (1991). Simulating the posterior distribution of loglinear contingency table models. Technical report.

Gelman, A., and Rubin, D. B. (1992a). A single sequence from the Gibbs sampler gives a false sense of security. In *Bayesian Statistics 4*, ed. J. M. Bernardo, J. O. Berger, A. P. Dawid, and A. F. M. Smith, 625–631. Oxford University Press.

Gelman, A., and Rubin, D. B. (1992b). Inference from iterative simulation using multiple sequences (with discussion). *Statistical Science* 7, 457–511.

Gelman, A., and Rubin, D. B. (1995). Avoiding model selection in Bayesian social research. Discussion of Raftery (1995b). In *Sociological Methodology 1995*, ed. P. V. Marsden, 165–173.

Gelman, A., Stevens, M., and Chan, V. (2003). Regression modeling and meta-analysis for decision making: a cost-benefit analysis of a incentives in telephone surveys. *Journal of Business and Economic Statistics*.

Gelman, A., and Tuerlinckx, F. (2000). Type S error rates for classical and Bayesian single and multiple comparison procedures. *Computational Statistics* 15, 373–390.

Gelman, A., Huang, Z., van Dyk, D. A., and Boscardin, W. J. (2007). Using redundant parameters to fit hierarchical models. *Journal of Computational and Graphical Statistics*.

Geman, S., and Geman, D. (1984). Stochastic relaxation, Gibbs distributions, and the Bayesian restoration of images. *IEEE Transactions on Pattern Analysis and Machine Intelligence* 6, 721–741.

Genovese, C. R. (2001). A Bayesian time-course model for functional magnetic resonance imaging data (with discussion). *Journal of the American Statistical Association* 95, 691–703.

George, E. I., and McCulloch, R. E. (1993). Variable selection via Gibbs sampling. *Journal of the American Statistical Association* 88, 881–889.

Geweke, J. (1989). Bayesian inference in econometric models using Monte Carlo integration. *Econometrica* 57, 1317–1339.

Geyer, C. J. (1991). Markov chain Monte Carlo maximum likelihood. *Computing Science and Statistics* 23, 156–163.

Geyer, C. J., and Thompson, E. A. (1992). Constrained Monte Carlo maximum likelihood for dependent data (with discussion). *Journal of the Royal Statistical Society B* 54, 657–699.

Geyer, C. J., and Thompson, E. A. (1993). Annealing Markov chain Monte Carlo with applications to pedigree analysis. Technical report, School of Statistics, University of Minnesota.

Gilks, W. R., and Berzuini, C. (2001). Following a moving target: Monte Carlo inference for dynamic Bayesian models. *Journal of the Royal Statistical Society B* **63**, 127–146.

Gilks, W. R., Best, N., and Tan, K. K. C. (1995). Adaptive rejection Metropolis sampling within Gibbs sampling. *Applied Statistics* **44**, 455–472.

Gilks, W. R., Clayton, D. G., Spiegelhalter, D. J., Best, N. G., McNeil, A. J., Sharples, L. D., and Kirby, A. J. (1993). Modelling complexity: applications of Gibbs sampling in medicine. *Journal of the Royal Statistical Society B* **55**, 39–102.

Gilks, W. R., Richardson, S., and Spiegelhalter, D., eds. (1996). *Practical Markov Chain Monte Carlo.* New York: Chapman & Hall.

Gilks, W. R., and Wild, P. (1992). Adaptive rejection sampling for Gibbs sampling. *Applied Statistics* **41**, 337–348.

Gill, J. (2002). *Bayesian Methods for the Social and Behavioral Sciences.* New York: Chapman & Hall.

Gill, P. E., Murray, W., and Wright, M. H. (1981). *Practical Optimization.* New York: Academic Press.

Gilovich, T., Griffin, D., and Kahneman, D. (2002). *Heuristics and Biases: The Psychology of Intuitive Judgment.* Cambridge University Press.

Giltinan, D., and Davidian, M. (1995). *Nonlinear Models for Repeated Measurement Data.* London: Chapman & Hall.

Glickman, M. E. (1993). Paired comparison models with time-varying parameters. Ph.D. thesis, Department of Statistics, Harvard University.

Glickman, M. E., and Normand, S. L. (2000). The derivation of a latent threshold instrumental variables model. *Statistica Sinica* **10**, 517–544.

Glickman, M. E. and Stern, H. S. (1998). A state-space model for National Football League scores. *Journal of the American Statistical Association* **93** 25–35.

Goldstein, H. (1995). *Multilevel Statistical Models,* second edition. London: Edward Arnold.

Goldstein, H., and Silver, R. (1989). Multilevel and multivariate models in survey analysis. In *Analysis of Complex Surveys,* ed. C. J. Skinner, D. Holt, and T. M. F. Smith, 221–235. New York: Wiley.

Goldstein, M. (1976). Bayesian analysis of regression problems. *Biometrika* **63**, 51–58.

Golub, G. H., and van Loan, C. F. (1983). *Matrix Computations.* Baltimore, Md.: Johns Hopkins University Press.

Good, I. J. (1950). *Probability and the Weighing of Evidence.* New York: Hafner.

Good, I. J. (1965). *The Estimation of Probabilities: An Essay on Modern Bayesian Methods.* Cambridge, Mass.: M.I.T. Press.

Goodman, L. A. (1952). Serial number analysis. *Journal of the American Statistical Association* **47**, 622–634.

Goodman, L. A. (1991). Measures, models, and graphical displays in the analysis of cross-classified data (with discussion). *Journal of the American Statistical Association* **86**, 1085–1111.

Goodman, S. N. (1999). Toward evidence-based medical statistics. 1: The p value fallacy. *Annals of Internal Medicine* **130**, 995–1013.

Goodman, S. N. (1999). Toward evidence-based medical statistics. 2: The Bayes factor. *Annals of Internal Medicine* **130**, 1019–1021.

Green, P. J. (1995). Reversible jump Markov chain Monte Carlo computation and Bayesian model determination. *Biometrika* **82**, 711–732.

Greenland, S., Robins, J. M., and Pearl, J. (1999). Confounding and collapsability in causal inference. *Statistical Science* **14**, 29–46.

Grenander, U. (1983). *Tutorial in Pattern Theory.* Division of Applied Mathematics, Brown University.

Groves, R. M. (1989). *Survey Errors and Survey Costs.* New York: Wiley.

Groves, R. M., Dillman, D. A., Eltinge, J. L., and Little, R. J. A., eds. (2002). *Survey Nonresponse.* New York: Wiley.

Gull, S. F. (1989a). Developments in maximum entropy data analysis. In *Maximum Entropy and Bayesian Methods*, ed. J. Skilling, 53–71. Dordrecht, Netherlands: Kluwer Academic Publishers.

Gull, S. F. (1989b). Bayesian data analysis: straight-line fitting. In *Maximum Entropy and Bayesian Methods*, ed. J. Skilling, 511–518. Dordrecht, Netherlands: Kluwer Academic Publishers.

Guttman, I. (1967). The use of the concept of a future observation in goodness-of-fit problems. *Journal of the Royal Statistical Society B* **29**, 83–100.

Hammersley, J. M., and Handscomb, D. C. (1964). *Monte Carlo Methods.* New York: Wiley.

Hansen, M., and Yu, B. (2001). Model selection and the principle of minimum description length. *Journal of the American Statistical Association* **96**, 746–774.

Hartigan, J. (1964). Invariant prior distributions. *Annals of Mathematical Statistics* **35**, 836–845.

Hartley, H. O., and Rao, J. N. K. (1967). Maximum likelihood estimation for the mixed analysis of variance model. *Biometrika* **54**, 93–108.

Harville, D. (1980). Predictions for NFL games with linear-model methodology. *Journal of the American Statistical Association* **75**, 516–524.

Hastie, T. J., and Tibshirani, R. J. (1990). *Generalized Additive Models.* New York: Chapman & Hall.

Hastings, W. K. (1970). Monte Carlo sampling methods using Markov chains and their applications. *Biometrika* **57**, 97–109.

Heckman, J. (1979). Sample selection bias as a specification error. *Econometrica* **47**, 153–161.

Heitjan, D. F. (1989). Inference from grouped continuous data: a review (with discussion). *Statistical Science* **4**, 164–183.

Heitjan, D. F., and Landis, J. R. (1994). Assessing secular trends in blood pressure: a multiple-imputation approach. *Journal of the American Statistical Association* **89**, 750–759.

Heitjan, D. F., Moskowitz, A. J., and Whang, W. (1999). Bayesian estimation of cost-effectiveness ratios from clinical trials. *Health Economics* **8**, 191–201.

Heitjan, D. F., and Rubin, D. B. (1990). Inference from coarse data via multiple imputation with application to age heaping. *Journal of the American Statistical Association* **85**, 304–314.

Heitjan, D. F., and Rubin, D. B. (1991). Ignorability and coarse data. *Annals of Statistics* **19**, 2244–2253.

Henderson, C. R., Kempthorne, O., Searle, S. R., and Von Krosigk, C. M. (1959). The estimation of environmental and genetic trends from records subject to culling. *Biometrics* **15**, 192–218.

Higdon, D. M. (1998). Auxiliary variable methods for Markov chain Monte Carlo with applications. *Journal of the American Statistical Association* **93**, 585–595.

Higgins, K. M., Davidian, M., Chew, G., and Burge, H. (1998). The effect of serial dilution error on calibration inference in immunoassay. *Biometrics* **54**, 19–32.

Hill, B. M. (1965). Inference about variance components in the one-way model. *Journal of the American Statistical Association* **60**, 806–825.

Hills, S. E., and Smith, A. F. M. (1992). Parameterization issues in Bayesian inference (with discussion). In *Bayesian Statistics 4*, ed. J. M. Bernardo, J. O. Berger, A. P. Dawid, and A. F. M. Smith, 227–246. Oxford University Press.

Hinde, J. (1982). Compound Poisson regression models. In *GLIM-82: Proceedings of the International Conference on Generalized Linear Models*, ed. R. Gilchrist (Lecture Notes in Statistics 14), 109–121. New York: Springer-Verlag.

Hinkley, D. V., and Runger, G. (1984). The analysis of transformed data (with discussion). *Journal of the American Statistical Association* **79**, 302–320.

Hirano, K., Imbens, G., Rubin, D. B., and Zhao, X. H. (2000). Estimating the effect of an influenza vaccine in an encouragement design. *Biostatistics* **1**, 69–88.

Hodges, J. S. (1998). Some algebra and geometry for hierarchical models, applied to diagnostics (with discussion). *Journal of the Royal Statistical Society B* **60**, 497–536.

Hodges, J. S., and Sargent, D. J. (2001). Counting degrees of freedom in hierarchical and other richly parameterized models. *Biometrika* **88**, 367–379.

Hoerl, A. E., and Kennard, R. W. (1970). Ridge regression: biased estimation for nonorthogonal problems. *Technometrics* **12**, 55–67.

Hoeting, J., Madigan, D., Raftery, A. E., and Volinsky, C. (1999). Bayesian model averaging (with discussion). *Statistical Science* **14**, 382–417.

Hogan, H. (1992). The 1990 post-enumeration survey: an overview. *American Statistician* **46**, 261–269.

Hui, S. L., and Berger, J. O. (1983). Empirical Bayes estimation of rates in longitudinal studies. *Journal of the American Statistical Association* **78**, 753–760.

Imai, K., and van Dyk, D. A. (2003). A Bayesian analysis of the multinomial probit model using marginal data augmentation. *Journal of Econometrics*.

Imbens, G., and Angrist, J. (1994). Identification and estimation of local average treatment effects. *Econometrica* **62**, 467–475.

Imbens, G., and Rubin, D. B. (1997). Bayesian inference for causal effects in randomized experiments with noncompliance. *Annals of Statistics* **25**, 305–327.

Jackman, S. (2001). Multidimensional analysis of roll call data via Bayesian simulation: identification, estimation, inference and model checking. *Political Analysis* **9**, 227–241.

James, W., and Stein, C. (1960). Estimation with quadratic loss. In *Proceedings of the Fourth Berkeley Symposium* **1**, ed. J. Neyman, 361–380. Berkeley: University of California Press.

James, W. H. (1987). The human sex ratio. Part 1: a review of the literature. *Human Biology* **59**, 721–752.

Jaynes, E. T. (1976). Confidence intervals vs. Bayesian intervals (with discussion). In *Foundations of Probability Theory, Statistical Inference, and Statistical Theories of Science*, ed. W. L. Harper and C. A. Hooker. Dordrecht, Netherlands: Reidel. Reprinted in Jaynes (1983).

Jaynes, E. T. (1980). Marginalization and prior probabilities. In *Bayesian Analysis in Econometrics and Statistics*, ed. A. Zellner, 43–87. Amsterdam: North-Holland. Reprinted in Jaynes (1983).

Jaynes, E. T. (1982). On the rationale of maximum-entropy methods. *Proceedings of the IEEE* **70**, 939–952.

Jaynes, E. T. (1983). *Papers on Probability, Statistics, and Statistical Physics*, ed. R. D. Rosenkrantz. Dordrecht, Netherlands: Reidel.

Jaynes, E. T. (1987). Bayesian spectrum and chirp analysis. In *Maximum-Entropy and Bayesian Spectral Analysis and Estimation Problems*, ed. C. R. Smith and G. J. Erickson, 1–37. Dordrecht, Netherlands: Reidel.

Jaynes, E. T. (1996). *Probability Theory: The Logic of Science*. bayes.wustl.edu/etj/prob.html

Jeffreys, H. (1961). *Theory of Probability*, third edition. Oxford University Press.

Johnson, N. L., and Kotz, S. (1972). *Distributions in Statistics*, 4 vols. New York: Wiley.

Johnson, V. E. (1996). On Bayesian analysis of multirater ordinal data: an application to automated essay grading. *Journal of the American Statistical Association* **91**, 42–51.

Johnson, V. E. (1997). An alternative to traditional GPA for evaluating student performance (with discussion). *Statistical Science* **12**, 251–278.

Johnson, V. E. (2002). A Bayesian χ^2 test for goodness-of-fit. Technical report, Department of Biostatistics, University of Michigan.

Kadane, J. B., and Seidenfeld, T. (1990). Randomization in a Bayesian perspective. *Journal of Statistical Planning and Inference* **25**, 329–345.

Kahneman, D., Slovic, P., and Tversky, A. (1982). *Judgment Under Uncertainty: Heuristics and Biases.* Cambridge University Press.

Kahneman, D., and Tversky, A. (1972). Subjective probability: a judgment of representativeness. *Cognitive Psychology* **3**, 430–454. Reprinted in *Judgment Under Uncertainty: Heuristics and Biases*, ed. Kahneman, D., Slovic, P., and Tversky, A., 32–47. Cambridge University Press (1984).

Karim, M. R., and Zeger, S. L. (1992). Generalized linear models with random effects; salamander mating revisited. *Biometrics* **48**, 631–644.

Kass, R. E., Carlin, B. P., Gelman, A., and Neal, R. (1998). Markov chain Monte Carlo in practice: a roundtable discussion. *American Statistician* **52**, 93–100.

Kass, R. E., and Raftery, A. E. (1995). Bayes factors and model uncertainty. *Journal of the American Statistical Association* **90**, 773–795.

Kass, R. E., Tierney, L., and Kadane, J. B. (1989). Approximate methods for assessing influence and sensitivity in Bayesian analysis. *Biometrika*, **76**, 663–674.

Kass, R. E., and Vaidyanathan, S. K. (1992). Approximate Bayes factors and orthogonal parameters, with application to testing equality of two binomial proportions. *Journal of the Royal Statistical Society B* **54**, 129–144.

Kass, R. E., and Wasserman, L. (1996). The selection of prior distributions by formal rules. *Journal of the American Statistical Association* **91**, 1343–1370.

Keller, J. B. (1986). The probability of heads. *American Mathematical Monthly* **93**, 191–197.

Kish, L. (1965). *Survey Sampling.* New York: Wiley.

Kong, A. (1992). A note on importance sampling using standardized weights. Technical Reauthor.ind yyport #348, Department of Statistics, University of Chicago.

Kong, A., McCullagh, P., Meng, X. L., Nicolae, D., and Tan, Z. (2003). A theory of statistical models for Monte Carlo integration (with discussion). *Journal of the Royal Statistical Society B.*

Knuiman, M. W., and Speed, T. P. (1988). Incorporating prior information into the analysis of contingency tables. *Biometrics* **44**, 1061–1071.

Krantz, D. H. (1999). The null hypothesis testing controversy in psychology. *Journal of the American Statistical Association* **94**, 1372–1381.

Kreft, I., and De Leeuw, J. (1998). *Introducing Multilevel Modeling.* London: Sage.

Kullback, S., and Leibler, R. A. (1951). On information and sufficiency. *Annals of Mathematical Statistics* **22**, 76–86.

Kunsch, H. R. (1987). Intrinsic autoregressions and related models on the two-dimensional lattice. *Biometrika* **74**, 517–524.

Laird, N. M., and Ware, J. H. (1982). Random-effects models for longitudinal data. *Biometrics* **38**, 963–974.

Landwehr, J. M., Pregibon, D., and Shoemaker, A. C. (1984). Graphical methods for assessing logistic regression models. *Journal of the American Statistical Association* **79**, 61–83.

Lange, K. L., Little, R. J. A., and Taylor, J. M. G. (1989). Robust statistical modeling using the *t* distribution. *Journal of the American Statistical Association* **84**, 881–896.

Lange, K., and Sinsheimer, J. S. (1993). Normal/independent distributions and their applications in robust regression. *Journal of Computational and Graphical Statistics* **2**, 175–198.

Laplace, P. S. (1785). Memoire sur les formules qui sont fonctions de tres grands nombres. In *Memoires de l'Academie Royale des Sciences*.

Laplace, P. S. (1810). Memoire sur les formules qui sont fonctions de tres grands nombres et sur leurs applications aux probabilites. In *Memoires de l'Academie des Sciences de Paris*.

Lavine, M. (1991). Problems in extrapolation illustrated with space shuttle O-ring data (with discussion). *Journal of the American Statistical Association* **86**, 919–923.

Laud, P. W., and Ibrahim, J. G. (1995). Predictive model selection. *Journal of the Royal Statistical Society B* **57**, 247–262.

Lauritzen, S. L., and Spiegelhalter, D. J. (1988). Local computations with probabilities on graphical structures and their application to expert systems (with discussion). *Journal of the Royal Statistical Society B* **50**, 157–224.

Le Cam, L. (1953). On some asymptotic properties of maximum likelihood estimates and related Bayes estimates. *University of California Publications in Statistics* **1** (11), 277–330.

Le Cam, L., and Yang, G. L. (1990). *Asymptotics in Statistics: Some Basic Concepts.* New York: Springer-Verlag.

Leamer, E. E. (1978a). Regression selection strategies and revealed priors. *Journal of the American Statistical Association* **73**, 580–587.

Leamer, E. E. (1978b). *Specification Searches: Ad Hoc Inference with Nonexperimental Data.* New York: Wiley.

Lee, P. M. (1989). *Bayesian Statistics: An Introduction.* Oxford University Press.

Lehmann, E. L. (1983). *Theory of Point Estimation.* New York: Wiley.

Lehmann, E. L. (1986). *Testing Statistical Hypotheses*, second edition. New York: Wiley.

REFERENCES

Leonard, T. (1972). Bayesian methods for binomial data. *Biometrika* **59**, 581–589.

Leonard, T., and Hsu, J. S. (1992). Bayesian inference for a covariance matrix. *Annals of Statistics* **20**, 1669–1696.

Leyland, A. H., and Goldstein, H., eds. (2001). *Multilevel Modelling of Health Statistics*. Chichester: Wiley.

Liang, K. Y., and McCullagh, P. (1993). Case studies in binary dispersion. *Biometrics* **49**, 623–630.

Lin, C. Y., Gelman, A., Price, P. N., and Krantz, D. H. (1999). Analysis of local decisions using hierarchical modeling, applied to home radon measurement and remediation (with discussion). *Statistical Science* **14**, 305–337.

Lindley, D. V. (1958). Fiducial distributions and Bayes' theorem. *Journal of the Royal Statistical Society B* **20**, 102–107.

Lindley, D. V. (1965). *Introduction to Probability and Statistics from a Bayesian Viewpoint*, 2 volumes. Cambridge University Press.

Lindley, D. V. (1971a). *Bayesian Statistics, A Review*. Philadelphia: Society for Industrial and Applied Mathematics.

Lindley, D. V. (1971b). The estimation of many parameters. In *Foundations of Statistical Science*, ed. V. P. Godambe and D. A. Sprott. Toronto: Holt, Rinehart, and Winston.

Lindley, D. V., and Novick, M. R. (1981). The role of exchangeability in inference. *Annals of Statistics* **9**, 45–58.

Lindley, D. V., and Smith, A. F. M. (1972). Bayes estimates for the linear model. *Journal of the Royal Statistical Society B* **34**, 1–41.

Little, R. J. A. (1991). Inference with survey weights. *Journal of Official Statistics* **7**, 405–424.

Little, R. J. A. (1993). Post-stratification: a modeler's perspective. *Journal of the American Statistical Association* **88**, 1001–1012.

Little, R. J. A., and Rubin, D. B. (1992). *Statistical Analysis with Missing Data*, second edition. New York: Wiley.

Liu, C. (1995). Missing data imputation using the multivariate *t* distribution. *Journal of Multivariate Analysis* **48**, 198–206.

Liu, C. (2002). Robit regression: a simple robust alternative to logistic and probit regression. Technical report, Bell Laboratories.

Liu, C. (2003). Alternating subspace-spanning resampling to accelerate Markov chain Monte Carlo simulation. *Journal of the American Statistical Association*.

Liu, C., and Rubin, D. B. (1994). The ECME algorithm: a simple extension of EM and ECM with faster monotone convergence. *Biometrika* **81**, 633–648.

Liu, C., and Rubin, D. B. (1995). ML estimation of the *t* distribution using EM and its extensions, ECM and ECME. *Statistica Sinica* **5**, 19–39.

Liu, C., Rubin, D. B., and Wu, Y. N. (1998). Parameter expansion to accelerate EM: the PX-EM algorithm. *Biometrika* **85**, 755–770.

Liu, J. (2002). *Monte Carlo Strategies in Scientific Computing*. New York: Springer-Verlag.

Liu, J., and Wu, Y. N. (1999). Parameter expansion for data augmentation. *Journal of the American Statistical Association* **94**, 1264–1274.

Lohr, S. (1999). *Sampling: Design and Analsis*. Pacific Grove, Calif.: Duxbury.

Longford, N. (1993). *Random Coefficient Models*. Oxford: Clarendon Press.

Louis, T. A. (1984). Estimating a population of parameter values using Bayes and empirical Bayes methods. *Journal of the American Statistical Association* **78**, 393–398.

Louis, T. A., and Shen, W. (1999). Innovations in Bayes and emprical Bayes methods: estimating parameters, populations and ranks. *Statistics in Medicine* **18**, 2493–2505.

Luce, R. D., and Raiffa, H. (1957). *Games and Decisions*. New York: Wiley.

Madigan, D., and Raftery, A. E. (1994). Model selection and accounting for model uncertainty in graphical models using Occam's window. *Journal of the American Statistical Association* **89**, 1535–1546.

Madow, W. G., Nisselson, H., Olkin, I., and Rubin, D. B. (1983). *Incomplete Data in Sample Surveys*, 3 vols. New York: Academic Press.

Malec, D., and Sedransk, J. (1992). Bayesian methodology for combining the results from different experiments when the specifications for pooling are uncertain. *Biometrika* **79**, 593–601.

Mallows, C. L. (1973). Some comments on C_p. *Technometrics* **15**, 661–675.

Manton, K. G., Woodbury, M. A., Stallard, E., Riggan, W. B., Creason, J. P., and Pellom, A. C. (1989). Empirical Bayes procedures for stabilizing maps of U.S. cancer mortality rates. *Journal of the American Statistical Association* **84**, 637–650.

Mardia, K. V., Kent, J. T., and Bibby, J. M. (1979). *Multivariate Analysis*. New York: Academic Press.

Marquardt, D. W., and Snee, R. D. (1975). Ridge regression in practice. *American Statistician* **29**, 3–19.

Martin, A. D., and Quinn, K. M. (2002a). Dynamic ideal point estimation via Markov chain Monte Carlo for the U.S. Supreme Court, 1953–1999. *Political Analysis* **10**, 134–153.

Martin, A. D., and Quinn, K. M. (2002b). MCMCpack. scythe.wustl.edu/mcmcpack.html

Martz, H. F., and Zimmer, W. J. (1992). The risk of catastrophic failure of the solid rocket boosters on the space shuttle. *American Statistician* **46**, 42–47.

McClellan, M., McNeil, B. J., and Newhouse, J. P. (1994). Does more intensive treatment of acute myocardial infarction reduce mortality? *Journal of the American Medical Association* **272**, 859–866.

McCullagh, P., and Nelder, J. A. (1989). *Generalized Linear Models*, second edition. New York: Chapman & Hall.

McCulloch, R. E. (1989). Local model influence. *Journal of the American Statistical Association* **84**, 473–478.

Meng, C. Y. K., and Dempster, A. P. (1987). A Bayesian approach to the multiplicity problem for significance testing with binomial data. *Biometrics* **43**, 301–311.

Meng, X. L. (1994a). On the rate of convergence of the ECM algorithm. *Annals of Statistics* **22**, 326–339.

Meng, X. L. (1994b). Multiple-imputation inferences with uncongenial sources of input (with discussion). *Statistical Science* **9**, 538–573.

Meng, X. L., and Pedlow, S. (1992). EM: a bibliographic review with missing articles. *Proceedings of the Statistical Computing Section, American Statistical Association*, 24–27.

Meng, X. L., Raghunathan, T. E., and Rubin, D. B. (1991). Significance levels from repeated p values with multiply-imputed data. *Statistica Sinica* **1**, 65–92.

Meng, X. L., and Rubin, D. B. (1991). Using EM to obtain asymptotic variance-covariance matrices: the SEM algorithm. *Journal of the American Statistical Association* **86**, 899–909.

Meng, X. L., and Rubin, D. B. (1992). Performing likelihood ratio tests with multiply imputed data sets. *Biometrika* **79**, 103–111.

Meng, X. L., and Rubin, D. B. (1993). Maximum likelihood estimation via the ECM algorithm: a general framework. *Biometrika* **80**, 267–278.

Meng, X. L., and Schilling, S. (1996). Fitting full-information item factor models and empirical investigation of bridge sampling. *Journal of the American Statistical Association* **91**, 1254–1267.

Meng, X. L., and van Dyk, D. A. (1997). The EM algorithm—an old folk-song sung to a fast new tune (with discussion). *Journal of the Royal Statistical Society B* **59**, 511–567.

Meng, X. L., and Wong, W. H. (1996). Simulating ratios of normalizing constants via a simple identity: a theoretical exploration. *Statistica Sinica* **6**, 831–860.

Metropolis, N., Rosenbluth, A. W., Rosenbluth, M. N., Teller, A. H., and Teller, E. (1953). Equation of state calculations by fast computing machines. *Journal of Chemical Physics* **21**, 1087–1092.

Metropolis, N., and Ulam, S. (1949). The Monte Carlo method. *Journal of the American Statistical Association* **44**, 335–341.

Meulders, M., Gelman, A., Van Mechelen, I., and De Boeck, P. (1998). Generalizing the probability matrix decomposition model: an example of Bayesian model checking and model expansion. In *Assumptions, Robustness, and Estimation Methods in Multivariate Modeling*, ed. J. Hox and E. D. de Leeuw, 1–19. Amsterdam: T-T Publikaties.

Mollie, A., and Richardson, S. (1991). Empirical Bayes estimates of cancer mortality rates using spatial models. *Statistics in Medicine* **10**, 95–112.

Morgan, J. P., Chaganty, N. R, Dahiya, R. C., and Doviak, M. J. (1991). Let's make a deal: the player's dilemma. *The American Statistician* **45**, 284–289.

Moroff, S. V., and Pauker, S. G. (1983). What to do when the patient outlives the literature, or DEALE-ing with a full deck. *Medical Decision Making* **3**, 313–338.

Morris, C. (1983). Parametric empirical Bayes inference: theory and applications (with discussion). *Journal of the American Statistical Association* **78**, 47–65.

Mosteller, F., and Wallace, D. L. (1964). *Applied Bayesian and Classical Inference: The Case of The Federalist Papers*. New York: Springer-Verlag. Reprinted 1984.

Mugglin, A. S., Carlin, B. P., and Gelfand, A. E. (2000). Fully model based approaches for spatially misaligned data. *Journal of the American Statistical Association* **95**, 877–887.

Mulligan, C. B., and Hunter, C. G. (2001). The empirical frequency of a pivotal vote. National Bureau of Economic Research Working Paper 8590.

Muller, P., and Rosner, G. L. (1997). A Bayesian population model with hierarchical mixture priors applied to blood count data. *Journal of the American Statistical Association* **92**, 1279–1292.

Mykland, P., Tierney, L., and Yu, B. (1994). Regeneration in Markov chain samplers. *Journal of the American Statistical Association* **90**, 233–241.

Nadaram, B., and Sedransk, J. (1993). Bayesian predictive inference for a finite population proportion: two-stage cluster sampling. *Journal of the Royal Statistical Society B* **55**, 399–408.

Neal, R. M. (1993). Probabilistic inference using Markov chain Monte Carlo methods. Technical Report CRG-TR-93-1, Department of Computer Science, University of Toronto.

Neal, R. M. (1994). An improved acceptance procedure for the hybrid Monte Carlo algorithm. *Journal of Computational Physics* **111**, 194–203.

Neal, R. M. (1996a). *Bayesian Learning for Neural Networks*. New York: Springer-Verlag.

Neal, R. M. (1996b). Sampling from multimodal distributions using tempered transitions. *Statistics and Computing* **6**, 353–366.

Neal, R. M. (2003). Slice sampling (with discussion). *Annals of Statistics*.

Nelder, J. A. (1977). A reformulation of linear models (with discussion). *Journal of the Royal Statistical Society A* **140**, 48–76.

Nelder, J. A. (1994). The statistics of linear models: back to basics. *Statistics and Computing* **4**, 221–234.

Nelder, J. A., and Wedderburn, R. W. M. (1972). Generalized linear models. *Journal of the Royal Statistical Society A* **135**, 370–384.

Neter, J., Kutner, M. H., Nachtsheim, C. J., and Wasserman, W. (1996). *Applied Linear Statistical Models*, fourth edition. Burr Ridge, Ill.: Richard D. Irwin, Inc.

Newhouse, J. P., and McClellan, M. (1998). Econometrics in outcomes research: the use of instrumental variables. *Annual Review of Public Health* **19**, 17–34.

Neyman, J. (1923). On the application of probability theory to agricultural experiments. Essay on principles. Section 9. Translated and edited by D. M. Dabrowska and T. P. Speed. *Statistical Science* **5**, 463–480 (1990).

Normand, S. L., Glickman, M. E., and Gatsonis, C. A. (1997). Statistical methods for profiling providers of medical care: issues and applications. *Journal of the American Statistical Association* **92**, 803–814.

Normand, S. L., and Tritchler, D. (1992). Parameter updating in a Bayes network. *Journal of the American Statistical Association* **87**, 1109–1115.

Novick, M. R., Jackson, P. H., Thayer, D. T., and Cole, N. S. (1972). Estimating multiple regressions in m-groups: a cross validation study. *British Journal of Mathematical and Statistical Psychology* **25**, 33–50.

Novick, M. R., Lewis, C., and Jackson, P. H. (1973). The estimation of proportions in m groups. *Psychometrika* **38**, 19–46.

O'Hagan, A. (1979). On outlier rejection phenomena in Bayes inference. *Journal of the Royal Statistical Society B* **41**, 358–367.

O'Hagan, A. (1988). *Probability: Methods and Measurement*. New York: Chapman & Hall.

O'Hagan, A. (1995). Fractional Bayes factors for model comparison (with discussion). *Journal of the Royal Statistical Society B* **57**, 99–138.

Orchard, T., and Woodbury, M. A. (1972). A missing information principle: theory and applications. In *Proceedings of the Sixth Berkeley Symposium*, ed. L. LeCam, J. Neyman, and E. L. Scott, 697–715. Berkeley: University of California Press.

Ott, J. (1979). Maximum likelihood estimation by counting methods under polygenic and mixed models in human pedigrees. *American Journal of Human Genetics* **31**, 161–175.

Pardoe, I. (2001). A Bayesian sampling approach to regression model checking. *Journal of Computational and Graphical Statistics* **10**, 617–627.

Pardoe, I., and Cook, R. D. (2002). A graphical method for assessing the fit of a logistic regression model. *American Statistician* **56**.

Park, D., Gelman, A., and Bafumi, J. (2003). State-level opinions from national surveys using poststratification and hierarchical logistic regression. Technical report, Department of Political Science, Columbia University.

Parmigiani, G. (2002). *Modeling in Medical Decision Making: A Bayesian Approach*. New York: Wiley.

Parmigiani, G. (2003). Uncertainty and the value of diagnostic information. *Statistics in Medicine*.

Parmigiani, G., Berry, D., Iversen, E. S., Muller, P., Schildkraut, J., and Winer, E. (1999). Modeling risk of breast cancer and decisions about genetic testing (with discussion). In *Case Studies in Bayesian Statistics*, volume 4, ed. C. Gatsonis, R. E. Kass, B. Carlin, A. Carriquiry, A. Gelman, I. Verdinelli, and M. West, 133–203. New York: Springer-Verlag.

Parmar, M. K. B., Griffiths, G. O., Spiegelhalter, D. J., Souhami, R. L., Altman, D. G., and van der Scheuren, E. (2001). Monitoring of large randomised clinical trials: a new approach with Bayesian methods. *Lancet* **358**, 375–381.

Pauler, D. K., Wakefield, J. C., and Kass, R. E. (1999). Bayes factors for variance component models. *Journal of the American Statistical Association* **94**, 1242–1253.

Pearl, J. (1988). *Probabilistic Reasoning in Intelligent Systems: Networks of Plausible Inference.* San Mateo, Calif.: Morgan Kaufmann.

Pearl, J. (2000). *Causality.* Cambridge University Press.

Pericchi, L. R. (1981). A Bayesian approach to transformations to normality. *Biometrika* **68**, 35–43.

Pettitt, A. N., Friel, N., and Reeves, R. (2003). Efficient calculation of the normalizing constant of the autologistic and related models on the cylinder and lattice. *Journal of the Royal Statistical Society B* **65**, 235–246.

Pinheiro, J. C., and Bates, D. M. (2000). *Mixed-Effects Models in S and S-Plus.* New York: Springer-Verlag.

Plackett, R. L. (1960). Models in the analysis of variance (with discussion). *Journal of the Royal Statistical Society B* **22**, 195–217.

Plummer, M. (2003). JAGS: a program for analysis of Bayesian graphical models using Gibbs sampling.

Pole, A., West, M., and Harrison, J. (1994). *Applied Bayesian Forecasting and Time Series Analysis.* New York: Chapman & Hall.

Pratt, J. W. (1965). Bayesian interpretation of standard inference statements (with discussion). *Journal of the Royal Statistical Society B* **27**, 169–203.

Press, W. H., Flannery, B. P., Teukolsky, S. A., and Vetterling, W. T. (1986). *Numerical Recipes: The Art of Scientific Computing.* Cambridge University Press.

Propp, J. G., and Wilson, D. B. (1996). Exact sampling with coupled Markov chains and applications to statistical mechanics. *Random Structures Algorithms* **9**, 223–252.

R Project (2002). The R project for statistical computing. www.r-project.org

Racine, A., Grieve, A. P., Fluhler, H., and Smith, A. F. M. (1986). Bayesian methods in practice: experiences in the pharmaceutical industry (with discussion). *Applied Statistics* **35**, 93–150.

Racine-Poon, A., Weihs, C., and Smith, A. F. M. (1991). Estimation of relative potency with sequential dilution errors in radioimmunoassay. *Biometrics* **47**, 1235–1246.

Raftery, A. E. (1988). Inference for the binomial N parameter: a hierarchical Bayes approach. *Biometrika* **75**, 223–228.

Raftery, A. E. (1995). Bayesian model selection in social research (with discussion). In *Sociological Methodology 1995*, ed. P. V. Marsden.

Raftery, A. E. (1996). Hypothesis testing and model selection via posterior simulation. In *Practical Markov Chain Monte Carlo*, ed. W. Gilks, S. Richardson, and D. Spiegelhalter, 163–187. New York: Chapman & Hall.

Raghunathan, T. E. (1994). Monte Carlo methods for exploring sensitivity to distributional assumptions in a Bayesian analysis of a series of 2×2 tables. *Statistics in Medicine* **13**, 1525–1538.

Raghunathan, T. E., Lepkowski, J. E., Solenberger, P. W., and Van Hoewyk, J. H. (2001). A multivariate technique for multiply imputing missing values using a sequence of regression models. *Survey Methodology*.

Raghunathan, T. E., and Rubin, D. B. (1990). An application of Bayesian statistics using sampling/importance resampling for a deceptively simple problem in quality control. In *Data Quality Control: Theory and Pragmatics*, ed. G. Liepins and V. R. R. Uppuluri, 229–243. New York: Marcel Dekker.

Raiffa, H., and Schlaifer, R. (1961). *Applied Statistical Decision Theory*. Boston, Mass.: Harvard Business School.

Raudenbush, S. W., and Bryk, A. S. (2002). *Hierarchical Linear Models*, second edition. Thousand Oaks, Calif.: Sage.

Reilly, C., Gelman, A., and Katz, J. N. (2001). Post-stratification without population level information on the post-stratifying variable, with application to political polling. *Journal of the American Statistical Association* **96**, 1–11.

Richardson, S., and Gilks, W. R. (1993). A Bayesian approach to measurement error problems in epidemiology using conditional independence models. *American Journal of Epidemiology* **138**, 430–442.

Richardson, S., and Green, P. J. (1997). On Bayesian analysis of mixtures with an unknown number of components. *Journal of the Royal Statistical Society B* **59**, 731–792.

Ripley, B. D. (1981). *Spatial Statistics*. New York: Wiley.

Ripley, B. D. (1987). *Stochastic Simulation*. New York: Wiley.

Ripley, B. D. (1988). *Statistical Inference for Spatial Processes*. Cambridge University Press.

Robbins, H. (1955). An empirical Bayes approach to statistics. In *Proceedings of the Third Berkeley Symposium* **1**, ed. J. Neyman, 157–164. Berkeley: University of California Press.

Robbins, H. (1964). The empirical Bayes approach to statistical decision problems. *Annals of Mathematical Statistics* **35**, 1–20.

Robert, C. P., and Casella, G. (1999). *Monte Carlo Statistical Methods*. New York: Springer-Verlag.

Roberts, G. O., and Rosenthal, J. S. (2001). Optimal scaling for various Metropolis-Hastings algorithms. *Statistical Science* **16**, 351–367.

Roberts, G. O., and Sahu, S. K. (1997). Updating schemes, correlation structure, blocking and parameterization for the Gibbs sampler. *Journal of the Royal Statistical Society B* **59**, 291–317.

Robins, J. M. (1998). Confidence intervals for causal parameters. *Statistics in Medicine* **7**, 773–785.

Robinson, G. K. (1991). That BLUP is a good thing: the estimation of random effects (with discussion). *Statistical Science* **6**, 15–51.

Rombola, F. (1984). *The Book on Bookmaking*. Pasadena, Calif.: Pacific Book and Printing.

Rosenbaum, P. R., and Rubin, D. B. (1983a). The central role of the propensity score in observational studies for causal effects. *Biometrika* **70**, 41–55.

Rosenbaum, P. R., and Rubin, D. B. (1983b). Assessing sensitivity to an unobserved binary covariate in an observational study with binary outcome. *Journal of the Royal Statistical Society B* **45**, 212–218.

Rosenbaum, P. R., and Rubin, D. B. (1984a). Sensitivity of Bayes inference with data-dependent stopping rules. *American Statistician* **38**, 106–109.

Rosenbaum, P. R., and Rubin, D. B. (1984b). Reducing bias in observational studies using subclassification on the propensity score. *Journal of the American Statistical Association* **79**, 516–524.

Rosenbaum, P. R., and Rubin, D. B. (1985). Constructing a control group using multivariate matched sampling methods that incorporate the propensity score. *American Statistician* **39**, 33–38.

Rosenkranz, S. L., and Raftery, A. E. (1994). Covariate selection in hierarchical models of hospital admission counts: a Bayes factor approach. Technical Report #268, Department of Statistics, University of Washington.

Rosenthal, J. S. (1995). Minorization conditions and convergence rates for Markov chain Monte Carlo. *Journal of the American Statistical Association* **90**, 558–566.

Ross, S. M. (1983). *Stochastic Processes*. New York: Wiley.

Rotnitzky, A., Robins, J. M., and Scharfstein, D. O. (1999). Adjusting for nonignorable dropout using semiparametric models. *Journal of the American Statistical Association* **94**, 1321–1339.

Royall, R. M. (1970). On finite population sampling theory under certain linear regression models. *Biometrika* **57**, 377–387.

Rubin, D. B. (1974a). Characterizing the estimation of parameters in incomplete data problems. *Journal of the American Statistical Association* **69**, 467–474.

Rubin, D. B. (1974b). Estimating causal effects of treatments in randomized and nonrandomized studies. *Journal of Educational Psychology* **66**, 688–701.

Rubin, D. B. (1976). Inference and missing data. *Biometrika* **63**, 581–592.

Rubin, D. B. (1977). Assignment to treatment group on the basis of a covariate. *Journal of Educational Statistics* **2**, 1–26.

Rubin, D. B. (1978a). Bayesian inference for causal effects: the role of randomization. *Annals of Statistics* **6**, 34–58.

Rubin, D. B. (1978b). Multiple imputations in sample surveys: a phenomenological Bayesian approach to nonresponse (with discussion). *Proceedings of the American Statistical Association, Survey Research Methods Section*, 20–34.

Rubin, D. B. (1980a). Discussion of 'Randomization analysis of experimental data: the Fisher randomization test,' by D. Basu. *Journal of the American Statistical Association* **75**, 591–593.

Rubin, D. B. (1980b). Using empirical Bayes techniques in the law school validity studies (with discussion). *Journal of the American Statistical Association* **75**, 801–827.

Rubin, D. B. (1981). Estimation in parallel randomized experiments. *Journal of Educational Statistics* **6**, 377–401.

Rubin, D. B. (1983a). A case study of the robustness of Bayesian methods of inference: estimating the total in a finite population using transformations to normality. In *Scientific Inference, Data Analysis, and Robustness*, ed. G. E. P. Box, T. Leonard, and C. F. Wu, 213–244. New York: Academic Press.

Rubin, D. B. (1983b). Iteratively reweighted least squares. In *Encyclopedia of Statistical Sciences*, Vol. 4, ed. S. Kotz, N. L. Johnson, and C. B. Read, 272–275. New York: Wiley.

Rubin, D. B. (1983c). Progress report on project for multiple imputation of 1980 codes. Manuscript delivered to the U.S. Bureau of the Census, the U.S. National Science Foundation, and the Social Science Research Foundation.

Rubin, D. B. (1984). Bayesianly justifiable and relevant frequency calculations for the applied statistician. *Annals of Statistics* **12**, 1151–1172.

Rubin, D. B. (1985). The use of propensity scores in applied Bayesian inference. In *Bayesian Statistics 2*, ed. J. M. Bernardo, M. H. DeGroot, D. V. Lindley, and A. F. M. Smith, 463–472. Amsterdam: Elsevier Science Publishers.

Rubin, D. B. (1987a). *Multiple Imputation for Nonresponse in Surveys*. New York: Wiley.

Rubin, D. B. (1987b). A noniterative sampling/importance resampling alternative to the data augmentation algorithm for creating a few imputations when fractions of missing information are modest: the SIR algorithm. Discussion of Tanner and Wong (1987). *Journal of the American Statistical Association* **82**, 543–546.

Rubin, D. B. (1989). A new perspective on meta-analysis. In *The Future of Meta-Analysis*, ed. K. W. Wachter and M. L. Straf. New York: Russell Sage Foundation.

Rubin, D. B. (1990). Discussion of 'On the application of probability theory to agricultural experiments. Essay on principles. Section 9,' by J. Neyman. *Statistical Science* **5**, 472–480.

Rubin, D. B. (1996). Multiple imputation after 18+ years (with discussion) *Journal of the American Statistical Association* **91**, 473–520.

Rubin, D. B. (1998). More powerful randomization-based p-values in double-blind trials with noncompliance (with discussion). *Statistics in Medicine* **17**, 371–385.

Rubin, D. B. (2000). Discussion of Dawid (2000). *Journal of the American Statistical Association* **95**, 435–438.

Rubin, D. B., and Schenker, N. (1987). Logit-based interval estimation for binomial data using the Jeffreys prior. *Sociological Methodology*, 131–144.

Rubin, D. B., and Stern, H. S. (1994). Testing in latent class models using a posterior predictive check distribution. In *Latent Variables Analysis: Applications for Developmental Research*, ed. A. Von Eye and C. C. Clogg, 420–438. Thousand Oaks, Calif.: Sage.

Rubin, D. B., Stern, H. S, and Vehovar, V. (1995). Handling 'Don't Know' survey responses: the case of the Slovenian plebiscite. *Journal of the American Statistical Association* **90**, 822–828.

Rubin, D. B., and Thomas, N. (1992). Affinely invariant matching methods with ellipsoidal distributions. *Annals of Statistics* **20**, 1079–93.

Rubin, D. B., and Thomas, N. (2000). Combining propensity score matching with additional adjustments for prognostic covariates. *Journal of the American Statistical Association* **95**, 573–585.

Rubin, D. B., and Wu, Y. (1997). Modeling schizophrenic behavior using general mixture components. *Biometrics* **53**, 243–261.

Sampson, R. J., Raudenbush, S. W., and Earls, F. (1997). Neighborhoods and violent crime: a multilevel study of collective efficacy. *Science* **277**, 918–924.

Satterthwaite, F. E. (1946). An approximate distribution of estimates of variance components. *Biometrics Bulletin* **2**, 110–114.

Savage, I. R. (1957). Nonparametric statistics. *Journal of the American Statistical Association* **52**, 331–344.

Savage, L. J. (1954). *The Foundations of Statistics*. New York: Dover.

Schafer, J. L. (1997). *Analysis of Incomplete Multivariate Data*. New York: Chapman & Hall.

Schmidt-Nielsen, K. (1984). *Scaling: Why is Animal Size So Important?* Cambridge University Press.

Scott, A., and Smith, T. M. F. (1969). Estimation in multi-stage surveys. *Journal of the American Statistical Association* **64**, 830–840.

Scott, A., and Smith, T. M. F. (1973). Survey design, symmetry and posterior distributions. *Journal of the Royal Statistical Society B* **55**, 57–60.

Searle, S. R., Casella, G., and McCulloch, C. E. (1992). *Variance Components*. New York: Wiley.

Seber, G. A. F. (1992). A review of estimating animal abundance II. *International Statistical Review* **60**, 129–166.

Selvin, S. (1975). Letter. *American Statistician* **29**, 67.

Shafer, G. (1982). Lindley's paradox (with discussion). *Journal of the American Statistical Association* **77**, 325–351.

Sheiner L. B., Rosenberg B., and Melmon K. L. (1972). Modelling of individual pharmacokinetics for computer-aided drug dosage. *Computers and Biomedical Research* **5**, 441–459.

Sheiner L. B., and Beal S. L. (1982). Bayesian individualization of pharmacokinetics: simple implementation and comparison with non-Bayesian methods. *Journal of Pharmaceutical Sciences* **71**, 1344–1348.

Shen, W., and Louis, T. A. (1998). Triple-goal estimates in two-stage hierarchical models. *Journal of the Royal Statistical Society B* **60**, 455–471.

Shen, X., and Wasserman, L. (2000). Rates of convergence of posterior distributions. *Annals of Statistics* **29**, 687–714.

Simoncelli, E. P. (1999). Bayesian denoising of visual images in the wavelet domain. In *Bayesian Inference in Wavelet Based Models*, ed. P. Muller, and B. Vidakovic (Lecture Notes in Statistics 141), 291–308. New York: Springer-Verlag.

Singer, E., Van Hoewyk, J., Gebler, N., Raghunathan, T., and McGonagle, K. (1999). The effects of incentives on response rates in interviewer-mediated surveys. *Journal of Official Statistics* **15**, 217–230.

Sinharay, S., and Stern, H. S. (2003). Posterior predictive model checking in hierarchical models. *Journal of Statistical Planning and Inference* **111**, 209–221.

Skene, A. M., and Wakefield, J. C. (1990). Hierarchical models for multicentre binary response studies. *Statistics in Medicine* **9**, 910–929.

Skilling, J. (1989). Classic maximum entropy. In *Maximum Entropy and Bayesian Methods*, ed. J. Skilling, 1–52. Dordrecht, Netherlands: Kluwer Academic Publishers.

Skinner, C. J., Holt, D., and Smith, T. M. F., eds. (1989). *The Analysis of Complex Surveys*. New York: Wiley.

Smith, A. F. M. (1983). Bayesian approaches to outliers and robustness. In *Specifying Statistical Models from Parametric to Nonparametric, Using Bayesian or Non-Bayesian Approaches*, ed. J. P. Florens, M. Mouchart, J. P. Raoult, L. Simar, and A. F. M. Smith (Lecture Notes in Statistics 16), 13–35. New York: Springer-Verlag.

Smith, A. F. M., and Gelfand, A. E. (1992). Bayesian statistics without tears. *American Statistician* **46**, 84–88.

Smith, A. F. M., and Roberts, G. O. (1993). Bayesian computation via the Gibbs sampler and related Markov chain Monte Carlo methods (with discussion). *Journal of the Royal Statistical Society B* **55**, 3–102.

Smith, A. F. M., Skene, A. M., Shaw, J. E. H., Naylor, J. C., and Dransfield, M. (1985). The implementation of the Bayesian paradigm. *Communications in Statistics* **14**, 1079–1102.

Smith, T. C., Spiegelhalter, D. J., and Thomas, A. (1995). Bayesian approaches to random-effects meta-analysis: a comparative study. *Statistics in Medicine* **14**, 2685–2699.

Smith, T. M. F. (1983). On the validity of inferences from non-random samples. *Journal of the Royal Statistical Society A* **146**, 394–403.

Snedecor, G. W., and Cochran, W. G. (1989). *Statistical Methods*, eighth edition. Ames: Iowa State University Press.

Snijders, T. A. B., and Bosker, R. J. (1999). *Multilevel Analysis*. London: Sage.

Snyder, J., with Herskowitz, M., and Perkins, S. (1975). *Jimmy the Greek, by Himself*. Chicago: Playboy Press.

Sommer, A., and Zeger, S. (1991). On estimating efficacy from clinical trials. *Statistics in Medicine* **10**, 45–52.

Speed, T. P. (1990). Introductory remarks on Neyman (1923). *Statistical Science* **5**, 463–464.

Spiegelhalter, D. J., Best, N. G., Carlin, B. P., and van der Linde, A. (2002). Bayesian measures of model complexity and fit (with discussion). *Journal of the Royal Statistical Society B*.

Spiegelhalter, D. J., and Smith, A. F. M. (1982). Bayes factors for linear and log-linear models with vague prior information. *Journal of the Royal Statistical Society B* **44**, 377–387.

Spiegelhalter, D., Thomas, A., Best, N., Gilks, W., and Lunn, D. (1994, 2003). BUGS: Bayesian inference using Gibbs sampling. MRC Biostatistics Unit, Cambridge, England. www.mrc-bsu.cam.ac.uk/bugs/

Spitzer, E. (1999). The New York City Police Department's "stop and frisk" practices. Office of the New York State Attorney General. www.oag.state.ny.us/press/reports/stop_frisk/stop_frisk.html

Stein, C. (1955). Inadmissibility of the usual estimator for the mean of a multivariate normal distribution. In *Proceedings of the Third Berkeley Symposium* **1**, ed. J. Neyman, 197–206. Berkeley: University of California Press.

Stephens, M. (2000). Bayesian analysis of mixture models with an unknown number of components: an alternative to reversible jump methods. *Annals of Statistics* **28**, 40–74.

Stern, H. S. (1990). A continuum of paired comparison models. *Biometrika* **77**, 265–273.

Stern, H. S. (1991). On the probability of winning a football game. *American Statistician* **45**, 179–183.

Stern, H. S. (1997). How accurately can sports outcomes be predicted? *Chance* **10** (4), 19–23.

Stern, H. S. (1998). How accurate are the posted odds? *Chance* **11** (4), 17–21.

Sterne, J. A. C., and Davey Smith, G. (2001). Sifting the evidence—what's wrong with significance tests? *British Medical Journal* **322**, 226–231.

Stigler, S. M. (1977). Do robust estimators work with real data? (with discussion). *Annals of Statistics* **5**, 1055–1098.

Stigler, S. M. (1983). Discussion of Morris (1983). *Journal of the American Statistical Association* **78**, 62–63.

Stigler, S. M. (1986). *The History of Statistics.* Cambridge, Mass.: Harvard University Press.

Stone, M. (1974). Cross-validatory choice and assessment of statistical predictions (with discussion). *Journal of the Royal Statistical Society B* **36**, 111–147.

Strenio, J. L. F., Weisberg, H. I., and Bryk, A. S. (1983). Empirical Bayes estimation of individual growth curve parameters and their relationship to covariates. *Biometrics* **39**, 71–86.

Tanner, M. A. (1993). *Tools for Statistical Inference: Methods for the Exploration of Posterior Distributions and Likelihood Functions,* third edition. New York: Springer-Verlag.

Tanner, M. A., and Wong, W. H. (1987). The calculation of posterior distributions by data augmentation (with discussion). *Journal of the American Statistical Association* **82**, 528–550.

Taplin, R. H., and Raftery, A. E. (1994). Analysis of agricultural field trials in the presence of outliers and fertility jumps. *Biometrics* **50**, 764–781.

Tarone, R. E. (1982). The use of historical control information in testing for a trend in proportions. *Biometrics* **38**, 215–220.

Thisted, R. (1988). *Elements of Statistical Computing: Numerical Computation.* New York: Chapman & Hall.

Thomas, A., Spiegelhalter, D. J., and Gilks, W. R. (1992). BUGS: a program to perform Bayesian inference using Gibbs sampling. In *Bayesian Statistics 4,* ed. J. M. Bernardo, J. O. Berger, A. P. Dawid, and A. F. M. Smith, 837–842. Oxford University Press.

Tiao, G. C., and Box, G. E. P. (1967). Bayesian analysis of a three-component hierarchical design model. *Biometrika* **54**, 109–125.

Tiao, G. C., and Tan, W. Y. (1965). Bayesian analysis of random-effect models in the analysis of variance. I: Posterior distribution of variance components. *Biometrika* **52**, 37–53.

Tiao, G. C., and Tan, W. Y. (1966). Bayesian analysis of random-effect models in the analysis of variance. II: Effect of autocorrelated errors. *Biometrika* **53**, 477–495.

Tierney, L., and Kadane, J. B. (1986). Accurate approximations for posterior moments and marginal densities. *Journal of the American Statistical Association* **81**, 82–86.

Tierney, L. (1998). A note on the Metropolis Hastings algorithm for general state spaces. *Annals of Applied Probability* **8**, 1–9.

Titterington, D. M. (1984). The maximum entropy method for data analysis (with discussion). *Nature* **312**, 381–382.

Titterington, D. M., Smith, A. F. M., and Makov, U. E. (1985). *Statistical Analysis of Finite Mixture Distributions.* New York: Wiley.

Tsui, K. W., and Weerahandi, S. (1989). Generalized p-values in significance testing of hypotheses in the presence of nuisance parameters. *Journal of the American Statistical Association* **84**, 602–607.

Tukey, J. W. (1977). *Exploratory Data Analysis*. Reading, Mass.: Addison-Wesley.

Tufte, E. R. (1983). *The Visual Display of Quantitative Information*. Cheshire, Conn.: Graphics Press.

Tufte, E. R. (1990). *Envisioning Information*. Cheshire, Conn.: Graphics Press.

Turner, D. A., and West, M. (1993). Bayesian analysis of mixtures applied to postsynaptic potential fluctuations. *Journal of Neuroscience Methods* **47**, 1–23.

Vaida, F., and Blanchard, S. (2002). Conditional Akaike informaion for mixed effects models. Technical report, Department of Biostatistics, Harvard University.

Vail, A., Hornbuckle, J., Spiegelhalter, D. J., and Thomas, J. G. (2001). Prospective application of Bayesian monitoring and analysis in an 'open' randomized clinical trial. *Statistics in Medicine* **20**, 3777–3787.

Van Buuren, S., Boshuizen, H. C., and Knook, D. L. (1999). Multiple imputation of missing blood pressure covariates in survival analysis. *Statistics in Medicine* **18**, 681–694.

Van Buuren, S., and Oudshoom, C. G. M. (2000). MICE: Multivariate imputation by chained equations (S software for missing-data imputation). `web.inter.nl.net/users/S.van.Buuren/mi/`

van Dyk, D. A., and Meng, X. L. (2001). The art of data augmentation (with discussion). *Journal of Computational and Graphical Statistics* **10**, 1–111.

van Dyk, D. A., Meng, X. L., and Rubin, D. B. (1995). Maximum likelihood estimation via the ECM algorithm: computing the asymptotic variance. Technical report, Department of Statistics, University of Chicago.

Venables, W. N., and Ripley, B. D. (2002). *Modern Applied Statistics with S*, fourth edition. New York: Springer-Verlag.

Verbeke, G., and Molenberghs, G. (2000). *Linear Mixed Models for Longitudinal Data*. New York: Springer-Verlag.

Wahba, G. (1978). Improper priors, spline smoothing and the problem of guarding against model errors in regression. *Journal of the Royal Statistical Society B* **40**, 364–372.

Wainer, H. (1984). How to display data badly. *The American Statistician* **38**, 137–147.

Wainer, H. (1997). *Visual Revelations*. New York: Springer-Verlag.

Wakefield, J. C. (1996). The Bayesian analysis of population pharmacokinetic models. *Journal of the American Statistical Association* **91**, 62–75.

Wakefield, J., Aarons, L., and Racine-Poon, A. (1999). The Bayesian approach to population pharmacokinetic/pharmacodynamic modeling (with discussion). In *Case Studies in Bayesian Statistics*, volume 4, ed. C. Gatsonis, R.

E. Kass, B. Carlin, A. Carriquiry, A. Gelman, I. Verdinelli, and M. West, 205–265. New York: Springer-Verlag.

Wakefield, J. C., Gelfand, A. E., and Smith, A. F. M. (1991). Efficient generation of random variates via the ratio-of-uniforms method. *Statistics and Computing* **1**, 129–133.

Waller, L. A., Carlin, B. P., Xia, H., and Gelfand, A. E. (1997). Hierarchical spatio-temporal mapping of disease rates. *Journal of the American Statistical Association* **92**, 607–617.

Waller, L. A., Louis, T. A., and Carlin, B. P. (1997). Bayes methods for combining disease and exposure data in assessing environmental justice. *Environmental and Ecological Statistics* **4**, 267–281.

Warnes, G. R. (2003). Hydra: a Java library for Markov chain Monte Carlo. software.biostat.washington.edu/statsoft/MCMC/Hydra

Wasserman, L. (1992). Recent methodological advances in robust Bayesian inference (with discussion). In *Bayesian Statistics 4*, ed. J. M. Bernardo, J. O. Berger, A. P. Dawid, and A. F. M. Smith, 438–502. Oxford University Press.

Wasserman, L. (2000). Asymptotic inference for mixture models using data dependent priors. *Journal of the Royal Statistical Society B* **62**, 159–180.

Weisberg, S. (1985). *Applied Linear Regression*, second edition. New York: Wiley.

Weiss, R. E. (1994). Pediatric pain, predictive inference, and sensitivity analysis. *Evaluation Review* **18**, 651–678.

Weiss, R. E. (1996). An approach to Bayesian sensitivity analysis. *Journal of the Royal Statistical Society B* **58**, 739–750.

Wermuth, N., and Lauritzen, S. L. (1990). On substantive research hypotheses, conditional independence graphs, and graphical chain models. *Journal of the Royal Statistical Society B* **52**, 21–50.

West, M. (1992). Modelling with mixtures. In *Bayesian Statistics 4*, ed. J. M. Bernardo, J. O. Berger, A. P. Dawid, and A. F. M. Smith, 503–524. Oxford University Press.

West, M. (2003). Bayesian factor regression models in the "large p, small n" paradigm. In *Bayesian Statistics 7*, ed. J. M. Bernardo, M. J. Bayarri, J. O. Berger, A. P. Dawid, D. Heckerman, A. F. M. Smith, and M. West, 733–742. Oxford University Press.

West, M., and Harrison, J. (1989). *Bayesian Forecasting and Dynamic Models*. New York: Springer-Verlag.

Wikle, C. K., Milliff, R. F., Nychka, D., and Berliner, L. M. (2001). Spatiotemporal hierarchical Bayesian modeling: tropical ocean surface winds. *Journal of the American Statistical Association* **96**, 382–397.

Wong, F., Carter, C., and Kohn, R. (2002). Efficient estimation of covariance selection models. Technical report, Australian Graduate School of Management.

Wong, W. H., and Li, B. (1992). Laplace expansion for posterior densities of nonlinear functions of parameters. *Biometrika* **79**, 393–398.

Yang, R., and Berger, J. O. (1994). Estimation of a covariance matrix using reference prior. *Annals of Statistics* **22**, 1195–1211.

Yates, F. (1967). A fresh look at the basic principles of the design and analysis of experiments. *Proceedings of the 5th Berkeley Symposium on Mathematical Statistics and Probability* **4**, 777–790.

Yusuf, S., Peto, R., Lewis, J., Collins, R., and Sleight, P. (1985). Beta blockade during and after myocardial infarction: an overview of the randomized trials. *Progress in Cardiovascular Diseases* **27**, 335–371.

Zaslavsky, A. M. (1993). Combining census, dual-system, and evaluation study data to estimate population shares. *Journal of the American Statistical Association* **88**, 1092–1105.

Zeger, S. L., and Karim, M. R. (1991). Generalized linear models with random effects; a Gibbs sampling approach. *Journal of the American Statistical Association* **86**, 79–86.

Zelen, M. (1979). A new design for randomized clinical trials. *New England Journal of Medicine* **300**, 1242–1245.

Zellner, A. (1971). *An Introduction to Bayesian Inference in Econometrics.* New York: Wiley.

Zellner, A. (1975). Bayesian analysis of regression error terms. *Journal of the American Statistical Association* **70**, 138–144.

Zellner, A. (1976). Bayesian and non-Bayesian analysis of the regression model with multivariate Student-*t* error terms. *Journal of the American Statistical Association* **71**, 400–405.

Zhang, J. (2002). Causal inference with principal stratification: some theory and application. Ph.D. thesis, Department of Statistics, Harvard University.

Zhao, L. H. (2000). Bayesian aspects of some nonparametric problems. *Annals of Statistics* **28**, 532–552.

Author index

Subject index